AF615402

High-Level Nuclear Waste Disposal

High-Level Nuclear Waste Disposal

Edited by

Harry C. Burkholder

Battelle Memorial Institute
Pacific Northwest Laboratory
Richland, Washington, USA

Associate Editors

Carl R. Cooley
U. S. Department of Energy
Washington, DC

H. Bernard Dietz
Rockwell Hanford Operations
Richland, Washington

George Jansen
Battelle Memorial Institute
Office of Nuclear Waste Isolation
Columbus, Ohio

Robert B. Laughon
Battelle Memorial Institute
Office of Crystalline Repository Deployment
Columbus, Ohio

Virginia M. Oversby
University of California
Lawrence Livermore National Laboratory
Livermore, California

Robert F. Williams
Electric Power Research Institute
Palo Alto, California

Donald E. Wood
Rockwell Hanford Operations
Richland, Washington

Editorial Assistants

Billie L. Neth
Battelle Memorial Institute
Pacific Northwest Laboratory
Richland, Washington

James W. Thielman
Battelle Memorial Institute
Pacific Northwest Laboratory
Richland, Washington

Columbus • Richland

A portion of the funding for this symposium was provided by the United States Department of Energy. However, the views and findings of the various papers are solely those of the authors and do not necessarily represent the policy of the agency. The agency encourages wide dissemination of the technical information contained herein, with due respect for the Publisher's rights regarding the complete volume.

Library of Congress Cataloging-in-Publication Data

High-level nuclear waste disposal.

Papers presented at the International Topical Meeting on High-Level Nuclear Waste Disposal—Technology and Engineering held in Pasco, Washington, 9/24–26, 1985. Sponsored by the American Nuclear Society (Fuel Cycle and Waste Management Division, Environmental Sciences Division, and the Richland, Washington Section). Co-sponsors were the Canadian Nuclear Society and other organizations.

1. Radioactive waste disposal—Congresses. I. Burkholder, Harry C. II. International Topical Meeting on High-Level Nuclear Waste Disposal—Technology and Engineering (1985: Pasco, Wash.) III. American Nuclear Society. Nuclear Fuel Cycle Waste Management Division. IV. American Nuclear Society. Environmental Sciences Division. V. Canadian Nuclear Society.
TD898.H54 1986 621.48'38 86-14015
ISBN 0-935470-29-8

Printed in the United States of America.

PREFACE

This volume contains the proceedings of the International Topical Meeting on High-Level Nuclear Waste Disposal--Technology and Engineering held in Pasco, Washington, USA on September 24-26, 1985. The meeting was sponsored by the American Nuclear Society (Fuel Cycle and Waste Management Division, Environmental Sciences Division, and the Richland, Washington Section). Cosponsors were the Canadian Nuclear Society, the European Nuclear Society, the Atomic Energy Society of Japan, and the U.S. Department of Energy. Corporate support was provided by Battelle Memorial Institute, Westinghouse Hanford Company, Rockwell Hanford Operations, and Stone and Webster Engineering.

The meeting was timely because many countries had begun their site selection processes and their engineering designs were becoming well-defined. The technology of nuclear waste disposal was maturing, and the institutional issues arising from the implementation of that technology were being confronted. Accordingly, the program was structured to consider both the technical and institutional aspects of the subject. The meeting started with a review of the status of the disposal programs in eight countries and three international nuclear waste management organizations. These invited presentations allowed listeners to understand the similarities and differences among the various national approaches to solving this very international problem. Then seven invited presentations describing nuclear waste disposal from different perspectives were made. These included: legal and judicial, electric utility, state governor, ethical, and technical perspectives. These invited presentations uncovered several issues that may need to be resolved before high-level nuclear wastes can be emplaced in a geologic repository in the United States. Finally, there were sixty-six contributed technical presentations organized in ten sessions around six general topics: site characterization and selection, repository design and in-situ testing, package design and testing, disposal system performance, disposal and storage system cost, and disposal in the overall waste management system context. These contributed presentations provided listeners with the results of recent applied R&D in each of the subject areas.

Papers for all eighty-four presentations are contained in this volume. The contributed technical papers were subjected to both refereed peer-review and editing; the invited papers were only edited. Thus, this volume allows the reader to explore the content of the meeting in detail, at his/her leisure, and with some assurance about the quality of the information.

The Technical Program was organized by a committee chaired by the Editor and composed of the Associate Editors and a team of international experts on nuclear waste disposal. The international members of the Technical Program Committee were:

A. Barthoux, French Radioactive Waste Management Agency, Paris, France
F. Girardi, Commission of European Communities/Ispra Joint Research Center, Ispra, Italy
R. Heremans, Belgium Radioactive Waste Management and Fissile Materials Company, Brussels, Belgium
K. Kuhn, German Radiation and Environmental Research Company, Institute for Underground Storage, Braunschweig, Federal Republic of Germany
C. McCombie, Swiss Radioactive Waste Disposal Company, Baden, Switzerland
E. L. J. Rosinger, Atomic Energy of Canada, Ltd., Whiteshell Nuclear Research Establishment, Pinawa, Manitoba
L. Werme, Swedish Nuclear Fuel and Waste Management Company, Stockholm, Sweden

The Organizing Committee for the meeting was composed of the following individuals:

M. J. Lawrence, U.S. Department of Energy; Honorary Chairman
L. R. Fitch, Rockwell Hanford Operations; General Chairman
H. C. Burkholder, Pacific Northwest Laboratory; Technical Chairman
D. E. Wood, Rockwell Hanford Operations; Assistant Technical Chairman
J. R. Kirkendall, Rockwell Hanford Operations; Finance Chairman
D. L. Borders, Rockwell Hanford Operations; Committee Member
R. J. Cash, Westinghouse Hanford Company; Technical Tours Chairman
B. C. K. Moravek, Rockwell Hanford Operations, Public Information Chairman
M. C. Brown, Rockwell Hanford Operations; Committee Member
H. Babad, Rockwell Hanford Operations; Publications Chairman
C. L. Parks, Pacific Northwest Laboratory; Registration Chairman
J. H. Jarrett, Pacific Northwest Laboratory; Arrangements Chairman
D. J. Forgette, Rockwell Hanford Operations; Committee Member
L. I. Homme, Pacific Northwest Laboratory; Committee Member
R. D. Peters, Pacific Northwest Laboratory; Committee Member
F. M. Rogers, Pacific Northwest Laboratory; Committee Member
D. J. Squires, U.S. Department of Energy; DOE Liaison
J. Graham, Rockwell Hanford Operations; ANS National Representative
B. A. Calicoat, Rockwell Hanford Operations; Graphics Support

The sessions were conducted by the Technical Program Committee, some members of the Organizing Committee, and other individuals from high-level nuclear waste disposal-related organizations in the United States. These other individuals who acted as session cochairman were:

D. H. Dahlem, U.S. Department of Energy
T. O. Hunter, Sandia National Laboratory
B. Maiden, Battelle Project Management Division
W. C. McClain, Weston

A. M. Platt, Pacific Northwest Laboratory
D. O. Provost, Washington Department of Ecology
P. Saget, U.S. Department of Energy
M. W. Shupe, U.S. Department of Energy

The preparation of these proceedings was made possible by financial support from the U.S. Department of Energy and the dedicated work of S. C. Cozad, K. L. Filsinger, H. R. Kern, R. A. Keefe, D. A. Parks, B. E. Roberts, and J. A. Smith, of the Word Processing Section, and my secretary, D. C. Larson, at the Pacific Northwest Laboratory.

H. C. Burkholder
Battelle Memorial Institute
Pacific Northwest Laboratory
Richland, Washington
March 1986

CONTENTS

PREFACE

STATUS OF INTERNATIONAL HIGH-LEVEL NUCLEAR WASTE DISPOSAL PROGRAMS

PERSPECTIVE ON NUCLEAR WASTE DISPOSAL

SITE CHARACTERIZATION AND SELECTION

WASTE PACKAGE DESIGN AND TESTING

DISPOSAL SYSTEM PERFORMANCE

DISPOSAL AND STORAGE SYSTEM COST

DISPOSAL IN THE OVERALL WASTE MANAGEMENT SYSTEM CONTEXT

Status of International High-Level Nuclear Waste Disposal Programs

STATUS OF THE NUCLEAR WASTE DISPOSAL PROGRAM IN BELGIUM

A. Bonne
G. Collard
P. Dejonghe
Study Center for Nuclear Energy SCK/CEN
Boeretang, 200
2400-Mol, Belgium

E. Detilleux
National Agency for Radioactive Waste and Fissile Materials
NIRAS/ONDRAF
Boulevard du Regent 54 - Boite 5
1000-Brussels, Belgium

I. INTRODUCTION

Since mid-1985, the installed nuclear power capacity in Belgium totals 5,450 MW(e) all of the PWR-type. Consequently, in 1986, about 65% of the generated electricity will be of nuclear origin. Belgium also takes part in the European Fast Breeder program and has taken a share in the SNRO-300 at Kalkar FRG. Two fuel fabrication plants are in operation: one for UO_2 fuel and one for MOX-fuel. With regard to reprocessing, the decision concerning the reopening of Eurochemic has not yet been taken but the plant has already accumulated some wastes from earlier operations. In the meantime, spent fuel is being reprocessed at La Hague-France.

The relative importance of this program gives a special emphasis on the radioactive waste management issue taking into account that the country has a high population density and that its natural characteristics offer only limited opportunities for disposal.

In this perspective a law, promulgated in 1980, created the National Agency for the Management of Radioactive Waste and Fissile Materials (ONDRAF-/NIRAS); the latter Agency is entrusted with the definition and application of the waste management policy in order to ensure the current and long term protection of the population and the environment. The law also provides that all the expenditures related to the activities developed or sponsored by the Agency, including research and development and the long term activities, are to be borne by the producers of radioactive wastes.

Table 1 gives an overview of the expected cumulative quantities of conditioned waste from the present nuclear power program in a period of 30 years. Taking into account the final destination of the conditioned wastes, two main groups of wastes are being considered:

1. nongeological (this includes low-level and short-lived wastes which, until 1982, were dumped in the Atlantic Ocean under international surveillance; for these wastes, a terrestrial solution is also being studied) and

TABLE 1. Conditioned Waste Volumes From the Present-Day Nuclear Program in Belgium (cubic meter) in a 30-Year Period

SOURCE			DISPOSAL			
			NONGEOLOGICAL		GEOLOGICAL	
			Operation	Decommissioning	Operation	Decommissioning
NUCLEAR ENERGY	Nuclear energy production		50,000	40,000	300	400
	Fuel cycle	Fuel production	1,000	2,000	300	50
		Reprocessing	17,500		12,500	
MISCELLANEOUS	Radioisotopes: production/utilization		13,000	7,000	400	50
	Research		9,500	10,000	400	100
TOTAL			91,000	59,000	13,900	600

- All spent fuel of the programme (4700tUeq) reprocessed
- Waste of the past Eurochemic plant operation excluded (3200 m^3)
- Decommissioning of the Eurochemic plant in present-day conditions corresponds to about 5,000 cubic meter waste

2. geological (this includes high-level wastes and various types of alpha-bearing wastes. Among the total of 14,500 m^3 about 500 m^3 will be high-level waste; the rest will be highly contaminated secondary and technical wastes).

Nonreprocessed fuel has not yet been considered as a potential waste form although it may be understood that, even with reprocessing being done in a national or foreign facility, a fraction of the spent fuel might eventually not be reprocessed and end up in the waste stream.

The large quantities of wastes in the "geological" group are, on the one hand, a strong incentive for a continued effort in volume reduction, mainly of the non-HLW (by more efficient segregation at the source and/or the application of more advanced conditioning technology) and, on the other hand, illustrate the magnitude of the problem of geologic disposal. It is only in the early 70's that it became evident that, in spite of increased international collaboration in R&D, Belgium would have to find, on its own, a solution for the disposal of the conditioned high level and other actinide containing wastes. Accordingly, the R&D program of the SCK/CEN has been focused on three major topics:

1. volume reduction and insolubilization of plutonium-contaminated wastes by high temperature slagging incineration (such a unit is operational at SCK/CEN-Mol; it has a capacity of about 50-kg material per hour; the end product is a highly insoluble glass-like slag),

2. characterization and study of the long-term behavior of conditioned wastes (bitumen, slags, borosilicate glass,...) and their behavior in the storage environment (e.g., clay), and

3. geological disposal of conditioned wastes in deep clay layers (this is the main subject of the present presentation).

The three mentioned programs have been integrated in the program of the Commission of the European Communities. Since 1983 the producers of radioactive wastes also support the program on geologic disposal, through the intermediary of the National Agency for Management of Radioactive Wastes, mentioned above. For the vitrification of HLW, adequate forms of co-operation are being negotiated with other programs in Europe.

II. GEOLOGICAL DISPOSAL INTO DEEP CLAY FORMATIONS

Detailed information about the Belgian program on disposal in clay of high-level and alpha-bearing wastes will be given in several papers during this conference. Therefore, this presentation is restricted to a few headlines.

Following a general survey with the help of the National Geological Service in Belgium and taking note of the absence of salt and appropriate crystalline rock formations at reasonable depth, an early decision was made to study clay and argillaceous rock as a possible host rock for geologic disposal. In particular, a thick and homogeneous clay layer, "Boom-clay", occurs beneath an area of nearly 5,000 km^2 in the northeast part of the country. This clay presents a compact and homogeneous facies. At the nuclear site of Mol, it is found at a depth between 160 and 270 meters and is covered by sand layers and aquifers. This clay layer is the subject of the study.

A. Present State

After more than 5 years of exploratory drilling, sampling, characterization (studies on retention mechanisms, measurement of permeability, etc.), a seismic campaign and the study of the hydrology in the considered area, it was decided, in 1980, to construct at Mol an underground laboratory in the clay, at a depth of 225 m. The construction of this laboratory was completed in 1984 and various in-situ experiments on corrosion, leaching, heat dissipation, water movement, mechanical and rheological properties, etc... have been or are being installed. The construction is composed of a single shaft, 227-m deep and a horizontal gallery 30-m long and 3.5-m in diameter. The gallery is sustained by modular cast iron liners provided with flanged openings giving access to the clay.

Later, starting from the floor of the main gallery, at a depth of 227 m, an additional shaft (20 x 2 m) and horizontal drift (7 x 2 m) were constructed in nonfrozen clay. The purpose was to prepare future works without using the freezing technique and to install various sensors in the intact clay. Part of this work has been done in co-operation with the French organization ANDRA. Figure 1 gives a schematic view of the existing underground laboratory. Table 2 gives a few important dates.

The results obtained after more than 10 years of laboratory research and field investigation are summarized as follows:

1. The "Boom-clay" formation at Mol offers a polyvalent and very efficient barrier against dispersion of radioactive material in the environment:

    - The average fluid flow rate in the clay is on the order of 1 to 10 picometers per second;

    - The clay offers an important cation exchange capacity (20 to 40 meq per 100 g);

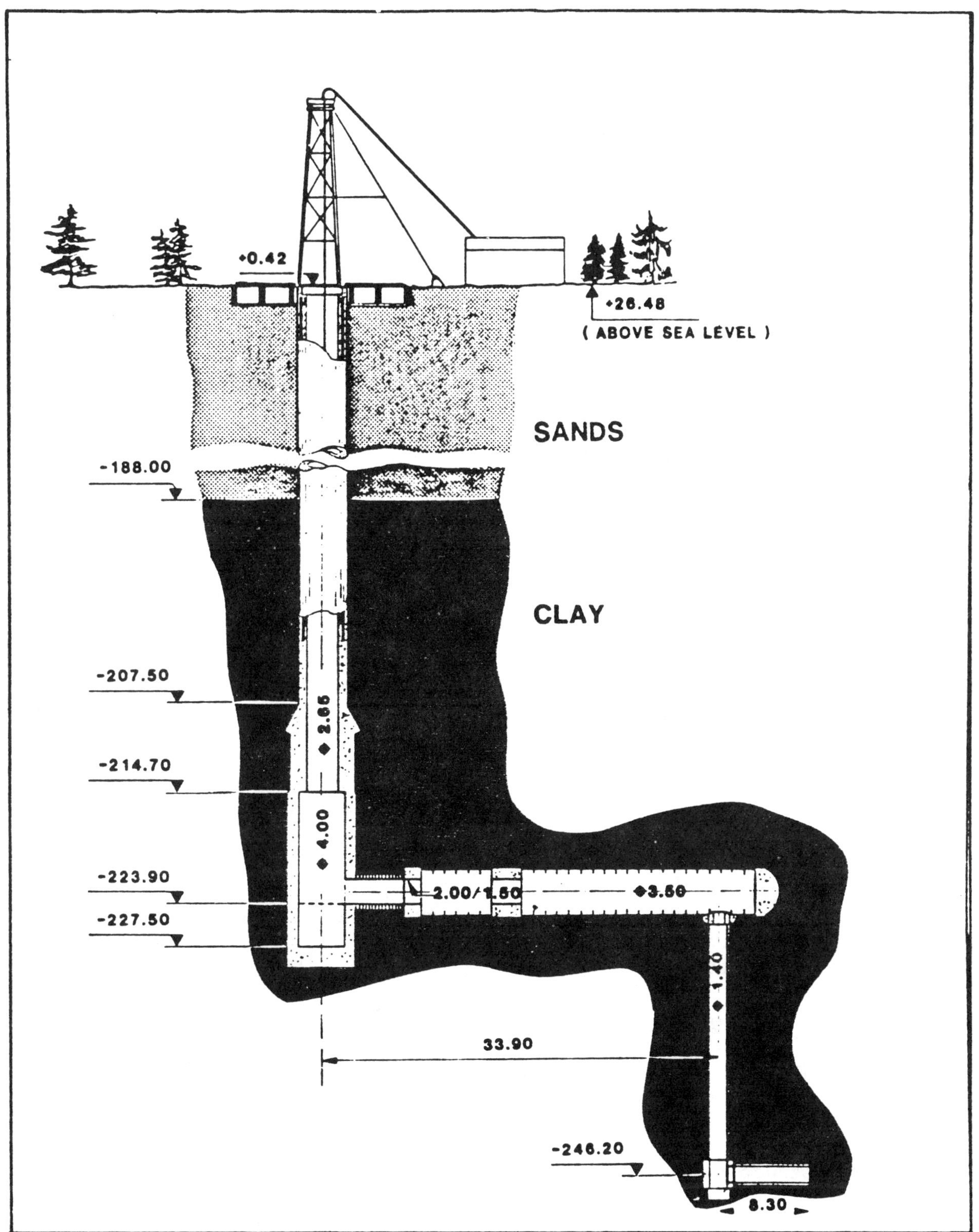

FIGURE 1. Underground Laboratory in Clay--As Built

TABLE 2. Belgian R&D Program Geological Disposal Main Achievements

1974	Project conceptualization
1974-1975	Reconnaissance drilling at Mol Site (Geology, Sampling, ...)
1975-1979	Positive conclusive results about homogeneity and geochemical barrier capacity of Boom-clay
	Repository design and concept developed
	Risk assessment methodology and tools developed and selected
	High resolution seismic campaign
1980-1984	Underground in situ experimental facility operational
	Risk assessment exercises completed
	Regional hydrological observation network (coverage 2500 sq km) installed

- Thanks to the physicochemical properties and the composition of the solutes in the interstitial fluid, the clay also behaves as a chemical "sink" for several radioelements. The interstitial fluid (15 to 19% by weight) is saturated with $CaCO_3$ and has reducing properties due to the presence of FeS_2 in the clay;

- The clay contains some residual organic matter (2 to 3%) which also contributes to the formation and/or trapping of chemical complexes and colloids involving radionuclides; and

- The global result is that not only characteristic cationic elements, Sr and Cs, are retained or precipitated, but also Pu and Np, in their valency state IV, Pu and Am in the valency III and even Tc.

2. Exposure of clay to air may lead to partial oxidation of FeS_2 and hence to acidification which, in turn, might enhance corrosion and reduce the ion exchange capacity of the clay. However, the amount of the FeS_2 (2 to 5%) is such that, after sealing and backfilling, the initial reducing state can be restored.

3. The construction of the infrastructure of the repository in frozen clay is technically feasible although economically prohibitive. Drifting shafts and tunnels in nonfrozen clay, even under the static pressure of 50 bar appears to be technically feasible with conventional digging equipment. However, for the latter type of construction in "Boom clay",

experience is still limited to a small shaft and drift. Therefore, this experience must still be confirmed at larger scale. In addition, whereas in the main parts of the underground laboratory a cast iron lining was installed, it is probable that, in future works, concrete lining will be used.

4. Taking into account the heat conductivity of the clay (1.7 W · m^{-1} · $^{o}C^{-1}$) heat dissipation is a limiting factor for the accumulation of high-level wastes in the repository. This problem can be avoided in a satisfactory manner by sufficient aging of the wastes before their final emplacement in the repository - in the Mol-study a reference cooling time of 50 years was adopted - and adequate design of the network of tunnels and disposal pits. For a 10 GW(e) program operating for 30 years, with 50 years cooling before disposal and a maximum temperature rise of 100°C in the clay in the immediate vicinity of the containers, the total underground surface area required has been estimated at 1 km^2.

5. Cost estimates have been made in co-operation with the CEC. The cost would be about 0.3 mill European Currency Unit (ECU) per kWh, paid at the time of production of the kWh. However, the study also acknowledges the influence of size of the repository and mode of financing upon the cost figure. Very similar figures were obtained for other disposal formations.

6. The safety and performance studies carried out up to now have to be understood as preliminary exercises aiming at 1) evaluating on a site specific basis the acceptability of the clay option from the safety point of view, 2) applying and validating available methodologies, codes and models, and 3) identifying areas which need further research or development.

The studies have shown that a radioactive waste repository in a clay formation, even at a depth of around 200 meters, is able to provide an adequate isolation of the buried waste under the circumstances of the natural degradation of the system. Also, under conditions of disruption of the geological barrier, the risks associated with these events can be shown to be acceptable.

In the case of the scenario of natural degradation, the results obtained for the consequences for a maximum-exposed individual of a critical group showed that the dose to man is principally determined by the geochemical barrier characteristic of the clay host rock. The results were demonstrated to be quite insensitive to the assumed parameters and hypotheses about the source term and near field alteration.

For the disruptive scenarios studied (tectonic faulting and glacial erosion) the risk assessment exercises reveal a higher sensitivity of the results to the characteristics of the artificial barriers and the geochemical barrier capacity of the overburden of the repository and to the characteristics of the repository debris (e.g., the degree of fracturing and degradation of the waste form in the faulting and glacial erosion scenario is an important parameter). For such scenarios however, uncertainty analyses are very important and place the results obtained in the right perspective.

Within the context of the PAGIS action (Performance Assessment of the Geological Isolation System) of the Commission of the European Communities, SCK/CEN is in charge of the co-ordination and realization of the performance studies to be made for various clay sites and repositories. For this action, the Mol site and its specific repository design have been retained as a reference case.

Simultaneously, and on national level, the conclusions from previous R&D and safety and feasibility studies will be drawn in the frame of a Safety Assessment and Feasibility Interim Report which will be prepared in a cooperation between ONDRAF/NIRAS and SCK/CEN. It is scheduled for end 1987. This report will consist of a presentation and analysis of the knowledge acquired in the past 12 years. The objectives are to allow the authorities to take position on the principle of the disposal of conditioned high-level and alpha-bearing wastes into the clay formation underneath the Mol-Dessel site and for the scientists and technicians to identify critical items which require further investigation.

B. Further Development

As was shown before, the results of previous studies are really promising. Nevertheless, in the technological field, many important items remain to be developed or demonstrated, e.g.,:

1. the possibility and/or technical problems in the digging of large diameter structures in non frozen clay,

2. the impact of heat load of the system on stresses and overall mechanical behavior of the structure,

3. backfilling and sealing,

4. costs for investment and operation, and

5. optimization of construction and operation of the repository and its components.

In the field of safety assessment more fundamental work remains to be done on the impact of heat and radiation upon the clay host rock and the retention of radioisotopes on the clay. It is also necessary to confirm present knowledge on retention and dispersion in longer term in-situ experiments. For these reasons a second phase of the project is being started. It is composed of two parts:

1. Construction of a 30-m long, up to 5-m outer diameter drift provided with storage pits, for the study of lining problems, feasibility of the digging technology in nonfrozen clay and, later on, emplacement of heat and radiation sources.

2. Construction of a longer section of tunnel, 200 m, provided with storage pits and handling equipment for the emplacement of conditioned wastes and backfilling; the latter tunnel will be accessible through the existing shaft of the underground laboratory and a second large shaft. This latter part is called demonstration.

Table 3 gives the timing for the second phase. This part of the program will also be covered by a contract with the CEC and coordinated within the framework of the Agency NIRAS/ONDRAF.

TABLE 3. Demonstration Program

Phase I: Experimental Drift	
1985-1987	Design, architecture, construction
1988-1992	Technological tests
Phase II: Desmonstration/pilot facility	
1986-1987	Definition of test program architecture
1988	Call for tenders and contact assignments
1989-1993	Construction
1993-1994	Dummy tests
1995	Operation of pilot facility

III. CONCLUSION

Developing and demonstrating a repository concept takes time and effort. However, work and knowledge has developed steadily and progressively. One may be confident today that the clay option is a reliable and safe one. However, several aspects remain to be demonstrated before final decisions as to the construction of a repository will be taken.

NUCLEAR FUEL WASTE DISPOSAL--THE CANADIAN PROGRAM

R. B. Lyon
P. D. R. Lisle
D. B. McConnell
Atomic Energy of Canada Limited
Whiteshell Nuclear Research Establishment
Pinawa, Manitoba ROE 1LO, Canada

ABSTRACT

The Canadian Nuclear Fuel Waste Management Program is in the fifth year of a ten-year generic research and development phase. The major objective of this phase of the program is to assess the basic safety and environmental aspects of the concept of isolating immobilized fuel waste by deep underground disposal in plutonic rock. The major scientific and engineering components of the program, namely immobilization studies, geoscience research, and environmental and safety assessment, are well established.

I. INTRODUCTION

In June of 1978, the governments of Canada and Ontario announced an agreement to cooperate in the development of technologies for the safe management and permanent disposal of Canada's nuclear fuel waste.[1] Under this agreement, the provincially owned utility, Ontario Hydro, is responsible for developing technologies for the interim storage and transportation of used fuel, while Atomic Energy of Canada Limited (AECL), a federal crown corporation, is responsible for coordinating and managing the research and development program for the immobilization and disposal of nuclear fuel waste.[2] The administrative structure, main research and development components, participating organizations, and international cooperation are described in the Program Guide.[3] An independent Technical Advisory Committee, established in 1979, provides an ongoing scientific review of the program.[4]

In 1981 April, the Canadian government approved a ten-year generic research and development program on nuclear fuel waste management. The objectives of this phase of the program are:

1. to develop the technology for storage, transportation, immobilization and disposal to the extent necessary to provide data for the assessment; to design facilities; to specify operating processes and procedures; and to demonstrate that practical technology is available for implementation of the concept;

2. to establish the requirements, equipment, and procedures for the site characterization and selection processes for the next phase of nuclear fuel waste management;

3. to assess the environmental and safety aspects of the concept of isolating immobilized fuel waste by deep underground disposal in plutonic rock; and

4. to develop the basis for public acceptance and support through scientific and regulatory review, and public information, interaction and participation.

II. RESEARCH AND DEVELOPMENT

A. Storage and Transportation of Used Fuel

Used CANDU fuel continues to be stored safely and economically in water-filled concrete storage bays at the nuclear generating stations. Ontario Hydro has been investigating several storage options for used fuel at reactor sites. Dry storage in concrete canisters has been demonstrated at the Whiteshell Nuclear Research Establishment (WNRE) in Manitoba. Experience with wet and dry storage of used CANDU fuel over the past 20 years provides confidence that interim storage is practicable for at least 50 years.[5]

Ontario Hydro is developing the technology for large-scale transportation of used fuel. The reference cask design has a two-module (192-bundle) payload, rectangular geometry, and monolithic stainless steel wall construction. The heat dissipation capabilities of the reference cask have been investigated in a full-scale simulation using electrically heated, simulated fuel bundles, and experiments have been conducted that show that little radioactivity would be released from the fuel in the event of sheath rupture during either normal or accident conditions of transport. Completion of design, construction and licensing of a full-size cask is scheduled for 1988.

B. Fuel Immobilization

Fuel immobilization studies involve the development of durable containment for the disposal of intact used-fuel bundles, and the characterization of used fuel as a waste form.[6] Studies have concentrated on simple cylindrical containers with a high-integrity corrosion-resistant metallic shell to isolate the fuel during its high toxicity phase. Containment systems that could offer substantially longer isolation, using materials such as ceramics, are also being studied.

Several container designs are being evaluated.[7] Prototypes of these containers were fabricated from stainless steel or grade-2 titanium and subjected to tests in a hydrostatic test facility at pressures up to 10 MPa and temperatures up to 150°C.[8,9] The long-term corrosion behavior of candidate container materials is also being studied.[10,11] Initial results indicate that grade-12 titanium is much more resistant to localized corrosion than grade-2 titanium,[12] as is Hastelloy C-276, a nickel-based alloy. Studies of the corrosion behavior of copper in simulated high-salinity groundwater have shown that copper is a suitable alternative to passive metals.[13]

C. Used-Fuel Characterization

Determination of the leaching and dissolution properties of used UO_2 fuel constitutes the major part of the fuel characterization program. During the past two years, emphasis has been on estimating the fractions of cesium-135 and iodine-129 that are released from the gap between the fuel and sheath during the early stages of used fuel dissolution. The gap inventory of cesium-135 and iodine-129 at the time of fuel discharge from the reactors is estimated to be about 2.2%. Recent studies have shown a correlation between fuel power history and fuel leaching properties.[14]

D. Waste Immobilization

Processes and products are being developed for immobilizing the waste that would arise if the fuel from CANDU reactors were recycled.[8,15] Glasses, ceramics and glass-ceramics are being evaluated as possible waste forms. A Waste Immobilization Process Experiment facility, consisting of a rotospray calciner and a ceramic electromelter, designed to produce 10 $kg{\cdot}h^{-1}$ of sodium borosilicate glass, is now operating at WNRE.

The behavior of glass waste forms and their durability in the hydrothermal environment anticipated in a disposal vault are being studied. A survey of borosilicate glasses[16] showed that durability increases with increasing SiO_2, Fe_2O_3 or Al_2O_3 content, but decreases with increasing Na_2O or K_2O content. Sodium aluminosilicate glasses[17] have a low, relatively constant, leach rate (less than 10^{-9} $kg{\cdot}m^{-2}{\cdot}s^{-1}$) within a wide composition range.

The ceramic waste forms being considered contain sphene ($CaTiSiO_5$). Calculations indicate that sphene should be stable in groundwater typical of the Canadian shield (high Ca^{2+}, low SO_4^{2-} and CO_3^{2-}) in the temperature range of 25 to 150°C. Three types of sphene-based matrices are being studied: natural minerals, ceramic pellets formed by pressing and sintering, and glass-ceramics formed by melting and controlled crystallization of the system Na_2O-Al_2O_3-CaO-TiO_2-SiO_2.[18]

E. Disposal Vault Sealing

Disposal-vault sealing studies involve the development of the buffer material (clay-sand mixture) to surround the containers, and other barriers to close the man-made openings to the surface: namely, the backfill and the plugs and grouts for shaft and borehole seals.[19]

A study of the physical and chemical properties of buffer and backfill clays[20] has provided information on the basic mineralogical, chemical and physical properties, and on the behavior of clays under wet-dry cycling. Bentonites are suitable as buffers because of their high swelling potentials, low hydraulic conductivities, low effective porosities, and high sorption capacities for radionuclides.[21] A compaction study of candidate buffer materials has been completed[22] and the hydraulic conductivities of two candidate materials have been measured.[23] Experimental studies have shown that shrinkage, long-term creep, drying and rewetting, and the removal of buffer material by groundwater are unlikely to prejudice the effectiveness of the buffer.[24-26] A major study on buffer and backfill engineering has provided information on procedures, schedules and costs.[27] Computer modeling studies have been performed to determine the effects of container and buffer geometry and the quality of the rock wall in the emplacement boreholes on the diffusional transport of radionuclides from failed containers.[28,29]

F. Immobilized Fuel Test Facility

The Immobilized Fuel Test Facility (IFTF)[30] at WNRE provides an environment for a wide range of multicomponent experiments[31] in radiation fields, under temperatures and pressures that simulate a vault environment. The experiments are designed to test active waste forms and materials proposed for engineered barriers. Preparation of long-term immersion experiments in passive canisters and of multicomponent-systems tests are underway. The first set of experiments were emplaced late in 1984. A typical set of experiments comprises 18 small titanium pressure vessels, each containing fuel waste, container material, buffer, groundwater and rock, loaded in one of the seven concrete canisters. The experiments are run for six months or more at temperatures up to 200°C and at pressures up to 8 MPa.

G. Geoscience Research

The emphasis of the geoscience research is on the evaluation of large plutonic rock masses in the Canadian shield as potential hosts for immobilized nuclear fuel waste.[32] Deep exploratory drilling, hydrogeological studies, and detailed surface mapping are being performed at designated field research areas in the Canadian Shield. The areas at Chalk River and Atikokan, Ontario, and Whiteshell, Manitoba, contain granite rocks, while that at East Bull Lake, Ontario, contains gabbro. Regional groundwater flow is being investigated in a major flow system study at Atikokan.

H. Underground Research Laboratory

The Whiteshell research area is situated on the Lac du Bonnet batholith, a large granitic body in southeastern Manitoba. This research area is the site of the Underground Research Laboratory (URL), which is being constructed below the water table in a previously undisturbed portion of the batholith. The objectives of the URL project are to study the correlation between surface and subsurface features, hydrogeological and geochemical systems in plutonic rock, excavation damage in rock, and the effect of heat on plutonic rock (including the effect on mass transport) and on buffer-backfill-rock interactions.[33]

Comprehensive geological, geophysical and hydrogeological investigations of the URL lease area are being done. Numerous geophysical surveys were performed in boreholes,[34] and three major subhorizontal fracture zones were identified. A network of instrumented boreholes has been established to provide baseline data on pre-construction hydrogeological conditions and to measure changes caused by the excavation. Groundwater levels are recorded continuously in about 75 groundwater monitoring locations. Predictions of changes in groundwater systems by several independent hydrogeological modelling groups are being compared with the groundwater system perturbations measured during and after excavation. Surface facilities are in place and shaft excavation was completed in March 1985. The URL will be ready for operation in 1986. The underground facilities have a 255-m deep, rectangular access shaft, a ventilation raise, and a test level with several experimental rooms.

I. Geochemistry and Applied Chemistry

The objective of the geochemistry and applied chemistry research is to quantify the chemical and physical interactions that occur between radionuclides and the geological materials lining water-bearing fractures in plutonic rock. These interactions can prevent, or retard, migration of radionuclides from the deep underground vault to the biosphere. Examinations of the geological record that exists in and along groundwater-bearing fractures in plutonic rock, and of geological analogues to a disposal vault, such as naturally occurring uranium deposits, are being used to assess the behavior of radionuclides in the geosphere. In two well-defined uranium deposits in northern Saskatchewan, the uranium has migrated less than 5 m into the clay surrounding the ore body over the last billion years.

III. ASSESSMENT OF THE CONCEPT

The objective of the environmental and safety assessment is to assess the impact of a disposal facility on man and the environment. The assessments are being published in a series of Concept Assessment Documents. The first interim concept Assessment document was published in 1981[35-37] and the second will be issued shortly. The formal Concept Assessment Document, scheduled for completion in 1988, will form the basis for concept evaluation by regulatory and environmental agencies and for subsequent review at a public hearing.

The environmental and safety assessment has two major components: pre-closure assessment and post-closure assessment. Pre-closure assessment covers the period up to and including vault backfilling, sealing and closure.[38,39] Post-closure assessment covers the period after the vault has been sealed and the surface facilities have been decommissioned.[40]

The post-closure assessment is being done with the SYVAC systems variability analysis code,[43] which allows for variability and uncertainty by representing input parameters as distributions rather than as single values. SYVAC contains a set of submodels, for the vault, the geosphere and the biosphere, that represent the components of the disposal system. The submodels and parameter distributions are derived by assimilating the results of the field and laboratory observations, which are usually interpreted by the use of detailed research models. SYVAC performs repeated deterministic calculations with parameter values sampled from their distributions in a Monte Carlo process. Figure 1(a) illustrates SYVAC results for a recently completed assessment, comparing risk versus time with a limit of 10^{-5} per annum suggested by the Nuclear Energy agency of the Organization for Economic Cooperation and Development (OECD/NEA).[42] The risk from used fuel disposal was estimated to be so low that it is almost indistinguishable from zero on the scale used. Figure 1(b) presents the probability of exceeding 30% natural background dose versus time. (This dose would be equivalent in risk to the OECD risk level.)

Validation of the assessment is achieved by a combination of quality assurance on software,[43] expert review, intercode comparison, and comparison with field and laboratory observations, the last applying mainly to the validation of the research models. An excellent example of the validation of a research model is the comparison between prediction and observation of the water-table drawdown during construction of the Underground Research Laboratory.[44]

IV. REVIEW PROCESS AND SCHEDULE

In August 1981, the governments of Canada and Ontario issued a statement describing the evaluation process, the roles and responsibilities of the environmental and regulatory agencies, and the involvement of the public.[45] The evaluation process, which will start in 1988, will involve a regulatory and environmental review and a full public hearing by 1991. A decision will then be made by the two governments on the acceptability of the concept. In the regulatory and environmental review, the Atomic Energy Control Board will then be made by the two governments on the acceptability of the concept.

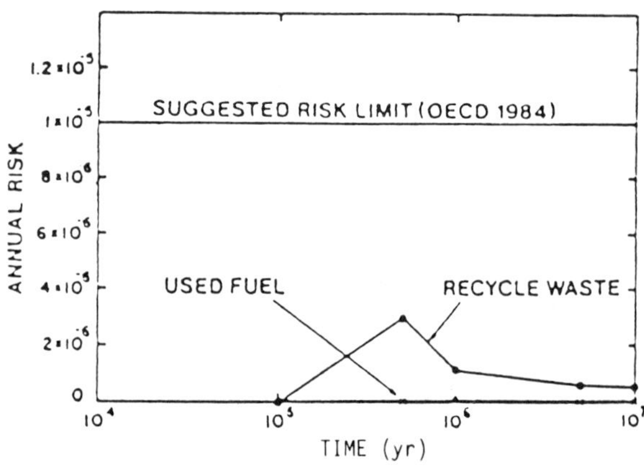

FIGURE 1(a). Predicted Annual Risk From Nuclear Fuel Waste

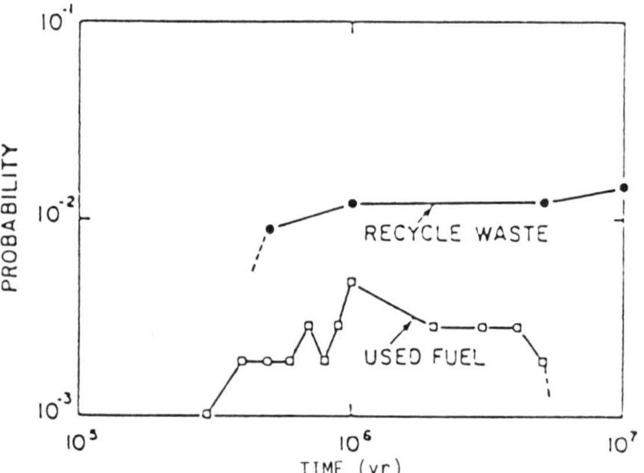

FIGURE 1(b). Probability of Exceeding 30% of Natural Background Dose

In the regulatory and environmental review, the Atomic Energy Control Board will act as the lead agency, assisted by the federal Department of the Environment and the Ontario Ministry of the Environment. The public hearing will be held under the auspices of the Canadian government.

REFERENCES

1. Minister of Energy, Mines and Resources Canada and the Ontario Energy Minister, "Joint Statement," June 5. Printing and Publishing Supply and Services Canada, Ottawa, Canada (1978).

2. D. B. McConnell, ed, The Canadian Nuclear Fuel Waste Management Program, 1984 Annual Report, Atomic Energy of Canada Limited Report (in preparation).

3. E. L. J. Rosinger, R. B. Lyon, P. Gillespie and J. Tamm, Guide to the Canadian Nuclear Fuel Waste Management Program, 2nd ed., Atomic Energy of Canada Limited Report, AECL-7790 (1983).

4. L. W. Shemilt (Chairman), Fifth Annual Report of the Technical Advisory Committee on the Nuclear Fuel Waste Management Program, Report TAC-5 (1987). Available from Dr. L. W. Shemilt, McMaster University.

5. J. A. Remington, R. C. Oberth and C. E. L. Hunt, "Twenty Years' Experience in Canada with Wet and Dry Storage of Irradiated CANDU Fuel," Paper No. IAEACN43/429, Proceedings of the IAEA International Conference on Radioactive Waste Management, Seattle, Washington, Vol. 3, 107; also available as AECL-8032 (1983).

6. Proceedings of the Nineteenth Information Meeting of the Nuclear Fuel Waste Management Program (1985 Topical Meeting), Atomic Energy of Canada Limited Technical Record*, TR- 350 (1985).

7. K. Nuttall, J. L. Crosthwaite, P. McKay, P. M. Mathew, B. P. M. Teper, P. Y. Y. Maak and M. D. C. Moles, "The Canadian Container Development Program for Fuel Isolation," Proceedings of the Materials Research Society Symposium, 15, 677 (1983).

8. K. J. Truss, "Container Development and Testing," Proceedings of the Eighteenth Information Meeting of the Nuclear Fuel Waste Management Program (1984 General Meeting), Atomic Energy of Canada Limited Technical Record*, TR-320, 121-130 (1985).

---

*Unrestricted, unpublished report available from SDDD, Atomic Energy of Canada Limited Research Company, Chalk River, Ontario KOJ 1J0.

9. B. P. M. Teper, Test Program of the Prototype of the Thin-Wall Packed Particulate Container, Part I: Hydrostatic Tests, Ontario Hydro Research Division Report 84-293-K (1984).

10. P. McKay and D. B. Mitton, "An Electrochemical Investigation of Localized Corrosion on titanium in chloride Environments," Corrosion, 52, 41 (1985).

11. B. M. Ikeda and P. McKay, "The Effect of Gamma Radiation on Electrochemical Processes Relevant to the Canadian Nuclear Fuel Waste Management Program," Proceedings of the 166th meeting of the Electrochemical Society, New Orleans, October 7-12, (1984).

12. J. Postlethwaite and R. J. Scouler, "Effect of Nickel and Molybdenum Additions on the Localized Corrosion Resistance of Titanium," Unpublished Contract Report No. 4 (1984).

13. P. J. King, K. W. Lam, and D. P. Dautovich, "The Corrosion Behavior of Copper Under Simulated Nuclear Waste Repository Conditions." Canadian Metall. Quarterly, 22, 125 (1983).

14. L. H. Johnson, K. I. Burns, H. H. Joling and C. J. Moore, "Leaching ^{137}Cs and ^{134}Cs and ^{129}I from Irradiated UO_2 Fuel," Nucl. Technol., 63, 470 (1983).

15. G. G. Strathdee and A. G. Wikjord, Waste Immobilization Research for the Canadian Nuclear Fuel Waste Management Program, Atomic Energy of Canada Limited Technical Record*, TR-148 (1982).

16. K. B. Harvey, "The Development of Borosilicate Glasses as Media for the Immobilization of High-Level Recycle Wastes I. Literature Survey," Atomic Energy of Canada Limited Technical Record*, TR-239 (1984).

17. J. C. Tait and D. L. Mandolesi, "The Chemical Durability of Alkali Aluminosilicate Glasses," Atomic Energy of Canada Limited Report, AECL-7803 (1983).

18. P. J. Hayward, W. E. Hocking, S. L. Mitchell and M. S. Stanchell, "Leaching Studies of Sphene-Based Glass-Ceramics," Nuclear and Chemical Waste Management, 5, 27 (1984).

19. G. W. Bird and D. J. Cameron, "Vault-Sealing Research for the Canadian Nuclear Fuel Waste Management Program," Atomic Energy of Canada Limited Technical Record*, TR-145 (1982).

---

*Unrestricted, unpublished report available from SDDD, Atomic Energy of Canada Limited Research Company, Chalk River, Ontario K0J 1J0.

20. McGill University Geotechnical Research Centre, Study of the Suitability of Various Candidate Backfill and Buffer Materials for a Deep Underground Nuclear Waste Disposal Vault: Final Report, Unpublished Contract Report (1982).

21. D. W. Oscarson and S. C. H. Cheung, Evaluation of Phyllosilicates as a Buffer Component in the Disposal of Nuclear Fuel Waste, Atomic Energy of Canada Limited Report, AECL-7812 (1983).

22. D. A. Dixon, M. N. Gray and A. W. Thomas, "A Study of the Compaction Properties of Potential Clay-Sand Buffer Mixtures for Use in Nuclear Fuel Waste Disposal," Proceedings of an International Symposium on Clay Barriers for Isolation of Toxic Chemical Wastes, Stockholm, Sweden, May 28-30, 1984, p. 53-62, Roland Pusch, editor, Elsevier Publishing Co., Amsterdam (1984).

23. H. S. Radhakrishna and H. T. Chan, "Strength and Hydraulic Conductivity of Clay-Based Buffer Materials," Atomic Energy of Canada Limited Technical Record*, TR-327 (1985).

24. A. P. S. Selvaduri, Influence of Non-linear Material Response on the Canister-Buffer-Rockmass Interaction in a Nuclear Waste Disposal Vault During Water Uptake, Carleton University, Department of Civil Engineering, Contract Report (1984).

25. McGill University Geotechnical Research Centre, Creep Behavior of Buffer Material in a Nuclear Waste Vault, Final Contract Report to Atomic Energy of Canada Limited (1984).

26. A. P. S. Selvadurai, R. S. Lopez and G. A. Hartley, "Geotechnical Modelling of Container-Buffer-Rockmass Interactions in a Nuclear Waste Disposal Vault," Proceedings of XI International Conference on Soil Mechanics and Foundation Engineering, San Francisco, August, 1985.

27. W. L. Wardrop and Associates Ltd., "Buffer and Backfilling Systems for a Nuclear Fuel Waste Disposal Vault," Atomic Energy of Canada Limited Technical Record*, TR-341 (1985).

28. S. C. H. Cheung, T. Chan and R. S. Lopez, "Numerical Analysis of Radionuclide Migration Through Engineering Barriers," Proceedings of the Materials Research Society Meeting on the Scientific Basis for Nuclear Waste Management, Boston, 1983 November, Elsevier Science Publishing Company (1984).

---

*Unrestricted, unpublished report available from SDDD, Atomic Energy of Canada Limited Research Company, Chalk River, Ontario KOJ 1J0.

29. S. C. H. Cheung and T. Chan, Effect of Localized Waste-Container Failure on Radionuclide Transport from an Underground Nuclear Waste Vault," Atomic Energy of Canada Limited Report, AECL-7796 (1983).

30. M. A. Ryz, "Whiteshell Nuclear Research Establishment Immobilized Fuel Test Facility," Trans. Amer. Nucl. Soc., 44, 590 (1983).

31. R. B. Heiman and L. H. Johnson, "Deign on Multicomponent Systems Test on High-Level Nuclear Waste Forms," Advances in Ceramics, 8, 337 (1984).

32. K. W. Dormuth and J. S. Scott, "Research and Development for a Plutonic Rock Radioactive Waste Disposal Vault," Paper No. IAEA-CN-43/166, Proceedings of IAEA International Conference on Radioactive Waste Management, Seattle, Washington, 1983 May, 221; also available as AECL-8087 (1983).

33. G. R. Simmons, A. Brown, C. C. Davison and G. L. Rigby, "The Canadian Underground Research Laboratory," Paper No. IAEA-CN-43/167, Proceedings of IAEA International Conference on Radioactive Waste Management, Seattle, Washington, 1983, 291; also available as AECL-7961 (1983).

34. N. M. Soonawala, "Geophysical Logging in Granites," Geoexploration, 21, 221 (1983), also available as AECL-7653.

35. R. B. Lyon, K. K. Mehta and T. Andres, "Environmental and Safety Assessment Studies for Nuclear Fuel Waste Management. Volume 1: Background," Atomic Energy of Canada Limited Technical Record*, TR-127-1 (1981).

36. K. Johansen, J. R. E. Harger and R. A. James, Environmental and Safety Assessment Studies for Nuclear Fuel Waste Management. Volume 2: Pre-Closure Assessment, Atomic Energy of Canada Limited Technical Record*, TR-127-2 (1981).

37. D. M. Wuschke, K. K. Mehta, K. W. Dormuth, J. A. K. Reid and R. B. Lyon, Environmental and Safety Assessment Studies for Nuclear Fuel Waste Management, Volume 3: Post-Closure Assessment, Atomic Energy of Canada Limited Technical Record*, TR-127-3 (1981).

38. J. H. Gee, K. L. Donnelly, B. J. Green, B. G. Rogers and M. A. Stevenson. Preliminary Environmental Assessment of the Canadian Nuclear Fuel Waste Management Concept: Pre-Closure Phase, Ontario Hydro Report No. 83137, Design and Development Division, Ontario Hydro, Toronto, Ontario (1983).

39. J. S. Nathwani, Nuclear Fuel Waste Management Concept: Preliminary Safety Assessment of the Pre-Closure Phase, Ontario Hydro Report No. 82175, Revision 1, Nuclear Studies and Safety Department, Ontario Hydro, Toronto, Ontario (1983).

---

*Unrestricted, unpublished report available from SDDD, Atomic Energy of Canada Limited Research Company, Chalk River, Ontario KOJ 1JO.

40. E. L. J. Rosinger, B. W. Goodwin, and D. M. Wuschke, "Performance Assessment for Nuclear Fuel Waste Disposal - The Canadian Approach," Proceedings of "Waste Management 85, Tucson, USA, March 1985 (to be published).

41. K. W. Dormuth and G. R. Sherman, SYVAC - A Computer Program for Assessment of Nuclear Fuel Waste ManaSement Systems, Incorporating Parameter Variability, Atomic Energy of Canada Limited Report, AECL-6814 (1981).

42. OECD, Long-Term Radiation Protection Objectives for Radioactive Waste Disposal, Report No. ISBN. 92-64-12604-X, OECD Nuclear Energy Agency, Paris (1984).

43. K. J. Hoffman and G. R. Sherman, "Standards for the Development of Scientific Research Software," Atomic Energy of Canada Limited Technical Record*, TR-291 (1985).

44. V. Guvanasen, J. A. K. Reid and B. W. Nakka, "Predictions of Hydrogeological Perturbations Due to Construction of the Underground Research Laboratory," Atomic Energy of Canada Limited Technical Record*, TR-344 (1985).

45. Minister of Energy, Mines and Resources Canada and the Ontario Energy Minister, Joint Statement August 4. Printing and Publishing Supply and Services Canada, Ottawa, Canada K1A 0S9 (1981).

---

*Unrestricted, unpublished report available from SDDO, Atomic Energy of Canada Limited Research Company, Chalk River, Ontario K0J 1J0.

STATUS OF THE COMMISSION OF THE EUROPEAN COMMUNITIES (CEC) RADIOACTIVE WASTE DISPOSAL PROGRAM

Serge M. Orlowski
Commission of the European Communities
Rue de la Loi, 200
B-1049 Bruxelles, Belgium

ABSTRACT

The two components of the third pluriannual R&D program of the Commission of the European Communities (CEC) on radioactive waste, namely the JRC action and the shared-cost-action with EC national laboratories, have been approved in 1984 and 1985 respectively. Research on the disposal of reprocessing waste will be pursued along the lines of the previous programs; it will regroup within coordinated activities and common projects most of the experiments on geological sites in the EC, migration studies (Project MIRAGE), performance and safety assessment studies (Project PAGIS), etc. In addition, the construction and/or operation of three underground facilities open to Community joint activities is included in the program; this represents a new step towards increased cooperation on waste disposal between EC Member States and towards the realization of future industrial underground repositories.

I. INTRODUCTION

On 22 November 1973, the European Council of Ministers approved, as part of the first European Community's program for the environment, the principle of a Community action concerning the management of radioactive waste. Since then, three pluriannual research programs succeeded each other, covering approximately the half decades 1975-1980, 1980-1985, and 1985-1990. As was the case for the first two programs, the third CEC program is being carried out partly by the Commission's Joint Research Center (JRC), mainly in its Ispra Establishment, and partly by research bodies of the European Community under the Commission's coordination by cost-sharing contracts.

The shared-cost action was approved by the EC Council of Ministers in March 1985 and is just starting; a call for R&D proposals has been issued in April and answers are under examination. The action's program comprises two parts: the first (part A) is a continuation of the previous R&D activities; the second (part B) represents a new step towards the realization of future industrial underground repositories.

A. Waste management studies and associated R&D actions:

- systems studies,
- improvement of radioactive waste treatment and conditioning technologies,
- evaluation of conditioned waste and qualification of engineered barriers,
- research in support of the development of disposal facilities; shallow burial and geological disposal studies,
- safety of geological disposal, and
- joint elaboration of radioactive waste management policies.

B. Construction and/or operation of underground facilities open to Community joint activities. The JRC action, initiated in 1984, comprises three projects:

- waste management and the fuel cycle,
- safety of waste disposal in continental geological formations, and
- feasibility and safety of waste disposal in deep oceanic sediments.

The action will be revised at the end of 1985 for possible updating and modification.

The JRC action and the shared-cost action together represent some 200 million ECUs, of which, in global terms, 25% fall on the first and 75% on the latter when the share of the contract partners is also taken into account. More than half of the budget will be devoted to waste disposal R&D; the wastes are high-level waste (HLW) and alpha waste resulting from spent-fuel reprocessing. Direct disposal of spent fuel will be considered only in the framework of system studies. I will now briefly review where we stand and what action is planned during the period 1985-1989, as far as HLW disposal is concerned.

II. R&D ON THE GEOLOGIC MEDIA

At the very beginning, in 1975, it was decided to limit the Community studies to rock types already under consideration in some member States such as clay (Mol project, Belgium), salt (Asse mine experiments, Federal Republic of Germany) and to add crystalline rocks; tasks were shared between countries, Belgium and Italy specializing in clays, the Federal Republic of Germany and the Netherlands in salt and France and the United Kingdom in granite; geological formations of these three types, relevant for waste disposal, were identified within the European Community; a number of data, related to the characterization of the various rocks and formations were obtained in laboratory and in situ by means of several deep drillings. Results of this first phase 1975-1980 have been extensively reported.[1]

During the second phase 1980-1985, the work continued in this particular field and was amplified mainly to obtain a better understanding of the properties of the host rock and of the long-term mechanical behavior of the geological formations.

To this effect, several sites were fitted up, inter alia:

1. Mol, in Belgium, where the construction of an experimental laboratory was completed in 1984 in an oligocene clay layer at a depth of 220 m, and where an important program of geomechanical measurements has already been going on for several years,

2. the Konrad iron mine (sedimentary environment) in the Federal Republic of Germany where advantage was taken of the existence of galleries to carry out measurements of pressures around underground cavities and of the resulting deformations,

3. the Asse salt mine, in the same country, where an extensive R&D program is being performed; accompanying flooding tests were carried out in the Hope salt mine, and

4. the Uranium mine of Fanay-Augeres in France, where a drift at 170 m depth is used as an underground laboratory to perform a "scale effect experiment" with a view to establish a possible relationship between dispersion tensor and fracture network in granite.

In addition, a preliminary investigation of an experimental gallery in clay was carried out at Pasquasia (Sicily, Italy) and complementary work done at other Italian clay sites (Monterotondo, Orte, Orciatico); in the UK, research on granitic sites (Altnabreac, Troon) was brought to completion. Finally, a coordinated intercomparison exercise of computer codes for geomechanical calculations was initiated (project COSA for salt).

Most of this Research has been reported at the 2nd European Community Conference on radioactive waste,[2] held at Luxembourg in April 1985, and is the subject of many CEC reports. Research on seabed disposal is being pursued, mainly at JRC Ispra, and will be reported elsewhere during this conference. During the same period, 1980-1985, new research on other aspects of geologic disposal was the subject of large coordinated efforts within the European Community. The most important may be the MIRAGE project, initiated in 1982. MIRAGE studies the MIgration of RAdionuclides in the GEosphere; as a pluridisciplinary project, it was built around several research themes: radionuclide chemistry, laboratory simulation of migration phenomena, hydrogeology, natural geological analogues, possible presence and roles of micro-organisms at depth, development and coupling of migration calculation tools. The state of progress of the project was presented in March 1984 and 1985 and widely reported.[3]

During the third period, 1985-1989, it is intended to carry out research on geological disposal along the same lines. Survey of suitable formations will be continued in order to increase the knowledge of their properties and most of the EC sites will be involved in this European program. MIRAGE project should continue taking stock of the lessons learned during its first phase. An increased contribution to this project is anticipated from the JRC Ispra, where two boreholes will be drilled at a depth of about 150 m in permeable sediments; they will be used as a field laboratory where methodology and technology of in situ migration investigations will be developed, and will help to validate the geochemical transport models derived from laboratory columns experiments.

III. WASTE PACKAGE AND REPOSITORY DESIGN

A. Vitrified Waste Overpack

Screening tests were started up in 1978 on a number of metal alloys which could be used as overpack material for the vitrified waste. As a result, three most promising reference materials (carbon steel, Ti-Pd and Hastelloy C-4) were selected in 1983 and are now subject to a detailed testing program, being performed in various laboratories under representative conditions for geological repositories. Conceptual designs of overpacks for a typical vitrified waste form (Cogema design), using the reference materials, have been developed. During the 1985-1989 period, the testing program of the selected materials will be completed, and model and full-sized containers will be tested and possible failure modes assessed.

B. Backfilling and Sealing

Work on backfilling and sealing of waste repositories and related openings began in 1982. The materials concerned were essentially clay-based mixtures for backfilling in plastic clay, new magnesium-based materials and concrete for hard rock, and crushed salt for backfilling in salt. In addition to this, the interface between backfill/sealing materials and the host rock was studied and some methods for monitoring and control as well as techniques for measurement were assessed.[4] During the 1985-1989 period, emphasis will be given to the engineering aspects of backfilling and sealing. Materials previously selected will be tested on scaled-down or full-size mock-ups, in galleries and boreholes, depending on the budget available.

C. Repository Design

Several repository variants have been considered during the past years for salt, clay and granite in relation to the progress of experimental research. These include deep dry drilling in salt, and the evolution of waste disposal strategies, such as mixed disposal of different waste categories. It is foreseen that researchers will update and possibly optimize the current designs.

IV. DISPOSAL SYSTEM PERFORMANCE AND SAFETY

An agreement was reached in 1981 to analyze in common, within the framework of the CEC program, the performances of confinement of clay, salt, granite and marine sediment repositories for vitrified high-level waste. The European PAGIS project was born. It will develop in three phases (see Table 1); the first one was closed in 1984 and the PAGIS I report was published at the end of 1984.[5]

The main feature of PAGIS is its intermediate character between the generic safety assessments which were made in the past and those which will be needed in the future for the licensing of specific repositories. Real sites or formations were selected as references for each option, although no site has yet been definitely chosen in any EC country as the site for an HLW repository. Other sites have also been included, as variants, in order to cover the variety of formation properties typical of the EC underground. Moreover, use was made of all data obtained in national and Community programs such as MIRAGE from laboratory and field experiments; these results and the corresponding models are applied within a common methodological framework, which constitutes

TABLE 1. The PAGIS Project

	First Phase	Second Phase	Third Phase
Period	1982-1984	1985-1986	1987- 1988
Goal	Collection of data, models and scenarios	Model and data validation	Safety assessment
	Establishment of an European methodology	Dose calculations	Guidelines for complementary R&D
		Sensitivity studies and uncertainty analysis	
Status	Published Dec. 1984	Underway	To be defined

the first European methodology in this field. In this context, an overall performance code (LISA), which has been developed by the JRC Ispra, will be presented at this Conference.

PAGIS is now in its second phase, the operational one, where dose and risk evaluations are made. Various sensitivity analyses are under way in order to show the influence of design parameters, the relevance of the single physico-chemical phenomena and of the input data as well as the impact of the formation and site characteristics on the doses to man. These analyses will result in both best estimates of the dose and risk curves for the normal evolution case and the most significant altered evolution scenarios.

V. STORAGE AND DISPOSAL SYSTEM COSTS

Since its beginning, the CEC program supported evaluations of storage and disposal cost; an evaluation performed at AERE Harwell, as part of a strategy study on HLW management and based on very preliminary data in 1978 showed that the cost of disposal (supposed to be in a granite repository) will not be detrimental to the economics of nuclear electricity production; it also stressed the economic incentive of solidifying HLW as early as reasonable for storage purposes. The variations in the cost of a repository in granite associated with practical alternatives was also studied by a French company and showed, for a given nuclear program, the large sensitivity of the repository cost to the HLW cooling time. Finally, a Dutch study was performed on a hypothetical case of interest for the Netherland's waste authorities, i.e., storing all waste categories in a salt dome.

In 1982, it was felt necessary to launch a much more comprehensive study of the HLW storage, transport, and disposal costs, taking into account representative cases of present tendencies in the European Community. The study has been carried out by four national organizations directly involved in waste management: ANDRA in France, CEN/SCK and ONDRAF/NIRAS in Belgium, and DBE in the Federal Republic of Germany with the support of two Belgian companies; the most recent technological concepts were therefore taken into consideration.

The study relates solely to the disposal of wastes from the reprocessing of irradiated fuels i.e., high-level wastes, hulls, medium-active wastes and technological alpha wastes, all conditioned. The wastes are disposed of in argillaceous, granitic or saline continental formations. Four reference scenarios are defined by the dates at which the HLW and alpha wastes are disposed of, simultaneously or not, either at 10, 30, or 50 years after discharge of spent fuel elements from the reactor. Three light-water nuclear power generating capacities of 10, 25 and 60 GWe, each in operation for 30 years, are being considered. This study, which has lasted almost 3 years, has been the subject of an anticipated presentation at the second EC Conference of radioactive waste[2] and is today near to completion. Here are some of the most interesting results:

1. At present, the geological disposal costs can be reasonably assessed. The total cost of a storage and disposal system for waste coming from a set of 25 nuclear power stations of a certain number of megawatts each, as defined in the study, will be of the order of two million ECUs. The variation of this cost with the size of the nuclear program is given in Figure 1.

2. Marked differences between the disposal costs in clay, granite and salt domes have not been ascertained.

3. Several ways of financing are possible. The formula of funds capitalized by paying a fee on the kWh leads to a fee of the order of 0.33 thousandths of ECU per kWh; this corresponds to about 1% of the price of the kWh produced.

In addition, a preliminary investigation of the seabed disposal costs, by means of penetrometers or by drilling into the marine sediments was performed by two British companies, Taylor Woodrow[6] and Ove Arup Partners, respectively. Costs seem to be of the same order of magnitude for the two options.

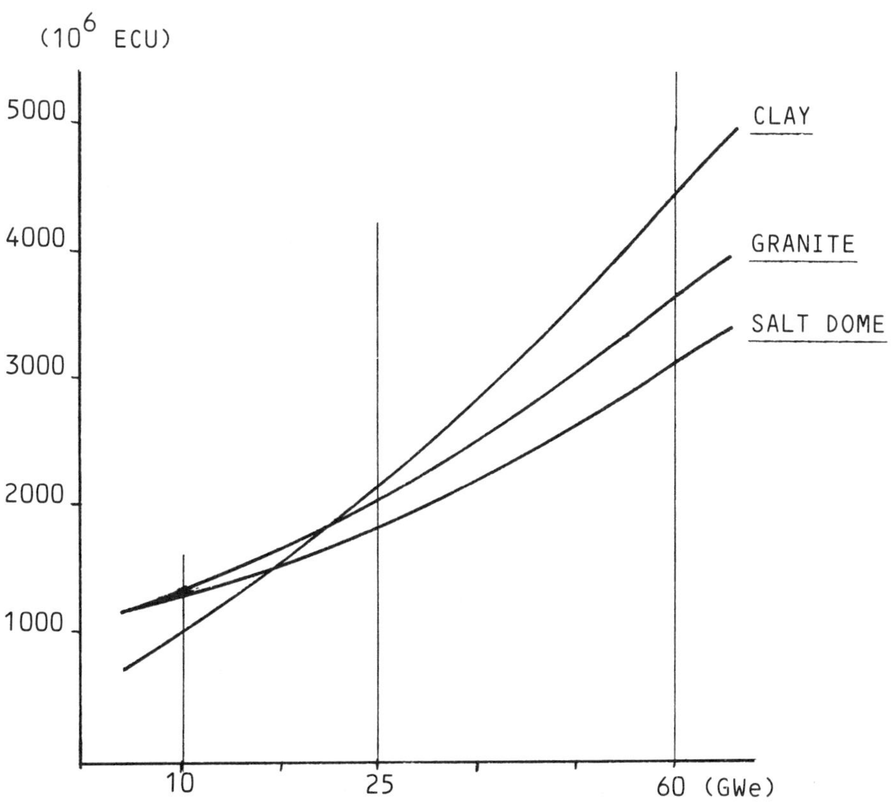

FIGURE 1. Variation of Disposal Costs with the Size of the Nuclear Program

VI. PILOT UNDERGROUND FACILITIES

On May 16th, 1983, the Commission, in a communication to the Council of ministers, invited "the member States to communicate, as from now, their possible projects concerning the study and realization of experimental or demonstration disposal installations". Belgium, the Federal Republic of Germany and France answered the invitation. Today, the CEC shared-cost action 1985-1989 includes (cf. Introduction) the following three specific projects, which are open to Community co-operation by the responsible national bodies:

1. pilot underground facility in the Asse salt mine (Federal Republic of Germany) - Project HAW,

2. pilot underground facility in the argillaceous layer located under the Mol nuclear site (Belgium) - Project HADES, and

3. experimental underground facility in France in a geological medium of complementary nature - Project ATLAS.

They deal with experimental and pilot facilities without industrial utilization. These facilities will make it possible to confirm on site the numerical values of the parameters to be taken into consideration for building industrial disposal facilities and to develop radioactive waste emplacement techniques. Radioactive waste or materials, which will be used in some projects for studying the operating conditions of an industrial facility, will be retrievable.

The co-operation includes, inter alia, participation of scientists of other Member States to the above-mentioned projects, especially by temporarily providing backup personnel, and the possibility of completing the programs with one's own specific activities, according to modalities to be specified on a case-by-case basis. The control and the responsibility of the projects will be ensured by the hosting bodies. Figure 2 shows the planning envisaged for the development of the three projects.

VII. CONCLUSION

The third radioactive waste R&D program of the CEC has already started and will continue to provide, throughout the present decade, a forum for intensive and practical cooperation between all the bodies of the European Community interested in waste disposal. The functioning of specialized working parties, which over the years have become genuine European research teams, the implementation of coordinated European projects involving many laboratories such as MIRAGE and PAGIS, and the setting-up of new ones as the need may be, are the basis of such a cooperation. The renewed interest shown by some EC member States for all disposal options, and especially clay, is making the clear focus of R&D specialization by rock types that was agreed upon ten years ago less sharp. On the other hand, it strengthens the need for cooperation

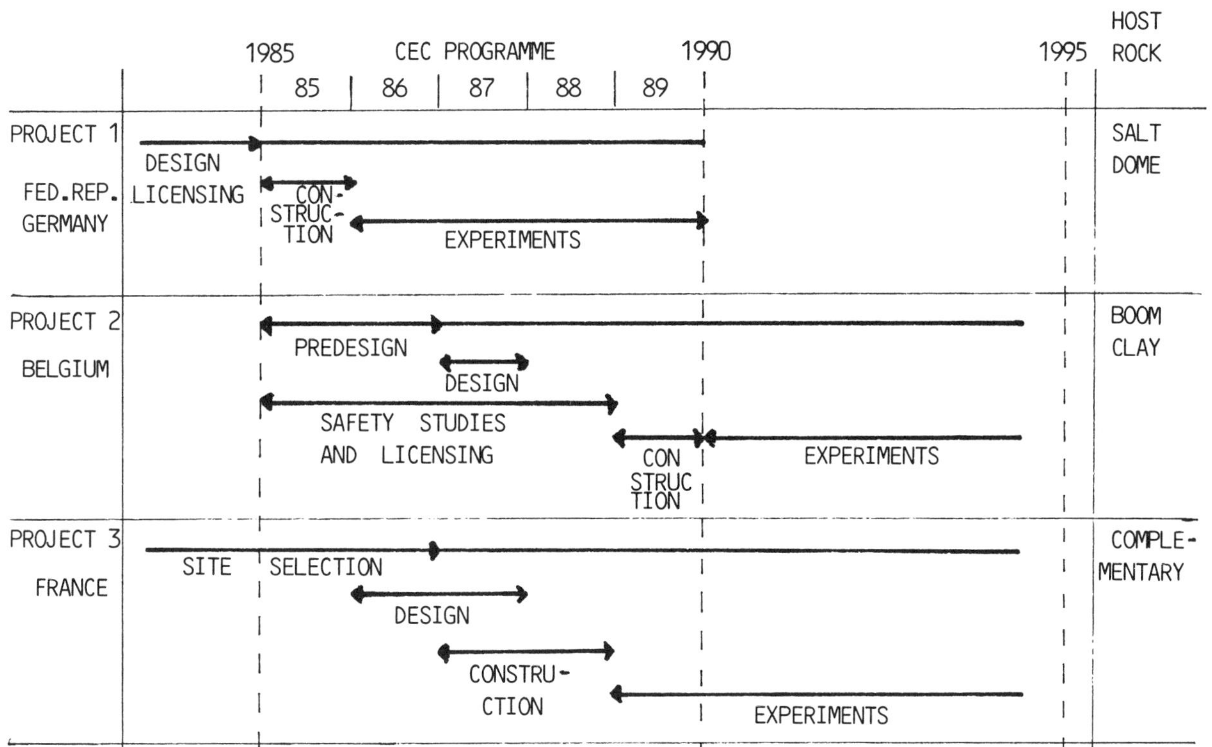

FIGURE 2. European Community Program on Radioactive Waste Management. Experimental underground facilities

at bilateral or community level between the various national laboratories; the Belgian-French agreement (CEN/SCK-ANDRA) to perform experiments in the Mol cavern is a good example.

The third CEC program is adding, however, a new and important dimension to the European Community joint effort on waste disposal, i.e., the implementation, within the framework of the CEC program, of the three projects of experimental underground facilities in Belgium, Federal Republic of Germany, and France.

REFERENCES

1. R. Simon and S. Orlowski, eds, Proceedings of the First European Community Conference on Radioactive Waste Management and Disposal, Luxembourg, May 1980, Harwood Academic Publishers, London, EUR 6871 (1980).

2. Proceedings of the Second European Community Conference on Radioactive Waste Management and Disposal, Luxembourg, April 1985, EUR 10163 (in print).

3. B. Come, ed., MIRAGE project, First Summary Report for 1983, CEC Report EUR 9143, 1985; Second Summary Report for 1984, CEC Report EUR 10023 (1985) (in press).

4. L. M. Lake et al, The Backfilling and Sealing of Radioactive Waste Repositories, CEC Report EUR 9115 (1985).

5. N. Cadelli et al, PAGIS, Performance Assessment of Geological Isolation Systems, Summary Report of Phase 1, CEC Report EUR 9220 (1984).

6. N. R. C. Bury, The Offshore Disposal of Radioactive Waste by Drilled Emplacement: a Feasibility Study, Graham and Trotman Ltd, London, EUR 9754 (1985).

STATUS OF THE NUCLEAR WASTE DISPOSAL PROGRAM IN THE FEDERAL REPUBLIC OF GERMANY

R. Ollig
Bundesministerium für Forschung
und Technologie
Heinemannstrasse 2
D-5300 Bonn 2
Federal Republic of Germany

H. Schneider
Physikalisch-Technische
Bundesanstalt
Bundesallee 100
D-3300 Braunschweig
Federal Republic of Germany

K. Kühn
Gesellschaft für Strahlen-
und Umweltforschung mbH München
Institut für Tieflagerung
Theodor-Heuss-Strasse 4
D-3300 Braunschweig
Fedederal Republic of Germany

ABSTRACT

Radioactive waste disposal in the Federal Republic of Germany is one cornerstone within the "integrated back end fuel cycle concept." The Federal Government decided in January 1985 that direct disposal of spent fuel elements without reprocessing can not serve as proof of spent fuel management from LWR's. Consequently, DWK decided on February 4, 1985, to build a reprocessing plant at the site of Wackersdorf. PTB, being responsible for construction and operation of repositories, is pursuing two projects: The abandoned iron ore mine Konrad for the disposal of non-heat-generating wastes and Gorleben for all types of waste. Related R&D programs are mainly performed by GSF in the Asse salt mine.

I. POLICY, DECISIONS AND ACHIEVEMENTS

The past years saw in 1979 a decisive turning point for the back-end fuel cycle policy in the Federal Republic of Germany, when the lower Saxony State Government declared the planned integrated back-end fuel cycle centre practically not feasible. However, the agreement between the Heads of Federal and States' Governments in the same year envisaged the possibility of tackling nuclear waste disposal nevertheless as a joint goal. This arrangement modified the "integrated back-end fuel cycle centre" to become an "integrated back-end fuel cycle concept." This concept envisages the possibility of implementing individual fuel-cycle services of the centre at different locations. In 1979, the following decisions were made:

1. Storage capacities for spent fuel elements at reactor sites should be expanded, and construction and operation of away-from-reactor-storage facilities should be undertaken as soon as possible.

2. Site selection for a reprocessing facility should be completed by 1985, and that facility should be in operation by 2000.

3. Evaluation of the Gorleben salt dome for disposal of all types of radioactive wastes should be completed by 1992, and operation of the repository should begin by 2000.

4. Assessment of the feasibility and safety-related aspects of direct disposal of spent fuel should be completed by 1985.

Since the decision of the Heads of Governments in 1979, the following actions in the field of radioactive waste management were realized:

1. Compact rack storage of irradiation fuel elements at reactors as a first step of the "integrated back-end fuel cycle-concept" has been licensed for most nuclear power stations in operation and is envisaged for all new power plants under construction.

2. Start of operation of a first external spent fuel interim storage facility near Gorleben was anticipated for early 1985 but could not yet be realized due to a court decision. Construction of another external interim storage facility at Ahaus (State of North-Rhine Westphalia) was begun but was also discontinued by court action.

3. On February 4, 1985, the German reprocessing company DWK (Deutsche Gesellschaft für Wiederaufarbeitung von Kernbrennstoffen) decided to construct a reprocessing plant with a capacity of 350 t/a at the site of Wackersdorf in the State of Bavaria.

4. The German pilot vitrification facility for high-level waste, "PAMELA" at the Mol (Belgium), went into "cold" operation in Autumn 1984 and into "hot" operation in August 1985.

5. At the Asse salt mine of the Gesellschaft für Strahlen- und Umweltforschung (GSF) a comprehensive R&D program for the disposal of radioactive wastes in salt formations is carried out with priority. Since December 1983 radioactive Co-60 sources are applied there within a joint U.S./German project. This will be terminated by the end of 1985. Another important project, namely the test disposal of vitrified high-level radioactive waste, is scheduled to start in early 1987. A bilateral U.S./German Project Agreement covering this test would be possible.

6. The exploration of the Gorleben site for the disposal of all types of radioactive wastes in the underlying salt dome is well in progress. Exploration from the surface is finished. The results of this program have led the Federal Government to allow for sinking of two shafts for exploration of the salt rock formation by mining. The necessary constructional measures were started in May 1984.

7. Since mid-1982 the licensing procedure for the abandoned iron ore mine, Konrad, is underway for the disposal of wastes with negligible thermal impact on the host rock.

8. The respective license for the Konrad repository is expected for 1987. It will also include the volume of waste which can be disposed of in that repository. In case there will be a surplus of wastes which cannot be disposed of in Konrad, the Federal Government will reconsider the option to use the Asse salt mine as an additional repository for certain categories of waste.

9. Finally, the R&D project on "Alternative Spent Fuel Management and Disposal Techniques" was implemented according to schedule.

The Federal Government, on the basis of a large-scale investigation led by the Karlsruhe Nuclear Research Center (KFK), stated in January 1985 that direct disposal of spent fuel elements without reprocessing does not have any decisive advantage with regard to safety and, for legal and technical reasons, cannot serve as proof of spent fuel management concerning light-water reactors from the present point of view. Therefore, the construction of a reprocessing plant is a necessary precondition for the further operation of German nuclear power plants.

A review of the last five years thus shows that the policies set out in the decision of the Heads of Governments in 1979 have been put into practice and that the schedules envisaged have been realized and in some cases even shortened.

II. REPOSITORY PROJECTS

A. Gorleben Site

Since 1979, the Gorleben salt dome has been investigated for its suitability as a repository for all types of radioactive wastes, including heat-generating ones. The objectives of these site investigations are to demonstrate that the site is suitable for the construction of a repository and to provide all necessary data for the site-specific planning, construction and safe operation of the repository.

Exploration from the surface is nearly completed.[1-3] This work comprises the determination of the kind and thickness of the overburden strata above the salt dome and of the hydrogeological situation in these overlying strata. In total, an area of about 300 km^2 was investigated by 129 prospective drillings, 275 well drillings, and 11 core drillings. Four deep drillings into the flanks of the salt dome and two shaft pilot drillings served to obtain a survey of the sequence of strata and the structural conditions within the salt dome. More than 40 drillings contributed to finding out the geological situation in the contact area between salt dome and overlying strata.

The latter drillings, carried out during the first half of 1985, were chiefly aimed at determining the effect of water from the overlying strata on the salt dome. According to the present state of knowledge, in a limited area, subrosion phenomena are to be investigated at a level about 100 m lower than the top of the diapir. The water-influenced zone in the upper portion of the salt dome lies beyond a protecting pillar of 150 m thickness - as defined by requirements of the Mining Authority - which separates the water-bearing strata from the mine workings. When establishing its own values, PTB increased this distance for repository-specific safety allowances to 300 m from the overlying strata and to 200 m from the edge of the salt dome.

Because the accidental inflow of water or brine cannot presently be precluded with 100% safety, PTB included such an inflow, a priori, in its safety analysis. Reasonable scenarios were devised on the basis of more than 100 years of experience from salt mining in Germany. From the safety point of view, the inflow of solution is only critical for a limited time period in the post-operational phase in repository tunnels which are sufficiently permeable even though they are backfilled.

To investigate further the hydrogeological conditions, pumping tests were carried out. They showed, among other things, that the hydraulic separation of the two main groundwater units in the overlying strata is more efficient than was assumed after completion of the drilling program. A comprehensive seismic measuring program served to determine the form and size of the salt dome and the structural situation of the overlying and neighboring strata.

The focus of the work carried out on the Gorleben site meanwhile shifted to preparatory measures for further investigations by an exploration mine. This mine will serve to obtain a complete picture of the internal structure of the salt dome by the means of galleries and of drillings originating from them. Only after all exploration results are available will it be possible to satisfactorily analyze the safety of the site and its overall geological situation.

Both planned shafts will be sunk by the deep freeze method in the area of the overlying strata. The refrigerated drillings were started in May 1984. They were drilled and cased in the meantime so that it will be possible to start shaft sinking in February or May 1986.

B. Konrad Repository

In contrast to the planned Gorleben repository, the Konrad abandoned iron ore mine is only intended to serve as a repository for waste with negligible thermal effects on the host rock formation. This corresponds, however, to about 95% by volume of the entire quantity of radioactive waste destined for disposal. Another difference between the Konrad and the Gorleben project is that the geological structure of the subsoil is already known to a large extent by existing mine information and the results of the investigations carried out by the Gesellschaft für Strahlen- und Umweltforschung GSF between 1975 and 1982.[4] The exploration of the site could therefore be limited to supplementary and more detailed work required for the licensing procedure. The respective licensing procedure, which is required for radioactive waste repositories by the German Atomic Act, is called "Planfeststellungsverfahren" (plan approval procedure). Within the scope of the exploration from the surface, seismic measurements to determine the stratification conditions in large areas are worth mentioning.

The current underground exploration data confirm that information obtained and the experience gained from the previous mining activities are also applicable to the future disposal areas. In this connection, galleries are driven to the boundaries and from these galleries, exploratory drillings are made.

When the long-term safety analysis was carried out in preparing documents for the plan approval procedure, it was necessary to drill a deep borehole to investigate the geological and hydrogeological situation of the overlying strata and of the Cornbrash zone situated underneath the ore deposit.

The activities were and are focused on the preparation of plan approval documents which can serve as a basis for the design. In the meantime it was possible to complete essential planning work. The acceptance requirements for the radioactive waste which will be disposed of in the Konrad repository have been defined. To achieve the objective of disposal, i.e., to limit the radiological effects of the repository to values specified by legislation and possibly also by the licensing authority, the proposed waste packages must be such that they meet the safety requirements for the repository in both the operational and post-operational phases. This performance is achieved by treating the waste and by appropriate packing. The requirements refer both to the waste products and the waste containers. Compliance with these requirements is checked by product control. In this connection a distinction is made between radioactive waste which has already been treated and radioactive waste which is to be treated in future.

1. For waste which has already been treated, compliance with the requirements is checked by testing the waste packages by nondestructive and, if necessary, destructive methods.

2. For waste to be treated in the future, qualifications are carried out for the different treatment methods and installations. Subsequent checks of these waste packages will then be unnecessary.

The type and scope of the checks depend on the radiological importance of the waste packages of interest.

When the decision was made to separately transport the waste packages and the rock tailings produced by mining during repository operations, the disposal concept was changed from individual drum emplacement in order

1. to reduce radiation exposure of the operating personnel,
2. to backfill the remaining openings as completely as possible,

3. to simplify and accelerate the emplacement process, and
4. to improve the utilization of the room volume.

The systematic coverage and characterization of the waste proposed for the disposal was completed.[5] The work carried out allows the various waste products to be subdivided into six groups of products:

1. solid matter,
2. metallic solid matter,
3. cemented/concrete waste,
4. compacted waste,
5. concentrates, and
6. bitumen and plastic products.

On the basis of the safety requirements for the waste packages with respect to their nuclide inventories as well as their product and container properties, two categories of waste could be defined. For the first category, only minimal packaging of the waste is needed, whereas the waste of the second category requires qualified packaging. The requirements were defined on the basis of the conditions prevailing in the planned repository and cover incidents which can occur during normal operation. If necessary, additional requirements defined on the basis of post-operational aspects must be set when the respective safety analyses have been carried out.[6-7]

Furthermore, the safety analyses showed that definite activity limits per waste package must be compiled with only for a limited number of nuclides. A limitation of activity results also from the requirements that the thermal effects of the emplaced radioactive waste on the host rock should be negligible. Here, a maximum increase in temperature of about 3° K at the sidewall is considered negligible. Criteria were derived for the resulting limitations of the activity of the waste packages. If these limitations are exceeded, these packages can only be disposed of together with packages of smaller thermal ratings.

During preparation of the plan for the licensing procedure pursuant to the Atomic Energy Act, all chapters which are not affected by the deep drilling program could be completed in time by March 31, 1985. The chapters for which the deep drilling is of importance will be completed by the end of 1985. PTB still assumes that the repository will start operation in 1989 because a large portion of the plan can be completed by the date originally provided and the licensing authority can start its work in good time.

III. RESEARCH AND DEVELOPMENT

Research and development for the disposal of radioactive wastes is naturally concentrating on the completion of the Gorleben and Konrad projects. Respective in-situ experiments are mainly performed by GSF in the Asse salt mine. An alternative concept for the disposal of low- and intermediate-level wastes which is called "in-situ solidification," is also presently tested there. Based on the agreement between the Heads of Federal and States' Governments in September 1979, a small program for studying the potential use of granite as alternative host rock formation compared to rock salt was recently started in the Federal Republic of Germany.

A. Asse Salt Mine

The Asse salt mine is well-known having served as a prototype repository for low- and intermediate-level wastes from 1967 until 1978.[8]. During this period, about 125,000 containers with LLW and about 1,300 two-hundred liter drums with ILW were successfully disposed of. Due to different judicial interpretations of the previously mentioned Fourth Amendment to the German Atomic Act and due to strong political discussions mainly between the Federal and the State Government of Lower Saxony, it was decided to stop radioactive waste emplacement into the Asse salt mine at the end of 1978. A reconsideration of this decision is scheduled by the Federal Government for 1987 if the following conditions have been met:

1. A license has been granted for the Konrad repository;
2. The capacity and the throughput of the Konrad repository are not large enough to dispose of those quantities of radioactive wastes which have accumulated and are expected to be produced; and
3. The presently ongoing site investigation at Asse has shown positive results.

Because a review of the present R&D activities was recently presented, only the most important projects shall be mentioned here.[9]

Since December 1983, radioactive sources in the form of Co-60 pins have been used for the first time in Asse to investigate simultaneously the influence of heat and radiation on rock salt. This is done in the so-called "Brine Migration Test" which is a bilateral U.S.-German project being performed jointly by ONWI and GSF.[10] The results hitherto achieved air in good agreement with calculated and predicted values so that the test can be terminated at the end of 1985 as originally scheduled. This brine migration test is the connecting link between a number of tests with electrical heaters and the planned HLW disposal test.

In this HLW disposal test, high-level radioactive glass blocks will be used for the first time in an underground laboratory. Also for this test a bilateral product agreement should be signed between BMFT and U.S./DOE. On behalf of DOE, Battelle Pacific Northwest Laboratories (PNL) will produce thirty borosilicate glass logs using only two radioactive isotopes, namely Sr-90 and Cs-137. The stainless steel canisters containing the high-level radioactive glass will be welded, leak-checked, and then transported to the Asse salt mine in Germany. One objective of this test is the checkout of a complete technical system for transportation, handling, and emplacement of HLW canisters. Other objectives are the thermomechanical behavior of rock salt under heat and radiation conditions of a repository, the production of gases by radiolysis, and the performance of borehole seals. It is also believed that public acceptance of waste disposal can be increased by performance of this test. The start is presently scheduled for early 1988.

B. In-Situ Solidification

An alternative method for the treatment and disposal of low- and intermediate-level radioactive waste was developed during recent years and is presently tested at the Asse salt mine on a technical scale using nonradioactive materials.[11]

A wide variety of low- and intermediate-level wastes from evaporator concentrates to ashes can be managed by this process. Even tritium effluents can be treated. In the first process step, the waste is mixed with a special blend of cement, water, and additives. These ingredients are fed into a mixer in which pellets or granules are formed which have a size between 0.3 and 5.0 mm. These pellets are allowed to cure. They can then be used at that location or transported elsewhere. At the site of the disposal cavity, which can be mined either conventionally or by solutioning, the pellets containing the waste are again mixed with a special blend of cement, water, and additives. This mixture is then fed through a pipe by free fall without any pumps into the cavity where a monolithic block of waste and cement is formed. After the successful technical demonstration presently underway at Asse, a management scheme for certain categories of radioactive wastes will be at hand which is cheaper and easier to handle than filling, transporting, and disposing of drums in a mined repository.

C. Granite Option

The Federal Republic of Germany, being a member state of the European Communities, is actively participating in the radioactive waste management R&D programs of the Commission of the European Communities and delivers its results to all other member states. Likewise, similar results are available to the Federal Republic of Germany from programs underway in France and in the United Kingdom. Consequently, it was decided not to start an R&D program for granite in Germany but instead to cooperate with Switzerland on a bilateral basis in their underground rock laboratory, Grimsel, in the Alps. This laboratory was constructed in 1983 by the Swiss company NAGRA. German R&D partners are BGR and GSF. The underground rock laboratory was officially dedicated on June 20, 1984. Meanwhile, the majority of tests, including geological, hydrogeological, seismic, rock mechanical, migration, and heater tests, have been installed. These tests are operating or will be operating shortly.

In addition to these R&D activities, the Federal Government asked BGR to conduct a paper study on granite formations occurring in the Federal Republic from available literature. This study will be finished shortly.

REFERENCES

1. Oesterle, F. P. "Status of Site Investigations at Gorleben," Proceedings of the ANS Annual Meeting, Detroit, June (1983).

2. Röthemeyer, H. "Die Geplanten Endlagerbergkwerke Gorleben und Konrad," 8. GRS-Fachgesträch, Köln, Nov. (1984).

3. "Zusammenfassender Zwischenbericht über bisherige Ergebnisse der Standortuntersunchungen in Gorleben," Physikalisch-Technische Bundesanstalt, Braundchweig, Mai (1983).

4. "Eignungsprüfung der Schachtanlage Konrad für die Endlagerung radioaktiver Abfälle," Abschlußbericht GSF-T 136 Gesellschaft für Strahlen- und Umweltforschung, münchen (1982).

5. Brennecke, P., Ehrlich, D., Illi, H., Viehl, E., Warnecke, E. "Radioaktive Abfälle endlagergerecht - Erfassung und Kategorisierung radioaktiver abfä lle," Energiewirtschaftliche Tagesfragen 34, Nr. 30 (1984).

6. Gründler, D., Wurtinger, W. "Sicherheitsanalyse Konrad - Bildung von Abfallklassen," 8. GRS-Fachgespräch, Köln, Nov. (1984).

7. Berg, H. P., Ehrlich, D., Illi, H., Thomauske, B. "Requirements on Radioactive Waste Derived From the Safety Analysis for the Repository in the Konrad Iron Ore Mine Under Normal and Incident Conditions During the Operational Phase," International Seminar on Radioactive Waste Products - Suitability for Final Disposal, Jülich, June 1985.

8. Kühn, K., Ollig, R. The German Approach for the Disposal of Low-Level Radioactive Waste - Part of the German 'Entsorgungs' Policy," Proceedings of Waste Management 1981, 529-546, University of Arizona, Tucson (1981).

9. Kühn, K. "Pilot Research Projects for Underground Disposal of Radioactive Wastes in the Federal Republic of Germany, Radioactive Waste Management," Vol. III, 351-369, Proc. of an International Conf., Seattle, 16-20 May 1983, Published by IAEA, Vienna (1984).

10. Asse Salt Mine Brine Migration Test, Gesellschaft für Strahlenund Umweltforschung, mbH and Office of Nuclear Waste Isolation, Battelle Columbus Laboratories, GSF-T 118, ONWI-245 (1981).

11. Kraemer, R., Kroebel, R. "In Situ Solidification of Low- and Medium-Level Wastes," Scientific Basis for Nuclear Waste Management, Proc. 5th Int. Symp. Berlin, 1982, 849-858, American Society for Materials Research, Boston (1982).

THE STATUS OF FRENCH HIGH-LEVEL NUCLEAR WASTE DISPOSAL

C. G. Sombret
Commissariat a l'Energie Atomique (CEA)
B.P. 171
30205 Bagnols Sur Ceze Cedex
Marcoule, France

ABSTRACT

French research on high-level waste processing has led to the development of industrial vitrification facilities. Borosilicate glass is still being investigated for its long-term storage properties, since it is itself a component of the containment system. The other constituents of this system, the engineered barriers, are also being actively investigated. The geological barrier is now being assessed using a methodology applicable to various types of geological formations, and final site qualification should be possible before the end of 1992.

I. INTRODUCTION

Research in France on high-level waste disposal dates from the end of the 1950s. Until recently, this work was centered around developing a conditioning method, involving only the confinement material and its manufacturing process. The major steps in this program leading to the commissioning of the Marcoule vitrification facility (which has vitrified 980 m^3 of high level liquid waste representing an activity of 170 MCi as of July 1, 1985) and to the construction of the R7 and T7 facilities at La Hague have already been described elsewhere, together with the interim surface storage facilities.[1]

Nevertheless, further research is necessary in this area to improve the material properties and to meet new requirements. At the same time, work must be done in other related areas involved in the concept of a long-term repository. These include: container handling and transport to the interim storage area and then to the geological repository, engineered barriers, modeling of radionuclide migration from the source term, and evaluation of likely disposal sites.

II. THE MATERIALS

Glass has been selected as a suitable material that combines radionuclide immobilizing capability with ease of manufacture. Sodium borosilicate glass is used, and the exact composition is adjusted according to the nature of the liquid to be solidified.[2] The formulation is determined according to the desired properties, the waste composition, the planned volume reduction, and in certain cases, the thermal power of the materials to be vitrified.

The following research areas are now being investigated: compatibility with the manufacturing process, material characterization, sensitivity to the variabilities of industrial production, and long-term behavior. This research is applicable to the previously mentioned borosilicates as well as to a new generation of high-melting-point glass formulations.

A. Manufacturing Process Compatibility

The most important aspects are the corrosiveness of the molten glass with regard to metals, the viscosity of the molten glass, and volatilization during glass fabrication. This research, which has already been described, is essential especially for its effect on storage equipment (material life and gas processing design).[2]

B. Material Characterization

Glass characterization consists in a series of operations providing a comprehensive assessment of material properties, some of which must comply with basic safety regulations stipulated by the French authorities for the safety of nuclear facilities. Other properties must be determined prior to approval by the French agency responsible for the long-term storage (ANDRA), and to provide basic data for repository design work by COGEMA clients. Glass characterization covers the following areas: physical and thermal properties, homogeneity, thermal stability, leaching resistance, high temperature volatility and the effects of irradiation and insoluble metal inclusions due to fuel dissolution. It should be noted that thermal stability and leach resistance data do not necessarily reflect the actual long-term behavior, but only the glass properties at the beginning of interim storage. This is not true, however, for the irradiation effects, which are determined by both research approaches.

C. Industrial Process Variations

Experience has shown that the characteristics of industrially produced solutions do not correspond exactly to those on which laboratory experiments are based, as determined by analysis of reprocessing plant flowsheets. Moreover, even if equipped with reliable process controllers, industrial equipment is inevitably subject to a certain setpoint variation range.

These two causes result in the following major effects:

1. variations in the percentage amounts or in the element composition of the solutions in storage,

2. composition variations in the raw materials required for vitrification,

3. fluctuations in the calcinate or process solution feed rate to the vitrification facility, and

4. temperature variations at vital points in the heating apparatus.

It is thus indispensible to investigate the acceptable variation limits for these parameters. Laboratory experiments and tests on industrial vitrification equipment are currently in progress at Marcoule.

D. Long-Term Behavior

During up to several decades of interim storage and subsequently in the geological repository, the radioactive glass will be submitted to highly specific conditions and phenomena. These include thermal effects, pressure, irradiation, and possible leaching.

Determining the effects of these factors requires very long investigation because of the complexity of certain phenomena involved in alteration of the vitreous medium and the multiple interactions between them. For example, irradiation may cause helium to be released, which may in turn modify the physical integrity of the material and therefore its mechanical properties, thus increasing the release rate because of the increased surface area exposed to leachants.

Investigations may be conducted on simulated glasses, but radioactive material should be implemented whenever possible, as is currently the case in France.

III. THE PROCESS

The French industrial reference process involves vitrification of fission product solutions in a metal furnace after rotary-tube calcining and has been implemented in the AVM facility at Marcoule since 1978.[2] The R7 and T7 facilities designed by SGN for operation by COGEMA are now under construction at La Hague (Figure 1). They will be used to vitrify the spent LWR fuel solutions reprocessed at the UP2 and UP3 plants, respectively. R7 is scheduled to begin operation early in 1987, and T7 at the end of 1988.

IV. INTERIM STORAGE

Interim storage is intended to maintain the glass at a temperature below a specified limit and to cool the glass containers to a temperature low enough to permit long-term disposal in a geological repository.

Interim storage sites are set up adjacent to the vitrification facilities (Figure 2). The containers are stacked inside vertical metal tubes suspended in an underground vault.

For LWR glass storage, each column includes a stack of nine containers made of refractory Z 15 CN 24-13 stainless steel, each holding 150 liters of

FIGURE 1. Construction Site at La Hague: The R7 Vitrification Facility

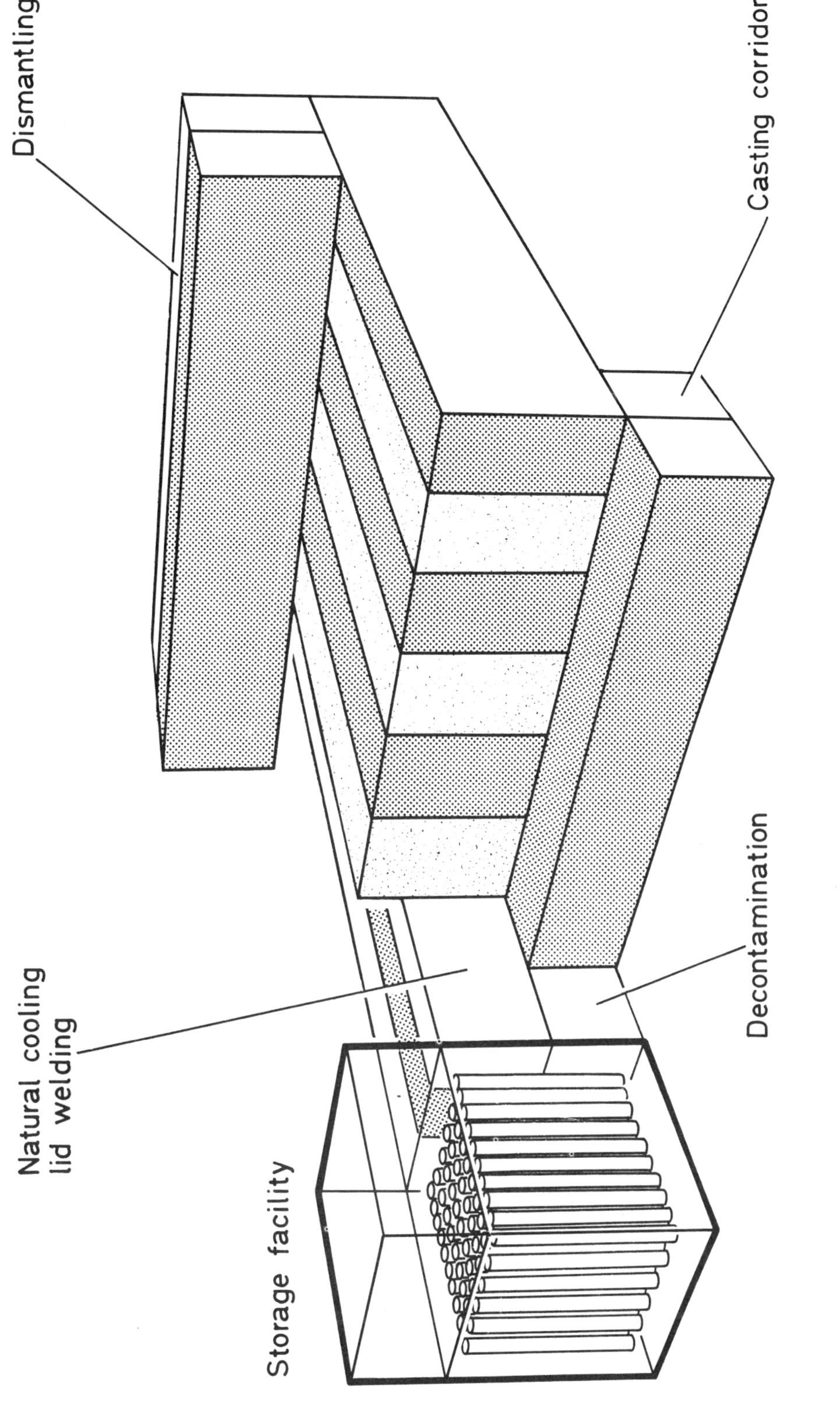

FIGURE 2. Layout of the La Hague Vitrification Plants (R_7 and T_7)

glass (Figure 3 and Table 1). On entry into the storage pit, their external contamination must not exceed 10^{-4} μCi·cm^{-2} for β,γ emitters, and 10^{-5} μCi·cm^{-2} for α emitters.

The containers are cooled by forced-air circulation, which may be followed by natural convection. The maximum specific power of the LWR glasses taken into consideration is 25 W·dm^{3}, and the glass core temperature must be held at least 100°C below the transformation point.

The storage facility is subject to a number of logical regulations covering the impact strength of the roof slab, the reliability of the handling equipment, the assurance of subcriticality even in the event of flooding, etc. The containers must also meet certain requirements. In particular, the topside distortion due to static pressure equivalent to the weight of 20 containers or to

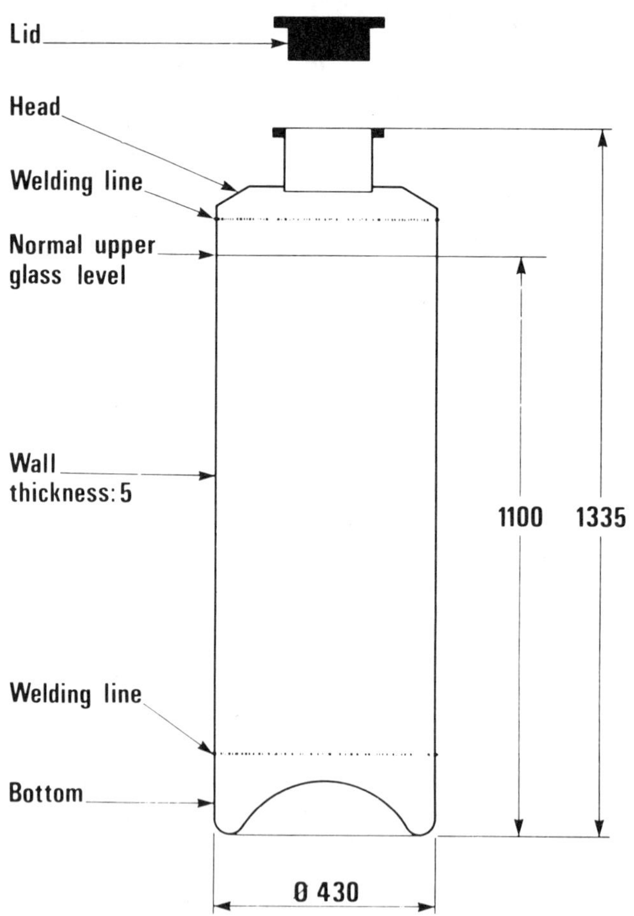

FIGURE 3. R_7/T_7 Glass Container - Vitrification of HLW in France (all dimensions in mm)

TABLE 1. Vitrified Waste Package Characteristics

Empty container weight	80 kg
Useful volume	170 liters
Overall height (incl. cover)	1334 mm
Glass volume	150 liters
Glass weight	400 kg
Number of glass castings per container	2
1 metric ton of reprocessed uranium produces approximately	120 liters of glass, (i.e. 0.8 container)

impact if dropped to the bottom of the pit must not prevent container recovery using standard equipment. The test results in this area have been satisfactory (Figure 4). The glass containers will remain in interim storage for several decades before being transfered to a permanent repository.

V. LONG-TERM DISPOSAL

Burial in deep underground formations has been adopted in France for long-term storage. No site has yet been selected, but the relevant agencies are in full agreement on the need for advancing research work to define the repository specifications.[3]

The conditions in which radionuclides could reach the biosphere from the stored waste material depend on the containment provisions and the geological barrier.

A. Containment Provisions

The containment consists of the waste package itself and the engineered barriers. Research is now in progress to assess the nature of the containment required. The waste package includes the glass and its container. Although current plans do not call for the use of the overpack, ongoing studies are evaluating the possibility of using different ceramic or metal overpack materials. Glass alteration studies are being conducted under realistic conditions allowing for underground water, temperature, pressure, backfill materials, the metal container, etc.[4]

There are two types of engineered barriers: near-field barriers, in which materials are placed between the waste package and the surrounding geological formation, and backfill barriers, consisting of the materials filling the access shaft and galleries. Both of these barriers must be sealed and ensure very low permeability. The near-field barrier must also provide adequate heat conduction and radionuclide retention properties.

FIGURE 4. Container Impact Test: Drop Test with the First Container Already in Place. (The Damper Stand may be seen at the bottom of the Figure.)

Research has been in progress for several years on these barriers. Investigation has revealed a number of potentially usable deposits of clays rich in smectites, an especially suitable swelling material. Characterization of these materials and certain illites is now being done as part of the research and development program that also covers the adaption of basic formulas, investigation of the barrier/package and barrier/environment interfaces, studies of implementation technology, mockup preparation, and physicochemical modeling.[5]

B. The Geological Barrier

Selecting a repository site consists in identifying a location in which the host rock, the barriers, and a portion of the biosphere all meet the specified technical safety options. Each site is assessed with reference to a number of guidelines designed to focus the search in order to limit the difficulties and cost involved. Most of the criteria are not absolute from a safety standpoint. A defect may be offset by other favorable qualities. France is large enough so that it contains a wide variety of formations. This means that many potential sites which would be difficult and costly to qualify may be eliminated from the outset.

The basic criteria include the following geotechnical, hydrogeological, geochemical, tectonic, geographic and socioeconomic factors:

1. mechanical properties of the host rock ensuring structural stability at depths ranging up to about 1000 meters,

2. conditions resulting in the slowest possible transfer to the biosphere (low permeability and low hydraulic gradient), implying hydrogeological behavior that can be modeled,

3. formations with high retention coefficients and sufficient thickness,

4. short-term stability (operation and surveillance) in normal and accident situations,

5. long-term stability,

6. allowance for existing or future resources (underground watersheet, possible human intrusion, etc.),

7. optimization of the waste transport cycle, and

8. cost.

The first step in the selection process was to identify potentially acceptable sites, i.e. sites meeting the above criteria. The preliminary selection included several dozen sites, classified by order of interest.

The next step, involving confirmation of the 12 most favorable sites (including both bedrock and sedimentary formations) is currently in progress. The objective of this phase is to provide all data required to characterize one or more sites, for the purpose of thoroughly evaluating them before undertaking costly field studies. The field studies are intended to confirm the assumptions on which the preliminary site selection phase was based. During this phase it is essential to check the assumptions and to confirm the absence of any latent defects. The work required depends not only on the geological formations themselves but also on the hydrogeological environment. A number of sites with major differences are being considered in this stage in order to keep a wide range of options open as long as possible.

A specific methodology has been developed to take into account individual characteristics of each site considered. Field work involving no detrimental effects and requiring no prior authorization has already been undertaken. Deep underground exploratory work is scheduled for the end of 1985. This second phase should lead to a preliminary evaluation of the potential repository site. Because of the wide variety of criteria and the different site characteristics, a simple sampling may be difficult; the advantages and drawbacks of each site will be established, however. A potential site should thus be designated in 1987.

The third phase, involving characterization and qualification of the selected site, can then be undertaken. This phase will have three primary objectives:

1. to confirm the structures identified during the preceding phases,
2. to acquire quantitative data on the site itself, and
3. to obtain field measurements on the dynamics of the geological environment.

An underground laboratory will be built for this purpose. If this field work shows that the site is unsuitable, the laboratory will be abandoned and another laboratory will be built on another site. If all goes according to schedule, the site should be qualified before the end of 1992.

REFERENCES

1. C. Sombret, "Vitrification Experience and Projects in France." Proceedings of International Meeting on Fuel Reprocessing and Waste Management, Jackson, Wyoming, August (1984).

2. C. Sombret, "The Vitrification of High Level Radioactive Wastes in France." Nuclear Energy, Journal of the British Nuclear Energy Society, Vol. 24, No. 2, April (1985).

3. A. Faussat, "La Politique Francaise de Gestion des Dechets Radioactifs." 2eme Colloque de Vincennes: Techniquees de Mise en Ouvre pour le Conditionnement, le Stockage et la Estion des Dechets Radioactifs. Vincennes (France) (1984).

4. J. L. Nogues, E. Vernaz, N. Jacquet-Francillon and S. Pasquini. "Interaction between a Nuclear Glass and Nearfield Materials in Various Rigs Simulating a Geological Repository." International Seminar on Radioactive Waste Products. Julich (FRG), June (1985).

5. R. M. Atabek, M. Jorga and R. Andre-Jehan. "Programs and Means Developed by the CEA for Clay Characterization." Engineering Geology No. 21. Elsevier Science Publishers, Amsterdam (1985).

UNDERGROUND DISPOSAL OF RADIOACTIVE WASTE--PROGRAM OF THE IAEA

K. T. Thomas
International Atomic Energy Agency
P. O. Box 100
A-1400
Vienna, Austria

I. INTRODUCTION

In the years since 1957, when the IAEA was established, radioactive waste management has grown in importance as an Agency activity keeping pace with the increasing role nuclear power has come to play in the energy economy of many countries.

The basic objective of the Agency's work in waste management is to assist member states in the safe management of radioactive wastes arising from the peaceful uses of atomic energy. This is achieved by exchange of information, the development of guidelines and international recommendations, the development and the exercising of responsibilities under international and regional conventions, the encouragement and sponsorship of research work, the provision of training, and the consideration of waste management and disposal in the frame of international cooperation.

The Agency's mechanisms for review, dissemination and exchange of information are by means of arranging conferences, symposia, seminars, technical committee and advisory group meetings, the use of recognized experts as consultants, sponsoring of research programs and training courses, study tours, etc. Support of research is carried out by organizing coordinated research programs and awarding contracts to laboratories in developing countries. The Agency awards fellowships, sponsors scientific visits and provides expert missions and field exports to assist member states in implementing their programs.

Guidelines and other technical documents in radioactive waste management are issued under the Agency's Safety Series or those issued as Technical Reports or Technical Documents. In the Safety Series Reports there are four categories, namely 1) safety standards and codes, 2) guides, 3) recommendations, and 4) procedures and data. In addition the Agency publishes the proceedings of its conferences and symposia.

The Agency is involved in giving assistance to a number of member states. This includes organizing training courses, study tours, scientific visits, awarding fellowships and providing expert missions and field experts, and also assisting in the implementation of specific projects.

II. UNDERGROUND DISPOSAL PROGRAM

In the program of the IAEA in the field of underground disposal, initially only general aspects of underground disposal were covered, which included publication of reports and holding of symposia and other meetings related to it. However, as the importance of underground disposal grew, efforts were increased for adequate programs and preparation of documents in the field. In 1977, a detailed questionnaire was sent to member states asking for information on national practices and programs in underground disposal, and requesting for indication of their interest on promoting IAEA activities in the field. The positive response from the member states resulted in the formulation of an underground disposal program based on the needs of the member states involved with this problem. Based on the interest shown by the member states and the recommendations of a special advisory group to advise the Agency on the subject, a technical review committee was established in 1978.

The Technical Review Committee on Underground Disposal of Radioactive Wastes (TRCUD) is a review and advisory group to periodically guide the Agency, and in particular to examine and make recommendations on the publications developed in this area. In making such recommendations, it looks into the need to ensure consistency between documents, quality of contents, and relevance in context of the overall objectives. National programs usually form the base from which committee members draw upon information while participating in the meetings.

The present program which was started in 1978 considers five major options of underground disposal of radioactive wastes. These options include disposal:

1. in shallow ground,
2. in rock cavities,
3. in deep geological formations,
4. by liquid injection, and
5. by hydraulic fracturing.

Major subject areas identified to be covered by a series of Technical and Safety Series Reports are:

1. general and regulatory activities,
2. investigations and selection of repository sites,
3. standards, criteria and safety assessments,
4. design and construction of repositories, and
5. operation, shutdown and surveillance of repositories.

Under these broad areas a number of sub-area topics have been identified. With experience and taking into consideration the national programs as they are developed, modifications in the subject matter and contents of the reports have been made since 1979.

In the disposal options, priority considerations have been given during the initial years on disposal of low- and intermediate-level waste in shallow ground and rock cavities due to the need to establish repositories for these wastes in many countries. Emphasis was later shifted to disposal of high-level and alpha-bearing wastes in deep geological formations.

Underground disposal is one of the important parameters in defining waste management systems for nuclear power development. A symposium on underground disposal held in 1979 reviewed the subject in great detail, including experiences of member states. The work of the underground disposal program was reported in the 1982 IAEA International Conference on Nuclear Power Experience and in the 1983 IAEA International Conference on Radioactive Waste Management. A seminar was held in 1984 in Sofia, Bulgaria on Site Investigation Techniques and Assessment Methods for Underground Disposal of Radioactive Wastes. In 1984 the first phase of the presently planned program was concluded and a review made. Table 1 gives details of the reports published and under publication.

In the future, emphasis will continue to be on disposal in deep geological formations, on development of reports on siting, design, construction and operation, and on criteria and codes of practice for such disposal. It is expected that the technical aspects of the topics to be covered under this phase would have matured to a sufficiently high level to prepare guidelines reports on deep geological repositories.

A report on "Standards and Criteria for Underground Disposal of High-Level Wastes" will be prepared. A code of practice on underground disposal with the necessary guides dealing with all the major options of disposal will also be prepared. The Secretariat is closely watching the development of the subjects in member states in these areas to make sure that the subjects are sufficiently developed for preparation of these reports. A symposium on Siting, Design and Construction of Underground Repositories for Radioactive Waste is scheduled for early 1986. Table 2 lists the reports which will be published before 1990.

III. CONCLUDING REMARKS

The program on underground disposal has been developed in an integrated manner. The setting of program goals, identification of topics to be covered, and review of the programs almost yearly, including examination of documents at different and final stages of their preparation, have been done by the Secretariat in close coordination with representatives of member states right from the beginning. Modifications have been made in the program contents as and when necessary, taking into consideration the importance and maturity of the subjects. An overview of the entire program is given in Table 3.

The whole field is still in the development stage. The disposal systems - whether in operation or under planning - being site-specific, the documents prepared on safety-oriented guidelines have to be of a relatively generic nature. However, some of the reports which primarily deal with technical aspects have to give more specific information. A few of the earlier reports are becoming obsolete and need revision.

The work on underground disposal of radioactive waste is of an interdisciplinary nature, and needs coordination with the other activities of the Agency, and the results achieved to date on underground disposal have been partly due to progress made in other programs of the Agency as well. The reports being prepared therefore take into consideration those being prepared under the Nuclear Safety Standards Program and on radiological safety and isotope hydrology. These are complementary activities within the Agency with different emphasis.

The Agency's work in the underground disposal program takes into consideration work being carried out nationally and internationally in the field. The Secretariat is closely associated with and attends the meetings of the CEC, NEA, and also meetings arranged by national authorities.

The reports published so far have been found to be useful. It is particularly important to note that deep geological disposal programs in many countries are just in planning and/or development stages, necessitating extensive multi-disciplinary technical investigations and information.

There are a number of special issues that require the attention of the Agency in particular. These include the needs of small countries, developing countries, and countries having small and large nuclear programs. This may require the attention of the Agency for cooperative activities as, for example, studies on the feasibility of setting up regional or international repositories for waste disposal. Other important issues include development of guidelines, standards and criteria for waste management operations including disposal. Multi-disciplinary efforts which are taking place at national level have to be taken note of and the Agency's efforts should stress work on finding commonalities and consensus in approaches for developing internationally accepted guidelines.

TABLE 1. Underground Disposal of Radioactive Wastes

DOCUMENTS PUBLISHED

DISPOSAL IN SHALLOW GROUND, ROCK CAVITY AND DEEP GEOLOGICAL REPOSITORIES

General Reports

Radioactive waste disposal into the ground	Safety Series No. 15 (1965)
Disposal of radioactive wastes into the ground (IAEA Symposium, Vienna, 1967)	Proceedings of Symposium (1967)
Regulatory aspects of underground disposal of radioactive wastes	IAEA-TECDOC-230 (1980)
Underground disposal of radioactive waste (IAEA-OECD/NEA Symposium, Otaniemi, Finland, 1979)	Proceedings of Symposium (1980)
Underground disposal of radioactive waste - Basic guidance	Safety Series No. 54 (1981)
Safety assessment methods for the underground disposal of radioactive waste	Safety Series No. 56 (1981)
Characteristics of radioactive waste forms conditioned for storage and disposal. Guidance for the development of waste acceptance criteria	IAEA-TECDOC-285 (1983)
Criteria for underground disposal of solid radioactive wastes	Safety Series No. 60 (1983)

Shallow ground and rock cavity disposal

Shallow ground disposal of radioactive waste: A guidebook	Safety Series No. 53 (1981)
Site investigations for repositories for solid radioactive waste in shallow ground	Technical Reports Series No. 216 (1982)
Disposal of low- and intermediate-level solid radioactive waste in rock cavities: A guidebook	Safety Series No. 59 (1983)
Site investigations, design, construction, operation, shutdown and surveillance of repositories for radioactive waste in rock cavities	Safety Series No. 62 (1984)

Underground Disposal of Radioactive Wastes

DOCUMENTS UNDER PUBLICATION

General reports

Performance assessment for underground disposal systems for radioactive wastes	Safety Series
Techniques for site investigations for underground disposal of radioactive wastes	Technical Reports Series
Management of wastes produced by radionuclide users in medicine and industry	Safety Series

Shallow ground and rock cavity disposal

Acceptance criteria for disposal of radioactive solid wastes in shallow ground and rock cavities	Safety Series
Operational experience in shallow ground disposal of radioactive waste	Technical Reports Series

Disposal of high-level radioactive waste

Deep underground disposal of radioactive waste - Near field effects	Technical Reports Series

TABLE 2. Underground Disposal of Radioactive Wastes

DOCUMENTS PLANNED FOR FUTURE PUBLICATION

General reports	
Siting, design and construction of underground repositories	Proceedings of Symposium
Standards and criteria for underground disposal of high-level wastes	Safety Series Category 1
Code of practice on underground disposal and guides to the code	Safety Series Category 1
Guidance for regulation of underground repositories for the disposal of solid radioactive waste	Safety Series
Disposal of high-level radioactive waste	
Waste acceptance criteria for solid waste in deep continental geological formations	Safety Series
In situ experiments for the disposal of radioactive waste in deep geological formations	Technical Reports Series
Borehole plugging and shaft sealing related to underground disposal of long-lived radioactive waste	Technical Reports Series
Siting, design and construction of geological repositories for high-level and alpha-bearing radioactive waste	Safety Series
Operation, shutdown and closing of deep geological repositories	Safety Series
Design, construction, operation, shutdown and surveillance of rapositories for solid radioactive waste in shallow ground	Safety Series No. 63 (1984)
Safety analysis methodology for radioactive waste repositories in shallow ground	Safety Series No. 64 (1984)

TABLE 2. (cont'd)

Disposal of high-level and transuranics-bearing waste in deep geological formations	
Site selection factors for repositories of solid high- level and alpha-bearing waste in geological formations	Technical Reports Series No. 177 (1977)
Development of regulatory procedures for the disposal of solid radioactive waste in deep, continental formations	Safety Series No. 51 (1980)
Site investigations for repositories for solid radioactive wastes in deep continental geological formations	Technical Reports Series No. 215 (1982)
Concepts and examples of safety analysis for radioactive waste repositories in deep continental geological formations	Safety Series No. 58 (1983)
Effects of heat from high-level waste on performance of deep geological repository components	IAEA-TECDOC-319 (1984)
Hydraulic fracturing	
Disposal of radioactive grouts into hydraulically fractured shale	Technical Reports Series No. 232 (1983)
Management of waste from uranium and thorium mining and milling	
Management of wastes from the mining and milling of uranium and thorium ores - A code of practice and guide to the code	Safety Series No. 44 (1976)
Current practices and options for confinement of uranium mill tailings	Technical Reports Series No. 209 (1981)
Management of wastes from uranium mining and milling (IAEA- OECD/NEA Symposium, Albuquerque, New Mexico, USA, 1982)	Proceedings of Symposium (1982)

TABLE 3. An Overview of the Entire Program

IAEA UNDERGROUND DISPOSAL PROGRAMME - OVERVIEW						
COMPONENT	OPTION		SHALLOW GROUND	ROCK CAVITIES	DEEP GEOLOGICAL REPOSITORIES	HYDRO-FRACTURING
Generic and Regulatory Activities, Standards, Criteria, Safety Assessments	General		SR No. 15 (1965)	SP (1967; 1980) SR No.Cat.I (1989)	TECDOC-285 (1983)	
	Basic Guidance		SR No. 54 (1981)			
	Regulation		TECDOC-230 (1980)			
	Performance and Safety Assessment		SR No. 56 (1981)		SR No. (1985)	
	Site Investigation Techniques		SE (1984) Tr No. (1985)			
	Standards, Criteria		SR No. 60 (1983)		SR No. Cat. I (1988)	
	For Specific Options	Guide to Disposal In	SR No. 53 (1981)	SR No. 59 (1983)		
		Safety Analysis	Sr No. 64 (1984)		SR No. 58 (1983)	
		Waste Acceptance Criteria	SR No. (1985)		SR No. (1988)	
		Regulatory Procedures, Guidance			SR No.51 (1980),SR No. (1988)	TR No. 232 (1983)
Site Selection and Investigations	SP (1986)		TR No. 216 (1982)		TR No. 177 (1977) TR NO. 215 (1982)	
Design and Construction			SR No. 63 (1984)	SR No. 62 (1984)	SR NO. (1989)	
Operation,Shutdown,Surveillance					SR NO. (1989)	
Near-Field Effects, Heat Effects In Situ Experiments, Borehole Plugging and Shaft Sealing					TR No.(1985); TECDOC No.(1985) TR No.(1987); TR No. (1988)	

LEGEND: SR Safety Series Report; SP Symposium; TR Technical Report; SE Seminar

STATUS OF THE HIGH-LEVEL NUCLEAR WASTE DISPOSAL PROGRAM IN JAPAN

Kunihiko Uematsu
Power Reactor and Nuclear Fuel Development Corporation
9-13, 1-chome, Akasaka Minato-ku
Tokyo, 107, Japan

ABSTRACT

The Japan Atomic Energy Commission (JAEC) initiated a high-level radioactive waste disposal program in 1976. Since then, the Advisory Committee on Radioactive Waste Management of JAEC has revised the program twice. The latest revision was issued in 1984. The committee recommended a four-phase program and the last phase calls for the beginning of emplacement of the high-level nuclear waste into a selected repository in the Year 2000. The first phase is already completed, and the second phase of this decade calls for the selection of a candidate disposal site and the conducting of the R&D of waste disposal in an underground research laboratory and in a hot test facility. This paper covers the current status of the high-level nuclear waste disposal program in Japan.

I. INTRODUCTION

The Advisory Committee on Radioactive Waste Management discussed the concept of disposal of high-level waste (HLW) in Japan in a 1980 report, which proposed that the waste be solidified into borosilicate glass and then be disposed in a deep geologic formation so as to minimize the influence of the waste on the human environment, with the aid of a multi-barrier system which is the combination of natural and engineered barriers. Recently the Committee revised the long-term program and developed new recommendations:

1. The Power Reactor and Nuclear Fuel Development Corporation (PNC) should lead the program in cooperation with governmental and industrial research organizations.

2. Vitrification technology should be demonstrated by the early 1990s through the construction and operation of the high-level liquid waste vitrification plant by PNC.

3. Research and development including surveys to select candidate disposal sites should be further accelerated to demonstrate the feasibility of geologic disposal in the Year 2000.

4. In order to get the public acceptance, which is ultimately needed for the siting procedure, public-information activities by the government and other organizations should be performed extensively.

In this recommendation, PNC is assigned as a leading organization to accomplish research and development on the management of HLW in Japan. Japan Atomic Energy Research Institute (JAERI) has a role to conduct research relating to safety evaluation of HLW management. The revised program is made up of four phases as follows:

1. survey on potentially suitable geologic formations for disposal (completed),

2. selection of candidate disposal site (possibly in about ten years),

3. demonstration of disposal technology, and

4. disposal of actual waste.

This four-phase program is illustrated in Figure 1.

II. RESEARCH AND SURVEY OF POTENTIAL FORMATIONS

Surveys of about thirty areas were conducted in Phase 1. The results showed that even in the same rock there are notable differences in characteristics, depending on the geologic bed situation. Geologic formations for disposal should be investigated without specifying the type of rock, except inherently inappropriate formations such as non-consolidated rock. The safety of geologic disposal can be assured by adopting a multi-barrier system, in which the engineered barrier can be designed appropriately depending on the geologic conditions.

III. RESEARCH AND DEVELOPMENT ON GEOLOGIC DISPOSAL

A. Geoscientific Study

In Phase 1, PNC has conducted two in-situ tests: the Shimokawa project for diabase and the Hosokura project for tuff breccia. PNC also participated in the international STRIPA project for granite. In these tests the rock characteristics, heat conductivity, water permeability and nuclide migration were studied to clarify the function of geological formations as natural barriers.

In Phase 2, hydrology in deep geologic formations and characteristics of rocks are to be studied. Effects from earthquake movements, drilling of rock and generation of decay heat on the retardation ability and integrity of geologic formation are also to be investigated. Existing rock caverns like the Hosokura mine are employed for these studies and for the collection and analyses of earthquake data. Basic studies on interactions of geologic media with radionuclides are also continuing.

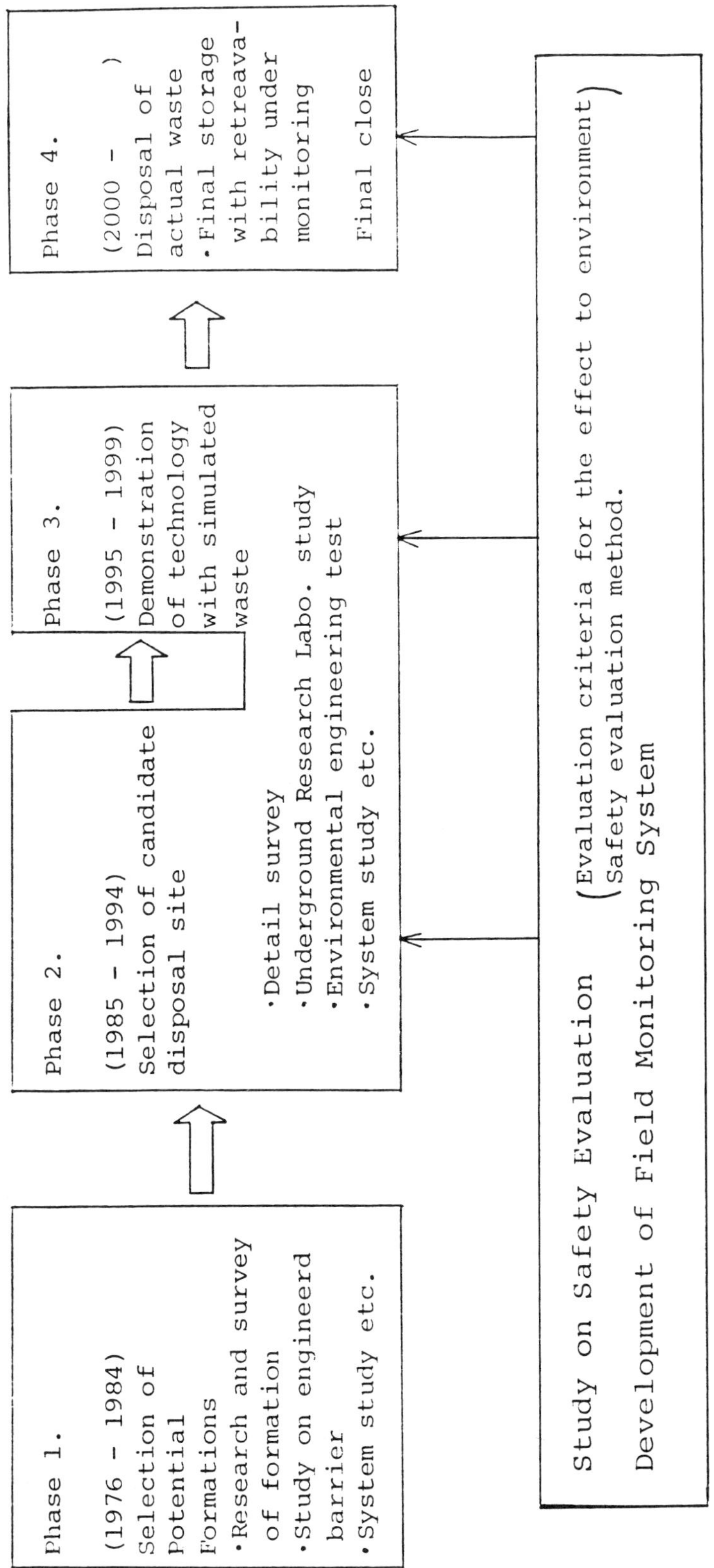

FIGURE 1. Four-Phase Program for the Development of Geologic Disposal in Japan

B. Engineered Barrier

Selection and development were conducted in Phase 1 and continue in Phase 2. The investigations focus on overpack materials, buffer materials, backfill materials and grout materials which are effective both for the protection from water intrusion into the repository and for retardation of radionuclide release after the water has intruded. In Phase 2, barrier construction technology with high reliability and stability is also developed.

The efficiency and integrity of the engineered barriers are also evaluated through the investigation of their chemical and thermal stabilities, mechanical properties and compatibility tests with the natural barrier. In addition, leaching mechanisms of radionuclides from wastes and the isolation effectiveness of both natural and engineered barriers are studied in cold and hot test facilities in situ.

C. Development of Disposal System and Performance Assessment

In Phase 1, radionuclide migration scenarios were studied on the basis of assumed migration routes through each barrier of the disposal system, and a fundamental evaluation model was developed that accounted for radionuclide retardation effects. A design study on a geologic disposal system was also made. Figure 2 shows the logic of the design study.

In Phase 2, geochemical evaluation models are to be developed to describe radionuclide behavior in the deep geological formations. The design of the disposal system will be made for the candidate disposal site. Comprehensive performance assessment models and computer codes are developed and improved. Data for the performance assessment will be provided by the field tests in underground research laboratories, hot laboratory tests on rock samples, and so on. These models and data will be used to establish disposal system safety standards.

In order to select the candidate disposal site, safety evaluations will be performed. Field monitoring technology is to be developed to measure the changes of the groundwater and rock properties, and to detect the radionuclides.

D. Underground Research Laboratory

One of the big activities in Phase 2 is the construction of an underground research laboratory (URL). Studies on both natural and engineered barriers are to be conducted in URL.

1. Study of the natural barrier--The hydrological mechanisms and characteristics of rock in deep geological formations are to be studied. The effects of earthquake movements, rock drilling, and heat generated by the waste on the retention ability and on the integrity of the natural barrier will be also determined.

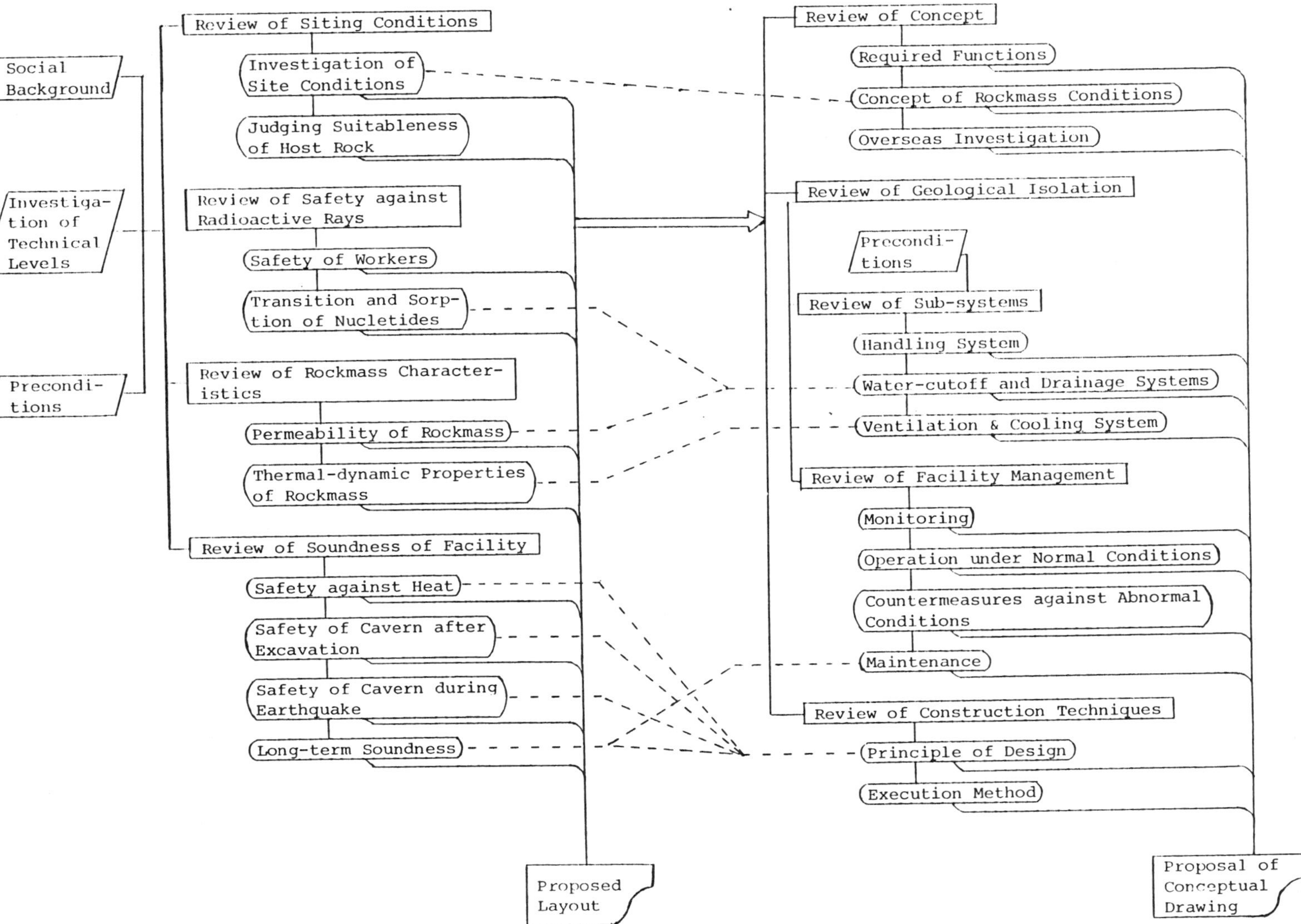

FIGURE 2. Flow of Design Study on Geologic Disposal System for High-Level Radioactive Waste

2. Study of the engineered barrier--Installation tests of engineered barriers in deep formations are to be performed. The effectiveness and integrity of installed barriers are to be evaluated.

E. Radwaste Isolation Research Facility

The radwaste isolation research facility will be constructed during Phase 2 for hot tests. The degradation, leaching and adsorption characteristics of natural and engineered barriers are evaluated under irradiation conditions. Nuclide migration tests are also carried out under the simulated repository conditions. Integrated tests will be made by using the small-sized and/or full-sized glass waste forms.

STATUS OF ACTIVITIES OF THE OECD NUCLEAR ENERGY AGENCY RELATED TO THE DISPOSAL OF HIGH-LEVEL RADIOACTIVE WASTE

S. G. Carlyle
OECD Nuclear Energy Agency
38 Boulevard Suchet
75016 Paris, France

ABSTRACT

Under the OECD/NEA Radioactive Waste Management Committee (RWMC), three main categories of work are carried out in support of the disposal of high-level radioactive wastes: first, the preparation of recommendations on appropriate radiation protection objectives for disposal, particularly in the long term; second, the development of systems performance assessment methodologies; and third, the provision of reliable field data and in depth understandings of field processes relevant to radioactive waste disposal. This paper briefly describes recent achievements of the RWMC, and possible future trends in the development and implementation of safe disposal routes for high-level radioactive wastes. Particular emphasis is given to current activities within established NEA projects such as the Stripa Project, the International Sorption Information Retrieval System (ISIRS), the User's Group for Systems Variability Analysis Codes, and work on Long Term Objectives for Radioactive Waste Disposal.

I. INTRODUCTION

At the most recent meetings of the NEA Radioactive Waste Management Committee, a review of its programme of work was carried out. As a result, it was decided that, in addition to addressing certain policy and topical issues, effort should be concentrated in the area of safety assessments - often referred to as system performance assessments - which could be used to judge the effectiveness of radioactive waste disposal, in particular the disposal of high-level waste (HLW). This really confirmed the recent trend in the NEA's activities towards the development of a more effective capability in the area of safety assessments. Nevertheless, this endorsement led the Agency to think deeply about its activities in the field of system performance assessments. In order to begin this process one must clearly understand what are the current trends and activities; what is their status and what still remains to be done. This paper gives some ideas on how the NEA views these questions, and identifies the main emphasis of the NEA's future activities in this area.

II. CURRENT TRENDS

Radioactive waste management thinking seems to have just passed a watershed. This is perhaps most clearly illustrated by the completion in 1984 of several activities which led to the publication of some notable reports.

The first activity that can be mentioned was a joint study by the NEA and the Commission of the European Communities on the current status of understanding and development of the geological disposal of radioactive wastes[1]. One of the main conclusions of this report states that, on the basis of available information, early disposal of either HLW or spent fuel is technically feasible, and that future research should concentrate on improving the understanding of some specific aspects and not on establishing the viability of any particular concept. Early engineering demonstrations of disposal are recommended in order to generate public understanding and confidence. The message from the report is clear: disposal options are available for HLW and spent fuel but further effort is needed to refine disposal systems and to provide for more realistic, that is, site specific performance assessments.

A similar report was also published giving the status of research on the disposal of high level waste into deep ocean sediments[2]. Although research is clearly less advanced than on equivalent disposal on land, early results indicate that this concept is also technically feasible. Again, refinement of the concept is still needed to give an increased understanding of the performance of successive barriers and engineering feasibility.

These general conclusions on the technical feasibility of primary radioactive waste disposal concepts coincided with the completion of a major study by an NEA group of experts on long-term radiation protection objectives radioactive waste disposal.[3] This report recommended that, in order that all events and processes can be taken into account on a rational basis, the design of objectives for the protection of the individual should be expressed in terms of risk limitation, where risk is defined as the product of the probability of exposure and the probability that the doses received will give rise to deleterious health effects. It was further suggested that the maximum risk objective for members of the public should be set at 10^{-5} per year, corresponding approximately to 1 mSv per year recommended by the ICRP for those scenarios where exposure is expected to persist for a decade or more in the lifetime of individuals in a critical group. These, together with other recommendations in the report, give further evidence that there now exists a basis for the rational evaluation of the safety of radioactive waste disposal in both the short and long term.

Early in 1983 the NEA Radioactive Waste Management committee evaluated the findings of these and other key reports in making a technical appraisal of the current situation in the field of radioactive waste management.[4] The fundamental conclusion of this "Collective Opinion" of the Committee was that detailed short- and long-term safety assessments can now be made which give confidence that radiation protection objectives can be met with currently available technology for most waste types, and at a cost which is only a small fraction of the overall cost of nuclear-generated power. In addition, it concludes that confidence and optimism prevails based on the large body of scientific and technical evidence available from past and ongoing studies. Finally, the report concludes that R&D will have to continue to fill remaining gaps for particular options in order to collect site-specific data and refine safety studies.

The inference from the RWMC Collective Opinion is that disposal of most waste types can now be carried out and that technical factors cannot be used as a reason for delaying disposal. This fact more than any other, means that a major milestone has been passed. Previously, technical feasibility governed most of the R&D; that is to say whether or not it was actually feasible to dispose of radioactive waste in a manner proposed as a result of conceptual studies and whether it was actually possible to carry out comprehensive safety assessments. Both of these are now technically feasible. This watershed being passed, a new era of how, when, and where disposal will take place has begun. This fact is evident in the widespread adoption of two activities; firstly, carrying out comprehensive assessments of the safety of proposed national solutions to waste management, and secondly, the development of site oriented research based on proposed or generic sites. The former is illustrated by three studies carried out by national organizations: the NAS Report[5] in the USA, the KBS-3 Study[6] in Sweden, and the Project Gewahr Study[7] in Switzerland. Evidence for the latter is found in the current OECD/NEA Stripa Project in Sweden, the Grimsel Test Site in Switzerland, the Mol Facility in Belgium, the Underground Research Laboratory (URL) Project in Canada, the Basalt Waste Isolation Project (BWIP) Project in USA and Gorleben in West Germany. In addition, several other nations intend to follow this approach, including Japan and France. Of course, these are not the only indicators of current trends, the Commission of the European Communities' Performance Assessment of Geological Isolation systems (PAGIS)[8] is an example of a concerted international effort to compare the performance of different geological disposal options.

III. CURRENT NEA ACTIVITIES

Largely through its Radioactive Waste Management Committee, the NEA seeks to respond to current needs in this area. On the above questions of how, when and where, the NEA is most actively involved in how waste disposal is to be implemented; the other two question marks, when and where, being largely national issues. How disposal is implemented will depend on many technical factors not the least of which will be the results of safety assessments. At the NEA three broad areas of activity are followed in order to help answer this question, i.e., i) the development of performance objectives; ii) the further development of performance assessment methodologies; and iii) the acquisition of field and laboratory data.

A. Performance Objectives

The development of a technical basis for establishing performance objectives is carried out by convening groups of experts to consider specific topics. Recent examples include developing long term radiation protection objectives for radioactive waste disposal[3] and the application of the ALARA radiation protection principle to uranium mining and milling waste.[9] An expert group is currently addressing acceptability criteria for shallow land burial of low-level waste.

The findings of the expert group on long-term objectives for radioactive waste disposal have already been mentioned. Its practical application has, however, raised some concern among the waste management community. This is well understood but it must be emphasized that, as with all guidance documents, their recommendations require interpretation in the context that they are made. The main objective in radioactive waste disposal is isolation of the radioactive materials from the biosphere for as long as it is considered necessary to protect man and the environment, which in practical terms means something of the order of 100,000 years or more for long-lived waste. The aim of geological disposal of HLW is isolation practically forever and performance assessments are used to check the degree of reliability of this isolation. For this purpose we use pessimistic scenarios and assumptions which usually tell us that in the worst possible cases, some 10,000 years or 100,000 years from now or even later, a few people may receive radiation exposures at levels of up to perhaps 1 mSv per year, which, as mentioned earlier, corresponds to present limits on persistent exposure to individual members of the public. It is this perspective which should be present in the minds of experts and authorities discussing the acceptability of waste disposal systems. It follows that a sense of perspective and an awareness of the probabilities of potentially disruptive scenarios is essential for the assessment of waste disposal systems. An illustration of this is given in Figure 1 which the Swiss experts have published recently. It compares the range of results of several safety assessments carried out using varying degrees of conservatism. Obviously the relevance of a very precise radiological criterion for high-level waste disposal seems somewhat academic, notably because most estimates are already well below any reasonable criterion.

B. System Performance Assessment

The high degree of sophistication reached in systems performance assessments, compared for example to performance assessment for the disposal of equivalent chemical wastes, has led to new challenges for those developing currently preferred methodologies. Recently, there has been a move away from exclusive reliance on deterministic calculations to the balanced use of deterministic and probabilistic models. This is in response to the need to take into account the many uncertainties in carrying out performance assessments covering the long term, due mainly to the intrinsic variability of natural processes. It is largely for this reason that the RWMC recently established a Users' Group for Systems Variability Analysis Codes. The objectives of the Users' Group is to provide an international forum for workers actively developing probabilistic codes to:

1. exchange codes, information and experience,
2. provide for mutual peer review,
3. discuss specific technical issues such as code quality assurance verification, validation, etc., and
4. identify aspects of code development and plan intercomparison of mutual benefit.

Nagra – Project Gewähr 1985

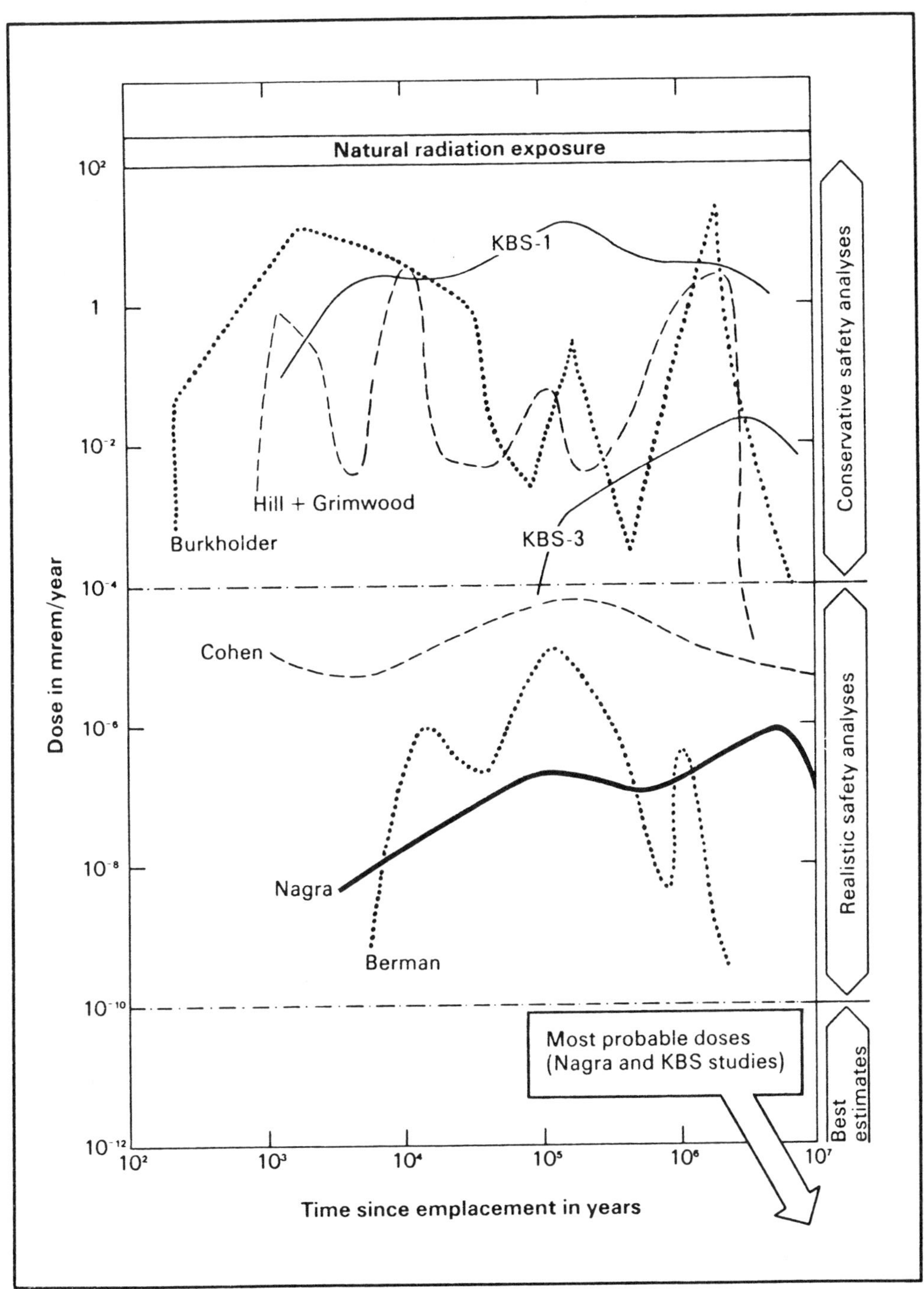

FIGURE 1. Annual Doses Predicted by a Number of Major Safety Analyses of HLW Disposal[7]

Specific technical issues are considered by ad hoc meetings of specialists drawn from Belgium, Canada, the CEC, Sweden, Switzerland, the United Kingdom, and the United States, to exchange experience, views and ideas on a topical subject designated by the main User's Group. Such meetings have already considered input data acquisition and handling and methods and procedures for sensitivity analysis and will in turn consider about a dozen other topics in the future.

Other activities of NEA involvement in this area concern the further development of deterministic codes with considerable effort being devoted recently to the comparison, verification, and validation of hydrological and geochemical codes. The NEA is currently taking an increasingly active role in the Swedish-initiated HYDROCOIN exercise.[10] The purpose of the study is to obtain improved knowledge of the influence of various strategies for groundwater flow modelling for the safety assessment of final repositories for nuclear waste. To this end calculations are made with different mathematical models used by a number of organizations. The study is intended to address:

1. the impact on the groundwater flow calculations of different solution algorithms,
2. the capabilities of different models to describe field measurements, and
3. the impact on the groundwater flow calculations of incorporating various physical phenomena.

This exercise illustrates the current quest for reliable tools for systems performance assessments and the increasing trend for scientists to share their experiences at an international level. The Agency is particularly well placed to respond to such developments because of the facilities provided by the NEA Data Bank which has an established international reputation in the exchange of scientific data and computer programs needed for nuclear energy applications. In the past, this has largely involved servicing the nuclear industry with codes and data for calculating the design and safety of plant and installations; now the service is being extended to embrace activities associated with carrying out performance assessments for radioactive waste disposal. Already, a number of codes have been loaded and tested on the NEA Data Bank computers and are available for distribution.

In addition, the International Sorption Information Retrieval System (ISIRS) has been established at the NEA Data Bank.[11] It is a data base system for the storage and handling of information related to the sorption of radioelements from solution onto geologic media. The objective of the project is to advance the understanding and prediction of the migration of radionuclides through geologic media in support of safety assessments of radioactive waste disposal. The system was initially created in July 1981 for an experimental period of two years. During this period, the data base management software was developed by Pacific Northwest Laboratories (PNL) in the United States. In 1983, the eleven participating countries agreed that the data base and computer software be transferred to the NEA Data Bank at Saclay, France, and that the Technical Committee of ISIRS should evaluate the performance of the system over the next two years, on the understanding that ISIRS was still in an experimental phase. No financial contribution was required for this period. The evaluation phase expired in July 1985 and ISIRS will now be operated as a service to interested countries within the framework of the NEA Data Bank. The NEA Secretariat and the NEA Data Bank are responsible for the administrative and technical support of the project.

A similar data base of chemical thermodynamic data is currently being developed to provide a comprehensive, internally consistent, and internationally recognized data base to meet the requirements of modelers developing advanced codes to predict radionuclide migration.[12] Currently, ten elements are being treated, i.e., iodine, lead, technetium, cesium, strontium, radium, uranium, neptunium, plutonium and americium. Data for ΔG_f^o, ΔH_f^o, S^o, C_p, and their temperature functions and uncertainties are being compiled and reviewed. Such data are necessary to establish constants for speciation, solubility and some leaching and retardation calculations. Emphasis is being placed on data for 298.15K, 10^5Pa, and zero ionic strength for solutes. In order to evaluate the data, teams of world experts are performing a seven-step critical review to develop a selected set of internally consistent data which is CODATA compatible.

C. Laboratory and Field Observations

The provision of data for use in performance assessments has already been mentioned as an activity involving the NEA Data Bank. However, one of the most difficult areas is the acquisition of reliable information on the processes which affect the performance of a waste repository. This is largely due to the very long time scales over which some of these processes operate, and which are currently considered necessary for inclusion in comprehensive performance assessments. It is clearly impossible to predict certain phenomena without having a large degree of uncertainty; examples include canister corrosion, hydrological flow regimes, and biosphere pathways. However, it is possible to gain an insight into likely future behaviour by examining current processes, using expert judgement, and the use of sensitivity and uncertainty analyses to guide the use and handling of data. Also probabilistic performance assessment methodologies can give a range of performance indicators, though these are subject to certain caveats. From the data acquisition standpoint, laboratory and field data are gathered and expert judgement used to place the data in context for use in various mathematical models. Primary in this search for data is the recent trend towards establishing field research centres, sometimes on proposed sites for repositories or at convenient sites which closely mirror the generic properties of a conceptual repository. The NEA is again actively involved in coordinating such research through the Coordinating Group on Geological Disposal. It was this group that was instrumental in producing the recent status report on Geological Disposal already mentioned.[1]

In addition, in 1980, the NEA International Stripa Project was established to carry out geochemical and hydrogeological studies, migration experiments, and studies of the function of engineered barriers for the disposal of heat-generating radioactive wastes in granite.[13] The first two phases of this project (1980-86) have concentrated on the development of methods and techniques for the investigation of the hydrogeochemistry of the Stripa granite, and in examining backfilling and sealing materials under simulated repository conditions. A Phase III is currently being considered, which would begin in mid-1986 to:

1. integrate different investigation tools and methods in order to validate predicted groundwater flow and radionuclide migration in a specific volume of Stripa granite and

2. investigate and verify the use of different materials and techniques for sealing groundwater flow paths in Stripa granite.

If Phase 3 is approved, the project will gradually progress from the development of techniques to investigate the characteristics of a granite repository site to the proving of these techniques by examining an undisturbed area of granite. It is important to note that a predictive model will be developed for the site and an attempt will be made to validate the model. It is only by such an integrated approach that public confidence will be generated. A similar approach is taken in most of the other underground research laboratories.

IV. FUTURE TRENDS

What are the likely future trends in radioactive waste management as seen from the NEA standpoint? It is clear that a firm scientific and technical basis for the implementation of safe radioactive waste management activities already exists in most OECD Member countries. This generates confidence, as illustrated by the RWMC Collective Opinion, that technical reasons cannot now be used to delay disposal. This is reflected in the firm plans by several countries for the investigation of sites and the implementation of disposal.

The results from safety assessments do indicate that proposed disposal solutions for most waste types are likely to be safe. However, more work is required if these broad indications are to be confirmed so that both the public and scientists have confidence in the results from safety assessments. But what remains to be done? This year the NEA has spent a great deal of effort in examining all the elements involved in carrying out comprehensive systems performance assessments for radioactive waste disposal. This has indicated that all too often specific activities are carried out in isolation from closely related topics. For example, those workers carrying out site-specific measurements may produce data to use in detailed research models without adequately considering the needs of those carrying out integrated assessments using simplified models. Similarly, those developing models may do so without being fully aware of the limitations on the acquisition of data. It is for this reason that, upon deciding to concentrate a considerable amount of effort in the area of safety assessments, a group of consultants was brought together by the NEA Secretariat to consider current needs in system performance assessments and prepare advice for the future orientation of activities in this field

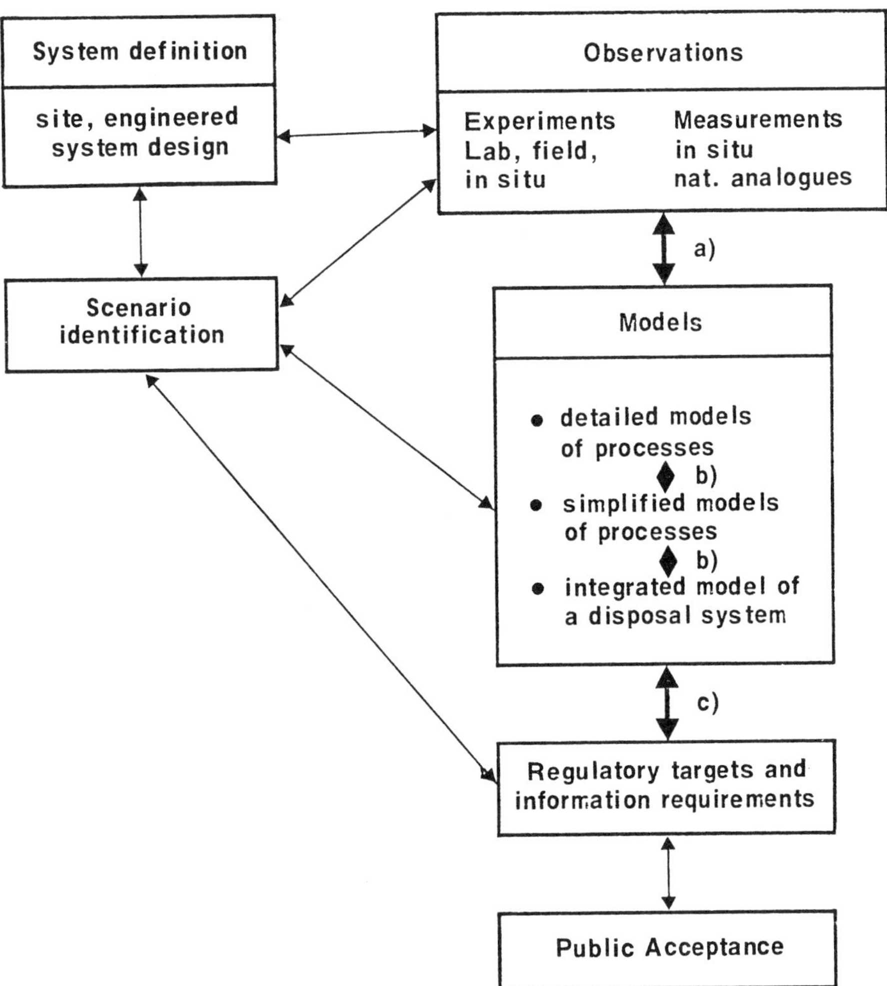

a) Link between the development of models and observations (validation)

b) Link between detailed models and simple models, and link between deparate models and an integrated system model

c) Link between the output of performance assessment and regulatory requirements.

FIGURE 2. Linkages in System Performance Assessments

with a view towards holding a workshop on performance assessments in October this year. The group considered that the increasing number of methodologies, models, and criteria has led to some concern among OECD member countries on how comprehensive assessments should be carried out. It was suggested that there exists one particular problem that requires resolution, i.e., how to rationalize all the various elements that combine together in system performance assessments. This rationalization is necessary in order to generate confidence that such assessments are realistic, reliable, and can provide the correct type of information on which the safety of disposal can be judged. The group considered that work being carried out on specific elements was progressing well but that the first step in rationalizing performance assessments was to examine the main links between these elements. Figure 2 outlines a simplified description of the main links that the group considered worthy of the most attention. Three broad links are emphasized:

1. the link between the development of models and observations,

2. the link between detailed models and simple models and between separate models and an integrated systems model, and

3. the link between the output of performance assessments and regulatory requirements.

Other links were discounted as being less important or, as in the links with scenario identification, because they are currently being debated in other fora. From the NEA viewpoint, the critical examination of these links may give an indication of those aspects of performance assessments that required most attention and where there is scope for further international collaboration.

With respect to the further development of long-term performance objectives and disposal criteria, it is probable that the trend will be towards broader perspectives in deciding the merits of disposal by one particular method over another or in optimizing disposal strategies. The consideration of a range of factors, not exclusively radiological protection, has already been advocated in a number of recent papers on this topic. One such study[14] inter alia advocated that policy makers should spell out broad criteria for the acceptability of waste disposal and that detailed criteria should be drawn-up by international consensus. Other authors make a plea for integrated approaches to the limitation of radiation risk from nuclear power.[15] A multi-attribute approach is recommended in order to systemize the decisionmaking process of those people who make decisions. Such analyses would be much more broad ranging than currently included in cost-benefit analysis. Recommendations such as these, when combined with those of the NEA Long-Term Objectives document, show that although we are confident that disposal in a properly located and engineered manner will be safe, a number of iterations of full-scale assessments such as the KBS-3 study and the Gewähr Project need to be carried through before there can be a realistic chance of producing widely acceptable criteria.

Another probable future trend is that there will continue to be emphasis placed on the search for more realistic data by establishing underground or surface experimental facilities to develop site investigation techniques and methodologies, acquire field data, and generally gain a detailed understanding of the likely behaviour of potential waste disposal sites. Finally, in the quest for public confidence, the use of natural analogues - an often misunderstood term - will be explored further as one of the few methods of validating parts of codes developed to predict the long-term behaviour of radionuclides in the environment.

Each of these ongoing activities when combined together will help the steady progress towards the establishment of actual disposal facilities. Already a number of low-level waste disposal sites are available, more will follow. Also, mined repository sites, such as Forsmark in Sweden, are under development and more will follow. In the long term, high-level waste or spent fuel repositories will be developed. However, to do so will take much effort to build-up public confidence. There is therefore a considerable need for further projects like the Canadian URL facility and also for a comprehensive and easily understood decision-making process, based on many factors, not the least of which will be the results of system performance assessments.

V. CONCLUDING REMARKS

The new era of radioactive waste management thinking described in this paper offers fresh challenges to those working in this field. The OECD Nuclear Energy Agency, through the expert guidance of its Radioactive Waste Management Committee, will continue to endeavour to address the most pressing issues, particularly in the area of safety assessments. Integral with this is the further development of models and codes to predict the performance of proposed disposal sites and the concurrent search for a detailed understanding of in situ behaviour and the acquisition of reliable data. Discussions are currently in hand to find the best way for NEA to help this process so that the current impetus towards the full implementation of disposal can be maintained.

ACKNOWLEDGMENTS

The author wishes to thank J. P. Olivier, O. Ilari, K. Bragg and A. B. Muller for their encouragement and helpful comments during the preparation of this paper.

REFERENCES

1. CEC and OECD/NEA, Geological Disposal of Radioactive Wastes: An Overview of the Current Status of Understanding and Development, Paris (1984).

2. OECD/NEA, "Seabed Disposal of High Level Radioactive Waste: A Status Report on the NEA Coordinated Investigation Programme," Paris (1984).

3. OECD/NEA, "Long Term Radiation Protection Objectives for Radioactive Waste Disposal," Paris (1984)

4. OECD/NEA, "Technical Appraisal of the Current Situation in the Field of Radioactive Waste Management - A Collective Opinion by the Radioactive Waste Management Committee," Paris (1985)

5. "A Study of the Isolation System for Geological Disposal of Radioactive Wastes," National Academy Press, Washington D.C. (1983)

6. "Svensk Karnbransleforsorjning Final Storage of Spent Nuclear Fuel KBS -3," 5 volumes, Stockholm (1983)

7. Nagra, "Question des dechets nucleaires en Suisse - Concept et apercu du Project Garantie," Baden (1985)

8. CEC, "PAGIS, Performance Assessment of Geological Isolation Systems Summary Report of Phase 1: A Common Methodological Approach Based on European Data and Models," Luxemburg (1984)

9. OECD/NEA, "Long-Term Radiological Aspects of Management of Wastes From Uranium Mining and Milling," Paris (1984)

10. "HYDROCOIN - An International Project For Studying Groundwater Hydrology Modelling Strategies," Progress Report No. 1, Sweden (1985)

11. OECD/NEA, "International Sorption Information Retrieval System (ISIRS)", "NEA Newsletter on Radionuclide Migration in the Geosphere," No. 10, Paris (1984)

12. A. B. Muller, "International Chemical Thermodynamic Data Base For Nuclear Applications - Radioactive Waste Management and the Nuclear Fuel Cycle," 6(2), pp. 131-141, June (1985)

13. OECD/NEA, "Geological Disposal of Radioactive Waste - In Situ Experiments in Granite," Paris (1983)

14. G. Bengtsson, "Judging the Long-Term Acceptability of Radioactive Waste Disposal Practices From the Radiation Protection Point of View and Other Perspectives," Proceedings of a Seminar on Interface Questions in Nuclear Health and Safety, OECD/NEA, Paris (1985) (in press)

15. G. A. H. Webb and H. J. Dunster, "An Integrated Approach to the Limitation of Radiation Risk for Nuclear Power," Proceedings of a Seminar on Interface Questions in Nuclear Health and Safety, OECD/NEA, Paris (1985) (in press).

CURRENT STATUS OF THE SWEDISH WASTE DISPOSAL PROGRAM

Per-Eric Ahlstrom
Swedish Nuclear Fuel and Waste Management Co (SKB)
Box 5864
S-10248 Stockholm, Sweden

ABSTRACT

Direct disposal without reprocessing is the main strategy for management of spent nuclear fuel in Sweden. Under the present circumstances this is the most economic route. Before final disposal the fuel will be stored for about 40 years in CLAB, a specially built underground facility which was put in operation in July, 1985. This will give flexibility in the choice of final disposal method. Wastes from reactor operation will be disposed of in SFR, another underground facility in hard rock with planned operation starting in 1988. The total cost for waste management in Sweden is estimated at 47 billion SEK ($5.5 billion US) or about 0.02 SEK/kWh (2.4 mills/kWh) including all types of waste and also the decommissioning of all twelve nuclear power facilities.

I. BACKGROUND

The nuclear power program in Sweden consists of 12 nuclear units. Their combined capacity is 9500 MWe. The last two units Forsmark 3 and Oskarshamn 3 reached full power for the first time in May 1985 and are now in their final startup and testing phase.

The Swedish legislation explicitly puts the primary responsibility for the management of radioactive wastes from nuclear facilities on the owners of these facilities. This means that the nuclear power utilities in Sweden are responsible for all necessary steps in the handling and final disposal of all radioactive wastes arising from the nuclear power program. The utilities have assigned the duty to perform these steps to their jointly owned Swedish Nuclear Fuel and Waste Management Co (SKB). The responsibilities of SKB in the nuclear waste management field thus include:

1. all necessary research and development work,

2. planning and cost calculations for the total nuclear waste management system (except handling and treatment at the reactor sites),

3. design, construction, and operation of all necessary facilities for storage and disposal of nuclear wastes, and

4. all transportation and handling of spent nuclear fuel outside the reactor sites.

There are mainly three national authorities in Sweden which deal with matters related to radioactive wastes from nuclear power plants:

1. The Swedish Nuclear Power Inspectorate (SKI) supervises and controls the safety of design, construction, and operation of nuclear facilities;

2. The National Institute of Radiation Protection (SSI) supervises plant owners to assure that appropriate measures for radiation protection are taken by the plant owner.

3. The National Board for Spent Nuclear Fuel (SKN) supervises the planning, research, and development for the waste management program. SKN also administers the funding for radioactive waste management, which is built up by a fee on nuclear power production.

II. WASTE MANAGEMENT STRATEGY

The planning of the Swedish waste management system is based on the operation of the 12 reactors up to the Year 2010. The main features of the planned system are shown in Figure 1, which also gives some data on energy and waste quantities produced to 2010.

The main strategy for handling of the spent fuel is direct disposal without reprocessing. Under the present circumstances in Sweden, this is the most economic way of handling the spent fuel. It is also at present the politically preferred option.

About 690 tonne of spent fuel have however been contracted for reprocessing at Sellafield in the United Kingdom and La Hague in France. An additional 178 tonne contracted for at La Hague were recently transferred from SKB to a Japanese company. SKB is also actively trying to transfer most of its remaining reprocessing contracts in order to limit the number of different wastes types as much as possible.

The spent fuel will be stored for some 40 years before final disposal according to present plans. This will allow the fuel residual power to decrease considerably and make the final disposal easier. It will also provide great flexibility to adjust to future developments in the area of spent fuel management. Further it will provide ample time for the research and development for the repository site selection and for the system's design and optimization.

After the interim storage period the spent fuel will be encapsulated in corrosion-resistant canisters and shipped to a final repository. The interim storage and the encapsulation facility will give rise to some additional quantities of waste which must be considered in the planning of a complete system. The total amount of conditioned wastes from the Swedish nuclear power program according to present plans will be about 245,000 m^3.

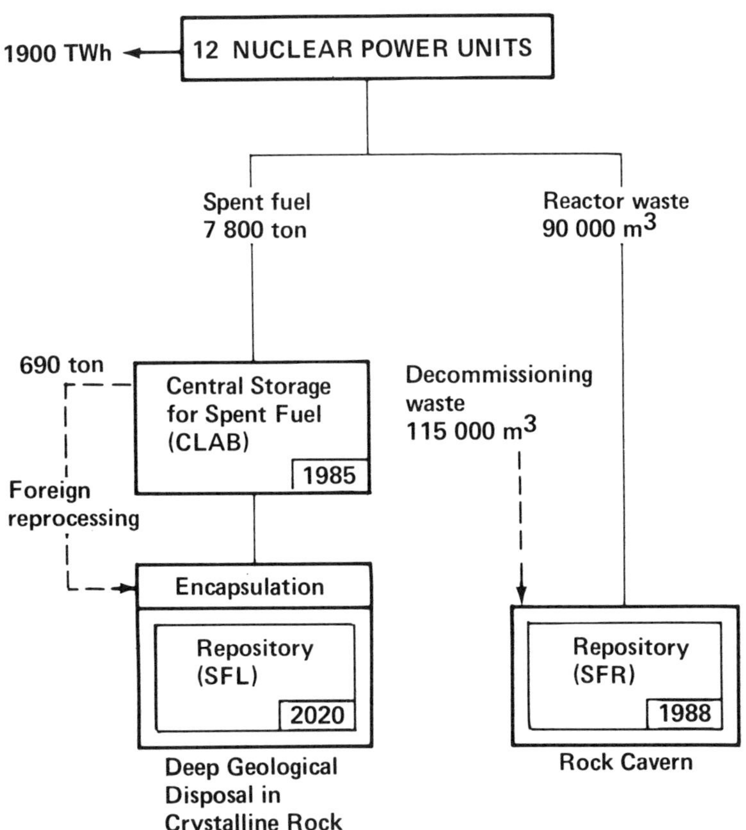

FIGURE 1. Main System for Management of Radioactive Waste in Sweden

III. THE INTERIM SPENT FUEL STORAGE FACILITY - CLAB

As mentioned above, interim storage of spent fuel plays a strategic role in the waste management system. Construction of CLAB - the central interim storage for spent fuel started in the spring of 1980 and the first spent fuel was shipped to CLAB in July 1985. The facility is located at the site of the Oskarshamn nuclear power plant on the Swedish east coast.

The fuel is stored under water in stainless steel-lined concrete pools in a large underground rock cavern with at least 25 m rock cover. There are 4 pools, with a total capacity of 3000 tonne. Additional storage pools can be added when needed up to a total of 9000 tonne. The fuel reception and unloading facilities are in a building above ground. The total receiving capacity is about 300 tonne per year or about 100 spent fuel shipping casks.

With the present plans CLAB would need to be expanded around 1995. It is also foreseen that used core components would be stored in a separate pool at CLAB.

IV. TRANSPORTATION OF SPENT FUEL AND RADIOACTIVE WASTE

All nuclear power plants in Sweden are located at the coast and the sea is used as the ultimate heat sink. The interim fuel storage, CLAB, and the final repository for LLW and MLW from reactor operation, SFR, are also on the coast. It has thus been decided to use sea transportation as the main method to ship spent fuel and radioactive waste. This is also quite favorable considering the large transport weights that must be handled with this type of cargo.

A specially built and equipped ship, M/S Sigyn, was put into operation in 1982. The ship can take 10 fuel casks of the type TN17 Mk2 each with a weight of about 80 tonne. Each cask can take 17 BWR or 7 PWR fuel assemblies. The fuel is transported dry and cooled by natural air convection. The casks are mounted on special transport frames which are handled by a specially built terminal vehicle.

The same equipment can also be used to transport MLW and LLW in containers of steel or concrete. Each container has a transport weight of up to about 120 tonne.

V. FINAL REPOSITORY FOR REACTOR OPERATIONAL WASTE: SFR

The first repository to be constructed in Sweden is a repository for LLW and MLW from the operation of the nuclear power plants. The repository -SFR- will be located at the Forsmark site 160 km north of Stockholm on the Swedish east coast. Construction started in 1983 and deposition of waste will start in 1988. The repository is built as rock caverns 50 m deep into the rock below the bottom of the sea about 1 km off the coast. The water depth at the site is about 5 km. The total capacity is planned for 90,000 m^3 of waste. SFR will be built in two phases - the first is under construction now and the second is planned for the end of the 1990s.

As the SFR is constructed in hard granitic rock it will give valuable experience for any future repository built in similar rock. A detailed hydrogeologic investigation program is carried through the whole excavation and construction phase. This program will give valuable field data for assessing mathematical models for groundwater movements.

VI. FINAL REPOSITORY FOR HIGH-LEVEL WASTE

The feasibility of final disposal of spent nuclear fuel and high-level waste have been demonstrated by the work reported in the KBS-1 and KBS-3 reports, which have been accepted by the Swedish government. These reports showed how long-lived wastes <u>can</u> be disposed of by present-day technology and within the geological conditions existing in Sweden.

However, considerable work remains to be done to show how these measures are to be realized in detail and in an optimum way.

The future work for realization and optimization of a safe system for final disposal of nuclear waste will comprise the following:

1. continued research and development work in order to further deepen the scientific knowledge, that constitutes the base for the performance and safety assessment,

2. studies and evaluation of alternatives to the methods and concepts investigated so far,

3. optimization of systems with respect to technology, economy, and use of resources against the improved scientific knowledge, and

4. investigations for site-selection.

A basis for the planning of the R&D work is the overall timetable for realization of a final repository for spent fuel, see Figure 2.

According to the "Act on Nuclear Activities" the R&D program shall be updated every three years and sent to the authorities for review. The first program subject to such review must be submitted in September 1986.

VII. RESEARCH AND DEVELOPMENT FOR SPENT FUEL AND HLW REPOSITORY

Important to the ongoing research program for a high-level waste repository in Sweden are the site selection studies. These are described in more detail in another paper to this meeting by Hans Carlsson.[1]

Site investigations started in Sweden in 1977. The investigations so far follow a standard program which is continuously updated as instruments and methods are further improved and developed. Eight sites have been fully investigated by this program. An additional three to four sites will be studied by 1990. Two or three sites will then be selected for more detailed studies during the 1990s. The final selection of a site for the repository will be made at the end of the century.

Besides the site selection studies the geoscientific R&D is directed towards three major areas:

1. comprehensive studies of fracture zones,
2. studies of bedrock stability, and
3. studies and modelling of groundwater movement and nuclide migration.

All these areas are of great importance for the final site selection and for the safety assessment.

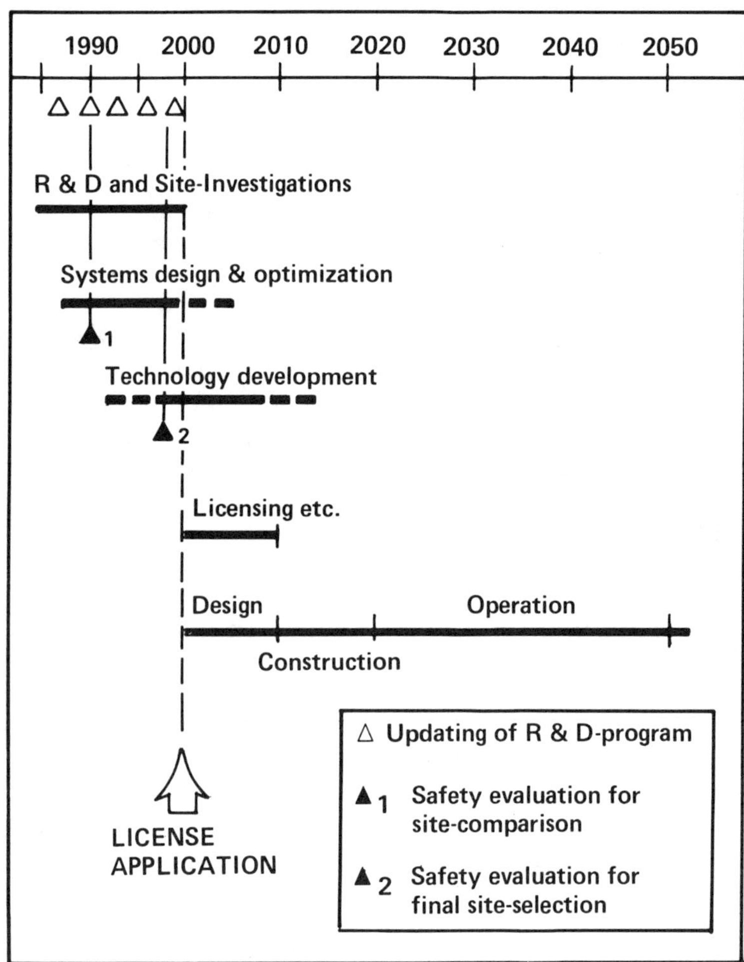

FIGURE 2. Overall Time-Schedule for Realization of HLW Repository

The studies on chemistry fall into four categories: ground water chemistry, radionuclide chemistry, release and transport model development, and in-situ tests and natural analogue studies. The very long time frame that has to be considered in the performance assessment of a nuclear waste repository makes it attractive to supplement laboratory and field experiments with observations in nature, which reflect similar processes over geological time spans. An international Workshop on natural analogues was organized by SKB and the US-DOE in Chicago in October 1984[3]. The main conclusion from the workshop was that the natural processes to be studied must be carefully selected and evaluated in close cooperation between the investigating geoscientist and the modeller of the corresponding processes in waste repository systems. The emphasis is on processes with well-defined boundary conditions rather than complete geological repository systems. SKB is supporting studies on uranium mobility in granitic rocks, on Morro de Ferro, and on evaluation of data from Oklo.

An important part of any R&D program for HLW disposal is the study of the waste form. The work in Sweden on spent fuel has continued for several years. Total contact times between spent fuel specimens and groundwater up to more than 900 days have now been achieved. Special studies of solubility constraints and the effects of radiolysis have been started. The measurements on uranium and plutonium indicate that a solubility limit is rapidly achieved and that this limit stays constant irrespective of contact time. Comparison between measured equilibrium concentrations and calculated solubilities show reasonably good agreement.[2]

The studies of canister materials and buffer and backfill materials are continued along the lines followed in the studies for the KBS reports. Future work will include investigations on alternatives to copper and bentonite, which were the main materials studied for KBS-3.

In the area of performance assessment and safety analysis, the main emphasis in recent years has been on the development of a computer code system called PROPER, which is based on the main ideas that were first implemented in the AECL-code SYVAC. The new code is mainly intended for intercomparison studies concerning R&D priorities, design optimization, site screening and selection, and safety assessments. In detailed safety analyses, the code must be backed up by more elaborate and comprehensive methods.

VIII. INTERNATIONAL COOPERATION

The Swedish waste management program includes an extensive international cooperative framework. Sweden is the host country and SKB the executive organization for the international OECD/NEA - Stripa project. Nine countries are cooperating in this project, which uses an abandoned iron-ore mine at Stripa some 250 km west of Stockholm. The project is now in phase II which will be completed in 1986. The Phase I and Phase II studies include the following items:

1. hydrogeological and hydrogeochemical characterization of the site and the groundwaters including application of new techniques and new sampling methods,

2. a major buffer mass test with electrically simulated heaters in half-scale deposition bore-holes,

3. development and demonstration of cross-hole techniques for detection and characterization of fracture zones,

4. 2D- and 3D-migration experiments, and

5. borehole, shaft, and tunnel sealing tests.

Results from the Phase I and the ongoing Phase II were presented and discussed at an NEA symposium in Stockholm in June this year. The planning of a tentative Phase III starting in 1986 is now in progress.

SKB also participates in the joint Japan - Switzerland - Sweden (JSS) project for studies of HLW glass. Radioactive glasses were obtained from Cogema in France and are being investigated at Studsvik in Sweden and at Wurenlingen in Switzerland. Supplementary investigations on inactive simulated HLW glass are being made at Stripa and in various laboratories.

SKB has bilateral information exchange agreements with US-DOE, CEA in France, AECL in Canada and NAGRA in Switzerland. An agreement with CEC-Euratom has been negotiated but not yet formally approved.

IX. COSTS AND FINANCING OF RADIOACTIVE WASTE MANAGEMENT IN SWEDEN

All costs for radioactive waste management and decommissioning of nuclear power plants in Sweden has to be carried by the owners of these plants. The costs are covered by a fee on the nuclear power production. The fee is set by the government and revised each year. For 1984 and 1985 the fee is 0.019 SEK/kWh (about 2.2 US mills/kWh). The fee is based on cost calculations made by SKB and submitted annually to SKN. This authority reviews the calculations and proposes a fee to the government. According to the law, the fees should be set individually for each reactor based on its individual projected costs. So far, however, there has not been a firm enough base for differentiating the fees, and therefore, the same fee applies to all units. In the future it is likely that the fees will be differentiated between the different utilities. The fees are collected in a special fund at the Bank of Sweden. The fund is administered by SKN, which also provides money from the fund to the various waste management activities performed by SKB. Excluded from this financing are all facilities needed for the handling and disposal of reactor operational wastes. These facilities are financed directly by the utilities through SKB. In particular, this is the case for the major part of SFR.

From the 1985 SKB report to SKN on cost calculations the following figures were obtained.[4] The expenditures were 3.4 GSEK for CLAB, SFR, reprocessing services, the R&D program, and the transport system. The expected future costs (in January 1985 funds) are 43.4 GSEK. It should be kept in mind that many of the costs will occur in the far future. Figure 3 gives a rough account of the distribution of these future costs with time. The relative distribution of the total costs, not accounting for interest or present value calculation, are roughly as follows:

Interim storage of spent fuel and other wastes	18%
Reprocessing of 690 MTU	10%
Final disposal of spent fuel and long-lived wastes	31%
Final disposal of operational and decommissioning wastes	4%
Transportation of wastes	6%
Decommissioning and dismantling of nuclear power plants	24%
Miscellaneous including R&D	7%

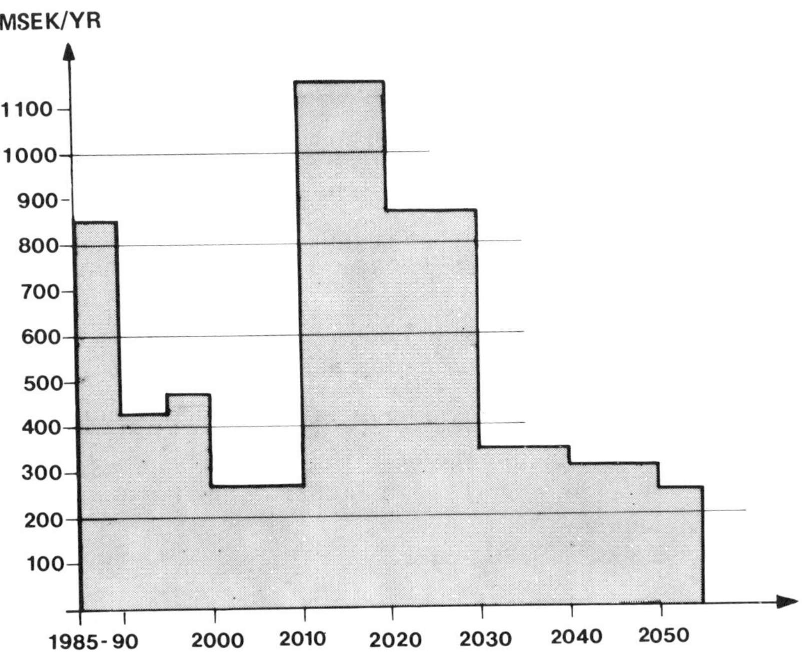

FIGURE 3. Approximate Distribution of Future Waste Management Costs for Sweden's Nuclear Power Program

The cost for direct disposal of spent nuclear fuel without reprocessing is calculated to be about 3100 SEK/kg uranium including associated R&D costs. Of this cost, 7% is attributed to transportation, 30% to intermediate storage in CLAB for 40 years, 37% to encapsulation in copper canister and 26% to final disposal according to the KBS-3 concept. A substantial part of the costs for spent fuel disposal are fixed costs. The marginal cost for additional quantities was calculated to be 1300 SEK/kg U.

X. SUMMARY

The current Swedish program for waste management is at present focused on

1. startup of the interim fuel storage facility, CLAB,
2. construction and completion by 1988 of the first phase of the repository for reactor operational wastes, SFR,
3. basic research and development on disposal of high-level waste with direct disposal of spent fuel as the preferred approach,
4. continued site investigations for finding a suitable site for a HLW repository (a major screening and selection of 2 to 3 sites for more detailed studies will be made around 1990),

5. studies of alternatives to the methods described in the KBS reports, and

6. extensive international cooperation in several areas in particular through the Stripa and JSS projects.

The total costs for waste management (including decommissioning and dismantling of the nuclear power plants) are about 0.020 SEK per kWh which is about 10% of the consumer cost for electricity. Under present conditions in Sweden, direct disposal of spent fuel is more economical than reprocessing.

REFERENCES

1. Hans Carlsson, SKB "Selection and Characterization of Potential Sites for a Spent Nuclear Fuel Repository in Sweden," in these proceedings (1986).

2. Annual Research and Development Report. SKB Technical Report 85-01 (1984).

3. John A T Smellie, Natural Analogues to the Conditions Around a Final Repository for High-Level Radioactive Waste, SKB/KBS Technical Report 84-18 (1984).

4. SKB-PLAN 85, "Kostnader for Karnkraftens radioaktiva restprodukter," Juni (1985).

STATUS OF THE HIGH-LEVEL NUCLEAR WASTE DISPOSAL PROGRAM IN SWITZERLAND

Rudolf Rometsch
Hans Issler
National Cooperative for the
Storage of Radioactive Waste (NAGRA)
Parkstrasse 23
5401 Baden, Switzerland

I. INTRODUCTION

In recent years, a quite far-reaching international consensus has emerged concerning the essential techniques for disposal of high-level radioactive waste. The presentation of a series of programs and results from different countries tends, therefore, to become monotonous and repetitive. To mitigate this effect, we will try to point out mainly the particularities of the Swiss program.

II. BOUNDARY CONDITIONS

The boundary conditions are somewhat unusual with regard to several aspects. The legal framework has been amended recently in the normal exercise of the procedures of direct democracy. A Federal Government Ruling of 1978 on the Atomic Energy Act had to be submitted to referendum and gained a large majority in 1979. It designates the guaranteeing of "permanent safe management and final disposal" of radioactive waste as a prerequisite to future development of the use of nuclear energy in Switzerland. For already existing nuclear power plants outside the scope of this ruling, the Federal Department dealing with energy matters demanded a project which offers a guarantee of feasibility and safety of final disposal as a condition to the extension of operating licenses beyond 1985.

The setting of such a task is novel and is not usual for other waste types. It was set for radioactive wastes, not because they presently pose a particular direct danger or might be released to the biosphere in an uncontrolled way, but because they are perceived by large sectors of the population as presenting a particular threat. This feeling is in the nature of a subjective expectation and is probably evoked by the socio-political controversy over nuclear energy and the memories of atom bombs ending the Second World War. Whatever its cause may be, it has resulted in the placing of extraordinarily strict requirements on the disposal of radioactive waste. The fulfillment of these may, however, provide guidance for disposal of other waste materials as well.

Final disposal of radioactive waste is the objective in most countries making use of nuclear energy. However, opinions differ widely on procedures and the timetable for realization. Especially in Switzerland there is the desire for rapid progress towards realization and the attempt by the authorities to provide a simple and clear definition of "final disposal". In guidelines issued in 1980 by the Federal Commission for Safety in Nuclear Installation (KSA) and the Nuclear Safety Division of the Federal Office of Energy (HSK), the protection objectives for final repositories are described, as follows:

1. A repository must be designed in such a way that it is capable of closure at any time within a few years. After closure of a repository, it must be possible to dispense with safety and supervision measures.

2. Radionuclides from a sealed repository which reach the biosphere as a result of realistically conceivable processes and events may at no time lead to individual doses which exceed 10 mrem per year.

The aim is to protect future generations. They should not have to concern themselves with our waste, at least not with radioactive wastes, and they must never be exposed because of that waste to a radiation risk higher than a small fraction of that given by the inevitable human radiation exposure from natural sources.

Within these societal boundary conditions, the technical ones are easy to set for Switzerland. At present 5 nuclear power plants are in operation with a total of 3,000 MW of installed electrical capacity. It is assumed that this will double in the next 25 years or so and that each power plant will have an operating lifetime of 40 years. The resulting 240 GW-years of electricity production will yield a total of 1,200 m^3 of vitrified high-level waste if the present policy of reprocessing all spent fuel is maintained.

Reprocessing and waste vitrification for the Swiss power plants is done abroad under contract. Should these operations, for economic or political reasons, be abandoned then the spent fuel elements would have to be disposed of directly. The volume of conditioned high-level waste would then triple and almost 2 t U and 30 kg Pu per m^3 would be added to the activity inventory.

Only about 6% of the total high-level waste quantity used as a basis for repository planning has been already produced. The treatment, conditioning, and planned intermediate storage time makes it necessary to have the disposal repository for high-level waste ready around 2020. This is taken as a target date.

III. RESPONSIBILITIES

Responsibilities for repository planning and construction in Switzerland differ from the common pattern in which the Government has to take over the last step of waste management. The responsibility in Switzerland remains with those producing and handling the radioactive waste. Naturally, there is special supervision by governmental authorities of waste disposal activities similar to that for the construction and safe operation of nuclear power plants.

For practical reasons, the six utility companies operating, constructing, or planning nuclear power plants and that branch of the Federal government responsible for handling the residues from radioisotope application in hospitals, industry, and research set up a common organization, Nagra, the National Cooperative for the Storage of Radioactive Waste. Nagra is currently responsible for planning the implementation of safe final disposal.

Anticipating the new legislation, Nagra began in 1977 to form a research, development, and project team to work on three programs: one to realize a high-level repository around the year 2020, one to realize a low- and intermediate-level repository towards the year 2000, and the third to establish by 1985 the required project demonstrating feasibility and safety of such repositories. In the following sections, the status of this work will be reviewed with regard to the high-level repository.

IV. REFERENCE PROJECT FOR HIGH-LEVEL WASTE DISPOSAL

Although a variety of engineering concepts have been reviewed, and a range of potential host rocks is available in Switzerland, one particular set of options was chosen to be followed up with priority and to be used as a basis for the warranty project meant to guarantee feasibility and safety of disposal. For the project, the data had to be as specific as possible.

The repository system proposed is a mined facility around 1,200 m below the surface in the crystalline bedrock which is covered by some hundreds of metres of sediments (see Figure 1). No definite site has been proposed but geological, hydrological, and geochemical data are based on field data gathered during a regional geologic program including a series of 6 deep drillings (1,500 - 2,500 m). The HLW to be disposed of in the repository consists of around 6,000 cylinders of borosilicate glass returned from reprocessing the fuel of the 240 MW(e)/year power program. A special feature of the planning is that flexibility in allocation of all nuclear wastes to the appropriate repository for disposal is ensured also for intermediate-level actinide wastes from reprocessing which may be unsuited for storage in the planned Swiss LLW/ILW repositories.

Engineering project studies have been performed to develop a reference repository design. The objectives were to assess the required technical expertise for engineering a repository system, and to provide a basis for safety analysis work. No technical optimization has been carried through. The repository consists of a series of tunnels 3.7 m in diameter lying in the interior of a tectonically stable block of crystalline bedrock. HLW canisters are

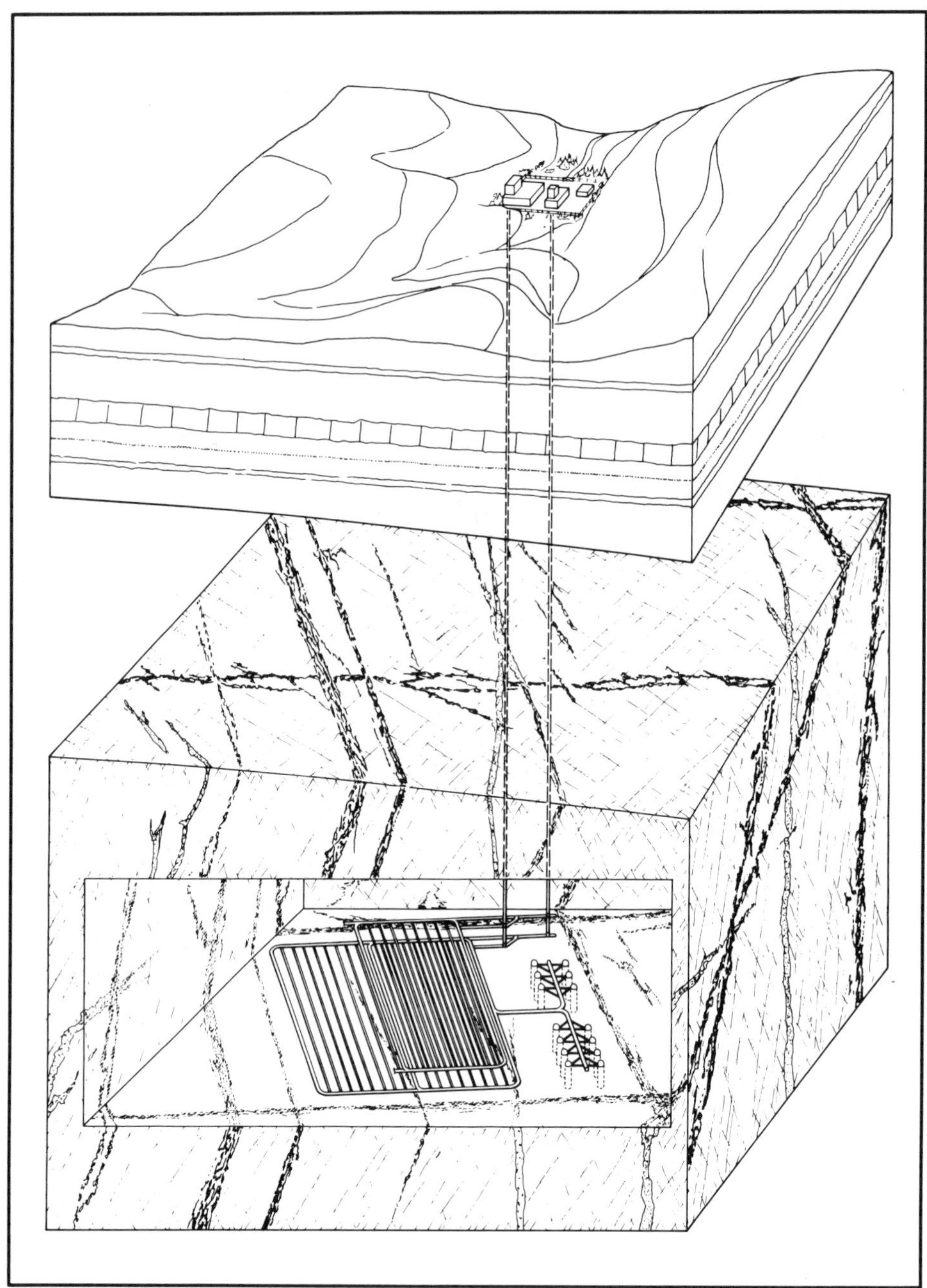

FIGURE 1. Perspective Overview of High-Level Waste Repository Installation. Example of management on two levels in a fissured granite block according to model data set.

delivered from an intermediate storage facility to the repository surface reception area where they are sealed into massive cast steel containers. The sealed packages are transferred in shielded containers down one of two 1,230-m shafts accessing the repository working level. They are then emplaced 5 m apart centrally in the tunnel axis surrounded by compacted bentonite blocks (see Figure 2).

The filling operations take place simultaneously with drilling of further storage tunnels. However, complete spatial separation of constructional and operational activities is maintained. The storage area for non-HLW consists of 10 m diameter silos all of which are excavated before the start of disposal operations. The total building costs and times are estimated at 600 million Swiss Francs and 15 years. The conclusion of engineering studies carried out within the framework of the Swiss warranty project is that the required technology for construction and operation of a deep HLW repository is already available today.

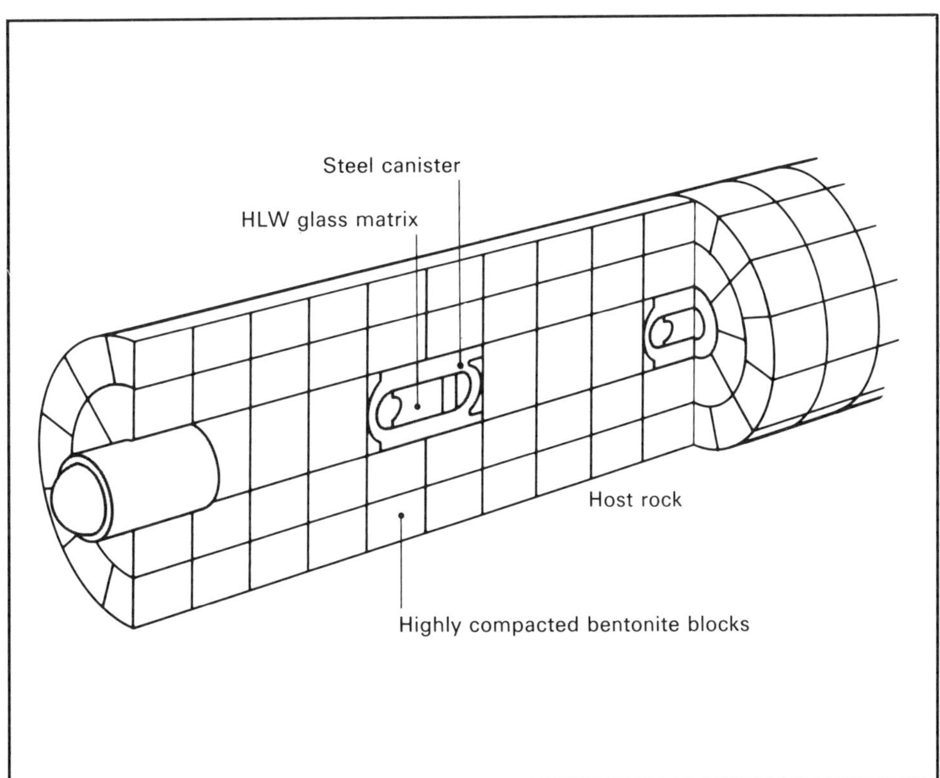

FIGURE 2. The Technical Barrier system. Gaps and cavities disappear when the bentonite swells with increasing water content.

V. ANALYZING THE SAFETY OF A HIGH-LEVEL REPOSITORY IN GRANITE

A broad program was aimed at understanding the behavior of all components of the safety barrier system, and thus at improving our confidence in long-term predictions. The characteristics of the host rock and surrounding geosphere can be investigated only by extensive geological field work, which in the Swiss case is particularly complex and costly. The crystalline bedrock (granite and gneiss) is overlain by varied sediment sequences, in total some hundreds of metres thick. Mapping the surface topography of the bedrock, and understanding its tectonic, hydrologic and geochemical characteristics require a combination of advanced seismologic methods, deep drillings with complex test methodology, and surface observations and measurements over a large region.

These approaches have been employed over several years to allow a picture of representative deep geologic conditions to be built up; for the warranty project the data are gathered into a model data set upon which construction and safety studies are based. The assumed host rock is granitic with extensive small-scale fracturing, almost all of which is, however, sealed by subsequent deposition of minerals so that the resultant hydraulic conductivity is low. Extensive borehole test sections have been shown with packer tests to have conductivities between 10^{-9} and 10^{-12} m/s. The small quantities of water which flow through such rocks (approx. 4 m^3/y for the whole repository area) are mainly confined to zones of relative weakness where tectonic movements have altered the rock structure. This has a huge impact upon system safety since the weak zones have a relatively high porosity (around 3%) and excellent sorption properties for radionuclides which diffuse out of the isolated water-carrying features within the altered zones.

The granite structure sketched here is found at considerable depths below the upper surface of the crystalline basement. The first few hundred metres of granite carry much more water and influence radionuclide release concentrations only by providing a significant dilution factor (around 10^4). Evidence for a qualitative difference between upper and lower granite is obtained not only from hydraulic testing but also from characterization of the chemistry of the rock itself and of the deep groundwater. At depth, reducing conditions are observed, which has an important influence on the solubilities of compounds of radionuclides and hence on potential release rates. Even with very pessimistic assumptions for the important safety parameters, the geosphere conditions described ensure that radionuclides released from the near field of a repository can reach the biosphere only much later at greatly reduced concentrations.

Moreover, the technical barriers within the repository system themselves greatly restrict releases of radionuclides from the near field. Extensive work in various countries has been devoted to demonstrating the corrosion resistance of borosilicate waste glasses; based on this and on results of a current joint Japanese/Swedish/Swiss active glass leaching study, a reference lifetime for the vitrified waste of 150,000 years has been conservatively adopted. The massive steel pressure-vessel around the waste delays the start of glass corrosion at least 1,000 years. Equally important, however, is the chemical buffering effect of residual iron and corrosion products which remain after mechanical failure of the container occurs. The resulting low solubilities for important radionuclides ensure that releases occur only at very low levels. Finally the compacted bentonite, which in the reference design is over 1-m thick, provides a diffusion barrier which can delay releases for tens of thousands of years and thus allow decay of all nuclides which do not have very long half-lives.

Furthermore, even at steady-state conditions, radionuclides can be transported through the bentonite by diffusion only up to a maximum rate which is independent of the water flow in the adjacent host rock so that a degree of decoupling is present between technical and geological safety barriers. Figure 3 gives a schematic view of the total repository system modeled in the safety analyses.

VI. THE WARRANTY PROJECT

The prime objective of the projects prepared for government review was to quantify the level of safety achievable for a repository system of the type described. Using appropriate calculational models the radiation doses which might arise due to releases of radionuclides from the HLW repository have been calculated. In choosing models and data, efforts were made to define a base case which should give a reasonably realistic picture of potential consequences, whilst ensuring that estimates remain on the high, i.e., conservative, side. For the chosen base case totally negligible doses were calculated (less than 10^{-7} mrem/year). Even when compounding pessimistic assumptions and data predicted doses remain very low (less than 10^{-2} mrem/year). It is clear that various issues remain to be further classified and that the predictions based on real, but geometrically limited, field data must be confirmed for a fully characterized site. However, the doses calculated are small, in absolute terms and relative to the low value of 10 mrem/y included in the Swiss safety authority guidelines, so that adequate safety is predicted for the high-level waste repository.

This prediction and the above-mentioned conclusion that the required technology for executing the reference project is already available, are the essential statements of the warranty project submitted to the Government in January 1985. The experts of the Government, in consultation with specialists from various countries, are now reviewing this project. The documentation contains, in 8 volumes, a summary of the analyses and data leading to the above statements, more detailed information is included in 150 reference reports on the work performed in the past 6 years. In the course of the next year the Government will, based on the findings of its experts, announce whether it judges that the warranty project provides that guarantee recognized as a political necessity for the further use of nuclear energy in Switzerland.

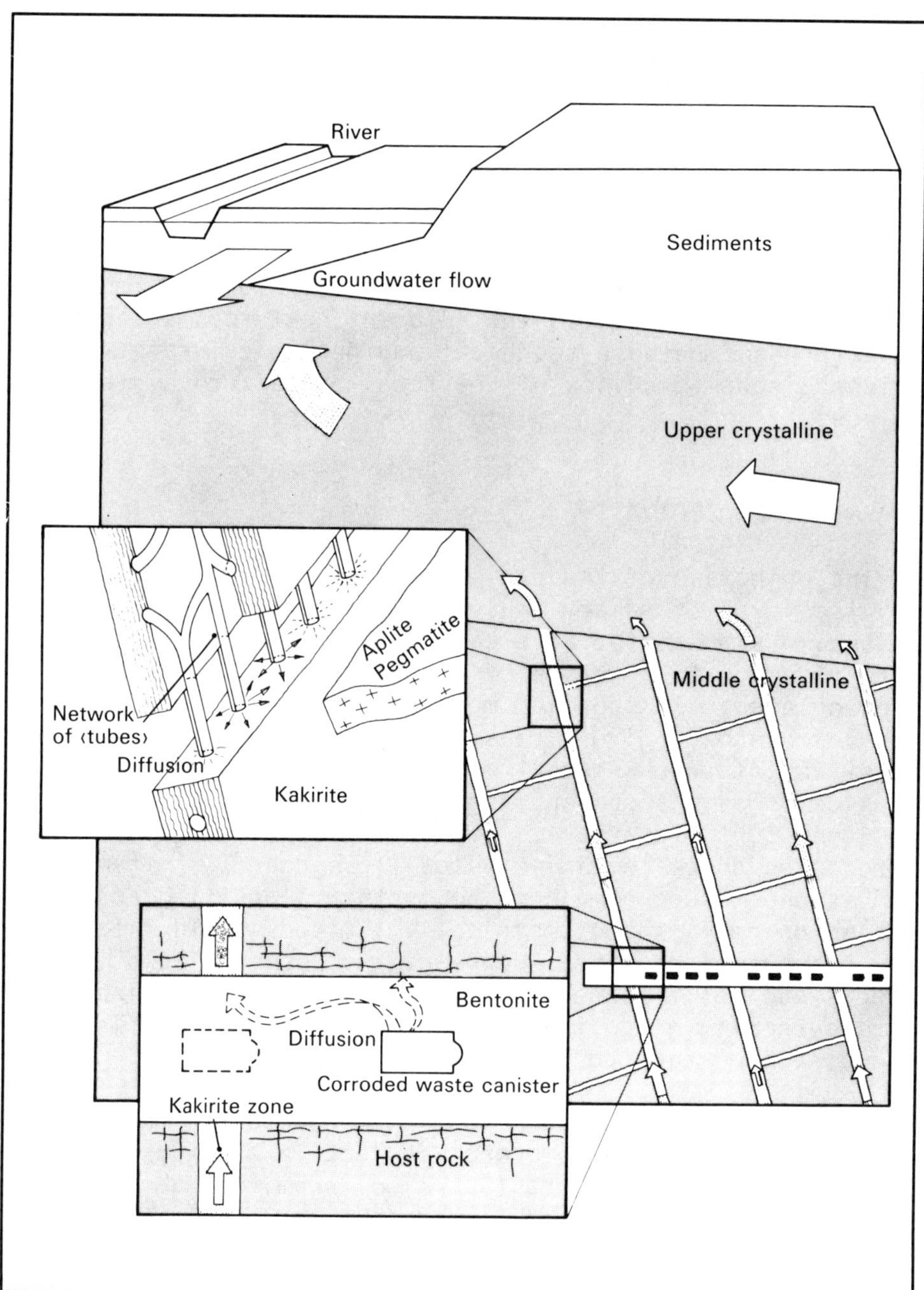

FIGURE 3. This gives a Schematic Overview of Those Features of the Total HLW Disposal System Which are Explicitly Modeled in the Safety Analyses

VII. CONTINUATION OF THE PROGRAM

Meanwhile work continues towards making the repository projects a reality. In the framework of Nagra activities, this involves (for the high-level waste repository) completion of the first phase of geological investigations on deep crystalline rock, operation of the Grimsel underground rock laboratory mainly to improve methods for non-destructive recognaissance of large rock masses, closer characterization of selected areas for repository construction and further improvement of data and of models, including their validation. This work should lead in about 10 years to definitive site selection and characterization and to application for a general permit for repository construction. For the latter, the Government will have to seek Parliamentary approval, which means that the project must find also a large degree of public acceptance. This is why Nagra attaches high importance to fully informing the public at all times.

Nagra will also maintain its cooperation with sister organizations in other countries. In fact, the work aiming at optimization of repository design involves also consideration of combining the efforts of several countries with a view to constructing a common repository for the relatively small volumes of high-level waste. Today such a desirable solution appears not to be feasible for political reasons. However, once national projects are accepted in several countries, joint ventures will become attractive again. Therefore we consider continuation of the Swiss national development program to be a basic scientific and political prerequisite to participation in any project for an international high-level waste repository - wherever this might become a reality.

STATUS OF THE UNITED STATES' HIGH-LEVEL NUCLEAR WASTE DISPOSAL PROGRAM

Ben C. Rusche
Office of Civilian Radioactive Waste Management
U.S. Department of Energy
Washington, D.C. 20585

I. BACKGROUND

The safe disposal of spent nuclear fuel and high-level radioactive waste in the United States has been a matter of national concern since the first civilian nuclear reactor began generating electricity in 1957. The spent fuel and high-level waste remain potentially hazardous for tens of thousands of years and it is important that they be permanently isolated from the environment until their radioactivity decays to levels that will pose no significant threat to people or the environment.

The passage of the Nuclear Waste Policy Act of 1982 was a major milestone in the Nations management of nuclear waste. The Act, which was passed by the United States Congress in December 1982 and signed into law by the President January 7, 1983, established a national policy for the safe and permanent disposal of spent nuclear fuel and high-level radioactive waste.

The Act established a schedule and a step-by-step process by which the President, the Congress, the States, affected Indian tribes, DOE and other Federal agencies can work together in the siting, design, construction and operation of deep, geologic repositories for disposal of spent nuclear fuel generated by civilian nuclear powerplants and high-level waste resulting from atomic energy defense activities.

DOE's activities to dispose of high-level waste and spent fuel are regulated by The Nuclear Regulatory Commission, the Environmental Protection Agency and the Department of Transportation. Incidentally, EPA, in a major step forward, recently issued its environmental regulations.

Before I discuss where we are in developing the waste disposal system, I would like to give you some background on the efforts, which resulted in the passage of the Nuclear Waste Policy Act.

II. THE ACT

In passing the Nuclear Waste Policy Act, the Congress realized that this waste creates potential risks and requires safe and environmentally acceptable methods of disposal. A national problem has been created by the accumulation of this waste and Federal efforts during the past 30 years to devise a permanent solution to the problem of waste disposal have not been adequate.

We currently have approximately 11,000 metric tons of spent fuel in storage pools at now more than 90 licensed commercial nuclear powerplants in 27 States and the equivalent of almost that much defense high-level waste in three or more States.

The question is clearly not one of do something or do nothing with the waste. The question is, rather, store in many locations (more than 100) or move to permanently isolate these wastes in one or more very secure and environmentally isolated sites.

The Federal Government has the responsibility to provide for the permanent disposal in order to protect health, safety and the environment. And the generators of the waste ought to pay the cost of disposal. Since April 7, 1983, utilities have been charged a fee of one-mill per kilowatt hour for nuclear-generated electricity.

The Act has provided a broad, flexible framework in which to conduct the necessary activities. It has authorized certain key facilities, set schedules and fees, articulated institutional interactions, provided the opportunity to analyze the desirability of enhancements to improved program performance, and recognized the need for flexibility and contingency planning in a large, complex, and controversial program that may span more than a century.

Specifically, the Act establishes:

1. a schedule for siting, construction, and operation of repositories to provide reasonable assurance that the public and the environment will be adequately protected,
2. the Federal responsibility and a definite National policy, for the disposal of such waste, and
3. the Nuclear Waste Fund, composed of payments made by the generators to ensure that the costs of carrying out activities relating to the disposal of such waste will be borne by the generators and beneficiaries.

III. IMPLEMENTATION APPROACH

The approach which we have taken, and which we have recently outlined in a document entitled, Mission Plan, involves the following goals:

1. We must protect the public health and safety and the environment.
2. The program must be credible to the public by virtue of its integrity and technical excellence.
3. The program must neither subsidize nor penalize nuclear power as an energy source.

4. The program must be conducted in a cost-effective manner, with full cost recovery.

Program objectives which evolved from these policy goals to implement the provisions of the Act are:

1. to site, obtain a license for, construct, and operate geologic repositories such that the transportation of radioactive waste to and disposal of the wastes in the repositories can be accomplished in a manner that is safe and environmentally acceptable,

2. to submit a proposal to Congress to develop one or more facilities for monitored retrievable storage,

3. to ensure the acceptance of waste for disposal by January 31, 1998, in accordance with the acceptance schedule provided for in DOE's standard disposal contracts with utilities and in conformance with the Act,

4. to assist utilities in providing adequate and safe at-reactor storage for spent fuel before transfer to DOE and to provide limited Federal interim storage for any utilities found by NRC to be eligible for such service, and

5. to manage the technical program and the funds collected for disposal and storage services, or otherwise provided through appropriation, in an effective, integrated, and efficient manner.

The strategy of the program is to ensure that the activities authorized by the Act are carried out in a vigorous manner; that potentials for system-performance improvements are analyzed and incorporated where useful, including requests for new Congressional authority as allowed by the Act, and that contingency plans are identified and evaluated to provide maximum confidence in implementation of the Act, notwithstanding uncertainties as to future Congressional decisions and technical and institutional matters. With these strategies in mind, let me give you a brief status report on the program.

IV. STATUS OF IMPLEMENTATION

When the Act was passed, DOE had underway field and laboratory testing at nine different sites. Typically, the field studies included the drilling of boreholes to investigate subsurface conditions and to determine whether a potentially suitable host rock existed.

The field studies were supported by laboratory studies that focused on the isolation and engineering characteristics of the rock. Measurements of groundwater characteristics were made. Also in progress were systems analysis, waste-package development and repository-design efforts.

In February 1983, in accordance with the Act, DOE formally identified those nine sites as being potentially acceptable. Those sites, as many of you know, are: one site in Louisiana, two sites in Mississippi, one site in Nevada, two sites in Texas, two sites in Utah, and one site in Washington.

Also, as required by the Act, siting guidelines were developed. After a long review process, including several public hearings and consultation with affected states, Indian tribes, and key Federal agencies, the NRC concurred with the guidelines. The siting guidelines were issued in final form in December 1984 and became effective January 7, 1985.

After issuance of the guidelines, we issued in December 1984, draft environmental assessments on each of the nine potential sites. These draft environmental assessments evaluated each site in terms of the siting guidelines and, when final, will be the basis for the nomination and recommendation of sites for site characterization.

V. FIRST REPOSITORY

In the draft environmental assessments, we announced the proposed sites for nomination and recommendation for site characterization. The proposed sites for nomination are:

- Deaf Smith County, Texas,
- Hanford, Washington,
- Yucca Mountain, Nevada,
- Davis Canyon, Utah, and
- Richton, Mississippi.

Of these five potential sites, we proposed to recommend to the President for site characterization:

- Deaf Smith County,
- Hanford, and
- Yucca Mountain.

Earlier this year, we conducted numerous public hearings and formal briefings in the six states containing the nine potentially acceptable sites. We are now reviewing more than 20,000 comments received on the draft environmental assessments.

After we have completed this review, we plan to finalize the environmental assessments, formally nominate sites suitable for characterization, and recommend three sites to the President for site characterization. We currently plan to do this around the end of this year.

VI. CONSULTATION AND COOPERATION

The decision to characterize particular sites formally triggers the Nuclear Waste Policy Act's Consultation and Cooperation (C&C) agreement provisions between the Department and affected tribes or states. However, well in advance of any legal requirement to do so, the State of Washington initiated Consultation and Cooperation agreement negotiations with DOE more than a year ago, as did the Yakima Indian Nation.

We have made considerable progress in reaching agreement with Washington on virtually all issues, the key exception being liability. Recently, the Umatilla Indians have requested the initiation of C&C agreement negotiations and discussions are underway.

A Consultation and Cooperation agreement has the advantage of regularizing DOE and state or tribe relations. As called for in the Act, site characterization activity may proceed whether or not a Consultation and Cooperation agreement has been signed between the parties. However, the Department of Energy is committed to negotiating agreements with each of the affected parties, and we will diligently work to conclude agreements.

VII. SITE CHARACTERIZATION

Site characterization, which I mentioned earlier, is geohydrological exploration, investigation, and evaluation.

To collect the subsurface data, construction of exploratory shafts at each of the three sites approved for characterization will be necessary. DOE plans to construct two shafts at each site. These shafts will be to the depth of a proposed repository--about 1,000 to 4,000 feet deep. Shaft construction at the three sites will take approximately two years with in-situ tests planned for FY 1988 through FY 1990.

Before proceeding to construct shafts at a site approved for characterization, the Act requires that DOE prepare a Site Characterization Plan. These plans will be submitted to NRC and the affected states and Indian tribes for review and comment and will be made available to the public. Public hearings will be held in the vicinity of each candidate site to inform the area residents of the plan and to receive their comments.

During site characterization, DOE will regularly report and consult with NRC and affected states and Indian tribes on the nature and extent of site characterization activities and the information developed from such activities.

In about 1990, and based on site characterization, DOE will evaluate each site and recommend one site to the President for the first repository. This recommendation will be accompanied by an Environmental Impact Statement which will have been prepared in accordance with Nuclear Waste Policy Act and the National Environmental Policy Act requirements which include public review and comment and public hearings.

When the President recommends to Congress the site for the Nation's first repository, which is estimated to be in 1991, the host State Governor or legislature or affected Indian tribe on whose reservation the repository is located may issue a notice of disapproval within 60 days of the President's recommendation. The disapproval can be overridden only by a resolution of both Houses of the U.S. Congress. Thus, Congress will then weigh our facts against the state's or tribe's objection.

If the disapproval is not overridden, the President must submit another repository site recommendation to Congress within 12 months. If no disapproval is submitted, or if the disapproval is overridden, then as prescribed by the Act, the site designation is effective and DOE will submit to the NRC a Construction Authorization Application within 90 days.

Under the Nuclear Waste Policy Act, the NRC has three years to review the application. NRC has indicated that three years is the minimum licensing review period required unless effective steps are taken to identify and resolve potential licensing issues during the next six years. We believe that licensing issues can be identified and resolved through the effective use of the on-going close and extensive interaction between DOE and NRC. With this schedule in mind, we believe we will be in a position to begin receiving waste for disposal by January 31, 1998, as called for in the Act.

VIII. SECOND REPOSITORY

While the Act does not authorize the construction of a second repository, it does require DOE to carry out the siting and development activities essential to preparation for such a facility. These activities trail those for the first repository by about five years.

For the second repository, DOE may consider:

1. sites identified as potentially acceptable but not nominated for the first repository,

2. sites characterized but not chosen for the first repository site, and

3. sites found potentially acceptable from rock formations not previously studied in the first repository selection process.

The screening process for the second repository is currently in the regional phase in which we have recently compiled open literature information on the geologic, environmental, and socioeconomic conditions regarding crystalline rocks in 17 states in the north central, southeastern, and northeastern regions of the country.

In April, we issued a document entitled, "Region-to-Area Screening Methodology for the Crystalline Repository Project." The screening methodology will be used to narrow geologic focus from large regions to smaller areas in studies to identify potential crystalline sites. Using this screening methodology, later this year, we expect to identify approximately 15-20 areas in four-to-six of those 17 states in which area phase field work will be conducted. To date, no field testing has been conducted in those 17 states.

IX. FINANCIAL ASSISTANCE

As required by the Act, financial assistance has been provided to States with potentially acceptable sites and affected Indian tribes to encourage participation in the analysis of technical information. To date, more than 20 million has been provided to states and national organizations for their participation.

X. SYSTEMS INTEGRATION

By considering all the elements of the program as part of a single system, we believe we can optimize them as a unit to best meet the program requirements. Beyond those activities specifically authorized by the Act, there are opportunities built into the Act to evaluate options for enhancing what is authorized. Careful analyses of the provisions of the Act and of programmatic options have shown that increased confidence and improved performance can be achieved by emphasizing systems integration.

This concept of optimizing the system by integrating the facilities and components applies not only to the authorized activities but also to any other waste management system that could be developed to meet the requirements of the Act. In particular, the Act requires DOE to complete a detailed study of the need for, and feasibility of, monitored retrievable storage (MRS) and submit a proposal to Congress for the construction of one or more MRS facilities.

XI. MONITORED RETRIEVABLE STORAGE

Analyses to date continue to reinforce the tentative conclusion that an MRS facility fully integrated into the overall waste management system can significantly enhance several important program objectives. And those improvements include the following:

1. improved transportation efficiency because spent fuel consolidation and packaging at the MRS facility would reduce the number of shipments to the repository with reduction in potential environmental impacts and risks to the public,

2. increased reliability and flexibility in operating the system in an integrated cost-effective manner by incorporating an additional facility that can regulate the flow of waste to the repository,

3. improved confidence in DOE's ability to meet schedules, particularly in beginning to accept quantities of waste no later than January 31, 1998. The integral MRS facility, if approved by Congress, would be scheduled for initial operation as early as 1996 or 1997,

4. an ability to accept significantly larger quantities of waste in the early years of operation, substantially reducing the added cost of providing increased at reactor storage capabilities, and

5. ability to focus repository licensing efforts on demonstrating the long-term isolation capability of the site because many of the operational functions, such as waste preparation, would be handled at the MRS facility.

In April, we identified three candidate sites expected to be included in a proposal to Congress. All three candidate sites are in Tennessee.

The studies and analyses necessary to fully describe the MRS facility and to define its potential costs and benefits are being prepared. The final results will be presented in a proposal to be submitted in January 1986 for Congressional consideration, as required by the Act.

XII. TRANSPORTATION

In implementing a waste disposal system, the Act places responsibility for the transportation of spent fuel and high-level waste on DOE but also states that nothing in the Act shall be construed to affect Federal, state, and local laws pertaining to the transportation of spent fuel and high-level waste. In addition, the Act directs that private industry be utilized to the fullest extent possible in performing the transportation functions.

Development of the transportation system is integral to the development and siting of repositories and in carrying out other activities within the total waste disposal system. Planning for the transportation system will provide for development and acquisition of the appropriate types and quantities of equipment and services as well as development of the appropriate institutional arrangements.

In August, we published a draft Transportation Business Plan, which delineates activities within the development of a transportation system. A preliminary draft was issued last year. Since that time, several meetings have been held with interested parties to discuss transportation issues and to obtain private sector participation in the formulation of DOE's transportation business strategies.

We also plan to issue a draft Transportation Institutional Plan around the end of this month. This plan will propose processes and schedules for working with potentially affected and interested groups in the implementation of the transportation aspects of the Act.

The Department of Transportation is responsible for routing and certain safety aspects of transportation, and NRC is responsible for certifying casks and for safeguards. Through its Ruling HM-164, the Department of Transportation has established the Federal-State relationship.

XIII. INTERIM STORAGE

Beyond these activities, the Act authorizes us to provide up to 1,900 metric tons of storage for utilities who run out of storage space prior to our accepting their spent fuel for disposal. And, further, we are conducting several cooperative agreements with utilities to demonstrate rod consolidation and dry storage technologies to make available new licensed technologies for more efficient storage capabilities at the reactor sites.

XIV. SUMMARY

All of us involved in and concerned about these activities recognize that this program is highly controversial. There are, indeed, many constituencies with widely varying views of how, when, and whether elements of the program should be carried out.

The Act is a remarkable piece of legislation in that there is general agreement on its key provisions. Nevertheless, this is a program intended to span more than a century, with some choices by Congress, states, Indian tribes and the nuclear power industry yet to be made.

The crafters of the Act clearly recognized this. And further, the crafters recognized, and I quote from the Act, "...that...state, Indian tribe and public participation in the planning and development of repositories is essential in order to promote public confidence in the safety of disposal of such waste and spent fuel... High-level radioactive waste and spent nuclear fuel have become major subjects of public concern, and appropriate precautions must be taken to ensure that such waste and spent fuel do not adversely affect the public health and safety and the environment for this or future generations."

XV. CONCLUSION

We realize the difficulty in predicting the future and that there are many uncertainties, but failure to solve our national problem of safe, permanent disposal of high-level waste would prolong the potential risks of spent fuel and high-level waste stored all around the country. It is important for this country for each state and for each citizen that we succeed.

Protection of health, safety, and the environment are paramount. To remain with the status quo is failure.

By passage of the Nuclear Waste Policy Act, the United States Congress clearly found that development and implementation of a waste disposal system for high-level waste is an important issue and that it is essential that we get on with it.

Other nations, I hope, will benefit by our experience and I plan to continue to learn as much as I can about the plans and problems of those other nations, so that we, too, will benefit in this most important issue.

REFERENCES

1. Public Law 97-425--January 7, 1983, 96 STAT 2201, "Nuclear Waste Policy Act of 1982." 42 USC 10101.

2. DOE (U.S. Department of Energy), Mission Plan for the Civilian Radioactive Waste Management Program, Volume I, DOE/RW-0005, Washington, D.C. (1985).

3. DOE (U.S. Department of Energy), The Monitored Retrievable Storage Proposal Research and Development Report, DOE/S-21, Washington, D.C. (1983)

4. DOE (U.S. Department of Energy), Selection of Concepts for Monitored Retrievable Storage of Spent Nuclear Fuel and High-Level Radioactive Wastes, DOE/RL-84-2, Richland Operations Office, Richland, Washington (1984).

5. DOE (U.S. Department of Energy), "Implementation of the Nuclear Waste Policy Act of 1982," DOE/RW-0008.

6. DOE (U.S. Department of Energy), The Integrated System for the Management of Spent Nuclear Fuel and High-Level Waste (1984).

7. DOE (U.S. Department of Energy), The Need for and Feasibility of Monitored Retrievable Storage--A Preliminary Analysis, DOE/RW-0022 (1985).

8. DOE (U.S. Department of Energy), Screening and Identification of Sites for a Proposed Monitored Retrievable Storage Facility, DOE/RW-0023 (1985).

9. DOE (U.S. Department of Energy), Office of Civilian Radioactive Waste Management, Transportation Business Plan: Strategy Options Document, DOE/RW-0007 (1984).

10. DOE (U.S. Department of Energy), Office of Civilian Radioactive Waste Management, Draft Environmental Assessment, 1984c,d,e,f,g,h,i,j,k DOE/RW-0009, 0010, 0011, 0012, 0013, 0013, 0014, 0015, 0016, and 0017, Washington D.C. (1984)

11. DOE (U.S. Department of Energy), Office of Civilian Radioactive Waste Management, "General Guidelines for the Recommendation of Sites for the Nuclear Waste Repositories; Final Siting Guidelines," 10 CFR Part 960 (1984).

12. EPA (U.S. Environmental Protection Agency), "Environmental Standards for the Management and Disposal of Spent Nuclear Fuel, High-Level and Transuranic Radioactive Waste," 40 CFR Part 191 (1985).

13. DOE (U.S. Department of Energy), Office of Civilian Radioactive Waste Management, Region-to-Area Screening Methodology for the Crystalline Repository Project, DOE/CH-1 (1985).

14. DOE (U.S. Department of Energy), Office of Civilian Radioactive Waste Management, "Draft Transportation Business Plan," DOE/RW-0026, Washington, D.C. (1985).

FIGURE A. Potentially Acceptable Sites for the First Repository

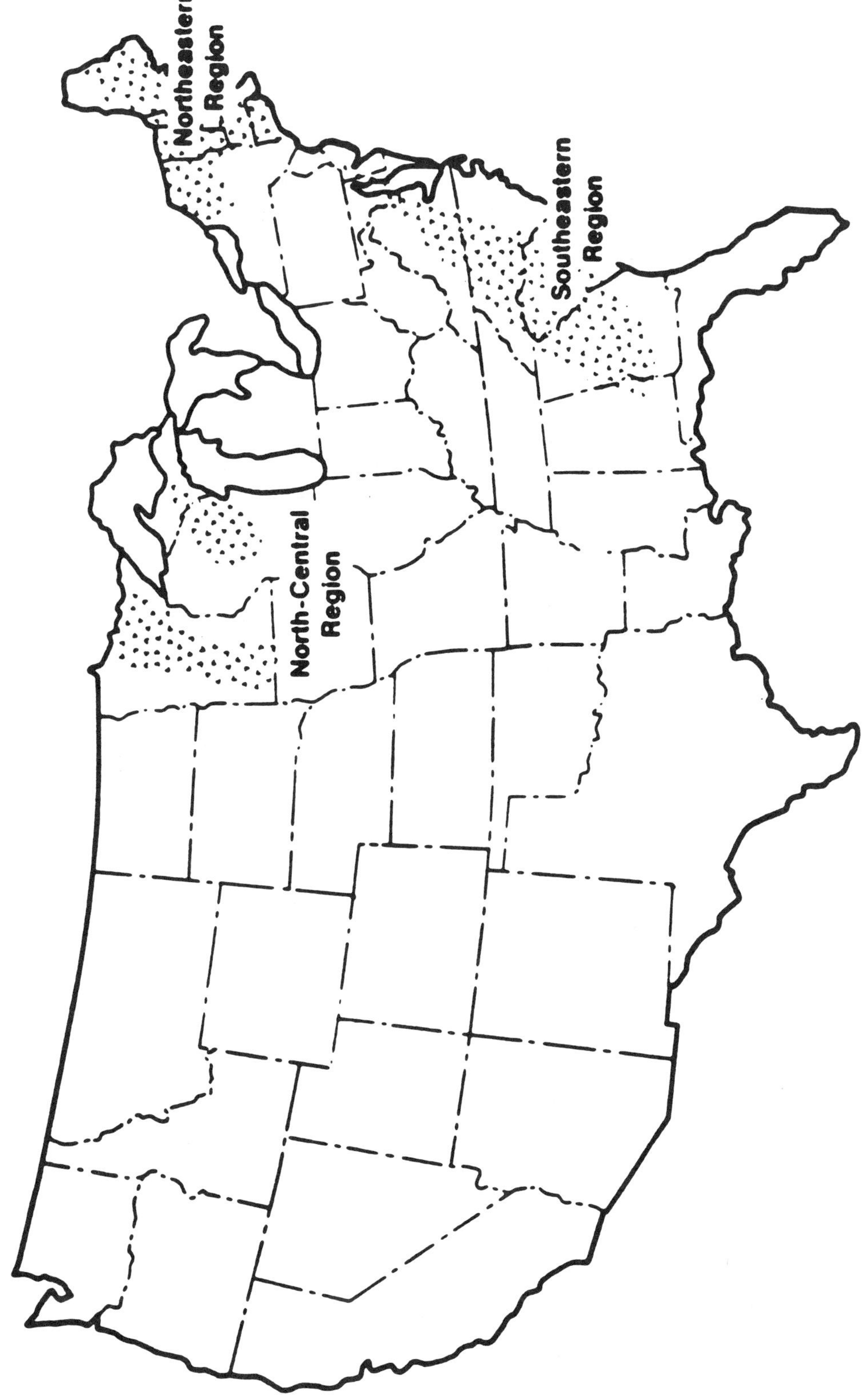

FIGURE B. Regions Being Considered for the Second Repository

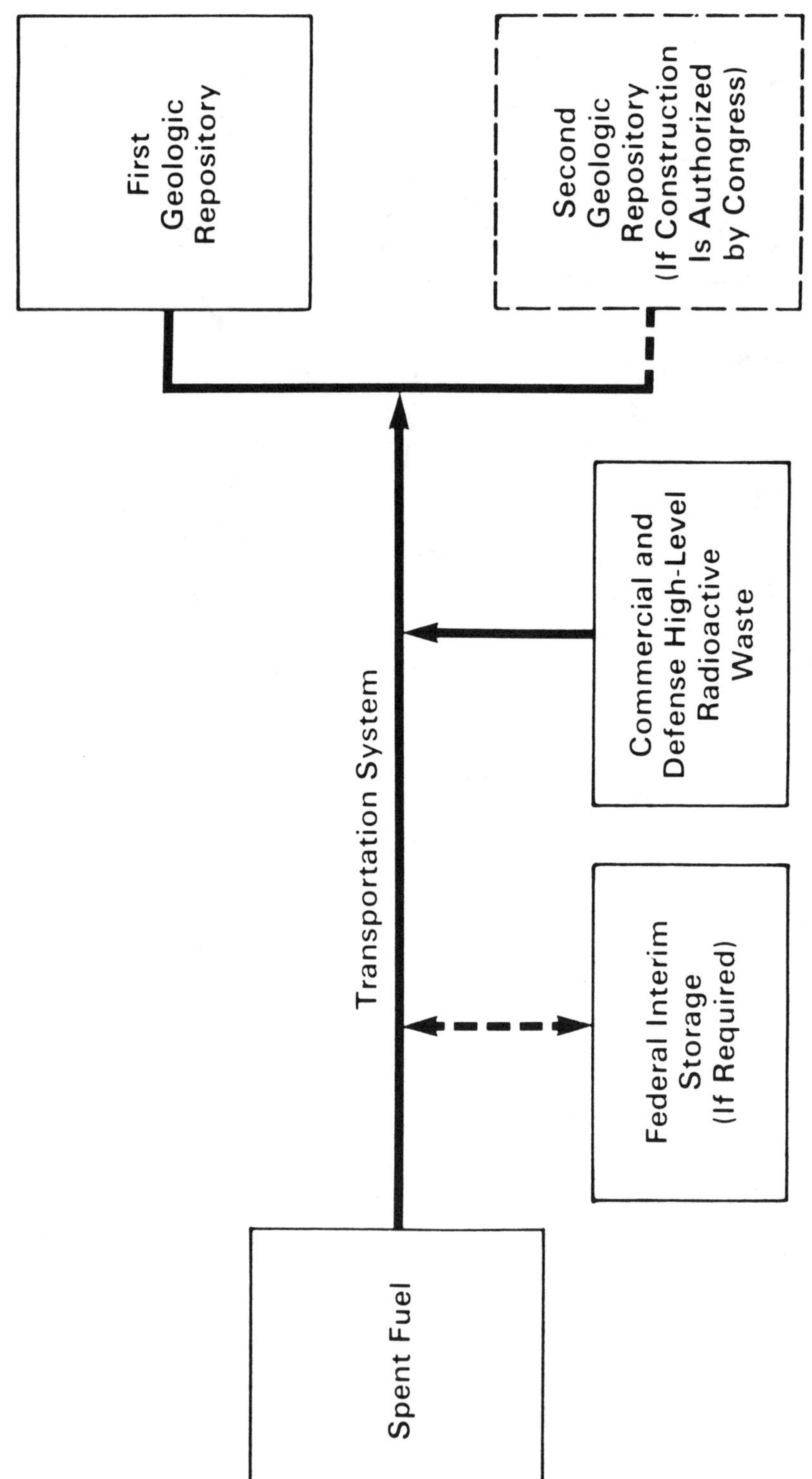

FIGURE C. Major Components of Activities Authorized Under the Nuclear Waste Act of 1982

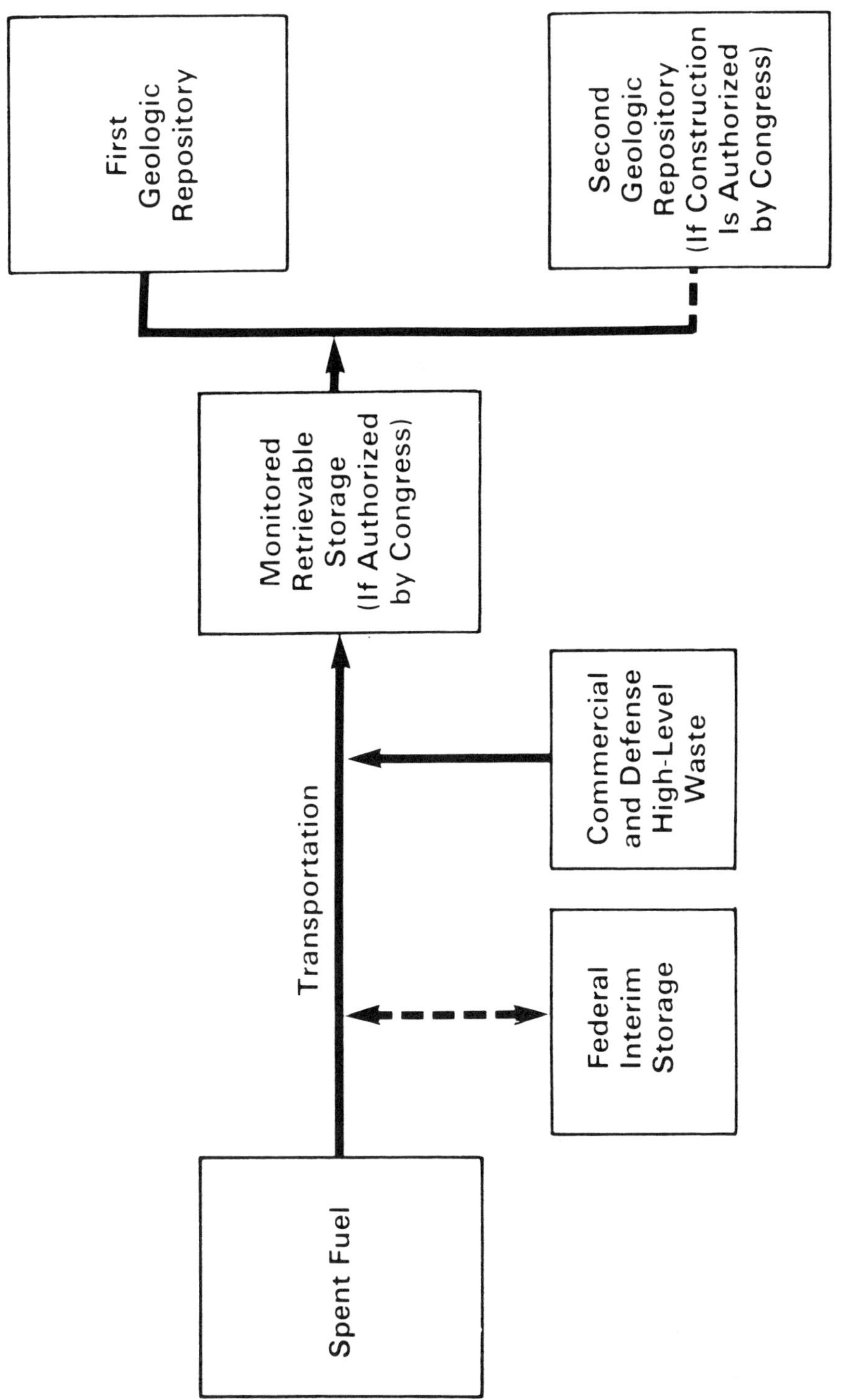

FIGURE D. Major Components of an "Improved (or Enhanced) Performance System"

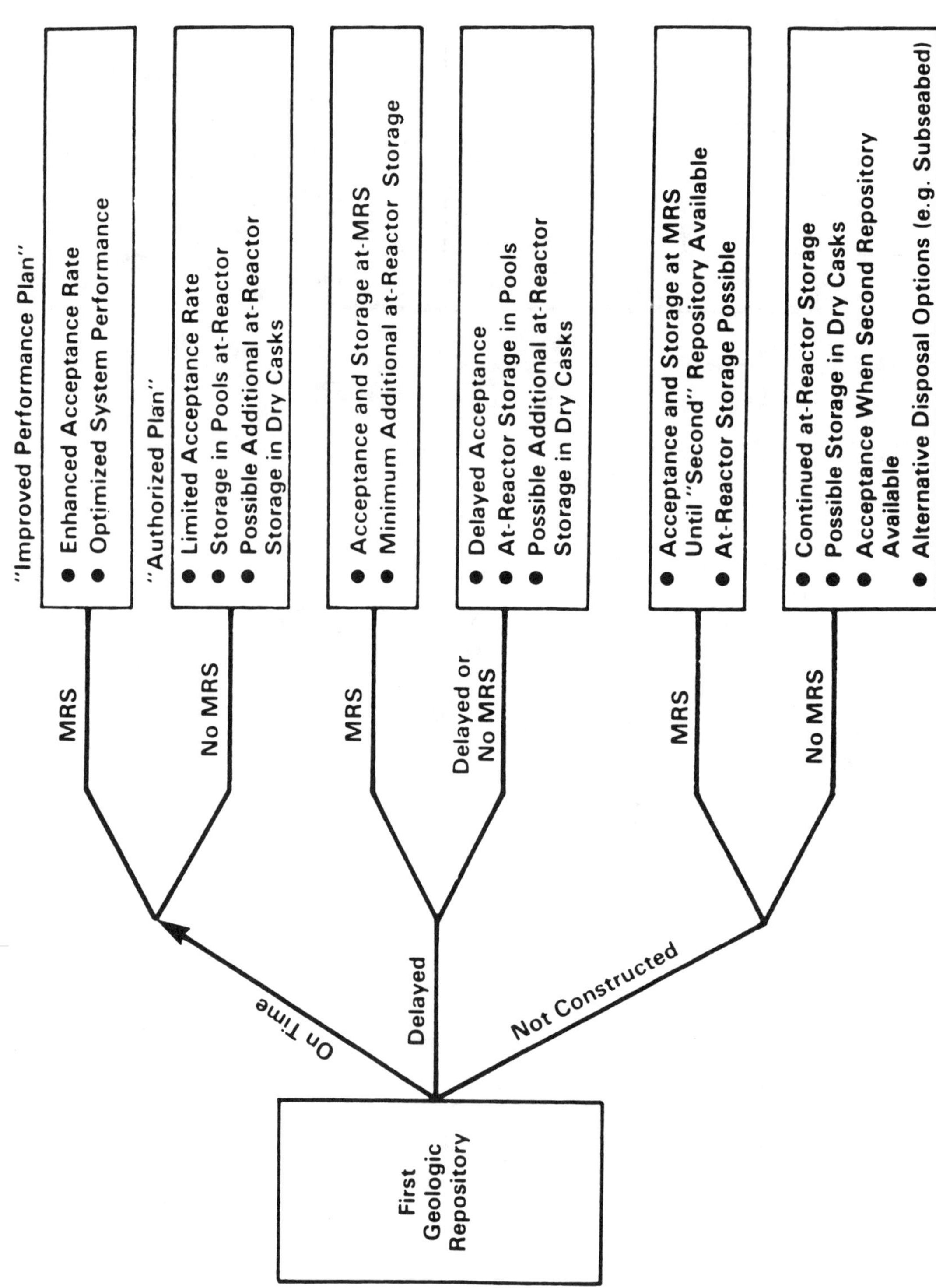

FIGURE E. Contingency Plans

Perspective on Nuclear Waste Disposal

DOE PERSPECTIVES ON INSTITUTIONAL INTERACTIONS

B. G. Gale
Siting Division
Office of Geologic Repositories
Office of Civilian Radioactive Waste Management
U.S. Department of Energy
Washington, D.C. 20585

D. J. Squires
Basalt Waste Isolation Division
U.S. Department of Energy
Richland Operations Office
Richland, Washington 99352

I. GENERAL

The Nuclear Waste Policy Act of 1982 (NWPA) (Public Law 97-425) states that "it is essential that host states, affected Indian tribes, and the public participate in the planning and development of geologic repositories, which are being evaluated by the U.S. Department of Energy." The NWPA places a great deal of emphasis on the U.S. Department of Energy's interactions with the affected and interested parties. The Office of Civilian Radioactive Waste Management's (OCRWM) public interactions include sharing information such as technical reports, plans, fact sheets, draft documents, and information and organizational brochures and pamphlets.

In addition to printed material, DOE-OCRWM shares information and seeks public participation through public meetings and hearings held in geographically affected areas and in central population centers. Because the NWPA authorizes and encourages the involvement of states and Indian tribes in the development of the Nation's waste disposal system, additional attention and planning have been directed toward intergovernmental and institutional activities.

A recent step toward assisting state and tribal involvement in the OCRWM process of managing the nation's nuclear waste is the establishment of an Office of Geologic Repositories (OGR) Desk Officer System. This system was established to provide a key headquarters liaison person for each affected state and tribe in the first and second repository programs. The Headquarters staff contact is available to answer questions and provide information about the OGR program on a daily basis. Updated public information, such as pamphlets, fact sheets and exhibits, are being developed for the program to better explain the OCRWM mission and the steps required by the NWPA.

This paper will focus on the institutional activities that the Basalt Waste Isolation Project has been involved with over the past several years and is involved with at the present time.

II. HISTORY

The Basalt Waste Isolation Project, or BWIP, is a component of the U.S. Department of Energy's (DOE's) Office of Civilian Radioactive Waste Management (OCRWM) Program, which is responsible for assessing the feasibility of using the basalts beneath the Hanford Site in south-central Washington State as a high-level nuclear waste repository. The prime contractor to DOE for BWIP is Rockwell International, Rockwell Hanford Operations. The BWIP activities are currently in the research and development stage and include studies in characterization of geology and hydrology of the basalts, geochemistry of the expected basalt environment, hydrothermal testing of waste forms and packages, conceptual design of a repository, heater and block tests at the Near Surface Test Facility, performance measures and models, and the design and construction of an exploratory shaft. The principal borehole at Hanford, which identified the proposed site for an exploratory shaft, was completed in May 1982, and the exploratory shaft starter hole was begun in November 1982. Further drilling of the exploratory shaft has been delayed since passage of the NWPA.

From 1977 to 1982, research and development activities were conducted as part of the BWIP. Results of these studies were documented in the BWIP Site Characterization Report (SCR) issued in November 1982, as required by the procedural rule of 10 CFR 60. The purpose of the SCR was to present the current status of the project and to identify future activities to be carried out to obtain the data needed to determine if the Hanford Site met the qualifications of a potential repository site.

In 1983, the U.S. Nuclear Regulatory Commission (NRC) prepared an analysis of the SCR, and documented for the DOE the issues the NRC considers important regarding the characterization program to assess the suitability of the basalt site. In addition, the DOE has received comments from the State of Washington, the U.S. Geological Survey (USGS), the Yakima Indian Nation, special interest groups, and members of the public. The BWIP evaluated and dispositioned the comments in consultation with the participating parties. The BWIP is presently conducting ongoing characterization activities and preparing a Site Characterization Plan (SCP) which is responsive to the resolution of these comments.

The NWPA provides for a number of procedural requirements (i.e., siting guidelines, a mission plan, environmental assessments, site characterization plans, etc.) that have resulted in the reprogramming of a number of BWIP activities. The siting guidelines and draft environmental assessments were issued in December 1984. The final EAs are scheduled to be released in December 1985.

III. STATE OF WASHINGTON INTERACTIONS

The State of Washington and the DOE have been discussing the management of high-level nuclear waste disposal at Hanford for more than five years. In 1979, the Intergovernmental Basalt Waste Isolation Project Working Group was formed to provide a formal mechanism to keep the State of Washington fully aware of the status of the BWIP. With the establishment of the State of Washington High-Level Nuclear Waste Management Task Force by Governor Spellman in August 1982, the objectives of the Working Group were transferred to the Task Force. In May 1983, the Governor signed legislation which established the Nuclear Waste Policy Review Board and Advisory Council, with responsibility for consultation covering the BWIP. The State of Washington Department of Ecology was the State organization responsible for implementing the provisions of the NWPA, and providing a status of the BWIP activities on a timely basis. In March 1984, legislation was passed by the State which transferred this overview function to the Nuclear Waste Board and Advisory Council. The Board remains the State of Washington organization that DOE interfaces with on BWIP activities, with assistance by the Department of Ecology. Day-to-day coordination and consultation activities associated with the BWIP continue to be handled by the Board with the assistance of the Department of Ecology.

The DOE has assigned priority to the implementation of the provisions of the NWPA relating to a participatory role by the State of Washington. The State has received a financial assistance grant so that it can conduct the overview of the BWIP activities in accordance with the NWPA. In addition to the financial assistance grant the DOE provided to the Nuclear Waste Board and Advisory Council, the State of Washington Legislature has been provided a separate grant by DOE. This grant provides funds for additional and independent review by the Legislature and its staff.

In July 1983, the State of Washington and the DOE identified respective negotiating teams to negotiate a Consultation and Cooperation (C&C) Agreement in accordance with requirements of the NWPA. Since July 1983, a number of meetings have been held by the teams, and a number of issues have been identified which the State believes should be resolved prior to consummation of the C&C Agreement. The State of Washington scheduled hearings throughout the State in February 1985 to describe what the C&C Agreement contains, and also to obtain from the public, comments and concerns which could be considered in the final C&C Agreement. These hearings were postponed and have not been held to date.

The State of Washington has been involved with the BWIP in an active role through the attendance at project briefings, meetings, etc., held with the NRC, USGS, and other organizations. In addition, the State also reviews project documentation and provides comments to DOE where appropriate. The Nuclear Waste Board and Advisory Council hold monthly meetings in Olympia, Washington, which BWIP staff attend and provide support whenever called upon to clarify or provide information covering project activities.

Many briefings and tours covering BWIP activities and facilities have been provided at the request of State agencies, commissions, departments, and legislative groups.

The Nuclear Waste Board and Advisory Council have acquired the services of a subcontractor who, among many other tasks, is assisting the State in its technical reviews of the program and developing a public information program by which the State plans to inform the public about BWIP activities and provide other information as requested by interested citizens. The DOE-BWIPO staff provides assistance to the State as requested in performing these consultation and public information programs.

The State of Washington operates a Reference Information Center in Lacey, Washington. The primary function of this center is to have and make available BWIP documents which the interested public may wish to review. With the assistance of the BWIP, essentially all BWIP documents are currently available, or are planned to be available, at the State of Washington Reference Information Center at Lacey.

IV. AFFECTED INDIAN TRIBES

A. Yakima Indian Nation

Pursuant to the NWPA, the U.S. Department of the Interior (DOI) notified DOE's Secretary Hodel (letter DOI to DOE, dated 3/30/83) that the Yakima Indian Nation was an affected Indian tribe and, therefore, full status as required by the NWPA should be afforded the Yakima Indian Nation by the DOE.

In July 1983, the Yakima Indian Nation and the DOE held their first meeting to discuss the potential for a financial assistance grant and the requirements for a C&C Agreement. In later discussions, the Yakimas decided not to negotiate a C&C Agreement until after the State of Washington and the DOE reached agreement. However, a financial assistance grant was established effective July 1983, and the Yakima Indian Nation has been involved with the BWIP in an active role through the attendance at project briefings, meetings, etc., held with the NRC, USGS, and other organizations. In addition, they also review project documentation and provide comments to DOE. To support the Yakima Indian Nation's review of BWIP activities, they have contracted for a number of consultants with expertise in the areas of hydrology, geochemistry, geohydrology, etc.

B. Umatilla Confederated Tribes

In July 1983, the DOI notified DOE Secretary Hodel (letter DOI to DOE, dated 7/13/83) that the Confederated Tribes of the Umatilla Indian Reservation were also an affected Indian tribe as set forth in the NWPA. In January 1984, representatives of the Umatilla Confederated Tribes were provided a briefing covering the BWIP activities, and preliminary discussions were held covering future proposals by the Umatillas. In May 1984, a financial assistance grant was established which provided funding to the Umatillas to identify what project activities they would be involved with and what level of overview they planned to conduct. Before the grant was in place, the Umatilla Indians' project participation was limited; however, since establishment of the grant, participation by them and their consultants has been more extensive. Consultation and Cooperations negotiations were initiated in August 1985.

C. Nez Perce Tribe

In September 1984, the DOI notified DOE Secretary Hodel (letter DOI to DOE, dated 9/17/84) that the Nez Perce Tribe of the Nez Perce Indian Reservation in Idaho was also an affected Indian tribe as set forth in the NWPA. In November 1984, representatives of the Nez Perce Tribe were provided a briefing covering the BWIP activities, and preliminary discussions were held covering future proposals by the Nez Perce Tribe. In December 1984, a financial assistance grant was established, which provided for the Nez Perce to identify what project activities they should be involved with and what level of overview they planned to conduct. Before the grant was in place, the Nez Perce Tribe's participation in the project was limited. The Nez Perce have not indicated a desire to begin C&C negotiations with the DOE.

V. OTHER FEDERAL AGENCIES

A. U.S. Geological Survey (USGS)

The USGS has provided varying degrees of support to the BWIP during the lifetime of the project. This support has included direct assistance to the project through interagency agreements covering the hydrologic and geologic activities during the initial years of the project.

The USGS also provided early reviews of the project through the DOE Technical Review Group. As noted earlier, the USGS was requested by the DOE to review the BWIP SCR, and provided comments primarily covering the geoscience disciplines. A number of meetings were held with the USGS to discuss these comments and, through a number of interactions, agreement was reached on disposition of the USGS comments. During the past two or three years, the USGS has been working with the BWIP through the Hydrologic Interagency Working Group to review and assist in development of models to be used in the characterization of the basalts in the Cold Creek Syncline, the area on the Hanford Site being considered for a potential repository. In addition, the USGS and DOE are currently holding meetings to review the planned program and results in specific technical disciplines such as hydrology, tectonics, geochemistry, geology, etc. The DOE plans to continue this exchange with the USGS and to work with them to receive their advice and resolve concerns covering the technical areas associated with the BWIP.

B. Nuclear Regulatory Commission (NRC)

The DOE has held meetings and discussions covering the BWIP activities with the NRC since 1978. During the past two or three years, these discussions have been more focused, and many more interactions have occurred. A number of meetings have been held with the NRC covering geology, tectonics, hydrology, performance assessment, waste package, geochemistry, engineering designs, quality assurance, etc. As noted earlier, the NRC provided many comments covering the BWIP SCR. The DOE and NRC have had numerous interactions to discuss the BWIP's disposition of the NRC comments. In addition, the NRC has assigned a full-time onsite representative to the BWIP to maintain day-to-day communications with the project staff. As required by the NWPA and 10 CFR 60, the DOE will continue to provide project information to the NRC during the site characterization and construction phases, if the BWIP is selected for these phases of the repository program.

C. U.S. Environmental Protection Agency (EPA)

Most all of the interactions that the DOE has with the EPA are performed by the DOE Headquarters (HQ) staff. The BWI Division provides support to DOE-HQ for review of and comment on EPA regulations.

D. Mine Safety and Health Administration (MSHA)

Support by MSHA included participation at a safety meeting to discuss mine safety, safety equipment operation, etc. The DOE has held discussions with MSHA staff covering levels of support that they might provide during construction of the BWIP underground facilities. However, no decision has been made in terms of future project support to be provided by MSHA.

E. U.S. Bureau of Mines

During the early years of the project, the Bureau of Mines provided an overview of the BWIP activities through the DOE Technical Review Group. Currently, the Bureau provides no support to BWIP.

VI. OTHER STATES

A. State of Oregon

In 1978 and 1979, the State of Oregon, under contract to Rockwell, provided support to the BWIP in limited geologic reconnaissance work associated with the basalts in the State of Oregon. Other activities have included tours and briefings for State of Oregon officials. In addition, the State of Oregon submitted a request for a financial assistance grant in 1983, which was not funded by DOE. However, recent discussions between the State of Washington and the State of Oregon indicate that the State of Oregon will be involved in the review of the BWIP through the State of Washington grant during the Site Characterization of the basalts, if the Hanford Site is selected for this phase of the program.

B. State of Idaho

In 1978 and 1979, the State of Idaho, under contract to Rockwell, provided support to the BWIP in limited geologic reconnaissance work associated with the basalts in the State of Idaho. Tours and briefings have been provided to State of Idaho citizens, when requested.

VII. PEER AND TECHNICAL REVIEWS

A. National Academy of Sciences (NAS)

In 1981, the NAS Waste Isolation Systems Panel held a number of meetings with the BWIP to discuss status of the project. The panel also obtained technical information covering BWIP which was used as part of the NAS report, issued in 1983, covering all the DOE geologic repository projects. Informal discussions continue to be held with some members of the panel.

B. Advisory Committee on Reactor Safeguards (ACRS) - Waste Management Subcommittee

In 1983, the DOE held meetings and discussions with the ACRS Waste Management Subcommittee in Washington, D.C., and at Richland, Washington. The purpose of the meetings was to provide the current status of the BWIP. The subcommittee provided a number of comments during the exchange, which are being considered by the BWIP in formulating the project objectives.

C. Technical Review Group (TRG)

Since early 1979, peer review groups have provided a review of the BWIP activities and assistance in establishing project objectives. During the past three or four years, membership and focus of the technical review groups have changed to provide more effective support covering most of the project activities. The current BWIP TRG participants include technical experts from universities, private consultants, industry, governmental agencies, etc. The TRG meets with DOE periodically (on a monthly basis) to provide support in establishing project goals.

VIII. PUBLIC

A. News Media

Tours and briefings are provided to the news media as requested. In addition, the BWIP makes information available about all project activities. Many television, radio, and newspaper interviews are provided when requested. The news media are also notified of all BWIP events. Examples include public information meetings, release of the draft Environmental Assessment document, draft EA public briefings, formal public hearings, DOE/NRC meetings, etc.

B. Public

The BWIP has conducted a number of public information meetings in Washington and Oregon, with the objective of providing the latest status of the project, identifying the future project activities, answering questions by the public during the meetings, and providing information that the public may desire. In addition, many tours and briefings of BWIP facilities and activities are provided as requested.

C. Reading Room

As part of the Hanford Science Center, a reading room containing most all of the BWIP documents has been established. The documents are available for public review. Many project documents such as Environmental Assessments, Site Characterization Reports, and many others are also provided upon request. In addition, and as noted earlier, the State of Washington's Reference Information Center in Lacey, Washington, makes available the same sort of BWIP documents.

IX. SUMMARY

Many interactions by the BWIP have been performed during the life of the project, with the organizations identified in this paper. These interactions have taken the form of reviews of documents and reports by the public, state and Federal agencies, affected Indian tribes, and many others. They have also included meetings, briefings, hearings, and tours of BWIP facilities by these groups and organizations. The objective of BWIP is to conduct an open information program and make available all data to those who wish to see it.

The extensive information process presently taking place is a major step in implementing the requirements of the NWPA for involvement by Federal agencies, states, affected Indian tribes, and the public. These actions will continue through the life of the project to allow for full participation in the Nation's repository selection process.

LEGAL AND JUDICIAL PERSPECTIVES ON THE DISPOSAL OF HIGH-LEVEL NUCLEAR WASTE

L. Manning Muntzing
Doub and Muntzing, Chartered
1875 Eye Street, N.W., Suite 775
Washington, D.C. 20006

ABSTRACT

Nuclear Waste Policy Act of 1982 creates a maze of legal requirements that are complex and subject to differing interpretations. The intervention of the courts should be anticipated as varying interests dispute the correct path through the maze to high-level waste disposal. A significant number of legal issues and unsettled questions exist that will need to be resolved. The challenge will be to minimize the delays that legal conflicts can produce. This can be helped by resolving uncertainties and conflicts early before they are on the critical path or negotiating resolutions, normally a difficult and costly endeavor.

I. INTRODUCTION

The Nuclear Waste Policy Act of 1982 was a major achievement in forging the institutional and technical forces that must deal with the solution of disposal of high-level nuclear waste in the United States. However, it was, as are all legislative solutions, a compromise. As a result, this legislative framework has inherent in it legal controversies that have already arisen and will continue to emerge. They will present roadblocks for the disposal program.

Siting waste repositories, which is the principal goal of the Act, will be opposed by state and local governments as well as other interested parties near a proposed site. One of the arrows in the opponents' quiver will be lawsuits. Opponents who choose to fight with lawsuits will look for arguments that the Department of Energy (DOE) has not complied with the Waste Act. Because the Act is complex, they will not need to look far.

This paper reviews the legal and judicial questions from two directions: first, it identifies some of the legal disputes that have arisen or can be anticipated, and, second, it discusses institutions that will be called upon to respond to or decide the disputes.

II. LEGAL DISPUTES THAT CAN BE ANTICIPATED

It is not possible to identify all of the legal problems that may arise, but we can identify the many issues which have already arisen and several we expect will arise. There have already been lawsuits filed concerning the method of establishing the fee for the utilities as well as contesting DOE's proposed siting of the first repository. These and other major issues that can be anticipated include the following:

- Legislative deadlines that compete with procedural requirements. Section 112 of the Act requires the Department of Energy to recommend to the President by January 1, 1985, three sites for technical characterization. However, DOE was not able to complete and get public comment on other Act requirements, such as environmental assessments, in time to meet that deadline.

- Role of the states and Indian nations in the approval process. Section 117 of the Act calls for DOE to enter into cooperative agreements with affected states. However, DOE has negotiated only with the State of Washington, and those negotiations have not produced an agreement.

- Siting issues involving complex procedures and technical analysis. DOE has published draft comprehensive environmental assessments, in excess of 1,000 pages each, on each of nine sites under consideration for a repository. Such extensive data requires considerable time for review by states and other parties affected. Some state officials have already claimed the assessments are insufficient compared to the Act's requirements in Section 112.

- Adequacy of DOE's repository siting guidelines. The Environmental Policy Institute (EPI) and several states have filed suit on this issue. EPI argues that DOE's guidelines do not meet the Act's requirements in Section 112. If this argument were to prevail, DOE would have to correct the guidelines, which would probably delay DOE's start of site characterization.

- Presidential and Congressional actions needed to reach final decisions. After characterization, Section 114 of the Act calls for DOE to recommend to the President a site to be approved for a repository. The President then is to recommend one site to Congress. Thereupon, a state or tribe may disapprove DOE's proposal. A state veto can be overridden only by a Congressional resolution. The state veto is a complex and unusual process, and no one can predict if it will work smoothly in the face of strong objection by the state selected.

- How many sites must survive characterization. DOE has interpreted the Waste Act to require three sites to be characterized, but after characterization, only one site needs to meet DOE's repository siting guidelines. Others argue that Congress intended that three sites must survive characterization. Therefore, DOE might characterize more than three sites to be sure that at least three survive. If after characterization less than three sites survive and if a court then finds that DOE must have three sites survive, DOE would have to characterize one or two more sites before it could proceed. This could delay DOE's program by the several years it would require to characterize these additional sites.

- Licensing requirements by the Nuclear Regulatory Commission and the regulatory role of other agencies. The Nuclear Regulatory Commission (NRC) has drafted licensing criteria for repositories as Part 60 of the NRC's rules and regulations. However, in licensing a repository, NRC will be attempting to license a first-of-a-kind facility, using NRC regulations that will not have yet been tested. Moreover, the Environmental Protection Agency (EPA) has authority to set generally applicable environmental standards for radiation to the accessible environment. EPA's standards may conflict with NRC's licensing criteria.

- Method of collecting revenues and handling disbursements including utility and public service commission accounting principles. Section 302 of the Act calls for funding of DOE's repository program through a fee charged to utilities of 1 mil per kWh. This provision sounds simple, but it was the subject of one of the first lawsuits under the Act. The suit was filed by utilities that wanted to pay based on a net generation to the busbar, rather than gross generation before subtracting electricity used in plant operation.

- Interim storage facilities. The Act provides at Section 141 that Congress can authorize a monitored retrievable storage (MRS) facility in addition to a repository. DOE has said it needs an MRS and will ask Congress to authorize one. However, DOE expects just as much opposition to siting an MRS facility as it expects in siting a repository.

III. INSTITUTIONS INVOLVED

The institutions that must deal with these issues are multiple. They include the courts, Congress, DOE, the states, industry, public interest groups, and the news media.

The final decisions on many of the issues will be made by the federal courts, potentially reaching the U.S. Supreme Court. Since state and local officials and organizations will oppose the repository, they will look very hard for legal errors on which to base lawsuits.

When it comes to matters of substance, if the decision makers have an adequate technical basis for their decisions, the courts will normally defer to those judgments and will not substitute their own. Also, with procedural matters where the decision makers have discretion, the courts will not substitute their judgment. However, if a decision maker is challenged on the basis of failing to implement the law, not following required procedures, or incorrectly interpreting the meaning of the statute, the courts will feel free to interpret the statute and decide what the Congress intended. The Waste Act is sufficiently complex that it will not be hard for opponents to argue that DOE did not follow the Act's provisions.

If the result is unacceptable because of the way the Waste Act was written, then it will be for the Congress to override the courts by writing new legislation. It seems clear that the leadership in the Congress that controls amendments or revisions to the Act does not favor rewriting or amending the law at this time. One reason why Congress is reluctant to amend the Waste Act is that it is becoming increasingly clear that the Waste Act was a delicate compromise. An attempted improvement on one provision could attract many other amendments.

The Department of Energy is in the middle of all the issues and must address some tough problems. The conflict between procedural requirements and Congressionally mandated deadlines remains. The issue of centralized or decentralized management of the program remains. A conflict exists over the amount of information required for licensing, at least of the first repository.

One of the important objectives of the Waste Act is to define the relationship between DOE vis-a-vis state governments and Indian tribes. The Act establishes an elaborate procedure to involve the states and tribes. However, to date, representatives of the state governments have been critical of DOE's repository guidelines and, in some instances, of DOE's failure to grant funds for the collection of site data.

DOE's reaction to the states and tribes is critical if DOE's repository program is to succeed. Some believe DOE's principal shortcoming is its inability so far to resolve conflicts with the states. DOE has announced which states contain its three preferred sites. All three states have filed lawsuits against DOE. Five other states have also filed lawsuits. This is not an auspicious beginning for such an important venture.

The states are also key players. So far, state governments have been adversaries, sometimes furious adversaries, of DOE. From each state's standpoint, an outsider -- DOE -- is trying to bury in the state nuclear waste that could, and thus should, be buried in some other state.

Industry played a significant role in aiding enactment of the Waste Act. Since then, industry has been less active. In order to balance the aggressive approach of other interested groups, industry needs to have a strong and vocal input. It is industry's waste being generated, and it is industry which will be most hurt if DOE's program fails. Moreover, it is industry which will pay DOE enormous sums to fund the program -- about 400 million per year and rising for new spent fuel, plus about 2 billion more for old spent fuel.

In view of the great importance to industry of permanent disposal of waste, industry should take a more active role with DOE. The attitude of many in industry seems to be that industry's role is to pay, and DOE's role is to solve the problem. That might be an acceptable arrangement if it were DOE's or the Nation's problem, but it is really industry's problem also.

Public interest groups are active and articulate. They believe that the Act will not work if the Act's deadlines are maintained. They may be correct that it is better for DOE to do it right than to do it fast. However, that idea should not become an excuse by DOE and its contractors to study the problem forever. After all, the Act's goal is to get the waste in the ground safely. Public interest groups are likely to be among the more active parties in filing lawsuits against DOE.

Part of the institutional framework is a regulatory function to be performed principally by the Nuclear Regulatory Commission (NRC) as well as by the Environmental Protection Agency. NRC's role is unusual. The Act gives NRC responsibility for licensing the repository and any MRS facility Congress authorizes. NRC and DOE both are new to their role. What if NRC and DOE reach an impasse?

In this network of institutions we should not forget the media, which in itself is one of the strongest institutions in America. In my opinion, most media descriptions of waste issues tend toward alarm. The message is -- the material is dangerous. If we bury it, there is no telling where it will go. For instance, the Portland _Oregonian_ ran a series of articles in May 1985, called "The Hanford Gamble." It began by saying that fundamental doubts about safety persist as the Federal Government pushes the concept of burying the nation's most dangerous radioactive waste four miles from the Columbia River, which flows through Portland. The article says the repository will leak, given enough time. If the nation's current amount of spent fuel could fill full-size pickups to their weight capacity, the resulting convoy would stretch from Portland to Salem, Oregon. But heat and radiation from unshielded spent fuel would quickly kill anyone who approached the trucks to drive them.

The point is not whether this article is objective or factual. The article is alarming. We must pay attention to such press reports because they carry a great deal of influence with the public.

With this complex web of issues and institutions, simple, direct, and expedited solutions are not likely. Disputes can be extensive, and only strong and resourceful leadership that is based upon sound technical assessments and data can minimize the legal battlefields that are potential sources of difficulty. However, early resolutions and negotiated agreements can help if they are sought with determination.

The Waste Act provides for DOE and states or tribes to enter into an agreement for consultation and cooperation. Section 117 (c) of the Act provides that soon after approval of a site for characterization, or upon the request of a state or tribe, DOE must seek to enter into a binding written agreement and shall begin negotiations. Unfortunately, the Act's agreement process has not gone well. DOE has had discussions with State of Washington officials which led to a draft agreement. However, higher state officials objected, arguing they were dissatisfied with DOE's position on liability and other issues. DOE and other candidate states, such as Nevada and Texas, have not yet begun discussions.

IV. CONCLUSION

In summary, the Waste Act of 1982 was an important step toward solving the nuclear waste problem in the United States. But the problem is difficult because the opposition to siting repositories is great. Opponents will file many lawsuits, which will both take advantage of existing legal controversies and create new controversies.

From a legal and judicial perspective, the many lawsuits expected will cause the courts to become the final arbiters of disputes, such as repository siting. The courts are both a good choice and a bad choice for this role. They are a good choice because most people respect the courts' authority, and the courts are relatively impartial. However, the courts are also a bad choice because they will have difficulty with technical arguments, and they are not the best arbiters of political issues, such as nuclear waste.

Surely this nation, however, with its technological abilities and institutional concepts, can rise above purely local objections and solve another national problem as we have so often done in the past so well.

A UTILITY'S PERSPECTIVE ON THE NUCLEAR WASTE POLICY ACT

William W. Berry, Chairman
Dominion Resources, Inc.
One James River Plaza
Richmond, Virginia 23219

ABSTRACT

The Nuclear Waste Policy Act is especially important to utilities because their customers pay for the disposal program, and the program is vital to nuclear operations and reconsideration of the nuclear option. DOE's accomplishments in implementing the Act are noteworthy, but we are concerned that some of them have been achieved later than specified by the schedule in the Act. We make recommendations regarding disposal fees, defense wastes, and shipping casks. Virginia Power has adopted a three-part strategy relying mainly on developing dry cask storage to solve the company's interim storage problems.

DISCUSSION

I'd like to give you a utility view of this country's progress in implementing the Nuclear Waste Policy Act. For several reasons, the Act is especially important to those of us who operate the nation's nuclear power stations.

First of all, the money in the Nuclear Waste Fund that finances programs under the Act comes not from government revenues, but from utility customers. Utilities are paying into the fund at the rate of over $300 million a year, and annual payments will increase as more nuclear units come on line. In addition, payments for past nuclear generation will amount to $2.3 billion. Building the high-level nuclear waste management system is the largest civil works project in American history. Utilities are responsible to their customers for seeing that the money is well spent, and our regulatory commissions hold us accountable. For that reason, we're paying close attention to the management of the Nuclear Waste Fund.

The second reason for utilities' keen interest in the Act is that the success of the nuclear waste program is vital to the continued operation of the nuclear units now in place. If any reactor reaches a point where the utility cannot ship its spent fuel or store it on site, then the unit must shut down. Loss of operating nuclear units would drive up electric rates and could have damaging effects on the utilities' ability to provide reliable service.

Virginia Power already faces a situation that could lead to reactor shutdown unless the measures we've initiated are fulfilled on time. The spent fuel pool at our Surry nuclear station could be the first in the country to run out of room. I'll discuss that situation and the development of new means of interim storage in a few minutes.

The third spur to utility interest in the Act is that creation of a permanent disposal system for nuclear waste is one of the many preconditions that must be met before nuclear power will again be considered for meeting future generation needs in the United States. By the year 2000, America is likely to need at least 100 to 200 gigawatts of generating capacity in addition to the power plants now in place and under construction. Last February, the Edison Electric Institute issued a special report on nuclear power. It declared that no utility would now choose nuclear power because of the risks and uncertainties involved, unless comprehensive reforms were implemented to remove the institutional and regulatory impediments. We believe it's in the national interest to take actions that will again make nuclear power an option American utilities can consider to meet future capacity needs. Timely actions to assure permanent waste disposal are essential to achieve that objective.

Since the difficulties in creating a repository are primarily institutional rather than technical, passage of the Nuclear Waste Policy Act went a long way toward establishing the confidence of industry in the country's waste program. The Act was a landmark achievement, culminating years of effort to develop a firm statutory basis for a successful program. But we recognized when it was signed that enactment was just the first step, and that successful implementation would prove equally challenging. The nuclear waste program tests the ability of government to carry out an exceedingly complex mission over a long period of time while dealing with forces that arise to oppose nearly every action.

Against that background, DOE's accomplishments are noteworthy. Last winter, DOE published the criteria for selecting a repository site and gave a preliminary indication of the five sites to be nominated for the first repository. The Department indicated the preferred three sites that are expected to be recommended for site characterization. Site selection is clearly both the backbone of the Act and its most politically troublesome aspect. We're pleased that DOE is proceeding with the difficult issues of permanent geological disposal. DOE must work diligently to complete the site selection phase so that site characterization can begin. I agree with DOE's plans to characterize three sites, as called for in the Act, and I see no need for the fourth site that's been suggested.

DOE is completing a proposal to be submitted to Congress for a Monitored Retrievable Storage facility, as required by the Act. In April, the Department announced that it would recommend a primary site and two alternate sites for the facility in Tennessee. If the MRS facility is approved by Congress, it may become an integral part of the final geological repository system, but of course, the MRS neither replaces a geological repository nor detracts from the need for one. DOE's recent concept of the MRS as part of an integrated waste handling system seems sensible. The Department envisions an MRS as a facility to receive and package spent fuel in preparation for final disposal in a geological repository. This arrangement could contribute to efficiency, particularly if it permits reduction in the volume of wastes and a consequent reduction in shipment-miles as spent fuel travels from reactors to repository.

We're pleased with last year's appointment of Ben Rusche as director of DOE's Office of Civilian Radioactive Waste Management. One of Mr. Rusche's first moves was to begin strengthening headquarters control of the policies and functions that are implemented in DOE's field offices. We welcome this action because the nuclear waste program needs firm central management.

DOE has proceeded diligently to implement the Nuclear Waste Policy Act, and has addressed many of the problems we've pointed to in the past. However, we continue to be concerned about the timely completion of the many steps that must be satisfied to achieve acceptance of spent fuel in 1998, which is the date specified by the Act.

Several of DOE's accomplishments have been achieved later than specified by the schedule in the Act. The recommendation of three sites for characterization was supposed to have been forwarded to the President by January 1 of this year. As I mentioned, DOE indicated its preferences for three sites in the draft environmental assessments that were issued for public review and comment on December 20 of last year. However, the official recommendation will not be made until this November. That will be almost a full year late. The Act called for a specific site recommendation for the first repository in March of 1987. DOE has announced that the recommendation won't be made until four years past the deadline. A full MRS proposal was to have been made to Congress by June 1 of this year, but DOE doesn't expect to complete it until January.

The question of how any particular overdue item affects the 1998 deadline can be argued at great length. But it's the cumulative effect of delays early in the program that worries us. If the early decision milestones are not reached more or less on schedule, then the goal of having an operating repository by January 1998 is not likely to be fulfilled.

The schedule is a significant component of the Act. Decision-making in nuclear waste management is the kind of activity that people like to put off, particularly because there's no immediate crisis that forces decisions to be made. Without the discipline of the schedule, the program would be subject to the forces of drift and delay that prevailed before the Act. The Act created both a process and a schedule, and they must be regarded as equally important. Since the Act was passed, DOE has been paying great attention to process and emphasizing extreme thoroughness. For example, there are more than ten thousand pages of environmental assessments for the candidate sites for repositories. Opponents will try to prolong the nuclear waste program by insisting on the preeminence of process and the subordinate role of schedule. They would delay the program to death. Those of us with a strong interest in making the program work must constantly press for adherence to a reasonable schedule.

We're pleased that DOE has repeatedly stated its obligation to take spent fuel beginning in 1998. But we'd rather not test the government's ability to keep that pledge. We'll sleep much easier as DOE shows its ability to meet critical dates.

We're encouraged by the fact that DOE and the Nuclear Regulatory Commission have begun to work together in a cooperative way to resolve licensing issues early in the program. NRC should work with DOE during site characterization to settle matters that otherwise would wait until the Commission's formal review of the license application. Much of this necessary review might thus be done even before DOE applies for a license.

Obviously, merely opening the repository will not be sufficient. We urge the government not only to hold to the promised date to begin accepting spent fuel, but to get repository operations up to speed and start clearing the spent fuel backlog as soon as possible.

Based on developments so far, we believe the program can be carried out without an increase in the program fee collected from utilities. The Fee Adequacy Report that DOE released in February indicates that there's no need for an increase for the time being. The industry will look very hard at any recommendation to raise the fee in the future, especially in view of the surplus funds generated by our one-time payments a few months ago. DOE gave utilities the option of choosing one of three schedules for paying the fee for nuclear electricity generated before April of 1983. To improve the cash flow of the nuclear waste program, the Department strongly encouraged utilities to choose the option of making a one-time payment this past June. So many companies responded favorably that DOE collected more than $1.4 billion, giving the program fund a significant surplus. Interest on the fund should defray cost increases due to inflation for some time to come. Since the President decided this year to include defense wastes in the repository, the defense effort should pay its fair share of the costs in a timely fashion, like the pay-as-you-go formula that industry is following. Deferral of defense payments would place an undue burden on the civilian program.

Developing storage facilities won't do much good unless we're ready and able to move spent fuel to them. We need to have approved spent fuel transportation casks that are designed to serve the variety of utility needs for different types of casks. For example, Virginia Power and a number of our sister utilities would prefer combined storage/transport casks. Other utilities could use simpler casks. While improved casks are being developed, the Federal Government should permit the use of existing serviceable casks to meet all the various needs. Early this year, DOE issued a report recommending maximum government involvement in development of shipping casks and transportation systems services, and creation of entirely new cask designs. But the Nuclear Waste Policy Act mandates that DOE utilize private industry to the fullest extent possible in each aspect of spent fuel transportation, and we at Virginia Power believe that a general policy of starting over on cask design is both unnecessary and unwise. Private industry has carried out cask development for many years here and abroad. Rather than start from scratch in every case and spend a planned $80 million, the Department should encourage continued development of improved cask designs. Certainly DOE involvement is essential. But we urge that the Department's primary role be in issuing performance specifications for cask procurement and shipping services.

While permanent disposal facilities are being developed, the Nuclear Waste Policy Act placed responsibility for interim storage on the utilities, and included provisions to help them fulfill that responsibility. Virginia Power's experience at our Surry station provides an example of how this is working out. There are two nuclear units at Surry. Unit 1 went into operation in 1972, and Unit 2 followed in 1973. As I mentioned, we're about to run out of spent fuel storage space in the Surry pool. I'll sketch the potential problems we face and then tell you how we're trying to solve them.

After we refuel Unit 1 early next year, there won't be room in the pool to hold all the assemblies from either unit. That's the situation we'll face unless we can add more storage outside the pool. This loss of full core discharge capability would not be a safety consideration. But it would prevent us from doing maintenance that requires unloading all the assemblies from either unit.

When we refuel the other Surry unit late next year, no room would be left in the pool for any more spent fuel. The units would have to shut down when the fuel in their core then is exhausted. This would happen to Unit 1 in the fall of 1987, and Unit 2 in the spring of 1988. Loss of the two Surry units would increase customer costs by over $200 million a year. Clearly, the spent fuel issue is a major concern for Virginia Power. We've already put in new racks that allow for closer packing of assemblies in the Surry pool, and we can't rerack again because the structure won't bear the added weight. A new pool would be the most expensive solution. And we can't wait the seven or eight years it would take to build and license a new pool. We developed a three-part strategy to meet the problems. First, we're developing the option of transshipment. We're taking the technical, legal, and licensing steps necessary to enable us to transship up to 500 assemblies to our North Anna nuclear station some 150 miles by road. Its pool has room to take Surry spent fuel until 1992. We applied for NRC permission for transshipment in 1982 and received the go-ahead just this month.

But transshipment obviously doesn't solve the fundamental problem -- it only postpones the day of reckoning. And it's the most politically sensitive option. The strategy we hope to adopt is dry cask storage, beginning next year. Dry storage casks are similar to spent fuel shipping casks and may eventually be used to ship the assemblies to the final repository. They'll be stored above ground on a specially constructed concrete pad.

In October, 1982, we applied to the NRC for a license for dry cask storage, which had never been used in the United States but was being carried out successfully abroad. We expect the NRC to issue our license soon -- it's essential if we're to avoid transshipment. With the permission of the NRC, we began construction of the storage facility in July, and we expect it to be ready in spring.

Last March we agreed to make our dry cask project part of the cooperative program of research on interim storage under Title II of the Nuclear Waste Policy Act. In addition to the first five casks at Surry, the demonstration involves DOE experiments on casks with our spent fuel at the Idaho National Engineering Laboratory. DOE will test the limits of the casks and try storage of consolidated fuel rods. Virginia Power, with partial support from the Electric Power Research Institute, is paying 75 percent of the costs for the entire project. DOE is funding the remainder with appropriated R&D funds.

DOE started removing our spent fuel in June. They'll take 121 assemblies, about equal to the number that goes into the pool from two refuelings. This is the third part of our strategy -- the project not only provides valuable research data, but also buys us time. If DOE continues to remove spent fuel at the planned rate, we'll be able to maintain a full core reserve until late next year, when our dry cask facility should be operational.

Other utilities will soon have to confront space problems. About half of them will have filled their pools before the government takes their spent fuel beginning in 1998. It's crucial to demonstrate new means of interim storage soon.

NRC can help through the procedures it establishes to carry out Section 134 of the Nuclear Waste Policy Act, which was expected to expedite the licensing of interim storage operations and facilities. This provision will be most important as utilities adopt new means to store spent fuel. NRC is deliberating now on how to discharge its Section 134 responsibilities. We hope that the outcome will be consistent with the spirit of the Act.

We commend DOE for its accomplishments in fulfilling the process of the Nuclear Waste Policy Act and urge the Department to give increased attention to the equally vital timetable. The technology for a geological repository is relatively simple; the Act allowed all of 15 years for it to be established; and adequate financing has been provided. It's certainly reasonable to expect that the statutory date for opening the repository can be met. We hope that everyone affected by the high level waste repository or involved in developing it will cooperate to assure that it meets the nation's urgent need for timely, permanent disposal.

THE COSTS AND RISKS OF DOING BUSINESS WITH THE PROPOSED MONITORED RETRIEVABLE STORAGE SYSTEM

Suzanne Hughes Rhodes
Governor's Division of Energy Policy
P.O. Box 11450
Columbia, South Carolina 29211

John J. Stucker, Ph.D.
Governor's Office
P.O. Box 11450
Columbia, South Carolina 29211

ABSTRACT

During the three and a half years that the Nuclear Waste Policy Act has been law, the Department of Energy (DOE) has been proceeding, albeit with some delays in its schedule, with its responsibility to establish a permanent repository for the disposal of this country's high-level nuclear wastes. As an adjunct to its responsibility, the DOE has recently proposed a major new program objective. A Monitored Retrievable Storage (MRS) facility has been suggested to relieve utility responsibility for spent fuel storage in 1996, two years prior to the most optimistic current plans for the opening of a permanent repository. The MRS would operate in conjunction with a proposed "integrated packaging and handling system" which would accomplish rod consolidation of most spent fuel before permanent emplacement in the repository. This large-scale MRS facility, if approved by Congress and successfully licensed, would mark the beginning of an unnecessary and expensive large-scale federally sponsored temporary storage program for civilian nuclear wastes. Furthermore, the MRS proposal could be expected to divert DOE attention from the permanent repository program which DOE is charged by law to implement.

I. SOUTH CAROLINA'S EXPERIENCE WITH NUCLEAR WASTE STORAGE

South Carolina first became concerned about federal plans for temporary storage of civilian spent fuel in the 1970s. One of the proposals put forward during that decade was that spent fuel be stored "away from reactors" at federally owned AFR sites; one of the proposed AFR sites was in South Carolina.

South Carolina has for 35 years been the host for temporarily stored liquid high-level defense nuclear wastes at the Savannah River Plant (SRP). An intense, bipartisan effort of Energy Secretary Edwards, Governor Riley, and the South Carolina Congressional Delegation was required to finally secure federal funds for the Defense Waste Production Facility (DWPF) which is now being designed and constructed at the SRP. This facility, the first of its kind in the United States and the largest in the world, will solidify defense wastes into a glass form and encase the wastes in special casks. These casks will ultimately be shipped to a federally operated repository, where the wastes will be buried thousands of feet underground.

Because of the decades-long problem in securing federal commitments to permanently care for the SRP defense wastes, South Carolina has repeatedly urged the DOE to cease emphasizing temporary storage of commercial wastes at any federal facility. South Carolina is among those fearing that DOE's reordering of priorities to emphasize temporary storage facilities would divert efforts away from the goal of permanent disposal. Governor Riley, as chairman of the South Carolina Nuclear Waste Consultation Committee, has signed six letters during the last two years to the Secretaries of Energy expressing this concern.

II. NUCLEAR WASTE POLICY ACT OVERVIEW

Section 111(a) of the NWPA states that Congress finds that a national problem has been created by the accumulation of radioactive wastes and that 30 years of federal efforts to devise a permanent solution have not been adequate. The Act clarifies that the federal government has the responsibility to provide for permanent disposal in an underground repository and that the nuclear power utilities are responsible for providing for spent fuel storage until the repository is available.

Sections 131 and 132 describe how utilities with demonstrated serious storage difficulties will have an opportunity to store spent fuel at a Federal Interim Storage facility prior to the availability of the repository.

Section 141 of the NWPA requires only the study of Monitored Retrievable Storage and a proposal for the need for and feasibility of MRS. In addition to a description of the need for an MRS, the proposal is to include a plan for funding "construction and operation of such facilities, which shall provide that the costs of such activities shall be borne by the generators and owners of the ..spent nuclear fuel.." (see Section 141(b)(2)(B) and Section 141(a)(4)). Section 141(a)(5) expresses Congress's intent that work on a repository should proceed regardless of what action is taken on an MRS.

As specified in Section 112, DOE is to consider each of several sites in the West, with each being a possible "first" site for a nuclear waste repository. These sites, particularly those in the states of Washington, Nevada, and Texas, will be closely studied for the next five years. At the same time, DOE is directed to examine other rock formations in other parts of the country as possible sites for a "second" nuclear waste repository. The Great Lakes region and the Appalachian Mountain ridge are now under consideration in this regard; several hundred areas will be narrowed to a group of approximately fifteen or twenty rock bodies in the next few months.

III. THE IMPORTANCE OF THE YEAR 1998

From the time of the earliest Congressional hearings, the leadership of the Office of Civilian Radioactive Waste Management has repeatedly stated that, although the schedule defined in the Nuclear Waste Policy Act (NWPA) is important, the quality of the work product is more important than the schedule. South Carolina has wholeheartedly supported that DOE position. However, the MRS program, a major programmatic diversion, is now under serious consideration; one of the reasons put forward by DOE for the MRS program is a legal commitment by DOE to receive spent fuel from utilities in 1998.

There are many dates specified in the NWPA. The repository siting guidelines, environmental assessments, defense waste plans, and numerous other projects are all to be developed, and each has a specific delivery date. The NWPA mentions the year 1998 only one time, in the Section which describes the Nuclear Waste Fund, not in the Section on MRS. The NWPA states that "in return for the payment of fees..beginning not later than January 31, 1998, (the DOE) will dispose of the high-level radioactive waste or spent fuel.." If the words had been "accept" or "store" wastes, the present MRS program proposal might be valid, but the NWPA clearly defines "disposal" as emplacement in the repository "with no foreseeable intent of recovery."

In fact, during the first 33 months of the NWPA no important dates have been met. Some of these dates were not met for very good technical reasons, others because of inter-agency problems, and some for the most partisan of political reasons. DOE has recently stated that the repository program as a whole is on schedule if there are no serious technical or legal challenges to DOE activities. Even if there is delay in the repository program, that is not necessarily sufficient basis for the establishment of an MRS. Thus, the MRS appears to be an unnecessary element of the DOE program which risks billions of dollars of the Waste Fund, would cause unnecessary DOE staff efforts, and could easily compromise the goal of establishing a permanent repository.

IV. CRITIQUE OF THE DOE MONITORED RETRIEVABLE STORAGE NEED AND FEASIBILITY REPORT

The Department of Energy is required by the NWPA to submit a report to Congress on the need for and feasibility of Monitored Retrievable Storage. Earlier this summer the DOE announced that this report will be submitted in draft form this month and that the final report will be submitted in January of 1986. Two Department of Energy reports were issued in April 1985: Screening and Identification of Sites for a Proposed Monitored Retreivable (sic) Storage Facility (DOE/RW-0023); and The Need for and Feasibility of Monitored Retrievable Storage - An Analysis (DOE/RW-0022). In addition, the July 1985 final Mission Plan for the Civilian Radioactive Waste Management Program (DOE/RW-0005) was released this July. In these three documents, DOE presents its case for the MRS; these documents are the basis for our comments.

South Carolina anticipated that the need for an MRS, as stated in the Need and Feasibility report, would be based on lack of affordable utility options to store spent fuel at reactor sites, as specified in the NWPA.

South Carolina anticipated that the feasibility of MRS would relate to political support or opposition during the licensing phase, which is likely to be tied to a public perception of whether or not the MRS would in fact be "temporary" and whether or not the repository would be available for spent fuel disposal early in the next century. In addition, the political feasibility of MRS was anticipated to be dependent upon its cost and the manner in which DOE would plan to charge utilities for the storage of their spent fuel. The NWPA clearly requires the utilities which benefit from MRS to pay for its costs. The technical feasibility of MRS was not anticipated by South Carolina to be at issue. As Governor Riley stated to Secretary Herrington last June, "The feasibility of temporary storage of nuclear wastes has never been questioned; temporary storage has always been this country's policy by default."[1]

The April '85 MRS Need and Feasibility report was unlike the report which South Carolina anticipated. It did not present substantive analytical information. Instead it presented a list of general strategies which could be characterized as "desirables"; the usefulness of the desirables was based on loose, self-serving concepts. The report suffered from lack of documentation, lack of scientific method, and lack of peer review. The usefulness of an MRS as presented in the Need and Feasibility report appears to hinge on several tentative and unfounded judgments. The report suggests an attempt to rush in prematurely with an unnecessary and expensive rescue based on a perception of utility helplessness in storing spent fuel.

The need for MRS, as presented in the April report, appears to be based solely on Energy Information Agency (EIA) mid-case electrical growth estimates.[2] Gross accumulations of spent fuel are put forward as reason enough for federal intervention into utility management.

In projecting the need for MRS, the DOE has neglected to consider any recently demonstrated utility management techniques such as extended burnup of fuel and on-site dry cask storage, even though these particular techniques are now undergoing licensing review. In addition, sophisticated new pool expansion opportunities have been explored and implemented by some utilities, on-site rod consolidation has been accomplished with mixed success, and expanded inter- or intra-utility transshipments have been negotiated. Other spent fuel management techniques cannot be ruled out for future potential use.

The DOE rationale of anticipating that none of these techniques will be of future use is puzzling. Each of these techniques increases utility opportunities to store spent fuel and minimizes the role of federal intervention in spent fuel storage. There are many indications that the proposed MRS option is more expensive and more complicated than the utility options which would be foreclosed by the MRS.

The April Need and Feasibility report did not address either the economic or the political feasibility of the MRS. Undocumented cost estimates were stated as 0.8 billion-1.2 billion; independent analyses range as high as three times this estimate[3]. Inappropriate cost comparisons are presented in the Need and Feasibility report. The cost of the demonstration dry cask storage program, for example, is compared with the estimated incremental storage costs of an operational full-scale MRS. The "cost savings" of several billions of dollars which could be gained through extended fuel burnup[4,5] (which might be substituted in place of large-scale rod consolidation) are not recognized. No economic consideration appears to be given to the costs of research, development, and construction of the rod-consolidation program, which is simply assumed as necessary to the "integrated system."

It is important that economic data in the final Need and Feasibility report be rigorous so that public policy planners and utilities can make an informed decision to chose or not chose the MRS storage option. A more complete document is expected in September 1985, and it may be more specific. At this time, however, it appears that the stronger utilities are expected to subsidize the weaker utilities, and there is absolutely no incentive for efficient or innovative utility management of spent fuel.

The "need" for rod consolidation is, inappropriately, the bulwark of the MRS proposal. Rod consolidation experience to date is with small quantities of rather old spent fuel. Nevertheless DOE has chosen to assume that there will be no problems in moving toward mass consolidation of fresh spent fuel. Upcoming demonstration projects may support a commitment to large-scale rod consolidation as a waste management tool, or may raise serious questions regarding technical design and operations, employee exposures, wastes generated, or economic viability. Experience is required before an analysis of rod consolidation costs can be weighed against its benefits to repository operations.

DOE claims that an MRS as part of an "integrated system" will eliminate the need for construction and licensing of packaging and handling facilities at the repository. However, this will be the case only if western fuel goes to the MRS before it goes to the repository (which would increase transportation) and only if there are no contingency packaging and handling facilities at the repository (for example, to repair fuel found to be damaged after shipment).

Peculiarly, the recently released final Mission Plan describes the MRS as the planned method for receiving spent fuel beginning in 1996, rather than the normally anticipated 1998. The Mission Plan states that after the MRS is full (apparently at approximately 15,000 metric tons, but perhaps at another capacity) spent fuel would be stored at utility sites. This new DOE plan is a reordering of the priorities established in the NWPA, which is that the utilities are to be responsible for storage of spent fuel until the repository is available, and the MRS is a back-up option in case of serious repository delays.

V. THE IMPORTANCE OF INNOVATION

The Need and Feasibility report introduces the new concept of unit trains to transport spent fuel. Such a transportation system, in conjunction with other measures, could provide extra safety and assurance to the concerned public, and should be applauded. However, this concept should not be limited to shipments from the MRS to the repository. Given the large quantity of spent fuel to be shipped from utilities, unit trains and truck convoys can and should be considered for all major shipments, including reactor-to-repository shipments.

There seems to be an underlying philosophy throughout the Waste Management program which discourages creative engineering and management at the utility level. While DOE presses to assume utility responsibilities, there is little reason for utilities to plan to store at reactor sites although this is directed in the NWPA. One way of eliminating the costs and risks associated with MRS construction and licensing would be for DOE to devise methods for rewarding aggressive utilities which minimize their spent fuel storage demands; for example, Waste Fund credits could be allocated. At the present time utilities which manage their own spent fuel may be penalized with extra charges rather than rewards.

The possibility of little or no near-term growth of the nuclear industry, in conjunction with innovative fuel management techniques, should be actively recognized. Anticipation of a "low-growth plus innovation" scenario could save tens of billions of dollars in MRS construction, repository, and utility operating costs. Few of us can imagine a strong nuclear industry which lacks a permanent repository for its wastes. Years ago the Congressional Office of Technology Assessment targeted public confidence as the most critical factor in establishing a permanent repository. Few of us would disagree with that conclusion. If the nuclear industry is to survive its present doldrums, it will have to develop public confidence through the design, sale, and efficient performance of light-water reactors; however, the establishment of a repository is equally important to public confidence.

VI. CONCLUSIONS

South Carolina is concerned with the in-house methods used in developing the early MRS documents and the related reference materials. The lack of peer review of these documents is apparent and inappropriate to both the National Environmental Policy Act and the spirit of the NWPA. South Carolina has formally requested to participate in a peer review process of the new MRS.

While it is important that the Office of Civilian Radioactive Waste Management programs maintain a rigorous schedule, the year 1998 is no more binding than any of the other dates stipulated in the NWPA - each of which has slipped. The DOE task is to establish a suitable repository, and the schedule specifies 1998 as the target date.

DOE is now 3-1/2 years into a 40-year program. DOE should continue permanent geologic repository siting activities with careful attention to rigorous technical processes and full communication with involved parties. At the same time, cooperative programs should be initiated with utilities to streamline waste management activities at reactors in conjunction with forthcoming repository operations. If the repository experiences serious delays, the MRS can be actively considered on the basis of specific factors of need, when more information is available concerning the location and geologic media of the first repository. At that time, we will have considerable experience with rod consolidation, extended fuel burnup, and various cask storage opportunities, as well as other reactor-site spent fuel management options.

At this time, insufficient information is available to justify a need for an MRS. In fact, the multi-billion dollar MRS as recently proposed may be irrelevant to repository operations. DOE should devote its resources to the permanent repository program, particularly improving the technical basis for its decisions and improving communications with the parties which must be involved to establish successful implementation of the Nuclear Waste Policy Act.

REFERENCES

1. R. W. Riley letter to Secretary Herrington, June 29, 1985.
2. Reactor-Specific Spent Fuel Discharge Projects, 1984-2020, PNL-5396 (1985).
3. Westinghouse Technology Services Division, Preliminary Cost Analysis of a Universal Waste Package Concept in the Spent Fuel Management System, Madison, Pennsylvania (1984).
4. W. A. Franks, L. Geller, The Benefit of Extended Burnup in Fuel Cycle Cost, S. M. Stoller Corporation (1985).
5. General Accounting Office letter report B-202377 to Richard L. Ottinger, November 18, 1981.

THE STATES AND THE NUCLEAR POWER INDUSTRY: ISSUES OF COMMON PERSPECTIVE IN NUCLEAR WASTE FACILITIES DEVELOPMENT

J. H. Davenport
Duryea, Murphy, Davenport & Van Winkle
Attorneys at Law
Evergreen Plaza Building
711 Capitol Way
Olympia, Washington 98501

I. INTRODUCTION

Before and during congressional consideration of the Nuclear Waste Policy Act and since it's enactment, the general assumption has been that the states and the nuclear power industry would be opponents regarding the Department of Energy's agenda in siting and developing nuclear waste facilities. That assumption is, like most, simplistic. When individual issues are analyzed, the states and the nuclear power industry can identify some common perspectives while retaining their independent objective.

The particular issues which this paper will analyze, seeking to identify the common or divergent points of view of the states and nuclear power industry include: 1) deliberateness and completeness of site investigation, including state independent monitoring, testing, and data collection; 2) transportation of nuclear waste; 3) collection, administration, and expenditure from the Nuclear Waste Fund; 4) liability for the nuclear waste program and the Price-Anderson Act; 5) equity between the states/multiple facilities; 6) licensability and the state role in licensing; and 7) public perception of and fairness objectivity of site selection.

One caveat should be remembered when considering the author's remarks here. Though I have sufficient daily contact with representatives of potential repository states to have, I think, a reasonably good understanding of the desires of those individuals, my interaction with individuals within the nuclear power utilities industry group is not that frequent. Consequently, my observations regarding their desires is based on some general knowledge about the industry itself.

II. REPOSITORY SITE INVESTIGATION

The Nuclear Waste Policy Act contemplates a site investigation process which contains a cursory site review and a complex site characterization. This two step investigation supports the Department's nomination and recommendation of sites for characterization and recommendation of a site to the President for development, respectively.

From the potential host state's perspective, the process which the Department should use in its site investigation, should be deliberate, scientifically sound, and objective. The issue of concern to the state is the health and safety of the state's residents and the protection of the quality of its environment. The state's desire (and presumably also DOE's desire) is that if a repository were developed within the state it would never have any off-site releases which would harm future generations of residents. Consequently, the state perspective favors a site investigation process which is complete and honest. It is important, therefore, that site investigators not hold their mission of building a repository above the objectiveness with which they are willing to treat the information which they may discover about the site.

Officials within state government also are concerned that a selection of a repository site within their state meet the political satisfaction of their constituent residents. Any waste facility siting raises the objections of nearby residents. But one of the basic policy compromises incorporated within the Nuclear Waste Policy Act is that the federal government would provide a site selection process which was so objective and fair that the residents of a host state selection process which was so objective and fair that the residents of a host state could be assured that a site within their state was selected because of its technical value and not because of its weakness of political representation. For this reason, state officials desire that the site investigation process conducted by the DOE be performed in a complete and objective manner with full disclosure of all known technical facts about the various available sites. It is to serve these respective state interests that the officials of host states desire the ability to do independent monitoring, testing, and data collection.

The interest of the utilities whose waste would be disposed under the DOE nuclear waste program is that their temporary waste storage problem be resolved. Their desire for resolution of that practical problem must be resolved in order to repair public confidence in the nuclear option. Utility spokesman have often stated that the nuclear waste problem has been technically resolved but that the political and institutional problems have not. Though the passage of the Nuclear Waste Policy Act two years ago prescribed resolution of the political and institutional deterrents to final nuclear waste disposal, the problem of the public's perception of nuclear waste continues. Unless the constituent population surrounding the prospective repository site can become confident of the technical soundness of the site, the public's confidence of the disposal methodology cannot be engendered. Consequently, the utility interest regarding objective and complete site investigation becomes identical to the perspective of state officials.

III. TRANSPORTATION

The interest of state officials, primarily of those states through which nuclear waste materials are proposed to be transported, but also of repository states is that they have full knowledge of all shipments of such materials so that they can plan and provide for the maximum public safety with regard to those shipments. Maximum public safety is, of course, enhanced by reducing the number of shipments and the number of road miles, using safe equipment (including casks), and detouring of transportation corridors from population centers. (Notice the countervailing effects of reducing road miles and detouring).

Though nuclear utilities are obviously as much concerned with public safety as are state officials (safety being morally important to them as persons and politically important to them in terms of the public's acceptance of the nuclear power industry), the quantifiable interest of nuclear utilities is the cost of transportation. The number of shipments, the adequacy of equipment, the number of road miles, and the location of routes are all relevant to the final cost of transportation. The more costly transportation becomes the more likely utilities face increases in the DOE's assessment to fund the Nuclear Waste Fund.

The interest of state officials and the nuclear utilities is not necessarily at odds on the transportation issue, however. For instance, a reduction of road miles traveled by nuclear waste shipments should reduce transportation costs as well as potential person exposures. However, selection of more circuitous routes, to avoid population centers may create greater transportation costs. The state interest in prior knowledge of shipments may be contrary to reduced costs of transportation. Ordinarily, prior notice means that more time is involved and that actual transportation takes longer. To the extent that time means money, transportation costs increase.

Once again, the primary objectives of nuclear utilities representatives is that institutional and political acceptance of the nuclear waste system be established and of the nuclear option be restored. Because safe transportation is an element of both, the interests of state officials and utilities representatives seem again to coincide assuming that an acceptable cost is established.

IV. THE NUCLEAR WASTE FUND

The interest of repository and transportation states is that their participation, as contemplated by the Nuclear Waste Policy Act, be fully funded. That includes, of course, funding for institutional interaction with the Department of Energy, the Nuclear Regulatory Commission, the Environmental Protection Agency, the Congress, and other states and Indian tribes. It includes full funding of the state's technical review of DOE site investigation, including independent testing and monitoring and data collection where necessary to verify or seek modification of DOE's investigative approach, and it includes full funding of informing the state's residents about the federal program. The Nuclear Waste Policy Act declares that the full costs of conducting the nuclear waste program, including state participation, should be funded by the nuclear utilities' rate payers, who are the beneficiaries of the federal program, rather than the citizens of the repository state. This is particularly true if the repository state has little or no nuclear power within the state.

The interest of such state officials is not constrained by the fact that other legitimate demands may be made upon the nuclear waste fund. The interest of the DOE and the contributing utilities is constrained by this fact. The reality is, however, that the states do desire to use the Waste Fund to fund full participation. At least as reflected in the state's request for funding in fiscal years 1984 and 1985, full funding of state participation is possible through a very small percentage demand upon the current funds available within the Nuclear Waste Fund.

Because the nuclear utilities have the interest of seeing the establishment of repositories that are accepted by the state's constituent population, and because the utilities are interested in the general public becoming more amenable to the nuclear option once the waste problem has been resolved, full state participation should be as desirable to the utilities as it is to state officials. However, as contributors to the Nuclear Waste Fund, utilities are realistically concerned that the Nuclear Waste Fund not be drained by purposes for which it was not intended. Obviously then utility officials have a more limited notion of what was intended. As each utility feels responsibility to its rate payers to keep its contributions to the Nuclear Waste Fund as low as possible, the interest of nuclear utilities can be seen to be as public spirited as the interests of repository state officials.

V. LIABILITY FOR THE NUCLEAR WASTE PROGRAM

The interest of state officials is that the Federal Government be wholly liable for any injury arising out of the Federal Government's nuclear waste program. This interest is dictated by a desire to guarantee its citizens that they will not be worse off because of the selection of a repository site within that state. The benefits of a Federal nuclear waste program fall to the public generally or to the users of power generated by nuclear facilities. It is conceivable, as in Nevada, that there is no such benefit, there being no nuclear utilities. The placement of liability in the Federal Government cures the possible disparity between benefit and burden for the nuclear waste program. The interest of nuclear utilities is that any liability for the injuries arising out of any handling of nuclear materials be dispossessed of them when the ability to control the materials has left them. Consequently when the Department of Energy takes title to radioactive waste, as prescribed by section 120 of the Nuclear Waste Policy Act, the interest of utilities is that their liabilities stop there.

Consequently, the interests of state officials and the nuclear utilities are identical with respect to federal liability for the nuclear waste program. However, nuclear utilities have a special interest in the question of the fiscal resources which may be utilized to pay claimants in the event of valid claims. The first obvious potential source is the Nuclear Waste Fund. However, since it is questionable that the nuclear utilities contemplated at the time that the Nuclear Waste Fund was created that they must bear the burden of potential errors and accidents caused by the Department of Energy itself or its contractors with whom the utilities have no corporate affiliation, the question arises whether general Federal funds should be exposed when the Federal Government is liable for waste accidents.

VI. INTERSTATE EQUITIES/MULTIPLE FACILITIES

Officials of potential repository and transportation states desire that the process of site and route selection will be objective and fair. One element of this equity is timing. State officials desire that the site screening and characterization process proceed at essentially even speed at all sites and that equivalent decisions in respective states be made at the same times with equivalent amounts of information. This interest is dictated by the political reality of repository siting. It is to a certain extent amazing that the repository states have worked so well together as of this time. The Nuclear Waste Policy Act recognizes this reality in requiring that a second repository be selected, contemplating the regional service potential of more than one repository. The location of a monitored retrievable storage facility, away from reactor interim storage facilities, within different states is a similar statutory gesture at equity between the states. Though nuclear utilities probably need not be concerned by this aspect of public acceptance of the Federal Government's nuclear waste program, the development of multiple facilities and the proximity of facilities to the generating nuclear reactors does have a cost feature primarily with regard to transportation.

The interest of nuclear utilities is, of course, that the DOE be able to receive waste on schedule. Consequently, utilities must favor forward motion at any site notwithstanding the lack of motion at other sites. However, a countervailing interest of utilities is the DOE's selection of the most licensable site. This is better guaranteed by deliberate speed with important, substantive choices between sites being made with the greatest amount of information, hopefully equivalent at each site for comparative purposes. There is a utility interest, therefore, in equity between the states as it concerns process.

VII. STATE ROLE IN FACILITIES LICENSING

The potential host states' interest in licensing is that the Nuclear Regulatory Commission perform its licensing function fully respecting its obligation to protect the public health and safety from the use of radioactive materials. State officials therefore desire that the standard of Department of Energy repository licensability be clearly established and that states be entitled to participate in NRC licensing proceedings in such a manner as to defend that standard. Clearly licensability is a higher standard than suitability as DOE has defined it in the repository siting guidelines (10 CFR 960). State officials desire a complete role in the licensing process including full prelicensing participation in discussions between the applicant Department of Energy and NRC staff, full party status within the licensing proceeding, full participation in discovery, examination of witnesses, introduction of witnesses, and of course, full rights of appeal of any NRC decision failing to apply the appropriate standard of licensability.

It is logical that the interest of nuclear utilities would be the same high standard for licensability of a high-level nuclear waste repository. This is because of the same utility interest, discussed throughout this paper, in the ultimate success of the Department of Energy's nuclear waste repository to fully and finally isolate radioactive waste. However, nuclear utilities would probably differ on the role which should be accorded to repository states in NRC licensing proceedings. A proceeding which involves the state as a full party will naturally be a longer proceeding than one in which the state did not participate. However, greater state participation should guarantee greater identification of conditions which should be imposed upon a DOE repository proposal in order that any licensed repository will be more apt to successfully isolate waste as originally contemplated by the Nuclear Waste Policy Act.

VIII. SUMMARY

As can be seen from the above discussion of several issues within the Department of Energy's nuclear waste program, potential host-states and utilities have some interests in common. Both desire the ultimate performance of the Department of Energy in identifying the best site in the nation for a first and second nuclear waste repository. Though state officials are obviously more conscious of the political receptiveness or lack of it of their own constituents that issue is important to nuclear utilities as well, because the public's willingness to regard the nuclear waste problem as resolved is important to the availability of the nuclear option within America's energy alternatives.

HIGH-LEVEL NUCLEAR WASTE DISPOSAL: ETHICAL CONSIDERATIONS

Margaret N. Maxey
Chair of Free Enterprise
ENS 103
The University of Texas at Austin
Austin, Texas 78712

ABSTRACT

Popular skepticism about, and moral objections to, recent legislation providing for the management and permanent disposal of high-level radioactive wastes have derived their credibility from two major sources: (1) government procrastination in enacting a waste disposal program, reinforcing public perceptions of their unprecedented danger and (2) the inflated rhetoric and pretensions to professional omnicompetence of influential scientists with nuclear expertise. Ethical considerations not only can but must provide a mediating framework for the resolution of such a polarized political controversy. Implicit in moral objections to proposals for permanent nuclear waste disposal are concerns about three ethical principles: fairness to individuals, equitable protection among diverse social groups, and informed consent through due process and participation. Now that the Nuclear Waste Policy Act of 1982 is the law of the land, the ethical task is twofold: (1) to examine how effectively the Act embodies these ethical principles and (2) to devise methods for monitoring the implementation of the Act to safeguard their observance.

> "It may well be said of man in the twentieth century that he spent most of his creative genius in trying to dispose of his own waste."
>
> (Anonymous)

Judging from public reactions to recent legislative proposals for the permanent disposal of high-level nuclear waste in the earth's crust, a majority of citizens nourish the popular belief that the state of nature is fundamentally benign and benevolent and should be protected from detrimental encroachments inflicted by "technological man." It is uncritically assumed that human transactions with nature, by definition, cause degradation of a pre-existing pristine condition and that biohazards threatening public health in the modern world are caused exclusively by industrial sources of pollution and toxic wastes.

It has come as a surprise to most citizens in our affluent, industrialized society to learn that we are virtually immersed in a sea of naturally occurring toxic substances and that the human diet contains a plethora of natural mutagens and carcinogens. Indeed, a storm of protest from the Epstein school of cancer etiology has greeted Bruce Ames' revealing and provocative essay in Science.[1] The Ames findings suggest that preoccupation with relatively small amounts of man-made effluents from various industrial processes diverts research attention away from the most health-effective and cost-effective ways to reduce the burden of cancer through dietary and life-style changes. By contrast, according to the Epstein view, cancer is basically a political problem since it is induced by the irresponsible policies and practices of corporate capitalism. The Epstein solution to the origins of "corporate cancer" does not require more scientific research since we already know that a new political agenda for corporate reforms will solve the problem.[2] The contrary argument, however, is that naive beliefs in a benign nature and in corporate causes of cancer are an implicit tribute to our conquest over communicable and degenerative diseases. This conquest is a direct result of technological advances in public sanitation practices and medical diagnoses. Research in natural carcinogens and mutagens would have ethical and political implications unrecognized by the Epstein school.

I. SOURCES OF OBJECTIONS TO NUCLEAR WASTE DISPOSAL

Situated in the context of a heated debate about the etiology of the dread disease of our age, cancer, our public controversy over siting nuclear waste repositories has understandably escalated into an exchange of ethical and moral arguments about the benefits vs. detriments of living in an industrial society, the protection vs. violation of individual rights, the distributive vs. aggregate consequences of guaranteeing fairness, and related issues. Controversies over the production and disposal of nuclear wastes have come to serve as a volatile instrument for debating divergent agendas for the future of our affluent technological society.

Moral objections to recent legislation providing for high-level nuclear waste disposal would not have gained such a high level of credibility were it not for two major sources of conflicting judgments. Forty years of government procrastination, accompanied by vacillations in technical choices and scientific research efforts, exacerbated by massive infusions of federal funding in pursuit of "a solution" to the radioactive waste problem, have made an indelible impression on the public psyche. It is now accepted as a "given" that the inherent dangers posed by nuclear wastes are so intractable that neither scientific nor engineering experts have been able to devise a solution that will guarantee protection of present and future generations from lethal exposure to radiation. Efforts have been made to counteract this impression--namely, arguments that the accumulated volume of wastes to date has not warranted expending funds for the construction of a repository; suggestions that a scientific breakthrough in transmutation of wastes would render the need for geological disposal moot; and claims that the decision to reprocess civilian nuclear wastes should be made prior to any disposal options. They have drowned in a sea of skepticism. Deeds and apparent inaction speak louder than words: if an achievable solution has existed, why has so much money been spent over so many years with no apparent result?

Reinforcing this procrastination has been the inflated rhetoric of influential scientists. Doubtless imbued with impeccable intentions and concern for publicly expressing their sense of moral responsibility toward society in introducing nuclear technologies, a few scientists with nuclear expertise have placed no boundaries to the range of their competence.

A striking case in point appears in papers privately circulated in 1975 and 1976 by John W. Gofman. In an essay titled "The Question of Law Versus Justice," he writes:

> There is no doubt in my mind that the promotion of nuclear power is a criminal act of the worst sort, given the constitution under which we live. The Constitution of the United States does not permit the taking of life without due process of law. Nuclear power, which begins its random killing of citizens of the United States even before the nuclear plant goes into operation, is an infringement of constitutional rights.[3]

According to Gofman, nuclear technologies begin their "random killing" of unconsenting, unsuspecting citizens from the moment "radioactive poisons"' e.g., radon and its daughter products, are brought to the earth's surface in the course of mining uranium. He rebukes the EPA and its standard-setting process, stating that:

> This is the old benefit: risk balancing, which justifies premeditated random murder of some individuals for the ostensible benefit [electric power] to society.[4]

The rhetorical persuasiveness of this line of reasoning is undeniable. Yet its selective attention ignores a more fundamental ethical question: Does not the same moral and legal indictment apply to the mining and dispersal of coal or petroleum or heavy metals, and does not ethical consistency require that prior moral attention be paid to technologies in more common commercial use, including autos, with known lethal effects presumably unjustified by putative "benefits"?[5] Gofman's myopic reasoning in the name of morality and due process trivializes the meaning of "justice."

A similar exercise in inflated rhetorical imagery remains unmatched in its persuasive power with the general public. Expostulating on the demands of nuclear energy in 1972, Alvin Weinberg questions the adequacy of social institutions "to deal with this marvelous new kind of fire." Invoking the image of Prometheus on the one hand, Weinberg on the other hand poses the problem of radioactive waste disposal as if it were a matter of striking a bargain with "unprecedentedly treacherous materials" which call up the image of Faust:

> We nuclear people have made a Faustian bargain with society. On the one hand we offer, in the catalytic nuclear burner, an inexhaustible source of energy. ...But the price that we demand of society for this magical energy source is both a vigilance and a longevity of our social institutions that we are quite unaccustomed to.[6]

With the fate of Prometheus and Faust firmly in mind, Weinberg embellishes his virtuous vision of the necessity for eternal vigilance and a permanently stable social order by recognizing the added necessity of a nuclear priesthood.

> The discovery of the bomb has imposed an additional demand on our social institutions. It has called forth this military priesthood upon which in a way we all depend for our survival. ...It seems to me ...that peaceful nuclear energy probably will make demands of the very same sort on our society and possibly of longer duration.[6]

Neither Weinberg's lament nor the analysis presupposed by it happens to coincide with the facts of the matter and the preponderance of scientific truth. Nonetheless, it has opened a deep and widening chasm separating public opinion from the professional judgment of technical experts who are in a position to distinguish actual facts and technical achievements from excursions of the imagination. As a consequence, public opposition to the production and disposal of nuclear wastes has been given its strongest expression in the form of moral objections based on ethical principles.

II. MORAL OBJECTIONS TO NUCLEAR WASTE DISPOSAL

Pretensions to professional omnicompetence may very well have created an unbridgeable chasm between public opinion and the judgments of technical experts. In doing so, they have also succeeded in attracting scholarly attention among professional philosophers and ethicists in the academic community.

To philosopher Robert Goodin, nuclear wastes are the prime example of "profound uncertainties" which render any form of decision-making based on a calculus of expected utility to distant future generations a "particularly poignant instance of a rather common phenomenon," namely "cheating our children."[7] Goodin concedes that his conclusions are derived from "the facts of the case as I have described them" and that, on such contentious issues, he "can offer only second-hand expertise." Nonetheless the major facts as he understands them are:

1. Nuclear wastes pose a unique threat of harm because they contain actinides with half-lives of millions of years, having toxic levels that could reach the biosphere in 1,700,000 years with concentrations 5 times the current MPC of international standards.

2. During this period, we can expect each of three proposed barriers (waste form, container, geological medium) designed to prevent reentry of radioactive poisons into the biosphere to be breached.

3. Such leakage can be expected to create grave hazards for human populations in the distant future since likelihood or leakage increases over a long time; hence, radiation exposure is inevitable.

Based on his set of accepted facts, Goodin avers that a conventional mode of learning by doing, or trial-and-error, violates ethical principles of fairness, equity, and participation, because any errors would entail exposure of entire populations to radiation with catastrophic consequences. Since any exposure is harmful and health effects are assumed to be latent, irreversible, and involuntary, they may not appear in time for changes in policy to be made. The magnitude of nuclear disaster makes this technology unique and incomparable to any other.[8]

It is noteworthy that Goodin's entire line of moral reasoning is predicated on a presumed set of empirical facts formulated in terms of an inventory and description of the physical properties of nuclear wastes: their length of half-life, number of lethal doses contained in small quantities, the no-safe-dose or zero-threshold or linear hypothesis of exposure to radiation.

Political philosophers Richard and Val Routley use a similar set of facts as unquestioned assumptions from which they derive moral conclusions.[9] They also focus on the philosophical problem of obligations to the future in their pattern of reasoning:

1. The risks and benefits of nuclear energy are distributively inequitable both in space and in time, so that a moral principle of justice and fairness imposes moral constraints on the deployment of a technology that can inflict widespread harm.

2. The turbulent record of human history, the fallibility of human beings, plus vast fluctuations in geological and climatic conditions for the past million years, eliminate any confidence that nuclear wastes can and will be safely stored for the millions of years required.

3. Leakage of even small quantities of nuclear wastes would result in catastrophe--namely, widespread disease, genetic damage, contamination of huge land masses, etc.[10]

Both Goodin and the Routleys base their moral assessments on the assumption that any error or accident entails catastrophic effects for unconsenting present and future generations. Nuclear technology is unique and incomparable precisely because there will inevitably be errors or accidents, and the effects on human health will be disastrous. This pattern of thought represents what Aaron Wildavsky terms "a new doctrine of risk aversion"--namely, "no trials without prior guarantees against error."[11]

Goodin insists that a revised set of criteria for policy choices must predetermine the moral acceptability of new technologies. There shall be no trial-and-error learning process unless there are guarantees that "errors will be small, immediately recognizable, and correctable."[12] Nuclear waste compels us to live "not merely with risk but also with *irresolvable* uncertainties."[13]

Again, it should be noted that selective attention apparently precludes the recognition that--far from being unique to nuclear technology--"irresolvable uncertainty" happens to be an inescapable condition of human life itself. Moreover, the flaw of selective attention appears to compound the prior ethical problem, namely the determination of a preponderance of scientific truth from which moral conclusions should be derived. Mere descriptions of physical properties, radioactive half-lives, lethal dose calculations, and hypothetical mechanisms for migration of waste inventories into the biosphere are not only ethically misleading but morally dubious. Half-life is no criterion for toxicity. Toxicity does not determine hazard. The mere existence of a toxic substance in the biosphere is no basis for a moral claim about its unacceptability. The only sound basis for evaluating the <u>ethical</u> significance of nuclear waste (or for that matter, any other toxic substance in the biosphere) is <u>likelihood of exposure and assimilation</u> of a harmful dose at a rate detrimental to human beings and other valued organisms.

Prior to examining whether ethical principles of fairness, equity, and participation are being protected or violated by proposed methods for nuclear waste disposal, the ethical task is to evaluate the scientific status of empirical evidence from which moral claims are derived.

III. SCIENTIFIC STATUS OF EMPIRICAL ASSUMPTIONS PRESUMED BY MORAL CLAIMS

Moral objections to proposals for nuclear waste disposal assume that catastrophic harm to present and future generations will result from exposure to radioactivity leaking from waste repositories, no matter how small the dose and rate. This moral claim derives from the linear or zero-threshold hypothesis adopted forty years ago by standard-setting agencies. It seemed "safer" then to assume that every radiation dose greater than zero entails some possibility of somatic and/or genetic harm. However, what philosophers and ethicists presume to be an established scientific "fact" is in reality only an untested theory, an extrapolated hypothesis, an ultra conservative and protective rule of prudence. Despite vast amounts of radiobiological data, there is no conclusive scientific evidence to prove the existence or absence of a threshold.

The absence of scientific evidence of harmful exposure to low levels of radiation is not due to incompetence or oversight or lack of attempts to find harmful effects. Lauriston Taylor's statement is ethically salient:

> No one has been identifiably injured by radiation while working within the first numerical standards set by the NCRP and then the ICRP in 1934. Let us stop arguing about the people who are being injured by exposure to radiation at the levels far below those where any effects can be found. The fact is, the effects are not found despite over forty years of trying to find them. The theories about people being injured have still not led to the demonstration of injury and though considered as facts by some, must only be looked upon as figments of the imagination.[14]

As for the presumed uniqueness of nuclear wastes and putative radiation exposures, it is certainly a scientific truth that, when delivered in sufficiently large amounts, ionizing radiation can cause determinable injuries to any biological system. Yet radiation experts repeatedly point out that any somatic or genetic effect caused by radiation can also be caused by at least 1,500 other agents, with no possibility of positively identifying the culprit. Besides not being unique in its biological effects, radiation dose-effects are not cumulative, since a process of repair or replacement of cells, both somatic and genetic in nature, has been demonstrated to occur in radiation therapy techniques.[15]

Furthermore there is a scientific and ethical basis for recognizing a practical threshold (or _de minimis_ dose) below which risks of radiation exposure are trivial and beyond regulatory concern--namely, a dose-level below which the latent period for the appearance of any likely effect is in excess of one's remaining life expectancy.[16]

As for expectations that dissolution and migration of nuclear wastes in a million-year period are inevitable, a geotoxicity survey quickly invalidates the assumption that nuclear wastes incorporated into the earth's crust would be its only toxic constituent. To the contrary, many naturally occurring toxic substances are continuously leached into water and food supplies, as Bruce Ames and others have pointed out. J. J. Cohen has considered only eight toxic substances: mercury, cadmium, chromium, lead, selenium, barium, arsenic, and uranium. He observes:

> It can be calculated that if our entire electrical supply came from nuclear power for 100 years (10^8 MWe-yr) and all the resulting wastes were buried, the resulting increase in the toxicity of the earth's crust would be one ten-millionth of one per cent.[17]

In response to the objection that other toxic substances are more uniformly distributed whereas nuclear wastes would be concentrated in a few repositories, Cohen writes:

> In fact nature has also concentrated toxic minerals into ore bodies. The toxicity of a typical high-level waste repository decreases in time due to radioactive decay. It can be shown that in a few hundred years the repository contents become relatively less toxic than typical mercury deposits and in about 1000 years it becomes less than the uranium ore body from which the nuclear fuel was originally derived. The ore body is at least as available to dissolution and transport as is the waste repository.[18]

To summarize: (1) The assertion that any and every radiation dose not only _can_ but _does_ cause somatic and genetic harm is due to a profound misunderstanding of the hypothetical status of the linear hypothesis. (2) There is neither a scientific nor an ethical justification for singling out radiation doses as a unique cause of cancer or genetic mutation since over 1000 other toxic agents are known to produce the same health effects at sufficiently high doses, with no way to single out the causal agent. (3) There is a scientific and ethical basis for recognizing a practical threshold or *de minimis* dose below which radiation exposures can be ignored. (4) The storage of nuclear wastes in technologically sophisticated containments, with multiple engineered barriers, represents such a trivial increase in geotoxicity levels already present that a moral claim based on uniqueness or catastrophic potential is seriously flawed.

The question remains: Now that the Nuclear Waste Policy Act of 1982 is the law of the land, how effectively does the Act embody the ethical principles which form the basis of major concerns expressed in moral objections to the very existence of nuclear wastes?

IV. THE NUCLEAR WASTE POLICY ACT AND ETHICAL PRINCIPLES

Three ethical principles form the basis of moral objections to proposals for permanently disposing of high-level wastes: (1) *Fairness* requires that no individual's material well-being, fiscal and physical, shall be made worse as the result of a permanent waste disposal program. (2) *Equity* requires that the aggregate economic vitality and environmental quality of a selected region, and of the diverse groups who may inhabit it, shall be preserved and protected. (3) *Informed consent* requires due process and participation of affected parties in the decisions taken.

The Nuclear Waste Policy Act of 1982[19] is vulnerable to criticisms which inevitably attend any legislation fraught with controversy and compromise.[20] Nonetheless, it provides not only for a comprehensive three-fold waste management program (i.e., interim on-site storage of spent fuel rods by utilities, monitored retrievable storage facilities for observing radiological behavior of wastes prior to disposal, and long-term geological repositories for permanent waste disposal); it also provides for strong institutional oversight by requiring a division of regulatory responsibility among three agencies: DOE is responsible for overall program management and site selections; EPA is charged with setting radiological performance standards; NRC is responsible for the development of licensing requirements and procedures.

As has been pointed out, moral objections assume as a given that fairness and equity to citizens in present and future generations will inevitably be violated because the long half-lives, concentrations of lethal doses, fallibility of human beings, and inevitable geological perturbations virtually guarantee leakage of wastes and catastrophic consequences of any exposure to radiation. This assumption is clearly recognized and placed in its proper context by the EPA in its proposed Environmental Standards for Waste Management and Disposal (40 CFR Part 191).[21]

The proposed EPA standards--predicated on the most pessimistic assumptions (including human intrusions) for a 10,000 year period--require that health risks posed by wastes during that time shall be no greater than the risks from an equivalent amount of unmined uranium ore. This requirement is based on the empirical fact that uranium ore, particularly uranium and radium, occurs in permeable geological strata through which ground water flows and transports radionuclides from the ore into the biosphere. Leaving the ore unmined over 10,000 years would result in excess cancer deaths ranging from 300 to more than 1 million. The standards proposed require that "very unlikely releases of waste to the accessible environment" shall be less than 10 times the quantities of radionuclides comprising the waste. Even using the linear, nonthreshold hypothesis, health risks of exposure would be no greater than, and indeed much less than, risks which future generations would face if no wastes had been created in the first place.

Moreover, NRC licensing requirements set forth rules whereby EPA performance standards shall be met with "reasonable assurance, making allowance for the time period, hazard, and uncertainties involved, that the outcome will be in conformance with these objectives and criteria."[22] For the first several hundred years, during which thermal and radiation levels are high, an engineered barrier system functions to confine any releases. But as radioactivity subsides, containment barriers are augmented by isolation methods of inhibiting transport of materials to an accessible environment in amounts and concentrations less than prescribed limits.[23] Intergenerational equity and fairness to present people are clearly preserved and protected by EPA and NRC requirements.

As for informed consent, due process, and participation, the Nuclear Waste Policy Act mandates an unprecedented and extensive process of "Consultation and Cooperation" with any affected states and/or Indian tribes, to the point of allowing the power to veto the final selection of a repository. Through its Secretary, the DOE is required to enter into binding written agreements with states selected for "site characterization." These agreements are subject to comments and recommendations by affected states or tribes as well as requests for impact assistance to local governments. They also specify procedures for independent monitoring of activities, for sharing technical and licensing information, and for resolving objections at any stage of siting, constructing, and operating a repository.

V. CONCLUSION

A skeptical and distrustful public has yet to recognize and be persuaded by the fact that the performance standards and licensing procedures required of EPA, NRC, and DOE by the Nuclear Waste Policy Act of 1982 are not only achievable, both technically and economically, but more than sufficient by ethical standards.

The thesis underlying this paper is that the burden of persuasion must finally be taken up by those who can make the case to the general public that the ethical principles which allow moral objections to waste disposal seem plausible do not require a nuclear priesthood or an eternal vigilance or a permanent stability of human institutions to achieve "safety." To the contrary, the ethical principles of fairness, equity, and informed consent simply require that we imitate the known behavior of natural processes, achievable with known engineering systems, in order to bequeath to future generations a level of risks and benefits neither greater nor less than those which we ourselves have inherited and accepted.

The key to persuasion is deeds, rather than words. The ethical task is to unlock the door to action by actually demonstrating that the art of political compromise is the touchstone of ethical consistency.

REFERENCES

1. B. N. Ames, "Dietary Carcinogens and Anticarcinogens," Science, 221, 1256-1264 (1983).

2. B. N. Ames, Reply to "Letters," Science, 224, 760 ff. (1984).

3. J. W. Gofman, "The Question of Law Versus Justice," quoted with commentary by S. J. Dundon, "Technical Evidence and Authority in an Anti-Nuclear Argument," in Values in the Law: Proceedings of the 14th Annual Conference of Value Inquiry, James Wilbur, ed. (1979).

4. J. W. Gofman, "Radiation Doses and Effects in a Nuclear Power Economy: Myths vs. Realities," Committee for Nuclear Responsibility Report, 1976-2, p. 3 (April 1976).

5. Simon Ramo, American's Technology Slip, pp. 147-128, New York: John Wiley (1980).

6. A. N. Weinberg, "Social Institutions and Nuclear Energy," Science, 177, 27-34 (1972).

7. R. E. Goodin, "Uncertainty as an Excuse for Cheating Our Children: The Case of Nuclear Wastes," Policy Sciences, 10, 25-43 (1978).

8. R. E. Goodin, "No Moral Nukes," Ethics, 90, 418-419 (1980).

9. R. and V. Routley, "Nuclear Energy and Obligations to the Future," Inquiry (Universitets-forlaget, Oslo), 21, 133-179 (1978).

10. Citations in support of these "scientific facts" include R. Nader and J. Abbotts, The Menace of Atomic Energy; A. B. Lovins and J. H. Price, Non-Nuclear Futures; J. W. Gofman and A. R. Tamplin, Poisoned Power.

11. A. Wildavsky, "Trial Without Error: Anticipation Versus Resilience as Strategies for Risk Reduction," in Regulatory Reform: New Vision or Old Curse?. M. N. Maxey and R. L. Kuhn, eds ., pp. 200-221, New York.

12. R. E. Goodin, "No Moral Nukes," Ethics, 90, 419 (1980).

13. Ibid., 421.

14. L. S. Taylor, "Some Nonscientific Influences on Radiation Protection Standards and Practice," Health Physics, 39, 841-874 (1980).

15. Ibid.

16. O. G. Raabe, S. A. Book and N. J. Harris , "Bone Cancer from Radium: Canine Dose Response Explains Data for Mice and Humans," Science, 208, 61-64 (1980).

17. J. J. Cohen, C. F. Smith and T. E. McKone, "A Hazard Index for Underground Toxic Material," UCRL-52889, Lawrence Livermore Laboratory (1980).

18. J. J. Cohen, "Statement to the Interagency Review Group on Nuclear Waste Management," San Francisco (July 1978).

19. Public Law 97-425 dated 7 January 1983.

20. S. Fred Singer, "Seeking Sane Ways to Store Nuclear Waste, Wall Street Journal (March 19, 1985).

21. Federal Register, 47/250, 58196-58206 (December 29, 1982).

22. 10 CFR 60.101(a)(2). Emphasis added.

23. 10 CFR 60.102(e)(1) and (2).

RELIABLE PREDICTIONS OF WASTE PERFORMANCE IN A GEOLOGIC REPOSITORY

Thomas H. Pigford
Paul L. Chambré
Department of Nuclear Engineering and Lawrence Berkeley Laboratory
University of California
Berkeley, California 94720

ABSTRACT

Establishing reliable estimates of long-term performance of a waste repository requires emphasis upon valid theories to predict performance. Predicting rates that radionuclides are released from waste packages cannot rest upon empirical extrapolations of laboratory leach data. Reliable predictions can be based on simple bounding theoretical models, such as solubility-limited bulk-flow, if the assumed parameters are reliably known or defensibly conservative. Wherever possible, performance analysis should proceed beyond simple bounding calculations to obtain more realistic--and usually more favorable--estimates of expected performance. Desire for greater realism must be balanced against increasing uncertainties in prediction and loss of reliability. Theoretical predictions of release rate based on mass-transfer analysis are bounding and the theory can be verified. Postulated repository analogues to simulate laboratory leach experiments introduce arbitrary and fictitious repository parameters and are shown not to agree with well-established theory.

I. INTRODUCTION

The Performance Assessment National Review Group[1] recommended that the U.S. geologic repository projects give greater emphasis to realistic and reliable predictions of long-term performance of repositories. Predictive reliability, the assurance that the actual performance will be as good or better than that stated by the performance prediction, is essential for any engineering project and particularly for a geologic repository, because real-time testing to confirm the repository design and to confirm the predictions of long-term performance is impossible. In any system design the use of well established and easily verified calculational techniques to establish the bounding values of predicted performance must be balanced with the desire to refine the performance prediction for greater realism. To predict what happens in tens of thousands of years in a repository we must emphasize sound theories of prediction, more so than in conventional, engineering design wherein performance can be predicted, validated, and remedied by real-time testing. Here we review the state of technology for predicting the long-term rate of dissolution of radionuclides from waste packages in a repository.

II. MECHANISTIC ANALYSIS OF RADIONUCLIDE RELEASE RATES

A. Release Estimates for Saturated Bulk Flow

Bounding analyses can be used to establish predictive reliability and to estimate limiting features of system behavior. In some instances, physically unrealistic assumptions or input values are used to obtain a conservative "bounding result."

1. Bounding estimates of repository release rates. To calculate radionuclide release rates from a repository, some projects have assumed unrealistically that all ground water flowing through the repository becomes saturated by the radioelements in the water or by the waste matrix. If the values chosen for the saturation concentrations and water flow rates are defensible, then the calculated releases are defensible as conservative upper bounds and are expected to be reliable.

   Wherever possible, performance analysis should proceed beyond simple bounding calculations to obtain more realistic--and usually more favorable--estimates of expected performance.[1] For example, because the emplaced waste packages are discrete and separated from each other, it is impossible for all water potentially flowing through a repository to become saturated with any waste constituent, assuming that it is not already saturated with that constituent before encountering the waste and assuming no large changes in saturation concentrations in the repository environment. Concentrations near saturation are expected only in the liquid immediately adjacent to the waste surface. All other water will be below the saturation concentration, and the average concentration in ground water leaving the repository will be below saturation. Thus, a more realistic calculation of repository release rates, if suitably reliable, is likely to be preferable to the extreme conservatism of saturated bulk flow. Although this simple theory of saturated bulk flow is conservatively bounding for estimating releases from a repository, it does not lead to conservative or bounding estimates of release rates from waste packages into the rock at the low flow rates predicted for the U.S. repository projects, as is explained below.

2. Nonbounding estimates of waste-package release rates. The tuff project[2] estimates waste-package release rates from a waste package by multiplying the saturation concentration of the waste matrix by the volume flow rate of ground water that is calculated to flow through rock equal in cross section to the cross sectional area of the waste canister. Whether partial or full repository flow rates are assumed, the results are not necessarily bounding or conservative for waste-package release rates, because zero waste-package release rate will be incorrectly predicted at zero flow. If there are pathways for molecular diffusion of dissolved species from a waste surface into surrounding stagnant ground water, a finite transient

release of dissolved species into the rock will occur. The ground water infiltration velocities predicted for tuff are so low, about 0.003 to 1 mm/year, that transient diffusional release will be more important than convective transport if the waste solid is conservatively assumed to be surrounded by ground water and moist tuff. Although the rate of diffusional release into the rock must eventually, at steady state, equal the repository release rate for long-lived species, steady state will not be approached for thousands of years in a low-flow repository, and the transient diffusional release into the rock can control the waste package release rates for long times. To develop a more reliable prediction that can be defined as conservative and bounding, a more mechanistic analysis of mass transfer of radionuclides to ground water could be adopted.

B. Mass-Transfer Analysis of Release Rate

Mass-transfer analysis is a general approach, highly refined in the field of chemical engineering,[3] to predict rates of transport of species within a phase and between phases, as affected by diffusion, convection, chemical reaction, adsorption, etc. It quantifies the actual mechanisms affecting the transport rate. An application by Chambré et al.[4-9] to a waste package surrounded by wet porous rock conservatively assumes that all the waste solid is suddenly exposed to ground water when the corrosion resistant barrier fails, and it conservatively assumes saturation concentration of dissolved radioelements in the liquid at the waste surface. Exact theoretical analysis of the diffusive-convective transport of the dissolved species from the waste surface into the surrounding porous rock results in an upper limit to the time-dependent dissolution rate that can occur.

Though more realistic than saturation bulk flow, Chambré's mass-transfer analysis for a bare waste solid is still conservatively unrealistic in neglecting the finite resistance to mass transfer presented by the partly failed waste canister and fuel cladding. It is further conservative in application by assuming saturation concentrations at the waste surface. More detailed mechanistic analysis[10] shows that saturation concentration is a close approximation to reality for borosilicate glass and for the spent-fuel matrix for all but the very early time of exposure in a repository. The theory contains no arbitrary adjustable parameters. For steady-state release rate it requires experimental data on saturation concentrations, diffusion coefficients, porosity, ground water approach velocity, as well as specifications of waste size and geometry. For transient releases, data on sorption retardation coefficients are also required. Because of the saturation boundary condition, information on degree of waste cracking and solid-liquid reaction rate does not enter the prediction. Even if solid-liquid reaction rate is included as a boundary condition, increased reaction surface from cracking of glass waste does not appreciably affect the dissolution rate after the first few days of exposure to ground water.

The important feature of this mass-transfer analysis is not that it predicts favorably low dissolution rates for most radionuclides, but that it is a mechanistic theory based on well-understood governing equations and conservatively bounding boundary conditions. The theory itself can be examined in detail. It can be subjected to verification in experimental real-time tests, as has been done at the Pacific Northwest Laboratory.[11] Such theories are the only reliable means of extrapolating into the future.

A similar theoretical mass-transfer analysis of time-dependent radionuclide release from a suddenly exposed waste form surrounded by backfill and rock is now used by the basalt project.[1] Although the salt project has based its waste-package release rates on a simple bounding calculation of the rate of brine inflow to an emplacement cavity, multiplied by the individual radioelement solubility, this approach is unrealistic and unnecessarily conservative because the salt will soon be consolidated against the waste package. More realistically, and now more reliably, one can calculate radionuclide release by calculating the diffusive transport from a brine layer at the waste surface into brine-filled grain boundaries in the surrounding salt.[12] The convective-diffusive mass transfer from the waste surface into interbed flows that may intersect the waste package can also be calculated, similar to the mass-transfer calculations used by Neretnieks[13] to predict container corrosion and waste-package release rates for the KBS project.

C. Some Results of Mass-Transfer Predictions of Release Rates

The profiles of ground water flow and concentration of a dissolved species for a sample waste solid surrounded by porous rock are shown in Figure 1. The diffusion-and-flow calculation by Chambré[4-9] uses the known distribution of ground water velocities around an infinite cylinder through pores in the surrounding rock. A general solution to the time-dependent dissolution rate of a radioelement with a constant boundary concentration N_i^* at the inner surface of the borehole has been given by Chambré.[7]

1. Steady-state diffusive-convective dissolution rates. The fractional rate of dissolution, f_i, of the elemental species i and its isotopes from a long water cylinder of radius R is calculated at steady state to be:

$$f_i = \frac{8 N_i^* \varepsilon \sqrt{DU}}{(\pi R)^{3/2} n_i}, \quad \frac{UR}{D} > 4 \tag{1}$$

where: D is the specie diffusion coefficient in pore liquid, U is the pore velocity of ground water before it comes near the waste, ε is the porosity of the surrounding rock, and n_i is the bulk density (g/cm^3) of the elemental species i in the waste. For a bounding calculation, N_i^* is chosen as the saturation concentration. For a waste cylinder of finite length, L, end effects are accounted for my multiplying the right-hand side of Equation (1) by a correction factor (1 + R/L).

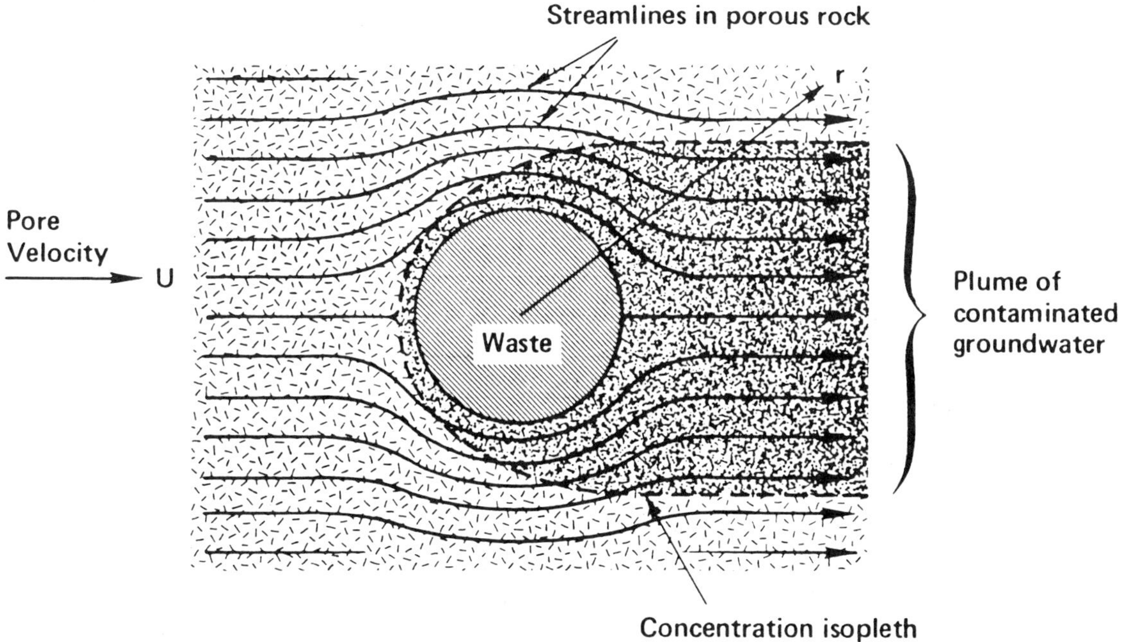

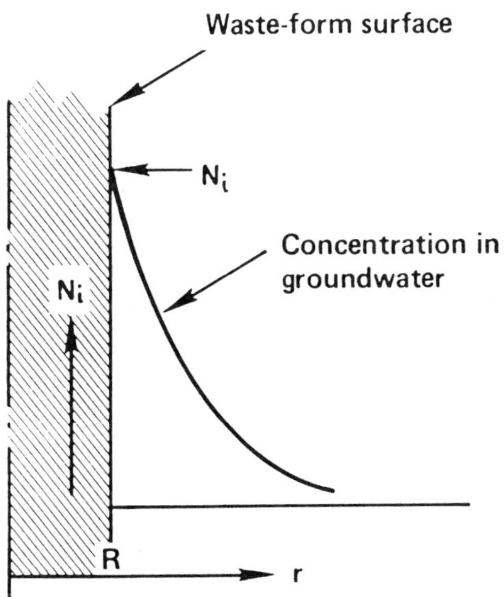

FIGURE 1. Flow of Water Past a Waste Cylinder; Concentration Profile

Table 1 gives estimated values[4,14] of the solubility of silica and the solubilities in water of radioelements in borosilicate glass waste. Also listed are the bulk densities and the calculated fractional release rates for a typical glass waste exposed to ground water at an approach velocity of 1 m/yr in rock of 1 percent equivalent porosity. The assumed diffusion coefficient of 10^{-5} cm^2/s is typical for an electrolyte in water. It conservatively neglects the effect of tortuosity[15], which in granite can result in more than a 100- to 1000-fold[16-18] reduction in D and more than a 10-fold to 30-fold reduction in estimated dissolution rates. Because the fractional dissolution rates of the low-solubility elements are lower than those of the silica matrix, these elements do not dissolve congruently with the matrix, forming precipitates as the matrix dissolves.

Table 1 gives values of fractional dissolution rates for silica and for various radioactive elements[19] in borosilicate glass, calculated from experimental data reported from IAEA-type laboratory experiments in which leachant is periodically replaced. For substances of limited solubility, the values of f_i computed by Equation (1) are smaller than those derived from laboratory leach tests, as is expected. For slightly soluble species in waste that has been embedded in a repository rock, the slow diffusion and slow movement of the liquid around the waste containers may be more significant in controlling the net rate of dissolution than the rate at which substances in the waste solid react with the surface liquid. If the solubility is very small, the rate of escape into ground water will be determined primarily by the solubility, the properties of the porous rock, and the ground water velocity; if the solubility is sufficiently large, the kinetics of the interaction between the solid waste constituents and water may dominate.

In Table 2 are calculated fractional dissolution rates at steady state for a waste package of commercial spent fuel. Because of the large inventory of uranium, its fractional dissolution rate is much lower than for borosilicate glass and is lower than that calculated for low-solubility fission products and other actinides. Therefore, unless there is a mechanism for waste constituents to be preferentially released from the UO_2 matrix more rapidly than the matrix dissolves, all of the listed species should dissolve congruently at the fractional dissolution rate of uranium.

2. <u>Steady-state diffusion-controlled dissolution rates</u>. In most of the repository designs the ground water velocity is so low that the

TABLE 1. Calculated Fractional Dissolution Rates for Commercial Borosilicate Waste

Waste cylinder radius: 0.15 m
length: 2.46 m
Amount of uranium[a] : 460 kg

Constituent	Solubility,[b] g/cm^3	Concentration in glass, g/cm^3	Calculated fractional dissolution rate[c] yr^{-1}	Observed fractional dissolution rate[d] yr^{-1}
SiO_2	5×10^{-5}	1.6	1×10^{-6}	2×10^{-3}
U	1×10^{-9}	1.2×10^{-2}	4×10^{-9}	2×10^{-6}
Np	1×10^{-9}	1.9×10^{-3}	2×10^{-8}	7×10^{-4}
Pu	1×10^{-9}	1.1×10^{-4}	4×10^{-7}	3×10^{-5}
Am	1×10^{-10}	3.6×10^{-4}	1×10^{-8}	3×10^{-5}
Se	1×10^{-9}	1.4×10^{-4}	3×10^{-7}	
Sn	1×10^{-9}	9.4×10^{-5}	5×10^{-7}	
Tc	1×10^{-9}	1.9×10^{-3}	2×10^{-8}	

a/ Amount of uranium initially in pressurized water-reactor fuel to produce the radionuclides contained in the waste.

b/ For amorphus SiO_2[14]. Other solubilities are from Krauskopf[4], at 20°C, moderately reducing conditions.

c/ Steady-state dissolution rates calculated from Equation (1) for diffusive-convective mass transfer. Ground water pore velocity = 1 m/yr. D = 3.2×10^{-2} m^2/yr.

d/ Date of McVay et al.[19] for IAEA-type leach tests, with periodic replacement of leachant.

TABLE 2. Calculated Fractional Dissolution Rates for Commercial Spent Fuel

Waste cylinder radius = 0.43 m

length = 2.46 m

Amount of initial uranium = 2770 kg

Constituent	Concentration in spent fuel, g/cm^3	Fractional dissolution rate,[a] yr^{-1}
U	1.2	2×10^{-12}
Np	6.2×10^{-4}	3×10^{-9}
Pu	1.2×10^{-2}	2×10^{-10}
Am	1.1×10^{-4}	2×10^{-9}
Se	6.4×10^{-5}	3×10^{-8}
Sn	1.5×10^{-4}	1×10^{-8}
Tc	9.5×10^{-3}	2×10^{-10}

a/ Steady-state dissolution rates calculated from Equation (1) for diffusive-convective mass transfer. Ground water pore velocity = 1 m/yr. D = 3.2×10^{-2} m^2/yr. Solubilities from TABLE 1.

convective component of the release from the waste form is negligible, Equation (1) is not applicable, and an alternate form developed by Chambré for a prolate spheroidal waste solid must be used:

$$f_i = \frac{\beta \varepsilon D N_i^*}{n_i}, \quad U \to 0 \tag{2}$$

where β is a geometrical parameter that can be calculated from the waste-form dimensions:

For a sphere of radius R

$$\beta = \frac{3}{R^2} \tag{3}$$

For a prolate spheroid waste of semiminor axis b and eccentricity e

$$\beta = \frac{3e}{b^2 \ln[\coth \frac{\alpha_s}{2}]}, \qquad \alpha_s = \cosh^{-1}\left(\frac{1}{e}\right) \tag{4}$$

Using the properties listed in Table 1, the limiting low-velocity fractional release rates are calculated to be about a fourth of those calculated for a pore velocity of 1 m/yr, which illustrates the nonconservatism (see II.A.2) in neglecting diffusional transport at pore velocities of the order of a few millimeters per year.

The time to reach steady state increases from the few years for a pore velocity of 1 m/yr to tens of thousands of years for near-zero velocities.[8] Sorption parameters do not enter these equations for steady-state release of long-lived radionuclides, but sorption increases the transient dissolution rate and the time to reach steady state.[8,9]

3. Extensions of the mass-transfer analysis. In subsequent studies, Chambré has extended the mass-transfer analyses to consider the effect of a backfill layer between the waste package and rock,[20] the increased dissolution rates that can result when the concentration profile is steepened by radioactive decay and sorption,[9] the effect of nonlinear sorption characteristics of backfill,[21] the effect of repository heating on mass transfer and release rate,[22] and the rate of release of species that have already diffused from the UO_2 matrix in spent fuel and are readily accessible as soluble constituents in the fuel voids and fuel-cladding gap.

4. Use of laboratory leach-rate data. If one wishes to include the solid-liquid reaction rate as part of a more comprehensive model of waste-package performance in a repository, a concentration-dependent reaction rate should be used as a boundary condition at the waste-form surface, with the concentration in the surface liquid determined by the calculated time-dependent rate of mass transfer into the exterior porous or fractured rock. Zavoshy et al.[10] show that when the boundary condition of constant surface concentration is replaced by an experimentally measured concentration-dependent solid-liquid reaction rate obtained from laboratory leach data for glass, the calculated dissolution rate approaches that from the simple model of saturation in surface liquid within a few years after emplacement, and the

surface concentration at steady state deviates in only a minor way from saturation for the low-solubility components. Therefore, the complication of a reaction-rate boundary condition is not necessary when the low-solubility elements approach saturation in surface liquid at times that are short compared to the times of interest in repository performance analysis, and the reliability of long-term prediction does not suffer from the uncertain extrapolation of laboratory leach-rate data.

5. Effect of borehole water. A recent mass transfer analysis by Chambré[23] shows that the volume of ground water trapped within the borehole and waste package introduces only a short time delay in the rise in concentration of dissolved species at the waste surface. Within the long times important in repository performance analysis, and for the low-solubility waste constituents, the concentration within the borehole liquid approaches saturation and the dissolution rate is controlled by the rate of mass transfer into the backfill or rock. Thus, assuming[30-33] that borehole water represents an equivalent volume of confined leachant to use in applying a laboratory leach correlation (cf. III B.1,2) ignores the important long-term release mechanisms in a repository, and it does not produce a conservative or bounding estimate of release rate. Because of the empirical extrapolation of real-time laboratory data, it cannot be a reliable technique to predict long-term releases.

6. Reliability and validation. Although these mass-transfer theories unrealistically assume that all waste solids are suddenly and completely exposed to ground water, these theories predict compliance with the numerical performance criteria for most of the radioelements in glass and spent fuel. Because the mass-transfer theories are mechanistic, mathematically formulated, exact, and require only a few directly measurable parameters, they are readily adapted to testing for validity and can be expected to result in reliable predictions of long-term performance, although still conservative and not realistic in all detail that might be desired. As the complexity increases, more phenomena, assumptions, and input data must be validated, and predictive reliability becomes more difficult, as illustrated below.

D. Effect of Partly Failed Protective Waste Containers

Further realism is usually expected to result in lower predicted releases from the waste package. For example, not all of the protective container is expected to fail at once, and releases from the waste solid will likely be reduced by the tortuous pathways through the partly failed outer layers and corrosion products. The multicomponent corrosion products can result in solid phases with low saturation concentration of contained radioelements. The protective features of these more realistic phenomena should be taken into account, where possible. However, in any predictive effort there is a compromise between the increased detail for realism as contrasted with the loss of predictive reliability, when the greater detail invokes additional physical parameters and requires more data and validation than may be possible within available resources and time.

The WAPPA code,[24] listed by all of the repository projects as one of their system codes for predicting waste-package performance, predicts release rates by finite-difference calculations of molecular diffusion through backfill. One of the several unjustified approximations[1] made in the WAPPA release-rate calculation is to assume that diffusional transport through holes of known area in the container is given by the diffusional transport from a bare waste solid multiplied by the ratio of hole area to total area. In attempting to be more realistic by taking into account partial container failure, predictive reliability suffers in two ways: (1) data are required on the time-dependent extent of container failure, including the number, size, and spacing of penetrations, and (2) the assumption of release rate proportional to hole area is incorrect when the holes are small.

Chambré's analytical solution[23] for diffusion through well-separated holes shows that for small holes the area proportionality assumed by WAPPA is not obeyed. As shown in Figure 2, if the equivalent hole radius is 1 mm and if the total hole area ia about 0.05 percent of the container area, the rate of diffusive transport through the holes is the same as if no container material were present.[25] This is a consequence of the large concentration gradients and large diffusive fluxes near the hole edges, and it may explain observations by Johnson et al.[26] of large releases of cesium through small apertures in Zircaloy cladding. Of course, the holes could become plugged with corrosion products, or the failure phenomena may be such that containers are penetrated by only a few openings, so that the net release rate could be appreciably lower than that of a bare waste solid. However, even if there are enough container holes to remove the container as an important barrier, the mass-transfer rates will remain low because of the slow diffusive-convective transport through surrounding backfill and rock.

Obtaining sufficient data to reliably predict the effect of partial failure of waste containers on release rate is a challenge to experiment and theory.

E. Effect of Statistically Distributed Container Failures

The Nuclear Regulatory Commission requires[27] that the yearly release rate of a radionuclide from engineered barriers in a geologic repository be no greater than 10^{-5} times the 1,000-year inventory of that radionuclide or 10^{-8} times the total curie inventory of all radionuclides at 1,000 years. For the simple analogue of a bare waste solid surrounded by rock: the release rate would equal the dissolution rate estimated by Equations (1) and (2) for steady state or by the analytical solutions for the transient dissolution.[7,8,10] If a backfill is present, the release rate would be the mass-transfer rate calculated at the backfill-rock interface.[8,9,20]

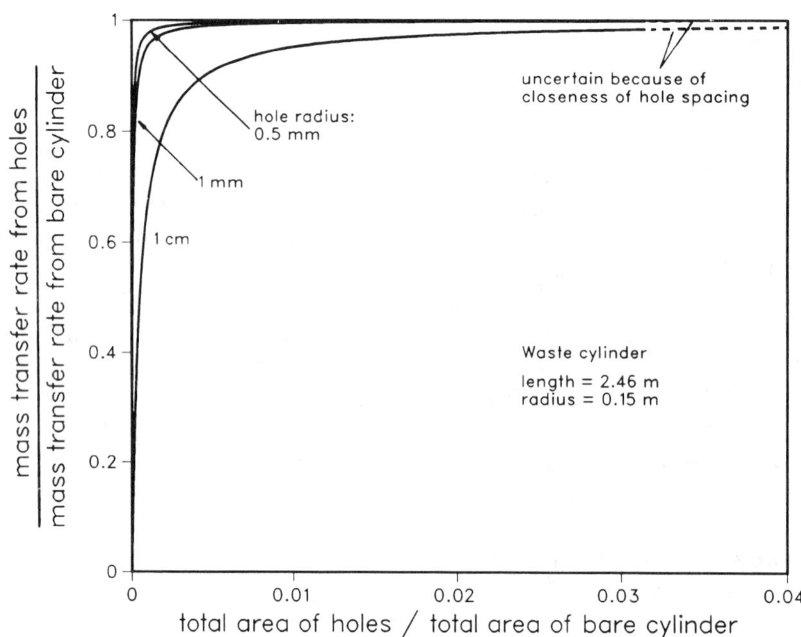

FIGURE 2. Diffusive Mass Transfer Through Holes in Waste Container

If the NRC release-rate criterion is to be applied to the entire ensemble of waste packages in a repository,[1] the statistical distribution of waste-package container failures can affect the average release rates for the repository.

At a given time t the average fractional release rate f(t) of the repository inventory of a radionuclide, based on the radionuclide inventory at 1,000 years, is a statistically weighted average of the fractional release rates from the waste packages failed up to time t. At time t the fractional release rate from a package whose container fails at time t', after emplacement if $f_p(t,t')$, and the failure probability per unit time at time t' is p(t'). The fraction of containers failing in the time span between t' and t' + dt' is p(t')dt', so the repository-average fractional release rate of the radionuclide is given by:

$$f(t) = \int_0^t f_p(t,t')p(t')dt' \tag{5}$$

The repository averaged fractional release rate will not differ much from the single-package fractional release rate for low-solubility long-lived species if waste dissolution continues after all containers have failed. It will be lower than the single-package release rate for soluble long-lived species that are available for rapid dissolution once the waste container fails, such as cesium-135 and iodine-129 in the gap activity in spent fuel.

1. Illustration for cesium-137. Statistically distributed container failures do not necessarily result in repository-average release rates lower than those for individual waste packages. To illustrate, we consider the release of cesium-137 from glass waste.[28] We assume that the cesium dissolves congruently with silica and apply Chambré's[23] analytical solution for the time-dependent fractional dissolution rate from a spherical waste surrounded by porous rock, for a silica solubility of 200 g/m^3, and for a waste package containing initially 270 kg of silica and 0.45 kg of cesium-137. The single-package fractional dissolution rate of cesium-137, for a container that fails at the time of emplacement, is shown in Figure 3. Because of the low 1,000-year inventory of 30-year cesium-137, the fractional dissolution rate is very large at early times before cesium-137 has decayed. However, if the container does not fall for 300 years, most of the cesium-137 will have decayed and the single-package fractional dissolution rate will be much smaller than the calculated equivalent fractional release limit of 0.02/year.

   Assuming that the container failure rate is governed by a log-normal distribution with mean time to failure of 300 years and a deviation of 300 years, we obtain the repository-average fractional dissolution rate of cesium-137 shown in Figure 3. The consequence of a statistical distribution of container failures is to allow earlier container failures and to increase the average normalized dissolution release rate of cesium-137. The calculated release rate of cesium-137 into rock will be much smaller if a sorbing backfill is present or if the low effective solubility (10^{-3} g/m^3) recently measured for cesium in defense glass[29] is considered.

2. Data needed for reliable prediction. Data on the probability distribution of container failures are necessary for reliable prediction of repository-average release rate, placing additional demands on the experiments and performance assessment to establish container failure modes.

III. NONMECHANISTIC ESTIMATES OF WASTE-PACKAGE RELEASE RATE

A. Use of Laboratory Leach Data to Predict Radionuclide Release Rates

Beginning over twenty years ago, laboratory leaching experiments have been performed on borosilicate glass and other candidate waste forms. Typically, a small sample of a waste-form material is exposed to a leachant liquid in a vial kept at constant temperature. The leachant is periodically analyzed for the concentration of dissolved constituents. Most data are reported for leach times of about one month, but some leach times of a few years have also been reported. Many publications propose correlations of the

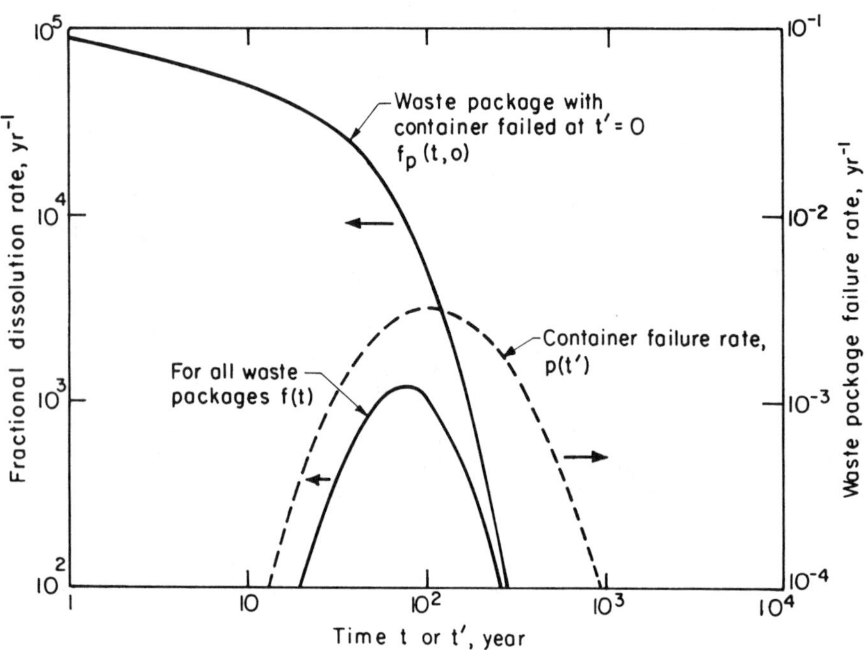

FIGURE 3. Normalized Release Rate of Cesium-137 From Borosilicate Glass Waste, Effect of a Lognormal Distribution of Container Failures (mean time to failure = 300 years, deviation 300 years)

rate of dissolution of silica from borosilicate glass, and a few extend the correlations to radioelements contained in glass waste. Some of the correlations are structured in a way to suggest possible mechanisms that control dissolution rate, such as surface films, sorption, etc., but all of the correlations are empirical and include several arbitrary and adjustable parameters that are determined by curve fitting to laboratory data.

In 1980, the Waste Isolation System Panel[14] (WISP) of the National Research Council began a three-year study that included an evaluation of the applicability of these laboratory leach data to predict release rates from waste solids in a geologic repository. In 1981, the panel concluded that (1) there is no reliable basis for extrapolating the empirical correlations of laboratory leach data to predict release rates at exposure times thousands of times longer than encountered in the experiments, and (2) the repository analogue proposed as a means of using the laboratory leach data is a nonmechanistic postulate and is not valid. The analogue problems are discussed in the WISP report and are summarized below.

B. Postulated Repository Analogues for Predicting Releases

The laboratory leach experiments measure the net rate of reaction between the leachant and the solid surface. In the liquid-continuum leachant no appreciable concentration gradients are expected, so the exterior-field diffusive-convective transport processes that control the dissolution rate in a repository environment are not present in the laboratory experiments. The buildup of corrosion products within the leachant can affect the rate of dissolution of the waste samples, so the ratio S/V of sample surface S to leachant volume V is used as a correlating parameter; a larger S/V results in more rapid increase in concentration of dissolution products in the liquid, more rapid approach to saturation, and more rapid reduction in leach rate.

1. Repository S/V based on borehole water. Various proposals have been made[30-33] to extrapolate the laboratory leach-rate data to repository conditions by adopting the laboratory data taken at the same S/V ratio that is presumed to exist for the waste package in a repository. Although it has been pointed out that there is no meaning to an equivalent volume of ground water in contact with the surface area of each waste package,[4] it is still assumed by some[31,32] that there is a repository analogue of the laboratory leach experiment. The assumption[31-33] that the equivalent repository water volume associated with a waste package is the volume of water in waste-package voids and in the borehole annulus leads to a prediction of zero steady-state release, because the concentration in this assumed confined liquid volume will reach saturation.

   However, because the borehole liquid is, in fact, not confined, dissolved species will transport into surrounding porous medium by diffusion and convection in pore water, the mechanisms considered in the mass-transfer analyses by Chambré et al. and others. Release will continue at a finite rate, the solute concentration in the borehole liquid will fall slightly below saturation, and solid-liquid reaction at the waste surface will proceed at a steady-state rate equal, for long-lived species, to the rate of diffusive-convective transport into the exterior pores.

2. Postulates of ground water residence time and volume. Macedo et al.[30-32] have obtained laboratory leach data for glass waste powder with periodic partial replacement of leachant, simulating a small and continuous leachant flow through a leach-test vial with well-mixed solid and liquid. At early times, the dissolution rates are controlled by the solid-liquid reaction rate, which decreases with increasing concentration of solute. At later times, the dissolution rate is found to be proportional to the volumetric rate of replacement of leachant, suggesting that an equivalent saturation concentration has been reached and that the release rate is given by the simple bulk-flow saturation-limited calculation discussed in II.A above. The diffusive-convective transport mechanisms that control the net release in a repository environment are not present in these experiments. Macedo et al. empirically correlate their leach data with the ratio S/V of powder surface to leachant volume and the average residence time T_r of leachant, the latter determined by dividing the leachant volume by the volumetric replacement rate of leachant.

   To apply their empirical correlation of laboratory leach data to predicting waste performance in a repository, Macedo et al.[31,32] propose a repository analogue that has waste fragments well stirred, with a specified volume of ground water associated with each waste package, and a specified volumetric flow of ground water through this well-stirred volume. For a waste solid surrounded by a large volume of wet rock, they propose that the leachant volume be identified as the volume of voids initially inside the waste container, which will become filled with water when the waste container fails. Here they do not include water that may be in the gap between the waste package and rock. They propose that the volumetric flow through this well-mixed container of waste fragments and void water be identified as the upstream Darcy velocity of ground water multiplied by the projected cross-sectional area of the solid waste. They do not consider the effect of any backfill between the waste package and rock, and they equate NRC's fractional release rate with the fractional dissolution rate. The expected fractional dissolution rate when liquid in contact with the waste is at concentration N_i is then estimated by Macedo et al. to be:

$$f_i = \frac{N_i V}{m_i T_r S} \qquad (6)$$

where m_i is the mass of species i in the waste solid per unit surface area of solid. It is clear that this analogue distorts the physical situation in the repository. Contrary to the assumptions of Macedo et al., fragmented waste is not well stirred with void water and with any water that may flow through the waste. If the waste solid has finite flow permeability, the actual rate of flow through the waste will depend on the ratio of waste solid permeability to the permeability of surrounding backfill and rock, and it can depart considerably, and either direction, from that estimated by Macedo et al.

More importantly, dissolved species can and will diffuse into surrounding backfill and rock and will be transported by diffusion and convection into a much larger volume of flowing water than estimated from the Macedo analogue. If the analogue were correct, Equation (6) would become identical with the diffusive-convective Equation (1) as the waste permeability becomes very small, resulting in no volumetric replacement of void water and an infinite void-water residence time. However, Equation (6) would incorrectly predict zero dissolution, whereas the dissolution rate from Equation (1) is finite. Furthermore, as the volume of void water goes to zero, Macedo's Equation (6) predicts a zero dissolution rate. In contrast, Equation (1) was derived for a waste solid in contact with wet rock, and it correctly predicts a finite dissolution rate whether or not void water is present. Because Equation (1) is for a steady-state dissolution rate, it applies also to a waste package with finite void water, provided that the radius R is taken as the radius of the borehole. Also, applying Macedo's Equation (6) to a repository would overlook the effect of rock porosity, an important parameter that affects the waste dissolution rate, as shown in Equation (1). The repository analogue proposed by Macedo et al. for finite convective flow is unrealistic and results in nonconservative predictions.

Recognizing that exterior-field diffusion can affect the dissolution rate in a repository, Macedo et al.[31,32] propose that at low ground water flow rates the equivalent contact time of ground water to be used in Equation (6) be calculated by:

$$T_r = \frac{d^2}{KD} \tag{7}$$

where: d is the waste diameter, K is the retardation coefficient, and D is the coefficient for molecular diffusion in ground water in the rock pores. If Equation (7) were corrected by moving the retardation coefficient to the numerator, T_r would be the time for a diffusion front to travel a distance d in a sorbing medium. However, it has not been explained why this transient diffusion time should have any connection with the mean residence time for steady-state dissolution, and why it should be related to the mean residence time for Macedo's laboratory experiments, for which Equation (6) was derived. It is not valid to adopt estimates of T_r based on exterior-field diffusion mechanisms to use in adapting correlations of laboratory data which represent dissolution mechanisms not affected by exterior-field diffusion. Different phenomena are involved, and the repository analogue is postulated without demonstrating its causal connection to the laboratory experiments.

The fallacy of the postulated analogue can also be demonstrated by comparing the results predicted by the technique of Macedo et al. with results predicted by Chambré's Equation (2), which is exact for steady-state dissolution at no flow and without decay. Macedo et al. state that, by using the diffusion estimate for T_r, Equation (6) will predict dissolution rates in a repository when convective effects on dissolution are negligible. In the absence of convection, there can be no finite flow through a fractured waste solid, so Macedo's Equations (6) and (7), derived for steady state dissolution, should agree with the exact steady-state solution in Equation (2). However, Macedo's incorrect repository analogue and his unjustified assumption of a diffusion-limited equivalent residence time introduce the sorption retardation coefficient K, even though sorption cannot affect steady-state dissolution of long-lived species. He fails to predict a functional dependence on rock porosity, which is shown in Equation (2) to be an important parameter in affecting dissolution rate. He incorrectly predicts a zero dissolution rate when there is no void water in the waste package, although as explained above (II.C.5), water in waste package voids and in the borehole annulus has no affect on the steady-state dissolution rate. Macedo's equations fail to predict the much greater transient dissolution rates that are shown by Chambré's exact solutions to occur over hundreds and thousands of years in a low-flow repository. Neglecting rock porosity and transient diffusion-controlled dissolution results in nonconservative estimates of dissolution rate.

As concluded by the WISP panel,[4] such empirical techniques are useful to correlate laboratory leach data, but postulating equivalent values of S/V, volume flow rate per waste package, and T_r in attempting to use these correlations to predict performance in a wet-rock repository is neither valid nor necessary. Predictive reliability is lost by such postulates. Proper mechanistic theories of repository performance exist, and these theories specify the kind of laboratory data, such as saturation concentrations, that are needed for valid and reliable predictions of waste dissolution rates.

3. Data on natural analogues. Macedo et al.[31] base their conclusion that solid-liquid reaction rates control dissolution in a repository on a limited number of observations by Berner,[34] who studied the dissolution rates of isolated grains of low-solubility minerals surrounded by a wet porous solid. By observing the temperature dependence of the dissolution rate, Berner concluded that dissolution was controlled by exterior-field diffusion for some mineral grains and by solid-liquid reaction rate for others. Ranking the few samples according to their solubility, he found that the dissolution of most of the low solubility grains was controlled by solid-liquid reaction rate. The observation was empirical and was limited to a small number of mineral samples. It does not justify the unqualified conclusion by Macedo et al.[31] that different materials, e.g., borosilicate glass, and enormously larger solid forms will follow the ranking observed by Berner.

Better insight into the fallacy of generalizing and extrapolating from Berner's data is provided by the analytical solution of Zavoshy et al.[10] for dissolution from a solid sphere of radius R surrounded by a saturated sorbing porous medium. The solid-liquid reaction rate, expressed by a simple first-order reaction dependent on the concentration of solute in the liquid at the solid surface, is used as a dissolution-flux boundary condition to connect with the mathematical analysis of exterior-field diffusion in the absence of convection. The analytical solution for the time-dependent dissolution rate contains a term ψ, the magnitude of which determines which phenomenon controls dissolution rate at steady state, where:

$$\psi = \frac{\text{forward reaction rate per unit area at R}}{\text{steady-state diffusive mass transfer rate at R}} = \frac{kR}{\varepsilon D} \tag{8}$$

where k is the forward reaction rate constant. When ψ is much larger than unity, steady-state dissolution is controlled by exterior-field diffusion; when ψ is much less than unity, solid-liquid reaction rate controls. Inferring k from early-time leach data for silica from borosilicate glass, assuming a waste sphere 0.44 m in radius, and with D = 7.7×10^{-2} m^2/yr for silica acid at 90^oC and ε = 0.01, we calculate ψ

(SiO_2) = 1240. For this large glass sphere, silica dissolution is controlled by exterior-field diffusion.

If we make the sphere radius small enough, Equation (8) will predict a small ψ and solid-liquid reaction rate will control. This is what one would expect from the physics of the problem. We know of no causal effect of curvature on the solid-liquid reaction rate, but the high curvature of a small-radius sphere promotes more rapid exterior-field diffusion and can eliminate it as a controlling phenomenon.

Equation (8) demonstrates that it is not valid to apply Berner's conclusions on small mineral grains to large waste solids.

The only physico-chemical property of the solid contained in Equation (8) is the forward reaction rate constant k. The theory shows that the dissolution rate of Berner's small mineral grains should have been ranked according to k, if it were known, instead of solubility. The ranking according to solubillty does not demonstrate a causal effect of solubility on first-order dissolution and should not be generalized. Berner observes that the solid-liquid reaction is usually complex, and a higher-order reaction could result in the observed ranking.

There is no valid basis for the conclusion by Macedo et al. that the dissolution rate of nuclear waste in a repository would be expected to be controlled by surface reaction mechanisms. Based on the foregoing analysis, and as demonstrated in Table 1, we conclude that for borosilicate glass the dissolution rate of silica and other low-solubility species in borosilicate glass waste will be controlled by exterior-field mass transfer and that solid-liquid reaction rate will not influence the dissolution rate except during the very early time of exposure to ground water. This early time is no more than a few days if the solid-liquid reaction rate is given by the parameters adopted by Zavoshy et al.[10]

4. Flow of ground water through fractured waste. In explaining his proposed model for predicting waste dissolution rate in a repository, Macedo[31,32] states that his model considers ground water flowing through a fractured waste solid, whereas the mass transfer analytical solutions presented by Chambré, assume an impermeable waste solid. Glass waste will be internally fractured from thermal stress. Water is likely to penetrate into fractured glass, but it will not alter the results of Chambré's mass transfer analysis if there is no net through flow of liquid through the waste. Solid-liquid reactions on internal surfaces will only increase the net solid-liquid reaction rate, which is already sufficiently rapid to maintain near-saturation concentrations of the low-solubility constituents in surface liquid in a repository environment.

Net through flow of ground water through the fractured waste is not included in the present mass-transfer analyses by Chambré, but it can be added. Because only a portion of the ground water flow can permeate the fractured waste, parameters appearing in the resulting mass transfer analysis must include the hydrodynamic permeabilities and porosities of the waste solid, backfill, and rock. Mass transfer from waste particles to through-flowing liquid will introduce dimensions of waste fragments and flow interstices, in addition to the other parameters already appearing in the mass transfer analysis. None of these additional parameters appear in Macedo's proposed method of predicting waste dissolution rate in a repository because Macedo assumes that all ground water flowing through rock of cross-sectional area equal to that of the waste solid will flow through the fractured waste, and he assumes that this permeating ground water is well stirred with powdered waste, as in his laboratory experiments. Therefore, the effect of any finite flow of ground water through fractured waste must rest on a more realistic and mechanistic analysis and cannot be predicted by the postulates of Macedo et al. The fact that Macedo's experiments include well-mixed flow through powder samples does not necessarily mean that the same flow and dissolution process will occur in repository waste.

Theoretical studies of the hydrodynamics and mass transfer in two-region porous media are under way by Chambré, and extension to include the effects of flow through fractured waste can be considered.

IV. SUMMARY AND CONCLUSIONS

Predicting rates that radionuclides are released from waste packages cannot rest upon empirical long-term extrapolations of laboratory leach data. Reliable predictions can be based upon simplified assumptions, such as solubility-limited bulk-flow, if the assumed parameters are reliably known or defensibly conservative, but assuming volumetric flow rates through a waste-package cross section is arbitrary and can be nonconservative.

Wherever possible, performance analysis should proceed beyond simple bounding calculations to obtain more realistic--and usually more favorable--estimates of expected performance. Desire for greater realism must be balanced against increasing uncertainties in prediction and loss of reliability. Theoretical predictions of release rate based on mass-transfer analysis are bounding, and the well-established theory is well adapted to verification. The results from the exact analytical solutions can be used to test predictions from numerical techniques and from less-mechanistic analogues.

Lower release rates are expected if less than complete failure of waste containers is considered, but data for reliable quantitative predictions are not yet available and will be difficult to obtain. Diffusive transport through small holes and cracks can be much greater than incorrectly predicted on the basis of area proportionality.

Repository-average release rates, taking into account statistical distribution of container failures, can be lower than individual-package release rates for some radionuclides and greater for others, depending upon mean-time to failure and the probability distribution of failures. Data are not yet sufficient for reliable prediction and will be difficult to obtain.

Several efforts to predict waste-package release rates in a repository, utilizing empirical correlations of laboratory leach-rate data, have invoked postulates of repository analogues to simulate the laboratory leach experiments. The postulated analogues are unrealistic, they introduce fictitious repository parameters, such as volume and volumetric flow rate of ground water associated with each waste package and ground water residence time, which are assigned arbitrary values for making predictions. They invoke functional dependence on parameters inconsistent with well-established mass-transfer theory, and they incorrectly assume that the dissolution mechanisms that control release rates observed in laboratory experiments are controlling or important in the repository.

The most useful experimental results from laboratory leach experiments are the saturation concentrations of radioelements released from the waste. Other parameters needed for reliable estimates of release rates in a repository can be directly measured, including rock and backfill porosity, diffusion coefficients, sorption, and ground water pore velocity upstream of the waste.

Effects on release rate due to colloids, radiolysis, possible flow through backfill and fractured waste, and grain-boundary diffusion and interbed flows in salt need to be resolved by theory and experiment.

V. ACKNOWLEDGMENT

This work was sponsored in part by the U.S. Department of Energy under Contract Number DE-AC03-76SF00098.

REFERENCES

1. J. A. Lieberman et al., "Performance Assessment National Review Group," Weston Report RFW-CRWM-85-01 (1985).

2. U.S. Department of Energy, Draft Environmental Assessment for the Yucca Mountain Site, DOE/RW-0012 (1984).

3. T. K. Sherwood, R. L. Pigford, and C. L. Wilke, Mass Transfer, McGraw Hill, New York (1975).

4. T. H. Pigford et al., "A Study of the Isolation System for Geologic Disposal of Radioactive Wastes," National Academy Press, Washington, D.C. (1983).

5. P. L. Chambré, T. H. Pigford, and S. Zavoshy, "Solubility-Limited Dissolution Rate in Groundwater," Trans. Amer. Nucl. Soc., 40, 153 (1982).

6. P. L. Chambré, S. Zavoshy, and T. H. Pigford, "Solubility-Limited Fractional Dissolution Rate of Vitrified Waste in Groundwater," Trans. Amer. Nucl. Soc., 43, 111 (1982).

7. P. L. Chambré et al., "Analytical Performance Models," LBL-14842 (1982).

8. P. L. Chambré, and T. H. Pigford, "Prediction of Waste Performance in a Geologic Repository," Proceedings of the Materials Research Society, The Scientific Basis for Nuclear Waste Management, Boston (1983).

9. P. L. Chambré et al., "Mass Transfer and Transport in a Geologic Environment," LBL-19430 (1985).

10. S. J. Zavoshy, P. L. Chambré, and T. H. Pigford, "Mass Transfer in a Geologic Environment," Scientific Basis for Nuclear Waste Management VIII, C. M. Jantzen, J. A. Stone, R. C. Ewing, eds., Materials Research Society Proceedings, 44, 311-322 (1985).

11. B. P. McGrail, L. A. Chick, and G. L. McVay, "Initial Results for the Experimental Validation of a Nuclear Waste Repository Source Term Model," Battelle Pacific Northwest Laboratory Report, PNL-SA-12015 (1984).

12. T. H. Pigford, and P. L. Chambré, Mass Transfer in a Salt Repository, Report LBL-19918 (1985).

13. I. Neretnieks, "Leach Rates of High Level Waste and Spent Fuel: Limiting Rates as Determined by Backfill and Bedrock Conditions," Scientific Basis for Nuclear Waste Management V, W. Lutze, ed., 559-568, Proceedings of the Materials Research Society, New York: Elsevier Science (1982).

14. R. O. Fournier, and J. J. Power, Amer. Minerol. 61, 1052-56 (1977).

15. I. Neretnieks, "Diffusion in the Rock Matrix: An Important Factor in Radionuclide Retardation," Journal of Geophysical Research, 85, 4379 (1980).

16. S. K. Neretnieks, "Diffusion in Crystalline Rocks," Scientific Basis for Nuclear Waste Management V, W. Lutze, ed., Proceedings of the Materials Research Society, Elsevier Science, New York (1982).

17. K. Skagius, and I. Neretnieks, "Diffusion in Crystalline Rocks," Scientific Basis for Nuclear Waste Management V, W. Lutze, ed., 509-518, Elsevier Science, New York (1982).

18. M. H. Bradbury, D. Lever, and D. Kinsey, "Aqueous Phase Diffusion in Crystalline Rock," Scientific Basis for Nuclear Waste Management V, W. Lutze, ed., 569-578, Elsevier Science, New York (1982).

19. G. L. McVay, D. J. Bradley, and J. F. Kircher, Elemental Release From Glass and Spent Fuel, ONWI-275 (1981).

20. P. L. Chambré, H. C. Lung, T. H. Pigford, "Mass Transport From a Waste Emplaced in Backfill and Rock," Trans. Amer. Nucl. Soc., 44, 112 (1983).

21. H. C. Lung, P. L. Chambré, and T. H. Pigford, "Nuclide Migration in Backfill With a Nonlinear Sorption Isotherm," Trans. Amer. Nucl. Soc., 45, 107 (1983).

22. P. L. Chambré, W. J. Williams, C. L. Kim, and T. H. Pigford, "Time-Temperature Dissolution and Radionuclide Transport," Trans. Amer. Nucl. Soc., 46, 131-132 (1984) (UCB-NE-4033).

23. P. L. Chambré, to be published.

24. L. M. Johnson, S. Stroes-Gascoyne, J. D. Chen, M. E. Attas, D. M. Sellinger, and H. G. Delaney, "Relationship Between Fuel Element Power and the Leaching of ^{137}Cs and ^{129}I from Irradiated UO_2 Fuel," Proceedings of the Topical Meeting on Fission Product Behavior and Source Term Research, Snowbird, Utah (1984).

25. U.S. Nuclear Regulatory Commission, "Disposal of High-Level Radioactive Wastes in Geologic Repositories - Technical Criteria," 10 CFR 60, Fed. Reg., 48, 120, 18194 (1983).

26. C. L. Kim, P. L. Chambré, and T. H. Pigford, Radionuclide Release Rates as Affected by Container Failure Probability, Report LBL-19851 (1985).

27. N. E. Bibler, E. I. du Pont de Nemours, Savannah River Laboratory, Private Communication (1985).

28. A. Barkatt, P. B. Macedo, W. Sousanpur, A. Barkatt, M. A. Boroomand, C. F. Fisher, J. J. Shirron, P. Szoke, and V. L. Rogers, "The Use of a Flow Test and a Flow Model in Evaluating the Durability of Various Nuclear Waste-Form Materials's Nuclear Waste-Form Materials," Nucl. Chem. Waste Management, 4, 153-169 (1983).

29. P. B. Macedo, "Phenomenological Models of Nuclear Waste Glass Leaching," Chapter 6, Final Report of the Defense High-Level Waste Leaching Mechanisms Program, J. E. Mendel, ed., Report PNL-5157 (1984).

30. A. Barkatt, P. B. Macedo, and B. C. Gibson, "Modelling of Waste Form Performance and System Release," Scientific Basis for Nuclear Waste Management VIII, Materials Research Society Symposium Proceedings, C. M. Jantzen, J. A. Stone, R. C. Ewing, eds., 44, 3-13 (1985).

31. M. J. Plodinec, G. G. Wicks, and N. E. Bibler, An Assessment of Savannah River Borosilicate Glass in the Repository Environment, DP-1629, E. I. du Pont de Nemours and Co., Aiken, South Carolina (1982).

32. R. A. Berner, "Rate Control of Mineral Dissolution Under Earth Surface Conditions," Am. J. Sci, 278, 1235-1252 (1978).

EDITORS NOTE: A portion of this invited paper by T. H. Pigford and P. L. Chambré provides a sometimes unfavorable review of efforts, principally by P. B. Macedo and his coworkers, to predict in situ waste-package release rates in a geologic repository using empirical correlations of laboratory leach-rate data. With the objective of clarifying and possibly resolving the issues raised by the Pigford-Chambré review, I have invited Professor Macedo to respond and Professors Pigford and Chambré to rebut that response.

Response to the Paper "Reliable Predictions of Waste Performance in a Geologic Repository"

P. B. Macedo
C. J. Montrose
Vitreous State Laboratory
The Catholic University of America
Washington, DC 20064

The paper "Reliable Prediction of Waste Performance in a Geologic Repository" by Thomas H. Pigford and Paul L. Chambré purports to give a realistic but conservative estimate of the long-term performance of radioactive waste packages in a geologic repository. The first half of the paper provides a summary of the diffusion-convection, mass-transfer analysis that they have previously presented. The rest of the paper is an attack on the usefulness of experimental studies of radioactive-waste leaching that have thus far provided a very reliable basis for predicting long-term durability of waste packages.

The repetition of the earlier theory that is given in Part II: "Mechanistic Analysis of Radionuclide Release Rates;" has been shown to be replete with shortcomings when used to describe dissolution of materials under geologic conditions.[1] Many of the shortcomings are traceable to the use of elemental solubilities rather than experimental studies to obtain a realistic source term that could be grafted onto the Pigford-Chambré transport calculation.

The authors' struggle to discredit the usefulness of laboratory leach tests and their interpretation is puzzling and disturbing. Pigford and Chambré disparge the importance of laboratory tests with a series of self-serving distortions. They contrast their mass-transfer analysis ("... a mechanistic theory based on well-understood governing equations ...") with measurement-based approaches ("empirical extrapolations" that depend upon "... arbitrary and adjustable parameters ..."). Mass-transfer theories are claimed as "... the only reliable means of extrapolating into the future ..." even though many laboratory experiments that simulate realistic conditions provide results that are consistent with water-glass interactions over millennia.[2-3]

This is asserted despite their own theory requiring the insertion of a parameter N_i^* which for a "... bounding calculation, ... is chosen as the saturation concentration." One must ask: saturation with respect to what? At what temperature and pH and under what redox conditions is this saturation concentration to be determined? And what is one to choose for the saturation concentration of an element such as boron? It is well-known, for instance, that interactions between the leachant and the waste package can produce rather substantial excursions in the ground-water pH upon which the solubilities of, for instance, silica, uranium, and technetium depend rather sensitively. Solubilities are also quite sensitive to the redox conditions;

of magnitude. The practical measurement techniques for determining appropriate N_i^* values are the pulsed-flow (or periodic replacement) leach tests that Pigford and Chambré have labeled as unnecessary and unreliable.

In Part III: Nonmechanistic Estimates of Waste-Package Release Rate, Pigford and Chambré intensify their attack. They refer to the study by Pigford et al.:[4]

> ... the Waste Isolation System Panel (WISP) of the National Research Council ... concluded that (1) there is no reliable basis for extrapolating the empirical correlations of laboratory leach data ... and (2) the repository analogue proposed as a means of using the laboratory leach data ... is not valid.

It would be interesting to discover whether WISP would support this conclusion in light of the extensive experimental programs of the last four years.[5] The authors' subsequent attempts to justify these rather extraordinary claims are seriously flawed. Responses to several of them are given below.

In Section B of Part III, Pigford and Chambré challenge the generic repository scenario that has been employed to connect laboratory measurements with repository performance. They assert that one cannot treat the volume of ground water in waste-package voids and in the borehole annulus as analogous to the volume of leachant in a laboratory leach test using simulated repository water. This is not correct. Under a reasonable set of generic repository conditions envisioned for the disposal of defense waste borosilicate glass, roughly 20% of the canister volume will be vacant; since the surrounding rock and backfill are characterized by a porosity on the order of 1%, the highly fractured nuclear waste glass encounters the flowing ground water within a quite well-defined reaction volume. Except under very low-flow-rate conditions where diffusion rather than flow is the dominant transport mechanism (see below), this corresponds almost excactly with the design of the laboratory-flow test. The discussion in the subsection entitled "Repository S/V based on borehole water" thus is of no relevance.

In the following subsection "Postulates of Ground-Water Residence Time and Volume" Pigford and Chambré criticize the approach of Macedo and his coworkers on a number of counts; however, except for the need to include a porosity factor in the numerator of Eq. (6), none of these survive careful scrutiny. Responsible engineering practice requires that no explicit account be taken of the effects of backfill between the waste package and the rock until the nature of the backfill is determined. Moreover, backfill materials will only be used so as to <u>decrease</u> the release rate. Using the fractional dissolution rate of the waste form as indicative of the fractional release rate is simply a conservative assumption that can only underestimate the durability of the waste isolation system. In the circumstances described above (where the fraction of void space within the borehole is much larger than the porosity of the surroundings), the upstream Darcy velocity multiplied by the cross-sectional area of the waste can correctly be taken as the volumetric flow rate. The authors' objection to substituting Eq. (7) in Eq. (6) in the case of very low flow rates, when diffusion is the chief means of transport, is puzzling, since doing so leads to their Eq. (2) apart from a trivial factor of order unity that depends on the geometry of the waste package.

In the paragraph headed "Data on Natural Analogues," Pigford and Chambré challenge the idea that surface-reaction mechanisms are influential in determining dissolution rates; rather, they assert that exterior-field, mass-transfer processes are completely dominant, and they cite an earlier theoretical paper of their own[6] and their Table 1 as demonstrating that this is so. Of course, the concentrations given in their Table 1 imply a ground-water composition that is unlike anything that might be encountered in a geologic repository.

Moreover, as pointed out in Reference 1, nothing in the Pigford-Chambré theory distinguishes the leaching of nuclear waste glass from natural minerals. Consequently one should be able to test the theory by analyzing the dissolution of natural materials that have undergone prolonged aqueous contact. For rock at a depth of 600 m, the theory overestimates the dissolution of SiO_2 by a factor of 10^7.

Conversely, when the laboratory leach test data of Macedo and his colleagues are scaled to describe the leaching of natural materials, the results are quite different. One finds order-of-magnitude agreement with ground water analyses in both basalt and granite rock formations.[7] The leaching of land tektites and deep-sea sediment microtektites are also predicted to similar accuracy by making use of scaled, leach-test results.[8-10] It would thus seem that laboratory studies of solid-liquid reaction mechanisms should not be discarded out of hand.

The final attack on Macedo's work is offered in the paragraphs entitled "Flow of Ground Water through Fractured Waste." Here Pigford and Chambré repeat their dissatisfaction with the lack of explicit accounting for backfill effects as well as with his method of calculating volumetric flow rates. They further dispute the assumption that the ground water will flow through the interstices of the fractured waste, and will be, in effect, well-stirred. That the water will thoroughly permeate the fracturede waste is to be expected since the porosity of the waste is so much higher than that of the surrounding rock. Contrary to the assertion of Pigford and Chambré, Macedo is correct in taking the ground water in the reaction volume with fragmented waste to be well-stirred; for repository relevant flows ($T_r \geq 1$ yr), diffusion distances over the time required for a ground water exchange within the reaction volume are on the order of or greater than 10 cm, a distance that is certainly large compared with the size of the voids. That is, slow flow in a repository implies rather thorough mixing; high flow rates are, of course, unacceptable in any repository context and need not be examined.

REFERENCES

1. Bernard L. Cohen, "Critique of the National Academy of Sciences Study of the Isolated System for Geologic Disposal of Radioactive Waste," Nucl. Technol., 70, 433 (1985).

2. T. M. El-Shamy, "The Chemical Durability of K_2O-CaO-MgO-SiO_2 Glasses," Phys. Chem. Glasses, 14, 1 (1973).

3. L. L. Hench, "Physical Chemistry of Glass Surfaces," J. Non-Crystalline Solids, 25, 343 (1977).

4. T. H. Pigford et al., A Study of the Isolation System for Geologic Disposal of Radioactive Wastes, National Academy Press, Washington, DC (1983).

5. See for example the Proceedings of the High-Level Waste Leaching Mechanism Symposium, Germantown, MD, September 18-20, 1984 (to be published in Nuclear Technology) (1986).

6. S. J. Zavoshy, P. L. Chambré and T. H. Pigford, "Mass Transfer in a Geologic Environment," in Scientific Basis for Nuclear Waste Management, Volume VIII, C. M. Jantzen, J. A. Stone and R. C. Ewing, eds., Mat. Res. Soc. Symp. Proc., 44, 311 (1985).

7. Aa. Barkatt et al., "Correlation Between Dynamic Leach Test Results and Geochemical Observations," Mat. Res. Soc. Symp, 15, 227 (1983).

8. Aa. Barkatt et al., "The Chemical Durability of Tektites--A Laboratory Study and Correlation with Long-term Corrosion Behavior," Geochim. Cosmochim. Acta, 48, 361 (1984).

9. B. P. Glass and J. Crosbie, The Rates of Solution of Certain Natural Glasses--Part I. Microtekties in Deep-sea Sediments, Technical Report NASA CR I66746, National Aeronautics and Space Administration, Goddard Space Flight Center, Greenbelt, Maryland (1981).

10. Aa. Barkatt et al., Chapter 1, "Mechanisms of Defense Waste Dissolution," in Final Report of the Defense High-Level Waste Leaching Mechanisms Program, J. E. Mendel, ec., Report PNL-5157, Pacific Northwest Laboratory, Richland, Washington (1984).

Response to Comments by P. B. Macedo and C. J. Montrose About Reliable Predictions of Waste Performance in a Geologic Repository

T. H. Pigford
P. L. Chambré

I. INTRODUCTION[(a)]

In a recent paper, we reviewed the need for reliable theory and experiment in making long-term predictions of waste performance in a geologic repository.[A] We discussed uncertainties in postulated techniques of applying closed-system laboratory leach data to predicting the dissolution of waste solids in a geologic repository, and we discussed the use of mass-transfer analysis to unify theory and experiment and to provide a clear theoretical basis for long-term prediction.

Comments on our recent paper by Drs. Macedo and Montrose provide a welcome opportunity to clarify several issues related to predicting waste performance.[B] Their comments help illustrate the need for reliable and sound theories for predicting waste performance in the long-term future, and they help focus the fundamental differences between waste dissolution in laboratory leach experiments and dissolution in a repository. To aid better understanding and resolution of the differences between mass transfer in the closed systems considered by Macedo et al. and others and mass transfer in the open systems of waste repositories considered in our mass transfer analyses, we comment here in some detail.

Our mass-transfer theory of waste-package performance has been developed with the objective of unifying theory and experiment to achieve reliable long-term predictions. A theory is adequate if it derives from well-understood and properly formulated governing equations, if it contains well-defined parameters that can be determined by experiment, if it survives tests and verification, and if it can relate experiment to the mechanistic processes that affect performance in a repository. The comments by Macedo and Montrose help illustrate these issues.

II. NEED FOR DATA ON SATURATION CONCENTRATIONS

Macedo and Montrose have correctly observed that the simplest and bounding application of mass-transfer theory requires, for low solubility species, data on saturation limits in the complicated chemical environment near a waste

---

(a) References listed in our earlier paper are shown as numerical superscripts. Additional references listed at the end of this response are indicated by letter superscripts.

package. Data on effective saturation concentrations obtained in laboratory leach experiments, including the valuable experiments of Macedo and Montrose are useful in applying the mass-transfer theory for predicting waste performance in a repository.

Macedo and Montrose question the saturation concentrations of soluble constituents, such as boron, to be used in applying mass-transfer theory. No such saturation concentrations are needed, other than to verify that the constituent is highly soluble. Our publications have stated that the theory of solubility-limited mass transfer does not apply to highly soluble species whose dissolution rate may not be limited by a solubility-limited dissolution rate of the waste matrix. Boron is evidently one such species in borosilicate glass waste. Here we would use the more general analysis, summarized in Section II-B-3 of our paper and derived in a separate paper,[10] wherein solid-liquid reaction rate is used as a boundary condition rather than saturation concentration.

In other publications, we apply mass transfer theory to the transient dissolution of highly soluble species, such as the "gap activity" of spent fuel, that can dissolve rapidly when exposed to ground water. Saturation concentrations are not required for that calculation.

III. WATER FLOW THROUGH WASTE FORMS IN A REPOSITORY

Macedo and Montrose correctly observe, as is also pointed out in our paper, that the particular form of the mass-transfer equations that we have published does not apply if there is flow of ground water through a highly fractured package.[A] Dissolution with flow through fractured waste and with diffusion and flow in the surrounding nondissolving media would require a reformulation of the mass-transfer analysis.

In the U.S. repository projects, the expected ground-water flow is so low that mass-transfer rates are evidently controlled by diffusion from the outer surfaces of the waste package into the surrounding media.[4] If any of the projects should expect appreciable flow through a fractured waste package in a repository, the mass-transfer analysis should be reformulated. We also note that in Sweden's waste disposal project, which predicts greater ground-water flows than in the U.S. projects, net flow through the waste package is not expected.

However, even if there is sufficient flow through the waste form to affect release rate from the waste form in a repository, there will remain open-system pathways for diffusion and convection from the waste form's exterior surface into the pore water in the surrounding media. The questions on the validity of the Macedo-Montrose proposed extrapolation of closed-system laboratory data to a waste package in the open system of a repository remain and are unanswered.

Also, if there is appreciable flow of ground water through a waste solid surrounded by porous or fractured rock, the actual flow rate through the waste must depend upon permeabilities of the waste solid and the surrounding rock, as pointed out in our paper. These parameters do not enter the equations proposed by Macedo. If the waste is more permeable than the rock, the volume flow rate through the waste can be greater than that estimated by Macedo's prescription of multiplying the approach Darcy velocity by the cross sectional area of the waste. The conclusion by Macedo and Montrose that their prescription gives the correct volumetric flow rate contradicts the known hydrodynamics of this problem.

IV. AN INVALID TEST OF THE PUBLISHED THEORY

Macedo and Montrose adopt a statement by B. L. Cohen that our published mass-transfer theory "... has been shown to be replete with shortcomings when used to describe dissolution under geologic conditions ...," and that "... for rock of a depth of 600 m the theory overestimates the dissolution of SiO_2 by a factor of 10^7 ...."[C] Evidently Macedo and Montrose were not aware of a 1984 publication by members of the National Research Council's Waste Isolation System Panel that had already analyzed each of Cohen's claims in detail and showed that Cohen has made an invalid test of the published equation for dissolution rate from a single dissolving object.[D]

Cohen argued that natural grains of siliceous rock dissolve much more slowly than predicted by our equations, even if it is assumed that the rock has long existed with water slowly flowing between the grains. Cohen made two fundamental errors: (1) he applied our equation derived for release from isolated dissolving objects spaced far from each other and surrounded by a nondissolving porous medium, to individual small "grains" of ore that touch each other, forming the pores through which ground water flows through ore bodies, and (2) he failed to allow for the rapid approach to saturation of ground water as it permeates an ore body. In Cohen's ore body all of the grains surrounding a given dissolving grain are also dissolving, whereas our published mass-transfer equations apply to a single dissolving object surrounded by a porous medium of nondissolving grains. The governing equations for diffusive-advective transport must apply also to the ore body, but they are of different form than those used in our analysis. The resulting analytical solutions will be different. We expect that in Cohen's ore body the water between dissolving grains would soon reach saturation, and the net rate of dissolution would be far less than predicted by Equation (1).

Cohen did not consider dissolution rates from the outer boundaries of a dissolving ore body into surrounding nondissolving porous media, so his proposed test should be applied to a closed-system analysis. Therefore, the data reported by Cohen are not a valid test of our mass-transfer equations.

Macedo and Montrose do not credit the dissolution tests conducted by McGrail et al.[12] in an open system, where they measured diffusive-convective mass transfer from a single dissolving object surrounded by a nondissolving porous medium, confirming Equation (1) of our paper.

V. VOLUME OF GROUND WATER ASSOCIATED WITH A WASTE PACKAGE

Part III-B-2 of our paper discusses proposals by Macedo and others to extrapolate leach-rate data, obtained in laboratory experiments with no continuous diffusive-convective transport across the container boundaries, to the open system of a waste package in a repository by assuming some finite volume of ground water associated with each waste package. The fallacies of assuming some ill-defined water volume were outlined in a National Research Council report.[4] Macedo and Montrose have not responded to the technical issues pointed out therein. Their statement that there is a definite void volume within the waste canister and that the surrounding rock and backfill have a specified porosity does not answer the question of what water volume per package is the analogue of the laboratory closed-system experiment. They have not shown what is the appropriate water volume associated with the rock and backfill porosity, nor have they shown that either the canister void water or the pore water, or both, is the appropriate analogue for the water volume in their closed-system experiment with well-mixed waste solid and water.

As a test, consider a limiting case of a nonporous solid waste form in direct contact with infinite porous rock through which ground water is flowing. Our exact analytical solution [Equation (1)] predicts a finite steady-state dissolution rate limited by diffusion and convection in pore water. However, the total ground water volume in rock pores associated with this single waste package is infinite. This, together with a finite surface area of waste, results in zero ratio of waste surface S to water volume V. If one now adopts zero as the proper S/V ratio to obtain appropriate dissolution rates from laboratory closed-system experiments, the predicted repository dissolution rate would be the forward rate of solid-liquid reaction measured in laboratory experiments, i.e., the rate of solid-liquid reaction in the absence of dissolved waste constituents in the reacting liquid. The forward reaction rate has been shown in both Table 1 of our paper and by our more detailed analysis to be far greater than the maximum value of the steady-state convection-diffusion-limited dissolution rate of low-solubility species such as silica that is predicted from the exact analytical solution [Equation (1)].[10] Therefore, the postulated S/V analogue fails. The S/V postulate fails other tests described in III-B-2 of our paper.

Macedo and Montrose's Reference (7) is an interesting application of their S/V postulate to connect closed-system laboratory leach data for powdered minerals mixed in water to the dissolution rate of rock into ground water in the rock interstices. Evidently there are no nearby outer surfaces of the rock formation for diffusion and convection of dissolving rock into surrounding nondissolving porous media. Therefore, their natural analogue appears to be a closed system similar to that in Macedo's laboratory experiments. Unfortunately, the successful comparison by Macedo et al. of their laboratory data with this natural analogue does not answer the questions of how to apply their closed-system laboratory data to the open system of a waste package surrounded by the exterior nondissolving medium of a repository.

For other closed-system natural analogues, we suggest that Macedo and Montrose consider Cohen's data on rock grains dissolving into interstitial ground water. There are as yet few natural-analogue data for testing the mass-transfer theories for open-system dissolution. We urge the test of open-system mass-transfer theories with open-system experiments, such as those of McGrail et al.[11]

VI. DISSOLUTION RATE AT LOW VELOCITIES

Our equations show that for very low groundwater velocities, characteristic of expected conditions in U.S. repositories, convection has little effect on dissolution rate, and the rate of mass transfer from waste to ground water can be analyzed by calculating molecular diffusion of dissolved species into pore water. In Equation (2), we presented the exact analytical solution for the steady-state mass transfer rate as an example of a reliable theory. Governing equations, boundary conditions, and derivations were referenced to earlier publications. We then examined the equations proposed by Macedo et al. for dissolution rates at low velocities, presented as Equations (6) and (7), having determined that they were indeed the equations proposed by Macedo et al. Macedo and Montrose state that "substituting Equation (7) in Equation (6) leads to Equation (2), apart from a trivial geometrical factor of order unity." We find that Equations (6) and (7) combine to yield:

$$f = \frac{NVKD}{md^2S} \quad \text{(after Macedo and Montrose)} \tag{9}$$

Functionally, Equation (9) is markedly different from the exact analytical solution [Equation (2) in our paper]. Equation (9) does not contain the rock porosity ε, which must be present because of the boundary condition that joins the waste dissolution rate to the exterior-field mass transfer rate. Omitting the porosity introduces an error of two to four orders of magnitude for some emplacement rocks in the U.S. program. Equation (9) incorrectly contains the retardation coefficient K. The governing transport equation shows that K cannot be a parameter affecting dissolution rate at steady-state for long-lived species. These two differences were pointed out in III-B-2 of our paper. Evidently, Macedo and Montrose have not recognized these points.

Equation (9) also contains the postulated ratio S/V of waste surface to ground water volume. The fallacy outlined above in attempting to define S/V for the open system of a repository also occurs here.

The view of Macedo and Montrose that ground water flows appreciably through the waste form does not affect the comparison of their Equations (6) and (7) to the exact solution, Equation (2), because there is no flow in this problem. Nonflowing permeation of a waste form by ground water does not affect the applicability of Equations (1) and (2).[4]

We understand from the publications of Macedo, and from the comments by Macedo and Montrose, that the surface area S in Equation (9) is intended to be the surface area of the individual grains of waste solid, i.e., the surface area of the particles of glass powder in their laboratory experiments and the surface area of particles of fractured waste in their repository analogue. The quantity m has been defined as the mass of reacting species per unit area of reacting surface of the waste particles. Macedo and Montrose consider their repository waste form to behave as waste fragments well stirred and mixed with water. Thus, the product mS is the total inventory of the reacting species in the waste. For a spherical waste form of overall radius R in a repository, the inventory of a dissolving species is also given by the product of the species density n_i and the waste volume, so that Equation (9) of Macedo and Montrose can be transformed to the form of Equation (2) by:

$$mS = 4\pi n_i R^3/3 \tag{10}$$

Also, from Macedo's qualitative description of Equation (7), we understand the diameter d to be 2R for a spherical waste in a repository.[31] Substituting Equation (10) and d = 2R into Equation (9) of Macedo and Montrose, we obtain:

$$f = \frac{3NVKD}{16\pi R^5 n_i} \quad \text{(after Macedo and Montrose)} \tag{11}$$

whereas Equation (2) for a sphere is:

$$f = \frac{3\varepsilon DN^*}{n_i R^3} \quad \text{(exact solution)} \tag{2}$$

If one postulates a repository volume of water given entirely by the waste-package void water volume, as in a closed system, then V would be proportional to the waste volume, and Equation (11) would given the same functional dependence on R as does Equation (2). However, Macedo and Montrose state that in this low-flow case the greatest contribution to V is from pore water in the rock surrounding the waste. For a waste form directly surrounded by a large amount of porous rock, the Macedo-Montrose prescription for V becomes the volume or pore water in the surrounding rock, in addition to the small amount of void water in the waste package. For a large amount of rock per waste package, V is mainly the pore water in the rock and is affected little by waste radius. The resulting incorrect fifth-power R^5 dependence in Equation (11) is a consequence of the unjustified assumption that the ratio S/V is a meaningful and correct parameter for the repository and the incorrect assumption that the size of individual waste particles affects net diffusion rates from the outer surface of the waste into the surrounding rock.

Macedo and Montrose state that in this low-flow condition there is "rather thorough mixing" of pore water in the surrounding rock with the solid waste, creating a well-stirred condition analogous to their closed-system laboratory experiments. The mechanisms of such mixing of pore water throughout the considerable amount of rock associated with each waste package have not been described, nor have the mechanisms of mixing of that pore water with the waste solid. Macedo and Montrose speak of rapid exchange of pore water with the reaction volume of the waste particles by diffusion, evidently visualizing the result equivalent to a well-stirred mixture of waste particles and pore water. However, the "exchange" is only outward from the waste package. It is the concentration gradients of dissolved species in ground water in rock pores, as shown in Figure 1 of our paper, that create the barriers to mass transfer that are so effective in a repository. The "diffusive mixing" envisioned by Macedo and Montrose to create the same concentration throughout rock pores as in the liquid at the waste surfaces, equivalent to the spatially uniform concentration in closed-system laboratory experiments, cannot occur.

Finally, Macedo and Montrose have not answered the test of their Equation (6) discussed in our paper. The principal point being made in our paper is that reliable theories must have a clear theoretical basis, and they must specify well-defined parameters for their application. No derivation of Macedo's Equation (7) has been presented. Since learning in 1984 of Macedo's proposed equation for predicting repository performance,[31] we have sought on several occasions his help in deriving Equation (7), but without success.[E] Regardless of numerical constants, Equation (7) does not produce the correct and physically necessary functional relationship for the parameters that are shown, by exact mathematical solutions, to control steady-state release rates in a repository.

The technical points at issue here can be best resolved by such tests, using the tools of mathematics and governing equations.

VII. WHEN DOES EXTERIOR-FIELD DIFFUSION CONTROL DISSOLUTION?

Part III-B-3 of our paper refers to a more detailed analytical solution of the mass transfer rate from a spherical waste form, wherein solid-liquid reaction rate is used as an interface boundary condition.[10] The analysis assumes a simple expression for the solid-liquid reaction rate, linearly dependent on the concentration of dissolved species and with experimentally determined rate constants, similar to a reaction-rate equation assumed by Macedo.[31] The results of the exact time-dependent solution tell us under what conditions mass-transfer rate is limited by chemical reaction rate and under what conditions it is limited by exterior-field diffusion.[10]

This analytical solution provides a tool for assessing Macedo's conclusion that Berner's data on dissolution of small mineral grains surrounded by a nondissolving porous medium show that solid-liquid reaction rates control waste dissolution in a repository.[31] Our exact analytical solutions show that whereas solid-liquid reaction rates may control dissolution of some small ore grains surrounded by nondissolving material, exerior-field mass transfer can control for the same species in large solids similar to the repository waste forms. We do not "assert that exterior-field mass transfer processes are completely dominant," for we show under what conditions solid-liquid reaction rate can control.

We cite our earlier paper to provide the full details and derivation and for a detailed numerical calculation, from the exact analytical solution, of the time-dependent concentration of silica in ground water adjacent to a glass waste package.[10] It is shown that the steady-state surface concentration of dissolved silica is within 0.1 percent of the saturation concentration, and it is shown that the solid-liquid reaction rate controls silica dissolution rates only during the first few days or months of exposure to ground water, depending on the retardation coefficient for dissolved silica.

The simple comparison in Table 1 of our paper between exterior-field mass transfer rates and solid-liquid reaction rates shows clearly that the latter do not control steady-state dissolution rates for low-solubility constituents in a repository waste package, a conclusion reached by the WISP panel and endorsed by its reviewers.[4] The more detailed analysis described in III-B-3 predicts the rapid build-up of dissolved species in surface liquid and the rapid transition from chemical reaction controlling, at early times, to exterior-field mass transfer controlling for low-solubility species such as silica.[10]

VIII. THE REPETITION OF FOUR-YEAR-OLD MASS-TRANSFER ANALYSES

Macedo and Montrose conclude that "the first half of the paper gives a summary version of what is essentially the same diffusion-convection mass transfer analysis that they (Pigford and Chambre) have been presenting for roughly the past four years." Mere repetition of our 1982 publication of the simple steady-state equations for mass-transfer in a geologic repository would be adequate for our paper, which is to emphasize the need for careful development and evaluation of means for predicting future performance. However, our paper goes further. It summarizes continued refinement and generalization of the mass-transfer theory during the last four years, and some new results are presented in our paper. Based on our more extensive work on mass-transfer in a repository, we caution readers from using the simple steady-state equation to predict waste-package performance. Steady-state is reached soon in some convective conditions, but our referenced work shows that the higher mass-transfer rates accompanying the approach to steady state are important for the U.S. repositories. We do not understand how Equations (6) and (7) postulated by Macedo et al. could be expected to predict mass-transfer rates in either steady-state or during the long-duration transient.

IX. MASS TRANSFER THROUGH BACKFILL

Macedo and Montrose criticize the analysis of mass transfer from a waste package through backfill and into surrounding rock, stating that the nature of the backfill has not been determined and that it is more conservative to determine the dissolution rate at the waste-package surface. Evidently they mean that release rates into the rock are always more conservatively predicted if it is assumed that no backfill is present and that the waste form is surrounded directly by rock. Their proposal to ignore backfill in predicting release rates into rock is indeed more conservative for radionuclides that are short-lived and that would otherwise decay in backfill before reaching the rock. However, for the more important long-lived radionuclides the steady-state release rates into rock with backfill present will be greater, and hence more conservative, if the backfill is of greater porosity than the rock that it replaces.[1,7,8,20] One of the U.S. projects now specifies backfill with porosity greater than that of the surrounding rock. Therefore, it is timely to consider the effect of backfill, and for the more important radionuclides, it would not always be more conservative to neglect backfill and consider a waste solid surrounded directly by rock.

One of our tasks is to develop correct analytical solutions and design equations to be used as tools by repository and waste-package designers in assessing what benefits can or cannot be achieved by incorporating backfill in waste packages. We urge that designers use these tools before decisions are made that determine the nature of the backfill to be used, if any.

X. THE USE OF MASS-TRANSFER ANALYSIS IN REPOSITORY PROGRAMS

Macedo and Montrose misinterpret our observation that mass-transfer theory is the only reliable means of extrapolating waste-package performance into the future. Mass-transfer theory relies on detailed and careful application of the governing equations that describe the mechanistic processes of transport of material between and within phases. It is applicable to all situations in which transport occurs. The form of the mass-transfer theories that are finally used may not be the same as the simple steady-state equations presented in this paper to illustrate results of mass-transfer analysis. However, a point of this paper is to illustrate the pitfalls of waste-performance predictions based on heuristic assumptions and plausibility.

Used correctly, mass transfer theory is a valuable and necessary tool for developing reliable predictions. Its use by Sweden's nuclear waste disposal project has been reviewed and endorsed by international peer review, including two reviews by panels of the National Research Council.

XI. ADEQUACY OF DATA ON SATURATION CONCENTRATIONS (SOLUBILITIES)

Macedo and Montrose conclude that the data on solubilities given in Table 1, used to illustrate the prediction of diffusive-convective-limited dissolution rates, "imply a ground water composition that is unlike anything to be encountered in a geologic repository." We understand this to mean that none of these solubilities is appropriate for any of the repository projects. Certainly the local chemical environment is likely to depend on many conditions that may not have been present in the laboratory experiments from which solubilities were derived. However, that all the solubility data are so unrealistic is not yet apparent. The solubility data on silica, neptunium, and plutonium from borosilicate glass were suggested by saturation data from leaching experiments at Battelle's Pacific Northwest Laboratory. The data on actinide and fission product solubilities were recommended in a 1982 study by Krauskopf,[4] from his analysis of the expected stable phases in reducing conditions.

The repository projects are seeking new and more realistic solubility data. Some new data have been published. We also find that one project is still using Krauskopf's data on an interim basis. We find that many--but not all--of the newer solubility data from the projects are similar to those listed in Table 1. For some species, such as uranium, higher values should be used in an oxidizing environment, but the higher values now suggested do not invalidate the applicability of the mass transfer equations to these constituents in the waste.

We do not know the basis for the Macedo-Montrose comment that none of the concentrations in Table 1 of our paper could occur in a repository. For those saturation concentrations in Table 1 that are appropriate to the repository environment, our detailed analysis shows that these limiting concentrations will be approached soon within a repository.[10] For example, the steady-state concentration of dissolved silica, from borosilicate glass, is predicted to be within 0.1 percent of saturation.

We continue to seek unification of theory and experiment for reliable prediction of waste performance. Macedo's measurements of saturation concentrations can contribute to that unification.

XII. ADDITIONAL COMMENTS

The term "nonmechanistic" used to characterize the postulates discussed in Part II-A,B of our paper was adopted from Macedo's published description of his own work.[31]

We thank Drs. Macedo and Montrose for their helpful and constructive comments, and we thank Dr. Harry Burkholder, technical editor of these proceedings, for providing the opportunity to resolve these issues and to bring them to the attention of the technical community. We welcome further comments.

REFERENCES

A. T. H. Pigford and P. L. Chambré, "Reliable Predictions of Waste Performance in a Geologic Repository," Proceedings of the ANS/CNS/AESJ/ENS International Topical Meeting on High-Level Nuclear Wate Disposal - Technology and Engineering, Pasco, Washington (1985).

B. P. B. Macedo and C. J. Montrose, "Response to the paper 'Reliable Prediction of Waste Performance in a Geologic Repository'" (March 1986).

C. B. I. Cohen, "Critique of the National Academy of Science Study of the Isolation Systems for Geologic Disposal of Radioactive Waste," Nucl. Tech., 70, 433 (1985).

D. T. H. Pigford, "Response to B. L. Cohen's Criticism of the Report: 'A Study of the Isolation System for Geologic Disposal of Radioactive Wastes,' Nat. Acad. Press, 1985," Report UCB-NE-4052 (September 18, 1984).

E. T. H. Pigford, Letter to P. B. Macedo, October 9, 1984.

Site Characterization and Selection

SELECTION AND CHARACTERIZATION OF POTENTIAL SITES FOR A SPENT NUCLEAR FUEL REPOSITORY IN SWEDEN

Hans S Carlsson
Division Research and Development
Swedish Nuclear Fuel and Waste Management Co. (SKB)
Box 5864
S-10248 Stockholm, Sweden

ABSTRACT

Geological, geochemical, geophysical, rock mechanical, hydrogeochemical and hydrogeological investigations are included in the site selection studies for a spent nuclear fuel repository in Sweden. The investigations started in 1977 and are carried out in crystalline rock. The final site will be selected around the year 2000 through a screening process of investigated sites. The field studies generate numerous data that are compiled into a descriptive model for each site. The descriptive model forms the basis for the numerical calculation of the groundwater flow within the site. This paper presents the major instruments and methods that are used in the ongoing program as well as some of the obtained results.

I. BACKGROUND

The Swedish legislation lays the primary responsibility for nuclear waste management on the reactor owners. They execute their duties in the waste field through the jointly owned Swedish Nuclear Fuel and Waste Management Co., SKB. Site investigations in Sweden for a spent nuclear fuel repository started in 1977 and will continue up to the year 2000, when the final site is expected to be selected (Figure 1). A screening process is planned in two steps, around 1990 and 2000 respectively. In the first step, two or three out of eight or more possible sites are selected for more detailed studies. In the second step, one of those sites is selected for the licensing procedure. According to the present concept, the final repository for spent nuclear fuel will be located at approximately 500 metres depth in crystalline rock. For a one-story repository, approximately 1 km^2 area is needed.

II. STANDARD PROGRAM

The site investigations up to year 1990 follow a "standard" program,[1] which however, is modified according to the special conditions at each individual site. The program is also continuously updated as instruments and methods are further developed. The standard program is divided into four phases as given below.

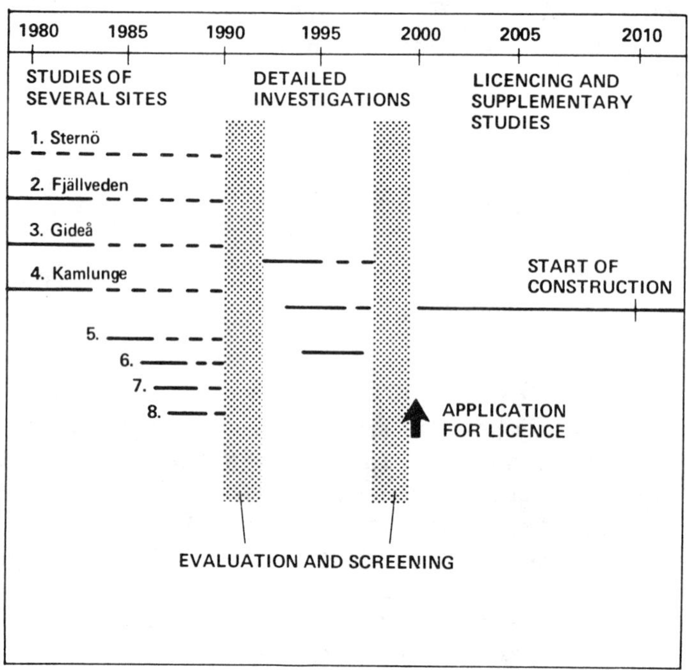

FIGURE 1. Time-Table for Site Selection Studies

A. Reconnaissance

In this phase, literature, maps and satellite images are studied and visits are made to the sites for preliminary judgements. The end result of this phase is a recommendation of one or more sites to be further investigated.

B. Surface Investigations

Detailed geological mapping and geophysical surface measurements are carried out within an area of 4-6 km^2. The main geological and tectonic features around this area are also mapped. The information on geological characteristics, fracture frequency and fracture zones constitute the base for the drilling program in the next phase.

C. Borehole Investigations

This phase normally starts with the drilling of a nearly vertical cored borehole down to 1000 m in order to get a first impression of the characteristics of the rock at depth. The deep drilling is followed by a series of inclined boreholes, normally 10-15 cored holes with a diameter of 56 mm down to 600-700 m. In addition, a considerable number of hammer drillings with a diameter of 110 mm down to a maximum depth of 250 m are carried out.

The core from the diamond drilled boreholes is logged and the geochemical characteristics are determined. A series of geophysical logs are made in the cored boreholes and the groundwater is sampled for hydrogeochemical analyses. Finally, the hydraulic conductivity and the piezometric pressure is measured in packed off sections of various length.

A minor geophysical program is conducted in the hammer drilled boreholes. All holes are directed to intersect major fracture zones as interpreted from the surface geological and geophysical investigations.

D. Evaluation of Results

In this last phase the collected data are processed and used for setting up a descriptive model of the investigated site. This model serves as the basis for calculation of groundwater flow and the migration of nuclides from repository depth to the biosphere.

III. INSTRUMENTS AND METHODS USED IN THE SITE INVESTIGATION

A. Geophysical Measurements

The surface and borehole geophysical methods are of a standard type. However, recently a new method called "tube-wave", developed within the Canadian program, has been introduced. By recording the response to a pressure pulse in a borehole, it may give information about water-bearing fractures.

B. Hydraulic Measurements

For downhole measurement of the hydraulic conductivity of the bedrock, special multihose equipment[2] has been developed. On the surface there are two trailers housing all the operating, regulating and computing devices, the reel for the 1000 m long multihose, and a diesel generator. The downhole probe consists of double packers, a pressure transducer and a test valve. The probe is connected to the surface by a multihose containing all the channels and conductors needed for inflation of the rubber packers, injection of water and for transfer of signals for maneuvers and recording.

By measuring flow and pressure of the water injected between the packers, the hydraulic conductivity of the rock in the packed-off section can be calculated. The computers in the trailer are programmed for this calculation and the results of the measurement in the form of a K-value is obtained almost immediately. As the computer is duplicated a new measurement can start in parallel to the data processing. The equipment is capable of measuring conductivities down to a value of $K = 10^{-11}$ m/s.

The multihose equipment has been shown to give reliable data and to allow for quick transport, mounting, recording and evaluation. One 1000-m borehole is measured, in 25-m sections, in one to two weeks.

Besides the hydraulic conductivity it is important to have information about the groundwater pressure and its variations within the investigated rock. For this purpose a special piece of equipment has been developed, the PIEZOMAC.[2] It is designed for continous recording of the piezometric pressure in five different boreholes, each packed off in five sections. A transducer records the pressure in each section sequentially and the values are stored in a central memory for a period of up to two weeks, when the memory has to be dumped. Alternatively the recorded values can be transferred by radio and telephone to a central office.

C. Hydrogeochemical Measurements

Information on the chemistry of the groundwater is of great importance for the estimate of canister corrosion as well as the behaviour of radionuclides escaping from a repository. Groundwater sampling and analysis are very sensitive operations with regard to possible contamination and changes during handling and transport. In order to improve the quality of gathered hydrogeochemical data a special field laboratory has been designed, where analyses can be conducted in-situ or close to the borehole.

The field laboratory is combined with multihose equipment similar to that used for conductivity measurements. In this case the double packers are used to seal off a water-bearing fracture from the rest of the borehole. Groundwater is then slowly pumped from the fracture and Eh, pH, sulphur and oxygen contents are measured in-situ by electrodes mounted between the packers. The measured data are transferred by conductors in the multihose to a data collection unit on the surface. For control purposes the same parameters are measured also on pumped-up water in the surface laboratory. A comparison between downhole and surface laboratory measurements of Eh have shown that the downhole measurements are much more stable and reliable, (Figure 2).

In addition, the concentration of all other major groundwater components are analysed in the field laboratory.

IV. RESULTS

So far, seven sites located in granitic and gneissic rocks have been investigated (Figure 3). In addition, information has been gained from several sites that have been abandoned after drilling and logging of one or two initial boreholes. In total, around 50,000 metres of cored boreholes have been drilled and extensive information has been gained from geological, geophysical, hydro- geochemical and hydrogeological characterization of the rock mass both in terms of actual measurements and interpretation of obtained data.

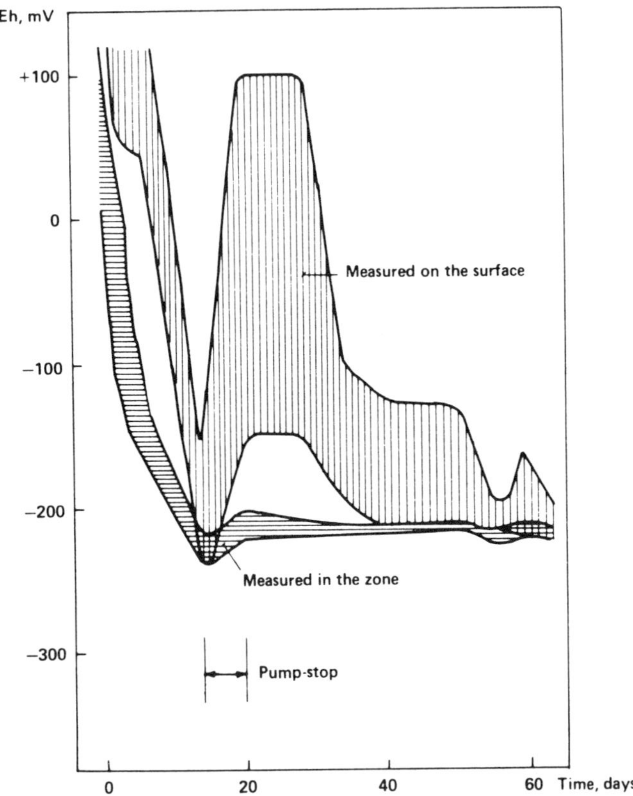

FIGURE 2. Comparison Between Eh Measurements In-Situ and in the Surface Laboratory

A. Descriptive Model

Existing fracture zones within a study site are considered to be the major flow paths for groundwater and the dominating mechanism for transportation of radionuclides from the repository to the biosphere. The site investigations have therefore been concentrated on the detection and verification of assumed fracture zones.

The work sequence for model calculations of the groundwater conditions within a study site is shown in Figure 4. The geological and geophysical investigations within a study site results in geological maps and maps of detected and verified fracture zones. The location, extent and width of the fracture zones, both on the surface and at the repository level, are presented. Figure 5 shows the study site Fjallveden[3] surrounded by large regional fracture zones. Figure 6 shows the detected local fracture zones at repository depth at the study site Kamlunge.[3] Figure 6 also shows the location of a conceived final repository. Hydraulic conductivity measurements have been performed in all diamond-drilled boreholes. The packed off sections normally vary in length between 2 metres and 25 metres. A decrease of the effective hydraulic conductivity versus depth has been observed[4] (Figure 7).

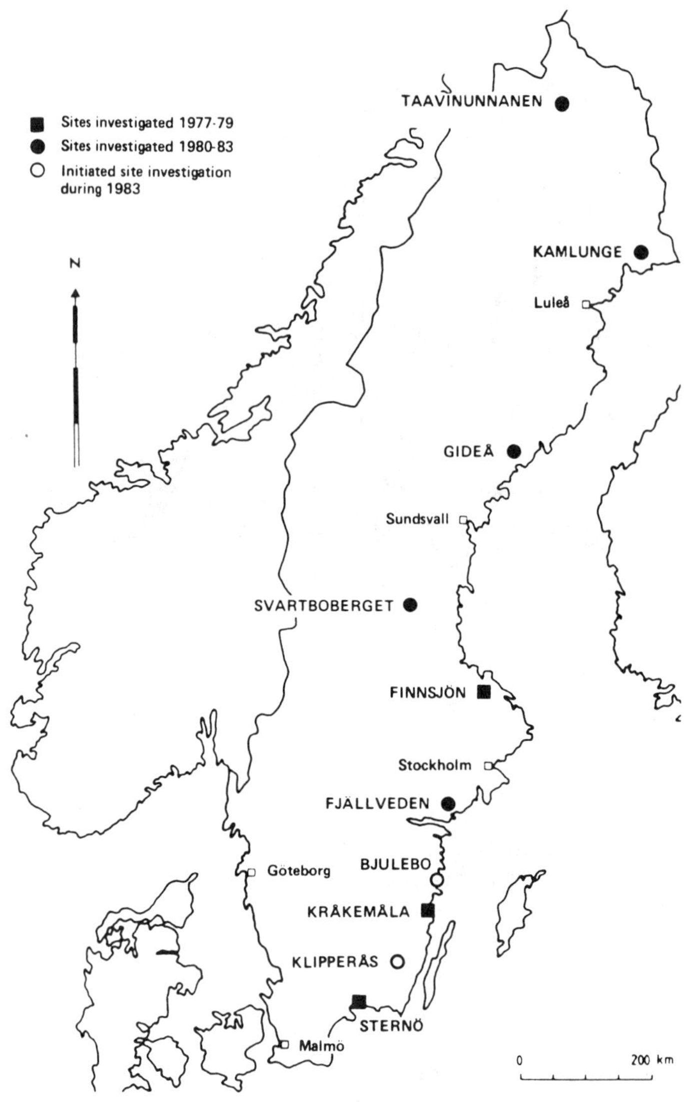

FIGURE 3. Location of Investigated Sites

At study sites Sternö and Finnsjön a "hydraulic fracture frequency" was calculated based on measurements made in two metre sections. Some 467 measurements at Finnsjon and 399 at Sternö have been used for the calculations. At both sites, the hydraulic fracture frequency decreases with depth. At levels below 300 m, the hydraulic fracture frequency is 0.2 - 0.3 fractures/metre at Finnsjön and 0.1 - 0.2 at Sternö[3] (Figure 8). The measured data comprise mean values from five boreholes per site. The measuring limit for measurements at Finnsjön was 2.5 x 10^{-9} m/s while it was 4 x 10^{-10} m/s at Sternö. Table 1 shows the results of chemical analyses on groundwater from the study site Gideå. The results are typical for other investigated study sites as well.[3]

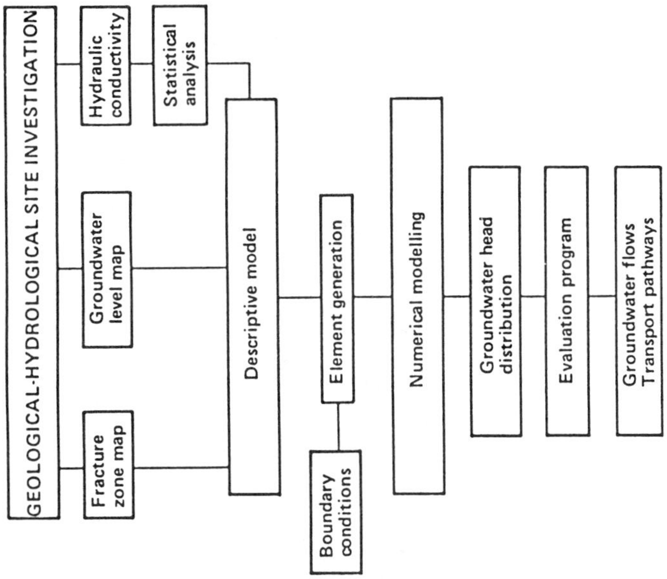

GEOLOGICAL-HYDROLOGICAL SITE INVESTIGATION
Fracture zone map
Groundwater level map
Hydraulic conductivity
Statistical analysis
Descriptive model
Boundary conditions
Element generation
Numerical modelling
Groundwater head distribution
Evaluation program
Groundwater flows
Transport pathways

FIGURE 4. Work Sequence for Model Calculations of the Groundwater Conditions Within a Site

LINEAMENT IN THE AREA AROUND FJALLVEDEN, FROM AIR-PHOTO INTERPRETATION
Well defined lineament
Poorly defined lineament
Site FJÄLLVEDEN
Topographical profile
Glacial striae
0
5 km

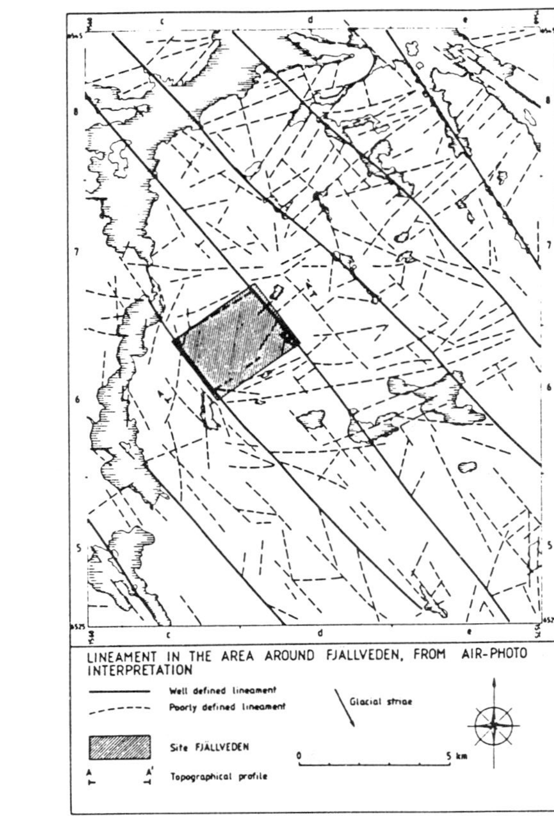

FIGURE 5. Interpreted Lineaments in the Area Around Study Site Fjällveden

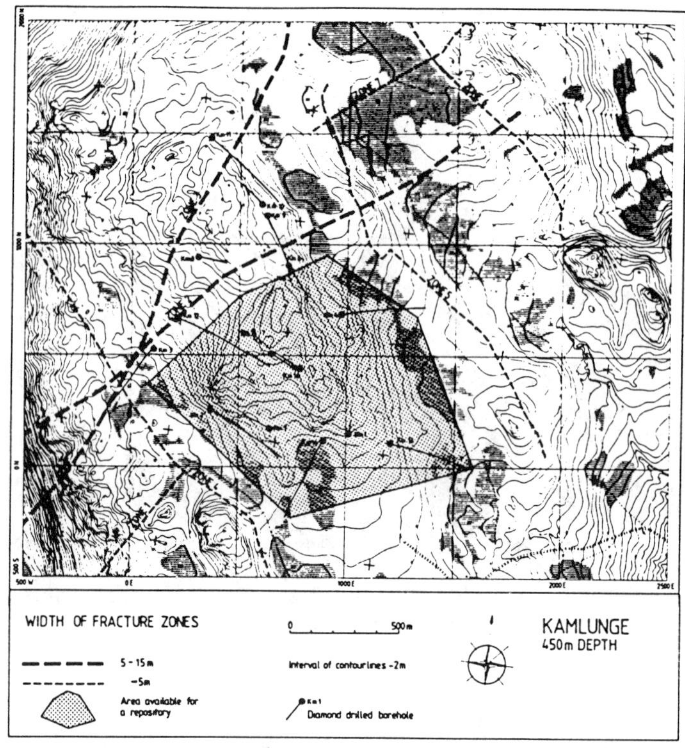

FIGURE 6. Detected Fracture Zones at Repository Depth at Study Site Kamlunge

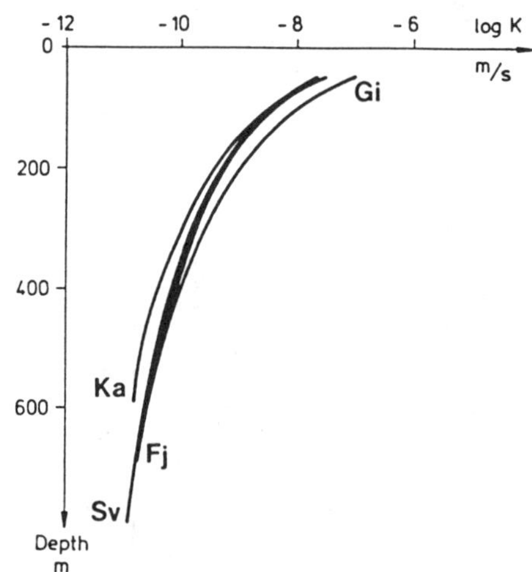

FIGURE 7. Effective Hydraulic Conductivity Versus Depth for the Study Sites Fjällveden (Fj), Gideå (Gi), Kamlunge (Km) and Svarboberget (Sv)

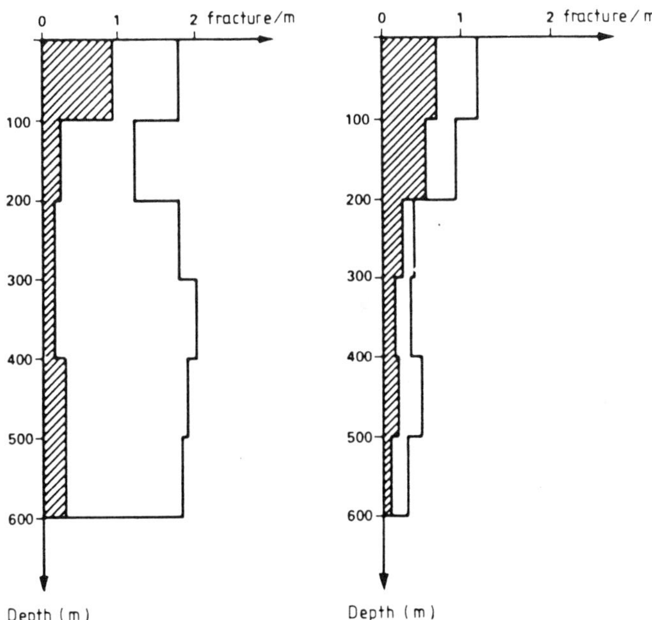

FIGURE 8. Total Fracture Frequency and Hydraulic Fracture Frequency in the Rock Mass Within the Study Sites Finnsjön (left) and Sternö (right)

TABLE 1. Results of Chemical Analyses on Groundwater from Gideå

Borehole	Depth m	pH	Eh v	HCO_3^- mg/l	Cl^- mg/l	HS^- mg/l	Na^+ mg/l	Ca^{2+} mg/l	Mg^{2+} mg/l	Fe^{2+} mg/l	TOC[a] mg/l
Gi 2	157	8.8	-0.10	161	4	0.02	50	10	3	0.1	4
	288	8.8	-0.09	163	5	0.03	49	10	3	0.6	5
	353	8.7	-0.09	162	5	0.04	52	10	2	0.5	3
	478	8.9	-0.09	160	5	0.02	50	10	2	0.2	3
	528	8.7	-0.10	159	5	0.03	50	10	2	-	2
Gi 4	91	8.0	-0.03	140	2	0.03	11	32	4	0.8	3
	212	9.0	-0.12	133	8	0.10	48	10	1	0.3	5
	385	9.3	-0.16	20	180	0.20	103	21	1	0.1	2
	498	8.5	-0.12	120	2	0.01	5	30	4	7.4	3
	596	8.9	-0.32	70	300	2.70	150	60	1	-	-

a Total organic carbon content of the water.

B. Model Calculations of Groundwater Flow

The groundwater calculations for each study site were performed utilizing a numerical three-dimensional FEM-model[4]. The calculations comprised the upper 1500 metres of the bedrock and were bounded by an assumed non-flow boundary at depth. The vertical boundaries were taken from regional modelling and the upper boundary constituted the groundwater head in the surficial rock. The size of the modelled areas varied between 2-5 km^2.

At study sites Fjällveden and Gideå, anisotropic hydraulic properties of the rock mass were encountered. The calculated groundwater flow at repository depth (appr. 500 metres) for study sites Gideå, Fjällveden and Kamlunge varies between 0.003 and 0.06 $l/m^2/yr$.[3]

C. Major Finding from the R&D within the Field of Geoscience

The major findings from the R&D within the field of geoscience including the site selection studies in Sweden to date can be summarized as follows:

- The Fennoscandian Shield is such a stable geological formation that substantial changes in its geohydrological characteristics or its ground water chemistry are not to be expected within time spans of the order of one million years. It is possible to select combinations of natural environ- ments and barrier materials in such a way that the function of the repository can be predicted with reasonable accuracy over periods of millions of years.

- Extensive site investigations have shown that the crystalline bedrock in Sweden can provide several areas of sufficient size, acceptable ground water chemistry and with groundwater flow low enough to meet the requirements for a final repository for spent nuclear fuel with a very high degree of safety.

REFERENCES

1. K. Ahlbom, L. Carlsson, N-Å. Larsson, "Final Disposal of Spent Nuclear Fuel - Standard Program for Site Investigations, Revision 1", SKB, Working Document 84-15, Stockholm (1984).

2. K. Almen, K. Hansson, B-E. Joansson, G. Nilsson, O. Andersson, P. Wikberg, H. Åhagen, Final Disposal of Spent Nuclear Fuel - Equipment for site Characterization, SKB, Technical Report 83-44, Stockholm (1983).

3. Final Storage of Spent Nuclear Fuel - KBS-3, Volume I-IV, SKB, Stockholm (1983).

4. L. Carlsson, A. Winberg, B. Grundfelt, "Hydraulic Properties and Modelling of Potential Repository Sites in Swedish Crystalline Rock," IAEA-seminar on Site Investigation Techniques and Assessment Methods for Underground Disposal of Radioactive Wastes, Sofia, Bulgaria (1984).

SITE INVESTIGATIONS FOR SPENT NUCLEAR FUEL DISPOSAL IN FINLAND

Veijo Ryhanen
Industrial Power Company Ltd
Fredrikinkatu 51-53 B
SF-00100 Helsinki, Finland

ABSTRACT

Site investigations related to the final disposal of spent nuclear fuel are being initiated in Finland. About five to ten areas will be selected for these investigations by the end of 1985. According to the decision in principle made by the Council of State (Government) the site for final disposal shall be selected by the end of the year 2000 and the repository shall be constructed by 2020. For bedrock investigations experience was gained from the tests in a deep borehole drilled in Lavia 1984. Hydrogeological and geophysical tests as well as hydrogeochemical sampling were carried out in the borehole. The aim of these studies was to develop and test methods required by deep borehole studies.

I. INTRODUCTION

Four nuclear power plant units are in operation in Finland: PWR plants in Loviisa [2 x 445 MWe(n)] owned by Imatra Power Company Ltd. (IVO) and BWR plants in Olkiluoto [2 x 710 MWe(n)] owned by Industrial Power Company Ltd. (TVO). The legislation concerning utilization of nuclear power stipulates that the producers of the waste, the power companies, are responsible for the spent nuclear fuel as well as for other nuclear waste.

The objectives and time schedule for spent fuel management are presented in the decision in principle made by the Council of State (Government) in 1983.[1] These objectives are included in the operating licenses of nuclear power plants; thus, even the license stipulations put the power companies under the obligation to observe the program and time schedule included in the decision in principle. The main principle is that the producer of waste shall study the possibilities of delivering the spent fuel irretrievably abroad, and as far as agreements for such arrangements do not exist, provisions shall be made for final disposal in Finland.

In practice the investigations and planning for the final disposal of spent fuel concern the spent fuel produced by Olkiluoto power plant as the spent fuel from Loviisa is transported to the Soviet Union to the supplier of the reactor. The total quantity of spent fuel from Olkiluoto power plant is estimated to about 1300 tU. Currently, an interim storage facility based on water basin techniques is being constructed on the Olkiluoto plant site and it will be completed by the end of 1987. The capacity of the first phase is

600 to 900 tU depending on the type of racks. It is possible to enlarge the capacity to the total quantity of spent fuel from the Olkiluoto power plant at a later stage.

A feasibility study was completed in 1982 on the final disposal of spent fuel in the Finnish bedrock.[2] The study surveyed the Finnish geological conditions from the viewpoint of final disposal, technical concepts suitable for these conditions, safety and costs. These results acted as a basis for the additional investigation program initiated in 1983. The bedrock investigations for the selection of a disposal site will constitute an essential part of these investigations.

II. SITE INVESTIGATION PROGRAM

All Finland lies within the Precambrian Baltic Shield.[3] Typical for shield areas is low seismic and tectonic activity. Thus, the emplacement of spent fuel deep into crystalline bedrock acts as a starting basis for the final disposal investigations in Finland. The decision in principle made by the Council of State requires that it shall be possible to complete the construction of the final repository by the year 2020, if necessary.

The time schedule for the investigations for the selection of the final disposal site is shown in Figure 1. The primary objective is selection of the final disposal site by the end of the year 2000. In possible disposal sites the field investigations will be carried out in two stages: preliminary site investigations during 1986 to 1992 and detailed studies during 1993 to 2000.

The program for the years 1986 to 1992 comprises five to ten investigation areas. The investigations of each area will be initiated by studies made from the surface (geological mapping, surface geophysics). The next stage will comprise drilling of one deep borehole (500 to 1000 m) and several

DRILLING OF TEST BOREHOLE AND SELECTION OF AREAS FOR SITE INVESTIGATIONS

PRELIMINARY SITE INVESTIGATIONS ON 5 - 10 AREAS

DETAILED SITE INVESTIGATIONS ON 2 - 3 AREAS

SITE SELECTION

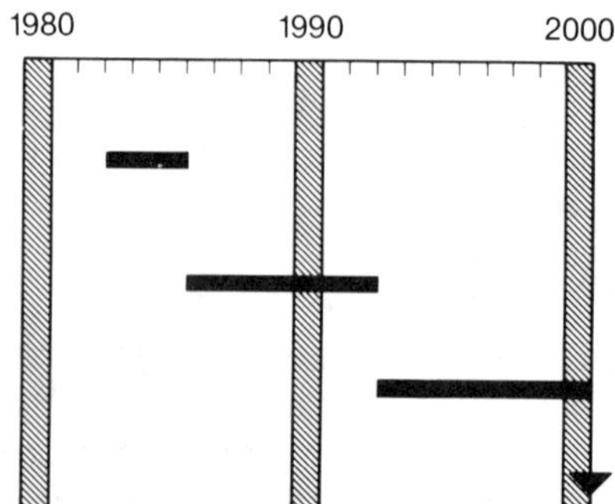

FIGURE 1. Time Schedule for the Investigations of Disposal Site for Spent Fuel.

shallow boreholes (100 to 300 m) in the area. Hydrogeological measurements, geophysical logging and groundwater sampling are made in the boreholes. On the basis of these results the suitability of the area for detailed investigations will be estimated.

The program for the years 1993 to 2000 comprises two to three areas selected on the basis of results obtained from the previous stage. These areas are studied in more detail, i.e., so that several deep boreholes are drilled for the tests in each area. On the basis of the results it will be possible to elucidate the suitability of the areas for final disposal of spent fuel and to select the final disposal site. After the site is selected the program provides ten years (2001 to 2010) for supplementary investigations on the selected site before the construction stage begins.

Concurrent with the site investigation program other studies for final disposal are carried out. Another focal point area comprises safety analyses, applying the investigation results of each area to them and acquiring data for analyses, i.e., by laboratory tests. The program consists, among others, of dissolution tests of uranium in Finnish ground-water conditions and sorption tests of radionuclides on species of rock typical in the site investigation program.

Two kinds of preparatory measures have been carried out to initiate the drilling program in 1986.

- Selection of investigation areas has been going on since 1983 (item III).
- Equipment and techniques required by field investigations were developed in 1984 to 1985 by the tests in a deep borehole drilled in Lavia (item IV).

III. SELECTION OF INVESTIGATION AREAS

The selection of areas for field investigations to be initiated in 1986 started in 1983. The main work phases are shown in in Figure 2. First, over 300 large bedrock areas were identified on a geological basis in Finland. The areas were identified on the basis of existing investigation material (satellite photographs, geological and geophysical maps, etc.). The size of these areas is 100 to 200 km^2 and they represent homogeneous bedrock blocks surrounded by fracture zones.

Then the suitability of the areas for site investigations were estimated on the basis of non-geological factors. These comprise above all demography, restrictions related to land use and existing road connections. On the basis of this survey, about 160 areas remained for further surveys.

FIGURE 2. Phases of Investigation Area Selection in 1983 to 1985

Based on more detailed geological and non-geological elucidations, areas suitable for site investigations are delimited to a size of 5 to 10 km^2. In this delimiting, the essential procedure is interpretation of aerial photographs. The number of potential investigation areas is estimated to be about 100 areas, in several parts of Finland.

By the end of 1985, from these crystalline rock formations, five to ten areas are selected for preliminary site investigations. An additional factor affecting the selection is land ownership of the areas, as the landowner's consent is needed for the investigations.

IV. DEEP BOREHOLE IN LAVIA

In April through May 1984 a cored borehole, 1001-m deep and 56 mm in diameter was drilled in the Lavia granite area in western Finland. This borehole is used in 1984 through 1985 to test and develop equipment and methods needed in deep boreholes.[4-7] Essential methods to be used in repository site investigations are included in the program in accordance with Figure 3. Several research organizations, universities and consultants from Finland and Sweden participate in the field work and in the interpretation of data.

The main species of rock at the drilling site are porphyritic granite and granodiorite. The borehole was drilled at an angle of 10° from vertical. Hydraulic tests were carried out in 30-m sections between depths of 73 and 973 m (Figure 4). Below the surface part of bedrock the sections having highest hydraulic conductivity were at depths between 400 to 430 m and 850 to 970 m. It was established that fracture frequency does not correlate directly with hydraulic conductivity.

Ground-water samples were taken at five different levels. The sampling levels were selected on the basis of results obtained by hydraulic tests and geophysical measurements (electrical, magnetic, radiometric and acoustic methods). Some results of the chemical analyses of ground-water samples are collected in Table 1. In hydrogeochemical sampling the relatively large share of flushing water used in drilling, at the most about 50%, constituted a problem.

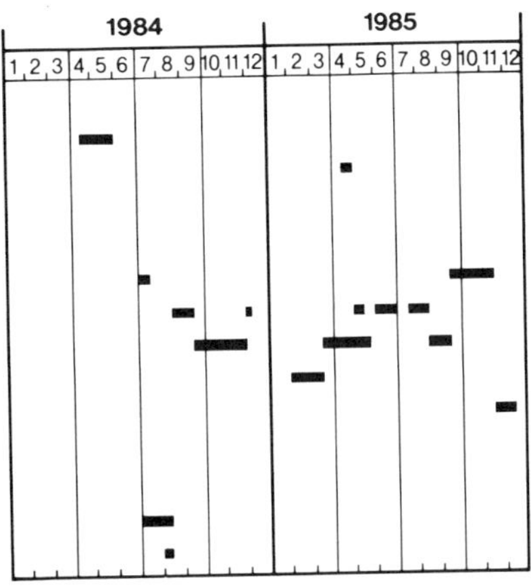

FIGURE 3. Time Schedule for Field Investigations of Lavia Borehole

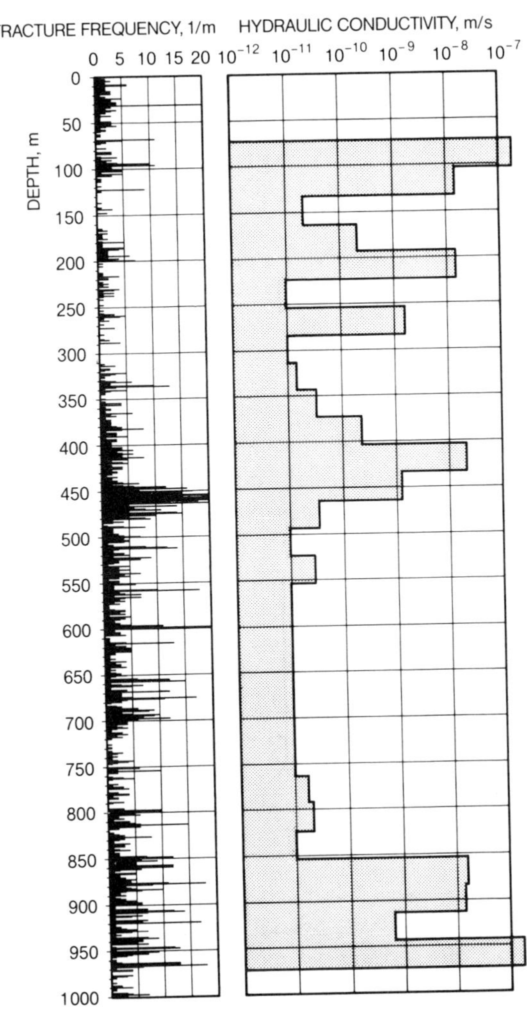

FIGURE 4. Lavia Borehole - Fracture Frequency and Hydraulic Conductivity

TABLE 1. Some Results from Lavia Ground-Water Analyses (Sampling of April through May 1985

Depth (m)	pH	Conductivity (mS/m, 25°C)	Na (mg/l)	K (mg/l)	Ca (mg/l)	Mg (mg/l)	Cl (mg/l)	Br (mg/l)
94-99	6.7	6.8	4.3	1.3	5.1	1.8	1.4	0.009
119-124	7.8	14.5	17	2.2	13.0	3.5	1.7	0.012
422-427	8.9	39	63	4.9	16.6	2.5	59	0.390
905-910	7.8	16.9	23	2.8	10.4	1.7	8.0	0.075
965-970	7.6	10.9	4.3	2.5	12.1	1.8	1.6	0,015

Together with the investigations of the borehole in Lavia, equipment development was initiated in Finland. A part of the Lavia borehole tests was carried out with new equipment constructed within the scope of the TVO nuclear waste research program. In 1984 equipment according to Figure 5 was constructed for ground-water sampling. The equipment comprises, besides the packers and pump located in the borehole, the flow-through cell on the surface, where the dissolved oxygen, pS, conductivity, pH and Eh are measured as soon as the sample reaches the surface.

V. EXPERIENCES AND CONCLUSIONS

In Finland the investigation program for final disposal of spent fuel has reached the phase when the field investigations for the selection of a repository site are about to begin. Stable bedrock is considered to offer favorable conditions for safe final disposal. Numerous formations suitable for field investigations can be found in the bedrock area in Finland.

The nuclear energy program in the background of the final disposal investigations is rather limited in Finland. This is the reason why the nuclear waste investigation program cannot be as large as in the leading nuclear power countries. The aim is to utilize in the Finnish nuclear waste studies as much as possible the results of the research and development work carried out abroad. The studies to be carried out in Finland are concentrated on sections where the Finnish work is most necessary. Such Finnish focal point areas comprise repository site investigations and safety studies.

Before starting more extensive field investigations, one deep borehole was drilled in Finland in 1984. Prior to this, except for drilling performed in connection with ore prospecting, there existed no other experience of bedrock investigations to be made in deep boreholes. During the last two years

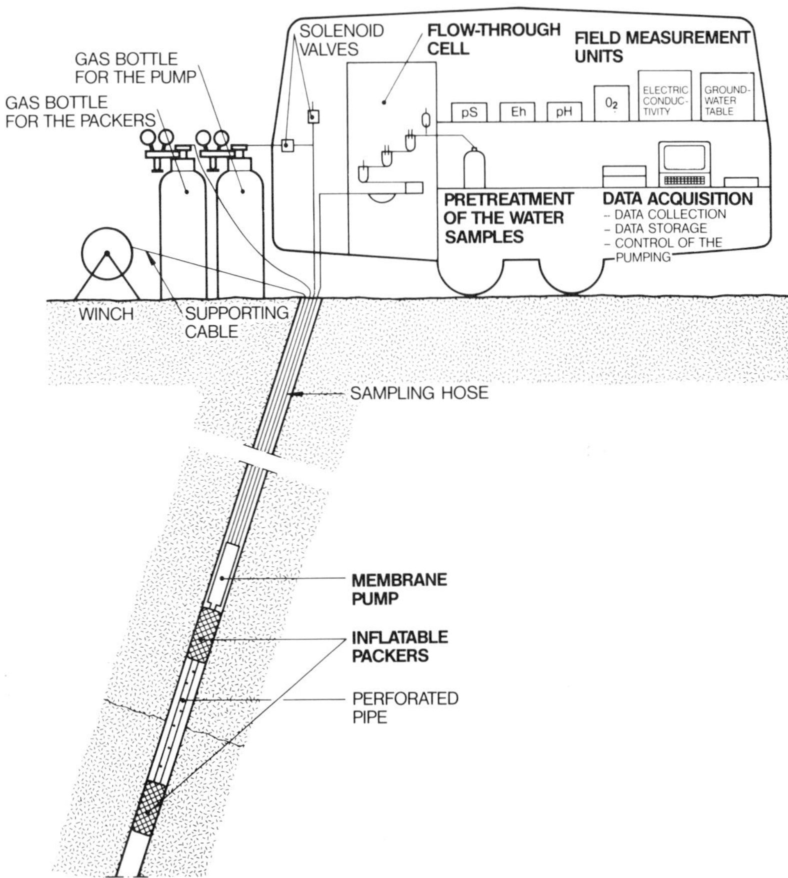

FIGURE 5. Equipment Constructed for Ground-Water Sampling in Deep Boreholes

the knowledge of investigation methods and data interpretation, as well as of the practical organizing of field investigations has advanced significantly. Thus, the borehole project in Lavia proved to be of great benefit.

Site selection investigations related to final disposal of spent nuclear fuel have already now received in Finland lively publicity which emphasizes the importance of public relations work in all phases of the site investigation program. The most important phases in the continued program in the immediate future comprises selection of investigation areas at the end of 1985 and starting of the drilling in the first area in 1986. Herein special attention shall be paid to the fact that sufficient information concerning the investigations should be available to the public already at an early stage.

REFERENCES

1. Decision in Principle of November 10, 1983, by the Council of State of Finland on the Objectives to be Observed in Carrying out Research, Surveys and Planning in the Field of Nuclear Waste Management.

2. Final disposal of spent nuclear fuel into the Finnish Bedrock, Report YJT-82-46, Industrial Power Company Ltd. (1982).

3. Simonen, A., "The Precambrian in Finland." Bulletin 304. Geological Survey of Finland (1980).

4. Aikas, T., A. Ohberg and V. Ryhanen, Lavian koereika - Yhteenveto vuoden 1984 tutkimuksista, Raportti YJT- 85-01, Saanio & Laine Oy, Teollisuuden Voima Oy. (In Finnish, Summary of the Lavia Borehole Investigations in 1984) (1985).

5. Almen, K-E., and O. Persson, Determination of Hydraulic Conductivity in Lavia Borehole, Finland, Report YJT-84-20, Swedish Geological Co., Uppsala (1984).

6. Saksa, P., Borehole Geophysical Investigations of Lavia Deep Testhole, Finland, Report YJT-85-06, Technical Research Centre of Finland, Geotechnical Laboratory (1985).

7. Anderson, P., and L. Stenberg, "Geophysical Borehole Logging in Lavia Borehole - Results and Interpretation of Sonic and Tube Wave Measurements," Report YJT-85-07, Swedish Geological Co., Lulea (1985).

THE SWISS SITE-SELECTION PROGRAM FOR HIGH-LEVEL RADIOACTIVE WASTE DISPOSAL

Marc Thury
National Cooperative for the Storage of
Radioactive Waste (Nagra)
Parkstrasse 23
5401 Baden, Switzerland

ABSTRACT

The Swiss National Cooperative for the Storage of Radioactive Waste, Nagra, started a site selection program for a high-level radioactive waste repository in 1980. The host rock of first priority is the crystalline basement in northern Switzerland which is covered by a few hundred meters of sediments. Phase 1 of the program, regional investigation of an area of about 1,000 km^2 and selection of one to three sites for further investigations is planned to be finished in 1989. At the moment, six deep boreholes of a total length of more than 10 km have been drilled and investigated. A regional geophysical program, a regional hydrogeological program and a neotectonic program are under way. A complete safety analysis was finished at the beginning of 1985. The present paper contains a summary of these programs and some results.

I. INTRODUCTION

The Swiss National Cooperative for the storage of Radioactive Waste, Nagra, initiated in 1980 a program for selecting and characterizing potential sites for a high-level radioactive waste repository. An investigation area of about 1,000 km^2 was selected in northern Switzerland. The host rock of first priority is the crystalline basement. This crystalline rock is covered in the investigation area by sediments which offer further potential host rock formations. The most investigated repository concept is disposal in galleries at a depth of up to 1,200 m with access by vertical shafts. An alternative concept involves disposal in large diameter deep boreholes down to 2,500 m.

At the moment, the major part of the field work of an extensive regional investigation program, including six deep boreholes with a total length of more than 10 km, geophysical surveys with more than 600 km of seismic lines and detailed hydrogeological and neotectonic studies, has been completed and detailed data analysis is being carried out. Planning for a few additional deep boreholes is going on.

The Federal Government Ruling of October 6, 1978, on the Atomic Act designates the guaranteeing of "permanent safe management and final disposal" of radioactive waste as a prerequisite to future development and use of nuclear energy in Switzerland. Nagra therefore presented to the Federal Government at the beginning of 1985 a report on feasibility and Safety Studies for final disposal of radioactive wastes in Switzerland, the so-called Project Gewahr 1985. A complete safety analysis had to be carried out, based on the data and on the results already available from the ongoing regional investigation program. This safety analysis allowed identification of the critical geological factors and emphasis can therefore be given to these factors during further data analysis and investigations.

II. SELECTION OF INVESTIGATION AREA AND GEOLOGICAL SITUATION

Switzerland can he divided into a southern part, the Swiss Alps with high mountains formed during the alpine orogeny, and a northern part with hills and with a few larger plains, only slightly affected by the alpine orogeny. The southern part has been excluded for repositories for high-level radioactive waste due to potential ongoing strong tectonic movements. The northern part has been carefully analyzed for potential host rocks for a repository. Some layers of rocksalt, anhydrite and clay of thicknesses rarely exceeding 100 m offer a limited potential as a host rock and it was therefore decided to investigate the crystalline basement as a first priority.

In the northern part of Switzerland the crystalline basement is covered by sediments up to several thousand meters thick. The surface of the crystalline basement generally rises towards the north, and north of the Swiss border it crops out in the Black Forest massif. The investigation area (Figure 1) was selected along the northern border of Switzerland in a region where the top of the crystalline basement was expected at maximum depth of 1,500 m. The investigation area lies between the two seismically active areas of Basel and Schaffhausen which have been excluded.

Figure 2 shows a geological north-south cross section through the investigation area, from the Schafisheim borehole. The section is based on an interpretation of a seismic reflection survey. A large Permocarboniferous trough which trends west-east through the investigation area has been discovered below the mesozoic sediments.

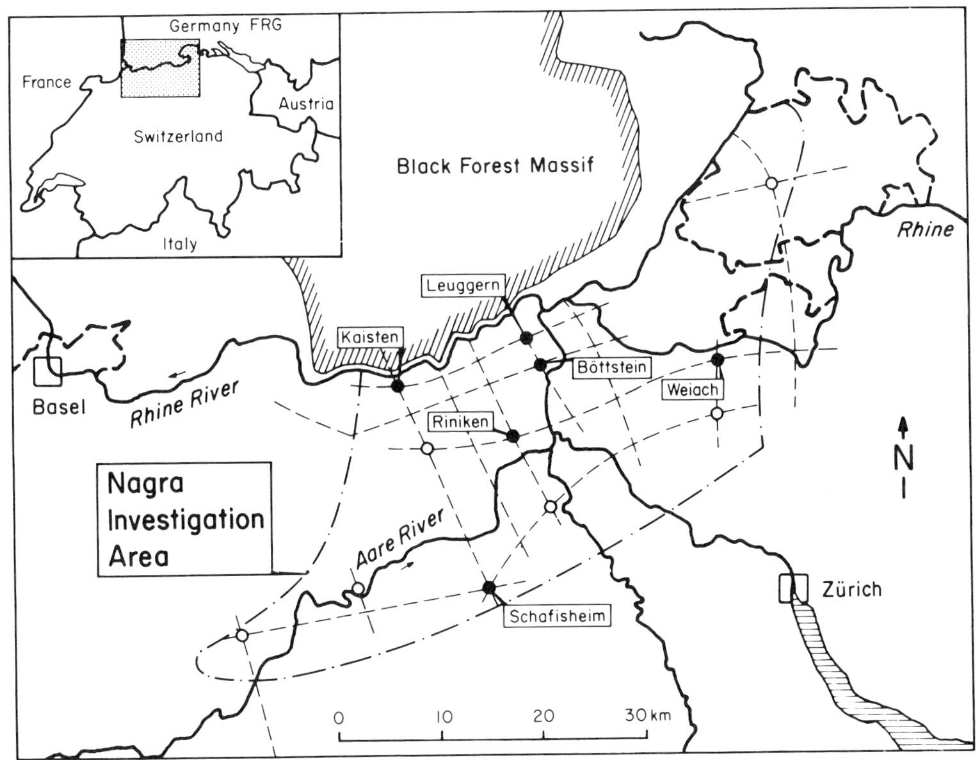

FIGURE 1. Nagra Investigation Area with Drilled Deep Boreholes (black circles) and Reflection Seismic Lines (dashed lines)

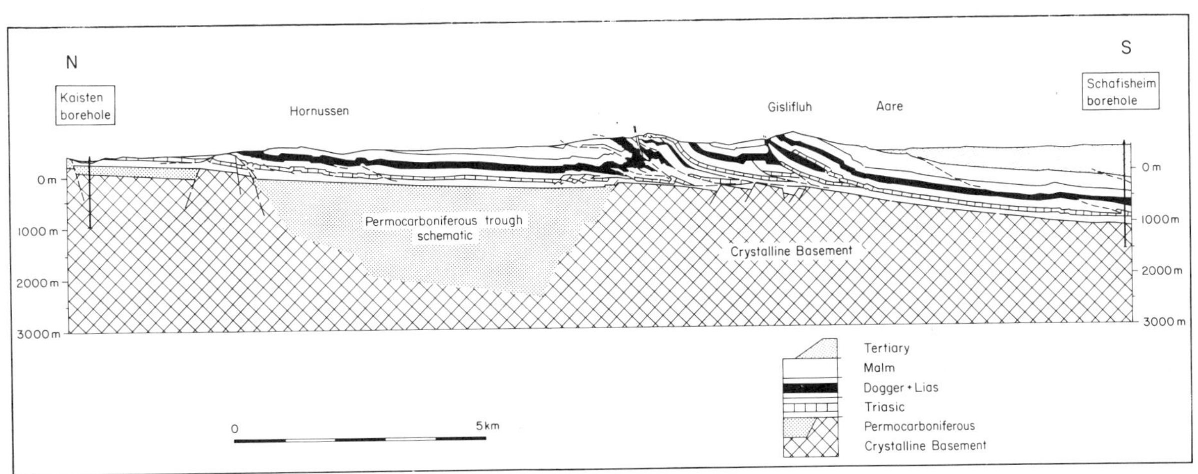

FIGURE 2. Geological Selection Across the Nagra Investigation Area

III. SITE-SELECTION PROGRAM

The aim of the program is to ensure that a Swiss repository, if needed, can be operational by the year 2020. The program is divided into 4 phases:

1980-1989 Phase 1	regional investigation and site selection,
1989-1995 Phase 2	detailed site investigations at one to three locations and selection of one site ,
1995-2005 Phase 3	detailed site investigation at one location with test shafts and underground laboratory, and
2005-2020 Phase 4	repository preparation.

The actual program of Phase I is divided into four programs, which are described below:

- deep drilling program,
- geophysical investigation program,
- regional hydrogeological program, and
- neotectonic program.

The program is managed by Nagra. More than 50 university institutes, consulting companies, consultants and contractors from 8 countries with more than 200 scientists work under contract for the program. The costs involved to date are approximately $50 Million (U.S.).

The results of the geological investigations are published in the Nagra Technical Report series, which can he ordered from Nagra, Baden, Switzerland. About 50 reports are presently available and about 20 more are in preparation.

IV. DEEP DRILLING PROGRAM

A series of 12 deep boreholes spread over the investigation area was planned to explore the geological structure of the subsurface with emphasis on detailed investigation of the practically unknown crystalline basement below the sedimentary cover. Investigation of water flow systems in the basement as well as their hydraulic characterization and water sampling should provide the input data needed for a hydrogeological model of the area.

To date, six boreholes have been drilled, with an average depth of about 1,600 m. Figure 3 shows an overview of the geology of these boreholes. The scientific investigation program was carried out as planned. The main parts of the program are:

Borehole:	BOETTSTEIN		WEIACH		RINIKEN		SCHAFISHEIM		KAISTEN		LEUGGERN	
Geology:	m	d	m	d	m	d	m	d	m	d	m	d
Quaternary	18	0	37	0	25	0	244	0	45	0	48	0
Tertiary	–		149	37	–		332	244	–		–	
Jurassic	–		518	186	~464	25	529	576	–		–	
Triassic	~297	18	285	704	327	489	385	1105	~80		~175	
Permian	–	–	461	989	985	816	–	–	172	125	–	–
Carboniferous	–	–	570	1450		1801	–	–	–	–	–	–
Crystalline Basement	1186	315 Granite 1501	462	2020 Gneis 2482			516	1490 Granite Syenite 2006	1009	297 Gneis 1306	1466	223 Gneis Granite 1689

▌ core   1501 (underlined) final depth   m: thickness (m)   d: depth of top formation (m)   ~ top of formation eroded

FIGURE 3. Geological Overview of Nagra Deep Boreholes

- Coring about 90% of the boreholes and detailed laboratory analysis of the cores.

- Core orientation (very successful with a sonic televiewer) and detailed fracture orientation statistics.

- Geophysical logging with practically all tools available today. For the granite of the Bottstein borehole, a synthetic electro-faciolog was elaborated, which corresponds remarkably well with the observed petrography.

- Fluid logging to detect water inflow zones. Very small inflow zones were detected with the logging of the electrical conductivity of the borehole fluid during pumping. In order to obtain a maximum conductivity contrast between borehole fluid and inflowing ground-water, the borehole fluid was replaced by deionized water.

- Hydraulic testing. Systematic double packer tests with intervals of 12 to 25 m have been carried out to obtain a continuous profile of the hydraulic conductivity of the crystalline basement. Figure 4 shows the conductivity profile of the Bottstein basement.

- Water sampling of ground-water of the sedimentary aquifers and of flow zones in the crystalline basement was carried out, mostly during the drilling operations which were interrupted when a promising flow zone was observed. The crystalline sections of the boreholes were drilled either with deionized water or inflowing deep ground-water, without any additives except two tracers, to allow in situ analysis of the remaining contamination of the water sample by borehole fluid during water sampling processes. Water samples were taken at the surface when the flow analysis

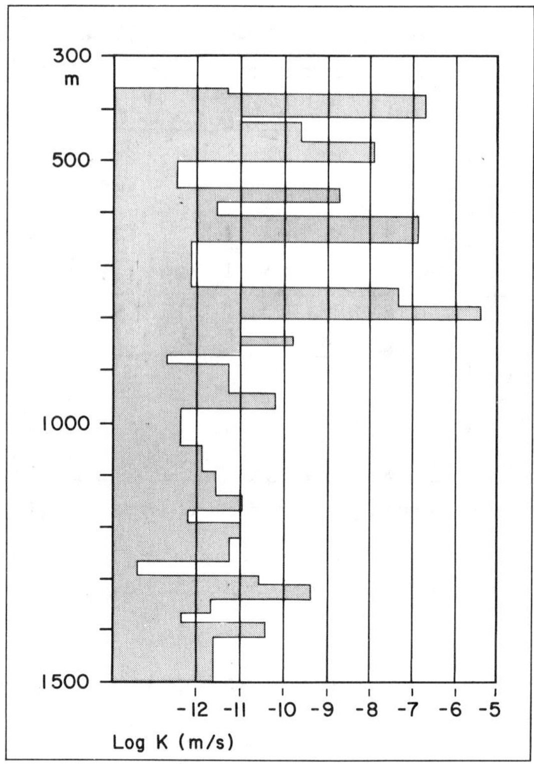

FIGURE 4. Profile of Hydraulic Conductivities of the Granite Section of Bottstein Borehole, Measured with Double Packer Tests

was sufficient and, in most cases, downhole with pressure vessels. The analysis program is described in the hydrogeology section below.

At the beginning of 1984, a multipacker system was installed in the Bottstein borehole with eight packers, allowing long-term observation of the hydraulic head of eight different zones of the crystalline section. The system has worked perfectly until now. During this year, a similar multipacker system was installed in four other boreholes. Some packer failures occurred and the long-term observation program is only partly carried out.

V. GEOPHYSICAL INVESTIGATION PROGRAM

A geophysical investigation program has been carried out over the whole investigation area with the aim of exploring the regional structural geology, together with the observations of the deep drilling program. The field work of the program was finished early this year. Interpretation work is still continuing. A summary of the work follows:

- In 1981/1982 a gravity survey with about 5,000 stations was carried out. A significant mass deficit was observed in the area of the Permocarboniferous trough.

- In 1981 an aeromagnetic survey with more than 6,000 km of lines was carried out. No significant magnetic anomalies have been detected.

- In 1981 a magnetotelluric survey was carried out along the Kaisten-Schafisheim line (Figure 2). The measurements have been disturbed by surface effects and are therefore not conclusive.

- In 1981/1982 a first seismic refraction survey with about 200 km of lines was carried out, giving the first indication of the existence of the Permocarboniferous trough.

- In 1984 an additional seismic refraction survey was carried out along the axis of the Permocarboniferous trough. Interpretation work is still going on.

- In 1981/1983 a first seismic reflection survey with about 180 km of lines was carried out. These lines allowed a detailed interpretation of the regional structural geology of the mesozoic sediments. In some parts of the area of the Permocarboniferous trough, a series of horizontal or inclined reflection bands indicate the presence of paleozoic sediments. However, the top of the basement below these paleozoic sediments does not reflect significantly and mapping of the top of the basement is therefore rather difficult. In the crystalline basement, some reflections have been observed down to a depth of about 7 km. Detailed interpretation of the structural geology below the mesozoic is still continuing.

- In 1983/1984 and 1984/1985, two additional seismic reflection surveys were carried out to provide a more detailed network of profiles. Interpretation is still going on.

VI. REGIONAL HYDROGEOLOGICAL PROGRAM

A clear understanding of the deep ground-water flow regime is needed to predict flow paths, travel times and dilution effects of ground-water between a potential repository and the biosphere. In addition, the assessment of the chemical composition of ground-water in the near-field of a potential repository is necessary. The flow regime in the Nagra investigation area is rather complicated: the mesozoic sedimentary cover contains some important limestone aquifers separated by impermeable formations of clay, anhydrite and rock salt. As shown on Figure 2, the mesozoic sediments are folded and overthrust in some areas, allowing potential water flow through fault zones and thrust zones connecting different sedimentary aquifers. The sedimentary filling of the Permocarboniferous trough is, in general, of fairly low permeability; however, some open fractures with large mud losses during drilling have been observed. The crystalline basement conducts ground-water along fracture systems. Based on the information from the first six deep boreholes, the upper 500 m of the basement have an average hydraulic conductivity of about 10^{-7} m/s. Below, the hydraulic conductivity is about 10^{-11} m/s, except in fault zones which can have high conductivities, as observed, down to a depth of at least 1,700 m.

The Nagra investigation concept is as follows:

- to prepare a hydrodynamic model for the assessment of the flow field,
- to investigate the hydrochemistry of the different aquifers and, as far as possible, to assess the model age of the water by isotope methods, and
- to validate the hydrodynamic model with the results of the hydrochemistry of the different water types and with the hydraulic head observation at different depths in the deep boreholes.

In 1981, the preparation of a regional hydrodynamicmodel was started. It covers a surface area of about 25,000 km^2 including the infiltration areas in the Swiss Alps and in the Black Forest massif in FRG. The model contains about 5,600 finite elements. Up to now more than 25 runs have been carried out, mainly with different sets of parameters for hydraulic conductivities of the various aquifer and aquitards.

In 1984, a local hydrodynamic model was prepared for the most interesting part of the investigation area, covering a surface of about 1,000 km^2. With about 6,000 elements, it allowed a more detailed representation of the geometry of the different aquifers. The lateral water inflow into the model was derived from the results of the regional model. More than 10 runs have been carried out with different sets of input parameters, resulting in a set of possible flow paths from a potential repository to the biosphere.

A first validation of the two hydrodynamic models has been carried out with the preliminary results of the hydrochemistry of the aquifers and hydraulic head observations of the deep boreholes. During the next year, improvements on the geometry of the aquifers of the models, as well as a detailed validation of the hydrodynamic models are planned.

The hydrochemical program started in 1981 with the collection of water from ahout 100 springs and wells of different aquifers in the investigation area and its surroundings. Very detailed hydrochemical and isotope analyses have been carried out. In the deep boreholes, about 60 water samples have been collected and analyzed. It was possible to define several different water types in the sedimentary aquifers as well as in the crystalline basement. A detailed study is currently being carried out for a better understanding of the rock-water interaction and the evolution of the ground-waters of the different aquifers.

VII. NEOTECTONIC PROGRAM

The main aim of the neotectonic program is to detect fault zones capable of tectonic reactivation in order to avoid these zones within repository sites. The program was started in 1981 with various approaches:

- Lineaments detected on satellite imagery have been checked on aerial photos and in the field without finding clear indications of active faults.

- Along the main river of the investigation area, a detailed analysis of morphology of bedrock below the quaternary alluvials was carried out, based on information from numerous shallow ground-water wells. Along the Rhine, which mainly forms the northern Swiss border, several fault zones seem to be active. These zones are at distances of 3-8 km; the movement rate is about 0.1 - 0.2 mm per year.

- The dip of the actual surface of quaternary alluvials of the main river plains has been analyzed to find dip anomalies which could be caused by recent fault movements. Some small anomalies were found.

- A detailed reanalysis of high precision leveling of the Federal topographical survey was carried out and two areas of relative uplift and subsidence were detected, with movement rates of 0.1 - 0.3 mm per year.

- The investigation area is well known as a seismically relatively quiet area, based on records from the 14th century. However, in 1983 Nagra installed a network of nine seismographs for a long-term microseismic survey. With this network, events of magnitude one can be detected with an accuracy of less than 1 km horizontally and vertically. During the last two years of records, three events of magnitude one have been registered in the investigation area.

- The neotectonic program continues with detailed analysis of the potentially active fault zones detected.

TEN YEARS OF SITE-SPECIFIC RESEARCH AND ITS RELEVANCE TO THE CONCEPT OF A WASTE DISPOSAL SYSTEM IN CLAY

P. Manfroy
R. Heremans
National Agency for Radioactive Waste and Fissile Materials
NIRAS/ONDRAF
Boulevard du Regent 54
B-1000 Brussels, Belgium

A. Bonne
Study Center for Nuclear Energy SCK/CEN
Boeretang 200
B-2400 Mol, Belgium

ABSTRACT

A R&D program on geological disposal of conditioned high level and alpha bearing radioactive waste was started in Belgium in 1975. The host formation being considered is a cainozoic weakly hardened clay layer (the Boom clay formation). Because mining technologies in such a material have not been developed in the same way as for hard rock or rock salt, special attention was to be given to investigations dealing with soil mechanics and mining problems to evaluate the technical feasibility of a deep disposal system. Intensive site specific research, including field reconnaissance, laboratory experiments, measurements on core samples and the construction of an underground laboratory led to the provisional conclusion that the Boom clay could be considered as an exploitable and acceptable medium for safe waste disposal.

I. INTRODUCTION

In 1975 the first nuclear power plants were put into operation in Belgium. At present their installed power reaches some 5,500 MW(e) which is almost 60% of the country's total electricity consumption. When, in 1974, we decided to start with research on disposal possibilities of radioactive waste arising from PWR spent fuel reprocessing operations, we took into consideration A production equivalent to an electronuclear program of 300 GW(e).

Our Nuclear Research Center (CEN/SCK) started with the inventory of deep underground geological formations which might be used for this kind of waste's safe and long term disposal. The area of Belgium's surface is only 30,500 km^2 and its hydrogeology at average depth is well-known, so that a first selection of a potential site could easily be made.

II. CHOICE OF A SITE AND INVESTIGATION PROGRAM

For several reasons we chose the nuclear site with a surface of some 700 ha, situated in the Northeast of the country above the Boom clay formation selected during the inventory.

Figure 1 shows the extension of the Oligocene (Rupelian) clay formation in Europe and its location in Northeast Belgium where it is named the "Boom clay". The Boom clay is well-known:

- at the SW of the site in the region of Boom where it is exploited for the production of bricks,
- at the NNW in the region of Antwerp where at regular intervals and at shallow depth it was met with and crossed by the construction of tunnels and docks of the port, and
- at the NE and the E where it was crossed by the deep underground mine prospection drillings in the coal basin of the Campine.

On the chosen site, in the region of Mol-Dessel (Figure 1), the interpolation of available data at that time indicated the presence of the clay layer with a thickness of at least 100 m and at an average depth of some 230 m, the layers above and underneath the clay being waterbearing and often glauconitic horizontal sand deposits.

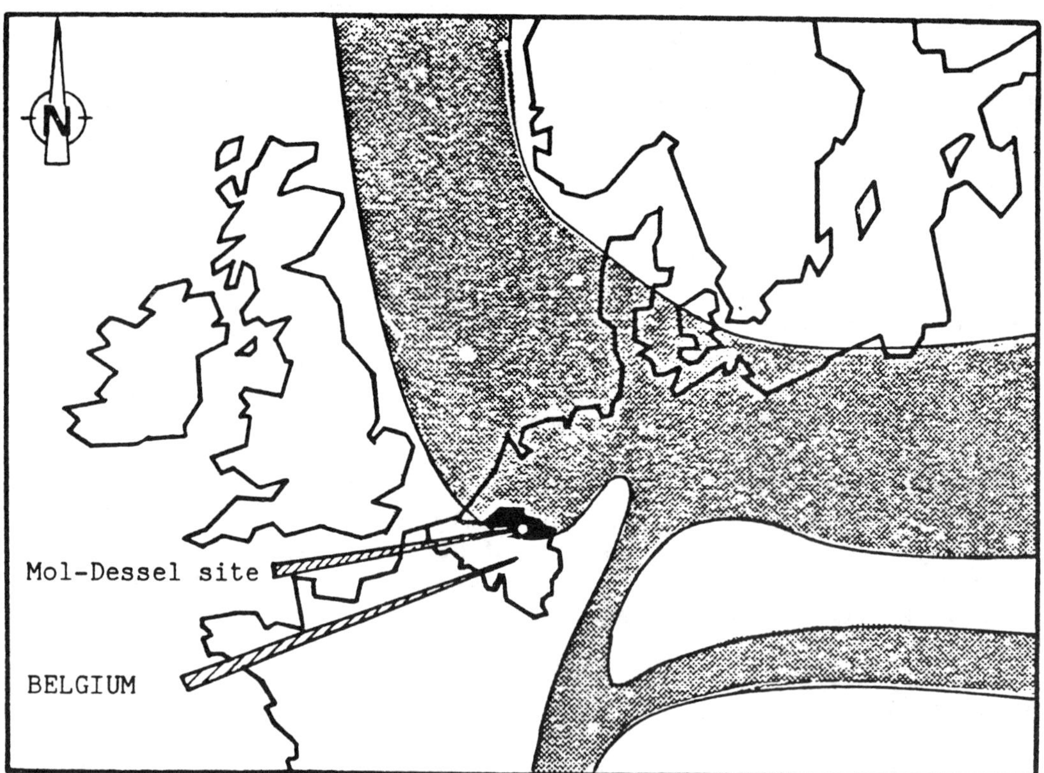

FIGURE 1. The Rupelian Clay in Europe (modified from POMEROL, 1973)

When establishing the objectives of the first five years R&D program for the period 1975-1979, we aimed:

- at an extensive investigation of the selected site,
- at a better knowledge of the characteristics of the clay considered as a host formation for several conditioned waste types,
- at the assessment of the technical possibilities with regard to the construction of underground facilities, and
- at a first assessment of the long term safety of the developed disposal system.

The R&D program and its realisations were focused on the mining aspect problems. Indeed, contrary to the other host rock types considered for waste disposal, such as rock salt and hard rocks, very little or no experience at all existed with regard to the clay behaviour during deep underground drilling operations.

III. FIELD WORKS

A. Drillings and bore-hole logging

With the first reconnaissance drillings on site, we aimed at defining more precisely the local litho-stratigraphy and hydrogeological parameters and also at taking core samples for a series of laboratory tests and experiments.

The geological drilling up to 570 m deep, the various hydrogeological drillings and the bore-hole logging made it possible to establish the stratigraphic section and to define precisely the piezometric reference levels in the different water tables, as presented schematically in Figure 2.

The bore-hole logging revealed, among other things, the presence of an important transition zone between the Mio Pliocene sands and the top of the "compact" clay. A very good agreement was found between the resistivity log and the grain size distribution determined on core samples.

During the geotechnical drillings, about 40 "nondisturbed" core samples in the clay layer were taken to perform geomechanical tests in laboratories that specialize in soil mechanics. Some first results are shown in Table 1.

B. Aerial photography and seismic campaign

Once it was proven that the clay under the site and at the drilling location was thick enough and sufficiently "homogeneous", we decided to continue the program with a more extensive field exploration. We proceeded to an interpretation of infrared pictures taken by the sattelite ERTS-LANDSAT 1, in order to determine and assess the possible presence of underground

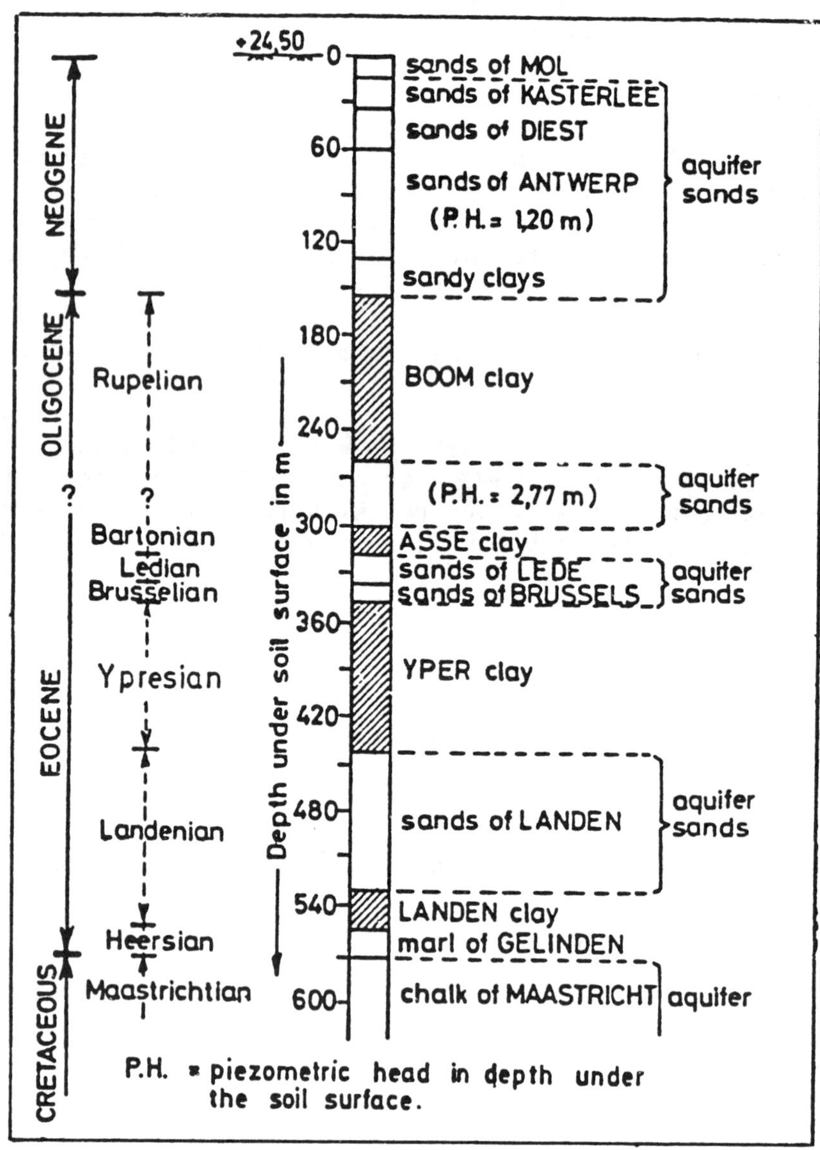

FIGURE 2. Simplified Stratigraphic Profile - MOL Site

TABLE 1. Some Geotechnical Characteristics of Boom Clay

- GRANULOMETRY (%) < 2 μ	53
2-60μ	42
> 60 μ	5
- ATTERBERG LIMITS (%) ω_p	26.5
ω_ℓ	77.5
I_p	51.0
- TRIAXIAL C.U	
C_{cu} (MN/M^2)	0.127
$\emptyset_{cu}$ (°)	17.5
- TRIAXIAL U.U. C_u (MN/M^2)	0.588

discontinuities by the detection of surface lineaments. We also evaluated the underground by geophysical surface measurements in order to verify the formation's lateral extension on and around the site. We opted for a tridimensional seismic reflexion campaign based on the principle of adjacent strips, rather than on the classical profiles. The range of the chosen parameters, namely a one millisecond sampling rate, two seconds recording time and order five cover, made it possible to obtain a vertical resolution of two to five meters for the top and basis of the clay body.

The interpretation of the results allowed us to know more of the details, the dimensions and forms of the top and the bottom of the clay body. It was proven that the deepest faults detected in the paleozoic and mesozoic formations had no influence at all on the cainozoic sand-clay cover including the Boom clay. It also confirmed that the dip of the clay layer was about one percent in NNE direction. In short, the general conclusion we were able to draw from this survey is that the clay layer under the chosen site is indeed "homogeneous" and has not been affected by any tectonic accident of the kind of a flexure or a fault able to interrupt the geometric continuity of the formation.

IV. SOME PRELIMINARY RESULTS FROM THE LABORATORY

Parallel with these field activities we continued our laboratory experiments and property determinations on core samples taken during the drillings. These laboratory activities concerned, of course, the determination of the geomechanical characteristics of the clay, but also its other physical, chemical and mineralogical characteristics, as well as its possible interactions with the conditioned waste to be buried on the site. Table 2 gives some other information on the Boom clay properties and composition.

As mentioned in the beginning of this paper, the clay layer outcrops in the region of Boom, situated at the SSW of the investigated site. This particularity was used to carry out in a clay quarry a series of experiments dealing with the heat transfer in the clay and related corrosion phenomena. The results from these tests were necessary to complete the information we used as a basis to initiate the first technical feasibility study. Using an electrical heater of 1.5 kW simulating a HLW canister we determined a thermal conductibility $K = 1.69\ W\ m^{-1}\ {}^{o}C^{-1}$ and a diffusivity $k = 18.8\ m^{2} \cdot y^{-1}$.

V. FEASIBILITY STUDY

Together with an engineering office and companies specialized in mining construction we started, in 1978, a technical feasibility study of a large scale facility for the disposal of conditioned waste in the Boom clay.

A. Basic Information

We had to take the following facts into account:

- a thickness of the "homogeneous" clay layer of some 100 m, covered by a sandy clay transition zone of about 10 m,

TABLE 2. Some Characteristics of the Boom Clay

- NATURAL WATER CONTENT (WEIGHT %)	≃26
- MINERALOGICAL COMPOSITION OF THE FRACTION < 2 μm (PARTS PER TEN)	. ILLITE (2-3) . SMECTITE (2) . VERMICULITE LIKE (3) ILLITE MONTMORILLONITE INTERSTRATIFIED (1-2) . CHLORITE + CHLORITE VERMICULITE LIKE INTERSTRATIFIED (1)
- ORGANIC MATTER (%)	BETWEEN 2.3 AND 5.5
- NATURAL VOLUME WEIGHT (TON.M^{-3})	≃ 1.93
- CATION EXCHANGE CAPACITY (MEQ/100 G DRY CLAY)	BETWEEN 20 AND 40

- a dip of one percent of the clay layer in the NNE direction, and

- a hydraulic gradient through the clay from the top to the bottom and the presence of an aquiclude waterbearing sand layer under the clay.

Moreover, the laboratory tests on core samples taken at different levels in the formation during drilling on site and the experiments carried out in the clay quarry caused us to consider:

- the typical rheological behaviour of a plastic clay and its consequences for the digging and lining of underground cavities and

- the weak thermal conductibility of the Boom clay.

B. Preconceptual Design

Taking all the above-mentioned parameters into account, a preconceptual design of the underground facility led to:

- a network of fully-lined circular tunnels with a useful diameter of 3 to 4 m with tunnels dug perpendicularly on the slope of the layer,

- a distance between the parallel tunnels of at least 30 m (to assure stability, taking into account the clay's rheological characteristics); in the case of heat emitting waste the distance between parallel galleries will be higher (taking into account the thermal characteristics of the clay and the temperature constraints imposed on the massif),

- pits for the burying of heat emitting waste dug at +/- 45° of the circular tunnels; the distance between these pits depends on the waste's thermal characteristics; depending on the thickness of the clay layer, the length of such a pit is limited to 20 m and its inclination was given to maximize the migration pathway through the clay layer; nevertheless, other concepts of the final disposal of heat emitting waste are possible and are presently being studied, and

- the implantation of the tunnels' axis in the center of the "homogeneous" clay layer; this central implantation with the burying pits situated in the second lower half of the clay layer is justified, the downward hydraulic gradient as well as the weak storing capcity of the aquifer underneath the clay being taken into account; however, the implantation is subject to revision according to results obtained from the experiments planned in the field of corrosion and migration.

To illustrate this, we present in Figure 3 a scheme of the reference concept established in 1978.

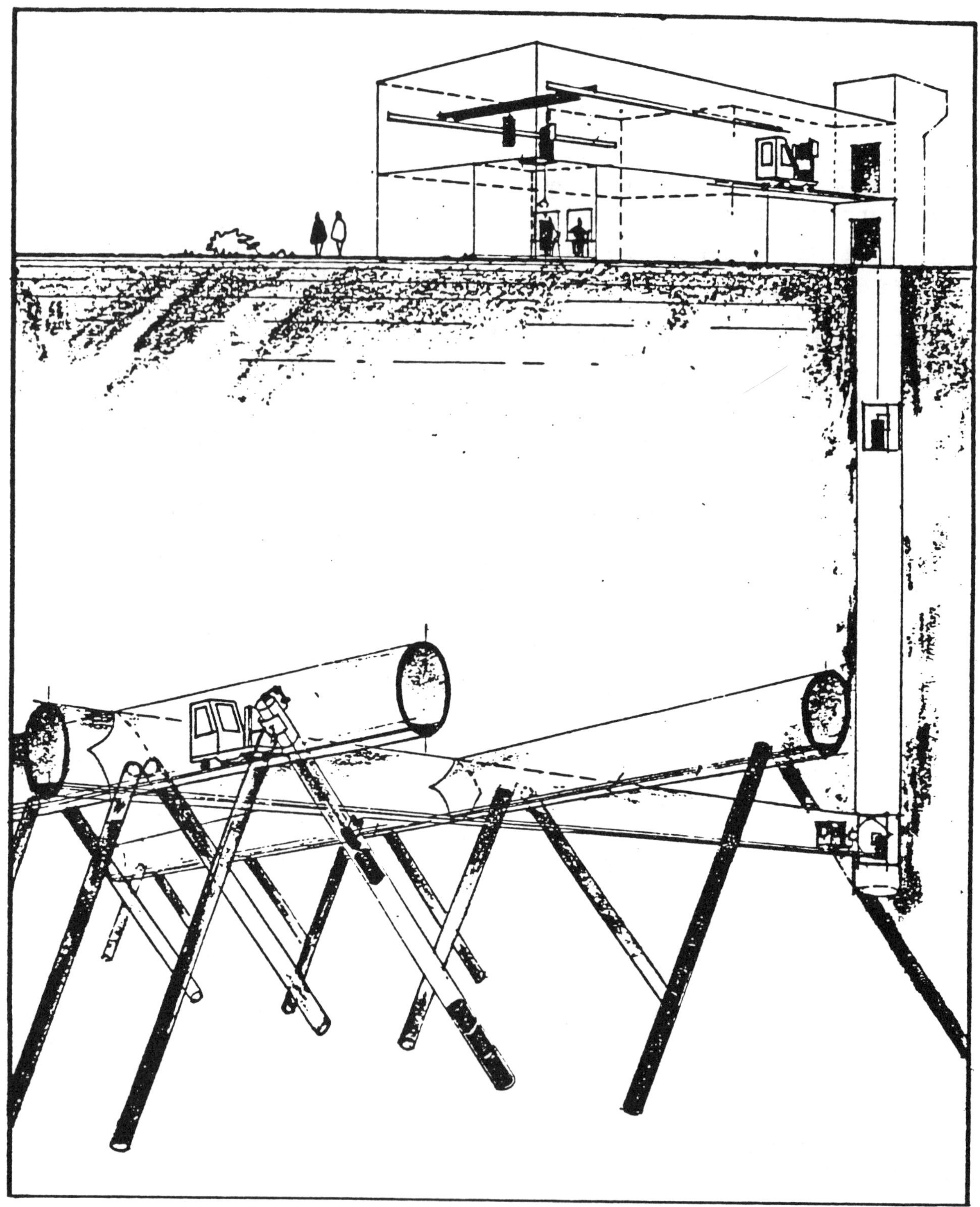

FIGURE 3. Underground Disposal in Clay - Artist View

VI. UNDERGROUND LABORATORY

Before the end of the R&D five year plan 1975-1979 and depending on the various results of the different activities which have already been carried out, we decided to examine the possiblity of building during the next five year plan an underground laboratory in the clay formation at the potential site.

A. Objectives

This underground laboratory had to enable us:

- to gather information on the clay's rheological behaviour at an average depth and on the technological possiblities to construct tunnels in it,
- to "visit" the clay massif and to "choose" core samples for the laboratory examination, and
- to carry out a series of measurements and experiments in situ under conditions representative for possible future disposal.

B. Realization

It is not possible to summarize in this paper the project's evolution from its conception till its final realization. Therefore we propose to give some basic information of the laboratory's construction and equipment in the light of Figure 4 which represents a scheme of the as built underground facility:

- The vertical shaft was built with a useful diameter of 2.65 m to a depth of -214.70 m, after having frozen the soil. Down to the clay layer it is covered with a double concrete lining (2 x 45 cm) with a polyethylene sheet of 3 mm in-between. Under the clay-sands limit, the concrete lining becomes monolithic and 0.90 m thick.
- From a depth of 214.7 m the shaft was enlarged to a useful diameter of 4.00 m in order to facilitate the hooking works of the horizontal gallery; the level of -277.50 m corresponds to the bottom of the vertical construction. For security reasons and taking into acount the lack of experience in this field we decided to also excavate the gallery by the aid of the freezing technique.

The whole of the gallery construction works has been scheduled in two phases (gallery at a depth of 222.9 m).

A first phase included the completion of:

- a rectangular first section, having a length of 4 m and an inner section of 1.5 x 2 m, with final support provided by IPN 450 adjacent rectangular steel frames and

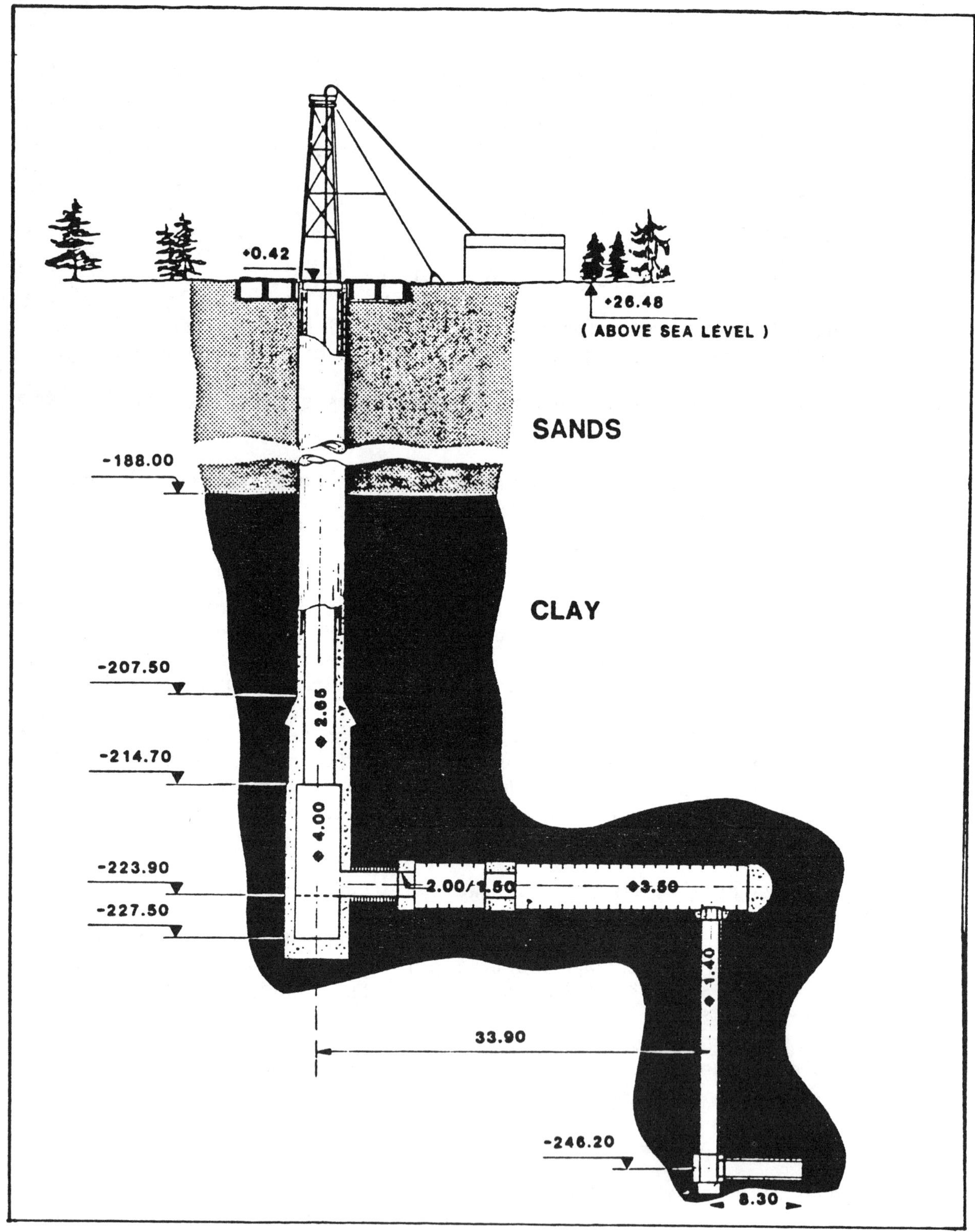

FIGURE 4. Underground Laboratory in Clay - As Built

- a reinforced concrete section, with a thickness of 1.4 m and an outer diameter of 4 m, allowing us to insure a transition between the first rectangular section and the next six meter long circular section with an inner diameter of 3.5 m lined with cast iron segments, and completed with a circular intermediate concrete plug.

The second phase consists of:

- after freezing the soil through the intermediate concrete plug, the digging and lining of the last twenty metres of the gallery in circular section with an inner diameter of 3.5 m and lined with cast-iron segments and

- concrete end plug poured against the clay front.

Additional excavations - a small shaft and a small gallery - dug at a two meter diameter and lined with concrete blocks have been made from the end of the main horizontal gallery. These reconnaissance cavities mined in unfrozen soil have given information about the behavior of the clay at depth during and after excavation owing to previous installation of extensometers and inclinometers.

Moreover, during the construction we placed at different levels (also indicated in Figure 4) along the shafts and along the galleries a whole set of instruments in order to follow the behaviour of the clay massif and to measure the constraints in the linings.

Alongside the two vertical shafts and the two horizontal galleries we displaced almost 400 measure points. Although, unfortunately, some of them have suffered quite a lot from the construction, enough data are available to allow good interpretation. During the excavation we could also verify "de visu" the perfect "homogeneity" of this clay, both in vertical and horizontal direction. We could also confirm the material's quality by the numerous "undisturbed" samples taken at different places and examined in the laboratory.

Preparation of the building on site began in January, 1980, and the facility as shown in the Figure 4 was completed mid-1984, including the lift, the ventilation, etc. From this moment our scientists and technicians have worked in the laboratory to carry out geotechnical, corrosion, heat transfer, migration and other tests.

VII. MAIN CONCLUSIONS AND FUTURE ACTIVITIES

The first conclusions of this large scale experiment are as follows:

- confirmed the "homogeneity" of the clay layer and its continuity in the prospected zone,

- allowed consideration of the construction of galleries in non-frozen clay at an average depth with existing technical means, provided that some adjustments are carried out,
- indicated that even in unfrozen soil, the build-up of external pressure on the lining, is until now, a very slow phenomenon which could make possible the reduction of the importance of the lining, such as used in the underground laboratory, and
- gained insite on a revision of the procedure for the construction of the access shaft and its lining, which could also lead to optimization of this part of the facility.

We can already affirm that considerable progress has been made through which the conception of a conditioned radioactive waste disposal facility in the Boom clay has become more realistic. The CEN/SCK, financially backed by the Commission of the European Communities and by ONDRAF/NIRAS, already considers the construction of a pilot unit in which excavation demonstration of a large diameter shaft in non-frozen clay will be possible, as well as demonstration of actual waste disposal operations, demonstrations of waste retrievability, gallery backfilling, etc. The next two or three decades will definitively open the way toward the realization of an industrial facility which will not only be acceptable economically, but also from the point of view of safety.

BIBLIOGRAPHY

P. Manfroy et al., "Conception d'une installation pour l'enfouissement dans l'argile de dechets radioactifs conditionnes," Proceedings of the IAEA/OECD-NEA International Symposium on the Underground Disposal of Radioactive Wastes, Otamiemi near Helsink 2-6 July, 1979, Paper: IAEA-SM-143/3.

Confinement geologique des dechets radioactifs dan la Communaute Europeenne - Catalogue europeen des formations geologiques faborables a l'evacuation des dechets radioactifs solidifies de haute activite et /ou de longue vie - EUR 6891 FR. (1980) - ISBN 92 - 825 - 1934 - 1.

P. Dejonghe et al., "General Perspectives in Radioactive Waste Management in Belgium," Transactio of the international ENS/ANS conference New Directions in Nuclear Energy with Emphasis of Fuel Cycles Proceedings ENC-3; pp. 216-228, Brussels, (Belgium), 26-30 April 1982, Ed. American Nuclear Society, Illinois, Volume 40, Tansac 40-273, 1982; ISBN - 0003 - 018X.

A. Bonne and R. Heremans, "Investigations on the Boom Clay, a Candidate Host Rock for Final Disposal of High-Level Solid Waste," Proceedings of the VII International Clay Conference 1981 Bologna and Pavia, Italy, 6-12 September, 1981, organized by the Italian group of AIPEA, pp. 799-818, Ed. Elsevier Scientific Publishing Company, (1982), H. Van Olphen, F. Veniale, ISBN-0-444-42096-7 (Volume 35); ISBN - 0-444-41238-7 (series).

R. Heremans, "Possibilites d'evacuation de dechets radioactifs dans une formation d'argile plastique," _Proceeding of an IAEA Conference on Radioactive Waste Management_, Seattle (USA), 16-20 May, 1983, pp. 243-259 (Volume 3); paper: IAEA-CN-43/56, Publ. IAEA, Vienna, 1984 (in five volumes); STI/PUB/649; ISBN 92-0-020484-8.

B. Dethy, B. Neerdael, "Correlations entre diagraphies et caracteristiques geotechniques d'une argile raide tertiaire," _Proceeding of the International Symposium Soil and Rock Investigations by In Situ Testing_, Paris, 18-20 May 1983, pp. 49-57, Publ: Bulletin of the International Association of Engineering Geology, Paris (Fr); n° 26-27, Dec. 1982, June 1983.

R. Funcken et al., "Construction of an Underground Laboratory in Deep Clay Formation," Papers of the "Eurotunnel 1983" Conference Basle (Switzerland) 22-24 June 1983, pp. 79-86, Ed. Acces Conference LTD 1983, U.K.

P. Manfroy, "Rejet des dechets radioactifs en formation argileuse profonde. Construction d'un laboratoire souterrain pour un programme experimental approfondi," _Proceeding of an IAEA Conference on Radioactive Waste Management_, Seattle (USA), 16-20 May, 1983, pp. 275-290 (Volume 3); paper IAEA-CN-43/54, Publ: IAEA, Vienna, 1984 (in Five volumes); STI/PUB649; ISBN 92-0-0202484-8.

B. Neerdael et al., "Field Measurements During the Construction of an Underground Laboratory in Deep Clay Formation," _Proceedings of the International Symposium Field Measurements in Geomechanics_, Zurich, 5-8 September 1983, pp. 1419-1430 (Volume 2) Publ, A. A. Balkema/Rotterdam 1984. Ed. K. Kovari; ISBN 37266 0005 1.

HYDROLOGICAL EVALUATION OF FIVE SEDIMENTARY ROCKS FOR HIGH-LEVEL WASTE DISPOSAL*

T. F. Lomenick
Chemical Technology Division
Oak Ridge National Laboratory
Post Office Box X
Oak Ridge, Tennessee 37831

B. Y. Kanehiro
Berkeley Hydrotechnique, Inc.
2150 Shattuck Ave.
Berkeley, California 94704

ABSTRACT

Utilizing performance criteria that are based upon siting guidelines issued by DOE for postclosure as well as preclosure conditions, a preliminary hydrologic evaluation and ranking is being conducted to determine the suitability of five sedimentary rocks as potential host rocks for a high-level radioactive waste repository. Based upon both quantitative and qualitative considerations, the hydrological ranking of the rocks in order of their potential as a host rock for the disposal of radioactive wastes would be shale, anhydrock, sandstone, chalk, and carbonates, with the first three rocks being significantly better than the remaining two types.

I. INTRODUCTION

A preliminary evaluation and ranking of the suitability of five sedimentary rock types as potential host media for the disposal of high-level radioactive wastes is being conducted under the Sedimentary Rock Program (SERP) at the Oak Ridge National Laboratory (ORNL) for the U.S. Department of Energy (DOE). The rocks under investigation are shale, anhydrock, chalk, carbonates (limestone and dolostone), and sandstone. Although the complete evaluation and ranking of these rock types includes consideration of the hydrologic, geologic, rock mechanical, geochemical, and thermal factors that would affect the construction, operation, and performance of a potential repository in each of these rocks, only the findings related to the hydrologic elements of the study are discussed here.

This study is confined to a generic, hydrologic consideration of anhydrock, chalk, carbonates, shale, and sandstone. It is specifically restricted to existing information and data, thus field and laboratory tests for purposes of developing new data were specifically excluded from consideration from the onset of the study. Because specific sites are purposefully not considered, reference site stratographics and characteristics were developed to allow for the analysis necessary to the evaluation and ranking of the rock types.

*Research sponsored by the Office of Nuclear and Chemical Waste Programs, U.S. Department of Energy under contract DE-AC05-84OR21400 with Martin Marietta Energy Systems, Inc.

II. EVALUATION AND RANKING CONSIDERATIONS

The performance criteria used to evaluate and rank the five rock types in this study were adopted from DOE's 10 CFR 960 Siting Guidelines (1984) and refer to the period of time before and during closure of the repository as "preclosure" and to the period after closing as "postclosure." For the preclosure guidelines in 10 CFR 960, it is clear that the only hydrologic disqualifying condition is any expected groundwater condition that would require engineering measures that are beyond the state of available technology to construct shafts and/or construct, operate, and close the repository. Since excessive groundwater inflow into the shafts and repository workings is probably the most likely condition that would lead to the need for such non-routine engineering measures, groundwater inflow was selected as the preclosure criterion for the hydrologic evaluation and ranking of the five rock types. With this approach, a single hydrologic performance measure (groundwater inflow rate) is used to integrate all of the geologic and hydrologic parameters that would have significant impact on the construction and operation of a repository.

Pre-waste-emplacement groundwater travel time was selected as the criterion for postclosure because it considers many of the hydrologic factors affecting the isolation capability of a potential repository and because it is the only disqualifying condition given in 10 CFR 960. Specifically, an expected pre-emplacement groundwater travel time to the accessible environment (10 km from the controlled area of the repository) of less than 1000 years would disqualify a potential repository site in any rock type from further consideration. Further a pre-emplacement groundwater travel time of more than 10,000 years is considered in 10 CFR 960 to be a favorable condition for siting. It is noted, however, that while pre-emplacement groundwater travel time incorporates many hydrologic parameters of concern into a single measure of performance, it does not consider all hydrologic factors related to the postclosure isolation potential of a waste repository. Notably, pre-emplacement travel time does not consider thermal driving forces, dispersion, or radionuclide sorption phenomena.

III. HYDROLOGIC PROPERTIES

In order to estimate the pre-waste-emplacement groundwater travel time (postclosure criterion) for the five rock types, it is first necessary to solve the steady-state flow equations to derive potential fields. Velocity fields are then developed from the potential fields and travel times are then estimated by solving purely advective equations for the geometries, boundary conditions, and material properties of the reference sites for analyses. Thus, hydraulic conductivity and effective porosity are the material properties required for the travel time analyses for each of the five rock types.

In this investigation, groundwater inflow is defined as the volume of water per meter of drift per unit of time that would enter the subsurface excavations of a repository as a function of time following excavation. Thus, the analyses for estimating the inflow of groundwater (preclosure criterion) into the repository workings for the five rock types involves solving the transient flow equations for the geometries, the boundary and initial conditions, and the material properties of the reference sites. Hydraulic conductivity and specific storage are therefore the material properties that must be obtained for the inflow analyses at the sites. Discussions on the development and utility of the data bases for these properties are given below. In all cases the data bases reflect those subsets of the five rock types that may reasonably be considered for waste disposal. It is noted that the literature search, while not exhaustive, did not indicate sufficient data for many of the hydrologic parameters to permit the derivation of typical probabilistic values and ranges. Thus, the "high" and "low" values that bracket the typical value are not intended to be statistically rigorous but are merely intended to account for the likely variations in properties.

A. Hydraulic Conductivity

Literature on the hydrologic properties of anhydrock was found to be the most difficult to find of all rock types studied. The limited laboratory and field results indicate that anhydrock has a very low hydraulic conductivity. Core results range from 4×10^{-13} to 1.6×10^{-8} m/s and are typically on the order of 1×10^{-11} m/s (Croff, 1985). Field test results indicate that horizontal conductivities range from 1.5×10^{-11} to 2.3×10^{-9} m/s and are typically on the order of 10^{-11} to 10^{-10} m/s (Croff, 1985). Field observations of anhydrock in outcrop indicate an intense pattern of very closely spaced, short fractures. The similarity between the core and field results is supported by these observations which indicate that the fracturing may be on a sufficiently small scale that the effects are seen in the core. In view of these similarities and of the reported inability to measure the lower conductivity horizons in some field tests, the typical data values are taken to be 1×10^{-11} m/s for horizontal hydraulic conductivity and 1×10^{-12} m/s for vertical hydraulic conductivity. Probable maximum and minimum values are assumed to be one order of magnitude higher and lower, respectively, than the typical values.

Although considerable information was found to be available from laboratory analyses of core samples, little information was found from in situ field tests for chalk. Both types of data were available from sites in England and Texas (Croff, 1985). The field hydraulic conductivity of chalk formations was found to vary greatly, depending upon the degree of fracturing. The English chalks for which data were available are used locally as major water supply aquifers, while the Texas chalks are much tighter and contain oil and gas. The field data for the Texas Austin chalk represents the more permeable, fractured portions of the formation, although some 50% of the formation has been estimated to be relatively unfractured (Croff, 1985).

The hydraulic conductivity values adopted for the data base represent the tighter end of the conductivity range, because only these chalks might be reasonably considered for nuclear waste disposal. The typical data base values adopted for this study are a horizontal conductivity of 1×10^{-7} m/s and a vertical conductivity of 5×10^{-8} m/s. The high and low values of the conductivity range are one order of magnitude greater and lower than these values. These values are lower than the lowest available field data and reflect the lower conductivities expected in the relatively unfractured portions of a chalk formation.

The literature on carbonates is relatively abundant for the more highly conductive members but is scarce for the tighter members of interest in waste disposal. Hydraulic conductivities for the more highly conductive carbonates can be very high, exceeding 1×10^{-2} m/s (Brace, 1980). These are among the highest conductivities observed in any rock type. Core data are generally orders of magnitude lower than field values, reflecting the importance of fractures in controlling hydraulic conductivity (Croff, 1985). A detailed review of the literature surveyed in this study indicated that the lower conductivity field data generally reflected either relatively thin beds of little interest to mined geologic disposal, or parts of thicker beds (Croff, 1985). Taking this into account, the typical data base values adopted for this study are a horizontal conductivity of 1×10^{-6} m/s and a vertical conductivity of 5×10^{-7} m/s. The high and low values of the conductivity range are again one order of magnitude greater and lower than these values.

Information on sandstone, like carbonates, is relatively abundant in the hydrology literature for the more highly conductive members, with data on the lower conductivity members originating primarily in the oil and gas literature. Representative hydraulic conductivity values for sandstone range from 1×10^{-5} m/s on the high end to 1×10^{-11} m/s on the low end (Croff, 1985). Core data were found to be similar in magnitude to field data, indicating a smaller influence of discrete fractures and a greater dependence of fluid movement on the bulk porosity (Croff, 1985).

Thick beds of very tight sandstones are found at many locations. The properties of these sandstones were therefore adopted for the data base. The typical data base values adopted for this study are a horizontal conductivity of 1×10^{-10} m/s and a vertical conductivity of 5×10^{-11} m/s. The high values of the conductivity range are two orders of magnitude higher than the higher values, and the low values are one-half an order of magnitude lower than the typical values. This asymmetric range was selected to reflect the large number of more permeable sandstones than might be considered for waste disposal.

The hydrologic properties of shale are not abundant in the literature because of its poor aquifer properties, and much of the available data originates from the oil and gas literature. Core data were found to be similar in magnitude to field data, indicating little influence of fractures on groundwater movement (Croff, 1985). Thick beds of very tight shales are found at many locations. The typical data base values adopted for these shales are a horizontal conductivity of 1×10^{-11} m/s and a vertical conductivity of 1×10^{-12} m/s. The high and low values of the conductivity range are one order of magnitude higher and lower than the typical values.

B. Effective Porosity

Effective porosity is traditionally considered to be that part of the total porosity that is actively involved in conducting fluid flow. In practice, effective porosity is measured in tracer tests and is equal to the ratio of the specific discharge (flow rate per unit area) across a surface to the linear particle velocity of groundwater movement across that same surface. As is evident from its measurement, "effective porosity" is determined purely from hydraulic factors and is in fact a hydraulic rather than a volumetric property of the rock.

Because of the lack of data on effective porosity in many rock types, it was usually necessary to estimate the magnitude of this parameter from total porosity data. Such data are available from laboratory tests on cores and from geophysical density logs in the field. The laboratory tests have the problem of small sample volumes, while the field tests have calibration uncertainties. Despite these difficulties, the laboratory values of total porosity are normally considered to be reasonably representative of field conditions, and the primary source of uncertainty is in approximating the effective porosity from the total porosity data. Because of the influence of fractures, the rock mass effective porosity of anhydrock would be expected to be lower than that observed in core measurements (Croff, 1985). The adopted data base typical values are 0.01 for total porosity and 0.001 for effective porosity. The ranges are approximately one-half order of magnitude higher and lower than the typical values.

Although the interstitial porosity of chalk is high, it has been found to be pressure and therefore depth sensitive. Because a repository would be expected to be relatively deep, the porosity data base was selected to represent the lower end of the range (Croff, 1985). In addition, the significant role of fractures in governing fluid movement suggests that effective porosity will be considerably lower than total porosity in chalk. In the adopted data base typical values are 0.10 for total porosity and 0.001 for effective porosity. The ranges are approximately one-half order of magnitude higher and lower than the typical values.

In view of the offsetting influences of the fracturing and of the relatively high matrix porosities locally observed in many carbonates, the effective porosity was assumed to be one order of magnitude lower than the total porosity (Croff, 1985). The adopted data base typical values are 0.10 for total porosity and 0.01 for effective porosity. The ranges are variable, extending to one order of magnitude on the low side for effective porosity.

Although sandstone porosities can be relatively high, the data base was selected to emphasize the tighter sandstones (Croff, 1985). The weak influence of fractures indicated by the hydraulic conductivity data suggests a smaller difference between total and effective porosity than was expected for the rock types previously discussed. The adopted data base typical values are

0.08 for total porosity and 0.01 for effective porosity. The high and low values are within an order of magnitude of the typical values, with the greatest range being assigned to the high value because of the potentially significant influence of flow in the unfractured matrix in the higher conductivity sandstones.

Because shale can be relatively compressible depending upon its stress history, the porosity of shale can be relatively high. The data base was selected to emphasize the compacted, older shales that would be expected to have been relatively deeply buried and therefore more representative of shales at typical repository depths (Croff, 1985). The difference between total and effective porosity is expected to be small in shale because of the generally low influence of individual fractures on groundwater movement. The adopted data base typical values are 0.03 for total porosity and 0.01 for effective porosity. The high and low values are within an order of magnitude of the typical values, and largely reflect uncertainties in degree of compaction.

C. Specific Storage

Specific storage is a measure of the volume of water that can be removed from or added to a volume of rock due to pressure changes within that volume. Specific storage in deep, confined strata results from the compressibility of the water and rock, and is orders of magnitude lower than specific storage in a water table aquifer where storage changes result primarily from emptying or filling unsaturated void space.

Accurate measurements of specific storage are obtained from multiple well interference tests in the field. However, no field data were found in the literature on specific storage for any of the low conductivity media considered in this study to be suitable for waste disposal. In addition, specific storage is not normally measured using core samples because of a lack of accuracy. Under such circumstances, the magnitude of specific storage is often approximated for deep, confined strata from compressibility data for the water and the sediments. The basic equation, as adapted from Freeze (1979), is:

$$S_s = \rho g \, (n_t C_w + C_r), \qquad (1)$$

where

S_s = specific storage
ρ = density of water
g = acceleration of gravity
n_t = total porosity
C_w = compressibility of water
C_r = compressibility of rock.

The compressibility of rock is defined for an isotropic, homogeneous elastic medium as the reciprocal of the bulk modulus, or:

$$C_r = 3(1 - 2u)/E, \tag{2}$$

where

u = Poisson's ratio
E = modulus of elasticity.

Specific storage values for this study were estimated using Equations (1) and (2).

The results show relatively little sensitivity to the ranges in geomechanical properties. The adopted data base values were generally assigned higher ranges than the computed results indicated, to account for additional uncertainties present in the computational method. With the exception of anhydrock and chalk, rounding of the computed results gave the same storage values to all rocks. Anhydrock and chalk varied from the general pattern because of their unusually high and low elastic moduli, respectively. The resulting typical data base specific storage values were 1×10^{-7} 1/m for anhydrock, 1×10^{-6} 1/m for carbonates, sandstone, shale, and salt, and 5×10^{-6} 1/m for chalk. The high and low values were taken to be approximately one-half order of magnitude larger and smaller than the typical value for each rock type.

D. Reference Stratigraphic Sections

Reference stratigraphic sections, that are believed to be representative of typical geologic repository environments, were prepared for each of the five rock types. In all cases the repository was placed at a depth of 700 m in the stratigraphic section.

Typical hydrologic properties (adopted properties) are required for each modeled stratum in the reference geologic sections for the five rock types being considered as potential repository host rocks. The hydrologic properties of the strata overlying and underlying the repository horizons in the reference sections were determined from the developed data bases for the repository horizons, and from general knowledge of hydrologic material properties of the various rock types and depositional environments as reported in standard works (DeWiest, 1965).

The reference stratiographic section for anhydrock shows that the uppermost 15 m of the section is made up of sand, silt, and gravel that is representative of recent surface deposits. Underlying this is a thick sequence of evaporites, alternating anhydrock and salt, with one shale unit. The reference repository horizon is an anhydrock unit 183-m thick. The evaporite sequence is assumed to continue below the anhydrock horizon.

In the reference stratigraphic section for chalk, the uppermost unit consists of shale and sandstone that is underlain by a unit of clay, shale, and marl. The next lower unit is composed of shale, marl, and chalk with some sandstone. The reference repository horizon is a 259-m-thick bedded chalk mixed with chalky carbonates and limey chalks.

The uppermost horizon for the carbonate section consists of a 488-m-thick shale unit, which is underlain by some 91 m of carbonates. A thin, 15-m-thick shale unit underlies this upper carbonate, followed by 77 m of sand and shale. The reference repository horizon is 275 m of fine- to coarse-grained crystalline carbonates (limestone/dolostone) with some sand grains and chert. This is underlain by a horizon of sandstone and conglomerate.

The reference stratographic section for sandstone consists of a 762-m-thick sandstone reference repository horizon that is overlain and underlain by some 600 m of shale.

The shale section consists of an upper 30-m-thick horizon of sand, silt, and clay followed by 122 m of shale with some silt and sand. This is underlain by the reference repository horizon, a 610-m-thick unit of gray to dark gray, low carbon, illitic shale with minor silt and limestone. A 122-m-thick carbonate horizon is assumed to underlie the shale.

The five analyses for travel time and groundwater inflow, performed to rank the five sedimentary rock types, are described below.

The pre-waste emplacement groundwater travel time is estimated by first solving the steady state groundwater flow equation:

$$(\partial/\partial x_i)(k_{ij}\rho g/\mu)(\partial\phi/\partial x_j) = 0 \tag{3}$$

where

k_{ij} = permeability tensor
ρ = density of the fluid
μ = viscosity of the fluid
ϕ = hydraulic potential
x_i = Cartesian coordinates
g = gravitation acceleration.

The relatively complex geometry coupled with the anisotropic hydraulic conductivities considered in these analyses necessitate the use of numerical solution techniques in the form of a computer code. The numerical code selected for this study is based on a weighted residual implementation of the finite element method. The code is documented in Baca (1981) and will not be described here.

The results of the pre-waste-emplacement groundwater travel time analyses are summarized in Table 1. The results are presented as distance traveled over 100,000 years. Groundwater velocities of the chalk and carbonate sections were found to be very significantly greater than the other rock types, with the velocities in carbonates being greater than those in chalk. Sandstone

TABLE 1. Pre-emplacement Distance Traveled by Groundwater in 100,000 Years, in m

Rock Type	Expected	High	Low
Anhydrock	46	317	20
Chalk	>15,000	>>15,000	>15,000
Carbonates	>>15,000	>>15,000	>15,000
Sandstone	30	3,150	15
Shale	6	55	<1

showed a greater range in velocities than anhydrock, resulting from the perceived possibility of higher conductivity sandstones. Shale and anhydrock uniformly showed the lowest groundwater velocities of all rock types studied. The ranking of the five rock types on the basis of groundwater travel time is thus: (1) shale and anhydrock, (2) sandstone, and (3) chalk and carbonates.

The first three rock types rank relatively closely on the basis of the typical cases. Because the sandstone in the reference section is surrounded by low permeability shale layers, it is suggested that the travel distances derived in the sandstone analyses may be nonconservatively low. The travel distances computed for the high and low cases for these three rock types show the same ranking as the typical cases. In view of the ranges and perceived degrees of relative conservatism associated with the various reference sections, it is believed that the ranking presented in the previous paragraph for the travel time analyses is valid for the purposes of the present study.

The groundwater inflow, for the purposes of ranking the five Sedimentary rocks addressed in this study, is estimated as the volume of groundwater per unit of time that may be expected to enter a very long drift of 5.5 m-diameter per meter of drift. Because the repository layouts have not been developed for the various rock types, the effects of flow field interference resulting from the multiple drifts of the final repository are not considered. Rather, it is believed that the inflow, estimated on the basis of a single drift, provides an adequate basis for comparison ranking of the five sedimentary rock types.

Inflow into a drift will be relatively large soon after excavation and decrease with time to approach a steady state value. Because the early time inflow can be significantly greater than the inflow at steady state, a transient finite element solution is used in this analysis.

The ranking for the five rock types on the basis of groundwater inflow is: (1) anhydrock and shale, (2) sandstone, (3) chalk, and (4) carbonates, which correspond to the travel time rankings.

IV. SUMMARY AND CONCLUSIONS

Results of the groundwater travel time analyses show that groundwater velocities for chalk and carbonate are significantly greater than the other rocks. Shale was found to have the lowest groundwater velocities, while sandstone and anhydrock rank behind but relatively close to shale. This is due to the lower conductivity of shale as compared to sandstone and the higher effective porosity of shale in comparison to anhydrock. In general, the ranking of the five rock types based on groundwater inflow analyses compares favorably to the travel time rankings. Although the chalk and carbonate rocks show relatively high inflow rates in comparison to sandstone, shale, and anhydrock, all values are considered to be below the point where engineering measures to remove the inflowing groundwaters would be problematical during the construction and operation of a deep repository.

REFERENCES

1. R. G. Baca and R. C. Arnett, Analysis of Fracture Flow and Transport in the Near-Field of a Nuclear Waste Repository, Rockwell Hanford Operations Basalt Waste Isolation Project Report, RHO-BWI-SA-81 (March 1981).

2. W. F. Brace, "Permeability of Crystalline and Argillaceous Rocks," Int. J. Rock Much. Min. Sci. and Geomech. Abstr. 17, 241-251 (1980).

3. A. G. Croff, T. F. Lomenick, R. S. Lowrie, and S. H. Stow, Evaluation of Five Sedimentary Rocks Other Than Salt for High-Level Waste Repository Siting Purposes, ORNL/CF-85/2/V1 (Draft), 1985.

4. R. DeWiest, Geohydrology, John Wiley and Sons, Inc., New York (1965).

5. R. A. Freeze and J. A. Cherry, Groundwater, Prentice-Hall, Inc., Englewood Cliffs, New Jersey (1979).

DIFFERENCES IN THE APPLICATION OF QA REQUIREMENTS TO NUCLEAR WASTE REPOSITORY SITE CHARACTERIZATION ACTIVITIES

Maxine A. Burgan
EBASCO Services Incorporated
160 Chubb Avenue
Lyndhurst, New Jersey 07071

ABSTRACT

Because of the direct involvement of technical personnel in performing vital functions to assure the quality of the results prior to and during site characterization for a nuclear waste repository, it is imperative that geoscientists and QA professionals understand the differences in application of 10 CFR 50, Appendix B, to their activities and to their typical roles. At this stage of NWR siting investigations, the mutual goal of scientists and QA professionals is to have confidence in the validity of the acquired data and resultant analyses, and the quality of the judgements which have led to the selection of a site and the proving of this site during the characterization processes.

I. INTRODUCTION

Because of two major areas of concern, siting work on nuclear waste repositories is under much closer scrutiny now, than was the siting work for nuclear power generation facilities during the 1960s and 70s. These areas of concern are: 1) the validity of the data and analyses, as well as the quality of the judgements which lead to site selection and characterization; and 2) the ability to satisfy the public sectors whose reluctance is due to suspected dangers of a nuclear waste repository in their back yard.

To provide confidence that these concerns are properly resolved with the proof of the resolution readily available, the imposition of the requirements for the assurance of quality contained in 10 CFR 50, Appendix B, has been mandated down to the lowest possible level of the hierarchy of sub-tiers who are involved in any contributary investigation or activity. Since the answer to the question of "who knows which contributing piece of evidence will be used during the licensing process? is unknown, all pieces must be treated as possible evidence. How do we assure the quality of these results of investigative activities?

II. DISCUSSION

In order to establish the proper perspective for this discussion, Figure 1 provides a description of the purposes of the siting activities, and a clarification of the terms, "investigations," and "site characterization."

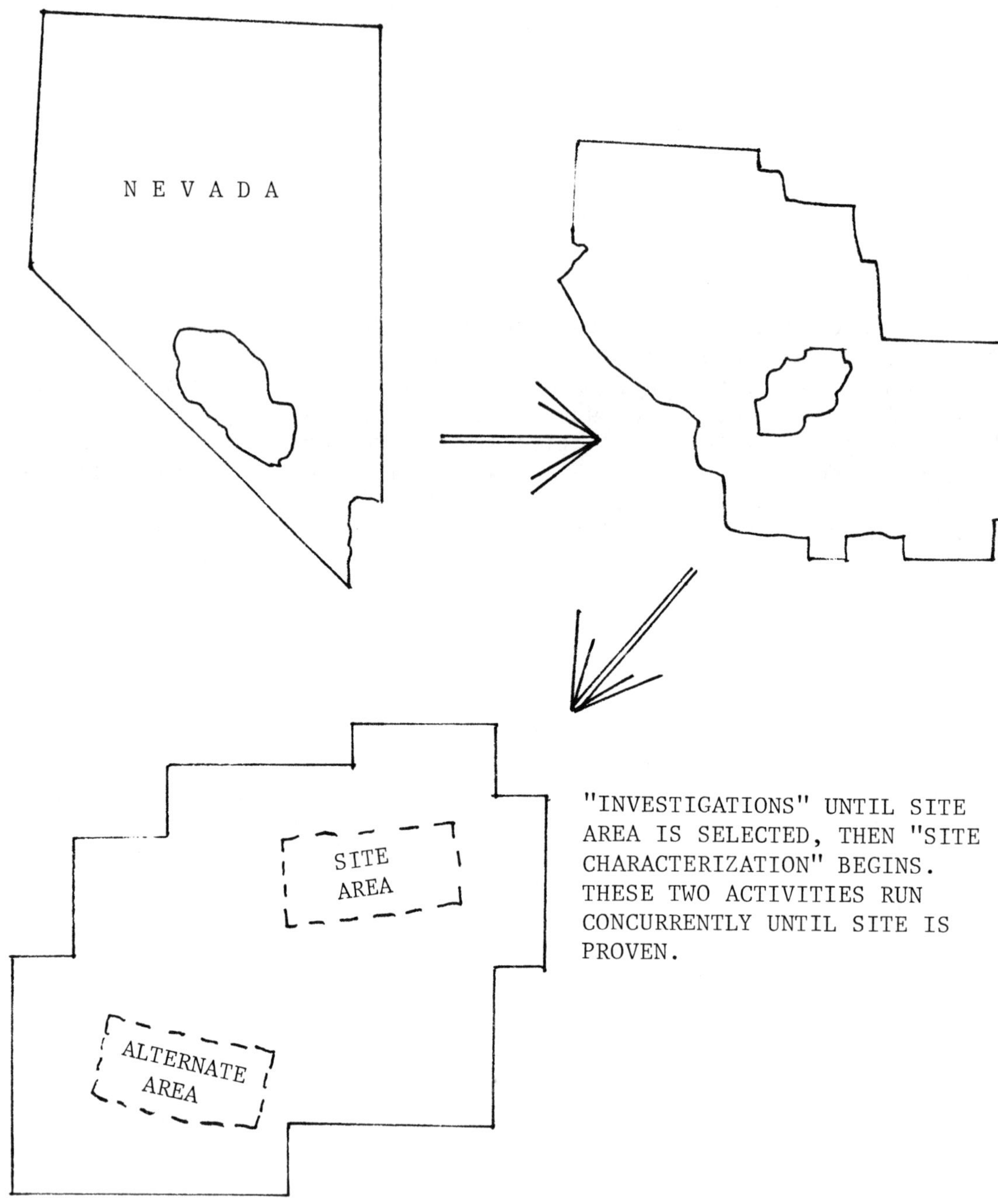

FIGURE 1. Selection of a Suitable Site

Several of the 18 Criteria of 10 CFR 50, Appendix B, demand specific actions applicable to a hardware product which can be measured, inspected, tested, or operated to assure its quality and conformance with prespecified specifications, or be rejected if it does not. Others of the Criteria provide flexibility.

The attempts to apply these well-established hardware concepts to the activity phases of siting investigations and site characterization, lead to major problems of misunderstanding. Hardware oriented mind-sets must be changed, the elements of 10 CFR 50, Appendix B, must be re-oriented toward the activities of data acquisition, data analyses, and data interpretation which are summed up in a report as the final product. During this phase, the concept of acceptance or rejection of acquired data is not applicable because all data gathered during the field work, laboratory testing, research, computer modeling, and analyses must be considered, whether favorable or adverse.

We may not be able to place a pair of calipers on a one-of-a-kind world-renowned authority's head to determine whether to accept, reject, or rework his conceptual results, but we must question whether his work encompassed a broad enough scope to adequately determine the characteristics. We must establish the validity of not only the authority's conclusions and recommendations, but also the validity of the data used during the analyses.

However, who is to perform the critical review, who will play devils advocate, who will resolve the differences of opinion, who will bless the use of unsubstantiated state-of-the-art results? The answer is a peer: A technically qualified individual having equivalent or better knowledge and capability for performing the task under review.

There may not be one peer capable of questioning or disputing the results, but a combination of many peers, and the documentation of comments and resolution of same can establish the current validity and acceptability of this state-of-the-art research.

This same principle of peer review applies to the field personnel who are acquiring data for input into the geoscientific analyses and computer modeling leading to site selection, as well as the engineering analyses leading to the design criteria for the waste repository. The reviewer questions whether the data acquisition and sampling were comprehensive enough to properly reflect the community as a whole.

Thus, peer review is an act of verification which is performed at all levels of activities such as a check for performance acceptance; and critical review and/or multiparty critical review is performed for acceptance of results.

III. APPLICATION OF CRITERIA ELEMENTS

The intent of this section is to address the criteria wherein the differences of application are meaningful. Those criteria which are not included in this section or those elements which are not included in the discussed criteria are directly applicable and create no major problems.

A. Criterion II - "Quality Assurance Program": Two elements of this criterion require the most important bases for establishing the validity of the data and research. The first base is the requirement for planning documents to assure that the scope of data acquisition activities provides a truly representative picture of the area under study. The second base is the interpretive requirements for qualified personnel. Because of peer review, establishing the criteria for qualifying specific reviewers has become crucial.

B. Criterion IV - "Procurement Document Control": It is acknowledged by industry and the NRC that many lower tier services are not required to have a complete 18-criteria, fully documented Quality Assurance Program. They must have adequate and proper controls over the pertinent areas such as sample control, test control, calibration controls, data controls, etc., but may work under the umnbrella of the contractor QAP. In this case, the contractor would invoke the required controls in the procurement documents and perform surveillance and/or audits during the lifetime of the activity on a schedule commensurate with the importance of the activity.

C. Criterion V - "Instructions Procedures": The NRC is concerned about the adequacy of detail in the technical procedures and instructions; will they truly define each step taken during an investigation? Their concerns are justified in view of the "state-of-the-art" aspects of determining site suitability, waste containment, etc. In most cases, the basis upon which the state-of-the-art research is done is little more than a concept. The methodology is refined as the research progresses. It is impossible to have a procedure or instruction which has been finalized and approved for use prior to the instigation of the activity. What is possible, and is mandatory, is to have a complete description of each step of the methodology, the assumptions, the variables, etc., which fully define the process after the fact (refer to Criterion XI). It is worth noting that the objective of having procedures or instructions is to provide the means for an equally qualified scientist to duplicate the process with reasonable duplication of the results.

D. Criterion X - "Inspection": "Inspections required to verify conformance of an item or activity to specified requirements." During the investigative activities, the specified requirements would be the activity planning documents and the procedure/instruction which defines the methodology for performing the activity. As long as it is clearly defined, the term "Inspection" can be replaced by "check," "peer check" or "peer review;" the objective being to verify that the performance is acceptable technically. This requires an equally qualified technically knowledgeable individual capable of making this objective judgement.

E. Criterion XI - "Test Control": Under Test Control, we must make it clear that we are testing naturally existing materials to determine their physical, chemical, thermal, etc., properties or characteristics. We also must make it clear exactly what procedures we are following in order to perform the tests:

- ASTM precisely,
- ASTM modified (documented and approved),
- published methodology precisely,
- published methodology modified (documented and approved), and
- state-of-the-art (after-the-fact documentation).

Additionally, consideration must be given to establishing adequate controls over storage and shelf-life of the samples and traceability between documents/records and samples.

F. Criterion XII - "Control of Measuring and Test Equipment": State-of-the-art testing may only require calibration of equipment prior to use, to an industry standard; a normal testing laboratory must have a complete calibration program.

G. Criterion XVII - "Quality Assurance Records": For Nuclear Waste Repositories, the needs require expansion of the requirements under Criterion XVII. Actually, it is an expansion into the unknown, such as "perpetual care" of records. What type of facilities will provide perpetual care? What type of media will retain its integrity perpetually? How long is perpetual?

H. Criterion XVIII - "Audits": After development of the QA Program, the prime function of the independent QA staff is to perform audits. These audits must be performed at all levels at least once, early in a short-lived activity and regularly in a long-lived activity. The objective of the audits is to correct deficiencies and to bring the system of controls into effective implementation. Secondary functions of QA are to perform ongoing QA indoctrination and training of all personnel involved in or impacted by the QA Program and to assure that technical training is provided by technical staff to all levels, as necesssary.

I. Another major factor which must be addressed, but which has no clear relationship to a single criterion is control over the objective evidence from the time of generation, through years of processing, until it is submitted to the DOE storage area as a QA record, or dispositioned otherwise.

From the outset, it is not clear what pieces of paper will be called upon to provide objective evidence in a future which may be several years down the road. As site characterization progresses into the licensing stage, each question and each resolution of a comment creates a change and usually expands the scope of the investigations. The objective evidence is in a dynamic state of change, modification, addenda, and/or expansion. During this licensing phase, past judgements and decisions will be questioned; objective evidence must be available to support a decision to use, or a decision not to use, regardless of the element of elapsed time.

The two elements of the factor for data control are:

1. The controls over the documents which describe the results of activity performance, in other words, the proof that the activity was performed. These controls must include the methods of recording the data, checks for completeness and acceptance, release and routing, and provide a closed loop feed-back system on changes, modifications, or conversions into analyses, graphics or software.

2. The controls over the collection, assembly, and temporary storage and maintenance of these working documents.

IV. SUMMARY

A. As a brief reference, Figure 2 lists some of the major differences in the application of QA elements and Figure 3 lists the responsibilities for assuring that the QA requirements are being met during the investigative phase. In its Review Plan, the NRC has recognized these differences, as well as most of the others herein discussed.

B. To combine the perspective and the application concepts, Figure 4 describes the progression from investigation activities versus facility hardware to in-situ site characterization activities which overlap with site preparation, construction and installation. This progression from the Quality Assurance role as the verifier that the system of controls is in place and functioning properly, to the full blown QA/QC application with which we are all too familiar in the design and construction of nuclear power plants, requires a readjustment in attitude.

C. Although the NRC has issued its Review Plan as applicable to the site characterization phase, and therefore includes many hardware concepts, the actual point of application is not clear in this document. Additionally, personal discussions with NRC have made it very clear that QA application is expected during the purely activities oriented earlier stages. This expectation could present us, as involved geoscientists, with a major problem unless we adequately and properly define the difference in application of the QA elements to our activities, prior to the start of the activity.

INVESTIGATIONS VS CONSTRUCTION

II	QUALIFIED GEOSCIENTISTS	VS	CONSTRUCTION OR MFG. LINE WORKERS
III	ANALYSES OF DATA GATHERED	VS	DESIGN TO SPECIFICATIONS, ETC.
	DATA & SAMPLE ACQUISITION	VS	CONSTRUCTION/FABRICATION TO SPECIFICATIONS
	DESCRIPTIVE GRAPHICS	VS	DESIGN/CONSTRUCTION/AS-BUILT DRAWINGS
	TECHNICAL CHANGE CONTROL	VS	DESIGN CHANGES
IV	PROCUREMENT OF UNIQUE EXPERTISE	VS	PROCUREMENT OF ITEMS TO DESIGN SPECIFICATIONS
V	STATE-OF-THE-ART	VS	STANDARD OPERATING PROCEDURES
X	PEER CHECKS, REVIEWS & VERIFICATIONS	VS	QC INSPECTIONS
XI	SAMPLE TESTING	VS	MEASURING & TEST OF ITEMS

BOTTOM LINE

RESOLUTION OF DIFFERENCES OF OPINION	VS	REJECT
REPORT	VS	PRODUCT OR FACILITY

FIGURE 2. Differences in Application of QA Elements

- RESPONSIBILITY FOR QUALITY PERFORMANCE?

  THE PERFORMER

- WHO IS RESPONSIBLE FOR PERFORMANCE ACCEPTANCE?

  PEER OR SUPERVISOR

- WHO IS RESPONSIBLE FOR EVALUATION OF COMPLIANCE?

  QUALITY ASSURANCE AND

  MANAGEMENT PERSONNEL

FIGURE 3. Responsibilities.

<u>SITE INVESTIGATIONS ACTIVITIES VS FACILITY HARDWARE (ITEMS)</u>

SITE INVESTIGATIONS ACTIVITIES

RESULT: REPORTS LEADING TO PSCR

MEANS: DATA ACQUISITION

- From existing material
- From research results
- From investigations
- From measurements of existing conditions
- From testing of existing samples

DATE ANALYSES

- Graphics
- Literature searches
- Geologic
- Engineering
- Hydrologic
- Seismologic
- Computer Modeling

VERIFICATIONS: Resolution of Differences

- PEER CHECKS
- PEER Reviews
- Multi-party reviews
- CLIENT Review
- DOE Review

RECORDS

- Reports
- Methodologies
- Back-up data
- Raw
- Processed

OBJECTIVE: To enable an equally qualified technical analyst to arrive at reasonable duplication of results by using the same raw data, but who may use alternate methodology.

FACILITY HARDWARE (ITEMS)

RESULT: PRODUCT

MEANS:

- Development of Design, Specifications and Drawings
- Procurement of Raw Materials/Parts
- Manufacture and/or fabrication to Specifications and Drawings

VERIFICATIONS: <u>ACCEPT</u> or <u>REJECT</u>

- In-process Inspections
- Final Inspections
- NDE as required
- Etc.

RECORDS

All of the above means and verifications produce records

CERTIFICATE OF COMPLIANCE

OBJECTIVE: To assure that the product was designed to meet established specifications, is reflected in as-built drawings; passed through rigid inspections to assure compliance; and is <u>Certified</u> as complying.

FIGURE 4. Site Investigations Activities vs Facility Hardware

SITE CHARACTERIZATION ACTIVITIES CROSS OVER THESE BOUNDARIES (AS DOES SITE PREPARATION FOR AN NPP). QA APPLICATION, THEREFORE, IS BASED ON WHICH ACTIVITIES ARE UNDER WHICH CATEGORY.

IN-SITU INVESTIGATION ACTIVITIES	IN-SITU PREPARATION/ CONSTRUCTION/INSTALLATION
RESULT: Modifications to PSCR	RESULT: LICENSE APPLICATION
SITE CHARACTERIZATION REPORT	
MEANS: DATA ACQUISITION	MEANS: Design Bases Analyses Design Bases Calculations
RAW DATA ANALYSES Geologic Mapping of Excavations and Shafts In-Situ Hydrologic Tests Down-Hole Geophysical Probes	Preliminary Design Drawings Engineering Specifications Procurement Specifications Seismologic, Hydrologic & Environmental Monitoring with In-Situ Arrays
IN-SITU ENGINEERING TESTS	
In-Situ Sampling/Testing Re-Analyses During Site Characterization Activities (Feed-Back Controls)	Design Bases Computer Modeling Fabrication of Exploratory Shaft Liners Shaft Electrical & I&C Installations Shaft Grouting/Sealing Procurement of Items to be Installed Etc.
VERIFICATIONS: Resolution of Differences PEER Checks PEER Reviews Multi-party reviews CLIENT Review DOE Review	VERIFICATIONS: ACCEPT or REJECT/ MODIFY Design Controls In-Process Inspections and Tests Final Inspections Receipt Inspections NDE Etc.
MULTI-PARTY TECHNICAL APPRAISALS OF EXPOSED SITE CHARACTERISTICS SITE CHARACTERIZATION REPORT REVIEW	Licensing Application Review
RECORDS: All	RECORDS: All

FIGURE 4 (contd)

THE EVALUATION OF LARGE-SCALE PERMEABILITY IN DEEP GRANITE

A. Barbreau
Commissariat a l'Energie Atomique
B.P. 6
92260 Fontenay-aux-Roses, France

G. de Marsilly
Ecole des Mines de Paris
35, rue Saint Honore
77305 Fontainebleau, France

P. Peaudecerf
E. Durand
Bureau de Recherches
Geologiques et Minieres
B.P. 6009
45060 Orleans Cedex, France

ABSTRACT

In order to extrapolate the results of hydraulic tests in fractured granite media and to optimize their conception, a significant experiment is being carried out on the evaluation of the consequences of scale effect on the value of hydraulic conductivity and dispersion coefficient. This experiment is being done in a 100-m long experimental drift in an underground laboratory at a depth of 170 m. The numerous structural and hydraulic data, which have been collected, should make it possible to build a 3-D model based on the fracture network.

I. INTRODUCTION

Within the framework of its program on the safety evaluation of radioactive waste disposal in crystalline formations the "Departement de Protection Technique de l'Institut de Protection et de Surete Nucleaire" (CEA, Atomic Energy Authority) is setting up a research program on the properties of fractured media. One of the studies currently underway is the evaluation of the consequences of the scale effect on the determination of permeability and dispersion coefficients. Its objective is to devise optimal hydraulic tests and improve the interpretation of their results, which may, in heterogeneous media, depend on the rock volume involved.

This experiment is being carried out in an underground laboratory at a depth of 170 m, in the drift of the Fanay-Augeres uranium mine, situated 30 km from the city of Limoges, in a granitic batholith in the Massif Central. It is a joint venture by the "Bureau de Recherches Geologiques et Minieres" (French Geological Survey) and the "Centre d'Informatique Geologique" of the Paris School of Mines. The program is funded through a cost-shared contract between the CEA (French Atomic Energy Authority) and the Commission of the European Communities.

II. OBJECTIVES AND PRINCIPLES OF THE RESEARCH WORK

Today, some of the most compelling problems in the field of hydraulics in fractured media are those concerning the determination of large-scale permeabilities in a formation and the ensuing identification of the pathways of the solutes. Several techniques can be used to obtain information relevent to this problem:

- direct observation of the fractures at the outcrops of the formation or in galleries or on cores from oriented exploration boreholes,
- measurements of the state of stress in the formation,
- injection tests between packers in exploration wells,
- hydraulic interference tests between exploration wells,
- large-scale water and tracer tests, and
- geophysical measurements in order to identify the properties of the rock (electric resistivity, sonic velocity) from which an estimate of the state of fracturing may be indirectly deduced.

From a technical point of view the methods which define the fractured medium stochastically appear to be the most suitable, e.g., in the framework of the percolation theory (References 1-10). These methods or parts of them have been used on certain sites with more or less satisfactory but always incomplete results. The main difficulty resides in the fact that local measurements (observations, water tests) do not make it possible to obtain a complete picture of the large-scale behavior of the formation. However, it is precisely the large-scale properties of the formation which influence the release of the radionuclides stored at great depths in a fractured medium. The problem of the scale effect must be solved.

It is, therefore, necessary to develop a methodology for estimating the "large-scale" permeabilities and pathways of a fractured rock formation, based on observations or tests which can be carried out on a real waste storage site, where measurements can be made in exploration shafts or deep experimental galleries. In order to achieve this, we have chosen to work on a rock volume whose three dimensions are close to 100 m. This "sample", necessarily at depth, is pierced by a gallery that constitutes a large-scale drain of the formation as the water flows more or less radially towards it. This then, by definition, gives the "large-scale" permeability of the formation when measurements are made simultaneously of the hydraulic head gradients at a fair distance from the gallery and of the rate of extracted flow. In reality, we are looking for the permeability tensor by measuring the pressures at a large number of points (70) in the formation around the gallery. A similar experiment has been conducted in Sweden at STRIPA.[11]

Similarly, the large-scale hydrodispersive properties of the formation will be investigated by means of tracer injections at various distances from the gallery and the observation of their migration towards (or their time of arrival at) the gallery.

In practice, these points of pressure measurement and tracer injection at various distances are provided by drilling radial boreholes from the gallery and equipping them with several successive packers that isolate a number of sections in each borehole, turning them into separate chambers which are linked to the gallery by tubes for the purpose of making the measurement and injecting the tracers.

Alongside of this basic program, all the "small-scale" measurements, which can be easily managed, are also made in order to find out how one can gain access to the "large-scale" properties by making "small-scale" observations. The theoretical tool used for this purpose is the stochastic description of the medium. To this end the following measurements are made:

- the geometric properties of the fracture network based on structural observations in the gallery and on the oriented core samples obtained by drilling the radial holes from the gallery,
- the state of stress of the formation, provided by stress measurements using flat jack, overcoring or, possibly, hydraulic fracturing techniques, and
- local permeabilities determined by water injection tests in the same radial exploration holes originating in the gallery.

III. EXPERIMENT

A. Experimental Site

The underground laboratory is located in a gallery of a working uranium mine belonging to the COGEMA of the CEA group. This mine is in the granitic formation of Saint-Sylvestre near the city of Limoges in the western part of the Massif Central, France. It is a leucogranite with two micas, with medium to large grains, and from the namurian period, i.e., around 320 million years old. This formation is intrusive into metamorphic rock formations, and is approximately 20 x 30 km in size. The massif is crossed by large N-W faults and it is in these faults that the uranium is found. This mine is well-documented through the work connected with the mining operation but also through studies already done on the structure and hydrogeology by the BRGM and the Paris School of Mines for the DGRST (General Delegation for Scientific and Technical Research) and on the fractures for the CEA-IPSN. The zone which was finally chosen for the experiment is around 100-m long and lies in a gallery situated at the level called 320, but actually around 170-m below the soil surface, which is very irregular in this area (Figure 1).

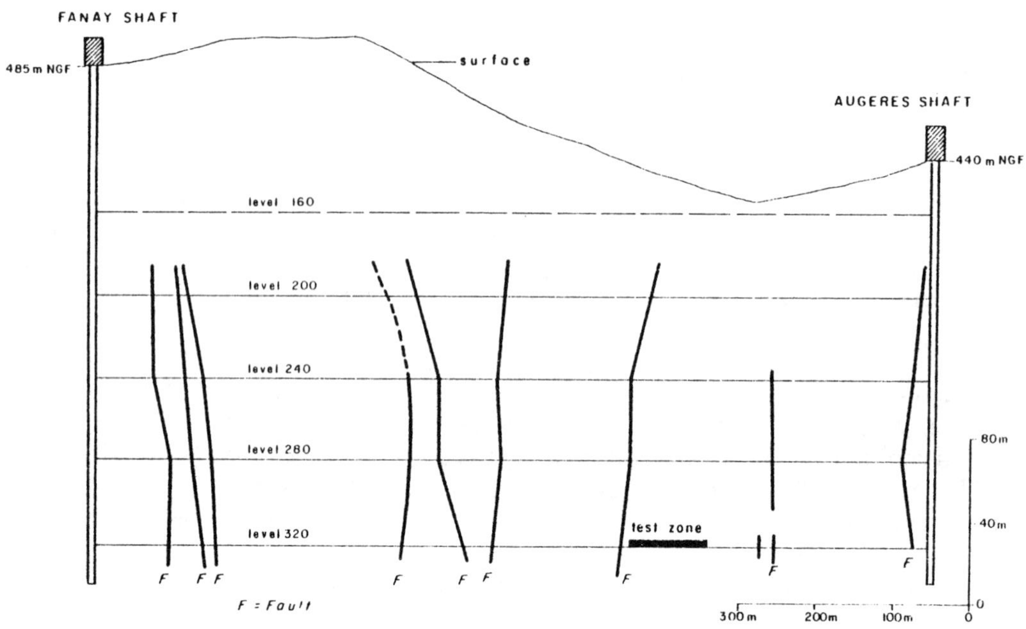

FIGURE 1. North-South Cross-Section of the Mine

B. Equipment of the Experimental Gallery

The onsite installations were made after an initial phase of site investigation including the collecting of structural data, core samples and measurements of pressure and flow rate. Cores were extracted from 10 exploration drillings, each 50-m long, in radial directions from the experimental gallery. The boreholes were distributed over three sections: a central one with four holes and two lateral ones with three holes each (Figures 2 and 3).

Basically, each borehole is equipped with seven packers, creating seven individual measurement chambers (Figure 4), the exceptions being boreholes F8 and F9 which only have six of a somewhat different length. Each of the latter are linked to the gallery by two rilsan tubes, which, through a complex network of hydraulic connections, make it possible to purge the system, measure the pressure and inject tracers. When the multiple packer devices are put in place and the pressure increase in the 68 measurement chambers monitored, the stabilized pressures and the flow rates are measured.

The gallery has, moreover, been fitted out in such a way that the water coming into the experimental section can be collected and its flow rate measured (Figure 5). For this purpose, small dams have been built above and below the experimental section as well as in between, and the measured global flow rate is close to 5 liters per minute.

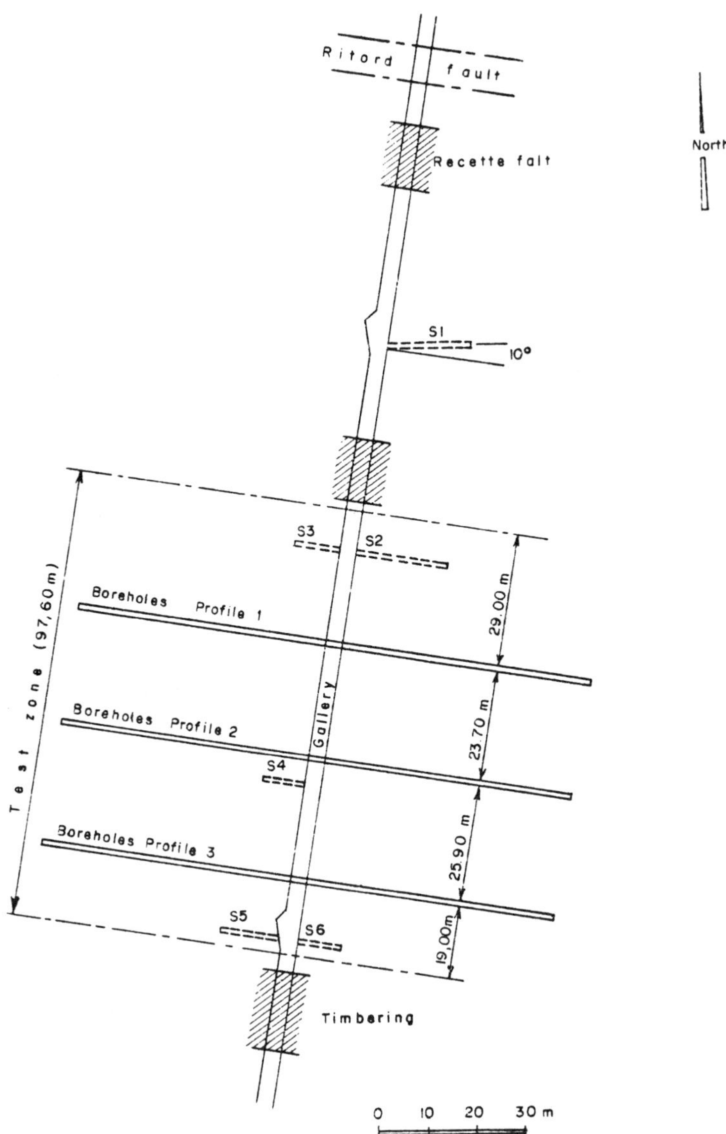

FIGURE 2. Plan View of the Test Zone Fanay-Augeres Mine, Level 320 (175 m below ground)

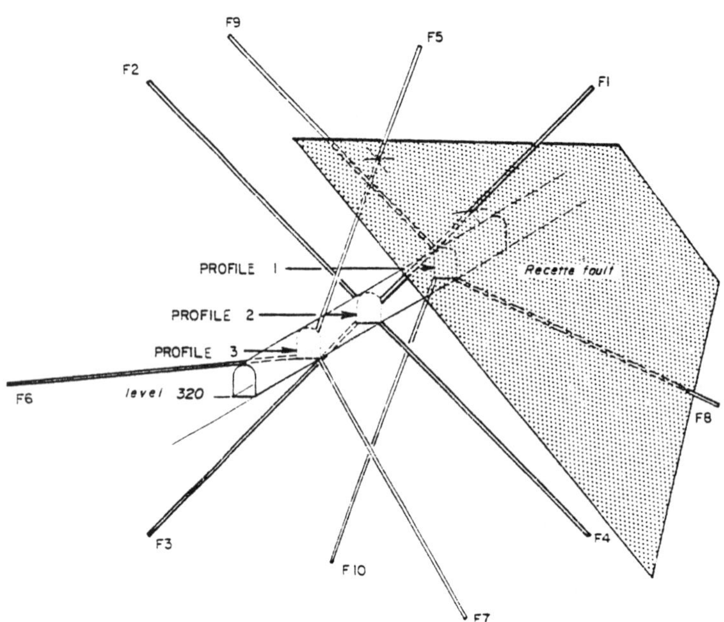

FIGURE 3. Perspective View of the Experimental Gallery at Fanay-Augeres and of the Three Sections of the Radial Exploration Borehole

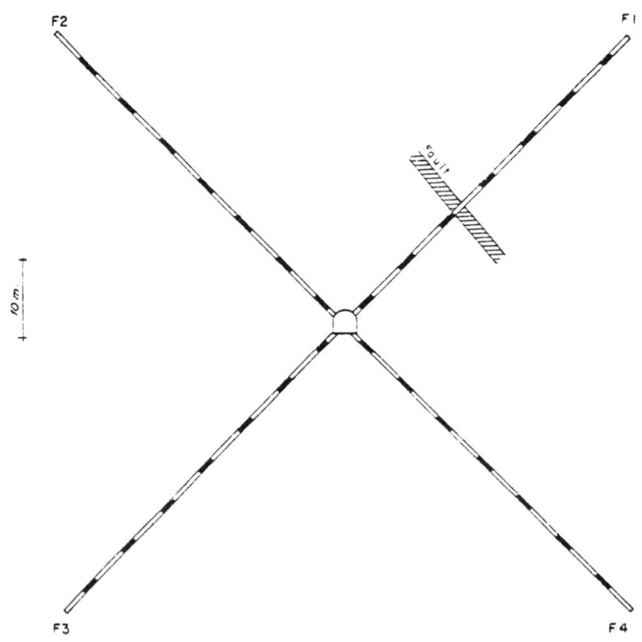

FIGURE 4. Diagram of the Location of the Packers and the Measurements Chambers

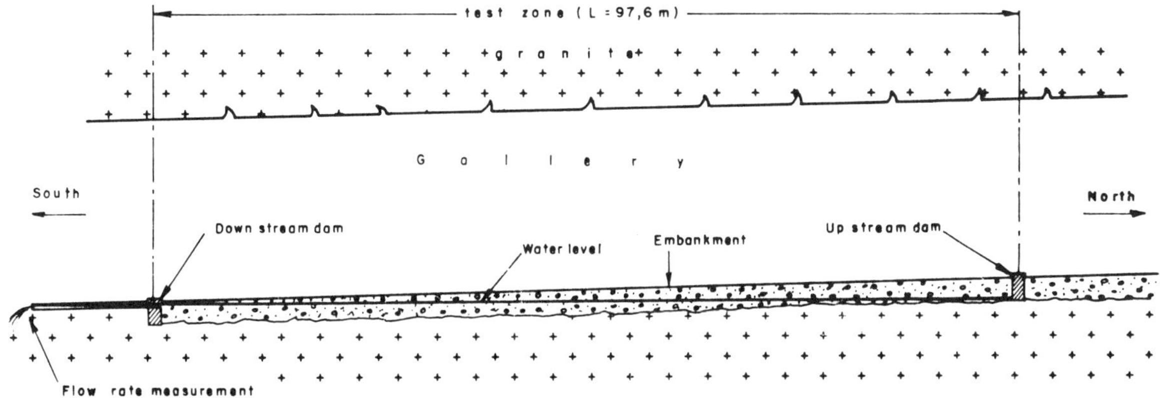

FIGURE 5. Longitudinal Cross-Section Along the Axis of the Experimental Gallery

C. Structural and Hydraulic Measurements and Results

The detailed structural examination consisted of a systematic recording of all the discontinuities on the eastern face of the gallery up to a height of 2 m. Several families of fractures were identified. A statistical study made on more than 1,000 fractures shows that, on the average, the fractures are fairly short (only 6% exceed 3.5 m). Two thirds of them do not have any noticeable aperture, 44% seem moist on observation.

The structural characteristics of all the discontinuities encountered by the exploration shaft were studied on core samples and their spatial orientation determined partly by a core orientator and partly by taking prints of the boreholes. The various data gathered in these different ways (orientation, distance between fractures,...) were treated statistically. The average fracture density is high: around 6 fractures per linear meter in the large-grain granite and 10 in the fine-grain one (see example, Figure 6).

The study of the inflow of water into the exploration wells was done with the help of a single packer moved in successive, 2 m steps and showed that the inflow points were generally localized in certain well-defined areas. The permeability was determined by injection of water either behind a single packer or between double packers. Three lengths were chosen for the chambers: around 50 m, 10 m and 2 m (or 2.5 m).

The permeability values obtained by these tests range from 1×10^{-5} to 3×10^{-9} m/s for the 50 m tests, from 8×10^{-4} and a value lower than 2×10^{-10} m/s (which corresponds to the lower limit of the capacity of the equipment) for the 10 m tests and from 5×10^{-5} to a value lower than 6×10^{-10} m/s for the 2 m and 2.5 m tests. Figure 7 shows the results obtained from the F10 borehole in the form of a permeability log. The formation presents a strong permeability heterogeneity, which suggests localized outflows.

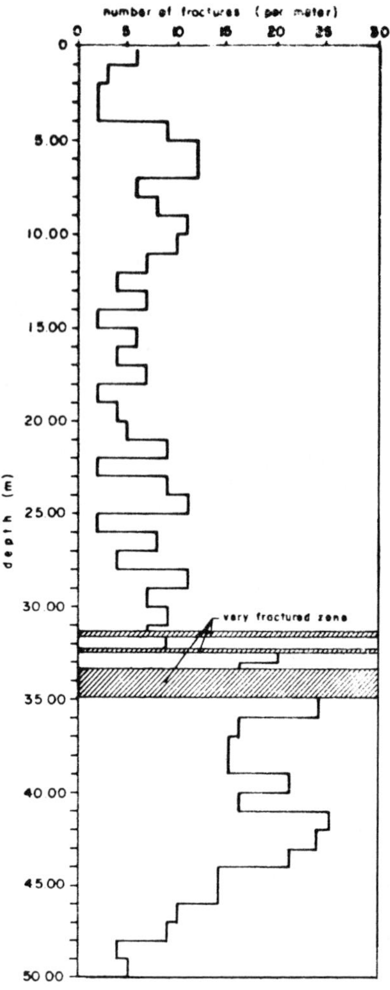

FIGURE 6. Fracture Density Along the F9 Borehole

From a structural point of view, there is no direct relation between the fracture density in the formation and the measured permeability. In particular, it should be noted that the most highly fractured zones do not have the highest permeability. The main factors governing the permeability are the degree of aperture of the fractures and of their interconnection.

Concerning the pressure field, it had not yet reached in July 1985 the same state as before the work started (June 1984) and one notices a strong dissymmetry between the east and the west sides of the gallery.

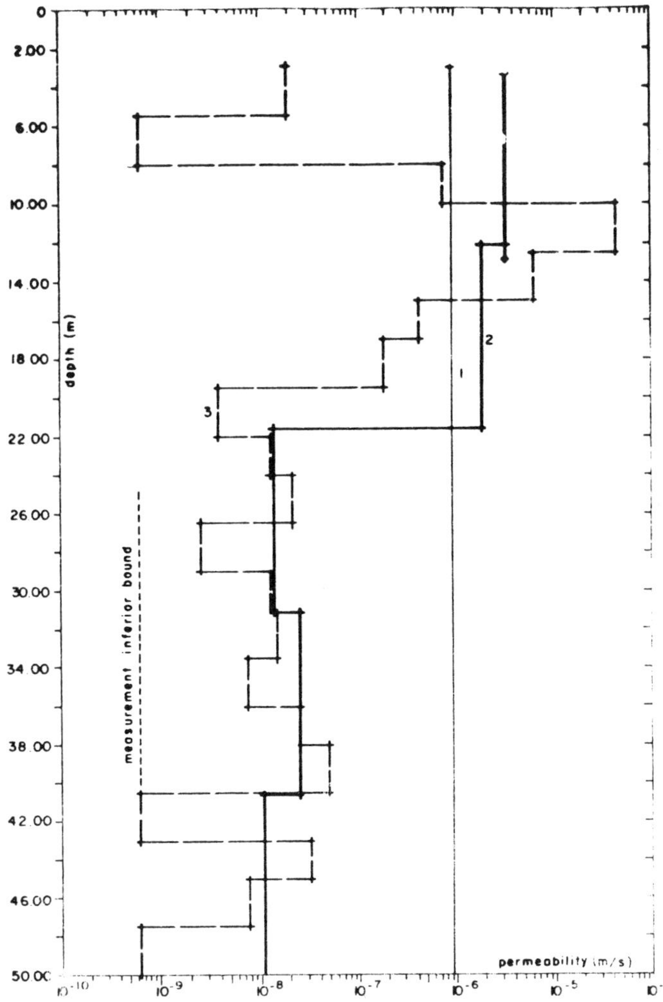

FIGURE 7. Permeability Log of Borehole F4 (curve 1 - global measurements; curve 2 - measurements on around 10 m chambers; curve 3 - measurements on 2 or 2.5 m chambers

IV. INTERPRETATION AND MODELING

A preliminary interpretation of the pressure distribution around the gallery was made using pressure measurements collected in an early phase of the program, using only 6 ten to twenty-meter-long horeholes with a single pressure measurement in the bottom of the hole. A 2-D finite element model was used to represent a single cross-section of the system, orthogonal to the gallery, and 300 x 800 m in size. The various drifts of the mine, parallel to the studied gallery, were represented, as well as the estimated position of the water table. Although all measurements could not be simultaneously interpreted, the general pattern of flow was reproduced (head distribution and flow rate) using an average constant isotropic hydraulic conductivity of

10^{-8} m/s (Figure 8). Attempts to use an anisotropic conductivity were not conclusive, for lack of pressure data.

At present, a similar type of interpretation is underway for the present pressure measurements, when a steady-state regime will be obtained. At first, each of the three sections orthogonal to the gallery where the boreholes have been located will be interpreted independently on a 2-D vertical model. Both an isotropic and anisotropic distribution of hydraulic conductivity will be used trying to fit, in the later case, both the orientation of the principal directions of the anisotropy tensor, and the magnitude of the anisotropy ratio. In a later stage, all three sections will be included in a 3-D model, if their difference in behavior is significant. The 3-D anisotropy tensor could then be analysed.

The next stage of the interpretation will be to relate the global hydraulic conductivity obtained from these pressure measurements to the geometric and hydraulic properties of the fractures. Each fracture will be represented by the coordinates (in 2 or 3 D) of its center; these are either observed on the walls of the gallery, along the cored holes, or generated randomly according

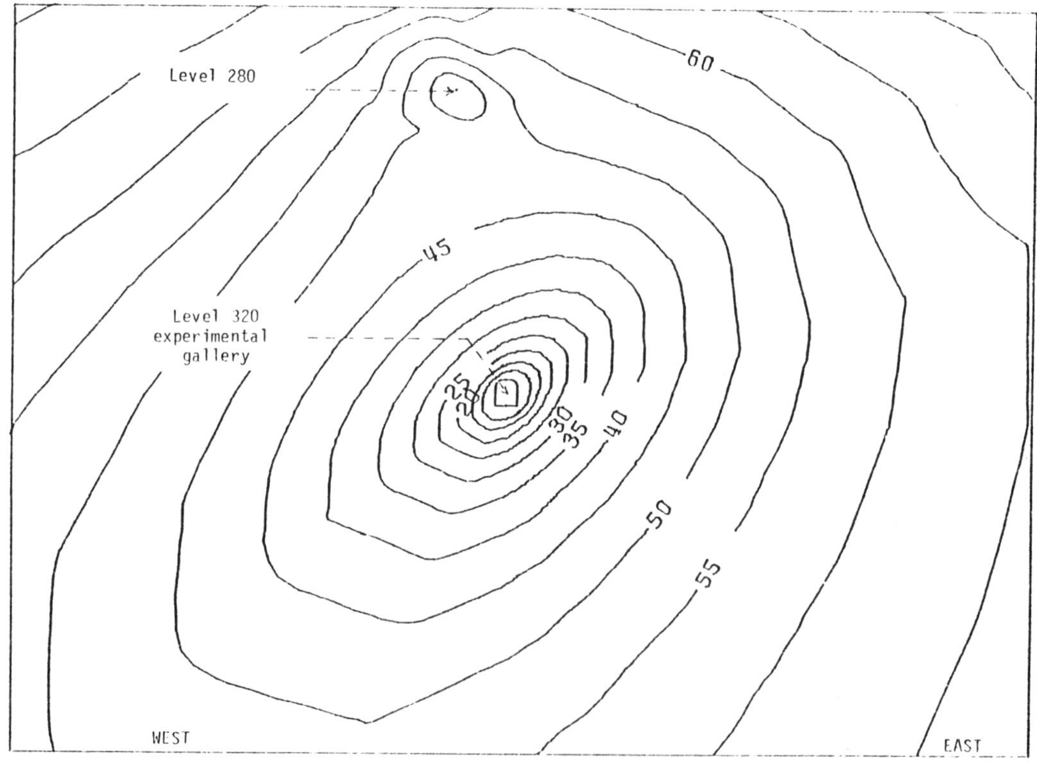

FIGURE 8. Calculated Heads (m) in the Vicinity of the Test Gallery

to the spatial probability distribution (determined from the observations) for each family of fractures. The shape of each fracture is assumed to be a circular disk, or a parallelogram; the orientation and extension of each fracture is generated randomly as in Long,[2] Robinson,[5] Rouleau,[12] or taking into account the spatial variability of the properties of the fractures, as in Long and Billaux.[10] Conditional simulations of the fracture network (conditioned by the observations, as in Anderson)[13] will also be considered.

From the geometrical description of the fracture system, a "connectivity" matrix will be determined. If two fractures intersect, their "centers" are connected; a non-zero entry will represent the direct link between the two nodes of the system. A zero value will be used otherwise.

The next step is to quantify the hydraulic resistance to flow between two nodes of the network. We intend to use here a different approach than what has been done so far; we believe that trying to solve the exact flow equations within each fracture plane is both too complex and unrealistic, since many totally arbitrary assumptions have to be made to correctly represent the flow: parallel plane, specified aperture, geometric shape of fracture (e.g., disks...), absence of fracture filling... We will therefore use here a Bayesian approach to determine the hydraulic resistance to flow between nodes in the connectivity matrix. The first step will be to produce an "a priori" estimate of this resistivity based on the ensemble of descriptive properties which have been collected on the fractures by observation on the wall of the gallery or on the cores of the boreholes. These are defined as averages for each family of fractures and are: (i) type of element (diaclase, fracture, fault, dyke, fractured zone); (ii) aperture; (iii) fracture filling; (iv) presence of water; (v) decompression; (vi) rugosity. An empirical relationship between the hydraulic resistivity and these parameters for each fracture plane will be established, based also on the length of the connection between two nodes in each fracture plane. The two resistivities forming the connection between two nodes (one for each fracture plane) are, of course, put in series to determine the unique resistivity between two nodes in the connectivity matrix.

The empirical relationship between fracture properties for each family of fractures and hydraulic resistivity will be based on a limited number of linear coefficients weighing the importance of each factor. The next step in the analysis will be to calibrate these coefficients on the hydraulic measurements, both the large scale permeability test and the small scale permeability tests by injection between packers. To do so, each steady state test will be defined by its boundary conditions, and the flow problem will be solved by inverting the "connectivity" matrix. This assumes that the flow problem is simply defined by applying Kirchoff's law at each node, i.e., the principle of mass balance in each fracture plane. Note in this respect that the precise determination of what is really the "center" or "node" of each fracture is unimportant. Since the flow equations are assumed linear, and since the resistivities between nodes are only parameters, any location within the fracture plane can be selected as a node.

A calibration criterion will be defined, e.g., based on the differences between calculated and observed pressures (for the large scale test) or calculated and observed flow rates (for the injection tests). The calibration will consist in minimizing the criterion with respect to the unknowns of the empirical relationship between hydraulic resistivity and fracture properties. A standard optimisation package will be used for that purpose, e.g., the code GRG. Given that a very limited number of unknowns will be determined in this way (4 or 5 per family of fractures) the computational time is expected to remain reasonable.

The calibration criterion being based on a large number of measurements, the objective of this approach is to determine an "average" behavior of each family of fracture, and thus the average large scale flow properties of the medium. But the use of the connectivity matrix, which is totally based on the geometric description of the fracture network, and which can be conditionned on small scale observations (wall of gallery, fractures of cores...), make it possible to also use local hydraulic measurements, like the injection tests between packers, for calibration in the same model. It represents both the large scale and the small scale features of fractured media, and nowhere makes any assumptions on the existence of a continuous equivalent medium or a representative elementary volume. Once it has been calibrated, it can be used on a predictive mode to define average properties (to be used on another continuous equivalent model) and the local variability of the flow in local fracture systems.

The expected proof of the validity of this approach will be that the model calibrated independently on the large scale pressure measurements and then on the small scale injection tests will yield the same results. Although preliminary tests have been started, this work is yet to be completed. It is worthwhile noting that this type of approach can easily be extended to include any new type of measurements which can be made on fractured systems, e.g., geophysical measurements, stress field, new tests, new parameters recorded, etc.

V. FURTHER WORK: STUDY OF THE INFLUENCE OF THE SCALE EFFECT ON THE DISPERSION

As indicated earlier, tracer tests will be carried out by injecting various tracers in several of the 58 injection chambers isolated in the ten boreholes of the test zone. The sampling of the water will be made in the flowing water arriving in the gallery, and not inside the formation. In a first phase, only conservative tracers will be used, but later on sorbing tracers or even colloids may be used.

The essence of the interpretation of this experiment will be very similar to what is planned for flow. Given the calibrated flow model, only a longitudinal dispersion mechanism will be determined, for each family of fractures. From the flow model, the flow path from any injection point will be determined, i.e., the proportion of the flow between each of the nodes of the connectivity matrix, from the injection point to the walls of the gallery. Between each connected node of the system, the flowpath will be decomposed into its two components, one in each fracture plane of the link. Inside one given plane, the transfer of solute will be represented by an empirical analytical transfer function, depending (through a limited number of linear coefficients) on the principal characteristics of each family of fractures. Between two nodes, the transfer of solute, for a pulse injection of tracer, will be given by the convolution of the two transfer functions for each of the two fractures which are interconnected. Between a series of interconnected nodes, the transfer is again given by the convolution of the transfer functions between two nodes. Finally, the arrival at the boundary (the walls of the gallery) is given by the weighted sum of all the transfer functions between the injection point and nodes at the gallery, all flowpaths being considered, the weights being the relative flow rates (from the flow model) along each flowpath. The calibration of the unknowns (the linear coefficients of the transfer functions, for each family of fractures) will be made by minimizing a criterion based on observed and calculated concentrations, for a number of tracer tests. Similarly, the validation of this approach will be obtained if it is possible to calibrate the model with a first set of tracer tests, and obtain the same degree of calibration for another set, independently of the distance from the gallery at which the injection was made. If this validation proves successful, the problem of the scale effect in fractured media will perhaps be better understood.

VI. CONCLUSION

This series of experiments under way at Fanay-Augeres is aimed at providing both a methodology for measuring the relevant properties of fractured media (for flow and transport problems), and a new method for numerically representing these media. The first phase of data collection on the flow problem is now almost complete, and the second phase on the transport problem should start in the very near future. The theoretical basis for the interpretation of these experiments has been defined in general terms, and the particular models which will be used are now under development. The chosen approach is based on a stochastic description of the geometry of the fracture system, as suggested by previous workers, but only to establish the connectivity of the network. The quantification of flow in this network is based on a Bayesian approach, defining a priori the resistance to flow in each branch of the network from all available observations or measurements on individual fractures, summarized into average properties for each family of fractures. The a posteriori quantification of the network is obtained by calibration of a very limited number of coefficients per family of fractures, calculated from the simultaneous fitting of a large number of hydraulic observations (pressure or flowrate). This spatial averaging of local measurements appears to be the only hope to use local measurements on a random network, by applying the principle of ergodicity and thus avoiding the need to use a large number of realisations of the intial random network.

The transport problem is treated in a very similar way, making use of the linearity of the transport equations to simply represent transport by a series of convolutions of prescribed linear functions through each flowpath of the system.

Further experimental research at Fanay-Augeres will include the use of nonconservative tracers in fractured systems, and later a coupled thermomechanico-hydrological experiment at the scale of 10 x 10 x 10 m.

REFERENCES

1. J. C. S. Long et al., "Porous Media Equivalents for Networks of Discontinuous Fractures," Water Resour. Res., 18, 3, p. 645-658 (1982).
2. J. C. S. Long, Investigation of Equivalent Porous Medium Permeability in Networks of Discontinuous Fractures, Ph.D thesis, Lawrence Berkeley Laboratory, Univ. of California (1983).
3. F. W. Schwartz, L. Smith, A. S. Crowe, "A Stochastic Analysis of Macroscopic Dispersion in Fractured Media," Water Resour. Res., 19, 5, p. 1253-1265 (1983).
4. R. Engelman, Y. Gur, Z. Jaeger, "Fluid Flow Through a Crack Network in Rocks," J. Appl. Mech., 50, P. 707-711 (1983).
5. P. C. Robinson, Connectivity, Flow and Transport in Network Models of Fractured Media, Ph.D. thesis, Oxford, Theoretical Physics Division, AERE Harwell TP 1072 (1984).
6. E. Guyon, J. P. Hulin, R. Lenormand, "Application de la Percolation a la Physique des Milieux Poreux," Annales des Mines, n° 5-6, p. 17-40 (1984).
7. E. Charlaix, E. Guyon, N. Rivier, "A Criterion for Percolation Threshold in a Random Array of Plates," Solid State Comm., 20, II, p. 999-1002 (1984).
8. S. Wilke, E. Guyon, G. de Marsily, "Water Penetration Through Fractured Rock: Tests of a Tridimensional Percolation Description," Mathem. Geology, 17, 1, p. 17-27 (1985).
9. G. de Marsily, "Flow and Transport in Fractured Rocks: Connectivity and Scale Effect," IAH International Symposium on the Hydrogeology of Rocks of Low Permeability, Tucson, USA, Jan 7-12 (1985).
10. J. C. S. Long, D. Billaux, "La generation aleatoire de champs de fractures: prise en compte de la variabilite spatiale de certains parametres," Symposium sur l'approche stochastique des ecoulements souterrains, AIRH, Montvillargenne (1985).

11. C. R. Wilson et al., "Large Scale Hydraulic Conductivity Measurements in Fractured Granite," Int. J. Rock Mech. Min. Sci. Geomech. Abst., 10, 6, p. 269-276 (1983).

12. A. Rouleau, Statistical Characterization and Numerical Simulation of a Fractured System. Application to a Groundwater Flow in the Stripa Granite, Ph.D. thesis, Univ. of Waterloo (1984).

13. J. Anderson, "Predicting Mass Transport in Fractured Rock With the Aid of Geometrical Field Data," Symposium international sur l'approche stochastique des ecoulements souterrains, AIRH, Montvillargenne (1985).

14. E. Durand et al., Etude de l'effet d'echelle en milieu fissure. Phase pilote: certification du site de Fanay-Augeres, Rapport CEA-BRGM-ENSMP, BRGM 84 SGN 237 STO (1984).

THE CHARACTERIZATION AND DISCRIMINATION OF FRACTURE ENVIRONMENTS WITHIN THE EYE-DASHWA PLUTON, ATIKOKAN, ONTARIO

P. A. Brown
Geological Survey of Canada
601 Booth Street
Ottowa, Ontario, K1A 0E7
Canada

N. A. C. Rey
Applied Geoscience Branch
Atomic Energy of Canada, Ltd
Whiteshell Nuclear Research Establishment
Pinawa, Manitoba, R0E 1L0
Canada

ABSTRACT

An integrated analysis of three scales of geological and geophysical data is used to identify, substantiate and characterize fracture environments within the Eye-Dashwa Pluton. Regional-scale data is used to identify the major structural discontinuities. Local-scale data substantiates these major structures and locates smaller scale features. Detailed-scale data rigorously identifies the location of fault-fracture zones and allows their fracture characteristics to be defined. Comparison of fracture orientation and fracture frequency of surface data and subsurface data, from a 1-km-deep borehole, indicates that the fracture characteristics defined from detailed surface data are compatible with fracture characteristics defined from subsurface data. Thus, the characteristics of subsurface fracture domains can be extrapolated from surface fracture data.

I. INTRODUCTION

Within plutonic rock environments the most important groundwater pathways are related to faults and fractures. Thus, the identification of these structures is of considerable importance to the Nuclear Fuel Waste Management Program. In this paper we outline a methodology for identifying and discriminating the potential flow paths within the Eye-Dashwa Pluton near Atikokan, Ontario and describe an analysis to determine if the surface characteristics can be extrapolated to depth.

Three scales of data are considered (Figure 1):

1. regional-scale data encompassing the pluton -- to identify the major structural discontinuities,

2. local-scale data covering a 4 km^2 region of the pluton -- to more precisely identify specific fracture environments, and

3. detailed-scale data incorporating both surface and borehole data -- to assess whether surface characteristics can be extrapolated to depth.

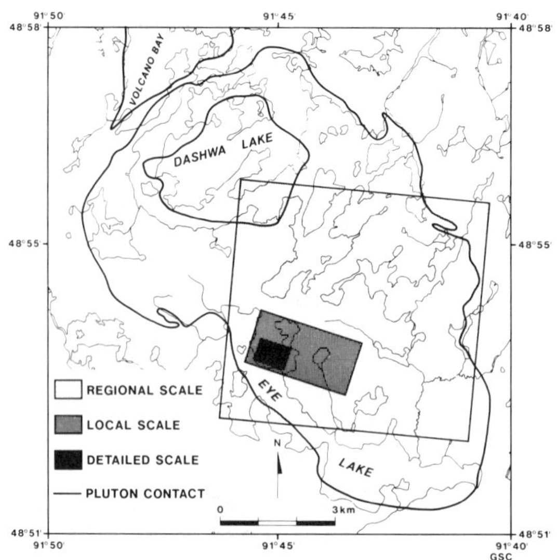

FIGURE 1. Location and Extent of Regional-, Local-, and Detailed-Scale Data Within the Eye-Dashwa Pluton.

II. IDENTIFICATION OF MAJOR STRUCTURAL DISCONTINUITIES USING REGIONAL-SCALE DATA

Regional data covering the Eye-Dashwa granitic pluton includes: (1) aerial photograph lineament analysis (at a scale of 1:50,000), (2) total field magnetic survey, (3) filtered total field magnetic data, (4) aeromagnetic gradiometer survey, (5) enhanced, filtered aeromagnetic data, and (6) outcrop fracture data. Unfortunately, none of these surveys yields an exact identification of discontinuities in the rock mass. Rather, each identifies certain anomalous conditions within the rock (or overburden) that often, but not always, correlate with structural discontinuities. For example, the majority of lineaments identified on aerial photographs are valleys, and hence, represent linear zones of enhanced erosion. As such, they are commonly interpreted as major discontinuities in the rock mass[1,2] on the assumption that erosion is enhanced along the surface trace of the discontinuity. However, other bedrock features such as geological contacts, variation in intensity of rock fabric, or degree of alteration may also contribute to enhanced erosion and thereby potentially form lineaments if they are aligned. Alternatively, major discontinuities may exist in the rock that do not exhibit enhanced erosion at the surface.

Similarly, negative linear anomalies defined by magnetic data may result from magnetite variation (often lithologically controlled) within the rock mass, or from variation in thickness or composition of overburden. Lapointe et al. (in press)[3] have shown that fracture zones within the Eye-Dashwa granite are strongly correlated with reduced magnetic susceptibility values. Thus, on a larger scale, linear magnetic anomalies may be correlated with fracture zones.

Even geological mapping, which has the capability of providing exact identification of discontinuities by direct observation, is restricted due to poor outcrop. The major discontinuities can be inferred from geological data only by comparison of fracture characteristics in adjacent outcrops and by searching for such things as linear zones of abundant fractures that may be associated with a major fault.[2,4,5,6] However, other factors such as lithological variations and bed thickness (in sedimentary sequences) may also affect fracture density.[5,7] Furthermore, Pohn[7] has noted a decrease in the frequency of some fractures adjacent to major structural discontinuities. Thus, the geological data must also be interpreted with caution.

The regional-scale data therefore provides, at best, a number of inferences concerning the occurrence and location of major structural discontinuities within the pluton. To assess and evaluate these inferences we established three classes of lineament:

Class 1: aerial photograph lineaments superimposed upon magnetic anomalies,

Class 2: aerial photograph lineaments showing only partial correlation with magnetic anomalies, and

Class 3: aerial photograph lineaments with no associated magnetic anomaly.

A total of 70 lineaments were defined with a 40-km^2 area of the pluton (Figure 2). Only 14 show a good correlation with magnetic anomalies (class 1), 35 show partial correlation (class 2), and 21 which were identified from aerial photographs, show no correspondence with magnetic anomalies (class 3). Class 1 lineaments should be more likely to represent large-scale structural discontinuities than class 3 lineaments. This premise is tested using the independently derived outcrop fracture data.

The outcrop fracture stations within 0.25 km of the surface trace of each lineament were identified, and the associated fracture frequency values defined as the 'lineament sub-sample'. Each lineament was then tested, using the outcrop fracture data, on the following premise: if a lineament represents a major structural discontinuity then the frequency of fractures, measured from outcrop in the vicinity of that lineament, should be significantly higher than the general level of fracture frequency throughout the pluton.

The performance testing is based on the nonparametric Mann-Whitney 'U' test with the significance of 'U' being calculated in the form of a Z score.[8] The random exceedence probability of this score can be quantified in relation

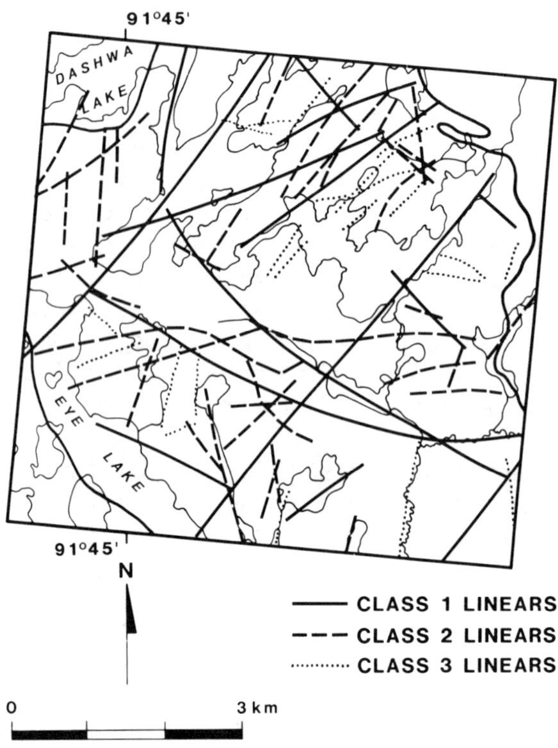

FIGURE 2. Location of Class 1, Class 2, and Class 3 Lineaments Within the Regional Area of the Eye-Dashwa Pluton

to the standard normal distribution, and expressed as a 'per cent level of confidence' (defined as 1 minus the exceedence probability, times 100).

This testing showed that (Figure 3) a total of 32 lineaments have higher than background fracture frequency at the 95% confidence level. A further 14 lineaments are associated with increased fracture frequency at the 80%-95% confidence level, and 24 lineaments show no associated increase in fracture frequency. Additional iterations failed to identify additional significant lineaments.

These two assessments, i.e. combined aerial photograph and magnetic anomaly, and fracture frequency derived from outcrop data, are completely independent. If the inferences regarding these lineaments are reasonably correct then there should be a good correlation between the two assessments. A standard Chi-square contingency table (Table 1) indicates that a correlation exists. Class 1 lineaments are preferentially associated with high frequency zones at the 95% confidence level. Conversely, class 3 lineaments

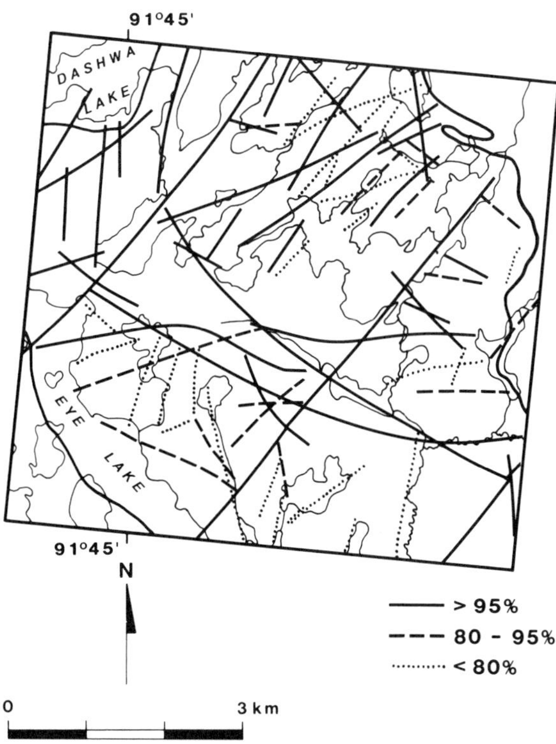

FIGURE 3. Location of Lineaments at 95%, 80%-95%, and Less Than 89% Confidence Within the Regional Area of the Eye-Dashwa Pluton

are preferentially associated with low frequency zones (less than 80% confidence). The random exceedence probability of this overall relationship (Q of chi-square) is 0.043 and is thus significant at the 95% confidence level.

This systematic correlation is a strong indicator that class 1 lineaments, which are associated with increased fracture frequency (at the 95% confidence level), have a high probability of representing major structural discontinuities within the rock mass and, as such, represent the dominant potential flow pathways (Figure 4). With decreasing class of lineament and decreasing fracture frequency confidence level there is increasing uncertainty in determining the nature of the underlying structure.

Additional, more detailed, data is required to:

1. define the characteristics of the major structural discontinuities and

2. evaluate the nature of the less significant structures.

TABLE 1. Chi-Square Contingency Table of Relationships Between the Two Independent Classifications of Regional Scale Lineaments.

LINEAMENT CLASS	CONFIDENCE LEVEL >95%	80 - 95%	<80%	ROW TOTAL
CLASS 1	(71.4) 10 (45.7)	(14.3) 2 (20.0)	(14.3) 2 (34.3)	14
CLASS 2	(48.6) 17 (45.7)	(22.9) 8 (20.0)	(28.5) 10 (34.3)	35
CLASS 3	(23.8) 5 (45.7)	(19.0) 4 (20.0)	(57.1) 12 (34.3)	21
COLUMN TOTAL	32	14	24	70

$Q(X^2) = 0.043$

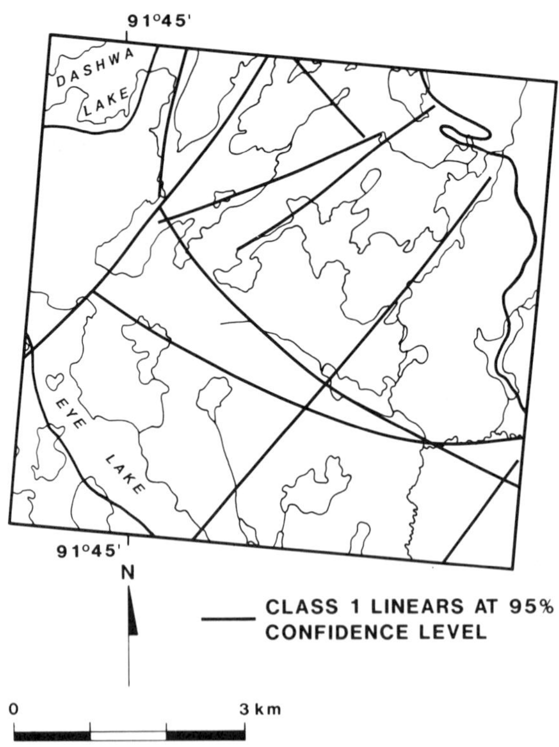

FIGURE 4. Location of Class 1 Linears Significant at 95% Confidence Within the Regional Area of Eye-Dashwa Pluton

III. IDENTIFICATION OF STRUCTURAL DISCONTINUITIES USING LOCAL-SCALE DATA

A 4 km^2 region of the pluton, which is bounded by, but does not contain, major structural discontinuities (as defined by the regional scale data), was selected for more detailed study.[9] Data include aerial photograph lineament analysis (at a scale of 1:17,000), and ground magnetic survey run on cut lines spaced at 100 m. Also run on the cut lines was a VLF-EM (very low frequency electromagnetic) survey. VLF-EM is sensitive to electrical conductors. In granitic plutons, linear anomalies are likely to correspond to zones of wet overburden associated with faults and large fracture zones. A combination of these data led to the identification of 45 linear anomalies, which may represent the prominent structural discontinuities within the area (Figure 5).

Additional outcrop fracture data were collected throughout the area and each lineament was tested to determine if it represents a zone of increased fracture frequency. A total of 8 lineaments showing higher than background fracture frequency (at the 95% confidence level) were defined. A further 15 lineaments showed increased values at the 80% to 95% confidence level, and 22 lineaments could not be distinguished from background values.

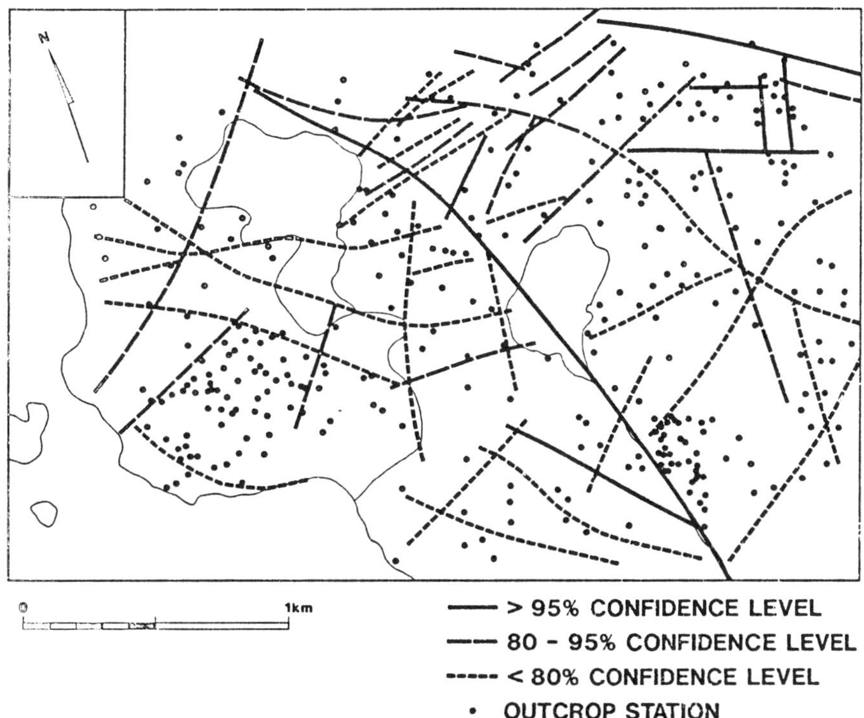

FIGURE 5. Local-Scale Area Showing Location of Outcrops and Lineament Confidence Level

These results suggest that either there are very few significant structural discontinuities in the area or the data (or data analysis) is deficient. The latter would appear to be more likely because:

1. the lineaments occupy, at this scale, narrow steep-sides valleys such that the available outcrop tends to occur on the wall rocks, i.e., the lineament core is rarely sampled at this scale and

2. the style of fracturing of small shear faults within the Eye-Dashwa granite is such that the wall rock adjacent to the central portion of small faults contains very few fractures.[10]

Thus, because of sampling problems and the style of fracturing, an analysis of fracture frequency is not particularly diagnostic at this scale of investigation. Fortunately, increased fracture frequency is not the only distinctive feature of small faults within the pluton. Small faults and fracture zones are characterized by increased alteration of the wall rock,[11,12,13] and increased abundance of fracture-filling materials.[11] Furthermore, recent data suggest that many of the fracture orientations adjacent to defined strike-slip faults conform to those predicted by a model of simple shear.[10] All of these parameters are presently being investigated to develop a rigorous method for identifying structural discontinuities at this scale of investigation.

IV. DETAILED DATA AND EXTRAPOLATION INTO THE SUBSURFACE

An area of 0.5 km^2 was selected for detailed work (Figure 1). Within this area additional outcrop fracture data, a ground magnetic survey and ground VLF-EM surveys were used to identify potential fault-fracture zones. The boundary faults (to the NW, NE and E) were defined by superimposed magnetic, VLF-EM and fracture anomalies (Figure 6). The smaller faults within the fault bounded block were identified by the geophysical surveys. To the south and southwest the block is bounded by a lithological contact between monzodiorite, which represents a marginal phase, and the central mass of the granitic pluton. The contact is a gradational zone with numerous mafic inclusions and mafic dykes.

Five fracture domains were defined:

1. northwest boundary fault,
2. northeast boundary fault,
3. eastern boundary fault,
4. contact zone, and
5. unfaulted 'intrablock' contained within these boundary conditions.

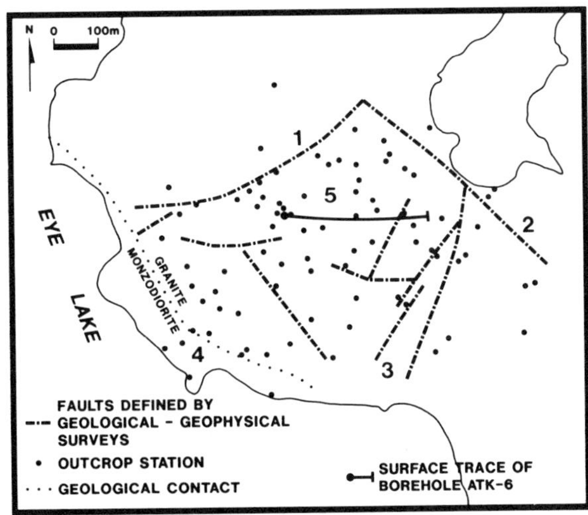

FIGURE 6. Detailed-Scale Area Showing Location of Faults Defined by Geological and Geophysical Surveys, Outcrop Stations, and Surface Trace of Borehole ATK-6. Numbers refer to text.

The orientation of fractures within each of these domains is shown in Figure 7. The three fault domains are dominated by vertical to subvertical fractures which, in each case, are within 20°-40° of the trend of the associated fault (Figures 7A, B, C). The contact zone is dominated by subhorizontal to moderate dipping fractures (Figure 7D) whilst the 'intrablock' contains both subhorizontal fractures and a prominant NW-SE trending subvertical set (Figure 7E).

The distribution of fracture frequency, within each environment, is shown in Table 2. The contact zone and intrablock have similar median and standard deviation values. The eastern and northwestern boundary faults are distinct and represent more highly fractured environments. The northeastern boundary fault is undersampled. Nonparametric, Kruskall-Wallis 'H'[14] testing of these sub populations indicates that both the northwestern and eastern boundary fault environments are significantly different at greater than 95% confidence.

Thus, the detailed geological and geophysical data has identified a small fault-bounded block of rock. The northwest and eastern boundary faults are characterized by significantly high fracture frequency values. The northeastern boundary fault locally shows high-fracture frequency but, in general, the observed distribution is no different from the 'intrablock' distribution. In terms of fracture orientation the bounding faults are characterized by subvertical fractures whose trend is within 20° to 40° of the trend of the fault. The 'intrablock' and contact zone both contain subhorizontal to moderate dipping fractures.

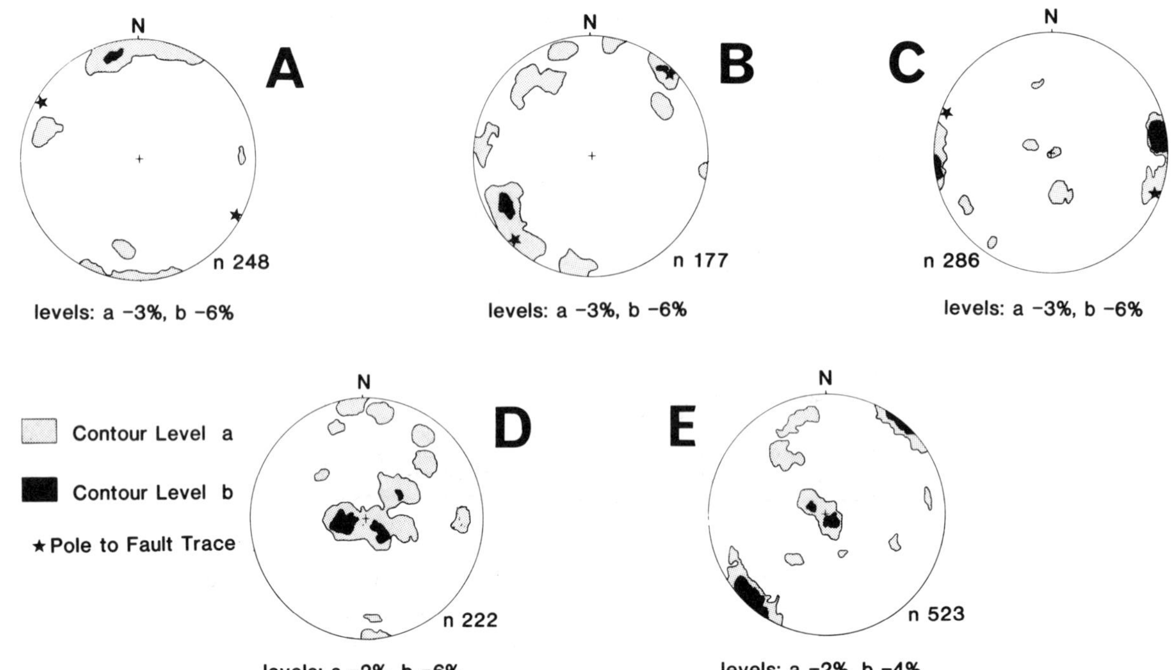

FIGURE 7. Orientation of Fractures Within the Five Fracture Domains Identified Within the Detailed Area. (a) Northwestern boundary fault; (b) northeastern boundary fault; (c) eastern boundary fault; (d) contact zone and (e) intrablock.

TABLE 2. Moments of Fracture Frequency Distribution Within the Five Fracture Domains Identified Within the Detailed Area

Environment	Arithmetic Median (fr/m)	Mean of Logs	Standard Deviation of Logs
Intrablock	2.73	1.435	2.246
Contact	2.61	1.485	1.352
E. Fault	4.14	1.617	1.921
NW Fault	6.26	1.797	3.006
NE Fault	3.16	1.500	6.152

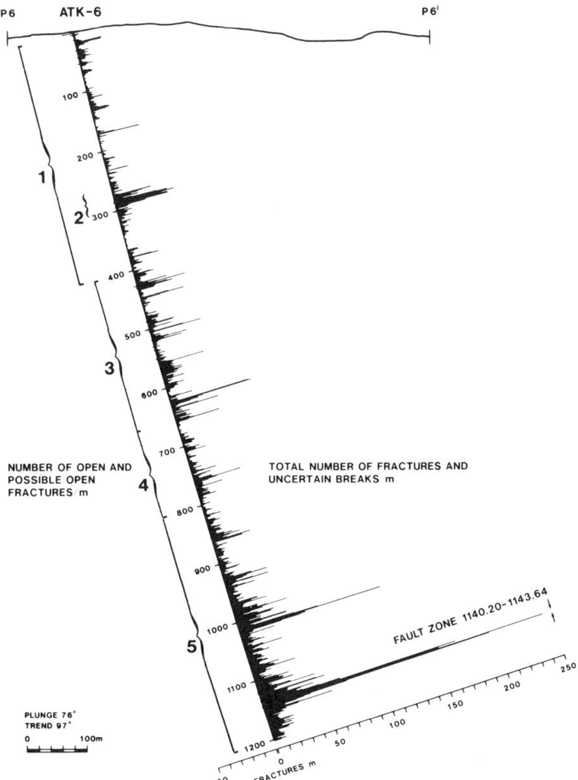

FIGURE 8. Fracture Frequency and Fracture Zones Identified in Borehole ATK-6: in all, a total of 9124 fractures were logged, of which 48 are open

A cored borehole (ATK No. 6) trending east-west and plunging 76°-72° to the east, was drilled to a vertical depth of 1 km (Figure 6) to assess whether the fracture environments, as defined at surface, could be extrapolated to depth. Results from the borehole are summarized in Figure 8 and indicate 5 distinct fracture environments:

1. few fractures from 0-400 m,

2. a fracture zone from 270 m - 300 m,

3. four small fracture zones from 400 m - 600 m,

4 a contact zone containing numerous mafic inclusions and dykes from 650 m - 800 m (below 750 m the rock type is predominantly monzodiorite), and

5. a highly fractured zone from 800 m - 1201 m containing a number of defined faults.

V. SURFACE-SUBSURFACE CORRELATIONS

The fracture orientation within each of the 5 environments defined in the borehole were compared with orientation data obtained from the fracture environments defined on surface. Comparison of the orientation of fractures within the 'intrablock' with the upper part of the borehole, i.e., 0-400 m (excluding the fracture zone at 270-300 m), indicates that the orientations in each domain are remarkably similar (Figure 9A).

The fracture zone between 270 m and 300 m is correlated with a small east-west trending fault which crops out 150 m south of the borehole and was identified by the VLF-EM surveys (see Figure 6). The fracture orientations are very similar (Figure 9B). Based on the similarity of fracture orientation the small fault at surface is tentatively correlated with the fracture zone at 270 m - 300 m.

The subsurface environment between 400 m and 600 m is characterized by four small fracture zones. A stereonet of the entire environment (Figure 9C) indicates that the dominant orientations in this segment of the borehole are subhorizontal and moderately dipping, with the moderate-dipping fractures trending ENE and dipping 20°-40° to the NNW. Fractures and fracture zones with this trend and dip were not observed within the detailed area and thus there is no direct correlation with this section of the borehole. However, a broad swamp-filled valley overlies the region where they should come to surface. The orientation of fractures in surface outcrops on the flanks of this valley (Figure 9C) show a remarkable similarity to the subsurface environment.

The contact zone, from 650 m to 800 m is directly comparable, lithologically, with the contact zone defined on surface. Both are mixed zones with abundant mafic inclusions and dykes. Spatial correlation of this environment indicates that the contact strikes 290°-300° and dips at 40°-60° to the northeast. Comparison of the surface and subsurface fracture orientations (Figure 9D) shows few similarities between the two data sets, and that there are no fracture sets parallel to the contact. It is suggested that the fracture environment within the contact zone cannot be correlated from surface to subsurface.

Below 800 m the borehole data indicates a highly faulted-fractured environment. Comparison of this with the surface data from the eastern boundary fault (Figure 9E) shows that the dominant fracture orientations are very similar and that there is a much greater scatter in the subsurface data. Furthermore, the dominant peak on surface is subvertical whereas in the subsurface it has a dip of 65° indicating that, if the two environments are correlated, then the fault is shallower at greater depths.

The northwestern and northeastern boundary faults showed no correlation, in terms of fracture orientation, with fracture zones intersected in the subsurface. It is concluded that these zones did not intersect the borehole.

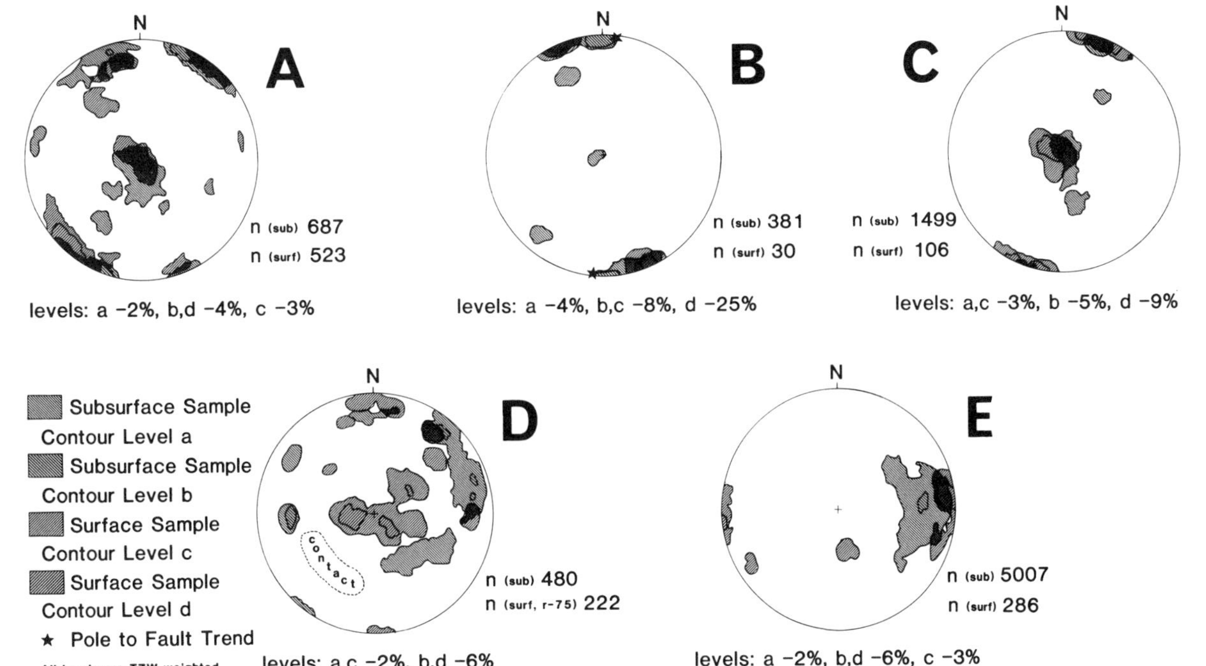

FIGURE 9. Comparison of Fracture Orientation, Surface-Subsurface, Within the Five Identified Fracture Domains; (a) intrablock is 0-400 m of borehole; (b) small E-W fault south of borehole is 270-300 m of borehole; (c) swamp-filled linear valley SE of detailed area is 400-650 m of borehole; (d) contact zone is 650-800 m of borehole; and (e) eastern boundary fault is 800-1201 m of borehole.

In terms of fracture frequency, a number of predictions were made before drilling the borehole.[15] The predictions were based on surface-subsurface relationships (defined from previously drilled boreholes,) which were extrapolated to the detailed area. Figure 10 shows the predicted distribution of fracture frequency within the top 150 m of the borehole and the actual distribution defined from the borehole data. The borehole data shows a bimodal distribution with each modal segment having a lower standard deviation than the predicted distribution. However, the actual and predicted median values are similar and, in general, the predicted distribution is a good approximation of the subsurface environment.

At greater depths, the frequency distribution was predicted to have a lower median value and slightly lower standard deviation than the near surface environment (Figure 11). Comparison of this predicted curve with the distribution of frequency values from 150 m to 400 m (excluding the fracture zone from 270 m to 300 m) shows, again, that the predicted distribution approximates the actual subsurface distribution (Figure 11). As in the case of the near surface data (0-150 m), the major difference between the two distributions is that the subsurface data shows a bimodal character. The median value of both populations is similar.

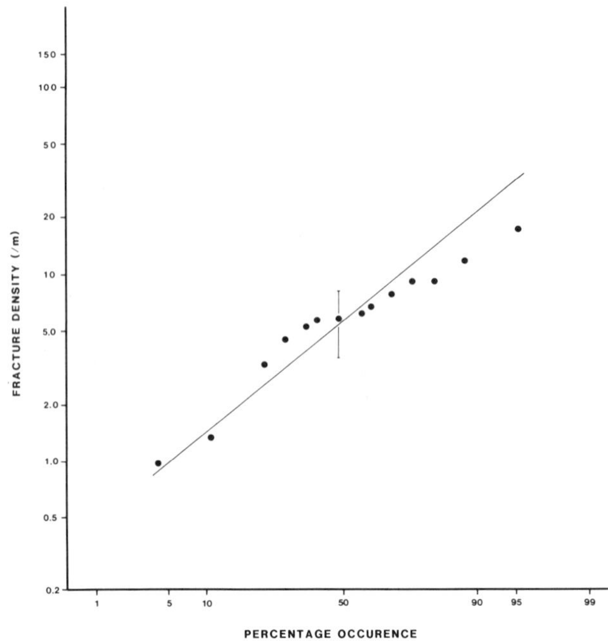

FIGURE 10. Distribution of Borehole Fracture Frequency Values from 0 to 150 m Versus Predicted Distribution Curve (solid line)

The fracture zone at 270 m to 300 m was not predicted. There is insufficient data on surface and in the subsurface to assess possible surface-subsurface relationships.

In addition to predictions regarding the intrablock, it was also suggested[15] that moderate dipping to sub-horizontal fracture zones may occur at depth, i.e. below 150 m. The predicted distribution, incorporating 30% moderate dipping to subhorizontal fracture zones is shown in Figure 12. Comparison of this predicted curve with the distribution of fracture frequency from 150 m to 650 m (excluding the fracture zone at 270 m - 300 m) indicates that the prediction is not as good as in the previous two cases. The subsurface data shows a bimodal character and the median value of the subsurface data is higher than predicted. This discrepancy in the median value may be attributed, at least in part, to the fact that the predicted value was computed assuming 30% moderate dipping fracture zones. The borehole section from 150 m to 650 m (excluding the fracture zone at 270-300 m) incorporates 4 moderate dipping fracture zones, which comprise approximately 40% of the total data.

The contact zone was expected to be subvertical and thus was not predicted to intersect the borehole. The eastern boundary fault was recognized before

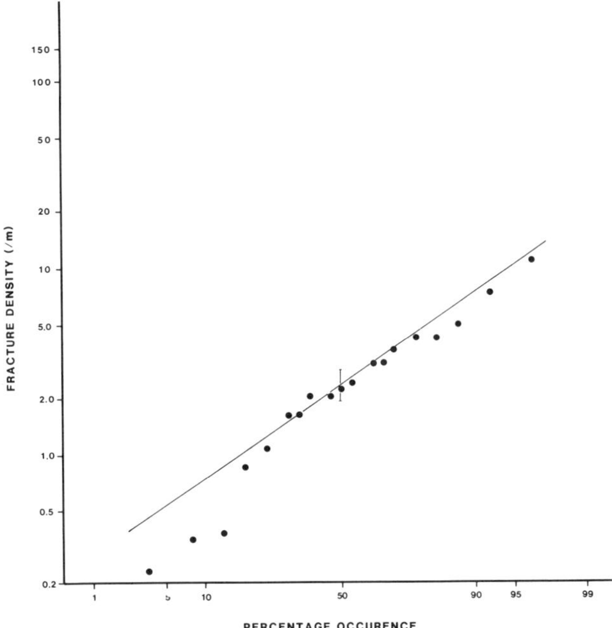

FIGURE 11. Distribution of Borehole Fracture Frequency Values from 150 m to 400 m (excluding the fracture zone at 270 m - 300 m) Versus Predicted Distribution Curve (solid line).

drilling and it was suggested (although not predicted) that it might intersect the borehole at approximately 800 - 1,100 m. No attempt was made to predict the subsurface frequency distribution of the fault.

Therefore, there appears to be good correspondence between the orientation of fractures within specific environments defined on surface and the orientation of fractures in these same environments in the subsurface. In addition, the distribution of fracture frequency, in the subsurface, was predicted with reasonable accuracy before drilling.

VI. CONCLUSION

Within the Eye-Dashwa Lakes Pluton the major structural discontinuities, which may represent the dominant flow pathways within a crystalline rock, are identified by a combination of aerial photograph lineament analysis, airborne geophysical surveys and outcrop fracture data. More detailed studies involving ground magnetic and VLF-EM surveys, as well as additional outcrop fracture data, are used to validate the major structures, and to identify smaller structures that are not apparent at the regional scale.

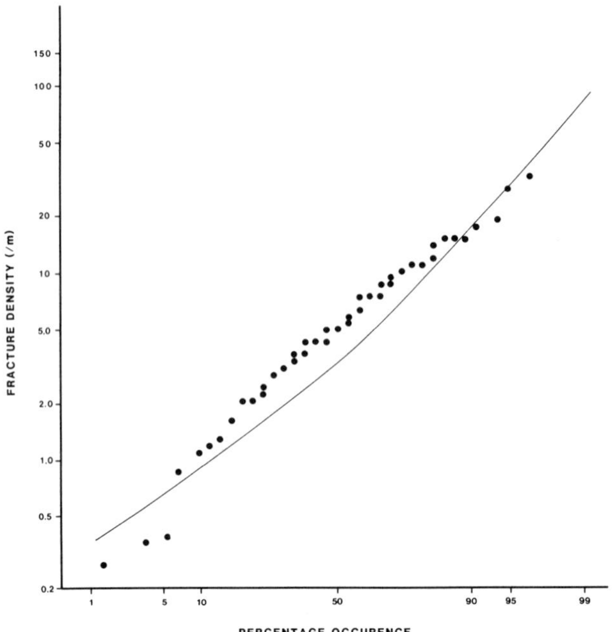

FIGURE 12. Distribution of Borehole Fracture Frequency Values from 150 m to 650 m (excluding the fracture zone at 270 m to 300 m) Versus Predicted Distribution Curve (solid line).

Finally, an extremely detailed evaluation of a 0.5-km^2 region of the pluton enabled an 'intrablock' environment and several fault environments to be identified and characterized in terms of fracture orientation and frequency. Comparison of these characteristics with subsurface data from borehole ATK No. 6 indicates that surface fracture domains can potentially be extrapolated to depth.

REFERENCES

1. R. M. Parkinson, "Operation Overthrust," in The Tectonics of the Canadian Shield, Royal Society of Canada, Special Paper, 4, p. 90-101 (1962).

2. A. Nur, "The Origin of Tensile Fracture Lineaments," J. of Structural Geol., 4. 4 (1982).

3. P. Lapointe et al., "Interpretation of Magnetic Susceptibility: A New Approach to Geophysical Evaluation of the Degree of Rock Alteration," Cna. Jour. Earth Sci. (in press).

4. N. Thomm, "Certain Methods of Mapping Faults in the Pretoria and Dolomite Series in West Witwatersand Areas, Ltd.," Geol. Soc. of South Africa, Trans., XLII (1939).

5. R. L. Wheeler and J. M. Dixon, "Intensity of Systematic Joints: Methods and Application," Geology, 3 (1980).

6. J. Sheperd et al., "Surface and Subsurface Geological Prediction of Bad Roof Conditions in the Western Coalfield, New South Wales, Australia," Institute of Mining and Metallurgy Trans., 90 (1981).

7. H. A. Pohn, "Joint Spacing as a Method of Locating Faults," Geology, 9 (1981).

8. S. Siegal, Non-Parametric Statistics for the Behavioural Sciences, McGraw Hill, New York (1956).

9. P. A. Brown, D. Stone and N. A. C. Rey, Criteria for Selecting New Areas for Geological Investigation Within the Eye-Dashwa Lakes Pluton, Atomic Energy of Canada Limited, Technical Record, unrestricted, unpublished reports, available from SDDO, Atomic Energy of Canada Limited, Research Company, Chalk River, Ontario KOJ 1JO, Canada, TR-237 (1984).

10. D. Stone, "Sub-Surface Fracture Maps Predicted From Borehole Data: An E0xample from the Eye-Dashwa Pluton, Atikokan, Canada," Int. J. Rock Mech. Min. Sci. and Geomech. Abstr., 21, 4 (1984).

11. J. J. Dugal et al., Well History Report for Borehole ATK No. 1 in the Atikokan Research Area, Atomic Energy of Canada Limited, Technical Record, unrestricted, unpublished reports, available from SDDO, Atomic Energy of Canada Limited, Research Company, Chalk River, Onartio KOJ 1JO, Canada, TR 292-1 (1985).

12. D. C. Kamineni and J. J. Dugal, "A Study of Rock Alteration in the Eye-Dashwa Lakes Pluton, Atikokan, Northwestern Ontario, Canada," Chem. Geol., 36 (1982).

13. D. C. Kamineni and D. Stone, "The Ages of Fractures in the Eye-Dashwa Pluton, Atikokan, Canada," Contrib. Mineral. Petro., 83 (1983).

14. W. H. Kruskall and W. A. Wallis, "Use of Ranks in One-Criteria Variance Analysis," J. Am. Statist. Assoc., 47 (1952).

15. P. A. Brown and N. A. C. Rey, "Prediction of Fracture Environments Within the Eye-Dashwa Pluton" (in prep).

VERIFICATION OF RADIONUCLIDE TRANSPORT MODELS USING NATURAL FRACTURE SYSTEMS

F. B. Walton
J. P. M. Ross
Atomic Energy of Canada, Ltd
Whiteshell Nuclear Research Establishment
Pinawa, Manitoba, ROE ILO
Canada

ABSTRACT

Four single and two multiple sorption site, kinetic models were tested on data for ^{137}Cs, ^{75}Se, ^{85}Sr, ^{144}Ce, ^{60}Co and ^{95m}Tc interactions with three natural-granite fracture systems. Special test sections were constructed for these experiments to eliminate dispersion and/or channeling in these fracture systems. The models were tested using a combination of model fitting, model extrapolation within the one fracture-cell data set, and prediction of data from two other fracture cells. A comparison of the models indicated that a model utilizing a linear combination of three first-order sorption mechanisms produced a better fit in the above test than the other models. The effect of variation in alteration mineral deposits on the fracture surface was also examined.

I. INTRODUCTION

The Canadian Nuclear Fuel Waste Management Program has two different radionuclide transport, modelling requirements, depending on the time scale being considered. For assessments on geological time scales, equilibrium models may be used. In contrast, laboratory and field-scale experiments require kinetic models. With respect to this latter requirement, six kinetic models - four single sorption site and two multiple sorption site - were tested for their ability to model radionuclide/rock interactions in natural-fracture systems in granite.

The unique physical structure of an intact fracture poses several experimental difficulties. Fracture apertures are typically of the order of tens of μm; hence, internal volumes of fracture-containing specimens used in laboratory experiments tend to be very small. To obtain fluid residence times long enough to allow reaction of radionuclides with the fracture surfaces, the flow rates may be too low to obtain with normal laboratory pumps. In addition, the irregularity of fracture surfaces may cause channeling and/or isolated areas of low or no flow. Hydraulic dispersion resulting from these flow irregularities and changes in flow geometry at the test section entrance and exit may prevent accurate assessment of the radionuclide/rock interaction data. Since the objective of this study was to obtain data to evaluate kinetic radionuclide/rock interaction models, new experimental procedures were required. Hydraulic dispersion was eliminated by opening the fracture and using each side of the fracture as one wall of a closed vessel or cell. The other walls of the cell were formed from pieces of Plexiglas, as illustrated in Figure 1. This configuration provides residence times in excess of three days using standard laboratory pumps. A teflon-coated magnetic stir bar was used to ensure homogeneity of the solution in the cells.

II. THEORY

The three most common isotherms used to correlate sorption data are the first-order reversible (SFO), the Langmuir (SLG) and the Freundlich (SFR). Kinetic versions of these isotherms have been developed[1] for single sorption sites and are given in Equations (1) to (3), respectively. The Dubinin-Radushkevich isotherm has also been used to account for non-linear, site-limiting sorption effects. A kinetic version of this isotherm (SDR) is given in Equation (4).

$$dS/dt = k_1 D - k_2 S \tag{1}$$

$$dS/dt = k_2 C(S_0 - S)/S_0 - k_2 S \tag{2}$$

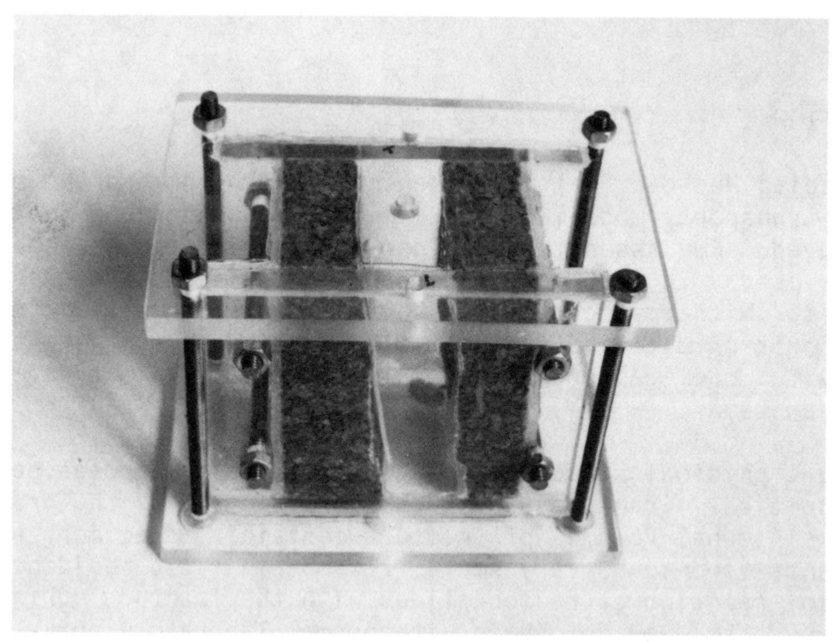

FIGURE 1. Photograph of Fracture-Cell Apparatus. The granite fracture surface is 9.9 cm wide by 5.2 cm high. The groundwater is stirred by a small Teflon-coated magnectic stirbar at the bottom center of the cell.

$$dS/dt = k_1C_0(C/C_0)^b - k_2S \tag{3}$$

$$dS/dt = - k_1 \ln(S/S_0) - k_2[\ln(1 + 1/C)]^2 \tag{4}$$

where

C is the concentration of the radionuclide in the aqueous phase,

C_0 is the initial concentration of the radionuclide in the aqueous phase,

S is the concentration of the radionuclide on the rock surface,

S_0 is the limiting concentration of the radionuclide on the rock surface,

k_1 and k_2 are rate constants, and

b is a constant.

We have shown, by selective chemical extraction procedures in a previous study,[2] that sorption of most radionuclides on granite surfaces is a function of two or more different mechanisms or sorption sites. The simplest multiple sorption site models consist of a linear combination of two or more single-site models. For two SFO sites reacting in parallel, the double first-order (DEO) model is given by

$$\begin{aligned} S &= S_1 + S_2 \\ dS_1/dt &= k_1C - k_2S_1 \\ dS_2/dt &= k_3C - k_4S_2 \end{aligned} \tag{5}$$

Similarly, for three sites reacting in parallel, the triple first-order (TFO) model is given by

$$S = S_1 + S_2 + S_3$$

with dS_1/dt and dS_2/dt defined as in Equation (5) above and

$$dS_3/dt = k_5C - k_6S_3 \tag{6}$$

where k_5 to k_6 are rate constants.

For a cell with volume V and initial concentration C_0 at $t = 0$, flushed with radionuclide-free water at a volumetric flow rate W, the mass balance is given by

$$VdC/dt = - WC - AdS/dt \tag{7}$$

where A is the fracture surface area.

For the special case of a non-reactive tracer, integration of Equation (7) gives

$$C = C_0 \exp(-tW/V) \tag{8}$$

III. EXPERIMENTAL

An intact fracture, about 0.2 m long and 0.1 m wide, was cut from Lac du Bonnet granite obtained from the Cold Spring Quarry, located on the Lac du Bonnet batholith near the Whiteshell Nuclear Research Establishment of Atomic Energy of Canada Ltd. The fracture was cut into four equal sections, which were used to fabricate fracture cells as described earlier (see Figure 1). Granite groundwater[2] (GGW) was fed from a reservoir to the fracture cells by a peristaltic pump. The effluent from the cells was collected every 90 minutes with a fraction collector. The various pieces of equipment were connected with small-bore Teflon tubing. A detailed analysis of all materials used in these experiments is given elsewhere.[3]

At T = 0 three cells were injected with GGW containing a mixture of six radionuclides: ^{137}Cs, ^{60}Co, ^{75}Se, ^{144}Ce, ^{85}Sr and ^{95m}Tc. The cells were then flushed with tracer-free GGW. In the first experiment, flow to the cell was maintained for 51 d and then halted for 2 d, to allow desorption to occur in the cell without elution. This procedure allowed the desorbing radionuclides to accumulate in the fracture cell, producing an increase in aqueous radionuclide concentrations. This discontinuity, in the otherwise smooth concentration-time curve for the cell, provided an excellent test of the predictive ability of the models. The pump was then restarted and elution continued for an additional 19 d.

The second and third experiments were stopped after 6 and 13 d, respectively, to allow determination of sorbed radionuclide inventories on the fracture surfaces as a function of time. These experiments were designed to examine the fact that multiple sorption site models predict different surface inventories associated with each mechanism as a function of time.

A fourth experiment was conducted with saline groundwater. The results of this experiment are described elsewhere.[3]

To test the models, a combination of model fitting and model extrapolation within the data from experiment 1, and model prediction of data from experiments 2 and 3, was employed.

IV. RESULTS AND DISCUSSION

Elution data for ^{95m}Tc in experiment 1 are plotted as a function of normalized time (tW/V), or cell volumes of GGW eluted, in Figure 2. As can be seen from this plot, the relative Tc concentration follows the elution

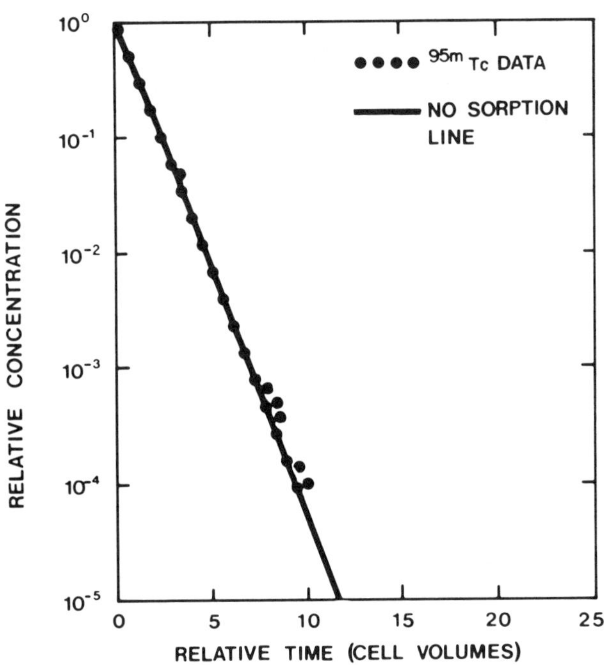

FIGURE 2. Plot of Normalized ^{95m}Tc Concentration Versus Relative Time in Experiment 1. The Tc concentration follows Equation (8) - the equation for the elution of a non-reactive tracer. Only every tenth data point is shown.

curve for a non-reactive tracer - Equation (8). The same observations were made for Tc elution in experiments 2 and 3.

The ability of a model to fit, or predict, a given set of data can be qualified by calculating the model residuals and the model variance. The model residual is defined as

$$\text{residual} = (C_{model} - C_{data})/C_{model} \tag{9}$$

and the model variance as

$$\text{variance} = (\text{residual})^2/n \tag{10}$$

where n is the number of data points. The residual is the relative deviation between the model and a single piece of data. The variance is a measure of the cumulative deviation between the model predictions and the whole data set.

The kinetic models of the four single sorption site isotherms given in Equations (1) to (4) and the multiple sorption site models given in Equations (5) and (6) were fit by regression to the elution data between 0 and 51 d (first 17 cell volumes) for experiment 1. Model fit variance were calculated and are given in Table 1. The models were then tested by extrapolating beyond this time period to predict desorption during the 2-d flow stoppage and the period between 53 and 72 d, when flow to the fracture cell was resumed. Model variances were calculated for this data (51 to 72 d) and are included in Table 1. Typical plots of model predictions and measured data, as a function of relative time, are given in Figures 3 and 4. These plots illustrate the superior fit and predictive capability of the TFO model over that of the SFO model for the ^{137}Cs data. Included in each figure is a plot of model residual as a function of relative time. A "residual plot" gives the relative deviation between the measured data and the model as a function of relative time. Models that fit the data will give a residual plot that is a horizontal line (i.e., the residual will be uniformly distributed about zero as a function of relative time, as shown for the TFO model in Figure 4). On the other hand, residual plots that show a systematic deviation from zero with relative time indicate an inappropriate choice of model to fit the data, as illustrated by the SFO model in Figure 3.

From Table 1 it can be seen that for ^{137}Cs, ^{85}Sr, ^{75}Se and ^{60}Co, the TFO model is superior to the other models in its ability to fit the data (0 to 51 d) and to predict the data (51 to 72 d) by extrapolation. A comparison of residual plots (not shown here) indicated that the single sorption site

TABLE 1. Single- and Multi-Site Model Variances for Data from Experiment 1

	Time Period (d)		Model Variances (x10^3)					
Tracer	Fit	Predict	SFO	SPR	SLG	SDR	DFO	TFO
^{60}Co	0 - 51	0 - 51	78.2	14.1	13.4	10.9	11.1	3.0
	0 - 51	51 - 72	30.0	225	25.4	183	109	12.3
	0 - 72	0 - 72	72.1	28.9	14.4	23.3	17.9	3.6
^{137}Cs	0 - 51	0 - 51	31.3	16.0	4.82	12.1	2.68	1.59
	0 - 51	51 - 72	1420	274	114	206	96.4	21.2
	0 - 72	0 - 72	59.6	34.2	8.59	25.6	6.13	3.54
^{85}Sr	0 - 51	0 - 51	49.0	2.55	32.5	22.9	2.02	1.26
	0 - 51	51 - 67.6*	360	183	1250	108	28.7	14.4
	0 - 67.7	0 - 67.7	65.5	6.93	49.3	63.4	2.70	1.66
^{75}Se	0 - 51	0 - 51	20.1	25.6	14.2	22.7	3.49	2.97
	0 - 51	51 - 72	1720	402	241	342	26.8	11.5
	0 - 72	0 - 72	30.9	47.1	19.4	41.4	5.80	4.98
^{144}Ce	0 - 51	0 - 51	102	27.3	78.2	27.2	27.3	29.2
	0 - 51	51 - 72	556	22.7	99.5	30.7	192	40.3
	0 - 72	0 - 72	177	26.9	78.8	26.7	31.4	29.6

Note: * - data below detection limit after 67.7 d.

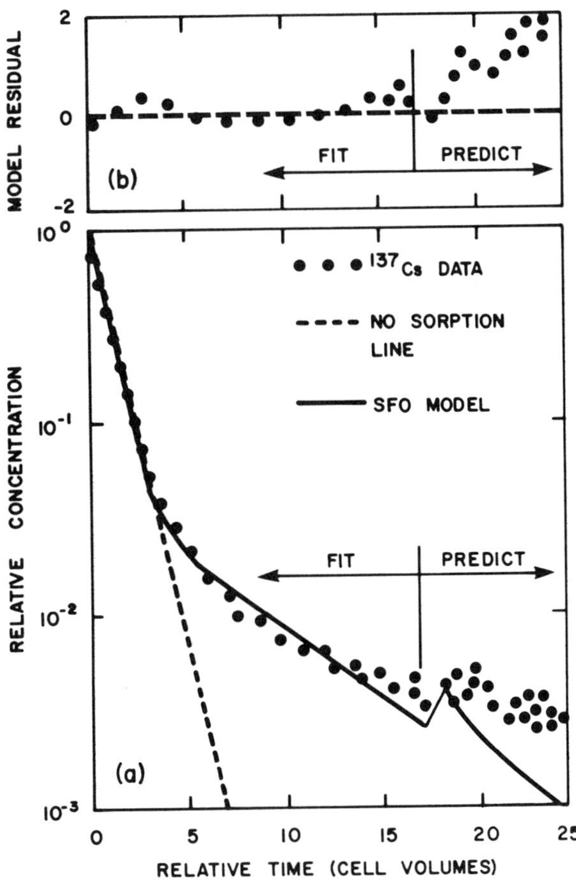

FIGURE 3. (a) Plot of SFO Model and ^{137}Cs Data as a Function of Relative Time or Cell Volumes (1 cell volume = 3 d). SFO model was fit to data from 0 to 17 cell volumes and extrapolated between 17 and 24 cell volumes. (b) Plot of Residual as a Function of Relative time for SFO Model for Both Fit and Extrapolated Portions of ^{137}Cs Data. Only every tenth data point is shown.

models (i.e., SFO, SFR, SLG and SDR) underpredict desorption from the granite fracture surface in 90% of the cases tested here. This indicates that the single sorption site models are not conservative in predicting radionuclide migration in granite fracture systems under kinetically controlled conditions typical of those tested here.

The models were fit to the whole data set in experiment 1 (i.e., 0 to 72 d) to determine whether the above observations were biased by the amount of data provided. These model variances are included in Table 1. The following observations were made from a comparison of the variances.

First, variances for the partial data set (0 to 51 d) were less than those for the full data set (0 to 72 d). If this had not been the case, this

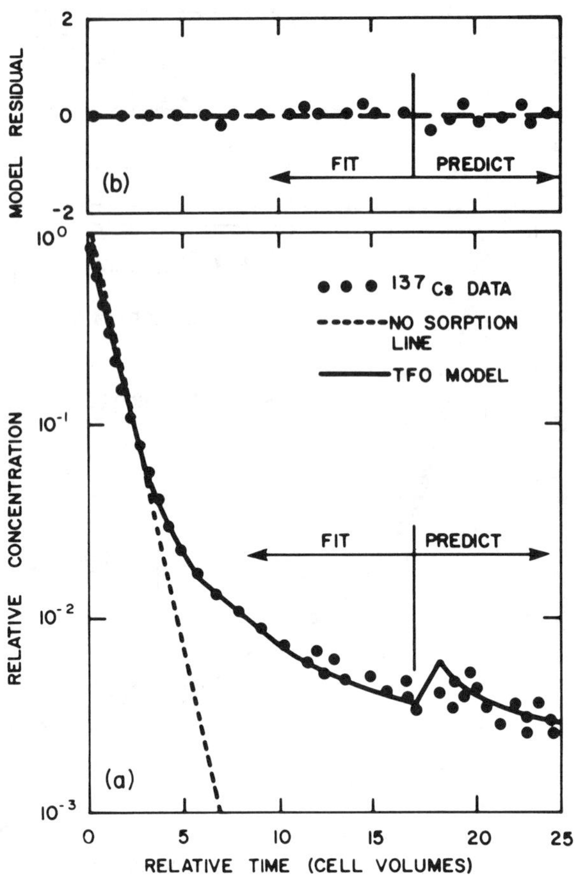

FIGURE 4. (a) Plot of TFO Model and ^{137}Cs Data as a Function of Relative Time (1 cell volume = 3 d). TFO model was fit to data from 0 to 17 cell volumes and extrapolated between 17 and 24 cell volumes. (b) Plot of Residual as a Function of Relative Time for TFO Model for Both Fit and Extrapolated Portions of ^{137}Cs data. Only every tenth data point is shown.

would have indicated that the data between 0 and 51 d were insufficient to allow a proper fit of the models. In most cases, the variances for the full data set were slightly larger than for the partial data set. This is, in part, a result of the higher gamma counting error (i.e., data scatter) at lower radionuclide concentrations for the data between 51 and 72 d.

Second, the relative ability of the models to fit the data did not change with the amount of data used to fit the model. This indicates that the model comparisons were unbiased; that is, the inclusion or exclusion of data did not change our conclusions on the ability of a particular model to fit or extrapolate the experimental data.

An apparent contradiction to the above observations is the SFO fit and extrapolation of the ^{60}Co data. In this case, variance for the period from 51 to 72 d was less than that from 0 to 51 d. Inspection of the model residual plots indicated that this was fortuitous. The model had a very poor fit to the data between 0 and 3 cell volumes and had a much better fit to the data from 3 to 17 cell volumes. The variance for the extrapolation was, therefore, less than that for the model fitting, due to the exclusion of the poorly fitting data. This indicates that both model variance and a residual plot are required to correctly assess model fit.

The GGW elution data for ^{144}Ce did not follow the same general pattern as those for the other radionuclides. First, the elution profile was more irregular or scattered. This same irregular behavior has been noted in similar mixing-cell experiments where filtration results suggested that the Ce was present as a colloid or particulate.[1,2] Second, the aqueous concentration of Ce did not appear to increase - at least to the same degree in GGW as the other radionuclides tested - during the 2-d flow stoppage period after 51 d. This was in contrast to its behavior in the fracture-cell experiment with saline groundwater[3], where the Ce concentration did increase for a similar flow-stoppage period, as in experiment 1. Since the starting concentration of Ce in both groundwaters was essentially the same, the observed differences can likely be attributed to differences in colloidal and/or particle behavior as a function of the ionic strength of the two groundwaters.

A final test was made by predicting the results of experiments 2 and 3 using the parameters derived from the fit of the data for 0 to 72 d in experiment 1. Only the TFO model was considered here because of its superior fit to the data in experiment 1. Appropriate adjustments were made in the model for fracture-cell geometric surface area, volume and flow rate. The variances are compared with those for experiment 1 in Table 2. Only in the case for Cs were approximately equal variances obtained for the three experiments. Several possible reasons for the poor prediction of other radionuclides were examined.

No direct measurement of actual fracture surface area is possible with present techniques. A large difference between actual fracture surface area and the geometric surface area used in these calculations would result in a

TABLE 2. TFO Model Variances for Prediction of Data from Experiments 1, 2 and 3

	Model Variances (x 10^3)				
Experiment	^{60}Co	^{137}Cs	^{85}Sr	^{75}Se	^{144}Ce
1	3.66	3.54	1.66	4.98	29.6
2	110	5.52	3.42	8.98	94.7
3	187	3.87	10.0	24.9	209

systematic over, or under, prediction of all radionuclide variances in a given experiment. Table 2 indicates no consistent pattern; hence, this hypothesis was rejected.

Surface mineral concentrations were examined by determining the amount of $Fe(OH)_3$ removed from the fracture surfaces using a hydroxylamine hydrochloride extraction procedure described in detail elsewhere.[3] The results of this analysis indicated a variation of a factor of 6 in $Fe(OH)_3$ surface deposits on the subsections of the fracture used to construct the fracture cells. With the exception of Cs, this large variation in surface mineralogy would explain, at least in part, the differences in fracture-cell sorption for experiments 1, 2 and 3. The variations in $Fe(OH)_3$ deposits apparently have little effect on Cs sorption. This is likely due to its apparent preference for clay minerals.[4]

Table 1 shows that the more complex the model, the better it's ability to fit the data (i.e., there are two parameters in the SFO model and six in the TFO model). From a purely mathematical point of view, this is to be expected. Both the complexity of the mineral structure of granite and previous selective extraction studies[2] give ample justification for the formulation of multiple sorption site mechanism models. The DFO and TFO are the simplest possible linear combinations. However, combinations of the other isotherms, although not tested here, are clearly possible. The variability of mineral deposits on portions of the same fracture used for these experiments did not allow a comparison to be made of measured and predicted multiple sorption site inventories as a function of time.

V. CONCLUSIONS

The following conclusions were reached from this study:

A. Six sorption models were fitted by regression to the experimental data form 0 to 51 d in experiment 1 for ^{137}Cs, ^{85}Sr, ^{144}Ce, ^{75}Se and ^{60}Co. A comparison of model variances indicated that the TFO model is superior to the other models tested in its ability to fit the data.

B. The TFO model was superior to the other models in its predictive ability in experiment 1. The single sorption sit models were found to be nonconservative in predicting radionuclide desorption from the fracture surfaces in 90% of the model extrapolation tests.

C. With the exception of Cs, the TFO model was not able to predict the results of experiments 2 and 3 using parameters derived from the data in experiment 1. Differences in iron oxyhydroxide concentration in the surface deposits in each of the fracture cells are believed to be responsible for significant differences in the sorptive capacity of the fracture surfaces. Cesium did not appear to be sensitive to variations in iron oxyhydoxide concentrations on the mineral surface.

D. No sorption of ^{95m}Tc was detected in these experiments.

E. Cerium solution concentrations in these experiments were very irregular with time, indicating colloidal or particulate behavior. Special models may be required to predict colloidal or particulate transport.

VI. FUTURE WORK

Test sections are currently under construction to perform radionuclide transport studies in large intact fracture systems. The kinetic parameters derived from the studies reported here will allow computer modelling and optimization of experimental operating parameters prior to actual tests. Further research is also under way to determine the variability in concentration of iron oxyhydroxide deposits found on granite fracture surfaces and its effect on radionuclide transport modelling.

REFERENCES

1. F. B. Walton et al., "The Determination of Nuclide-Geologic Media Reaction Kinetics Using Mixing-Cell Contractors," Chem. Geo., 36, 155 (1982).

2. F. B. Walton et al., "Determination of Radionuclide Sorption Mechanisms and Rates on Granitic Rock by Selective Chemical Extraction Techniques", A. C. S. Symposium Series, 246, 45 (1984).

3. F. B. Walton and J. P. M. Ross, Modelling of Radionuclide Interactions in Laboratory Scale Experiments with Granite Fracture Systems, Atomic Energy of Canada Report (in preparation).

4. R. D. Koons, P. A. Helmke, and M. L. Jackson, "Association of Trace Elements with Iron Oxides During Rock Weathering," Soil Sci. Soc. Amer. J., 44, 155 (1980).

STUDY OF THE BOOM CLAY LAYER AS A GEOCHEMICAL BARRIER FOR LONG-LIVED RADIONUCLIDES

L. H. Baetsle
P. Henrion
M. Put
Studiecentrum voor Kernenergie
Centre d'Etude de l'Energie Nucleaire
Boeretang, 200
B-2400 Mol, Belgium

A. Cremers
Laboratorium voor Colloidale Scheikunde
Katholieke Universiteit Leuven
Kardinaal Mercier laan, 92
B-3030 Leuven, Belgium

ABSTRACT

The Boom clay layer below the nuclear site of Mol, Belgium has been thoroughly investigated on its geohydrologic and physicochemical characteristics as well by laboratory experiments and in situ tests in the underground laboratory. Hydraulic permeabilities have been measured in situ; the chemical composition of the interstitial clay water is related to the mineralogical composition. Radionuclide sorption data and sorption mechanisms are given for Cs, Sr, Eu, Tc, Am, Pu and Np; experimental diffusion coefficients were determined by clay plug migration tests in representative conditions. Results of model calculations for the migration of radionuclides in dense porous media are given for Cs, Sr, Pu and Np.

I. GEOLOGY AND HYDROGEOLOGY

The Boom clay layer underneath the Nuclear site at Mol, Belgium was selected as one of the most promising locations for undertaking extensive geotechnical and geochemical investigations in order to assess its capacity as a host formation for the disposal of high level and TRU Waste. The Boom clay layer occurs in the Mol area at depths ranging from 160-180 m at its top surface and to 260-280 m at its lower interface. The 100 m to 120 m thick clay layer is tilting about 1% in the NNE direction and consists of very compact plastic clay of the Oligocene period. The clay is surrounded above and below by a sandwich type succession of sandy and clayey layers.

It is particularly important to mention that the sandy layers superposing the clay (-130 to -160 m) are very glauconitic (up to 80%) and constitute an additional secondary ion exchange barrier between the Diestian aquifer (-32m to -130m) and the Boom clay. The Diestian aquifer is used as a regular drinking water source. The regional flow in the sandy aquifer above as well as below the Boom clay amounts to 10 to 50 m per year and drains in the WSW direction towards the Scheldt river.

A hydrologic network consisting of 132 piezometric observation wells over an area of 2500 Km^2 permitted the measurement of the water level fluctuations over several years. A computer modelling study (NEWSAM model) carried out by

Patyn[1] in the frame work of a contract with Ledoux of the Ecole des Mines de Paris, has shown that the most probable fit of the experimental water level data implies a vertical seepage through the Boom clay layer which can be up-or downwards according to the location in the hydrogeologic system. At the Mol site the piezometric head is only 2 m and seepage of the upper aquifer theoretically penetrates the clay body and discharges into the underlying sandy aquifer (Sands of Berg) with higher salinity and small water transmissivity ($T = 10^{-5}$ m^2/s). According to extensive laboratory permeability studies on the Boom clay, a range of permeabilities of 10^{-11} to 10^{-9} m/s was determined on non disturbed samples. The global value for the clay formation amounts to 10^{-10} m/s and this value was found during in situ experiments in the underground laboratory. It is difficult to state positively that Darcy's law is valid in such impermeable media but some preferential waterflow is not to be excluded and therefore a seepage rate of 0.1 to 1 l/m^2/year is to be expected.

If no preferential pathways exist, the renewal of a interstitial water column of 100 m thickness at 40% void fraction would take 630,000 years. However, if only 1 to 3% of the interstitial water participates to the drainage phenomenon, the renewal period is reduced to a time interval ranging from 15,000 to 47,000 years. Measurements of C14 have been performed on interstitial claywater samples and on water samples of the underlying aquifer. The order of magnitude of the interstitial clay water ages deduced from these ^{14}C measurements are at least 15,000 years and of the underlying aquifer of about 45,000 years (10 half-lives of ^{14}C).

II. GEOCHEMISTERY OF THE CLAY LAYER

A. Solid Phase Composition of Clay

The Boom clay (BC) deposit is from the mid-Oligocene age (40 to 60 million years) and of marine origin; it has been compressed throughout the geologic ages by the weight of the overlying sedimentary formations. The total lithologic and hydrostatic pressure amounts to 40 bars. The Boom clay formation is composed of clay minerals (smectite, illite and vermiculite), particles of iron oxide, and quartz. It contains, furthermore, organic matter (humic acids), pyrite (FeS_2), and calcite ($CaCO_3$), which are in dynamic chemical equilibrium with the interstitial groundwater. The clay mineral fraction (≤ 2 µm 0) represents about 40 weight %.

The formation is saturated and the water content varies around 18 wt%. The maximum value of the porosity is therefore 35%. The cation exchange capacity determined by saturation with Ca^{++} and Sr^{++} ions lies around 0.24 meq/g while an exhaustive cation exchange with Ag thiourea, leads to a value of 0.3 meq/g. The surface area of the fresh material varies between 150 and 200 m^2/g.[2]

The organic carbon content of Boom clay consists essentially of humines and humic acids: a statistical average content of 1.23% (±0.41%) was found for a large number of clay samples representing the entire cross section of the formation. The organic matter contribution to the total ion exchange capacity is about 0.05 to 0.1 meq/g. The picture emerging from a recent investigation of the organic material in BC is that of a true humus-clay complex where the organic material is uniformly distributed and firmly attached to the mineral surfaces through coordinated bonds with metallic atoms (FeII, Al).

The $CaCO_3$ content is variable and sometimes locally accumulated in concretions; an overall mean value varies between 0.5 and 1% (5-10 mg/g clay). Finally the framboidal variety of pyrite was found in all clay samples with microcrystals of 0.5 μm diameter.

B. Interstitial Groundwater in the Clay Formation

Taking advantage of the existence of the underground laboratory, a series of wells were installed in the clay formation. All precautions were taken to avoid contamination by air or foreign matter (boring aids). Water samples are collected regularly under strict anaerobic conditions. The composition of this clay water as well as the compositions of the groundwaters in the surrounding aquifers are given in Table 1 and show the over-all relation in the hydrogeologic system. The interstitial solution consists essentially of $NaHCO_3$. Measured under in situ CO_2 partial pressure (3.4×10^{-3} atm; 17°C) the pH values range between 8.4-8.7.

The analogy between the chemical composition of the interstitial clay water with that of the underlying sandy aquifer is obvious and supports the theory of the vertical downward movement of the water through the Boom clay.

Not shown in Table 1 is the organic material (O.M.) content which amounts to about 120 ppm in the clay water and 20 ppm (max) in the underlying aquifer, indicating the restricted mobility of the O.M. in the compacted clay.

C. Redox Potential of the Clay Medium[3]

Pyrite is the main reason for the very reducing character of Boom clay and was formed over geologic time by reduction of sulphate through the action of sulphate reducing bacteria under anaerobic conditions; FeS_2 determined the redox potential in the formation which can be evaluated by the method of Garrels. At pH 8.7, one finds Eh = -0.262 V in good agreement with the measured values.

As a consequence BC is very sensitive to air and its study requires elaborate precautions. When reexposed, the reverse reaction occurs, namely

$$2FeS_2 + 7.5\ O_2 + H_2O \quad -> \quad Fe_2(SO_4)_3 + SO_4^{--} + 2H^+$$

H_2SO_4 reacts with $CaCO_3$ and, under extreme conditions, the final pH depends only on the relative proportions of these two substances.

TABLE 1. Composition of Interstitial Clay Water and Surrounding Groundwater

Element	Na	K	Mg	Ca	Fe	Si	SO_4	Cl	CO_3	pH units	Eh mV
	in mg/l										
Interstitial Clay water	405	11	2.6	5.5	0.74	4.26	14.3	36.7	927	8.68	-260
Aquifer underlying the Boom clay	387	9.9	2.05	6.9	-	-	6.3	29.9	942	-	-
Lower Diestian aquifer overlying the Boom clay	9.4	2.3	2.8	33	0.01	-	17.8	-	39.8	8.3	0

The pH may drop to as low as 2. It is therefore not envisaged to store BC in direct contact with the atmosphere if it is to be used to subsequently refill the galleries. In addition, the formation of sulfate in exposed clay must be taken into account during the construction and exploitation of the galleries network.

The humic acids under alkaline condition are also very reducing and their ability to reduce even TcO_4^- in the clay water has been conclusively demonstrated.[4]

III. SORPTION PROPERTIES OF BOOM CLAY

A. Sorption of Sr and Cs

The sorption of Sr and Cs on the Boom clay can be described on the basis of ion exchange mechanisms.[5] From selectivity coefficients determinations with Mg^{++} and K^+, the distribution coefficients of Sr and Cs can be derived, and are in good agreement with values measured under in situ conditions. The corresponding retardation factors,

$$R = 1 + \frac{\rho\,(1-\eta)}{4}\,Kd$$

where ρ is the density of the solid phase and η the porosity factor, were found to be in fair agreement with the R value derived from direct diffusion measurements in clay gels (6). On the basis of this study, the <u>in situ</u> diffusion coefficients of Sr and Cs are expected to be 3.2×10^{-6} and 3.2×10^{-3} m^2/y^{-1}, respectively. Direct measurements in clay plugs at near in situ moisture levels give rather dispersed values about 3.2×10^{-4} and 3.2×10^{-5} m^2/y^{-1}, respectively.[4] This sharp difference in the sense of more safety is interpreted in terms of hindered accessibility of the sorption sites when the clay particles are close together. The difference might be explained by supposing a continuous organic coating.

In relation to waste disposal in the BC formation, and because of their short lives, the migration of Sr and Cs is considered only in connection with accidental release from leaking waste containers.

B. Sorption of Actinides and Tc

Sorption tests were carried out under anaerobic conditions with very high ratios clay/liquid (to be closer to in situ conditions), the solution having a composition very similar to the percolation water except for the O.M. (which is absent in the synthetic water). The fundamental role of the O.M. in the clay in controlling the sorptions of Pu, Am, Np and Tc was immediately apparent.

It follows that the Kd values depend directly on the method used to separate the phases since, according to the method, more or less O.M. and, correspondingly, more or less activity is removed from the solution. In all cases also, larger Kd values are found at larger values of the liquid/solid ratio. Except for Tc, which is critically dependent on Eh, a limited degree of oxidation of the clay rather increases the sorption of the other elements.

Attempts to extrapolate the Kd values down to moisture levels comparable to in situ conditions are so far only tentative because they require assumptions to be made regarding the molecular dimensions of the mobile O.M. in the clay, regarding its actual concentration in the pore network and its intrinsic mobility. Therefore, some Kd values are presented in Table 2 as indicative values.

1. Sorption of Pu

The Kd values obtained with Pu show some scattering; they also tend to increase somewhat with time. If, on the basis of the molecular size distribution of O.M. found in the percolation clay water one agrees to consider 40,000 as the maximum molecular weight for the mobile material, fairly large values of Kd are indeed observed (column 5). The presence of colloidal particles resulting from the polymerisation of $Pu(OH)_4$ was suggested by B. Skyte Jenssen.[7] The occurrence of inorganic radiocolloids cannot be excluded but, recently, in studying the formation of O.M. complexes in percolation clay water, striking similarities were obtained for Am, Eu, and Pu, suggesting a considerable amount of reduction to Pu III and an organic matter controlling behaviour.

TABLE 2. Kd Values (ml/g) for Pu, Am, Eu, Np and Tc in Boom Clay[(a)]

1	2	PHASE SEPARATION			
		3	4	5	6
RADIOELEMENT	Sol/liq	Centr	HSC	UF $4x10^4$	UF 10^3
Pu	1/1 1/2 1/5	-	318 342 670	2084 2731 4912	
Am	1/1 1/2 1/5	- 33 93	73 142 370	8.1×10^4 1.9×10^5 3.1×10^5	
Np	1/1 1/2 1/5	 211 349	- 177 249	- 600 807	
Tc	1/1 1/2 1/5		39 60 116	89 147 254	 164 304

(a) In most cases these values are averaged over several identical or near-identical experiments. The largest variations are observed for Pu, e.g., for Sol/Liq = 1/2 and HSC Kd ranges between 160 and 400.

(b) Centr = centrifugation 2,300 g; HSC = centrif 30,000 g; UF 4.0E4 = Amicon membrane filtration 40,000 cut-off.

The extent to which carbonate complexation contributed to keep traces of Pu in solution even after a filtration to 10^3 MWt units deserves further examination.

2. Sorption of neptunium

Table 2 shows for the system Np-Boom clay Kd values quite comparable to those observed for Pu and suggests very similar retardation factors in the clay formation. The presence of trivalent elements in the system, even at a concentration initially 10^{-5} M in the solution does not modify the Np distribution. Speciation work in the interstitial clay water is under way.

3. Sorption of trivalent elements (Eu, Am)

Here again, Kd values, very similar to those for Pu and Np are obtained after centrifugation (column 4). However, the complexation of Am and Eu with the O.M. in the solution is virtually complete because, a subsequent filtration quantitatively removes the trivalent element (very high Kd values, column 5).

It seems therefore that the O.M. in the clay completely displaces Eu from its carbonate complexes. In fact, as suggested by Cremers[7], the O.M. in the clay appears to be the exclusive sorption sink for Eu or Am. Ultimately trivalent species may be regarded as immobile in the clay since their association with the O.M. takes place preferentially with larger molecules which cannot migrate in the tiny clay pores.

4. Sorption of Tc

Kd values in Table 2 indicate a considerable sorption of Tc, though smaller than for the above mentioned elements. This system is mainly under the control of the redox potential but the study of the Tc-clay water system has clearly demonstrated the complexation of the reduced Tc (TcO^{2+}) species by the O.M.

Kd values increase only slightly by membrane filtration indicating very limited organic complexation of the Tc remaining in solution; the latter identified by Gel filtration chromatography as TcO_4^-. The least exposure of the clay system to air causes a drop in Kd or a complete absence of sorption. It is believed that the ratio Tc VII/Tc IV is under the control of the Eh and that, as TcO-migrates, the retardation of Tc will ultimately be controlled by the rate of desorption of Tc IV from the clay-humus complex. Work is under way to test this hypothesis.

IV. MIGRATION OF RADIONUCLIDES IN HOMOGENEOUS CLAY

To describe quantitatively the migration of radionuclides in a clay formation, a unidirectional mathematical model has been developed by considering the mass transfer to occur by simultaneous diffusion and convection.

The clay formation is considered to be homogeneous, porous and saturated with water. Analytical solutions have been obtained by PUT for three important time dependent source conditions, namely, concentration source, flux source and impulse source.[8] A program package named MICOFP has been written in FORTRAN 77. The package is fully interactive and equipped with input-, output-, help- and graphic-panels.

To use the program, the model parameters; the apparent dispersion constant, the apparent convection rate and the ratio of the bulk concentration to the concentration in the liquid phase have to be determined. For these purposes, clay plugs were loaded with specific radioactive tracers and kept under equivalent in situ pressures during long time periods. The radionuclides diffuse into the clay plug and the resulting concentration profiles are analyzed to obtain numerical values for the model parameters.

Diffusion coefficients of 0.0126 m^2/y (4.0E-6 cm^2/s) were found for iodine, and this value serves as a reference level with which the other less soluble or less mobile nuclides are compared. Extremely slow diffusing species (e.g. Pu, Np) display diffusion coefficients of less than 3.2E-6 m^2/y (1.0E-9 cm^2/s). Table 3 gives the values of the model parameters used for the migration calculations of the radionuclides ^{90}Sr, ^{137}Cs, ^{239}Pu and ^{237}Np.

Some of the results of the calculations with MICOFP are displayed in Figures 1 to 4. These figures show the concentration of the radionuclides considered in the liquid phase as a function of time, for various distances from the source, and clearly locate the penetration depth, defined as the

TABLE 3. Data of Considered Radionuclides for Migration Calculations, and Reference Element Iodium

SPECIES	I	^{90}Sr	^{137}Cs	^{239}Pu	^{237}Np
Half-live (y)	1.57 E7	28	30	24400	210000
Appar.dispersion constant (m2/y)	0.0126	6.3E-4	3.2E-5	3.2E-6	3.2E-6
Appar.convection rate (m/y)	7.0E-5	3.5E-6	5.9E-7	1.7E-8	1.7E-8
Disposed quant. (Bq/m2)	-	2.3E+13	3.0E/13	1.0E+9	1.0E+8
C(bulk)/c (liquid)	0.4	8	160	1600	1600

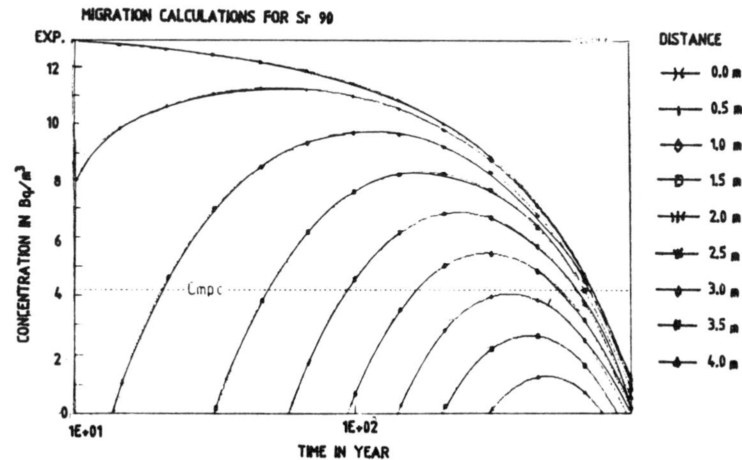

FIGURE 1. Concentration in Bq/m3 as a Function of Time and Distance

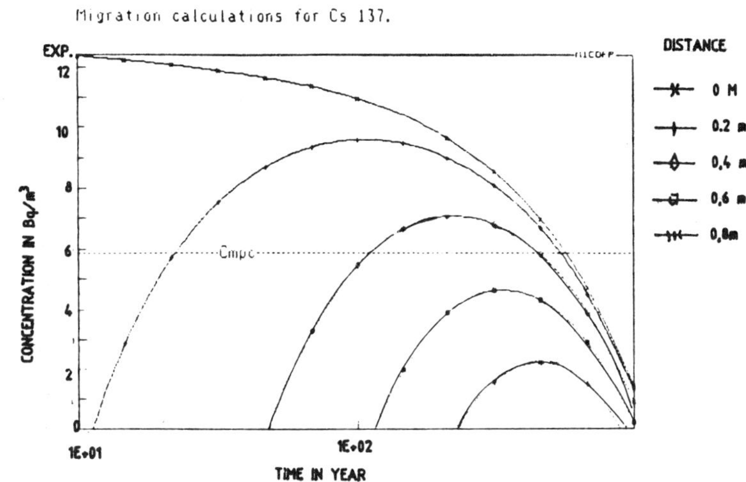

FIGURE 2. Concentration in Bq/m3 as a Function of Time and Distance

distance where the maximum concentration stays always below the maximum permissible concentration for drinking water (on the figures noted as Cmpc). For the model calculations the very pessimistic hypothesis was made that the dissolution of the conditioned waste starts at time of disposal.

Figure 1 gives the results for ^{90}Sr and shows a penetration depth of 3 meters. Figure 2 shows a penetration depth for ^{137}Cs of about 0.5 m. The long lived radionuclides ^{239}Pu and ^{237}Np show penetration depths of 0.4 and 6 m respectively (Figures 3 and 4), indicating that the Boom clay layer, under investigation, is a very effective barrier for the concerned radionuclides.

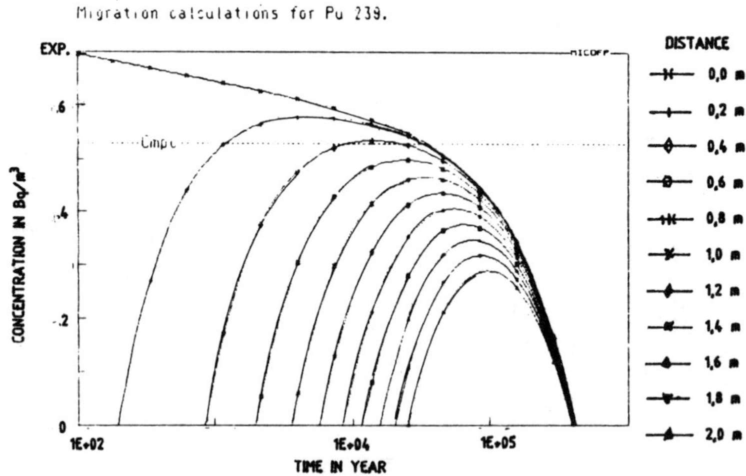

FIGURE 3. Concentration in Bq/m3 as a Function of Time and Distance.

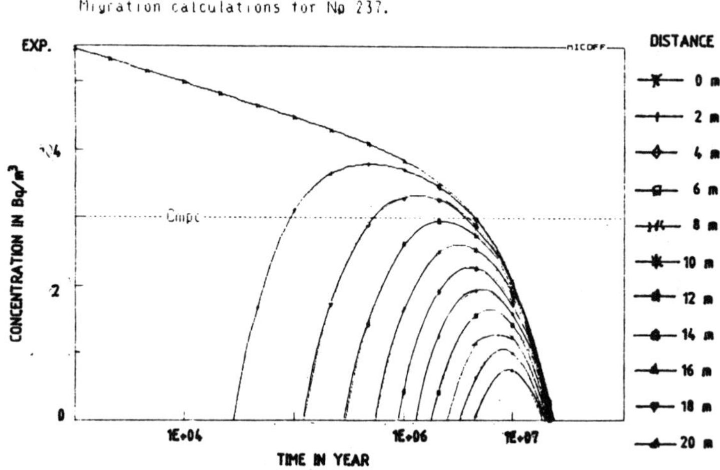

FIGURE 4. Concentration in Bq/m3 as a function of Time and Distance.

V. CONCLUSION

The hydrogeological investigations on the Boom clay layer at Mol have shown that the water movement within the formation is extremely slow with renewal times of 15,000 to 45,000 years. The surrounding aquifers impose a downward movement of the interstitial water towards the underlying saline aquifers.

The physicochemical studies on Boom clay have shown that the interstitial solution is constituted of $NaHCO_3$ in equilibrium with the solid phase compounds at pH 8.5 and Eh = -260 mV. The clay suspension is very sensitive to air oxidation.

Extensive investigations on the sorption behaviour of Boom clay have shown that the long-lived actinides (Pu, Am) are sorbed by the humic acid materials present in the interstitial pore water. The organic material is immobile due to its high molecular weight.

A computer programme based on the analytic solution of the diffusion equations permits the calculation of expected migration distance of some important radionuclides (Cs, Sr, Np, Pu) in the Boom clay.

REFERENCES

1. J. Patyn and E. Ledoux, Contribution a la Recherche Hydrogeologique liee a l'Evacuation de Dechets Radioactifs dans une Formation Aargileuse, These de doctorat, Ecole Nationale des Mines de Paris et l'Universite P&M Curie, Paris VI, 13.9 (1985).

2. B. Baeyens et al., "In Situ Physicochemical characterization of Boom Clay," J. of Waste Management and the Nuclear Fuel Cycle (Belgian National Issue) (in press 1985).

3. B. Baeyens et al., "Ageing effects in Boom Clay," J. of Waste Management and the Nuclear Fuel Cycle (Belgian National Issue) (in press 1985).

4. P. N. H. Henrion et al., "Migration of Radionuclides in Clay," J. of Waste Management and the Nuclear Fuel Cycle (Belgian National Issue) (in press 1985).

5. D. Verheyden et al., "In Situ Kd Values for Cs and Sr in Boom Clay at Trace Levels," Progress Report, Centrum voor Oppervlakte Scheikunde en Colloidale Scheikunde KUL Leuven (1982).

6. B. Baeyens et al., "Diffusion Behaviour of Cs in Boom Clay," J. Colloid Interface Science.

7. B. Skyte Jenssen, Migration Phenomena of Radionuclides into the Geosphere, Harwood acad Publishers, EUR 7676, Brussels & Luxemburg (1982).

8. M. J. Put, "A Unidirectional Analytical Model for the Calculation of the Migration of Radionuclides in a Porous Geological Medium," J. of Waste Management and the Nuclear Fuel Cycle (Belgian National Issue) (in press 1985).

HYDROLOGIC TEST PLANS FOR LARGE-SCALE, MULTIPLE-WELL TESTS IN SUPPORT OF SITE CHARACTERIZATION AT HANFORD, WASHINGTON

Phillip M. Rogers
Randolph Stone
Allen H. Lu
Rockwell Hanford Operations
P.O. Box 800
Richland, Washington 99352

ABSTRACT

The Basalt Waste Isolation Project is preparing plans for tests and has begun work on some tests that will provide the data necessary for the hydrogeologic characterization of a site located on a United States government reservation at Hanford, Washington. This site is being considered for the Nation's first geologic repository of high level nuclear waste. Hydrogeologic characterization of this site requires several lines of investigation which include: surface-based small-scale tests, testing performed at depth from an exploratory shaft, geochemistry investigations, regional studies, and site-specific investigations using large-scale, multiple-well hydraulic tests. The large-scale multiple-well tests are planned for several locations in and around the site. These tests are being designed to provide estimates of hydraulic parameter values of the geologic media, chemical properties of the groundwater, and hydrogeologic boundary conditions at a scale appropriate for evaluating repository performance with respect to potential radionuclide transport.

I. INTRODUCTION

The Basalt Waste Isolation Project (BWIP) is chartered by the U.S. Department of Energy with the geologic and hydrogeologic characterization of a site at Hanford, Washington where a high level radioactive waste repository may be constructed. The site is located on a federal reservation in south-central Washington (Figure 1). If a repository is constructed at this site, the main body of the repository would be at a depth of about 900 m (3000 ft) below ground surface.

A thorough understanding of the hydrogeology of this site is very important because the movement of radionuclides from a repository here would most likely be with the flow of groundwater. Hydrogeologic characterization of the site will draw on several parallel efforts that are currently ongoing or planned. These efforts include: piezometric baseline monitoring, experiments performed from an exploratory shaft (ES), geochemistry investigations, regional studies, and site specific investigations using large-scale, multiple-well hydraulic tests. The scope of this paper is limited to plans for and objectives of the

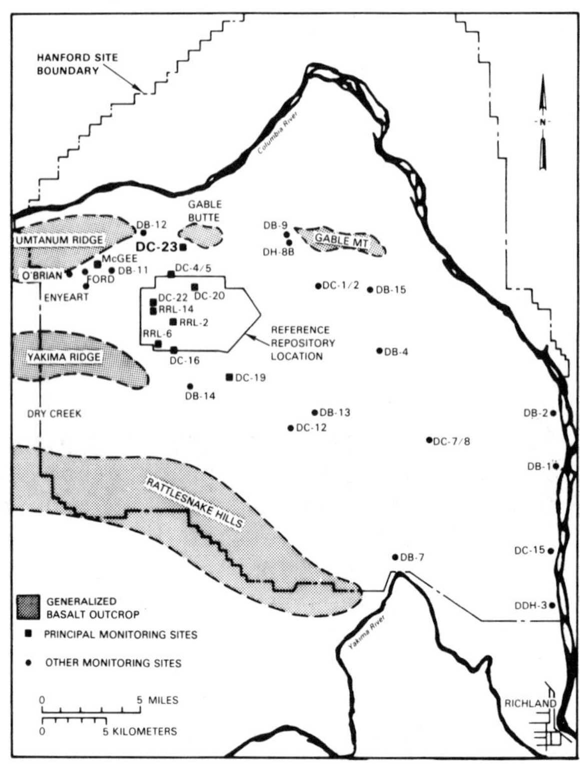

FIGURE 1. Hanford Site with Locations of Wells, Piezometers, and Boreholes

site specific investigations. Particularly, it is limited to a description of tests that will be used to evaluate the characteristic hydraulic parameters and natural boundary conditions of selected hydrogeologic units in and around an area known as the reference repository location (RRL) (Figure 1).

Large-scale, multiple-well hydraulic tests are required to supplement the small-scale information hitherto collected. Some parameters such as aquifer storativity, porosity, and dispersivity are best estimated from multiple-well tests. Some of the predictive models employed in repository performance assessment deal with the heterogeneous nature of hydrogeologic characteristics. It is important that the characteristic parameter values input to the models be representative of the actual average site conditions on the scale of the models. Large-scale multiple-well hydraulic tests will help assure that hydraulic parameter values used in modeling will meet this requirement.

A strategy has been developed for the site-specific hydrogeologic characterization of the site and surrounding areas. This strategy is a staged process (Figure 2) with four broad stages which are: piezometric baseline monitoring (PBM) program (stage 1) and a three-stage testing program (stages 2, 3, and

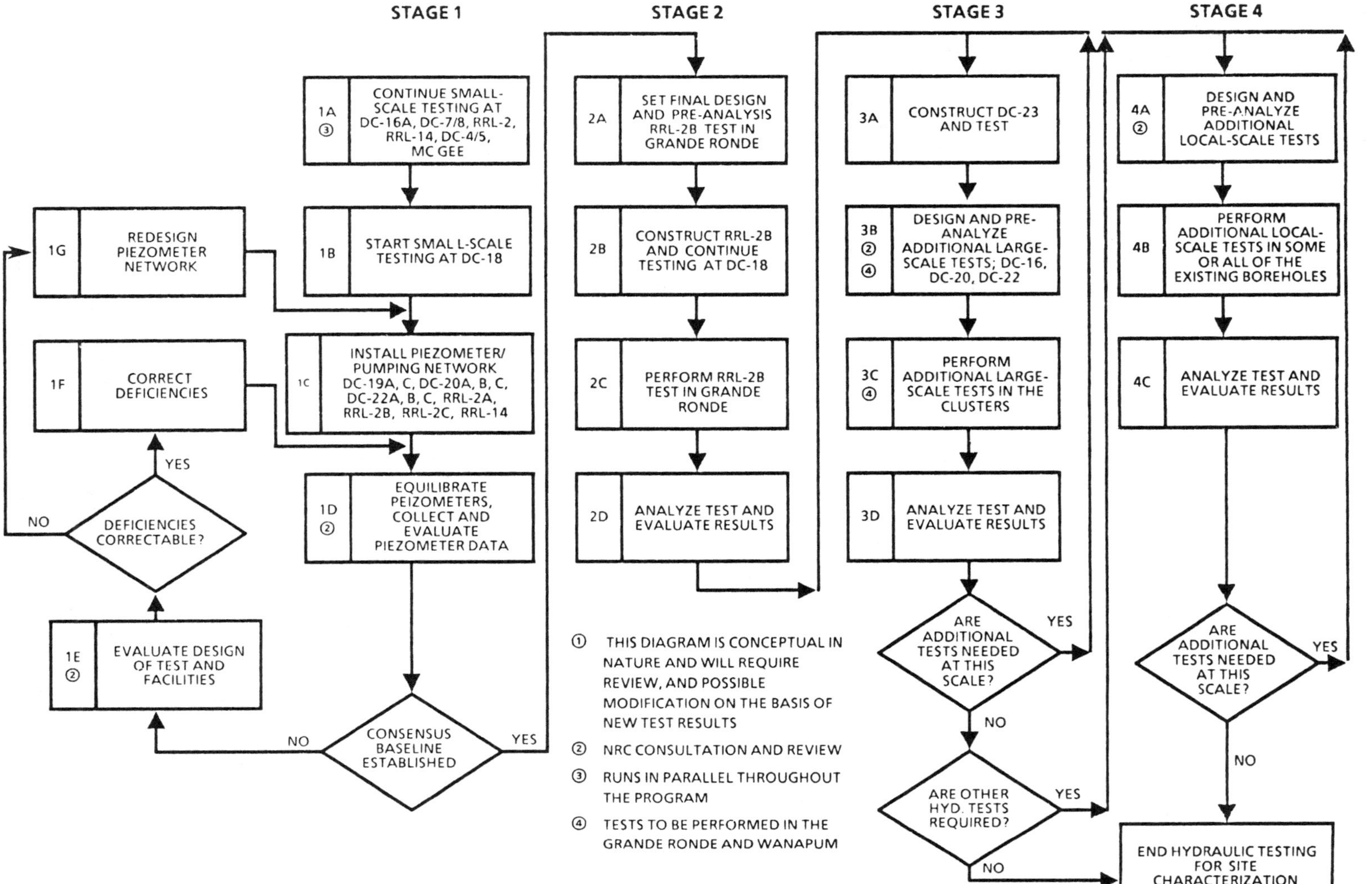

FIGURE 2. Site Hydrologic Testing Sequence

4).[1] The PBM program (stage 1) is designed to support the hydrologic testing in stages 2-4 as well as other regional and local studies. The PBM program will provide high-resolution, time series hydraulic head data for establishing a baseline, estimating groundwater flow gradients, and enhancing and extending the analyses of data from subsequent testing (stages 2-4). Data from this activity may also be useful for the inference of hydraulic boundaries. These data are being collected and criteria are being developed for their evaluation. The monitoring network will be expanded with the addition of several monitoring wells to be constructed over the next two years. They will be located on the Hanford Site but outside of the RRL.

The hydrologic testing program (stages 2-4) is designed to quantify hydraulic parameters of the geologic media, chemical properties of the groundwater, and hydrogeologic boundary conditions at a scale appropriate for evaluating repository performance with respect to potential radionuclide transport. Large-scale, multiple-well pumping tests are an essential aspect of the testing program. The initial testing stage (second stage of the strategy) is designed to obtain data for the estimation of the parameters and characteristics noted above at a location near the center of the site and at the depth of the candidate repository horizon (Cohassett flow) which is about 900 m (3000 ft) below ground surface. Stage two testing will be centered around a test well designated RRL-2B, centrally located in the RRL (Figure 1). Testing at this location will consist of a series of four discrete tests, stressing the candidate horizon and adjacent horizons.

The third stage is a logical extension of the second stage. The tests will be centered around test wells at DC-16, DC-20, and DC-22 (Figure 1). Overlapping areas of test influence are expected for some of the hydrogeologic units tested, thus allowing for the complete coverage of the RRL plus parts of the surrounding area.

The fourth stage lends considerable flexibility to the site investigation with ad hoc testing as necessary. Depending on the results of the first three stages of investigation, additional large-and/or small-scale tests may be performed.

II. Site Description

The RRL lies within the central portion of the Cold Creek syncline Figure 3). This syncline is located in the Pasco Basin, one of several structural and topographic basins within the Yakima fold belt subprovince of the Columbia Plateau. Within the Pasco Basin were deposited many flows of the Miocene Columbia River Basalt Group.

The Hanford Site is underlain by at least 50 basalt flows with accumulative thickness greater than 3000 m (9800 ft).[2] The basalt flow identified as the candidate repository horizon (Cohassett flow) lies about 900 m (3000 ft) below ground surface in the Grande Ronde Basalt and is about 75m (250 ft) thick. Average basalt flow thickness is between 30 to 40 m (100 to 130 ft).[3] The

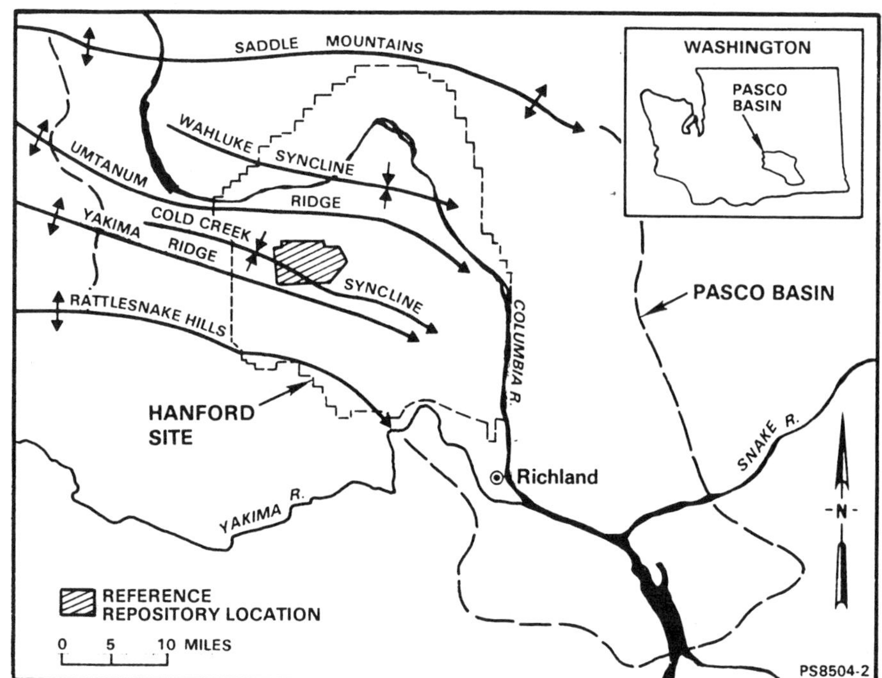

FIGURE 3. Principal Fold Structures of the Pasco Basin

basalt flows generally consist of an upper vesicular and/or brocciated flow top overlying a jointed, dense interior. The flow top typically accounts for about 15 percent of the total flow thickness.

Groundwater beneath the Hanford Site occurs in 1) fluvial and lacustrine sediments lying atop the basalts, generally under unconfined conditions, and 2) within interflows (composition flow bottom and flow top) and interbeds of the basalt sequence, under confined conditions.

The main groundwater occurrence and movement in basalt formations is within flow tops and sedimentary interbeds. The basalt flow interiors that separate individual flow tops appear to act as aquitards through which some degree of vertical leakage may occur along cooling fractures.

III. TEST OBJECTIVES

The first suite of multiple-well hydraulic tests has its own set of objectives. These objectives are based on several considerations, which include: information needs of the project, proximity of the test well to the ES and its location in the RRL, anticipated hydraulic characteristics of the hydrogeologic units to be tested, and the state-of-the-art for hydrologic testing at depths in excess of 900 m (3000 ft). Several general objectives of the large-scale hydraulic test of the Rocky Coulee flow top (the first to be tested at RRL-2) have been identified. These include:

- the facilitation of the design of subsequent tests,
- evaluation of lateral hydraulic conductivity and storativity of interflow zones near the Exploratory Shaft for input into ES construction and operation,
- evaluation of vertical hydraulic conductivity of flow interiors using parameter variation and analytical analysis techniques,
- assessment of the degree of vertical leakage into the test interflow zones from adjacent flow interiors,
- evaluation of effective porosity and longitudinal dispersivity of interflow zones using radial convergent tracer tests concurrent with the pumping test,
- assessment of the areal representativeness of hydraulic characteristic values obtained by previous singlewell testing,
- characterization of the dissolved substances in groundwater removed from the Grande Ronde Basalt at the RRL-2 site, and
- identification and classification of hydraulic boundaries.

The objectives and strategy are intended to evolve as more data and information are collected and evaluated. Thus the following discussion will focus on the ways and means of accomplishing the objectives of the first test that will involve pumping from the Rocky Coulee flow top in test well RRL-2B.

IV. TEST DESCRIPTION

The multiple-well hydraulic tests are dynamic tests that purposefully will cause groundwater-level fluctuations within a large area. Assuming a constant rate pumping test is the type of test employed, response to the test would be in the form of groundwater-level decline within the test area of influence during the period of pumping, followed by a period of gradual groundwater-level recovery.

At the RRL-2 site, four multiple-well hydraulic tests are planned. These would all be in the Grande Ronde Basalt beginning with the Rocky Coulee flow top and sequentially progressing to the Cohassett, Grande Ronde #5, and the Umtanum flow tops. In all four tests, assuming a constant discharge type of test, pumping would be from a well designated RRL-2B, which at the scale of Figure 1 is located at the site designated RRL-2. If a response from pumping at RRL-2B can be propagated to the existing monitoring facilities, DC-20, and DC-22, then hydraulic parameter values averaged over a large area of the RRL may be obtained.

For these tests, a large area of influence will be achieved by applying a large stress, on the order of 230-300 m (750-1000 ft) of drawdown at the pumping well, for a period of time, up to 30-50 d. Discharge rates necessary for this type of test will range from what may seem surprisingly small, possibly less than 0.004 m^3/min (1 gal/min) for the relatively impermeable hydrogeologic units (Cohassett flow top), to over 0.4 m^3/min (100 gal/min) for the hydrogeologic units with the largest values of transmissivity (Grande Ronde #5 and Umtanum flow tops).

Based on observations made during the drilling and development of RRL-2B and RRL-2C, the transmissivity of the Rocky Coulee flow top in the vicinity of RRL-2B and RRL-2C appears to be near the geometric mean value of the small-scale test values of the Rocky Coulee flow top in the RRL area or about 0.25 m^2/d (2.6 ft^2/d).[4] Numerical simulation using this value and the large-scale stress scenario (30-50 d of pumping with a drawdown of 230-300 (750-1000 ft) indicates that, with a discharge rate of 0.03 m^3/min (8 gal/min), there should be a measurable water-level response to pumping at DC-22 and DC-20 but probably not at DC-19.

A. Test Facilities

Large-scale hydraulic tests conducted in the Grande Ronde Basalt at the RRL-2 site will utilize one pumping well, designated RHL-2B, and two observation wells close to the pumping well. These observation wells are designated RRL-2C and RRL-2A, located 75 m (250 ft) east and 150 m (500 ft) south of the pumping well, respectively. Tracers will be injected at these two facilities as discussed in a subsequent section. The other wells, piezometers, and boreholes completed in the Grande Ronde Basalt Formation that will be used as observation points during the tests appear on Figure 1 and include piezometer clusters DC-19, DC-20, and DC-22; boreholes RRL-6, RRL-14, and DC-4; and the McGee Well.

The two most recently completed facilities, RRL-2B and RRL-2C, have some rather special features that are discussed briefly below. Details of the other facilities have been published.[5,6,7]

1. Well RRL- 2B - The pumping well, designated RRL-2B, will be advanced incrementally through the Grande Ronde Basalt so that each hydrogeologic unit to be tested can be investigated individually and then sealed before proceeding to the next hydrogeologic unit. Upon completion of testing in each hydrogeologic unit, the well will be lined with steel casing with the annulus between the casing and the borehole cemented before drilling to the next test horizon.[8] The lower most test horizon will be left open after testing for future monitoring (Figure 4).

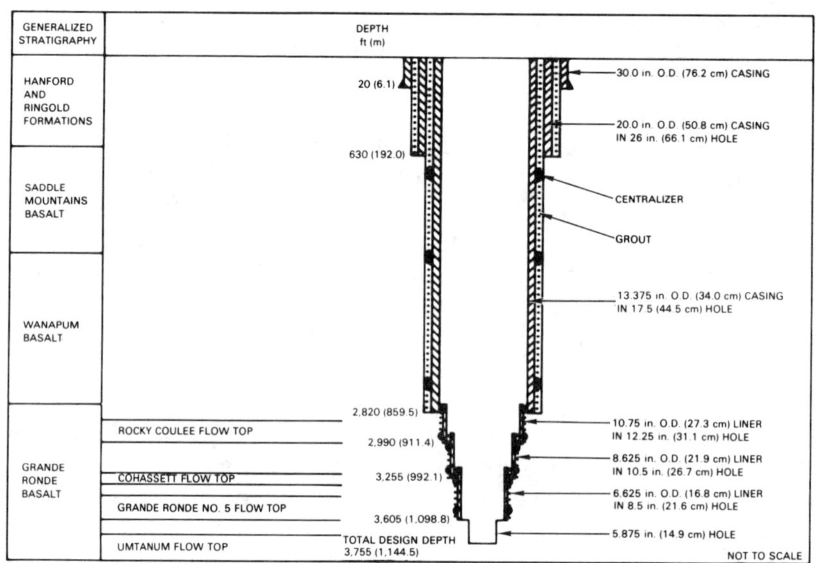

FIGURE 4. Design of Well RRL-2B

2. Well RRL-2C - Observation well RRL-2C will provide the means to measure head and formation pressure in throe flow tops in the Grande Rondo Basalt (Rocky Coulee, Cohassett, and Grande Ronde #5) and formation pressure in three flow inferiors (Rocky Coulee, Cohassett, and Grande Ronde #5). This multiple-level piezometer differs from others on the Hanford Site in that the RRL-2C design permits pressure monitoring of basalt flow interiors as well as flow tops.[8] The general design of this facility is shown in Figure 5. Formation pressure in the flow interiors will be measured using a transducer which will be shut-in with a packer set at formation depth. Well RRL-2C, with its piezometers placed within flow interiors and flow tops, provides an opportunity to estimate vertical hydraulic conductivity of the flow interiors using the ratio method.[9]

B. Test Design

Test design to date has been focused upon the first suite of tests which will be at the RRL-2 site. Test design as discussed here deals specifically with the first test which will be of the Rocky Coulee flow top. There are several experiments that are being attempted in the first test, namely:

- radial convergent pulse tracer tests for estimation of porosity and longitudinal dispersivity of the interflow,

- the so-called ratio test[9] for estimation of diffusivity of the flow interior subadjacent to the Rocky Coulee flow, and

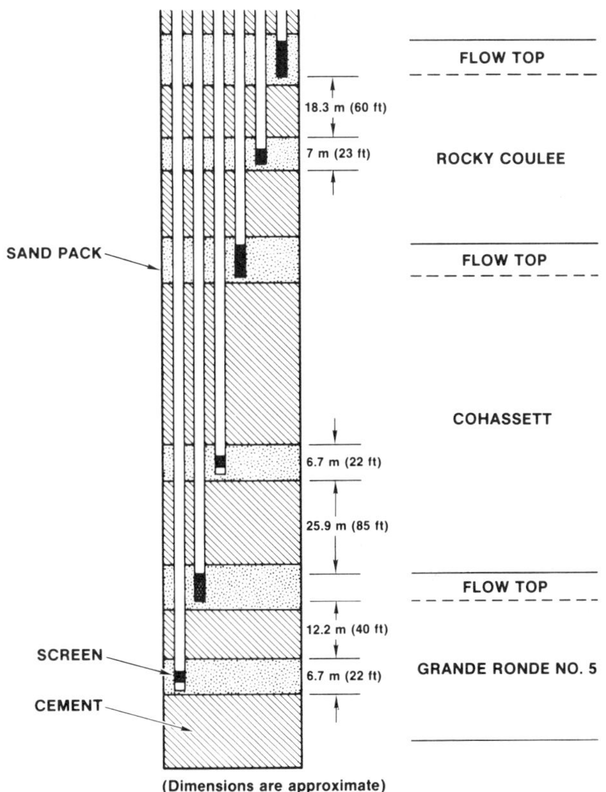

FIGURE 5. Piezometer Completions in Well RRL-2C

- constant rate pumping test for estimation of interflow transmissivity, storativity, and boundary conditions; and leakance from adjacent flow interiors.

These planned tests have required two major design efforts: the design of the ratio test and constant rate pumping test which will be referred to as the hydraulic test design and the design of the radial convergent tracer test. The approach to the hydraulic test design required the use of two numerical models. A quasi 3-dimensional finite difference model was used to estimate the areal response in the pumped interflow (Rocky Coulee interflow) and adjacent interflows.[10] With this model test response was simulated with a variety of potential boundary conditions and also assumed heterogeneities. However, it was decided that complex boundary conditions and assumptions about heterogeneities were not commensurate with the available data. So, for design, it is assumed that homogeneous, isotropic conditions exist, and boundaries are assumed to be either no-flow or set to infinite.

An axisymmetrical, finite element model was used to simulate response in both interflow and intervening basalt flow interiors for various ranges of assumed hydraulic parameter values and testing scenarios (discharge rate and test duration).[11] The simulations were useful in the design of the interior piezometers in monitoring well RRL-2C, specifically, the vertical distance from the piezometer to the basalt interflow-interior interface.

Several ranges of hydraulic parameter values were used in the initial test design (sensitivity studies). But obviously, final design must rely on one value for each of the pertinent parameters which are interflow transmissivity and storativity and interior vertical hydraulic conductivity. At this time, the assumed values of these parameters for the Rocky Coulee, for the purpose of test design, are:

- transmissivity, $T = 0.25\ m^2/d$ ($2.6\ ft^2/d$),
- storativity, $S = 10^{-5}$ (dimensionless), and
- vertical hydraulic conductivity, $k_v' = 1.5 \times 10^{-6}$ m/d (5×10^{-6} ft/d).

The design flow rate based on these assumed parameter values is 0.03 m^3/d (8 gal/min) resulting in a water level drawdown at the pumping well, RRL-2B, of about 270 m (875 ft) after 50 d pumping. Response should be evident in the Rocky Coulee interior piezometers at RRL-2C, which will allow the use of the ratio method, as well as at the monitoring facilities DC-20 and DC-22 (Figure 1).

The tracer test will be performed by injecting tracers into RRL-2A and RRL-2C and then observing the change in concentration of tracers with time at pumping well RRL-2B. Tracers have been selected based on background concentration present in the Grande Ronde groundwater, detection limit of the tracer, and the anticipated tracer behavior in basalt. At RRL-2A, LiBr and deuterium will be injected. At RRL-2C, NH_4SCN will be used.

Expected results from the evaluation of tracer test data are estimates of effective porosity and longitudinal dispersivity of the Rocky Coulee interflow zone. Knowledge of the thickness of the hydraulically active portion of the interflow zone is necessary for estimating the effective porosity. This thickness can be estimated from such data sources as borehole geophysical logs and core samples.

The objectives of tracer test design are to 1) select a discharge rate that will result in a tracer velocity such that the travel time between injection well and pumping well is within the scope of the overall test (30-50 d), and 2) select a tracer mass that is great enough to be detected at the pumping well but can yet be injected at reasonable concentrations within a reasonable time. The approach to tracer test design is based on an approximate solution for radial convergent flow from an injection well to a pumping well of an instantaneously injected tracer. This approximate solution is given by Lenda and Zuber (1970).

Several ranges of parameter values were used in the initial tracer test design. Although final design has not yet been completed, the results of the initial design (sensitivity studies) indicated that the mean transit time should be on the order of 1 d from RRL-2C to RRL-2B, 75 m (250 ft) separation and 5 d from RRL-2A to RRL-2B, 150 m (500 ft) separation. These results are based upon assumed parameter values of 0.84 m (2.75 ft) for longitudinal dispersivity, 0.003 m (.01 ft) for the product of interflow thickness and effective porosity, and a flow rate of 0.03 m^3/min (8 gal/min) from hydraulic design.

REFERENCES

1. Nuclear Regulatory Commission, "BWIP Site Technical Position No. 1.1: Hydrogeologic Testing Strategy for the BWIP Site," Division of Waste Management, Washington, D.C. (1983).

2. S. P. Reidel et al., "New Evidence for Greater than 3.2 Kilometers of Columbia River Basalt Beneath the Central Columbia Plateau," RHO-SA-162A, Rockwell Hanford Operations, Richland, WA (1981).

3. D. A. Swanson and T. L. Wright, "Guide to Field Trip Between Pasco and Pullman, Washington, Emphasizing Stratigraphy and Vent Areas and Inter-canyon Flows of the Yakima Belt," Proceedings, Geological Society of America, Cordillan Section Meeting, Pullman, Washington, Field Guide 1, p. 33 (1976).

4. S. R. Strait and R. B. Mercer, Hydrologic Property Data from Boreholes on the Hanford Site, SD-BWI-DP-051, Rockwell Hanford Operations, Richland, WA (1985).

5. R. L. Jackson et al., Piezometer Completion Report for Borehole Cluster Sites DC- 19, DC-20, and DC-22, SD-BWI-TI-226, Rockwell Hanford Operations, Richland, WA (1984).

6. L. C. Swanson and B. A. Leventhal, Water-level Data and Borehole Description for Monitoring Wells Used by the Basalt Waste Isolation Project, SB-BWI-DP-042, Rockwell Hanford Operations, Richland WA (1984).

7. T. M. Wintczak, Principle Borehole Report RRL-2, SD-BWI-TI-113, Rockwell Hanford Operations, Richland, WA (1984).

8. R. L. Jackson and R. L. Jones, "Drilling and Completion Specifications for Boreholes RRL-2B (Test Well) and RRL-2C (Multi-level Piezometer Nest)," SD-BWI-TC-023. Rockwell Hanford Operations, Richland, WA (1985).

9. S. P. Neuman and P. A. Witherspoon, "Field Determination of the Hydraulic Properties of Leaky Multiple Aquifer Systems," Water Resources Research, Vol. 8, No. 5 (1972).

10. M. G. McDonald and A. W. Harbaugh, A Modular Three-Dimensional Finite-Difference Ground-Water Flow Model, U.S. Department of the Interior, U.S. Geological Survey, Reston, VA (1984).

11. Golder Associates, Golder Groundwater Computer Package, Bellevue, WA (1983).

ROCKSALT SOLUTION AND NUCLEAR WASTE DISPOSAL - THE CONCEPT OF OVERBURDEN

P. Berest
M. Ghoreychi
L.M.S., Ecole Polytechnique
21128 Palaiseau Cedex, France

R. Andre Jehan
ANDRA 31,
33, rue de la Federation
75015 Paris, France

M. Kiener
GRECO 52,
Museum d'Histoire Nationelle
43, rue Buffon
75005 Paris, France

ABSTRACT

Ground water flow is a major danger for nuclear waste disposal, particularly in the case of rocksalt. In the case of bedded salt, the geological situation is often favorable. Salt is separated from ground water by thick layers of relatively impervious stratas (overburden). However, carrying out a disposal may disturb this favorable situation. Particularly the most frequently considered option is a "compact" design of a disposal in salt; the result being an important rise in temperature; the salt thermal expansion leads to significant mechanical effects on the salt and its overburden. An analysis of these effects and the study of the concrete properties of the overburden of a French salt deposit are discussed.

I. INTRODUCTION

The host rocks for high level waste disposal contemplated within the European community are rocksalt, clay and granite. In fact, these media hold little water and/or only allow low water flow rates. Rocksalt has these two properties. However rocksalt raises a particular problem due to its solubility. The water flow can, in some extreme conditions, lead to the destruction of the deposit.[1]

This risk must not be exaggerated; rocksalt has remained unchanged for millions of years in numerous deposits whose total surface makes up a significant part of the surface of immerged lands.[2] Nevertheless in the carrying out of a disposal, the heat emission from the wastes can perceptibly modify the natural conditions. The solution risk is thus an essential factor in the choice of a site and for the design of a disposal system.

II. SALT DISSOLUTION BY UNDERGROUND WATER

A. Water Flow Characteristics

Salt mine flooding is an extreme example of the action of water. This kind of accident is not interesting as such from the point of view of disposal, for deposits (and mining technology) make this accident most unlikely. However, the study of flooding brings in a very useful "model" of dissolution phenomena. The Ronnenberg accident W.G. is a classical example (Figure 1). The water inflow takes place in a zone where salt is divided into two distinct masses by almost vertical marl/anhydrite layers. The cumulative water inflow over the period 1905-1973 amounted to approximately 200,000 m^3; then 240,000 m^3 for the year 1974. Later on, the brine flow rate into the mine reached 1 m^3/mn in April 1975; 1.5 m^3/mn on June 27, 1975; and 3.7 m^3/mn on June 29. In spite of the efforts put forth in order to reduce the flooding, the flow rate reached 15 m^3/mn on June 30. The mine was then abandonned. The flow rate reached 60 m^3/mn on July 10.

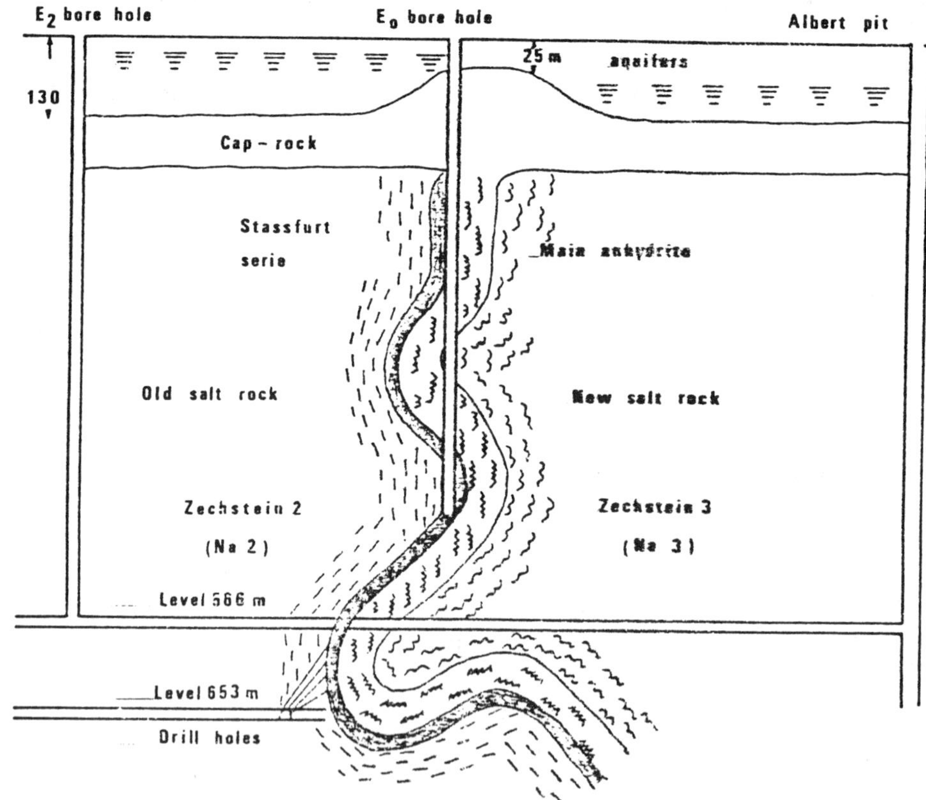

FIGURE 1. Ronnenberg Salt Mine

Some aspects concerning this type of accident should be underlined:

1. The open galleries of a mine build up a hydraulic "potential pit"; they offer an exsurgence to the water flow. This allows rapid and active dissolution. Indeed, when the salt and the overlying ground water come into contact, a saturated brine layer appears in contact with salt and tends to stagnate because of its own weight. It thus protects the salt. Rapid leaching requires a situation where the brine can flow.

2. When non-saturated brine flow is possible, its flow rate may increase very rapidly. Indeed, the brine enlarges its own flow channel in salt through dissolution.

3. The geological context is unfavorable. The salt is separated by a discontinuity which allows the water flow from the very beginning. Generally speaking, many European salt mines can be found in districts where the presence of salt has been known for centuries owing to the existence of salt springs. These are often fringing deposits, shallow and in lateral or vertical contact with underground water.[3] The deposit is not well protected in these districts; it leads to natural solution and particular risks for the mines. Accidents due to solution mining,[4] or even to natural deposit solution,[5] bring in similar information. They originate from an unfavorable geologic context or the destruction of natural protection by man-made works.

B. Concept of Geological Overburden

Many desposits remain undamaged from any solution for millions of years. This fact proves the efficiency of the natural barriers which separate the deposits from the ground water. The deposit of rocksalt is often followed by a deposit of very thick layers of sediments in which the prevailing materials are those which change into impervious rocks (for example clays). The French subsurface contains rather large quantities of rocksalt.[6,7] This salt seldom gave rise to salt domes. In most cases, it remained protected from ground water by impervious layers. The whole of these layers build up the "overburden" of the salt deposit. From our view point, the existence of such an overburden is a favorable element to be considered in the design of a disposal system.

This concept suits the particular geological conditions in France. It is different from the one considered in other countries (Figure 2). For example, in West Germany, the Gorleben site[8] is made of a salt dome whose top reaches within 300 meters of the ground surface; the salt is in contact with ground water to a depth of 8 km^2; the overburden concept does not exist. But in this case it should be possible to carry out the disposal under a very thick layer of rocksalt which would then become a geological barrier. In French sites the roof of the salt is deeper; the thickness of the salt over the disposal will be smaller; in these conditions the overburden must be part of the geological barrier.

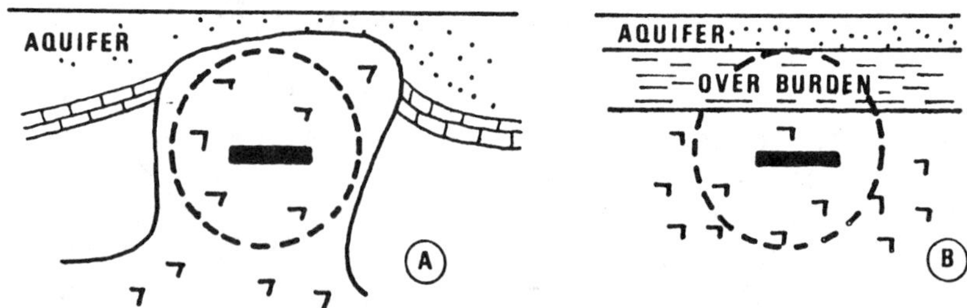

FIGURE 2. The Geological Barrier (A. Salt Dome; B. Salt Layer)

C. Modification of Natural Conditions by Carrying Out Disposal

The first point has not yet been enough considered. The rocksalt seems to have stress relaxation and autosealing properties which makes it more favorable than granite from this point of view. The second point is being carefully studied and results of numerous surveys will be presented in the following paragraphs.

A natural favorable context can be changed by the existence of a disposal system:

1. The drilling of galleries and pits destroys the natural protections of the salt deposit. Even after a crushed salt backfilling, the galleries will retain a significant proportion of empty space which may be an exsurgence for a water flow. The access pits which vertically cross the layers may remain a weakness for a long time, as they sometimes are in oil or gas storage cavities.[9]

2. The rise in temperature due to heat output from the waste can result in irreversible modifications of the salt or its overburden.

III. EFFECTS OF THE THERMAL LOAD

A. Mechanical Behavior of the Rocksalt

The studies linked to waste disposal have allowed significant progress in the study of the rheological behavior of rocksalt.[10,11] A few results have the agreement of most laboratories they involved:

1. Under deviatoric stresses, even moderate ones, the salt creeps without reaching equilibrium even after several months. The behavior of the salt thus resembles that of a highly viscous liquid.

2. A rise in temperature strongly increases the creep rate.

3. The creep rate seems to depend mainly of the second invariant of the deviatoric stress tensor[12] (or of the gap between extreme main stresses).

Our own uniaxial and triaxial creep tests[13] confirm these results. Other controversial questions remain:

1. Creep tests must be carried out over a period of several months. How can an extrapolation for longer periods be made?[14] The BGR of Hanover[15,16] proposes the concept of stationary creep (the creep rate reaches a constant value after a transient phase) but other laboratories wish to show evidence of a time hardening of rocksalt.[17]

2. Is salt a liquid or does it show a certain yield criterion? The latter, if it exists, is probably weak so that its determination in laboratory is difficult. Nevertheless, the existence of such a yield criterion can have significant consequences from a practical point of view when long-term behavior is considered.

3. It seems most unlikely that a universal law on rocksalt creep can be found. A very good fit with a simple law is sometimes possible for a pure and homogeneous salt.[15,16] Our experience is that French salts are relatively impure and show a great variety of crystal size and water content, proving on the contrary a wide dispersion.

B. Rocksalt Failure

The problem of rocksalt failure is one of the most difficult ones in rock mechanics. It includes two aspects: the research of a local failure criterion and the debate over the emergence and propagation conditions of failures in a structure (i.e. for a given geometry and loading conditions).

1. The search for a local criterion is carried out on samples in the laboratory. The results obtained are often questionable; some factors such as the geometry of the sample play a significant role in the conditions of emergence of the failure. Thus, even at a laboratory scale, the local and general aspects (i.e. concerning a particular structure) are closely linked.

2. Nevertheless we have a few results. As the creep rate seems to depend only on the gap between the main stresses, the emergence of the failure depends on other factors. The uniaxial compression, the triaxial compression, and the triaxial extension cause failures of different nature. This difference can be compared to in situ observations: the spalling of the walls of the salt mine pillars is frequent (simple compression or triaxial extension) whereas the emergence of discontinuities within the rocksalt massif is more seldom than for most other rocks (triaxial compression).

3. If we do not consider the galleries, the studied structure is an infinite semispace loaded by deep heat sources of decreasing power. The medium is thus subjected to expansion followed by contraction. The expansion effects will be debated in the following paragraphs. The contraction phase may also cause problems which have been less studied until now.[18] Let us

consider a sample placed in a stiff frame (Figure 3a); its behavior is elastoviscoplastic (Figure 3b). It is subjected to heating followed by cooling (Figure 3c). Considering the viscosity of the material, the thermal loading rate has a major influence on the moment and on the nature of the failure (Figure 3d).

C. Nature and Importance of the Thermal Loading

In the case of rocksalt, the thermal loading is particularly significant because of the very "compact" concept generally considered. A very simple comparison can be carried out by looking into the concepts presented in the report. "Admissible thermal load" published in 1982 by the European community (report EUR 7620 EN/FR).[19] This report deals with clay, rocksalt and granite. Its interest is only instructive; nevertheless it clearly explains some of the differences.

1. Estimate of the Maximal Temperature in Disposal

It is assumed that the disposal system has the shape of a not very thick horizontal disc (Figure 4A). The report suggests a very useful concept, that of the average thermal output (m^2) obtained by the division of the total thermal output laid in by the horizontal surface of the disposal system. This output decreases with time. For a period of about one century, the approximate following formula is adequate:

$$q = q_o \exp(- t/t_o)$$

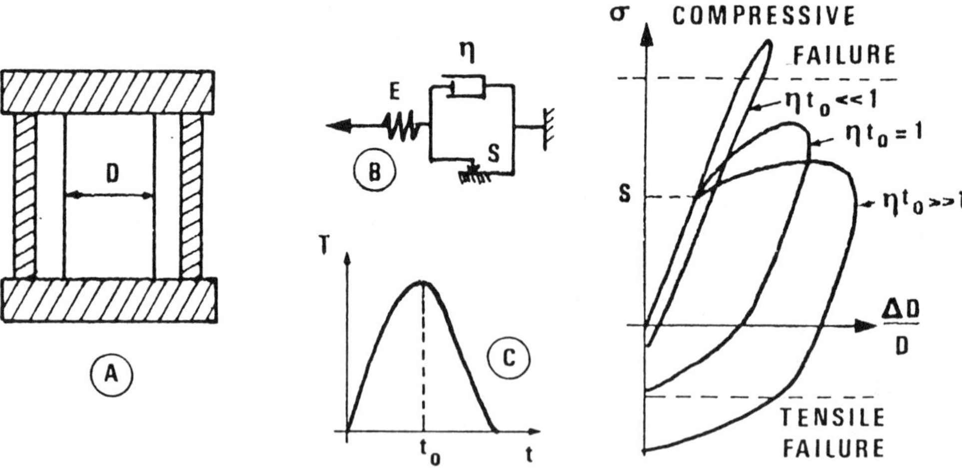

FIGURE 3. Failure Modes Under Thermal Loading

where t_0 is on the order of 40 years. The heat produced is discharged by thermal conduction,

$$\rho C \frac{\partial T}{\partial t} = K \Delta T$$

The solution of this problem is easy. For an estimate of the scale of sizes,[20] the evolution of the temperature in the centre of the disposal is adequate.

$$T(t) = q_o \left(\frac{t_o}{K \rho C}\right)^{1/2} \theta(t/t_o); \quad \dot{\theta}(u) + \theta(u) = \frac{1}{(\pi u)^{1/2}}$$

and

$$\theta(o) = o$$

The function $\theta(t/t_o)$ (Dawson function) accepts a maximum of 0.315 reached for $t = 0.85\ t_o$ (34 years). The "thermal climax" is thus characterized by the maximal temperature:

$$T_{max} = 0{,}315\ q_o \quad (t_o/K \rho C)^{1/2}$$

This simple formula serves to check the computer calculations; it also serves to easily debate the influence of the various parameters for t_o which are characteristic of the wastes; $K \rho C$ depends on the media considered. On the contrary, the parameter q_o (average thermal load emitted by horizontal surface unit) is not prescribed; it is the result of the choice of a certain design of the disposal system.

	$\rho \times C$ (J/m²/°C)	K(N/m/°C)	$(t_o/\rho C K)^{1/2}$	q_o(N/m²)	Tmax
CLAY	2040 x 940	1,5	27	2,5	16,5°C
GRANITE	2700 x 800	2,5	15	6,67	32°C
ROCKSALT	2160 x 850	5,2	11,5	> 50	> 181°C

2. Consequences of the Heating During the Thermal Climax

Considering clay and granite, disposal must be spread because the thermal diffusivity properties ($\sqrt{t_o/\rho ck}$) are weak and particularly because severe rules concerning the maximal temperature have been accepted in the quoted document, in order to avoid dessication (clay) and thermoconvection (granite).

In salt, on the contrary, the disposal is compact. This is allowed by favorable thermal properties and less severe requirements concerning the maximal temperature: the problems already quoted for clay and granite do not have the same importance in the case of salt. It is important to emphasize that this compact concept is not necessary. It has some advantages: a compact disposal system is less vulnerable to certain kinds of accidents. But this concept gives a certain acuteness to the problems of rock mechanics. The rise in temperature has two distinct effects:

- It modifies the mechanical characteristics of the rocks, generally by making them more ductile and, in the case of salt, in considerably reducing the viscosity (in simple terms, the hot salt will rapidly creep to resorb the shear stress as a liquid tending towards an equilibrium position). This circumstance should be favorable to the creation of an adequate sealing of the hottest zone in which the containers are placed.

- It brings about a thermal expansion which is due to the mechanical loading and whose effects will be noticeable up to the ground surface. It can be easily proven (Figure 4B). Let us imagine that the expansion is free. The land cylinder surrounding the disposal is free from any contacts with the massif (this assumption is unrealistic; it only serves to assess a scale of sizes). Thus, the heat emitted during time T by 1 m^2 of the horizontal surface of the disposal system, can be found in the vertical cylinder leaning on this specimen.

$$\int_o^T q\ (t)\ dt = \int \rho CT\ (z)\ dz$$

If this cylinder is free to expand vertically, it will extend by $\Delta H = \int \alpha T\ (z)\ dz$. Thus, after a period of $T = t_o = 40$ years the result will be:

$$H = \alpha/\rho\ C \int_o^{t_o} q\ (t)\ dt = \alpha q_o t_o/\rho\ C\ (1 - \frac{1}{e}) = 8 \times 10^8\ \alpha q_o/\rho C$$

with the resumption of the previous numerical values.

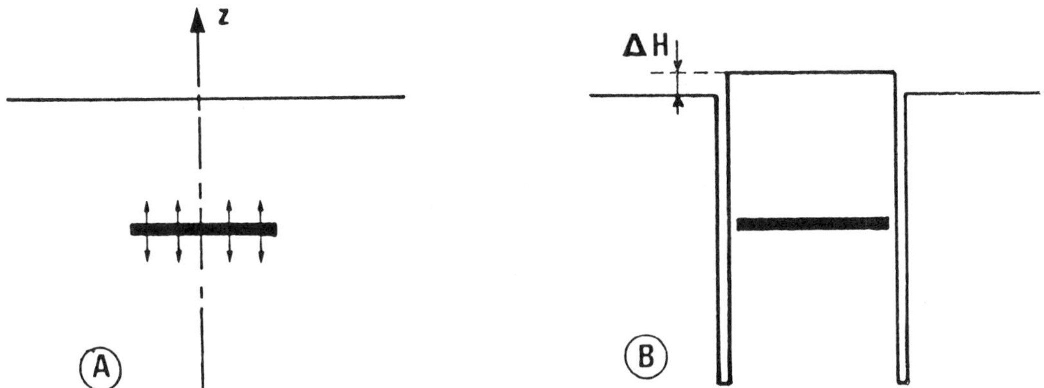

FIGURE 4. A - Heat Output; B - Thermal Expansion

	$\alpha(o/c)^{-1}$	q_0 (W/m^2)	Δ H
CLAY	10×10^{-6}	2,5	1 cm
GRANITE	8×10^{-6}	6,67	2 cm
ROCKSALT	43×10^{-6}	50	90 cm

From a mechanical point of view, the physical properties of the rock-salt (thermal expansion coefficient) develop the effects of a high thermal output per surface unit. This leads to mechanical stability problems at the scale of the whole massif. These problems appear far more serious in the given concept than those concerning granite and clay. In order to indicate clearly these problems, the previous rough description must give way to computer calculations.

3. Mechanical Effects of Heating in the Overburden

In the following calculation, salt is considered as a non-linear viscoelastic medium showing a strong influence of the temperature on the viscosity. The remaining part of the massif is elastic[21]. The emitted thermal output corresponds to that contemplated in the previous paragraphs. The situation 40 years after the deposit of the containers is discussed.

- Figure 5 shows the values of the second invariant of the deviatoric stress tensor (often quoted as J2). This invariant represents a variation compared with a hydrostatic distribution of stresses (i.e. the intensity of the shear stresses). There is no variation far from the disposal, which is a normal situation. The variation

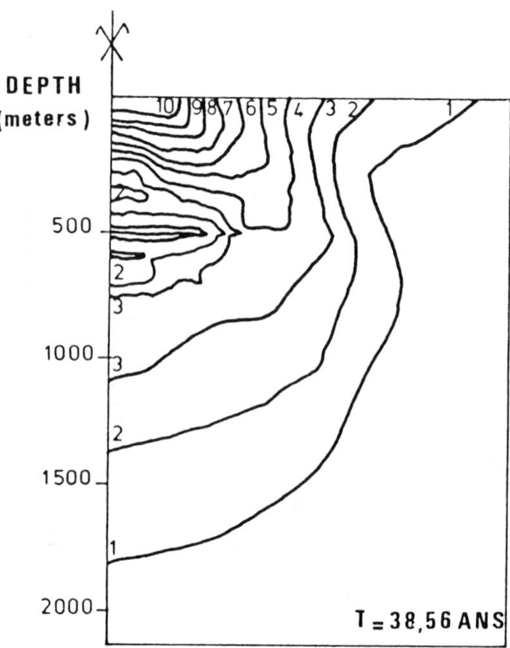

FIGURE 5. Second Invariant of the Deviatoric Stress

increases with the proximity of the disposal system. It approaches zero again inside the disposal system in the hottest zone. In this zone the viscosity is poor so that the stress relaxation rate is high. This is very characteristic of the behavior of rocksalt. If this behavior was only elastic, there would be on the contrary, noticeable variations compared with a hydrostatic distribution of stresses near the disposal system.

- The first invariant of the stress tensor which is the mean pressure (often quoted as I1) has not been represented. At an initial state, it increases with depth in the proposed calculation. After 40 years, it varies perceptibly (from about 10 MPa) from the natural distribution near the disposal system. There can be no free thermal expansion. It is partly prevented by the stiffness of the overlying ground (cold salt and overburden). More precisely, the vertical thrust caused by the expansion of the heated ground makes the thick plate made up by the overlying ground, bend, which opposes the complete upheaval assessed in the previous paragraphs. It leads to an upheaval of the ground surface, 1.20 meters directly above the disposal center (but which rapidly decreases when moving off the center), and the emergence of tensile stresses near the ground surface which is normal in the case of a bent plate.

It is obvious that the given numerical values are only indicative. The ground situated above the disposal system makes up a sole medium in the calculation, continuously homogeneous and isotropic. In nature, the general behavior of such ground is far more complex; their general stiffness must be inferior to the one assessed. (For example, if the ground is made of a piling of layers likely to slide at their interfaces). The tensile stresses appearing at the surface are difficult to assess. However, they do not a priori show any serious risk for the geological barrier which does not include the first ten meters of ground below the ground surface. The shear stresses which appear at the periphery of the ground cylinder above the disposal system are more interesting. On the average and on the total height of the cylinder, they must balance the thrust of the expanded hot salt.

Taking into account the size and the depth of the disposal system, the scale of the mean vertical shear stress in the cold part of the salt and in the overburden will be on the scale of the increase in the mean pressure in the disposal system, that is to say, 10 MPa. This is an average value. The exact distribution greatly depends of the rheological assumptions considered for the different materials. The compact design of the disposal system in salt thus leads to a strong stress of the ground above the disposal system. The properties of the ground must be carefully assessed in order to study an eventual failure emergence.

IV. OVERBURDEN OF THE BRESSE DEPOSIT

A study of the Bresse deposit, to the north of Lyons, has been undertaken. In the drill hole studied (Etrez 08, May 83) the overburden lies between a roof of Stampian salt at 714 m depth and a tortonian water-bearing stratum at 436 m depth. Results of creep tests on Stampian salt samples are referred to above (see II A). The lower part of the overburden is clayey and the upper part limestone. There are also anhydrite insertions. Cores have been taken from about 50 m of rocks, spread in three characteristic areas of the overburden. A test program has been carried out from these samples.

Testing included:

1. petrographic and sedimentological analyses undertaken by the Geological Department of the Museum of Natural History,[22]

2. study of hydrogeological properties (porosity, permeability), and

3. study of mechanical properties (uniaxial compression, Brazilian test, deflection test, triaxial test, creep test).

These tests have revealed a moderate permeability and considerable deformability of the overburden. The limestone and marl exhibit a ductile behavior, when samples are submitted to even a moderate confining pressure. The rigid materials (anhydride) are present as layers of only slight thickness, often broken up in the natural state under the effect of differential settlement. On the whole, their mechanical role seems slight. The large scale behavior of the overburden is still to be investigated, as for those results to be obtained from laboratory testing of core-samples. But from this first study it is possible to conclude that, in the site investigated, the overburden presents relatively favourable properties.

V. SUMMARY

The confinement of high level radioactive wastes must be secured over a very long period. It is therefore most interesting to make good use of the favorable natural geological circumstances whose longevity seems assured. From this point of view, rocksalt shows creep and self-healing properties which are a definite advantage.

The protection created by nature between a salt deposit and ground water can also be used. We must then check that the thermal load is consistent with the preservation of the wholeness of this protection. These considerations will play an important part in the selection of a favorable site and in the design of the disposal system.

REFERENCES

1. J. Goguel, Application de la Géologie aux Travaux de l'Ingeniéur, 2nd Ed., Chap. 7, p. 152, Masson et Cie, Paris (1967).

2. E. C. Pendery, "Distribution of Salt and Potash Deposits: Present and Potential Effect on Potash Economics and Explorations," Vol. 2, Proceedings of the 3rd Symposium on Salt (1966).

3. J. Bonvallet, P. Duffaut, Historique de l'Exploitation du Sel Gemme en Lorraine, Service Geologique Regional de Lorraine, BRGM (1979).

4. P. Combes, E. Ledoux, G. de Marsily, "Etude des Dissolutions Naturelles du Sel en Couche," in Journée du Sel, p. 151, P. Habib and P. Berest Editions, Ecole Polytechnique (1984).

5. Davies, "Structural Characteristics of a Deep Seated Dissolution-Subsidence Chimney in Bedded Salt," Sixth Symposium on Salt, May 25-27 (1982).

6. A. Autran, A. L'Homer, M.J. Lienhardt, R.-André Jehan, "Inventaire des Formations Saliféres en France," in Journée du Sel, p. 19, P. Habib and P. Berest Editons, Ecole Polytechnique (1984).

7. G. Fries, B. Beaudoin, P. Berest, "Eléments d'un Inventaire des Gisements de Sel Francais," Annales des Mines, p. 39, n° 5/6, May-June 1983.

8. M. Langer, "La Géologie de l'Ingénieur Peut-elle Aider a Resoudre le Probleme du Stockage des Dechets Radioactifs," Engineering Geology, p. 5, no 28, December 1983.

9. P. Berest, J. Goguel, "Accidents de Stockages Souterrains d'Hydrocarbures," Annales des Mines, p. 21, no 5/6 (May-June 1983).

10. M. Langer, "Rheological Behaviour of Rock Masses," Proceedings of the 4th International Congress on Rock Mechanics, Montreux, Vol. 3, p. 29, Balkema, Rotterdam (1979).

11. W. R. Wawersick, D.S. Preece, "Creep Testing of Salt-Procedures Problems and Suggestions," Proceedings First Conference on the Mechanical Behaviour of Salt, p. 421, Pennsylvania State University, November 1981, Trans Tech Publications, Clausthal, Germany (1981).

12. S. Heusermann, "Kritische Gegenuberstellung und Bewertung von Stoffgesetzen zur Beschreibung des Kriechverhaltens von Steinsalz auf der Grundlage von Laboruntersuchung un in situ-Messungen," Forschungsergebnisse aus dem Tunnel- und Kavernenbau, Heft 6, Universitat Hannover (1982).

13. J.P. Charpentier, P. Berest, "Fluage du Sel Gemme en Temperature. Moyens d'Essais et Resultats," Revue Generale de Thermique, p. 282-283 (June-July 1985).

14. C.J. Spiers and al., The Influence of Fluid-Rock Interaction on the Rheology of Salt Rock on Ionic Transport in the Salt," Periodic Report, contract n° WAS-153-80-7N(N), European Energy Community (1983).

15. M. Langer, "The Rheological Behaviour of Rock Salt," Annales des Mines, p. 201, no. 516 (May-June 1983).

16. U. Hunsche, "Results and Interpretation of Creep Experiments on Rock salt," Annales des Mines, p. 159, no. 516 (May-June 1983).

17. G. Vouille and al., "Experimental Determination of the Rheological Behaviour of Tersanne Rock Salt," Annales des Mines, p. 407, no. 516 (May-June 1983).

18. P. Berest, "Elastoplastic closed from solution for a spherical distribution of temperature in an infinite medium," submitted to Journal of Thermal Stresses.

19. Proost, Charge Thermique Admissible en Formations Geologiques - Conséquences sur les methodes d'Évacuation dés Déchets Radioactifs. Rapport CCE n° EUR 8179, Luxembourg (1983).

20. P. Berest, "Modéles Alytiques Pour les Contraintes Thermiques dans le Cas des Milieux Granitiques," in Journée sur le Granite, p. 249, documents du BRGM, n° 84.

21. H. Rolnik, "Dimensionner Aujourd'hui un Stockage de Dechets de Haute Activité dans le Sel Gemme : Quelle Rhéologie?" Thèse de Docteur Ingénieur a l'ENSMP, Ecole Polytechnique, Palaiseau (September 1984).

22. M. Kiener, "Etudes Sédimentologiques et Corrélations par Diagraphies: le sommet de la Série Salifère et sa Couverture dans le Sondage Ez 08," in Journée sur le Sel, p. 207, P. Habib and P. Berest Editons, Ecole Polytechnique (1984).

BOREHOLE GEOPHYSICAL INVESTIGATIONS OF A DEEP TESTHOLE IN FINLAND

Pauli Saksa
Technical Research Centre of Finland
Geotechnical Laboratory
SF-02150 Espoo, Finland

ABSTRACT

According to the Government's decision in principle in 1983, Industrial Power Company Ltd (TVO) is making preparations for all the steps of final disposal of the spent fuel produced by its power plants. Before the actual site investigation phase, TVO drilled a deep borehole in Lavia, western Finland. The borehole was used during 1984-85 for testing investigation techniques and methods used for bedrock characterization. Borehole geophysical logging performed in Lavia consisted of galvanic electrical, electromagnetic, radiometric, temperature, seismic and magnetic measurements. This composite survey provided both lithological and structural information on the rock mass. The neutron-neutron, density, natural gamma radiation and susceptibility methods characterized the rock type. Fracturing and its type could be interpreted most effectively with resistivity, P-wave velocity and density logs. Temperature and tube-wave measurements revealed several fractured zones relating to possible water flow in rock. Lavia investigations indicated that a high quality of instrumentation and careful calibration are necessary for this type of site investigations.

I. INTRODUCTION

In the high-level nuclear waste management program of the Industrial Power Company, which owns the Olkiluoto power plant, investigations are being made for final disposal of spent fuel in Finland. According to the schedule for the site investigations, the first stage took place in 1984-85 when an extensive research project was carried out by the Industrial Power Company and Finnish research organizations. The purpose of the project was to gain experience of different site characterization methods and their applicability.

One essential part of this project was drilling of a 1001 m deep testhole in Lavia in a granitic massif and borehole investigations in the hole. Hydraulic constant head injection tests[1] and water sampling have been performed in the hole. Geological mapping[9] and ground geophysics were conducted at the borehole site. The geographic location of the hole and geological setting in the area are shown in the site map in Figure 1.

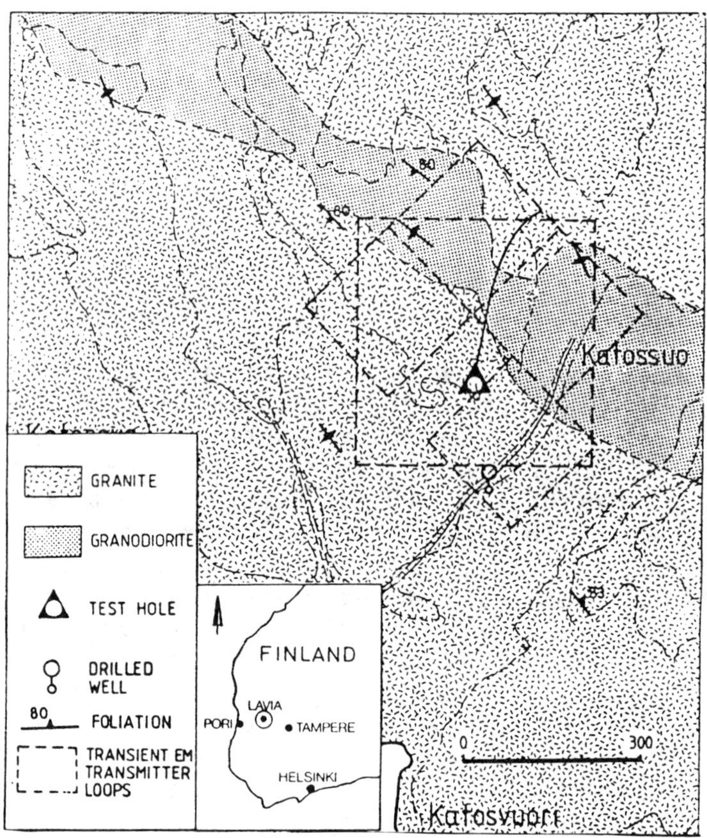

FIGURE 1. Lavia Site Map[9]

II. ROLE OF BOREHOLE GEOPHYSICS

Borehole geophysical methods belong to the class of near-field investigations. The information achieved consists of many physical rock parameters. These parameters can be further related to formation geology and its structure[3]. In crystalline bedrock with very low average-hydraulic conductivity, the fractures and fractured zones are assumed to form the main potential pathways for radionuclide movement. Thus, the mapping of fractures and their characteristics is essential in evaluating crystalline bedrock formations. Borehole loggings map fracturing efficiently and in natural undisturbed conditions. They indicate suitable test sections for hydraulic testing and water sampling. Determinations of groundwater flow directions, possible permeable zones, and elastic-mechanical parameters, can be used in hydrogeological modelling and in simulation of the mechanical behaviour of a rock massif with time.

Borehole geophysical methods can be divided into two subclasses within this site investigation application. The single-hole methods record the geological and structural variations in the vicinity of the borehole, generally within a few metres at best. The purpose of single-hole methods is to provide accurate small-scale knowledge of formation conditions. The constructed structural model can be further tested and extended by crosshole investigations. Both of these methodological types have been used in Lavia investigations during 1984-85.

III. INSTRUMENTATION AND MEASUREMENTS

The logging program in the Lavia borehole consisted of galvanic electrical, transient electromagnetic, radiometric, temperature, seismic and magnetic measurements. The principal logging was performed by the Swedish Geological Company (SGAB) in accordance with an agreement for co-operation with the Swedish Nuclear Fuel and Waste Management Company (SKB)[2]. Finnexploration Company[7] and the Geotechnical Laboratory of the Technical Research Centre of Finland (VTT/GEO) also performed measurements in the borehole. The data processing and interpretation have been carried out by VTT/GEO and SKB/SGAB[5,2]. A list of all the methods tested and their geometric configurations are given in Table 1.

TABLE 1. Borehole Methods Tested in Lavia Borehole 1984[5]

METHOD	ORGANIZATION
* Resistivity logging	
- Single-point	SKB/SGAB
- Short normal 16"	SKB/SGAB
- Long normal 64"	SKB/SGAB
- Long lateral 18'	VTT/GEO
- Wenner-geometry a = 0.318 m	FINNEXPLORATION
- Groundwater resistivity	SKB/SGAB
* Radiometric logging	
- Density (γ - γ)	SKB/SGAB
- Neutron-neutron	SKB/SGAB
- Natural gamma	SKB/SGAB
* Seismic logging	
- Acoustic P-wave	SKB/SGAB
- Tube-wave	SKB/SGAB
* Self-potential	SKB/SGAB
* Temperature	SKB/SGAB
* Susceptibility	FINNEXPLORATION
* Transient electromagnetic (SIROTEM)	FINNEXPLORATION
* Caliper (one arm type)	SKB/SGAB

The experience gained from logging showed that special attention should be given to proper instrument calibration in crystalline rocks and accurate recording of hole distance coordinates[5]. On-site digitization of measurements is also necessary.

IV. RESULTS AND INTERPRETATION

A. General

A comprehensive set of logs is important when evaluating the serviceability of different methods, sources of anomalies recorded, and effects of measurement geometries. One purpose of the Lavia investigations was to study whether separation of lithological and structural property variations is possible by borehole logging. This is important when using percussion drilled holes in site investigations. One has to be able to resolve both lithology and structural variations from borehole logs.

Some essential Lavia logging results are depicted in Figure 2. All the available material, such as drilling and core logging reports[6], reports of core sample petrography and mineralogy[4], laboratory determinations of core sample parameters[8], was of great help in interpretation.

B. Log Analysis

The borehole measurements in granitic massifs each describe characteristic properties. They can be classified into methods most sensitive to fracturing and its intensity, rock type and mineralogy, and those methods sensitive to variations in hydrogeological conditions. Resistivity and acoustic logging are sensitive to fracturing. The general relationship noticed between fracturing intensity and resistivity is presented in Figure 3.

The general correlation between sample values is good. However, it is evident that resistivity values or anomaly amplitudes are not directly related to the degree of fracturing or the permeability of the rock.

Density, susceptibility, natural gamma radiation, and neutron-neutron measurements accurately characterize mineralogical changes in rock, as along the granodioritic rock sections with more abundant biotite and hornblende contents. Granitic rock sections have higher natural gamma radiation levels but lower n-n values, density and susceptibility. Variations in magnetization intensity were in the marginal range--less than 100×10^{-5} SI units[7]. Density log values between 2.6 to 2.8 g/cm^3 agreed also within 1 to 2% with the determinations of core samples made in the laboratory[8]. It was also found that density logging was sensitive to fracturing.

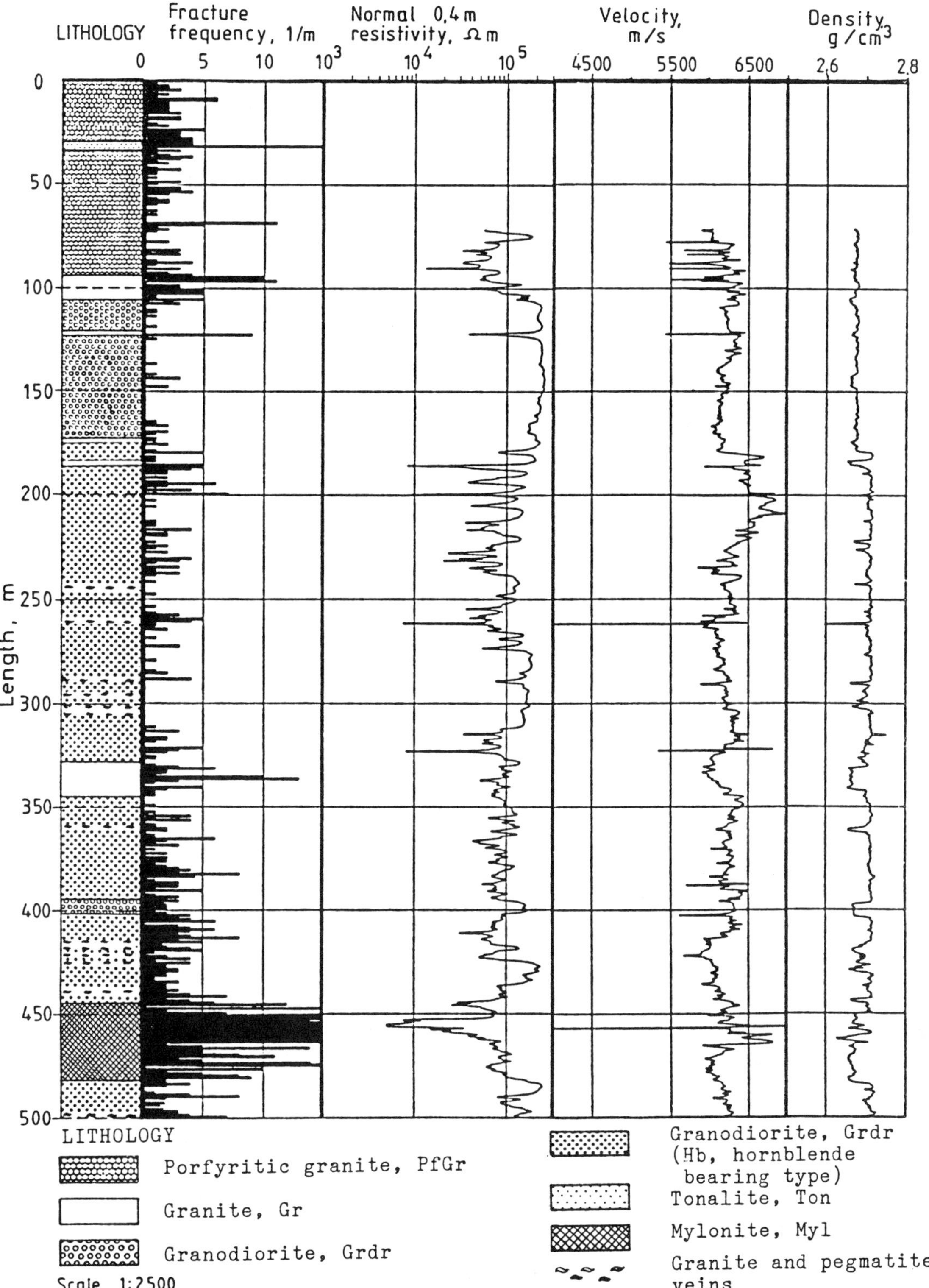

FIGURE 2. Set of Logging Results from the Lavia Borehole (0...500 m)[5]

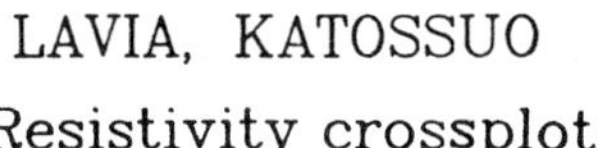

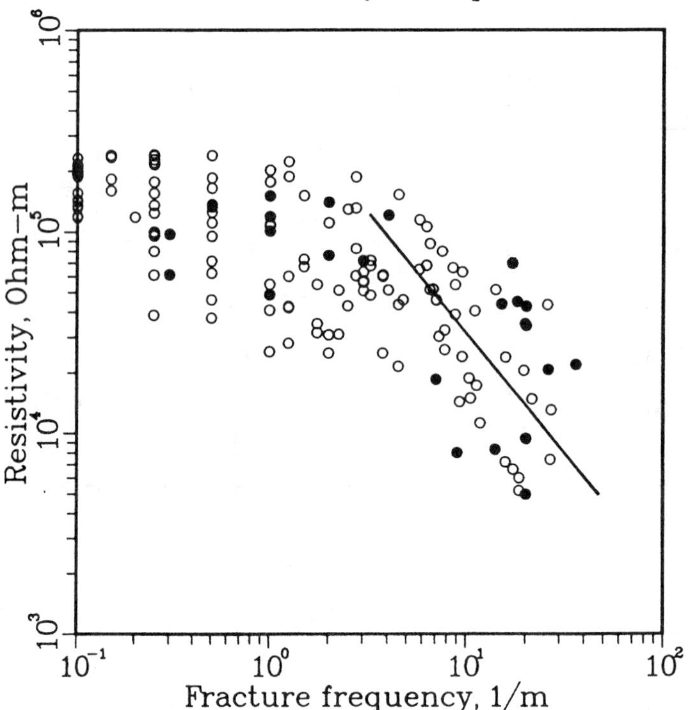

FIGURE 3. Resistivity Versus Fracture Frequency Crossplot[5]

Other methods, related either to groundwater flow in fractures or possibly to hydraulic conducting zones, were temperature, groundwater salinity, self-potential, and tube-wave loggings. The temperature profile along the hole was smooth; the average vertical temperature gradient was 13.5°C/km. Small deviations in temperature and groundwater salinity profiles were recorded in upper 100 m of the hole and near hole bottom. Tube-wave recording was successful in the Lavia hole. Hydraulically conductive or water filled fractured zones were mapped in several places. The tube-wave generating points agree well with recorded hydraulic conductivity values and indicate probable pathways for water flow[2].

When dealing with the problem of mapping fracturing in crystalline bedrock, the most applicable methods are galvanic resistivity and density logging, and acoustic velocity recordings. The usefulness of the above-mentioned methods in crystalline rock is based simply on the fact that the highest contrast between fractured and intact rock exists between electrical conductivity, ability to support propagation of seismic waves, and density. The results of these logs can be further converted into the forms of semi-quantitative apparent-fracture porosity profiles[5]. The basic model of the conversion is a thin fracture plane model embedded in intact rock. The recorded electrical conductivity, density, and transit time values are sums of corresponding rock matrix and fracture property portions between the source and the receiver. In modelling, the fracture parameters are assumed to be the same as for the groundwater.

In spite of the idealizations related to the model, and without defining the physical relationships necessary for actual quantitative modelling, we have found that these conversions are useful. Fracturing effects are more visible from converted profiles, the converted logs are more comparable with each other, and logging results from different rock formations are more comparable with each other.

C. Integrated Use of Logging Results

It is always advantageous to use borehole logging as a suite in structural investigations. Such a composite survey gives additional information about the physical properties of the subsurface[3]. It also helps in resolving geological structure and reduces the uncertainty inherent in the interpretation of borehole logs. As an example of rock-type characterization, Figure 4 shows the general range of natural gamma intensity and susceptibility in the Lava hole. Granodiorite rich in hornblende and biotite differs from the granite. Also tonalite and mylonite have certain parameter ranges in the

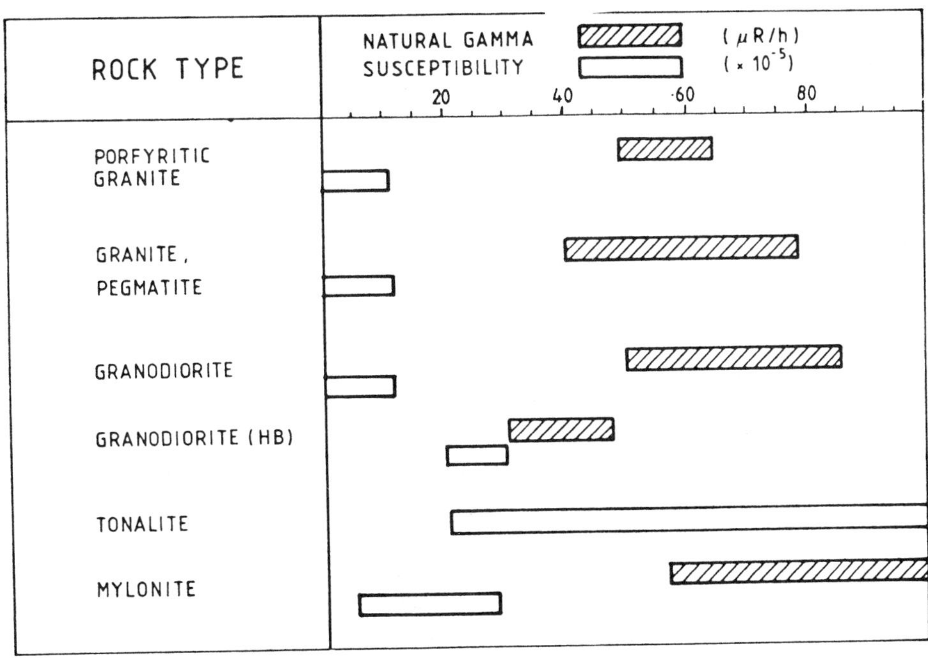

FIGURE 4. Natural Gamma and Susceptibility Variation of Lavia Hole Rock Types

figure. Fractured mylonite zone is associated with low susceptibility and gamma radiation increase, which is probably due to more abundantly occurring fractures filled by hematite, calcite and possibly pyrite[4,6].

Some analysis of apparent-porosity profiles and their ability to infer the type of fracturing can be made with crossplotting. For the evaluation of crossplotting results we selected idealized fractures with infilling materials: electrical conductors, clay minerals, water, non-conductive mafic and felsic minerals, and tight fractures. These characteristic properties of fractures should be visible, for example in resistivity vs. density and resistivity vs. P-wave velocity crossplots.

To test the validity of presented crossplot models, the responses of fractures and fractured zones with known characteristics were selected from resistivity, velocity, and density apparent-porosity profiles. The samples were chosen on the basis that they have homogeneous and confirmed properties. The crossplots of selected water-filled, conductive mineral-filled (conductors, clays), and tight fracturing examples are shown in Figure 5. In

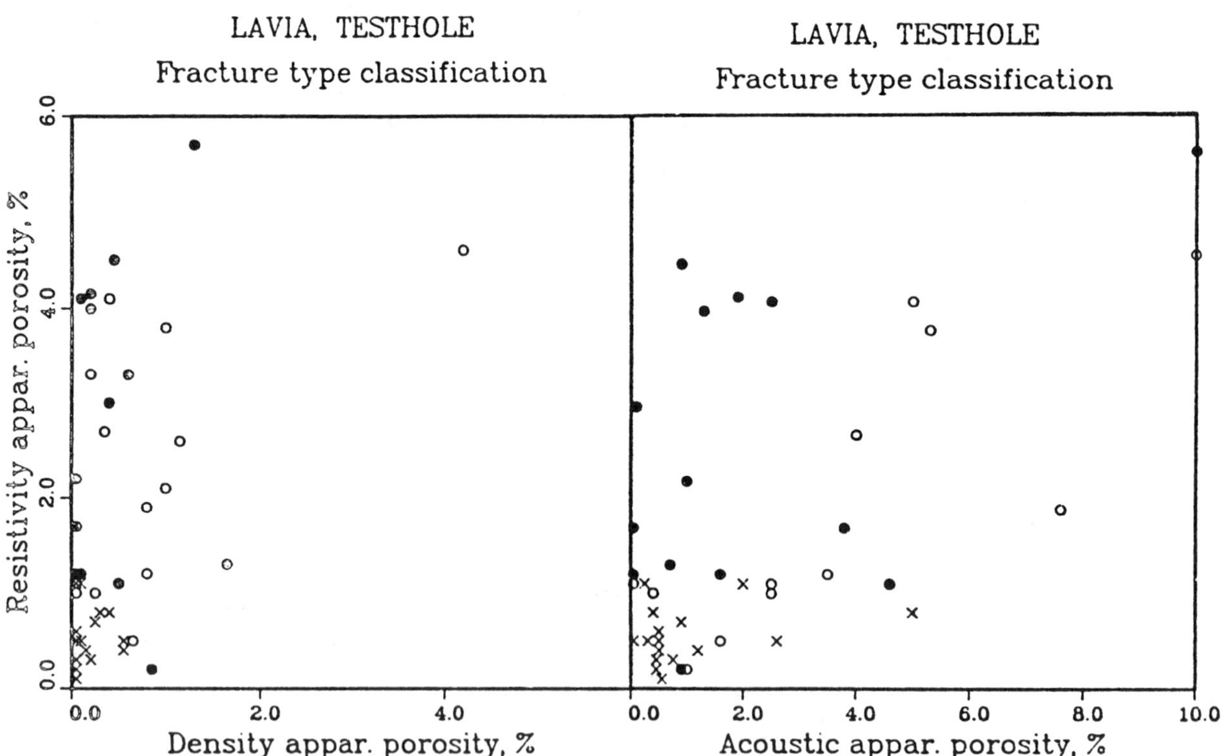

FIGURE 5. Crossplot Results.[5] Fracturing Samples Classified as Filled with Water; Conducting Minerals and Tight Ones are Marked by Open and Shaded Dots and Crosses.

both crossplots the tight fractures are grouped near axis origins, which means low resistivity, velocity, and density apparent-fracture porosity. Even fractures filled with good conductors have higher resistivity-based fracture porosities than those filled with water. Water bearing fractures seem to have the highest porosities derived from acoustic results[5].

V. CONCLUSIONS

The logging results from the Lavia borehole give convincing evidence that borehole geophysics can infer the rock mass in several detailed ways. Both lithological and structural properties can be mapped even in minor scales. The most serviceable logs in lithology studies have been density, susceptibility, neutron-neutron, and natural gamma, in this order. The temperature, groundwater resistivity, and tube-wave measurements were successful also for direct or indirect investigations of hydraulically conductive zones.

Fracturing of rock mass can be studied efficiently with the help of resistivity and acoustic P-wave velocity logs. Density logging, too, was useful in these fracturing studies. Fractures and fractured zones can be accurately positioned. However, the importance of an anomaly cannot be evaluated only with its size, or how large an amplitude it has. As an example, a fracture filled with massive pyrite would give high apparent-resistivity porosity values, but probably not anomalous low-velocity values. Thus, the integrated use of logs and careful study of core samples gives the most reliable description of rock properties. The comprehensive set ofborehole measurements will also allow rock type and fracturing classification in percussion drilled holes, if formation specific petrophysical calibration can be done in a diamond drilled borehole, or by ground geophysical measurements.

VI. ACKNOWLEDGMENTS

Industrial Power Company Ltd (TVO) permitted the publication of the material. Mr. Pertti Hassinen, MSc Tech., critically read this paper. Mrs. Tarja Pyymaki drew the illustrations and Ms. Yeasmine Blomqvist typed the manuscript. I thank also Mrs. Marjatta Tommila, MA, who revised the English of the paper.

REFERENCES

1. K-E. Almen and O. Persson, Determination of Hydraulic Conductivity in Lavia Borehole, Finland, Report YJT-84-20 (1984).

2. P. Andersson and L. Stenberg, Geophysical Borehole Logging in Lavia Borehole. Results and Interpretation of Sonic and Tube Wave Measurements, Swedish Geological Company. Report YJT-85-07 (1985).

3. J. K. Hallenburg, Geophysical Logging for Mineral and Engineering Applications, Tulsa, PennWell Publishing Company (1984).

4. A. Lindberg, "Lavian Koereian Kivilajien Petrografinen ja Kallioraon-taytteiden Mineraloginen Tutkimus" (The Petrography and Fracture Mineralogy of Lavia Borehole Core Samples), Geologian Tutkimuskeskus, Ydinjatteiden Sijoitustutkimukset, TVO/Koereika, Tyoraportti 84-12 (1984).

5. P. Saksa, Borehole Geophysical Investigations of Lavia Deep Testhole, Finland, Report YJT-85-06 (1985).

6. Suomen Malmi Oy (Finnexploration Company), Kallionaytekairaus, Koereika Lavia 1984 Loppuraportti (Lavia borehole drilling report), TVO/Koereika, Tyoraportti 84-06 (1984).

7. Suomen Malmi Oy, Geofysikaaliset Mittaukset Lavian Koereiassa (Borehole geophysical loggings in Lavia borehole), TVO/Koereika, Tyoraportti 84-11 (1984).

8. Valtion Teknillinen Tutkimuskeskus, Geotekniikan Laboratorio, Lavian Koereian Kallionaytteiden Laboratoriotutkimukset (Laboratory investigations of Lavia borehole core samples), TVO/Koereika, Tyoraportti 84-15 (1984).

9. P. Vuorela et al., Lavian Koereian Ymparistonin Kivilaji- ja Rakokartoitus (The lithology and fracturing of Lavia borehole site), Geologian Tutkimuskeskus, Ydinjatteiden Sijoitustutkimukset, TVO/Koereika, Tyoraportti 84-10 (1984).

SITE INVESTIGATION METHODS USED IN CANADA'S NUCLEAR FUEL WASTE MANAGEMENT PROGRAM TO DETERMINE THE HYDROGEOLOGICAL CONDITIONS OF PLUTONIC ROCK

C. C. Davison
Atomic Energy of Canada Limited
Whiteshell Nuclear Research Establishment
Pinawa, Manitoba R0E 1L0, Canada

ABSTRACT

Atomic Energy of Canada Limited (AECL) is investigating the concept of disposing of Canada's nuclear fuel wastes in a mined vault at a depth of 500 m to 1000 m within a plutonic rock body. Much effort has been directed at developing site investigation methods that can be used to determine the hydrogeological conditions of plutonic rock bodies. The primary objective of this research is to define the physical and chemical characteristics of groundwater flow systems at the various scales that are relevant to the prediction of potential radionuclide migration from a disposal vault. Groundwater movement through plutonic rock is largely controlled by fractures within the rock, and the hydrogeological parameters of fractured geological media are extremely scale dependent. AECL's hydrogeology research program began in 1978 and was initially aimed at developing methods to measure physical and chemical hydrogeological parameters at a small scale (within individual boreholes and fractures). The program has since moved toward developing methods for determining these parameters at the larger regional scale, which is relevant to assessing the safety of a nuclear fuel waste vault.

I. INTRODUCTION

The primary objective of AECL's hydrogeological research is to define the physical and chemical characteristics of groundwater flow systems in plutonic rock bodies of the Canadian Shield at the various scales that are relevant to the prediction of potential radionuclide migration from nuclear fuel waste disposal vault constructed at a depth of 500 m to 1000 m. To achieve this objective, realistic estimates must be obtained for the distribution of physical and chemical hydrogeological parameters, such as hydraulic conductivity, hydraulic potential, storage, hydraulic boundaries, redox potential, total dissolved ions, and ion distribution, that govern the interaction between any escaping radionuclides and the fluid/solid portions of the geosphere. Groundwater movement through plutonic rocks is largely controlled by fractures within the rock, and it is well known that most estimates of the hydrogeological parameters of fractured geological media are extremely scale dependent. Therefore, special care must be taken to ensure that a parameter value measured at one scale can be transposed to a different scale. At different scales of investigation and at different sites, a different study approach may be required, depending on the geometric and hydraulic characteristics of the various fractures controlling the hydrogeological conditions within the domain of interest. Figure 1 illustrates three different scales of investigation, which are important in understanding the hydrogeological behaviour of a site being considered for a nuclear fuel waste disposal vault. At the near-field scale, changes in fracturing caused by stress adjustments due to excavation or blast damage will be important, along with the natural jointing and fracturing in the vicinity of the vault. At the regional scale, features such as fault zones, shear zones, and major lithological units will largely control the groundwater flow regimes.

AECL's hydrogeological research program has gradually evolved from one of measurements made at a very small scale (within individual boreholes and fractures) to measurements made at the site and regional scales that are most relevant to assessing the safety of a rock body for the location of a nuclear fuel waste vault. Investigations of the near-field stress-related and blast-related hydrogeology aspects are currently being studied as part of AECL's Underground Research Laboratory (URL) experimental plan,[1] and will not be discussed any further in this paper. Currently, there are three main interrelated aspects of AECL's hydrogeological research program: developmental research, research area characterization and theoretical modelling.

II. DEVELOPMENTAL RESEARCH

Developmental Research includes designing, fabricating and evaluating instrumentation for measuring the physical and chemical hydrogeological parameters of fractured plutonic rock bodies. These involve mostly techniques that are carried out within small diameter (76-156 mm) boreholes drilled to depths of up to 1200 m into the rock. When AECL's research program began in 1978 very few methods were available to make these borehole measurements, and much of the initial development was aimed at producing the required borehole equipment and techniques. During the past several years, AECL has established many procedures to measure hydrogeological parameters and to attempt to collect uncontaminated groundwater samples from deep exploratory boreholes in fractured plutonic rock. The methods involve equipment employed during and after drilling in open boreholes, and also in borehole-monitoring systems used to seal borehole sections for long-term observation.

Boreholes used for the initial characterization of a research area are continuously cored using NQ-size (76 mm dia.), triple-tube-wireline, diamond-drilling techniques. Detailed logging of the recovered core during drilling provides immediate information on fracture location and character. Often, bottom-hole single-packer tests are performed through the drill bit at selected depth intervals during the diamond-drilling operations to measure hydraulic potential conditions and to determine the hydraulic conductivity. If other boreholes have been previously drilled in the vicinity, monitoring the water level in these boreholes during the drilling of a new borehole in the area often yields useful information about the location of fractures and fracture zones that are hydraulically interconnected between the boreholes.

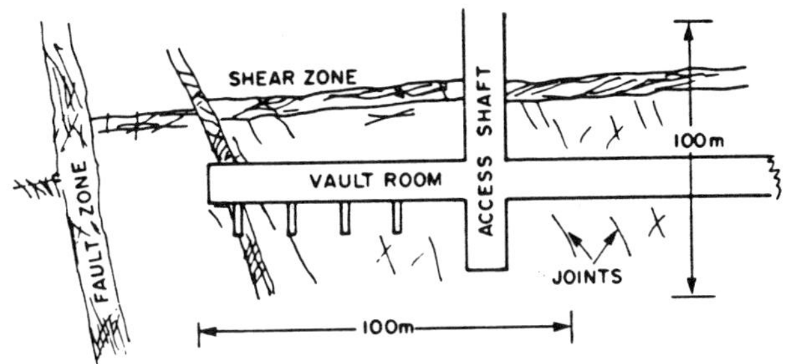

(a) Near Field or Vault Scale

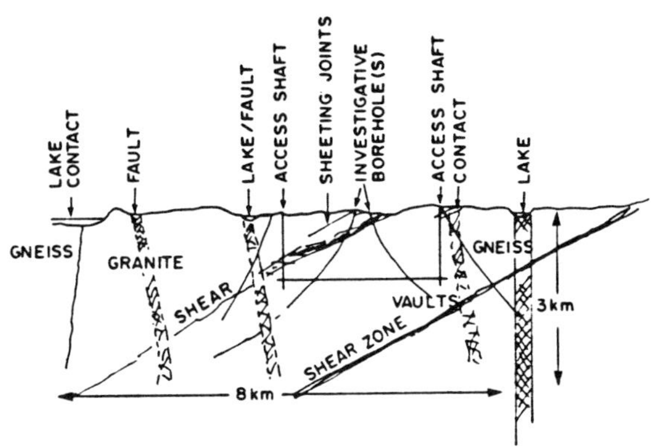

(b) Local or Site Scale

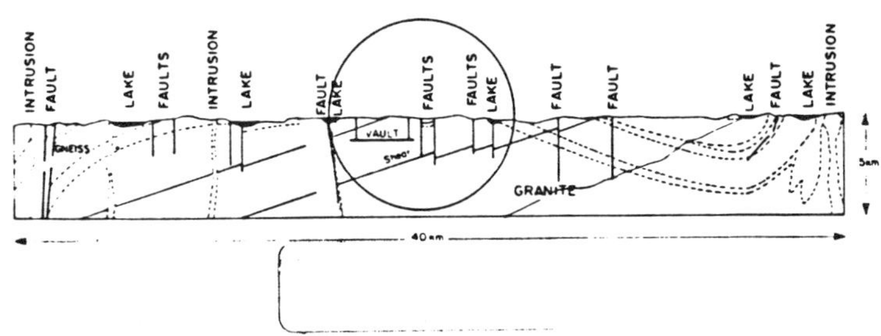

(c) Regional Scale

FIGURE 1. Scales Relevant to Hydrogeologic Studies of Nuclear Fuel Waste Disposal Site

It has been found to be particularly valuable to monitor the water levels in nearby boreholes during the drilling of deep percussion boreholes because the drilling method induces large water-level declines in the drilled borehole.

Upon completion of drilling, accurate borehole deviation surveys are conducted, and various geophysical surveys are run in the boreholes to identify fractures and lithological variations. Single-borehole logging methods include acoustic televiewer, sidewall television scanning, high resolution caliper, full waveform acoustic velocity, surface seismic tube-wave, fluid temperature and conductance, electrical, and nuclear logs. The logs provide useful information about the location, orientation and potential hydrological character of fractures that intersect the borehole.[2] Multi-hole geophysical techniques are also carried out where boreholes are located within 500 m to 1000 m of each other. Cross-hole seismic surveys show good potential to resolve gross fracture characteristics between boreholes up to 500 m apart in plutonic rocks.

Detailed hydrogeological tests are performed after the suite of geophysical logs are obtained from the borehole, the borehole cleaned thoroughly, and test intervals selected on the basis of a complete set of borehole information. The hydrogeological tests are usually conducted using a straddle-packer test tool, which is lowered into position by means of either a solid tubing string or a flexible multi-line umbilical cable. Shut-in pressures are used to estimate the hydraulic potential conditions, and the hydraulic conductivity is determined by a variety of methods including transient pressure-pulse response tests, fluid injection tests, rising or falling fluid-level recovery tests, or pumping withdrawal tests.[3] Water samples are recovered by either pumping or swabbing water from the straddle-packer zone, during which time the chemical quality is continuously monitored until stability is achieved. A downhole sample cylinder, which retains a fluid sample at in situ pressure conditions until laboratory analyses are performed, is subsequently lowered to the test zone. AECL has developed a straddle-packer geochemical sampling probe, which monitors downhole measurements of chemical parameters during pumping and sampling activities.[4]

Immediately following completion of the detailed hydrogeological testing and sampling program, retrievable single-packer elements are placed at various depths within the borehole to eliminate undesirable vertical fluid flows that would otherwise occur. Such fluid movements can create extensive disturbances to both the hydraulic and chemical regimes at a site. For example, recent observations made at one of AECL's research sites showed that one open borehole caused a significant hydraulic disturbance to a large surrounding area (several tens of square kilometers). Upon removal of the temporary single-packer elements, fluid samples can be obtained from the sealed-off zones. These samples are extremely valuable because they are often the first representative groundwater samples that can be obtained from low hydraulic-conductivity regions of a borehole.[5]

Ultimately, long-term hydrogeological access must be provided at isolated zones in each borehole so that no hydrogeological disturbance is transmitted along the borehole to other zones when a measurement is made. Monitoring systems consisting of either a multiple-interval packer/casing system or a multiple-packer standpipe piezometer system are installed in the boreholes, which enable hydrogeological measurements to be made at numerous isolated intervals with minimal disturbance. Figure 2 illustrates these two types of borehole monitoring systems. Once the monitoring systems are installed, long-term pumping tests, pressure pulse response tests, and tracer experiments can be performed between boreholes to assess the hydrogeological characteristics of large volumes of rock. These systems also allow the long-term monitoring of piezometric pressure trends and collection of representative water samples at numerous depths and locations within the rock mass. When this information is combined with the major structural and lithological information an understanding of three-dimensional regional groundwater flow systems can be obtained for a site.

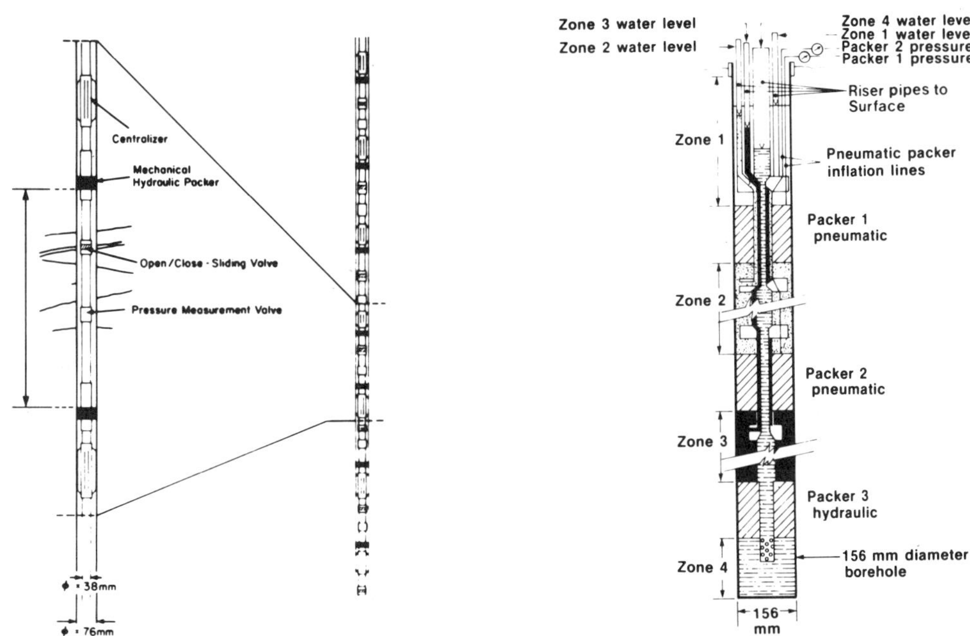

FIGURE 2. Examples of Hydrogeological Monitoring Equipment Installed in 76-mm and 156-mm Diameter Boreholes

III. RESEARCH AREA CHARACTERIZATION

AECL's Research Area Characterization involves the application of these methods at specific plutonic rock sites on Canada's Precambrian Shield. Hydrogeological studies are currently underway at four research areas: Chalk River and East Bull Lake in Eastern Ontario, Whiteshell in Eastern Manitoba, and Atikokan in Northwestern Ontario (Figure 3). The scale of investigation

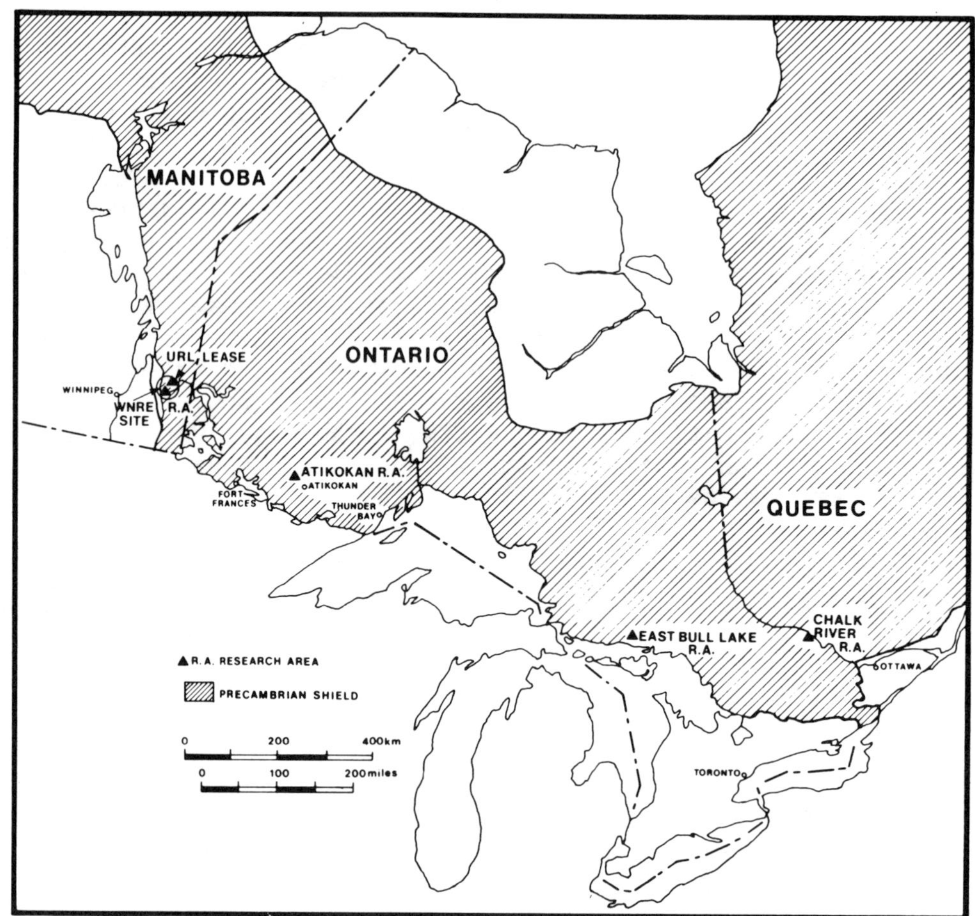

FIGURE 3. Location of CNFWPM Research Areas

at each of these areas is being increased to move from an understanding of local hydrogeological conditions toward an understanding of regional hydrogeological conditions.

The Chalk River research area, located on highly fractured granitic gneiss, has been used mainly for methodology development and groundwater tracer studies, but it is also the location of a detailed hydrogeological study of a block of rock, 100-m square by 75-m deep. The results from this block study are being used to assess how well a porous-media equivalent model can be applied to a highly fractured plutonic-rock mass.[7]

Hydrogeological investigations examining the characteristics of the Eye-Dashwa Lake granite pluton have been ongoing at the Atikokan research area since 1979. The studies have recently been expanded to analyze aspects of regional groundwater flow in a large area (> 200 km^2) of the granite pluton.

The East Bull Lake research area encompasses a pluton of gabbroic composition. Investigations are being carried out on a region 4 km^2 in area at a level sufficient to compare the hydrogeological conditions on the East Bull Lake research area with other plutons having granitic composition.

The Whiteshell research area involves two sites: the Whiteshell Nuclear Research Establishment (WNRE) and the site of the URL, located approximately 12 km apart within the same granitic pluton, the Lac du Bonnet granite batholith. WNRE has been used extensively, both as a site for methodology development and as a site for large-scale hydrogeological profile 2 km long.

Canada's URL is being constructed to provide researchers with representative geological environments in which to carry out a variety of in situ, near-field scale geotechnical experiments.[1] One unique feature of the URL project compared to similar projects in other countries, is that hydrogeologists are monitoring the hydrogeological conditions within a volume of previously undisturbed rock (approximately 2.5 10^9 m^3) surrounding the URL excavation site prior to, during and after the excavation of the access shaft, ventilation shaft and underground workings. Work began at the URL site in 1980 to characterize the hydrogeological conditions within the unconsolidated overburden deposits, the shallow bedrock (0-100 m) and the deeper bedrock (100-500 m). This involved an extensive program of borehole drilling, logging, testing, instrumentation, sampling and monitoring using the methods discussed previously in this paper. Over 130 boreholes have been drilled at the URL site, and the locations of these are shown on Figure 4. Twenty-five of these boreholes were drilled to depths ranging from 160 m to 1090 m to investigate the deep portions of the rock mass. Many of the deep boreholes were drilled close to the site chosen for the construction of the URL facility to provide a comprehensive network of boreholes with which to monitor the groundwater perturbations caused by the excavation (Figure 4).

Assimilation of all the fracture information obtained from logging, testing and monitoring the entire network of boreholes revealed that three major extensive low-dipping fracture zones exist in the URL rock mass. These three zones have been found to largely control the movement of groundwater at the study area. Figure 5 presents geologic and hydrogeologic cross sections of the study area along line 1-1', which illustrates this. The hydrogeology of fracture zones 2 and 3 has been studied in considerable detail, and the results indicate that a complex pattern of permeability exists within each of the fracture zones. These permeability distributions have been found to control the patterns of hydraulic head and groundwater chemistry at the URL site.[8]

The information obtained from the detailed hydrogeological investigations of the URL research area that were done prior to any excavation has been incorporated into various numerical computer models describing the piezometric drawdown that will occur in the surrounding rock mass during and after the underground excavation.[9] The drawdown predictions are presently

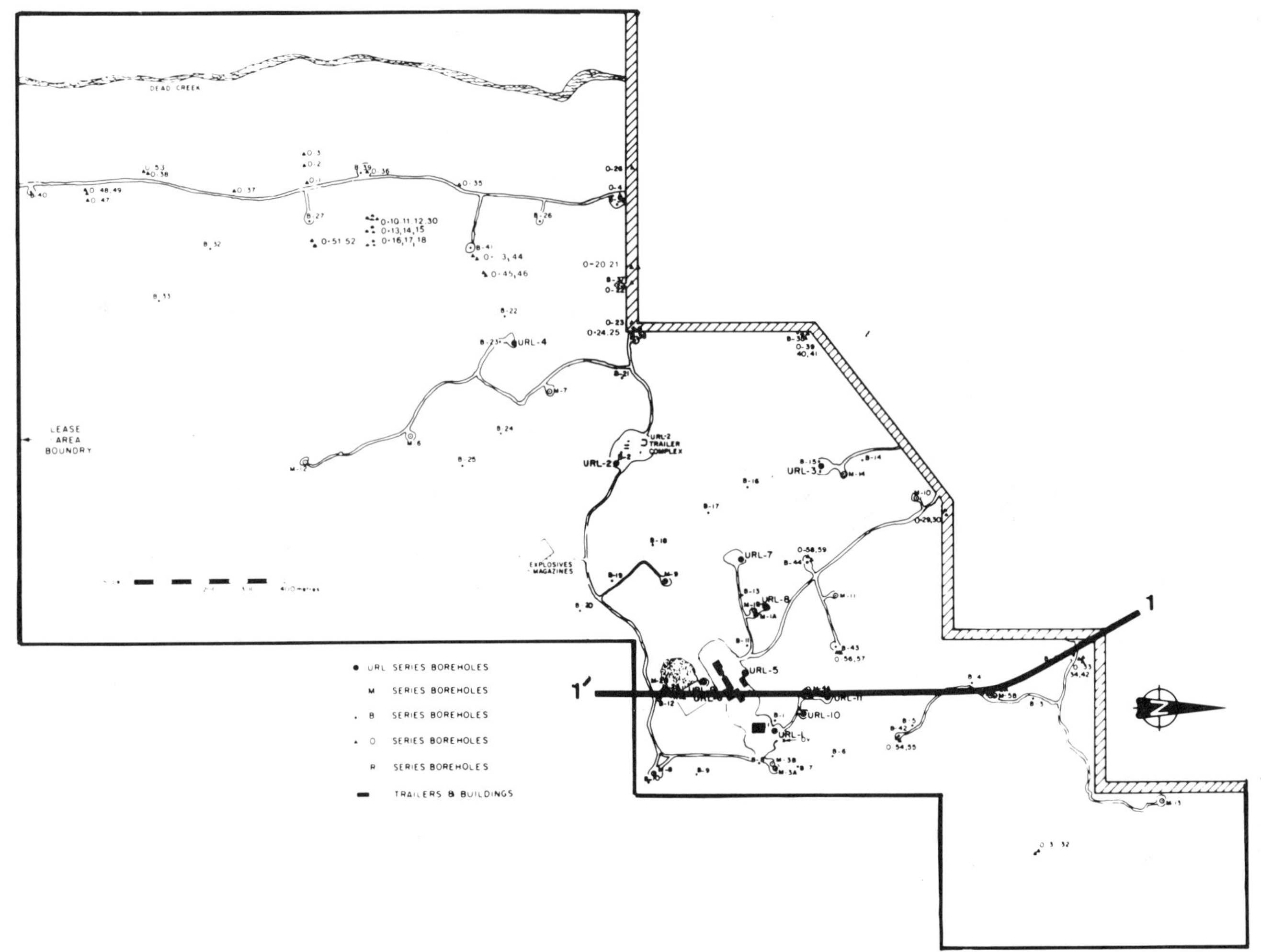

FIGURE 4. Borehole location plan of URL Site.

being compared with results of the continuous monitoring of the physical and chemical hydrogeological conditions within the rock mass surrounding the URL excavation site to assess how well the models have represented the hydrogeological conditions.

Currently, AECL is planning to conduct a detailed regional groundwater study, referred to as the Flow System Study, at either the Atikokan or Whiteshell research areas. Geophysical, geological and hydrogeological methods proven to be useful in understanding the large-scale hydrogeological conditions of plutonic rocks will be applied in this study. Particularly, many of the approaches that have been developed during the past years to successfully investigate the hydrogeology of the 2.5×10^9 m^3 volume of

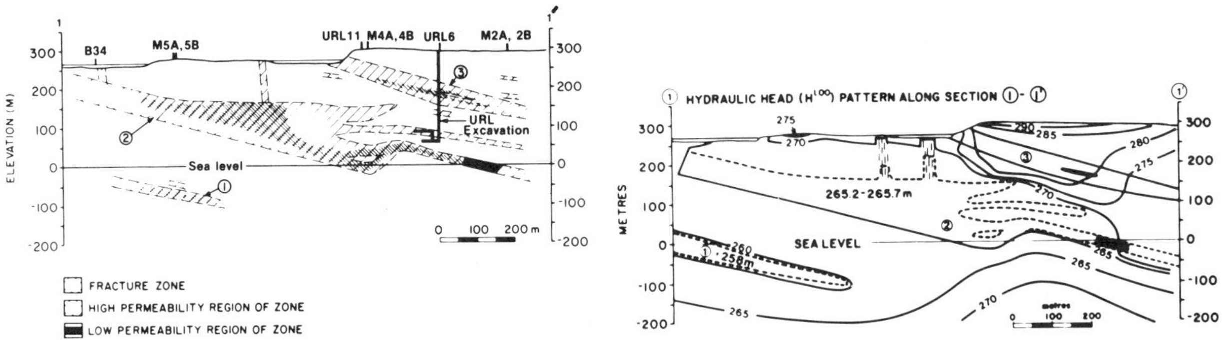

FIGURE 5. Cross Sections of Typical Geologic and Hydrogeologic Conditions at the URL Site

rock at the URL site will be used to study a rock volume of approximately 3 × 10^{11} m^3. Such a study will provide valuable data regarding the nature of regional groundwater movement through plutonic rock(s) and is thus essential for the development of realistic radionuclide release models of a potential nuclear fuel waste disposal site.

IV. THEORETICAL MODELLING

The third aspect of AECL's hydrogeological research program is the Theoretical Modelling of groundwater flow and radionuclide transport. Research is aimed both at evaluating existing mathematical models and at developing new or modified models that will be applicable to the three-dimensional analysis of radionuclide transport by groundwater through fractured plutonic rocks. A major modelling experiment currently being conducted predicts (by various modelling techniques) the hydrogeological response produced within the rock mass at the URL site by the excavation of the shaft and underground facilities. The various model predictions are being compared with the actual response, as measured by the extensive hydrogeological monitoring system. Both finite-difference and finite-element modelling methods have been used. Another aim of the modelling program is to derive representative geosphere models of three of AECL's research areas: Whiteshell, East Bull Lake and Atikokan. These models will be based on field and laboratory data from the specific research areas, and will be used to ensure that the analysis of radionuclide transport from a hypothetical vault is related to experimental data obtained from natural systems. All of the models will be based on varying levels of detailed field data from the research areas. However, aspects of regional groundwater movement will be determined in detail at one of the research areas to ensure that the regional models are representative.

REFERENCES

1. G. R. Simmons and N. Soonawala, Underground Research Laboratory Experiment Program, Atomic Energy of Canada Limited Technical Record TR-153 (1982).

2. C. C. Davison, W. S. Keys and F. L. Paillet, The Use of Borehole Geophysical Logs and Hydrologic Tests to Characterize Crystalline Rocks for Nuclear Waste Storage, Office of Nuclear Waste Isolation Report, ONWI-418 (1982).

3. C. C. Davison, Physical Hydrogeologic Measurements in Fractured Crystalline Rocks - Summary of 1979 Research Program at WNRE and CRNL, Atomic Energy of Canada Limited Technical Record, TR-161 (1981).

4. D. G. Bottomley, J. D. Ross and B. W. Graham, A Borehole Methodology for Hydrogeochemical Investigations in Fractured Rock, Wat. Res. Res. (1984).

5. R. Dickin et al., "Groundwater Chemistry to Depths of 1000 m in Low Permeability Granitic Rocks of the Canadian Shield," in Proceedings of IAH International Symposium on Groundwater Resource Evaluation and Contaminant Hydrogeology, Montreal Canada (1984).

6. C. C. Davison, "Monitoring Hydrological Conditions in Fractured Rock at the Site of Canada's Underground Research Laboratory," Groundwater Monitoring Review, pp. 94-102 (1984).

7. K. G. Raven et al., "Field Investigations of a Small Ground Water Flow System in Fractured Magnetic Gneiss," in Proceedings 17th Inter. Assoc. of Hydrogeologists Meeting, Hydrogeology of Rocks of Low Permeability, Tucson, Arizona (1985).

8. C. C. Davison, "Hydrogeological Characterization of the Site of Canada's Underground Research Laboratory," in Proceedings of IAH International Symposium on Groundwater Resource Utilization and Contaminant Hydrogeology, Montreal, Canada (1984).

9. V. Guvanasen, J. A. K. Reid and B. W. Nakka, Predictions of Hydrogeological Perturbations Due to the Construction of the Underground Research Laboratory, Atomic Energy of Canada Limited Technical Record, TR-344 (1985).

EXPERIMENTAL APPROACHES TO DETERMINING THE EXTENT OF EQUILIBRIUM IN A BASALTIC HYDROTHERMAL SYSTEM

C. C. Allen
R. B. Kasper
D. L. Lane
R. G. Johnston
S. A. Rawson
Rockwell Hanford Operations
P.O. Box 800
Richland, Washington 99352

ABSTRACT

Experiments in a basaltic hydrothermal system were conducted at 300°C to assess the extent to which equilibrium can be approached on a laboratory time scale. In these tests, the primary reacting phase in the basalt was mesostasis glass. In closed systems a metastable suite of secondary minerals was formed, which included cristobalite, smectite clay, and high silica/alumina zeolites. In open-system experiments, the secondary minerals more closely approximated natural high-temperature assemblages than did those from the closed-system tests. These experiments provide input to source term and modeling calculations of the geochemical behavior of a nuclear waste repository in basalt.

I. INTRODUCTION

The licensing of a nuclear waste repository will require that any proposed combination of engineered barriers and site geology be capable of minimizing the release of radionuclides to the environment for thousands of years. Laboratory experiments are being conducted by the Basalt Waste Isolation Project (BWIP) to investigate the behavior of specific design elements under the anticipated physical and chemical conditions to be found in basalt. The results of these experiments are useful as tests of geochemical models that will be used to assess engineering and site performance.

The varying of experimental parameters, including temperature, fluid/rock ratio, and fluid flow rate, has been proposed to enhance the approach to equilibrium in hydrothermal reactions. The present study addresses the effects of fluid/rock ratio and flow rate in the geochemical system of basalt and groundwater.

Laboratory hydrothermal experiments are often strongly influenced by reaction kinetics. Metastable phases may be formed initially, and then persist for months or years. This can lead to problems in relating laboratory results to long-term repository behavior. Experiments are therefore conducted to test the short-term evolution of systems under carefully controlled conditions. It is also important to identify natural analogs in which equilibrium conditions can reasonably be assumed. Mineral assemblages found in the geothermal fields of Iceland are used here as standards of comparison for the minerals produced in the laboratory.

II. EXPERIMENTAL

Six hydrothermal experiments conducted in closed Dickson-type autoclaves[1] and six others conducted in open "flowthrough" columns[2] are described in Table 1. All of the experiments were run at 300°C, and used crushed basalt and synthetic basalt groundwater. Several of the closed-system experiments have been discussed previously.[3]

The starting basalt in all of the present experiments consisted of crushed and sieved grains that had been washed in ultrapure deionized water to remove adhering fines. The major phases in this basalt were calcic plagioclase, pyroxene, titaniferous magnetite, and a glassy mesostasis.[4] The mesostasis, consisting of a mixture of silica-rich glass and immiscible iron-rich blebs, constituted approximately half of the basalt's volume.[5]

Table 1. Experimental Parameters

Experiment	Duration (h)	Fluid/Rock Mass Ratio	Flow Rate (mL/min)
DICKSON			
D3-5	1,337	2	--
D3-43	831	5	--
D1-9	616	10	--
D4-5	1,674	10	--
D4-64	2,278	100	--
D2-73	2,107	200	--
FLOWTHROUGH			
FT-100	2,100	--	0.004
FT-62	2,161	--	0.004
FT-30	691	--	0.005
FT-113	337	--	0.024
FT-131	531	--	0.024
FT-144	531	--	0.024

NOTE. Basalt: Reference Umtanum Entablature (RUE)[4,5] -120+230 mesh (D3-5, D3-43, D1-9, FT-62, FT-30, FT-113, FT-131, FT-144) -16+60 mesh (D4-5, D4-64, D2-73, FT-100)

Fluid: Synthetic basalt groundwater (GR-3)[6]

Temperature: 300°C

Pressure: 30 MPa

Porosity (flowthrough columns): 45%-50%

The starting solution was GR-3, a synthetic basalt groundwater.[6] Initial solution concentrations were reproducible to within 10% of the nominal values. The fluid composition was dominated by sodium (370 mg/L), chloride (313 mg/L), and sulfate (189 mg/L). It also contained silicon (33 mg/L), fluoride (32 mg/L), and traces of potassium, calcium, and magnesium. The room temperature pH of the starting fluid was 9.7 ± 0.1.

Dickson-type autoclaves were closed experimental systems wherein the starting materials were contained in externally pressurized gold bags. Flow-through columns, on the other hand, were titanium tubes (30.5-cm long, 0.64-cm inside diameter) containing the solid sample, through which GR-3 was continually pumped. Both systems allowed for sampling of the fluid without quenching the test.

Experimental solution samples were analyzed by inductively coupled plasma-atomic emission spectroscopy and ion chromatography. The pH values of the solutions were measured at room temperature using a combination electrode. The solid reaction products were vacuum dried and then analyzed by x-ray diffraction (XRD), optical microscopy, and scanning transmission electron microscopy (STEM). The STEM analyses were complemented by energy-dispersive spectrometry and electron diffraction.

III. RESULTS

Previous Dickson-type autoclave experiments utilizing basalt and synthetic groundwater[3] have demonstrated the tendency for this system to reach "steady state" after several hundred hours. This is indicated by solution concentrations of major elements, which arrive at essentially constant values. The reactants and products in such a state may be metastable, but they are observed to persist for the duration of experiments lasting for many months. This appears to be the case in many of the present experiments.

A. Solution Analyses

1. Low fluid/rock ratio tests. Data for several Dickson autoclave experiments employing low fluid/rock ratios have been previously reported.[3,7] These data, as well as data from the present study, indicate that solution chemistry is relatively insensitive to changes in initial fluid/rock ratios between 2 and 50. Concentrations of silicon, aluminum, and potassium in the present low fluid/rock tests initially rose rapidly and, within a few hundred hours, decreased to approximate steady-state values. The steady-state concentration of silicon in these experiments was approximately 530 mg/L (Figure 1). Final concentrations were measured below 10 mg/L for aluminum and 30-80 mg/L for potassium. Sodium decreased from its initial concentration of 370 mg/L to 250 mg/L, and then rose slowly to a value near 300 mg/L. The maximum solution concentrations for calcium and magnesium measured in these experiments were <6 mg/L

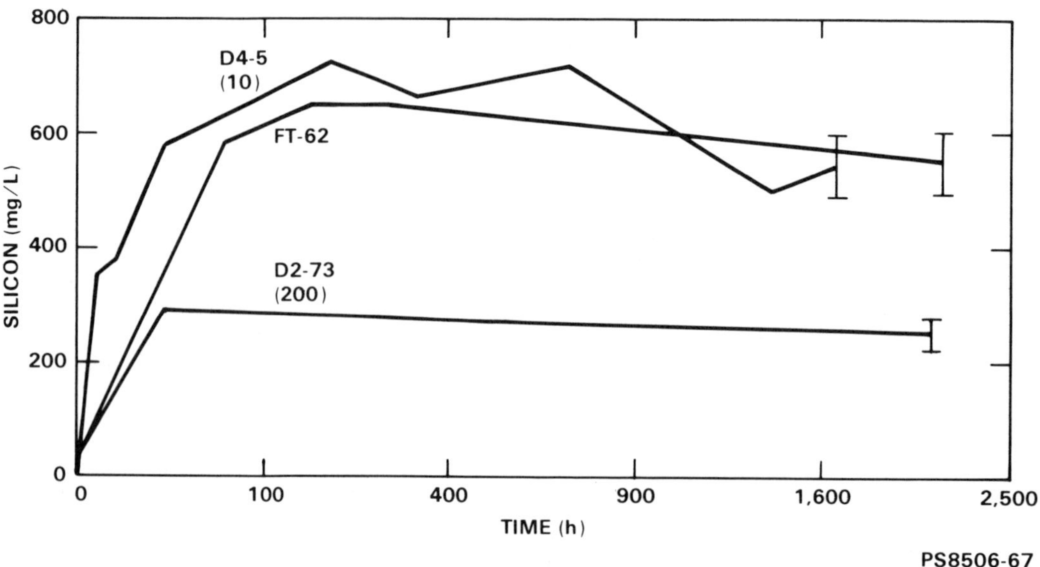

FIGURE 1. Silicon Concentration in Solution Versus Time (Solutions filtered through 0.4 micron membrane; 10% error bars include sample preparation and analysis effects. Numbers in parentheses represent initial fluid/rock ratios).

and <0.1 mg/L, respectively. Fluoride, chloride, and sulfate concentrations all remained within 30% of their values in the starting groundwater.

The solution pH in the low fluid/rock experiments dropped from its initial value of 9.7 to 5.6-6.0 after several hundred hours. Subsequent pH values rose slowly to 6.5-7.0 after approximately 1,500 hours (Figure 2).

2. High fluid/rock ratio tests. The major differences in solution composition between the experiments with high fluid/rock ratios and those with low ratios were in the concentration of silicon and in the pH (Figures 1 and 2). The silicon concentration for experiment D4-64 (fluid/rock of 100) rose to 400-450 mg/L, while that for experiment D2-73 fluid/rock of 200) reached a steady value of 250 mg/L. The steady-state pH value in experiment D4-64 was approximately 7.2, and the pH for experiment D2-73 never fell below 8.0.

3. Flowthrough tests. Solution chemistry data for the flow-through experiments are incomplete, due to complexities in the sampling system. Limited data sets are available for tests FT-62, FT-113, FT-131, and FT-144.

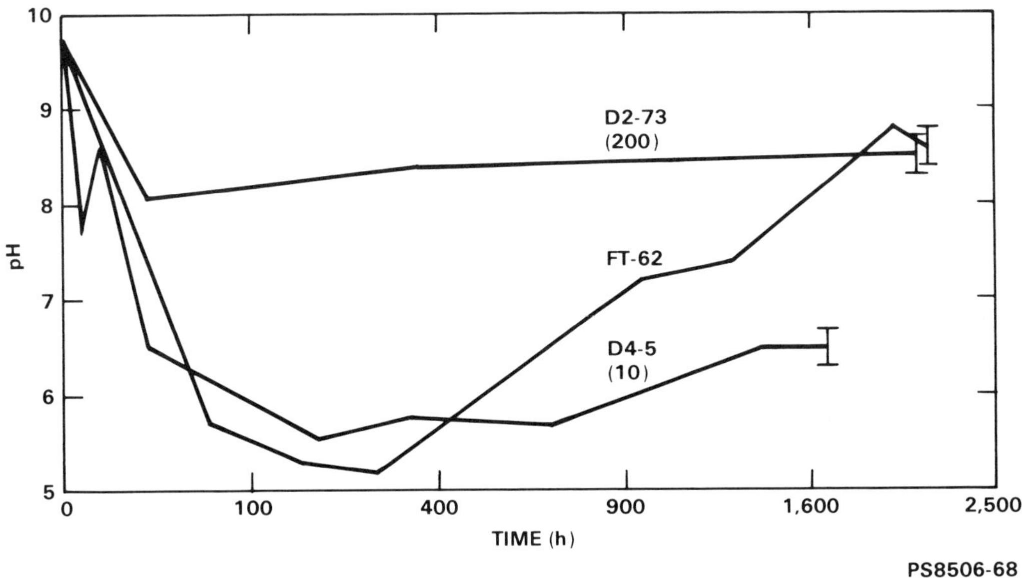

FIGURE 2. Solution pH Value Versus Time. (Error bars of 0.2 pH unit include sample preparation and analysis effects. Numbers in parentheses represent initial fluid/rock ratios).

Solution concentrations of silicon rose rapidly, and attained final concentrations in the range of 500-600 mg/L (Figure 1). Aluminum, absent in the starting groundwater, showed concentrations at the column outlets which did not exceed 5 mg/L. Potassium concentrations rose rapidly and reached final values of approximately 110-130 mg/L for all tests except FT-62. In this case the initial rise was followed by a continuing decrease to a final value of 34 mg/L. The ending concentration of sodium was in the range of 320-380 mg/L. The behavior of other analyzed components was similar to those of the Dickson tests. The pH values in the flowthrough experiments dropped rapidly from 9.7 to approximately 5.0-6.0, and then increased to the range of 8.5-9.5 (Figure 2). The rate of increase was a function of flow rate, taking approximately 300 hours and 1,800 hours for flow rates of 0.024 mL/min and 0.004 mL/min, respectively.

B. Solids Analyses

The solids recovered at the termination of every test consisted primarily of partially reacted basalt grains. This basalt showed evidence of preferential dissolution of the mesostasis and only limited attack on the mineral phases.

1. Low fluid/rock ratio tests. The reaction products identified at the end of these experiments varied as a function of fluid/rock

ratio, and they also reflected the influence of experiment duration.[3] The experiment (D3-5) with the lowest fluid/rock ratio (2) produced significant quantities of crystalline silica (cristobalite plus tridymite) and potassium feldspar, along with smaller quantities of the zeolite mineral potassium ferrierite and an iron-rich phyllosilicate. Similar products were obtained in experiment D3-43 with a fluid/rock ratio of 5. The XRD results indicated that ferrierite was the dominant reaction product phase, with silica as a minor constituent. A small quantity of a phyllosilicate that was different from that seen in experiment D3-5 was also produced. In the shorter of the two experiments with a fluid/rock ratio of 10 (D1-9), little evidence of reaction products was found. Traces of potassium feldspar, crystalline silica, and a clay rich in iron and potassium were detected by XRD and/or STEM. Experiment D4-5 (fluid/rock ratio of 10) produced, after 1,674 hours, an assemblage dominated by a mixture of cristobalite and tridymite, with lesser quantities of ferrierite, an iron-rich smectite clay tentatively identified as iron saponite, and minor amounts of potassium feldspar. All of these samples contained traces of magnetite, which was not evident in the XRD patterns of the starting basalt.

2. High fluid/rock ratio tests. Experiments conducted using high fluid/rock ratios yielded somewhat different reaction products from those discussed above. Silica and zeolites were either not present or barely detectable. Iron-rich, magnesium-bearing smectite (saponite ?) and iron-rich biotite were the dominant reaction products in test D4-64, along with minor potassium feldspar, magnetite, and hematite. Reaction products in experiment D2-73 were scarce with respect to basalt. The only secondary phases detected were iron-rich smectite and biotite, plus hematite.

3. Flowthrough tests. These experiments yielded a variable alteration assemblage, with the exact phases somewhat dependent upon the sample's position in the column. Secondary magnetite was ubiquitous throughout the column in every test. Potassium feldspar was generally present, although this mineral was occasionally absent near a column inlet and increased in abundance in the "downstream" direction. Basalt grains among the reaction products were more strongly attacked than those in the Dickson experiments.

   The major secondary minerals produced in the flowthrough experiments were phyllosilicates. Iron-rich biotite was identified by XRD and/or STEM analysis in all six samples. Iron-rich, magnesium-bearing smectites were also commonly produced. Iron-rich chlorite or septechlorite, either interlayered with the smectite or present as a separate phase, was detected in four of the solid samples. Vermiculite was identified in the reaction products of FT-100 and interlayered chlorite-vermiculite was tentatively identified in FT-62.

Minor amounts of zeolites were identified in all of the flowthrough solid samples. Potassium ferrierite and analcime were the most abundant, while clinoptilolite and possibly mordenite were occasionally encountered. Additional minor reaction products identified in individual flowthrough samples included hematite, apatite, and possibly ferro-actinolite. Silica was not detected in five of the six experiments, and only trace amounts were found in FT-100.

IV. DISCUSSION

The geothermal fields of Iceland have been proposed as being petrologic, hydrologic, and thermal analogs to the near-field environment of a nuclear waste repository in basalt (NWRB).[8] The host rocks are basaltic, and the groundwater composition in the low-salinity areas approximates that at the proposed NWRB location. Temperatures in the Icelandic fields span the range predicted in thermal models of a basalt repository. Furthermore, hydrothermal conditions in Iceland are believed to have persisted long enough for the systems to have attained equilibrium among the basalt, the groundwater, and the secondary mineral assemblages.

Extensive studies of hydrothermal alteration in Iceland[9] have demonstrated that temperature is the dominant variable that has controlled the formation of particular alteration phases. Factors such as pressure, detailed rock composition, and even fluid composition have much less effect. Thus, a comparison of the 300°C experiments discussed above to the Icelandic analog provides a useful indication of equilibrium attainment in the laboratory.

Table 2 is a summary, based on the work of Browne[10], of the secondary minerals found in low-salinity Icelandic geothermal wells. The table is divided according to the presently measured hydrothermal fluid temperature.

The low fluid/rock Dickson tests yielded secondary mineral assemblages more similar to those observed in the low-temperature Icelandic geothermal wells than to those in the high-temperature wells. Iron-rich smectites were the only clays produced experimentally, and in Iceland these minerals are stable only to 200°C.[9] High silica/alumina zeolites such as clinoptilolite and mordenite, which are structurally similar to ferrierite, are common in the low-temperature wells in Iceland. Ferrierite and the low-temperature form of cristobalite are common products of the Dickson experiments with low fluid/rock ratios. Thus, even after many months, the chemistry of these closed system tests remained controlled by reaction products that are metastable at the experimental temperature of 300°C.

Dickson autoclave experiments using high fluid/rock ratios produced small amounts of smectite and possibly biotite, but little or no silica phase or zeolites. The presence of magnesium in the smectites, not seen in the low fluid/rock tests, indicated the partial dissolution of pyroxene crystals. The dearth of solid reaction products, combined with fluid compositions that approached those of the starting groundwater, showed that both experiments

TABLE 2. Icelandic Geothermal Minerals

Mineral	Temperature <150°C	Temperature >150°C
Silica	Cristobalite	Quartz
Feldspar		Sodium feldspar
Zeolites	Clinoptilolite Heulandite Analcime	Analcime
Phyllosilicates	Saponite Illite Illite/Smectite	Illite/Smectite Chlorite Chlorite/Smectite Biotite
Iron Oxides	Hematite	
Iron Sulfides	Pyrite	Pyrite
Epidotes		Epidote
Amphiboles		Actinolite

were strongly fluid-dominated. These experiments also produced hematite (Fe_2O_3) or hematite plus magnetite (Fe_3O_4), as opposed to the magnetite alone produced in tests with lower fluid/rock ratios. Thus, at high fluid/rock ratios, secondary iron became progressively more oxidized.

The flowthrough experiments produced mica, smectite clay, vermiculite, chlorite or septechlorite, and possible mixed-layer clays. Comparison to the Icelandic geothermal minerals (Table 2) indicates that this assemblage is characteristic of temperatures above 150°C, in contrast to the assemblages produced in the Dickson experiments. Similarly, the production of analcime, the possible presence of actinolite, and the lack of cristobalite in the flowthrough experiments are indicative of higher temperatures than those that characterized the closed-system tests.

None of the present experiments produced mineral assemblages that exactly matched those reported from the high-temperature geothermal wells of Iceland. Much of this discrepancy can be traced to the glassy mesostasis that is a dominant phase in the basalts used in these experiments. The glass is considerably richer in silicon, iron, sodium, and potassium, and poorer in calcium and magnesium than the bulk rock. Observations of reacted basalt grains reveal that the mesostasis is preferentially attacked in the experimental systems, and compositions of the reaction products reflect its chemistry. Conversely, some natural hydrothermal systems, with lifetimes of centuries, involve extensive alteration of both glassy and crystalline phases in the basalt.

The results of the present experiments support the use of flow rate as an experimental parameter to facilitate the approach to equilibrium in a basaltic hydrothermal system. This is consistent with modeling results which predict that equilibrium can be achieved more rapidly in open experimental systems than in closed systems.[11] These experiments provide input data to source term and modeling calculations of the performance of a NWRB.

V. ACKNOWLEDGMENTS

The authors thank the BWIP Materials Testing Group staff, notably K. Cooley, C. Hobbick, S. Kunkler, G. McKeon, D. Schatz, S. Showalter, and J. Smith, for valuable scientific and technical support, and J. Burnell for careful peer review.

REFERENCES

1. W. E. Seyfried et al, "A New Reaction Cell for Hydrothermal Solution Equipment," Amer. Mineral., 64, 646 (1979).

2. W. E. Dibble, Jr. and J. M. Potter, "Effects of Fluid Flow Rates on Geochemical Processes," Soc. Petrol. Eng. Paper 10994, 57th Annual Fall Tech. Conf. and Exhibition (1982).

3. D. L. Lane et al, "The Basalt/Water System: Considerations for a Nuclear Waste Repository," Mat. Res. Soc. Symp. Proc., 26, 95 (1984).

4. A. F. Noonan et al, "Phase Chemistry of the Umtanum Basalt, A Reference Repository Host in the Columbia Plateau," Mat. Res. Soc. Symp. Proc., 6, 51 (1982).

5. C. C. Allen and M. B. Strope, "Microcharacterization of Basalt -- Considerations for a Nuclear Waste Repository," Microbeam Analysis -- 1983, 51 (1983).

6. T. E. Jones, Reference Material Chemistry Synthetic Ground Water Formulation, RHO-BW-ST-37 P, Rockwell Hanford Operations, Richland, WA (1982).

7. E. L. Moore et al, Hydrothermal Interactions of Columbia Plateau Basalt from the Umtanum Flow with its Coexisting Groundwater, SD-BWI-TI-228, Rockwell Hanford Operations, Richland, WA (1984).

8. G. G. Ulmer and D. E. Grandstaff, Icelandic Basaltic Geothermal Field: A Natural Analog for Nuclear Waste Isolation in Basalt, SD-BWI-TI-257, Rockwell Hanford Operations, Richland, WA (1984).

9. H. Kristmannsdottir and J. Tommasson, "Nesjasvellier -- Hydrothermal Alteration in a High Temperature Area," Proc. Int. Symp. on Water-Rock Interaction, Geol. Surv. Czechoslovakia, 170 (1976).

10. P. L. Browne, "Hydrothermal Alteration in Active Geothermal Fields," Ann. Rev. Earth Planet. Sci., 6, 229 (1978).

11. H. C. Helgeson and W. M. Murphy, "Calculations of Mass Transfer Among Minerals and Aqueous Solutions as a Function of Time and Surface Area in Geochemical Processes. Part I. Computational Approach," Math.Geol., 15, 109 (1983).

^{36}Cl MEASUREMENTS OF THE UNSATURATED ZONE FLUX AT YUCCA MOUNTAIN

A. E. Norris
K. Wolfsberg
Los Alamos National Laboratory
P.O. Box 1663
Los Alamos, New Mexico 87545

S. K. Gifford
Hydro Geo Chem, Incorporated
1430 North Sixth Avenue
Tucson, Arizona 85705

ABSTRACT

Determining the unsaturated zone percolation rate, or flux, is an extremely important site characterization issue for the proposed Yucca Mountain nuclear waste repository. A new technique that measures the ^{36}Cl content of tuff from the Exploratory Shaft will be used to calculate flux through the unsaturated zone over longer times than could be measured by the more conventional ^{14}C method. Measurements of the ^{36}Cl "bomb pulse" in soil samples from Yucca Mountain have been used to confirm that infiltration is not an important recharge mechanism.

I. INTRODUCTION

The Yucca Mountain site being studied by the Nevada Nuclear Waste Storage Investigations Project is the only one of the nine potentially acceptable candidate sites identified by the US Department of Energy in 1984 in which a high-level nuclear waste repository would be located in unsaturated rock. A performance evaluation of this site requires knowledge of the rate of water movement through the unsaturated zone because water-borne transport is a possible means for nuclear wastes to travel from the repository to the accessible environment. This paper discusses the use of ^{36}Cl to measure the groundwater velocity and calculate the flux in the unsaturated zone. The ^{36}Cl half-life is 3.01×10^5 years, which permits measurement of water movements from 5×10^4 to 2×10^6 years. This range is significantly longer than that afforded by the decay of ^{14}C. That nuclide's 5,730-year half-life precludes measurements of travel times longer than 4×10^4 years.

P. Montazer and W. E. Wilson[1] studied the available information on the percolation rate, or flux, in the unsaturated zone at Yucca Mountain. They report that recharge estimates range from 4.5 to 0.5 mm/yr based on regional water budget studies. Geothermal heat-flux methods yield estimates of vertical recharge from 10 mm/yr downward to 150 mm/yr upward. Analyses of the hydraulic properties of the volcanic rock, named tuff, from three strata at Yucca Mountain result in the following flux estimates:

1. 0.1 to 98.6 mm/yr in the Paintbrush nonwelded unit,
2. 10^{-7} to 0.2 mm/yr in the matrix of the Topopah Spring welded unit, and
3. a maximum flux of 0.006 mm/yr in the Calico Hills nonwelded unit.

Direct measurements of flux in Yucca Mountain have not been made. We calculate the water travel time through the repository host rock to be approximately

$$\frac{356\ m \times 0.14 \times 0.65}{0.0002\ m/yr} = 162{,}000\ yr,$$

where 356 meters is the thickness of the Topopah Spring Member host rock at the Exploratory Shaft site,[2] the porosity and the saturation are estimated[1] to be 14% and 65%, respectively, and the flux of 0.0002 m/yr is the maximum value calculated from the hydraulic properties of the host rock, as indicated in the preceding paragraph. The purpose of the ^{36}Cl measurements is to measure the travel time of water moving through the unsaturated zone, which may involve very long times, as this calculation shows. Measurement of the travel times in the unsaturated zone will permit calculation of the flux in a direct manner.

The determination of hydrologic flux through the Yucca Mountain unsaturated zone is the objective of more traditional techniques that are part of the site characterization work,[3] as well as the ^{36}Cl measurements. The other techniques include calculations based on the hydraulic properties of large-rock samples from the Exploratory Shaft, _in situ_ fluid flow experiments, and radioisotopic dating of pore water with ^{3}H and ^{14}C analyses. The data from all of these flux measurements will be used to evaluate the site conditions with respect to the US Nuclear Regulatory Commission requirement that the pre-waste-emplacement groundwater travel time along the fastest path of likely radionuclide travel from the disturbed zone to the accessible environment be at least 1,000 years. Additionally, the data will permit accurate calculations of radionuclide releases from the nuclear waste repository.

II. THE USE OF ^{36}Cl AS A GROUNDWATER TRACER

The geochemical properties of chlorine make it a useful tracer of subterranean water movements.[4] Chlorine is so electronegative that it exists as the chloride ion under most geohydrologic conditions. Chloride ions are nonvolatile. They do not react chemically with the tuffs in the unsaturated zone at Yucca mountain, nor are they sorbed. Thus, the movement of chloride ions will trace the movement of water in the liquid phase through the unsaturated zone.

Chlorine is deposited globally both in rainfall and in dry fallout. The source of most of the chlorine in this fallout is sea salt lofted into the troposphere by surface winds. A very small fraction of the chlorine atoms in the fallout consists of ^{36}Cl, which results from cosmic ray reactions in the atmosphere. The dominant reaction is cosmic ray-induced spallation of ^{40}Ar. The reaction can be represented as $^{40}Ar(x,x'\alpha)^{36}Cl$. A less important source of ^{36}Cl in the atmosphere arises from the reaction $^{36}Ar(n,p)^{36}Cl$. The neutrons that induce this reaction are cosmic ray secondaries.

The specific activity of ^{36}Cl in the environment is so low that measurements of the ^{36}Cl isotopic abundance have been infeasible with a few exceptions until the recent development of tandem accelerator mass spectrometric analyses. The sensitivity of this technique currently[5] is approximately one atom of ^{36}Cl in 10^{15} atoms of chlorine. The sample size required for these measurements is 10 mg of chloride or more. The background $^{36}Cl/Cl$ ratio in surficial deposits of alluvium at Yucca Mountain has been measured in this work to be ~5×10^{-13}, which is more than a hundred times greater than the limit of this technique. Any samples collected in which ^{36}Cl has undergone detectable radioactive decay will result in $^{36}Cl/Cl$ ratios lower than the cosmogenic background.

III. ^{36}Cl DATA FROM THE EXPLORATORY SHAFT

This site characterization work requires collection of samples as the Exploratory Shaft is mined, preparation of AgCl precipitates from those samples, measurements of the $^{36}Cl/Cl$ ratios in the precipitates using tandem accelerator mass spectrometry techniques, and interpretation of the data. The Yucca Mountain Exploratory Shaft will be mined by a drill and blast procedure that will deepen the shaft about two meters per blasting round. Approximately 225 blasting rounds will be required to mine the Exploratory Shaft. The rubble from some thirty rounds will be collected for ^{36}Cl analysis. Each measurement requires ~50 kg of rubble. The blasting rounds to be sampled have been selected on the basis of lithology and stratigraphy that might affect hydrologic characteristics. The stratigraphic data come the USW G-4 core hole,[2] which is 80 meters from the Exploratory Shaft site. The locations selected for sampling are indicated by triangles in Figure 1. These locations may be changed, depending on the geologic features observed during mining.

The $^{36}Cl/Cl$ ratios to be measured in this work are subject to error caused by the adventitious introduction of chlorine. The primary sources of chlorine not inherently in the samples are the ~10 mg/l chloride content of the water that will be used during the mining operations and the chlorine content of the explosives. The water used underground will be traced with NaBr to permit correction for this source of chlorine. In addition, the rubble pieces chosen for analysis will be the largest from the blasting round, to minimize the surface-to-volume ratio. This selection process should mitigate the effects of chlorine introduced by the blasting operations.

The preparation of each sample for tandem accelerator mass spectrometric analysis involves crushing the rubble collected from the Exploratory Shaft, then leaching the crushed rubble with chloride-free water. The bromide content of the resulting solution is determined to allow correction for chloride introduced with the water used for mining. A silver nitrate solution is added to the leachate, and the resulting AgCl precipitate is removed by filtration. This precipitate is the source material for the tandem accelerator mass spectrometric analysis. Details of that instrumental technique have been published by D. Elmore, et al.[5]

DEPTH, m (ft)	LITHOLOGY
	ALLUVIUM
TIVA CANYON MEMBER	WELDED, DEVITRIFIED TUFF
35.9 (118.0)	PARTIALLY WELDED VITRIC — WEATHERED
BEDDED TUFF	BEDDED TUFF, VITRIC
YUCCA MT. 51.3(168.2)	NONWELDED, VITRIC (PAH C.)
AND PAH 57.3(188.0) CANYON MEMBERS	BEDDED TUFF, VITRIC (POORLY CONSOLIDATED)
71.1 (233.4)	
80.9 (265.5)	CAPROCK ZONE, WELDED, CRYSTAL RICH
	WELDED, DEVITRIFIED TUFF, VAPOR-PHASE CRYSTALLIZATION ZONE
121.9 (400.0)	WELDED, >10% LITHOPHYSAL CAVITIES
128.0 (420.0)	WELDED, SPARSE LITHOPHYSAE
143.2 (470.0)	
TOPOPAH SPRING MEMBER	WELDED, DEVITRIFIED TUFF, COMMON TO ABUNDANT LITHOPHYSAL CAVITIES (UP TO 40%)
207.2 (680.0)	WELDED, DEVITRIFIED TUFF, RARE LITHOPHYSAE; SOME HIGH-ANGLE FRACTURES PRESENT
235.0 (770.0)	WELDED, DEVITRIFIED TUFF, LITHOPHYSAL CAVITIES 2-30%; SPLINTERY AND CONCHOIDAL-TYPE PARTING COMMON THROUGHOUT INTERVAL. LOWER CONTACT GRADATIONAL.
343.7 (1127.9)	
381.0 (1250.0)	WELDED, DEVITRIFIED TUFF; LITHOPHYSAL CAVITIES EXTREMELY RARE TO ABSENT
394.0 (1293.0)	WELDED, PARTLY VITRIC, ZEOLITIC AND ARGILLIC
401.2 (1316.5)	WELDED TUFF VITROPHYRE
410.0 (1345.4)	MOD. WELDED TO NONWELDED, VITRIC
420.0 (1377.8)	NON- TO PARTLY WELDED, PARTLY ZEOLITIC
428.7 (1406.8)	BEDDED AND NONWELDED TUFF, SLIGHTLY TO MOD. ZEOLITIC
434.7 (1426.3)	BEDDED & NONWELDED TUFF ZEOLITE
443.4 (1456.5)	
TUFFS OF CALICO HILLS	NONWELDED TUFF, ZEOLITIC
451.1 (1480.0)	NOT TO SCALE

FIGURE 1. USW G-4 Stratigraphic Column with Exploratory Shaft Sampling Locations for ^{36}Cl Shown as Triangles

The $^{36}Cl/Cl$ ratios, when plotted against sample depth, provide a measure of the flux. The accuracy of this measurement depends on the velocity of water movement through the unsaturated zone. The accuracy is high for uniform velocities less than ~5 mm/yr. The accuracy decreases for greater water velocities until the practical limit of the measurement is reached at water velocities of ~10 mm/yr. The following statistical analysis shows the sensitivity of this technique.

Power curves were used to estimate the probability of detecting small water movement velocities from $^{36}Cl/Cl$ measurements at the depths indicated in Figure 1. The analysis is based on a constant water flow rate throughout the volume penetrated by the Exploratory Shaft. A null hypothesis was tested, namely that the rate of water flow through the unsaturated zone is equal to or greater than 15 mm/yr. Figure 2 plots velocity versus power. Power is the probability of rejecting the null hypothesis when the null hypothesis is false. The significance level of 0.05 implies a 5% chance of an erroneous rejection of the null hypothesis. The three curves in the figure are labelled with the one standard deviation values as a percentage of the individual data points. We expect the data points in this work to have standard deviation values of ~10%. An example of the interpretation of the information in Figure 2 is the following. If the actual water velocity is 10 mm/yr, there is a 38% chance that the data will indicate a velocity equal to or greater than 15 mm/yr. However, when the actual velocity is less than ~5 mm/yr, there is essentially no chance that the data will indicate a velocity greater than 15 mm/yr.

IV. THE USE OF ^{36}Cl TO MEASURE INFILTRATION AT YUCCA MOUNTAIN

We have used ^{36}Cl analyses of soil samples already to measure infiltration. The source of ^{36}Cl for this study is not the cosmogenic fallout that will be used to trace water movements in the unsaturated zone. It is the pulse of ^{36}Cl that was deposited globally after high-yield nuclear weapons tests conducted at sea level in the Pacific Ocean between 1952 and 1959. This bomb pulse was carried beneath the surface during the past thirty years by local precipitation.

Samples of alluvium for this study were collected at measured depths from two trenches in the vicinity of Yucca Mountain. One trench was bulldozed at the site of the Exploratory Shaft. The other was at Yucca Wash, about four kilometers east of the Exploratory Shaft site. Samples were prepared for tandem accelerator mass spectrometry by the method described in Section III. The results of the $^{36}Cl/Cl$ measurements are shown in Figure 3. The uncertainties indicated are one standard deviation based on counting statistics.

The data from the two sites are qualitatively different. The distribution of ^{36}Cl with depth at Yucca Wash shows one well-defined peak. The data in the other set show two peaks. The ^{36}Cl data and the measured chloride content of each sample were used to calculate the integrals under the peaks

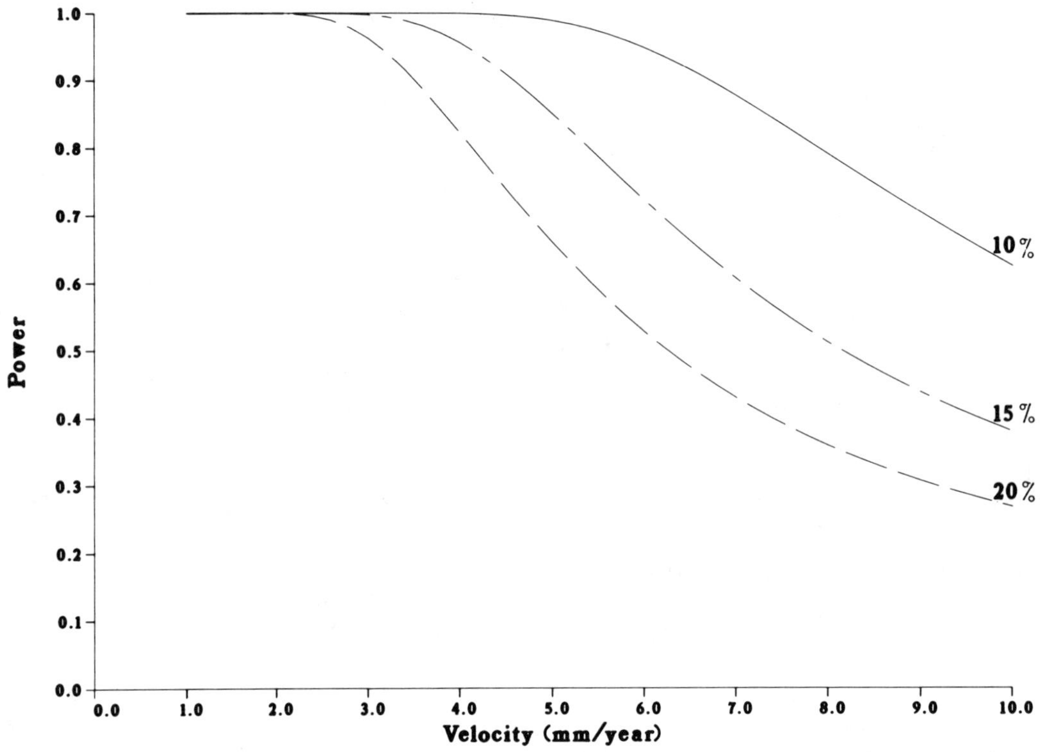

FIGURE 2. Power Curves with Significant Levels of 0.05 for Exploratory Shaft $^{36}Cl/Cl$ Measurement. The Curves are Labelled with the One Standard Deviation Values in Percent of the Individual Data Points.

for a quantitative comparison. The preliminary value for the integral under the peak at the Yucca Wash site is 6.0×10^{12} atoms/m^2. The preliminary value for the integral under the two peaks at the Exploratory Shaft site is 4.5×10^{10} atoms/m^2, or 0.8% of that at the Yucca Wash location.

The Yucca Wash data indicate that infiltration has proceeded vertically downward during the thirty years since the deposition of the ^{36}Cl bomb pulse. There is no evidence for recharge below 1.7 m. The data from the Exploratory Shaft site are more difficult to interpret. One explanation is that vertical percolation was followed by subsurface lateral water flow. The location of the "missing" ^{36}Cl is not known.

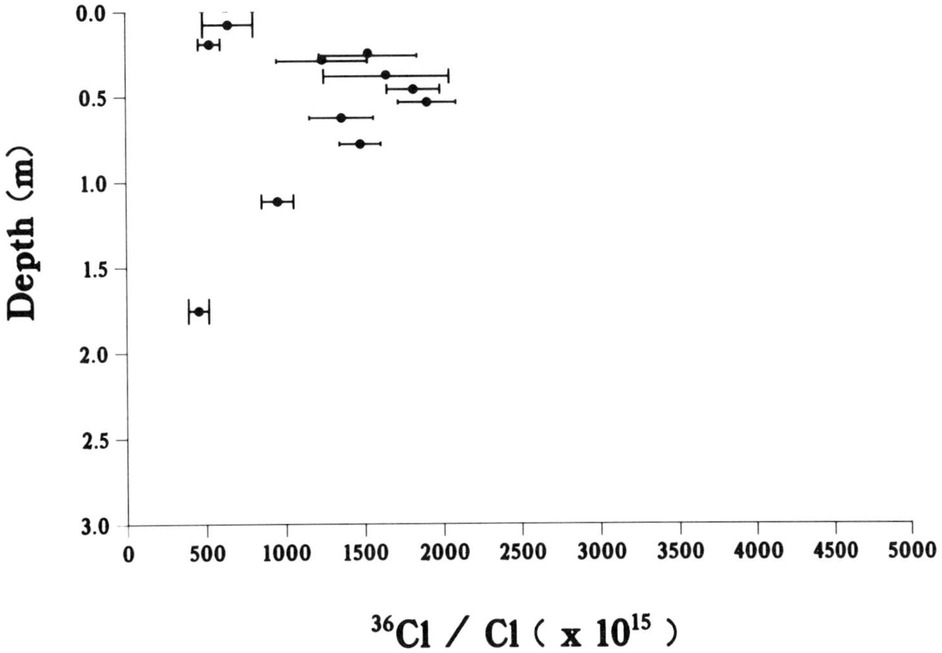

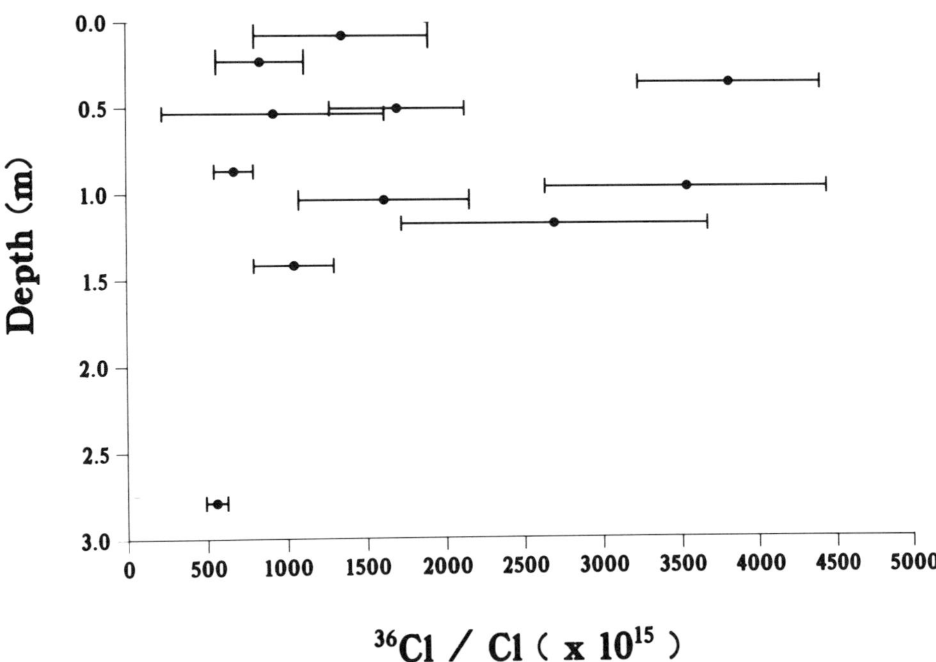

FIGURE 3. $^{36}Cl/Cl$ as a Function of Depth (a) at the Yucca Wash Site and (b) at the Exploratory Shaft Site

V. ACKNOWLEDGMENTS

Funding for this work was provided by the U.S. Department of Energy through the Nevada Nuclear Waste Storage Investigations project. The authors extend their thanks both to H. W. Bentley, Hydro Geo Chem, Inc., for discussions concerning the interpretation of the ^{36}Cl bomb pulse data and to K. Campbell, Los Alamos National Laboratory, for the ^{36}Cl sensitivity analysis. Helpful suggestions from C. W. Myers and G. L. DePoorter are appreciated.

REFERENCES

1. P. Montazer and W. E. Wilson, Conceptual Hydrologic Model of Flow in the Unsaturated Zone, Yucca Mountain, Nevada, Water-Resources Investigation Report 84-4345, US Geological Survey (1984).

2. R. W. Spengler et al., Stratigraphic and Structural Characteristics of Volcanic Rocks in Core Hole USW G-4, Yucca Mountain, Nye County, Nevada, Open File Report 84-789, US Geological Survey (1984).

3. Department of Energy (DOE), Test Plan for Exploratory Shaft at Yucca Mountain, Nevada, NVO-244 (Rev. 1), U.S. Department of Energy (in preparation).

4. H. W. Bentley, F. M. Phillips, and S. N. Davis, "Chlorine-36 in the Terrestrial Environment," in Handbook of Environmental Isotope Geochemistry, Vol. II, Elsevier (in press).

5. D. Elmore et al., "The Rochester Tandem Accelerator Mass Spectrometry Program," Nucl. Inst. and Meth. in Phys. Res., B5, 109 (1984).

ELIMINATION OF FREQUENCY NOISE FROM GROUNDWATER MEASUREMENTS

Y. M. Chien
R. W. Bryce
S. R. Strait
R. A. Yeatman
Rockwell Hanford Operations
P.O. Box 800
Richland, Washington 99352

I. INTRODUCTION

The Basalt Waste Isolation Project (BWIP) was established to assess the suitability of basalt as a repository medium for the long-term disposal of commercial high level radioactive waste beneath the Hanford Site in south-central Washington State. This project is sponsored by the U.S. Department of Energy (DOE) in conjunction with the Office of Civilian Radioactive Waste Management as mandated by the Nuclear Waste Policy Act of 1982.[1] Rockwell Hanford Operations (Division of Rockwell International) serves as the prime contractor to the DOE for operating the BWIP. In support of this work, long-term hydraulic stress tests are being run to determine areal aquifer characteristics in selected basalt formations.

Groundwater response to hydraulic stresses imposed during aquifer testing may be obscured by responses to other stresses on the aquifer. Formation pressure responses during a large-scale stress test, for example, may be affected by responses to atmospheric pressure variations, earth-tides, river stage fluctuations, or irrigation pumping cycles. This paper describes two techniques that may be used to remove these extraneous transients: a systematic method and a frequency analysis method. The two potentially extraneous stress responses removed in this study were response to atmospheric pressure fluctuation and response to earth tides. The other influences mentioned above do not appear to be important for the data set evaluated in this study.

II. SYSTEMATIC METHOD

Systematic removal of external stresses from downhole-pressure data requires either monitoring or synthesizing the external stress. A numerical relationship between the external stress and the downhole-pressure then can be established. Using this relationship, the groundwater response to the external stress may be systematically removed from the downhole-pressure record.

Formation pressure was monitored with a 0- to 3000-lbf/in^2 digiquartz pressure transducer at formation depth. Atmospheric pressure was monitored with a 0- to 15-lbf/in^2 transducer at the well head. Pressures were recorded at 1-h intervals by a desktop computer. The data set used to demonstrate the

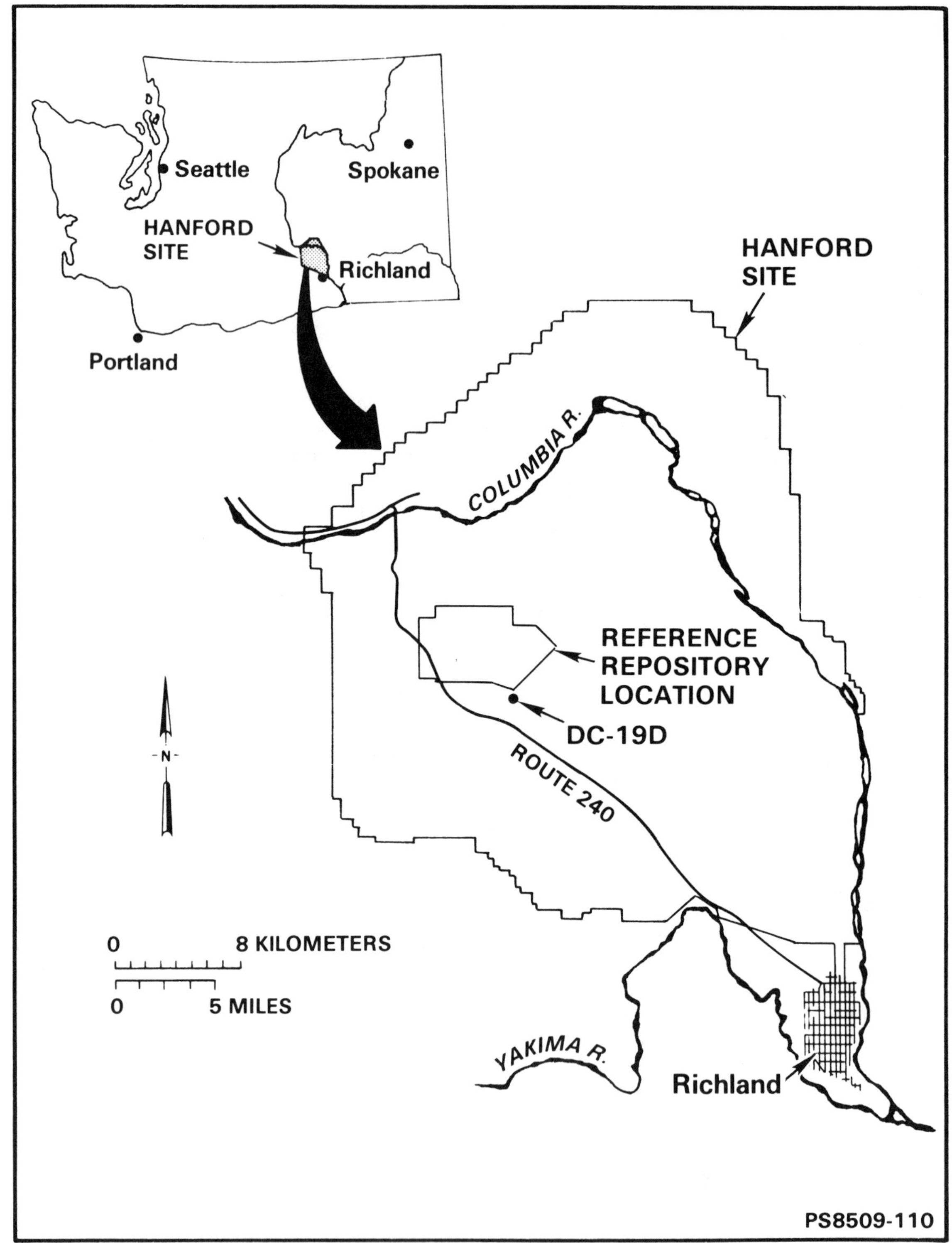

FIGURE 1. Location of the Hanford Site and Borehole CD-19D

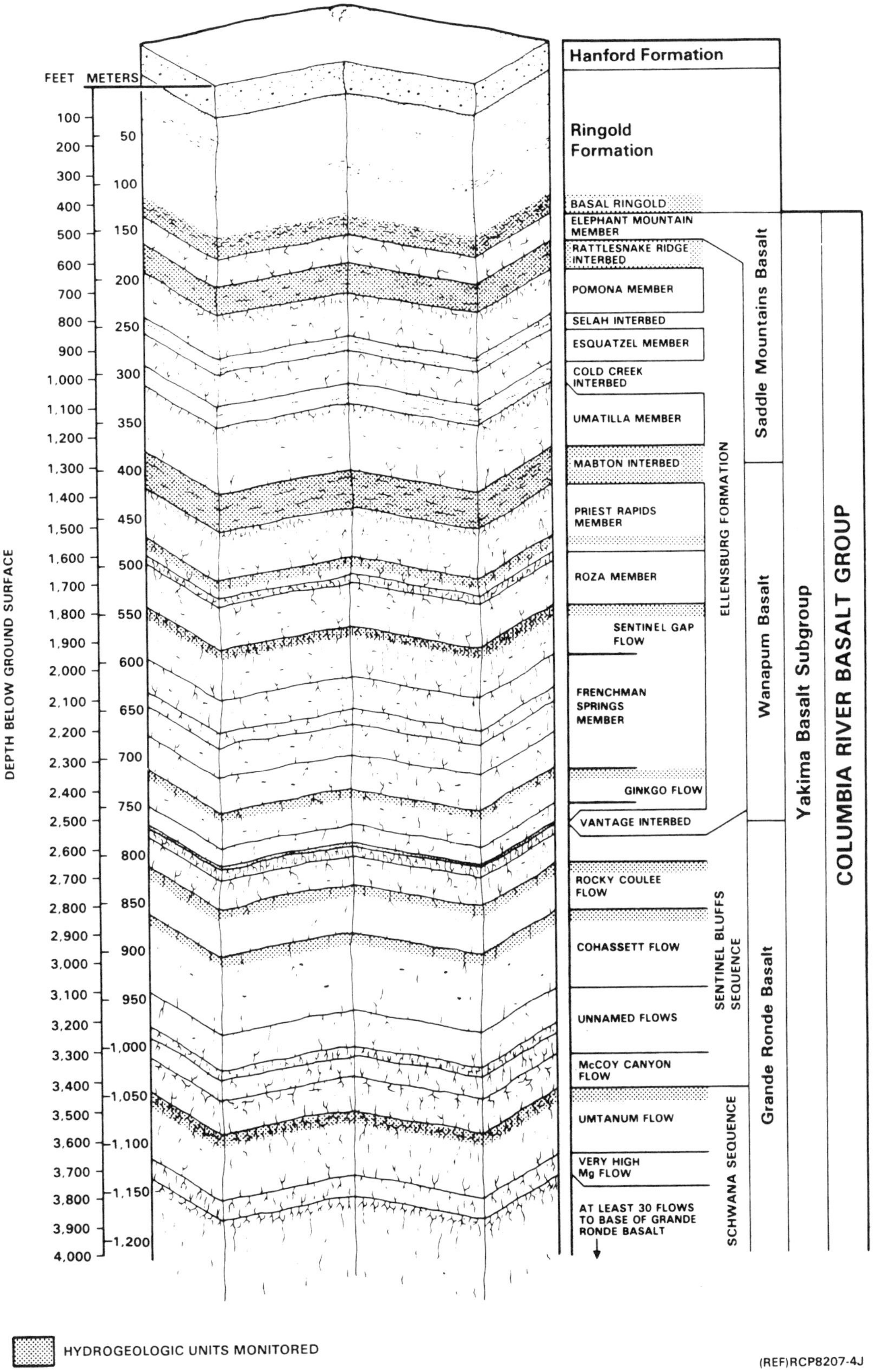

FIGURE 2. Stratigraphc Column Showing Location of Mabton Interbed

systematic technique is from the Mabton interbed at borehole DC-19D (Figure 1 and Figure 2) for the month of September 1984. A longer period of record (June 1984 to January 1985) was required for the application of frequency analysis.

A plot of downhole-pressure for the Mabton interbed in borehole DC-19D is shown in Figure 3. A visual examination of this figure and a comparison to a plot of atmospheric pressure at DC-19D (Figure 4) indicates that the major pressure fluctuations observed are caused by atmospheric pressure changes.

A. Removal of Atmospheric Effects

The ratio between the change in atmospheric pressure and the change in water level in a borehole is known as the barometric efficiency.[2]

$$\text{barometric efficiency} = \frac{\gamma \Delta h}{\Delta p}$$

where

γ = the specific weight of water
Δh = the change in piezometric head
Δp = the change in atmospheric pressure.

This barometric efficiency is related to the characteristics (i.e., compressibility) of the well aquifer system. A method described by Clark[3] was used to calculate barometric efficiency because it compensates for long-term trends in the potentiometric data.

Once the barometric efficiency of the aquifer has been calculated, the difference between a standard atmospheric pressure and the atmospheric pressure corresponding to a formation pressure is determined. The average monthly atmospheric pressure recorded during the last 30 years at the meteorological station located on the Hanford Site was used as the standard atmospheric pressure. Atmospheric pressure was not always recorded coincident with formation pressure, so a linear interpolation of the atmospheric pressure data was used to derive a data set corresponding by time with the downhole-pressure data.

The difference between the standard atmospheric pressure and the atmospheric pressure corresponding to a downhole-pressure reading is multiplied by the complement of the barometric efficiency and added to the downhole-pressure reading. This value is the downhole-pressure adjusted to the standard atmospheric pressure. A graph of the resultant data for September 1984 is shown in Figure 5.

Periodic fluctuations with an amplitude of approximately 0.03 lbf/in^2 remain in the plot of atmospherically adjusted downhole-pressure data. A visual comparison of this plot with the plot of the theoretical earth-tide potential (Figure 6) for the borehole DC-19D location suggests a causative relationship.

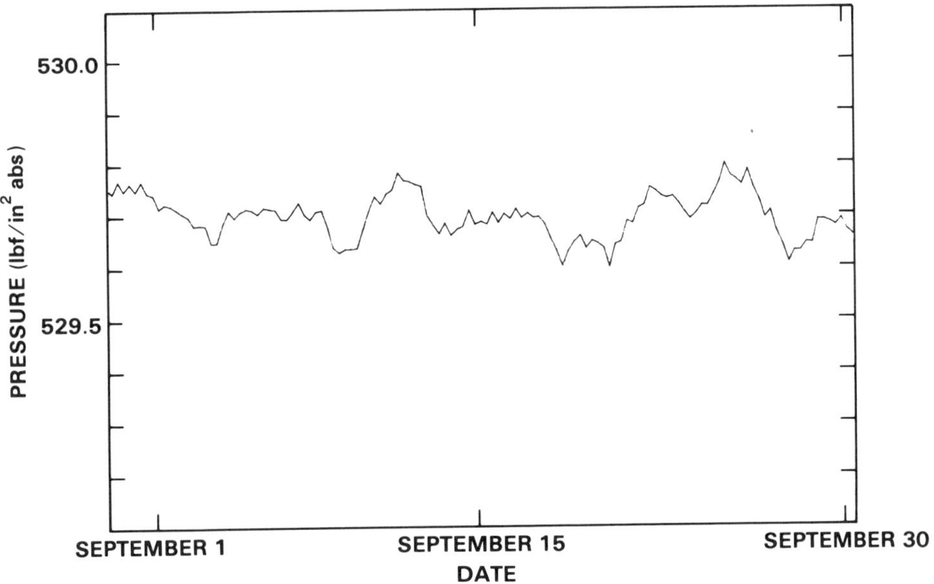

FIGURE 3. Downhole Pressure Data for the Mabton Interbed in Borehole DC-19D

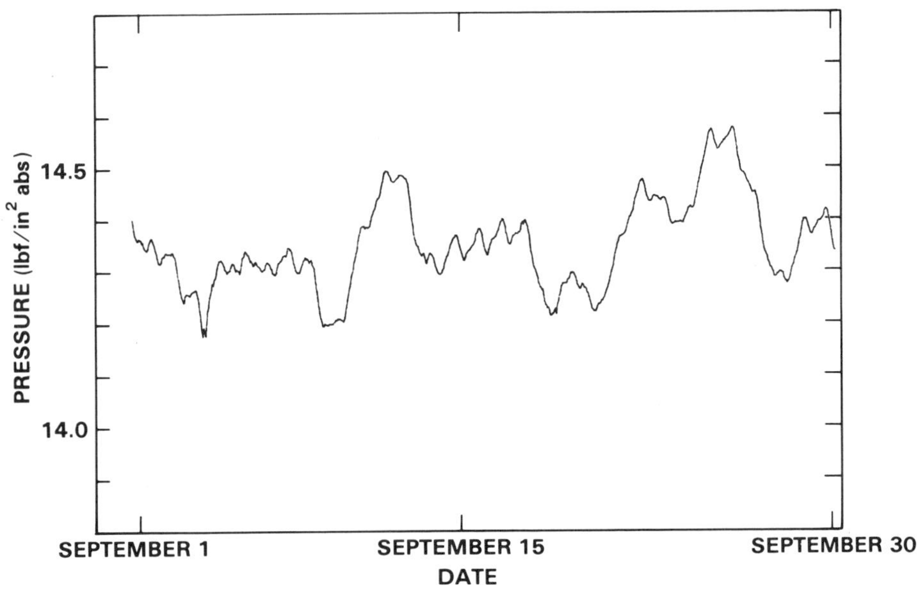

FIGURE 4. Atmospheric Pressure Data for the Mabton Interbed in Borehole DC-19D

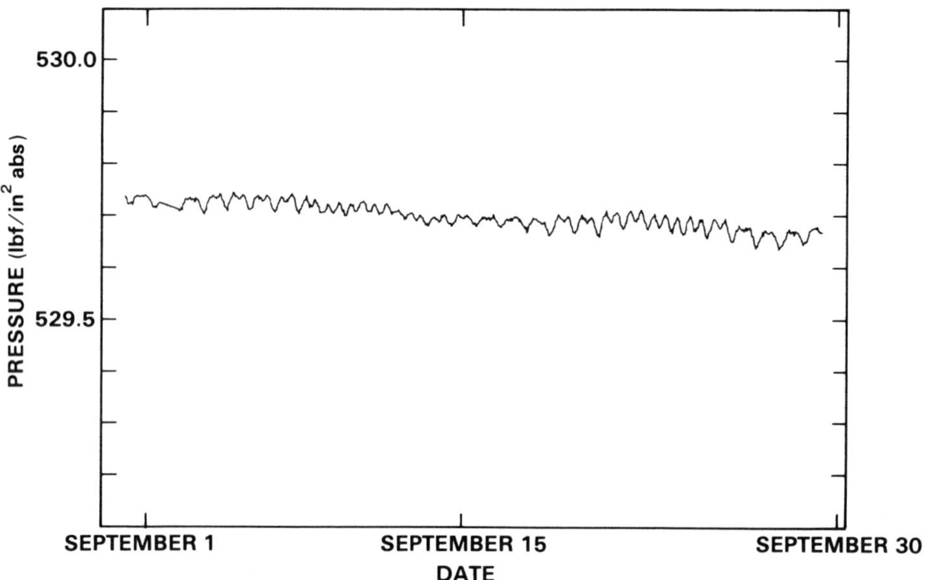

FIGURE 5. Downhole Pressure Data with Barometric Effects Removed

B. Removal of Earth-Tide Effects

The relationship between theoretical earth-tide potentials and downhole-pressure (earth-tide efficiency) was evaluated using a technique similar to the procedure proposed by Clark[3] for the evaluation of barometric efficiency. Earth-tide potentials were generated using the theory described by Melchior[4] (Figure 6). Response to earth tides was subtracted from the data using the earth-tide efficiency method and a procedure similar to that out-lined above for barometric effects. A plot of the resultant data is shown in Figure 7.

The removal of atmospheric and earth-tide responses from groundwater potentiometric data yields a fairly smooth graph. Some low amplitude ($\pm$0.01 lbf/in^2) variations remain. These may be judged to be inconsequential or may be removed by further attempts to identify the cause of the fluctuations.

III. FREQUENCY ANALYSIS

The theory of frequency analysis is based on the assumption that all dynamic responses can be approximated by a combination of sines and cosines (or any other periodic functions) of different frequencies, including frequencies of infinitely long periods. Fourier transform is a process of transforming a response from the time domain into the frequency domain or mathematically mapping an event from the time space into the frequency space.

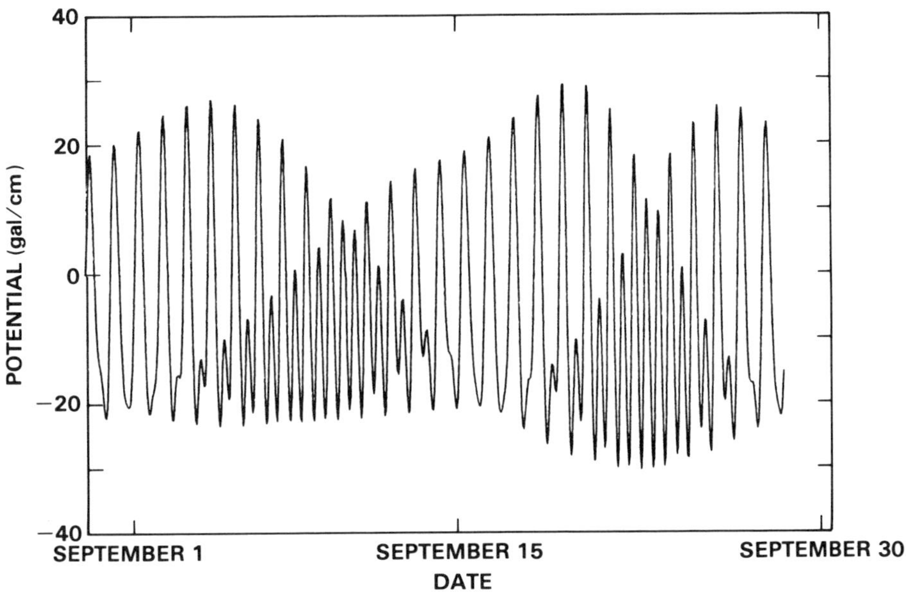

FIGURE 6. Theoretical Tidal Potentials (latitude)

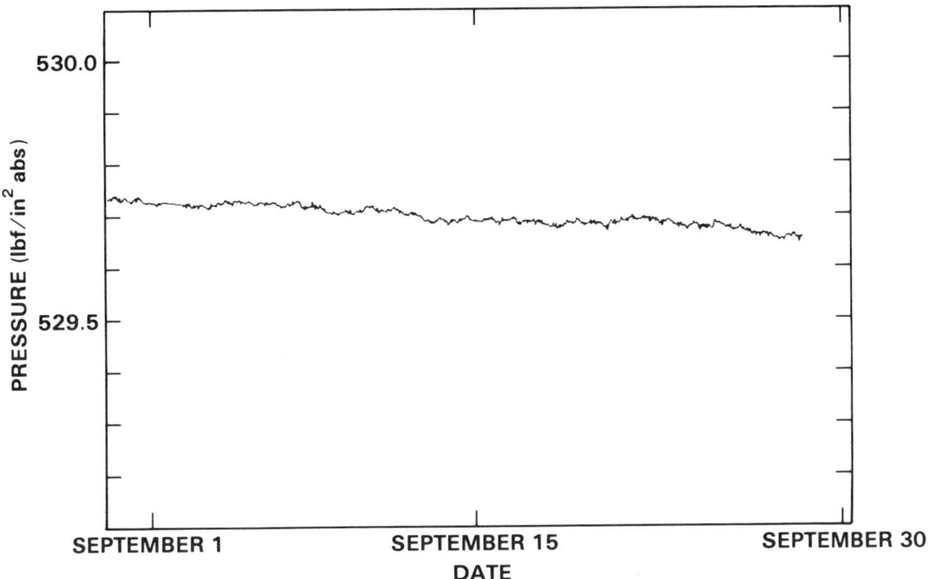

FIGURE 7. Downhole Pressure Data with Barometric and Tidal Effects Removed

Frequency analysis is the analysis performed in the frequency domain. Some physical phenomena that are obscured in the time domain may become clear in the frequency domain.

A. Data Preparation for the Frequency Analysis

A linear interpolation, like that used in the removal of atmospheric effects, is used to provide a set of downhole-pressure data at regular time intervals. Fourier transform requires regular data sampling intervals to preserve the frequency. To verify that the processed data do not alter the response characteristics, the processed data are compared with the raw data to make sure that the hydrologic trends have not been altered.

B. Fast Fourier Transform

Fast Fourier transform (a special method of Fourier transform) is applied to the interpolated data. To minimize the extraneous noise caused by abrupt signal rise or fall at the ends of a data record, a 'cosine tapering' technique is used. 'Tapering' is generally used to avoid generation of erroneous frequencies. A typical frequency spectrum obtained from the downhole-pressure data is shown in Figure 8. Note the high magnitudes of the low frequencies (long periods) near the origin of the frequency axis and the the distinct peaks around .04 and .08 cycles per hour (c/h). The .04 c/h corresponds to a 24-h period, and the .08 c/h corresponds to a 12-h period. The very low frequencies are believed to be the hydrologic recovery trends, and the .04- and .08-c/h frequencies correspond to the major earth-tide cycles.

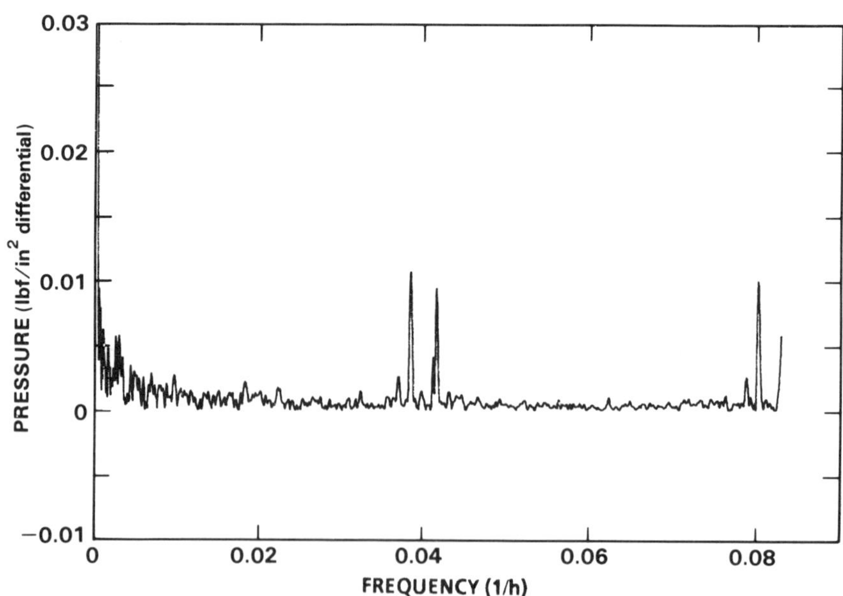

FIGURE 8. Downhole Frequency Spectrum

C. Data Set Frequency Averaging

Frequency analysis as applied to hydrology is quite different from its conventional applications to the electrical transients or structural vibrations. Conventional frequency analysis generally acquires many periods of data per data set, transforms the sets into a frequency domain by Fourier transform, and then averages the frequency sets, so that the random noises are canceled while the real signals are enhanced. Hydrologic cycles on the other hand are natural cycles of very long periods. There are rarely enough data sets to be averaged because the natural cycles (seasons, etc.) are very slow. Therefore, conventional frequency techniques cannot generally be applied to hydrology.

To circumvent this difficulty and to minimize the statistical uncertainty in the frequency spectrum, several data sets are derived from the original data set. A typical scheme is to pick data points 1, 7, 13, 19,... as Data Set 1; data points 2, 8, 14, 20,... as Data Set 2; and data points 6, 12, 18, 24,... as Data Set 6. These sets are then Fourier transformed and the corresponding frequencies are averaged among the different sets. This scheme has the advantage of preserving the low-frequency components corresponding to the slow recoveries and seasonal variations, as well as gaining confidence in the spectrum. The disadvantage is that it cannot resolve frequencies higher than 1/6 of an hour. Exactly 2048 frequencies are used in the Fourier analysis program, and all 2048 frequencies are averaged separately.

D. Evaluation of the Earth-Tide Effects

The influence of the earth-tide upon the downhole-pressure can be evaluated by applying an appropriate filter to the pressure data set. In the frequency band of .03 to .09 c/h (Figure 8), only the earth-tides cause significant responses (.04 c/h and .08 c/h). If a 'band pass' filter of this range is applied to the data set, then only the frequencies of the data set between .03 to .09 c/h can pass through. The resulting pressure history is shown in Figure 9. Since the frequency band consists of the earth-tide frequencies, this method allows the measurement of the earth-tide effects in a specific location. The effects of earth-tides acting on an aquifer at a specific location are difficult to measure because the aquifer expands neither homogeneously nor isotropically with gravity change. Such earth-tides, once known, can be removed from the downhole-pressure using the earth-tide efficiency method described above. An easier and straight forward alternative is the application of a 'band stop' filter of this frequency range on the pressure data. The band stop filter in this case is approximately equal to a 'low pass' filter.

E. Evaluation of Long-Term Trends

Low frequencies near the origin of the frequency axis in Figure 8 represent the signals with very long periods. These frequencies, whether artificially induced by aquifer pumping tests or naturally occurring as a result of seasonal variations, are the long-term aquifer recovery trends. Filtering

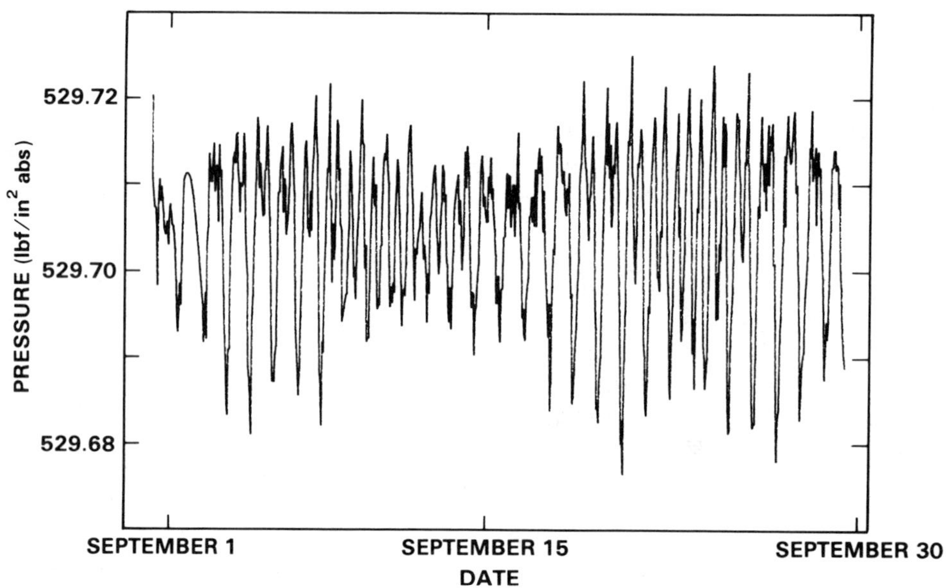

FIGURE 9. Band-Pass-Filtered Pressure Data for 0.03 to 0.09 Cycles per Hour

out all higher frequencies and letting only the low frequencies pass will yield such recovery trends. When such a 'low pass' filter is applied, the resulting pressure is shown in Figure 10.

F. Removal of Atmospheric Effects

Theoretically, all frequencies can be identified with Fast Fourier transform and each component can then be filtered out by successive application of the band pass filter technique. The atmospheric pressure and its corresponding frequency spectrum are shown in Figure 4 and Figure 11, respectively. The close resemblance of the atmospheric frequency spectrum (Figure 11) to the downhole frequency spectrum (see Figure 8) makes the successive filtering technique both cumbersome and impractical. If all the frequency components from the atmospheric spectrum are successively removed from the downhole-pressure, there may be very little left in the downhole-pressure curve other than a horizontal line. This is due to the broadband characteristics (a wide range of frequencies all with significant magnitudes) of the atmospheric pressure itself and the lack of sufficient data to precisely identify each atmospheric frequency. The systematic removal of atmospheric fluctuation from the downhole-pressure data mentioned above is more applicable.

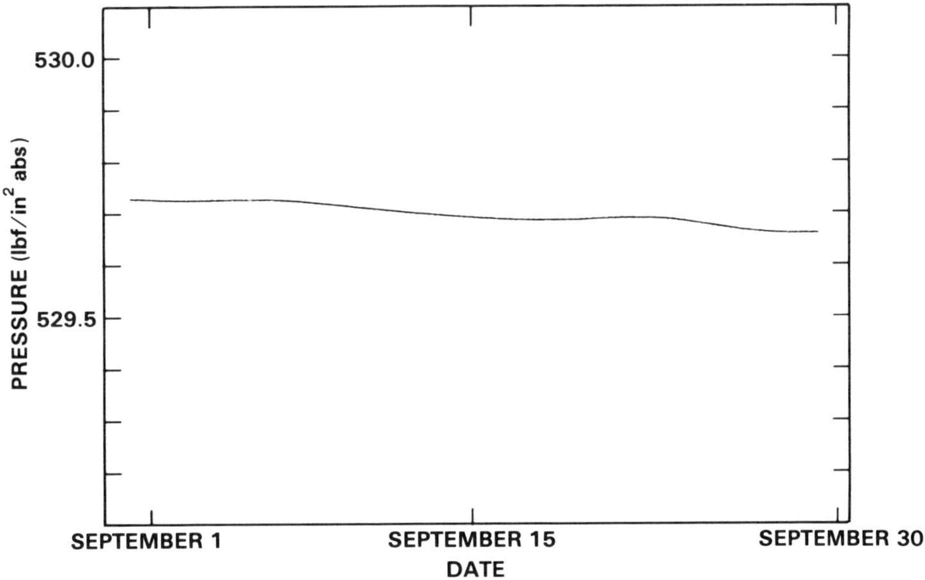

FIGURE 10. Low-Pass-Filtered Pressure Data for 0.005 Cycles per Hour

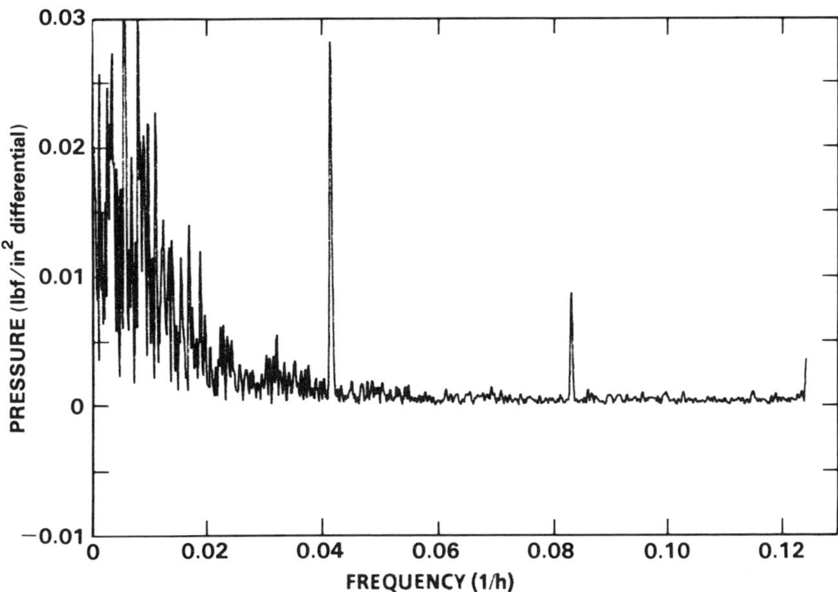

FIGURE 11. Atmospheric Frequency Spectrum

IV. SUMMARY

Groundwater response to atmospheric fluctuation can be effectively removed from downhole-pressure records using the systematic approach. The technique is not as successful for removal of earth-tides, due to a probable discrepancy between the actual earth-tide and the theoretical earth-tide. The advantage of the systematic technique is that a causative relationship is established for each component of the pressure response removed. This concept of data reduction is easily understood and well accepted. The disadvantage is that a record of the stress causing the pressure fluctuation must be obtained. This may be done by monitoring or synthesizing the stress.

Frequency analysis offers a simpler way to eliminate the undesirable hydrologic fluctuations from the downhole-pressure. Frequency analysis may prove to be impractical, if the fluctuations being removed have broadband characteristics. A combination of the two techniques, such as eliminating the atmospheric effect with the systematic method and the earth-tide fluctuations with the frequency method, is the most effective and efficient approach.

REFERENCES

1. "Nuclear Waste Policy Act of 1982," Public Law 97-425.

2. D. K. Todd, Groundwater Hydrology, John Wiley & Sons, New York (1980).

3. W. E. Clark, "Computing the Barometric Efficiency of a Well," Proc. ASCE, J. Hyd. Div., pp. 92-98, (July 1967).

4. Melchior, P., 1966, The Tides of the Planet Earth, Pergamon Press, London, England, (1966)

DISEQUILIBRIUM OF NATURAL RADIONUCLIDES IN HANFORD SITE GROUNDWATERS

J. C. Laul
M. R. Smith
Pacific Northwest Laboratory
P.O. Box 999
Richland, Washington 99352

V. G. Johnson
R. M. Smith
Rockwell Hanford Operations
P.O. Box 800
Richland, Washington 99352

ABSTRACT

Concentrations of U, Th, Ra, Pb and Po radionuclides are extremely low in well DC-14 groundwater of the Grande Ronde Formation. Uranium, Th and Ra appear to be highly sorbed. Relative to Rn, the retardation factors are: 1200 for U, 12000 for Th, and 1500 for Ra in DC-14 water. The $^{234}U/^{230}Th$ ratio is about 10 in the DC-14 well water and 240 in Columbia River water indicating that U is largely in the +4 oxidation state in the DC-14 water and implies a reducing environment. Colloids appear to be important for Pb and Po but not for U, Th and Ra in the DC-14 well water.

I. INTRODUCTION

All groundwaters and geologic materials contain measurable concentrations of natural radionuclides of the uranium and thorium decay series. Consequently, site specific investigations of these radionuclides in groundwater can provide direct and analog information on the expected behavior of radwaste radionuclides in these aquifers. In addition, measurements of natural radionuclide concentrations can provide a baseline for future monitoring of any possible releases of radwaste.

Figure 1 lists the natural radionuclides and half-lives of the ^{238}U and ^{232}Th decay series measured in this investigation. Since these radionuclides have widely different geochemical properties, their behavior may provide indications of the in-situ chemical behavior of other nuclides (tracers) injected into the system. Here the results of a feasibility study of natural radionuclides are reported. These radionuclides were measured in groundwater during May-July, 1984, from borehole DC-14 completed in a basalt flow-top aquifer in the the Grande Ronde Formation (994-1017 m below ground surface) at the Hanford site. This zone lies stratigraphically below the candidate horizon for the proposed repository. Columbia River water adjacent to the DC-14 well was also sampled about the same time. These preliminary results permit evaluation of retardation factors and sorption-desorption characteristics of various radionuclides in site specific waters. In addition, the question of solution versus colloid transport of the radionuclides is addressed.

^{238}U Chain

^{238}U ⟶ ^{234}Th - - - - ^{234}U ⟶ ^{230}Th ⟶ ^{226}Ra ⟶
4.4×10^9 yr, 24.1 d, 2.44×10^5 yr, 7.5×10^4 yr, 1600 yr

^{222}Rn - - - - - ^{210}Pb ⟶ ^{210}Bi ⟶ ^{210}Po
3.82 d, 22.3 yr, 5.0 d, 138.4 d

^{232}Th Chain

^{232}Th ⟶ ^{228}Ra - - - - ^{228}Th ⟶ ^{224}Ra
1.4×10^{10} yr, 5.76 yr, 1.91 yr, 3.66 d

FIGURE 1. Radionuclides measured in ^{238}U and ^{232}Th chains

A. Water Sampling and Analysis

The concentrations of natural radionuclides (Figure 1) in groundwater are extremely low. Consequently their measurements require pre-concentration and ultraclean sample handling and analysis techniques. The details of field sampling and analysis of groundwaters for U, Th, Ra, Rn, Pb, Bi and Po nuclides are described elsewhere by Laul et al.[1] In brief, to measure Ra radionuclides, several hundred gallons of water were passed through a Battelle large volume water sampler containing several filters (0.8 μm) and two quarter inch thick Al_2O_3-$BaSO_4$ adsorption beds. The adsorption beds were then dried in the laboratory and counted by gamma spectrometry for ^{228}Ra, ^{226}Ra and ^{224}Ra radionuclides. A direct counting of an aliquot was made to measure ^{222}Rn and ^{226}Ra via ^{222}Rn by an alpha scintillation method. For Th, U, Pb, Bi and Po nuclides, unfiltered and filtered waters were collected and acidified with HCl on site. Four to 50 liters of each were subsequently analyzed in the laboratory using carrier-free radiochemical separations[1] and ^{233}U, ^{230}Th and ^{208}Po as tracers. Uranium-238, ^{235}U and ^{234}U were measured by mass spectrometry; ^{232}Th, ^{230}Th, ^{228}Th, ^{210}Pb and ^{210}Po by alpha spectroscopy; ^{234}Th and ^{210}Bi by beta counting.

The DC-14 well water was free of gross amounts of particulates and was collected at the surface under artesian flow conditions. For the colloidal study, unfiltered and filtered (0.10 μm nucleopore filter) waters were collected. The Columbia River water contained gross amounts of particulates (sediments) and algae. The Columbia River water was filtered through a prefilter (25-2 μm) and then through a ~1 μm filter. For colloidal study, three samples were collected. One was filtered through a prefilter (~1 μm), one through a 0.40 μm and one through a 0.10 μm nucleopore filter. Where possible, the measurements were made in duplicate.

II. RESULTS AND DISCUSSION

The concentrations (dpm/L or ng/L) of U, Th, Ra, Pb, Bi and Po radionuclides in the ^{238}U chain for unfiltered water from well DC-14 (Grande Ronde Formation) and prefiltered (1.0 μm) water from the Columbia River are shown in Table 1. Analytical uncertainties associated with the measurements are either due to counting statistics or precision of duplicate analyses, whichever is larger. Typically, the errors in the analysis are: ±5-10% for ^{238}U, ^{234}U and ^{222}Rn; ±10-20% for ^{234}Th, ^{230}Th, ^{226}Ra, ^{210}Pb, ^{210}Bi, and ^{210}Po. The radionuclide concentration limits[2] for drinking water approved by the Environmental Protection Agency (EPA) are also included in Table 1 for comparison.

The radionuclide concentrations in DC-14 well water are extremely low, far below their laboratory solubility limits from a basalt substrate.[3] When compared with the Columbia River water the ^{238}U and ^{226}Ra concentrations in DC-14 are low by factors of 30 and 3 whereas ^{222}Rn, ^{210}Pb and ^{210}Po concentrations in DC-14 are high by factors of 18, 6 and 65 respectively. Thorium radionuclides are about the same concentration in both waters. However, these concentrations in both waters are below the EPA drinking water limits[2] (Table 1).

The ^{222}Rn activity is the highest in the ^{238}U chain, and thus the radionuclide data are normalized to ^{222}Rn. The activity ratios for DC-14 and Columbia River waters are shown as histograms in Figure 2. The interpretations of these data are as follows:

A. Retardation Factors

If all the radionuclides in the ^{238}U decay series are in radiochemical equilibrium, then the activity ratio (AR) of each daughter/parent will be unity (secular equilibrium). Considering the different chemical properties of U, Th, Ra, Pb, Bi and Po in groundwaters, a significant disequilibrium among these radionuclides is expected as shown in Figure 2. Activity of ^{222}Rn in the groundwater is the highest in the ^{238}U chain. Because Rn is a noble gas and chemically inert, it is not (or weakly) sorbed on the substrate. Thus, Rn is soluble in groundwater and its movement not retarded by processes such as sorption, membrane filtration and precipitation. The ^{222}Rn activity as a result of recoil from ^{226}Ra adsorbed on a substrate in a given aquifer is taken as an indicator of the effective supply rate relative to its more reactive radionuclide members (U, Th, Ra, Pb, Bi and Po) in a chain. The retardation factor (RF) relative to Rn is just the inverse of activity ratio (RF = 1/AR), assuming the same recoil supply factor (ε) of other radionuclides relative to ^{222}Rn. This approach has been used by Krishnaswami et al.,[4] to compute adsorption/desorption constants for various nuclides in an aquifer. They estimated ε values of 0.96 for ^{226}Ra and 0.90 for ^{230}Th.

Table 1. Radionuclide Concentrations (dpm/L or ng/L) in Borehole DC-14 and Columbia River Water

Radionuclide	DC-14 Unfiltered (dpm/L)	DC-14 Unfiltered (ng/L)	Columbia River Prefiltered (1.0 µm) (dpm/L)	Columbia River Prefiltered (1.0 µm) (ng/L)	EPA Drinking Water Limit (dpm/L)
U-238	0.040 ± .005(2)[a]	54	1.10 ± .05	1480	33[b]
Th-234	0.2	4×10^{-7}	<0.2	$<4 \times 10^{-8}$	
U-234	0.057 ± .005(2)	4.2×10^{-3}	1.38 ± .08(2)	1.0×10^{-1}	
Th-230	0.0051 ± .0010(2)	1.1×10^{-4}	0.0057 ± .0017	1.2×10^{-4}	
Ra-226	0.048 ± .007(3)	1.6×10^{-5}	0.12 ± .03	6.4×10^{-5}	11[c]
Rn-222	72 ± 5(2)	2.1×10^{-7}	4.1 ± .2	1.2×10^{-8}	
Pb-210	1.7 ± 1.0(2)	1.0×10^{-5}	0.26 ± .11	1.5×10^{-6}	
Bi-210	3.5 ± .5	1.3×10^{-8}	<1.2	$<4 \times 10^{-9}$	
Po-210	17.0 ± 0.9	1.7×10^{-6}	0.26 ± .11 0.11 ± .02	2.6×10^{-8}	

(a) Values in parenthesis are the number of analyses

(b) This represents gross alpha activity of ^{238}U and its daughters except ^{234}Th, ^{222}Rn, ^{210}Pb and ^{210}Bi

(c) ^{226}Ra and ^{228}Ra activity

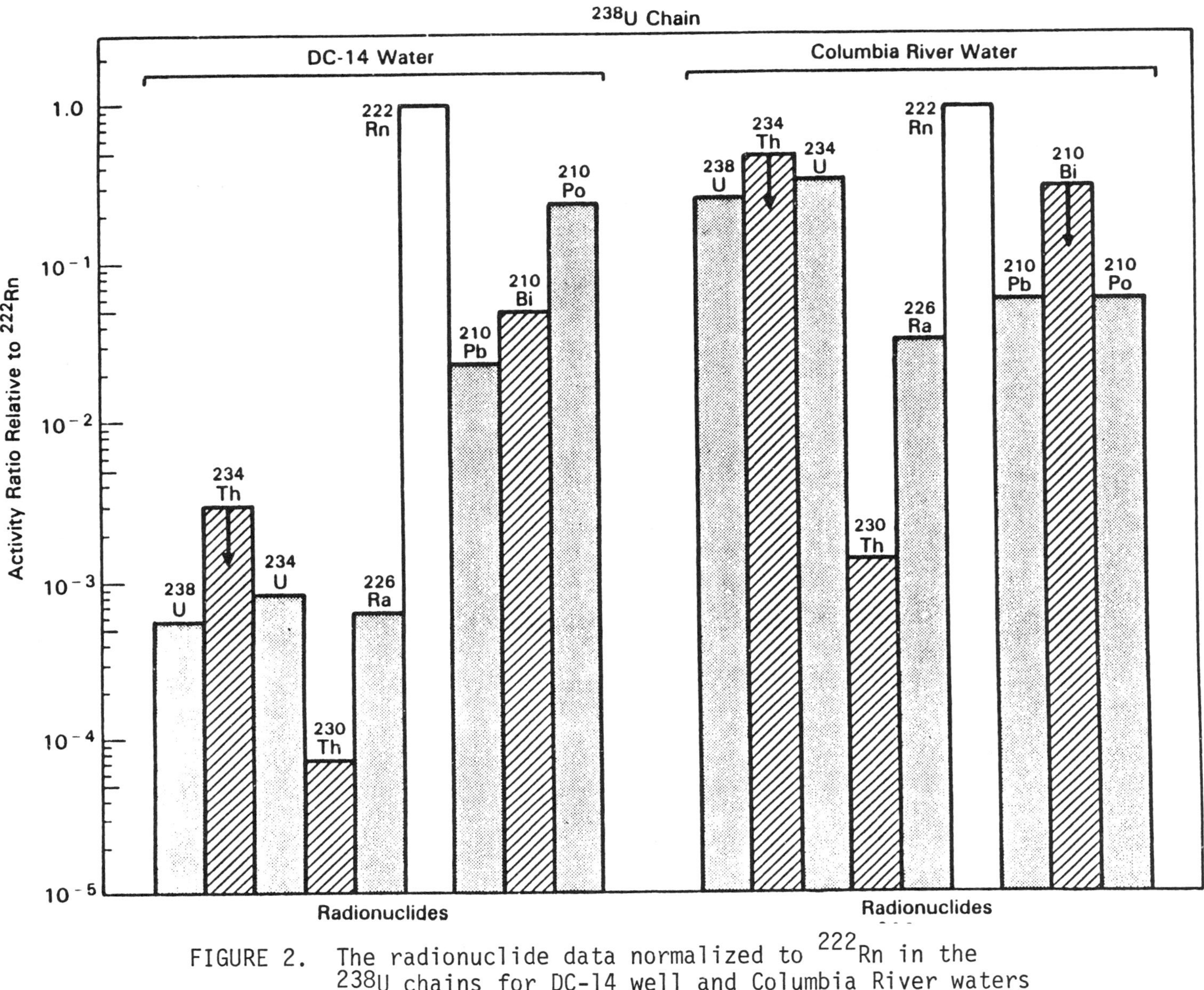

FIGURE 2. The radionuclide data normalized to ^{222}Rn in the ^{238}U chains for DC-14 well and Columbia River waters

Rama and Moore (1984)[5] pointed out that the observed Rn in groundwater (use of Rn for normalization) may be an overestimate of the supply rate (except for ^{210}Pb, ^{210}Bi and ^{210}Po) since ^{222}Rn can be supplied by diffusion from nanopore fluids into the intergranular water as well as by direct in-situ recoil from ^{226}Ra adsorbed on grain surfaces. They showed this in laboratory experiments using a coarse grained (200-800 μm) granite saprolite in contact with water. However, laboratory conditions (static) based on short time scales are often far different from physical and chemical environments found in actual aquifers where a steady state is reached over a long time period. If Rama and Moore's assertion is correct then large excesses of Rn would always prevail, with the result that the parent/daughter ratio such as $^{226}Ra/^{222}Rn$ would always be significantly lower than unity. Hubbard et al.[6] and Laul et al.[7] consistently observed $^{226}Ra/^{222}Rn$ activity ratios of near unity in nine individual deep briney aquifers in the Palo Duro Basin, Texas, confirming radiochemical equilibrium and suggesting that the addition of Rn by diffusion from other sources is not likely. In an open flowing system, loss of Rn can occur. However, in deep or confined aquifers this does not seem to be the case.

From Figure 2, the RFs (RF = 1/AR) range from 4 to 10^4 in DC-14 well water. Accounting for the recoil supply rate (ε) for ^{226}Ra and ^{230}Th, the retardation factors (relative to Rn) for various radionuclides are: 1200 for ^{234}U, 12000 for ^{230}Th, 1500 for ^{226}Ra, 44 for ^{210}Pb, and 4 for Po, respectively. For Columbia River water, the data are again normalized to ^{222}Rn. However, the Columbia River is an open system and ^{222}Rn might be continuously outgassing. Thus, these RF values are most likely lower limits ranging from 4 to 700.

B. Reducing/Oxidizing Environment

The ^{234}U-^{230}Th pair can reveal information on the oxidation state of U and thus on the reducing/oxidizing conditions found in a site specific environment. Because quadravalent U and Th have similar chemical properties (i.e. same +4 charge and ionic radii of 1.07 A°) and solubility products[9,10] the low $^{234}U/^{230}Th$ activity ratio (10) as compared to Columbia River water (240) indicates that uranium may be present as U (IV) and thus suggests a reducing environment. Under oxidizing conditions, U (VI) will predominate. Since U (VI) is far more soluble than U (IV) because of its ability to form soluble complexes, the $^{234}U/^{230}Th$ ratio is expected to be much higher than unity.

Because the Columbia River is an open system it is in an oxidizing condition. The $^{234}U/^{230}Th$ ratio in the Columbia River is 240 and 10 in DC-14 well water. By relative comparison of the ratios, U appears to be predominately in the +6 oxidation state in Columbia River water and in the +4 oxidation state in DC-14 well water; this +4 oxidation state of U suggests a reducing environment in the DC-14 well water (Grande Ronde Formation).

C. Solution versus Colloidal Transport

The presence of colloids in groundwater can increase the concentrations of radionuclides transported by the ground water.[11] However, very little is known about the occurrence, nature and abundance of natural colloids in groundwater and the possible affinity of radionuclides may have for these sub-micron particulates.[11] This state of knowledge is especially true for basalt groundwaters at the Hanford site. To evaluate the effects of potential colloidal phases in Hanford groundwater, we analyzed uranium and thorium series radionuclides (Figure 1) in filtered and unfiltered DC-14 well water. To test the effect of colloids, we analyzed unfiltered and filtered (<0.10 μm) DC-14 well water for the radionuclides. The data are plotted in Figure 3. Uranium-238 and ^{234}U show about a 30% increase in the unfiltered versus filtered water. However, within a 2σ error ^{238}U and ^{234}U concentrations between unfiltered and filtered water overlap. The ^{232}Th, ^{230}Th, and ^{228}Th concentrations are the same within the experimental (counting statistics) error. On the other hand, ^{210}Pb and ^{210}Po concentrations are higher in the unfiltered water than in filtered water by factors of 5 and 3. Correspondingly, the RF values for Pb and Po would be 5 and 3 times higher in the filtered than unfiltered water. Thus, colloids appear important for the transport of Pb and Po but not for U and Th in the DC-14 well water. For Ra nuclides, no analysis was made on the <0.10 μm fraction.

The radionuclide data for prefiltered (1 μm) and filtered (<0.40 μm and <0.10 μm) Columbia River water are also shown in Figure 3. Based on the comparison, the concentrations of ^{238}U, ^{234}U, ^{228}Ra, ^{226}Ra and ^{224}Ra are about the same between the prefiltered and filtered samples. The Th radionuclide concentrations are extremely low and thus our values contain large counting statistical errors. Within the experimental error ($\pm 2\sigma$), ^{232}Th, ^{230}Th, and ^{228}Th activities are identical among prefiltered, <0.40 μm and <0.10 μm size fractions. Only ^{210}Po in DC-14 well water shows an increase by a factor of 3 in the unfiltered versus filtered water. Thus, except for Po, colloids are not important for the transport of U, Th and Ra in Columbia River water.

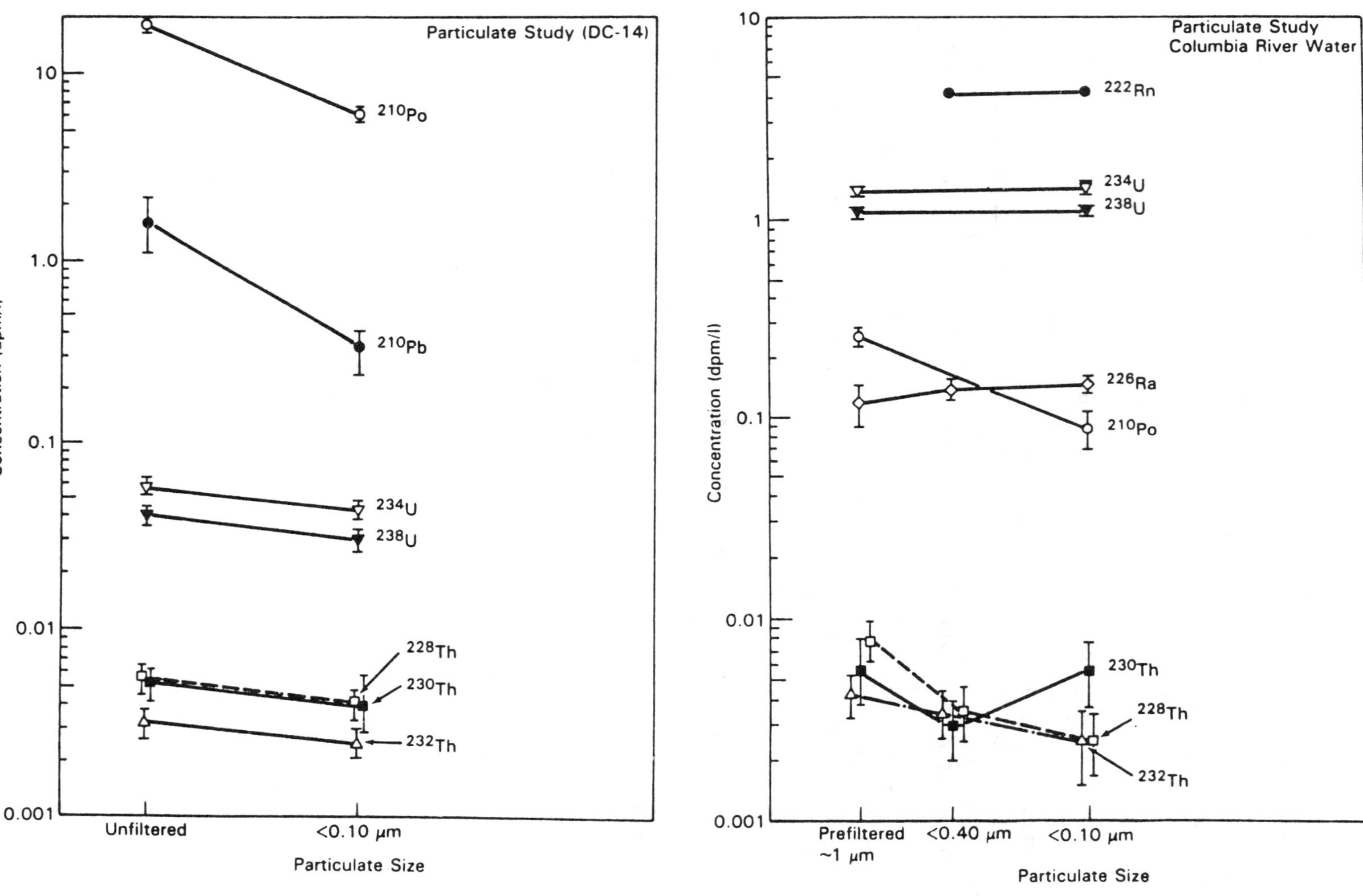

FIGURE 3. Particulate size study of radionuclides for confined zone of DC-14 water and Columbia River water

III. CONCLUSIONS

This feasibility study demonstrates that in spite of the extremely low radionuclide concentrations in the DC-14 well water, the ^{238}U and ^{232}Th decay chain radionuclide measurements can be used to evaluate in situ radionuclide transport characteristics and redox states of the environment in basalt aquifers at the BWIP site.

Based on the radionuclide data, specific conclusions are as follows:

- Concentrations of U, Th, Ra, Pb and Po radionuclides are extremely low in DC-14 well water from the Grande Ronde Formation. Relative to the DC-14 well water, Columbia River water is high in U and Ra content by factors of 30 and 3, and low in Rn, Pb and Po content by factors of 18, 6, 65, respectively. The Th nuclides are about at the same concentration in both waters. However, and these concentrations are far below the EPA drinking water limits.

- Uranium, Th, Ra, Pb, Bi and Po are highly sorbed by the solid phases in DC-14 and Columbia River waters.

- Relative to ^{222}Rn, the retardation factors (RF) for the various radionuclides range from 4 to 12000 in DC-14 and from 4 to 700 in Columbia River. The RF values for the DC-14 well water are: 1200 for U, 12000 for Th (^{230}Th), 1500 for Ra, 44 for Pb and 4 for Po.

- The Columbia River, being an open flowing system, is an oxidizing environment. The $^{234}U/^{230}Th$ ratio of 240 suggests U largely in the +6 state. This ratio is far less (10) in DC-14 well water indicating that U is largely in the +4 state, which suggests a reducing environment.

- Colloids do not appear to play an important role in the occurrence of U, Th, and Ra in DC-14 well water. Lead-210 and ^{210}Po on the other hand, appear to be associated with a submicron particulate phase. The nature and abundance of the apparent solid phase is unknown.

IV. ACKNOWLEDGMENTS

We thank J.S. Schmitt and J.G. Pratt for their assistance in this experiment. This work was supported by Basalt Waste Isolation Project under the U.S. Department of Energy, Contract DE-AC06-76RLO 1830.

REFERENCES

1. J. C. Laul et al., "Analysis of Natural Radionuclides from Uranium and Thorium Series in Briney Groundwaters," J. of Radioanalytical Chemistry (in press).

2. Environmental Protection Agency: Part 141 - Interim Primary Drinking Water Regulations, EPA-570/9-76-003 (1976).

3. G. S. Barney et al., Radionuclide Sorption Kinetics and Column Sorption Studies with Columbia River Basalts, SD-BWI-T1-108, Rockwell Hanford Co., Richland, WA (1983).

4. S. Krishnaswami et al., "Ra, Th and Radioactive Pb Isotopes in Groundwaters: Application to the In-Situ Determination of Adsorption-Desorption Rate Constants and Retardation Factors," Water Resources Research, 18, No. 6, 1633-1675 (1984).

5. Rama and W. S. Moore, "Mechanism of Transport of U-Th Series Radioisotopes from Solids into Groundwater," Geochemica. et Cosmochimica Acta., 48, 395-399 (1984).

6. N. Hubbard, J. C. Laul and R. W. Perkins, "The Use of Natural Radionuclides to Predict the Behavior of Radwaste Radionuclides in Far-Field Aquifers," Scientific Basis for Radioactive Waste Management VII, Mat. Res. Soc. Symp. Proc., 26, 891-897 (1984).

7. J. C. Laul, M. R. Smith and N. Hubbard, "Behavior of Natural Uranium, Thorium and Radium Isotopes in the Wolfcamp Brine Aquifers, Palo Duro Basin, Texas," Scientific Basis for Radioactive Waste Management VII, Mat. Res. Soc. Smp. Proc., 44, 475-482 (1985).

8. R. J. Serne and J. F. Relyea, The Status of Radionuclide Sorption-Desorption Studies Performed by the WRIT Program, PNL-3997, Pacific Northwest Laboratory, Richland, WA.

9. J. E. Grindler, "The Radiochemistry of Uranium," Nuclear Science Series, NAS-NS, 3050 (1960).

10. E. K. Hyde, "The Radiochemistry of Thorium," Nuclear Science Series, NAS-NS, 3004 (1960).

11. J. A. Apps et al., Status of Geochemical Problems Relating to the Burial of High-Level Radioactive Waste, LBL-15103, NUREG/CR-3062 (1982).

Repository Design and In-Situ Testing

THE HIGH-LEVEL WASTE REPOSITORY CONCEPT FOR THE 1985 GEWAEHR PROJECT IN SWITZERLAND

A. L. Nold
R. Gassner
National Cooperative for the Storage of Radioactive Waste (Nagra)
Parkstrasse 23
5401 Baden, Switzerland

ABSTRACT

The main aim of the 1985 Project Gewaehr is to demonstrate the long-term safety of final disposal of radioactive wastes in Switzerland. An engineering project study aimed at demonstrating the feasibility of constructing and operating a repository for high-level waste on a mining concept basis has been carried out. The study is based on a model data-set representing typical geological and rock mechanical conditions as found in northern Switzerland. The project led to the conclusion that construction and operation of a repository for high-level radioactive waste is feasible with present-day technology.

I. INTRODUCTION

The main aim of the 1985 Project Gewaehr is to demonstrate the long-term safety of a final repository for radioactive waste. However, this aspect of long-term safety is closely related to that of technical feasibility: all construction options being considered as a basis for safety analysis of the repository must also be capable of realization on a technical and organizational basis.

An engineering project study aimed at demonstrating the feasibility of constructing a deep repository for high-level waste HLW (Type C repository) has been carried out, the study is based on a model data-set representing typical geological and rock mechanical conditions as found outside the so-called permocarboniferous basin in northern Switzerland.

The Type C repository accomodates HLW and those types of alpha-containing intermediate-level waste whose radionuclide concentrations exceed the maximum permissible concentration values for the planned Swiss repository for intermediate- and low-level waste (ILW/LLW). According to the Swiss nuclear waste management concept, disposal is to be in suitable geological formations. The protection of man and the environment is ensured by a system of multiple safety barriers.

II. BASIC DATA

A. Choice of Concept, Host Rock

Analysis of possible scenarios for transport of radioactive waste materials from the repository to the biosphere shows that water is the only critical transport medium. The safety barriers are therefore selected on the basis of how they restrict penetration of water into the storage area, how they limit the removal of radioactive material from the waste into water, how they hinder or slow down the transport of possibly contaminated water from the repository to the earth's surface and how they can retard radioactive material in the water on its way to the surface.

The principal stipulations of the waste management concept initially leave various realization alternatives open for a repository. These can be characterized by two largely independent parameters:

- the construction concept and
- the host rock (the overall situation given by the host rock and its overburden).

From a construction point of view the repository could be designed as, for example, a mined system of tunnels and silos; as an arrangement of deep boreholes from the earth's surface into the host rock, which can take several waste canisters at a time and can be individually sealed; or as a combination of both systems with underground tunnels in stable overlying rock, from which boreholes are sunk into a host rock, which is suitable on a safety basis but shows rock-mechanical problems. Other possibilities are also conceivable.

A priori, low permeable clays, anhydrite formations and the crystalline basement can be considered as host rocks for HLW in Switzerland and a further degree of freedom - albeit limited by the choice of host rock - exists with regard to the selection of a region for the repository location.

The repository concept selected for Project Gewaehr 1985 was a system of mined tunnels and silos at a depth of about 1200 m in the crystalline basement of northern Switzerland. This concept was selected for evaluation with regard to its long-term safety and constructional feasibility and was found to be safe and feasible. However, the choice of concept in no way prejudices later planning of a final repository project with regard to either the host rock, the region of location or the engineering design.

B. Geological Situation

The design studies for the repository concept in Project Gewaehr 1985 were based on a representative geological situation (in the form of a model data-set) as can be found adjacent to the so-called permocarboniferous trough in Nagra's region of investigation in the area of the Cantons Aargau, Schaffhausen, Solothurn and Zurich. The repository was assumed to be located in a several kilometer-wide granite block between two major faults in the crystalline basement. The data-set is used for calculating safety analysis models and for the construction project study. Further information is given in Reference 1.

C. Radioactive Waste

The repository is intended to accommodate a total of about 6,000 vitrified HLW-cylinders (gross volume of about 1,200 m^3) and about 10,000 m^3 of ILW with concrete and bitumen solidification matrix.

D. Safety Barriers

The safety barrier system for HLW is presented in Figure 1. The technical barriers comprise the waste glass matrix, the corrosion-resistant steel canister and the surrounding bentonite clay. The function of the technical barriers and of the geosphere is described in more detail in Reference 2.

The technical barriers for alpha-containing ILW differ from those for HLW. Cement or bitumen is foreseen as the solidification matrix and a thick-walled steel canister is not required. The barriers correspond to those in a Type B repository with the exception that the concrete walls of the waste silos are additionally surrounded with a bentonite layer (Figure 5).

III. THE PROJECT STUDY

A. Main Aspects Concerning Design and Operation

The Type C repository for HLW described in the construction study in Project Gewaehr 1985 is characterized by the following essential aspects with respect to design and operation:[3]

- For this study, the repository is intended for disposal of HLW and some ILW but can also be used for the disposal of nonreprocessed fuel elements.

- Final disposal of HLW is effected in horizontally mined tunnels while ILW is disposed of in vertical silos. The common entrance to the repository area is through two vertical shafts and surface reception facilities are foreseen. The tunnel and silo system is conceived in such a way that it can be adapted to take account of the geometry of potential disturbed zones in the host rock at repository depth without compromising long-term safety. The silo and tunnel areas are spatially separated in order to avoid any undesirable influence (particularly chemical) from either side.

- Systematic, multi-phase investigations of the host rock in the repository area are carried out before detailed planning of the layout of the storage tunnels and silos to ensure that zones of greater disturbance potential are avoided by observing a sufficient safety clearance. Minor disturbed

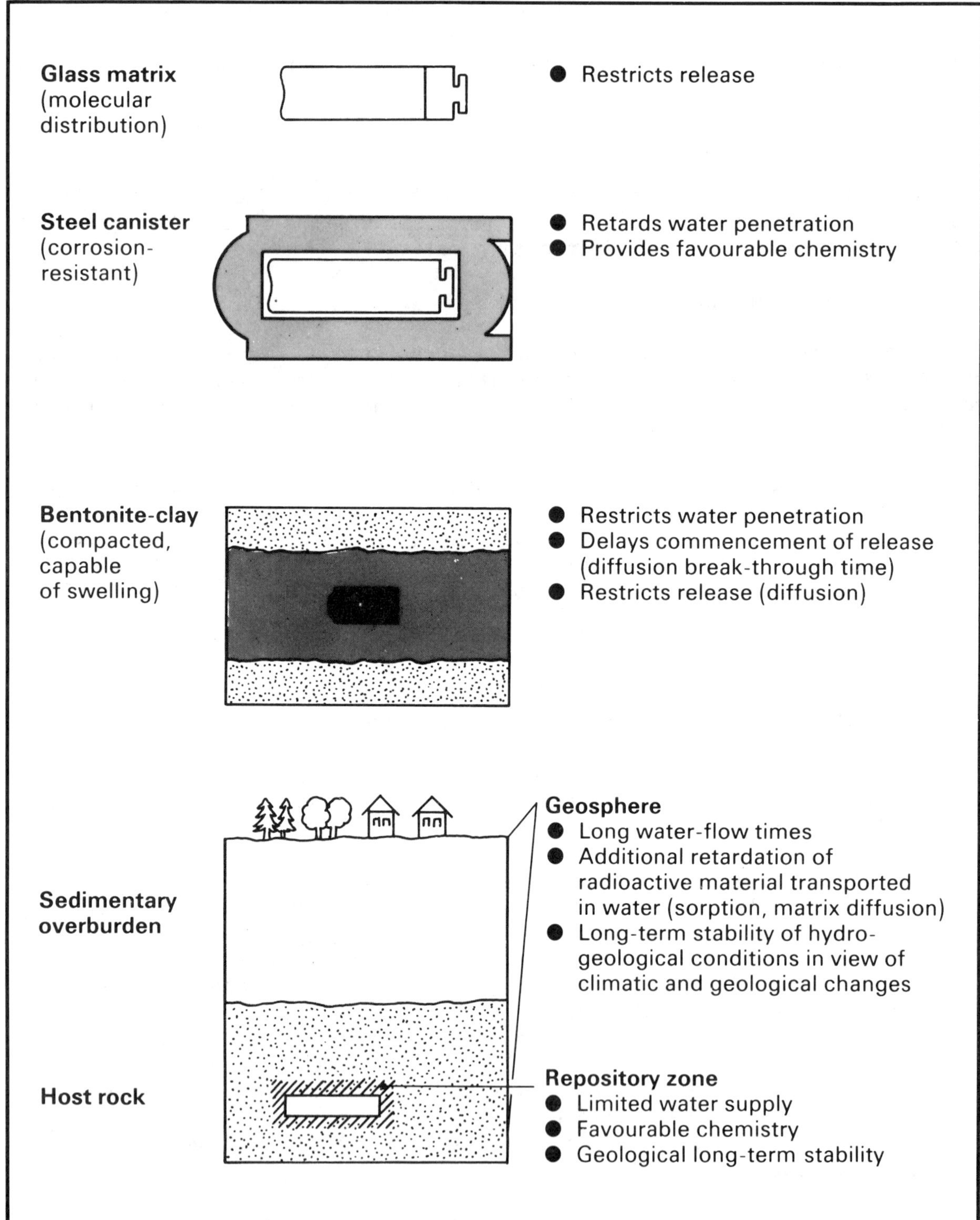

FIGURE 1. The System of Safety Barriers for High-Level Waste

zones which intersect the tunnel system are dealt with by storing no waste in their vicinity and sealing the relevant section of the tunnel with backfill material.

- Waste emplacement and backfilling of the storage tunnels is carefully supervised and controlled. Before final closure of the repository, technical safety evaluation of long-term, in-situ experiments will take place (observation of storage materials over several decades).

- The repository is conceived in such a way that, after closure, maintenance and supervision can be dispensed with and a high level of long-term safety can nevertheless be ensured. Retrieval of the waste is not planned; this would be technically costly but is not completely impossible.

- The proposed safety barrier system conforms to the basic strategy of the waste management concept. The waste is delivered in a conditioned form (in the solidification matrix) and all remaining technical barriers are provided during construction, operation and closure of the repository.

- The waste is prepared for emplacement in the surface reception area, the HLW in particular being encapsulated in corrosion-resistant repository canisters. All waste - the encapsulated HLW and the ILW as delivered - is conveyed to the repository in additional transport shielding. Work in zones with higher local radiation level is done by remote handling.

- The delivered waste containers and the encapsulated waste for final disposal undergo quality control. Facilities for remedial work, for replacement of defective HLW canisters, for decontamination of container surfaces etc. are foreseen, as well as measuring equipment for monitoring the radioactivity of the waste containers.

- Organizational measures ensure that records are kept of all delivered and stored waste. From these it is possible to determine which waste containers have been stored in which particular position and the waste specifications also provide a continuous overview of the accumulated amounts of significant radionuclides in the waste already stored.

- The construction and operational phases of the repository run simultaneously but all work involving the use of explosives will be concluded before the beginning of emplacement of radioactive waste. Appropriate arrangement of the repository installations ensures that subsequent mechanical excavation with tunnelling machines is spatially separated from the emplacement operations.

B. Description of Main Structural Parts

Figure 2 gives a perspective overview of the installations of a Type C repository using the example of arrangement of the repository area on two levels in the fissured granite block of the model data-set. Figure 3 gives

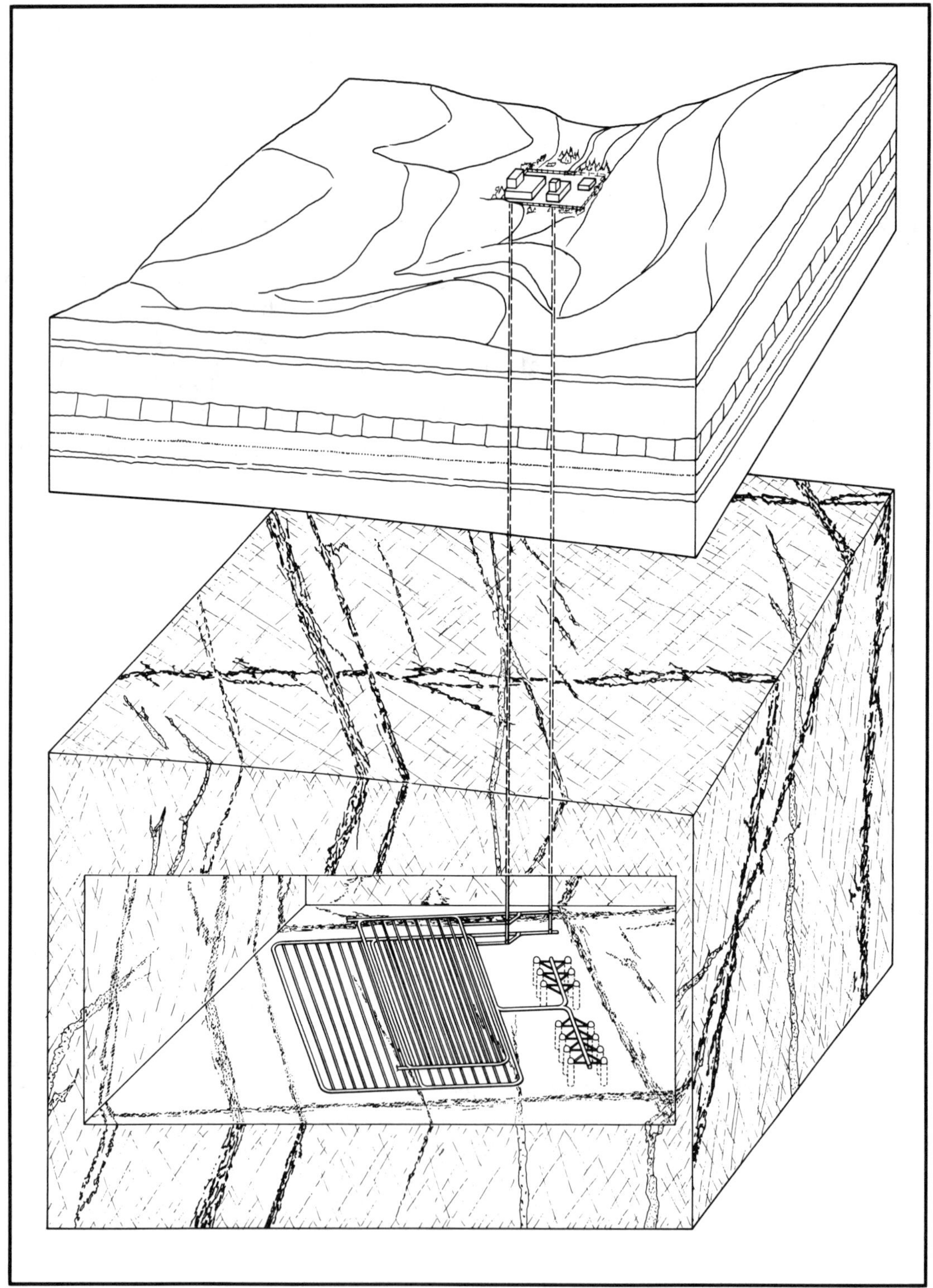

FIGURE 2. Perspective Overview of Type C Repository Installations. Example of arrangement on two levels in a fissured granite block.

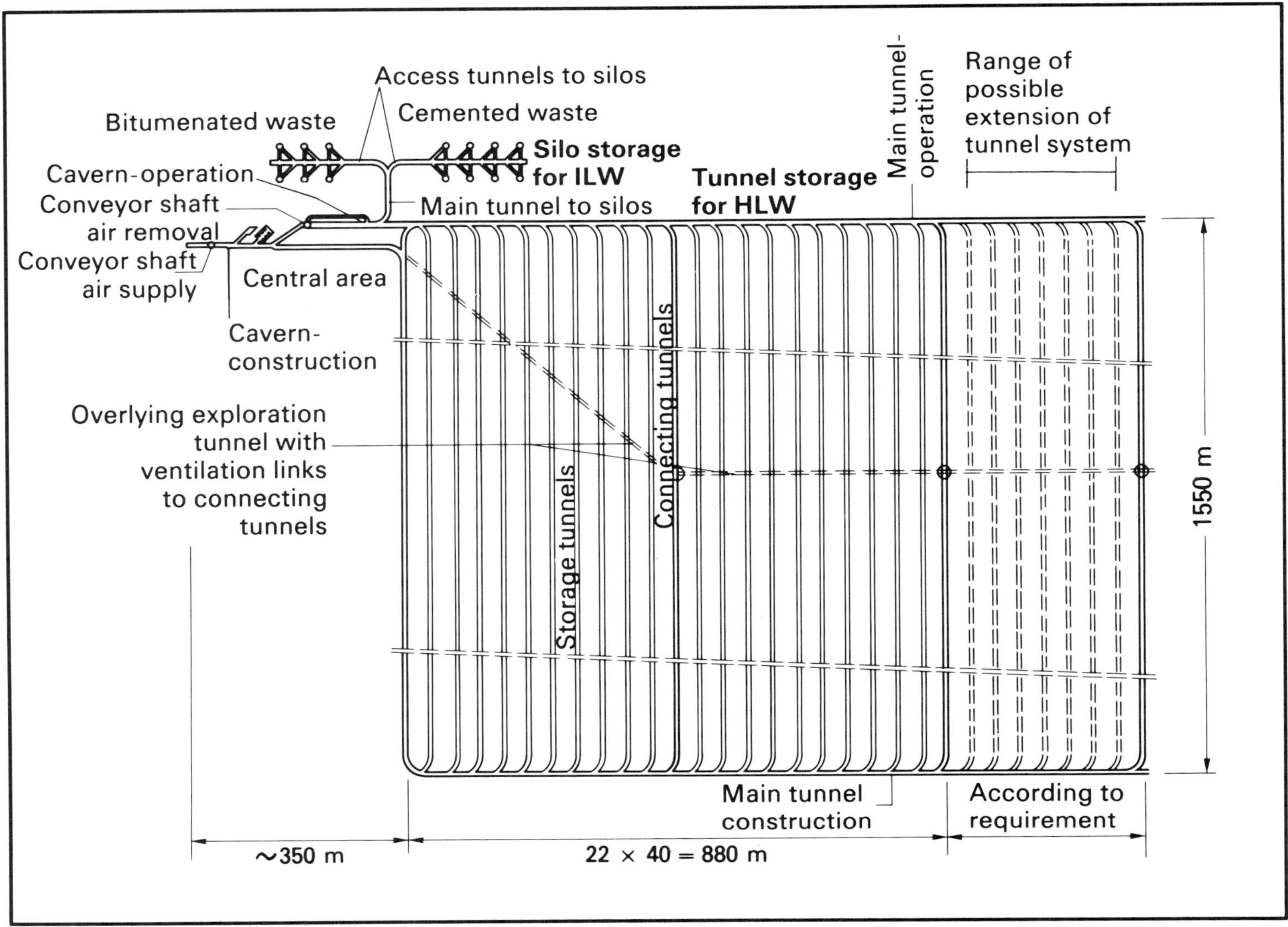

FIGURE 3. Schematic Representation of Repository Layout. Arrangement can be adapted to the specific geometric structure of the host rock.

the schematic layout of the repository - this has to be adapted to the geometric realities of the host rock in the repository zone. The repository consists of the following units:

- the surface reception facilities for receiving the delivered waste and some additional surface buildings (here, the HLW cylinders receive their corrosion-resistant overpack in the form of a thick-walled steel canister),

- two independent vertical shafts with practically identical conveying arrangements (transportation for the construction operations is housed in the air inlet shaft and the air outlet shaft is intended for the conveying of radioactive waste),

- the underground central area with two large caverns, for organizing construction activities and emplacement operations,

- the two main tunnels with a cross-sectional area, around 60 m^2 which surround the repository area for HLW (these tunnels will be completed before the commencement of waste emplacement),

- the horizontal storage tunnels with a diameter of approximately 3.7 m, to be arranged according to geologic situations (twenty-two of them with an individual length at 1550 m are shown in Figure 3; Figure 4 indicates the emplacement procedure of an encapsulated HLW-Canister and the highly compacted bentonite blocks in the circular, full-face machine-excavated storage tunnel), and

- the repository silos which are in a separated area for the disposal of cemented or bitumenized ILW (the silo caverns have a depth of more than 50 m and a diameter of around 10 m; in each silo is a concrete structure standing free from the rock, in which the waste is emplaced and immobilised with special concrete; the empty space between the rock and the silo wall is backfilled with bentonite granulate, as shown in Figure 5).

C. Construction Time, Cost

The total excavated volume of the underground constructions is about 1,100,000 m^3; the construction time is estimated at 15 years (until commencement of emplacement operations, without geological investigations); the building costs will be around 600 million Swiss francs. At a first guess the emplacement operations and the simultaneous tunnel excavation requires a team of around 200 people.

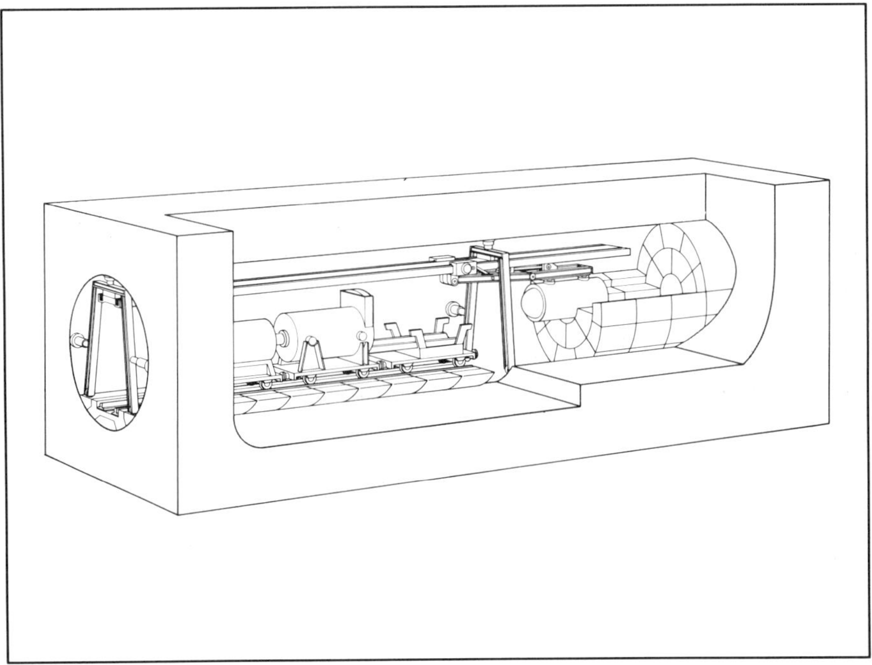

FIGURE 4. Emplacement Procedure for HLW Canisters in the Storage Tunnels of the Type C Repository with Highly Compacted Bentonite Blocks

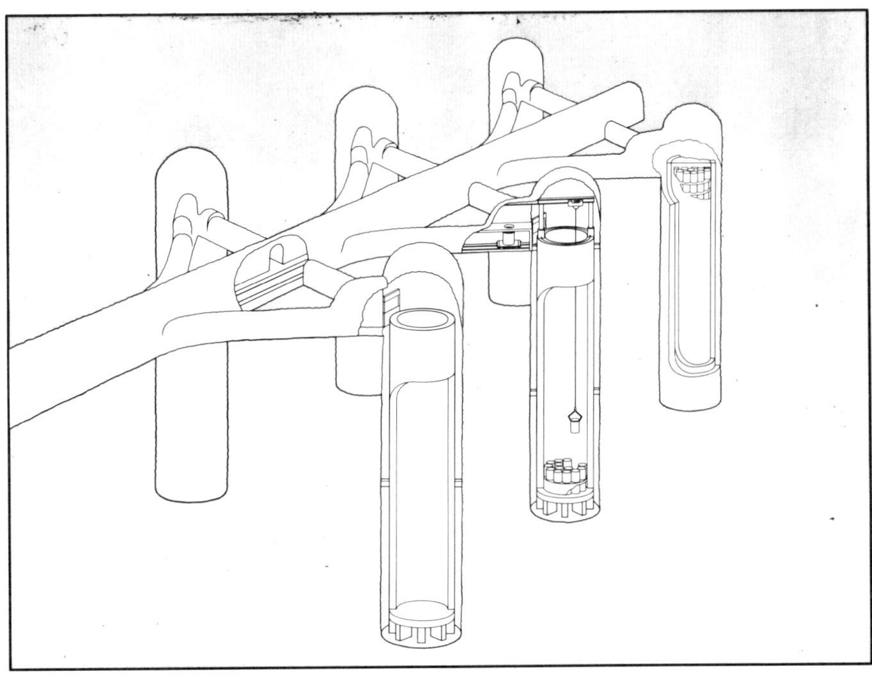

FIGURE 5. Silos for Alpha-Containing Intermediate-Level Waste in the Type C Repository, Shown in Different Stages of Emplacement and Sealing

IV. SUMMARY

The project permits the conclusion that construction and operation of a repository for high-level radioactive waste and, if necessary, spent fuel rods is feasible with present-day technology.

Although the project study was not economically optimised, it is apparent that final disposal of high-level waste will be feasible from a financial point of view.

As described in Reference 2, quantitative evaluation of realistic release scenarios give radiation dose predictions which lie far below the protection objective of 10 mrem per year, which is specified by the Swiss nuclear safety authorities.

REFERENCES

1. M. Thury, "The Swiss Site Selection Program for High-Level Nuclear Waste Disposal," in these proceedings (1985).

2. Ch. McCombie, "Predicting the Safety Performance of a High-Level Nuclear Waste Repository," in these proceedings (1985).

3. Projekt Gewaehr, Project Reports NGB 85-01 to 85-08, Nagra Parkstr. 23, CH-5401 Baden, Switzerland.

LOCATION RECOMMENDATION FOR SURFACE FACILITIES FOR THE PROSPECTIVE YUCCA MOUNTAIN NUCLEAR WASTE REPOSITORY*

James T. Neal
Nuclear Waste Engineering Projects Division
Sandia National Laboratories**
P. O. Box 5800
Albuquerque, New Mexico 87185

ABSTRACT

Six locations east of Yucca Mountain were evaluated for use as the site for repository surface facilities. Fourteen factors in three major categories were considered. These categories, in order of importance were: 1) preclosure radiological safety, 2) environmental quality, and 3) ease and cost of construction, operation, and closure. The evaluation factors and process closely followed the DOE siting guidelines (10 CFR Part 960). Site-specific factors also were considered. The results indicate that the location immediately east of the northern portion of Exile Hill is preferred. We suggest the location be termed the "reference conceptual site" for use during the conceptual design phase of repository development.

I. INTRODUCTION

To guide the siting of nuclear waste repositories, the Department of Energy (DOE) has developed guidelines (10 CFR Part 960) specifying technical factors to be used in making siting decisions. These guidelines are compatible with the technical criteria promulgated by the Nuclear Regulatory Commission and standards developed by the Environmental Protection Agency. The DOE guidelines dictate that factors relating to preclosure radiological safety weigh most heavily in the evaluation process for siting surface facilities, followed in order of importance by environmental quality, socioeconomic impact, transportation, and the ease and cost of construction, operation and closure. The latter category includes most geotechnical factors.

The candidate horizon for the underground facility is in tuff at Yucca Mountain, under an area on and adjacent to the southwestern part of the Nevada Test Site. Yucca Mountain is an area of high relief (some 250 m) and difficult

---

*This work was supported by the U.S. Department of Energy (DOE), under Contract DE-AC04-DP000789.
**A U.S. DOE facility.

access. Consequently, the relatively flat alluvial portions along the eastern side of the site are most suitable for the surface facilities, but do not lie directly above the underground facility.

The Nevada Nuclear Waste Storage Investigations Project Office adopted fundamental criteria for potential surface facility sites. Such sites must include at least 30 hectares and have foundation conditions capable of supporting reasonable seismic design of waste handling facilities.

Of engineering interest in siting surface facilities at Yucca Mountain are numerous 12-million-year old normal faults, none of which display definite Holocene movement, and varying thicknesses of alluvial cover. These attributes may significantly affect the seismic suitability of the site; ground motion from earthquakes and underground nuclear testing may vary for the prospective sites. Susceptibility to flash flooding was also found to be a discriminating factor among the candidate sites. The site must also have reasonable proximity to portal access to the underground facility, because an inclined ramp is planned for waste emplacement.

Parcels of land meeting the fundamental requirements were evaluated further by a three-member site-evaluation team applying the DOE Siting Guidelines. Although the process was more arduous than that commonly used in site selection, the nature of the facility, the close coupling required with the underground facility, and the concern for public safety warranted the effort.

The evaluation of six candidate sites (Table 1; Figure 1) for repository surface facilities showed that the location east of the northern half of Exile Hill (Site 3) is highly preferred, followed in order by sites to the north (Site 2) and south (Site 4) of Site 3. The evaluation is based on current understanding of the factors as they relate to Yucca Mountain; most are specified in the DOE Siting Guidelines (10 CFR Part 960). As new information becomes available, such as modification of the subsurface facility area, revisions in site evaluation could occur and conceivably change the preferred location. For these reasons, it is suggested that the recommended location be termed a "reference conceptual site," consistent with its intended use during repository conceptual design.[1]

REFERENCE

1. J. T. Neal, Location Recommendation for Surface Facilities for the Prospective Yucca Mountain Nuclear Waste Repository, Sandia National Laboratories, Albuquerque, New Mexico, SAND84-2015 (1985).

TABLE 1. Yucca Mountain Nuclear Waste Repository Surface Facility Site Evaluation Factors and Scores[(a)]

			Comparative Performance Measurement						FIGURE OF MERIT					
PRECLOSURE RADIOLOGICAL SAFETY (48%)		WEIGHT	S_2	S_3	S_4	S_5	S_7	S_8	S_2	S_3	S_4	S_5	S_7	S_8
o Air transport and diffusion	meteorological factors	0.132	5	5	3	3.67	3	2.33	.66	.66	.396	.48	.396	.308
o Flash flooding		.05	5	4.33	2.33	1	1.67	1	.25	.217	.117	.05	.084	.05
o Ramp inclination to centroid	operating safety factors	.078	5	5	5	5	3	2.33	.39	.39	.39	.39	.234	.182
o Ramp inclination to exploratory shaft		.10	3	3	3	5	1	1	.30	.30	.30	.50	.10	.10
o Tectonic/fault displacement		.12	1	5	3	5	3	3	.12	.60	.36	.60	.36	.36
ENVIRONMENTAL QUALITY (31%)														
o Protection of botanical/faunal species		.155	5	5	3	3	3	3	.775	.775	.465	.465	.465	.465
o Archaeologic/cultural suitability		.155	3	3	5	1	5	5	.465	.465	.775	.155	.775	.775
EASE AND COST OF CONSTRUCTION, OPERATION AND CLOSURE (21%)														
o Average slope	surface characteristics	.04	3	5	3	3.67	1	3	.12	.20	.12	.147	.04	.12
o Area availability/contiguity		.018	3.67	5	5	4.33	1.67	2.33	.066	.09	.09	.078	.030	.042
o Structural complexity/faults	rock characteristics (tectonic factors)	.023	3	3	3	3	5	5	.069	.069	.069	.069	.115	.115
o Ground motion		.066	1	5	3	5	3	3	.066	.33	.198	.33	.198	.198
o Portal access		.015	1	5	3	5	5	5	.015	.075	.045	.075	.075	.075
o Ramp length to centroid	operational factors	.015	2.33	3	3	1	5	5	.035	.045	.045	.015	.075	.075
o Proximity to northern block		.033	5	5	3	1	1	3	.165	.165	.099	.033	.033	.099
							TOTAL		3.496	4.381	3.469	3.387	2.98	2.964
							RANK		2	1	3	4	5	6
							%		69.9	87.6	69.4	67.5	59.6	59.3

(a) Site evaluation factors, weights and composite scoring. Weights of major categories are averages of four individual's judgments. The 5, 3, and 1 scoring connotes judgment indicating "favorable," "average," and "less favorable," respectively; values other than whole numbers reveal differences in opinion among the three-person scoring team.

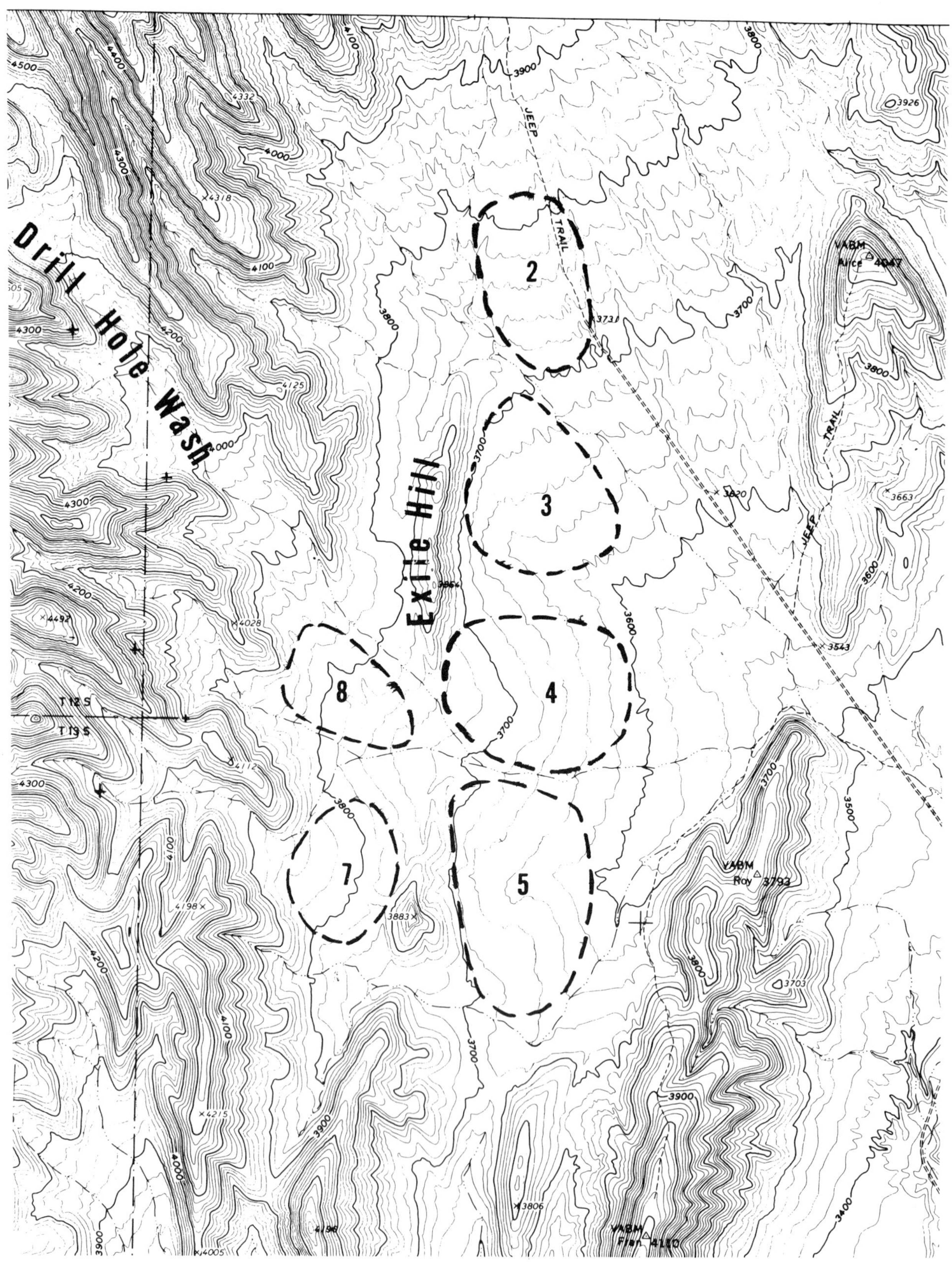

FIGURE 1. Location of Six Candidate Sites on East Flank of Yucca Mountain, Nevada

ASPECTS OF RADIOLOGICAL SAFETY AS ONE DESIGN CRITERION FOR THE OPERATION FOR A SPENT FUEL REPOSITORY IN SALT FORMATIONS

Hans J. Engelmann
Bernward Hartje
Deutsche Gesellschaft zum Bau
und Betrieb von Endlagern für
Abfallstoffe mbH (DBE)
Woltorfer Strasse 74
D-3150 Peine
Federal Republic of Germany

ABSTRACT

In the Federal Republic of Germany the R&D-program "Alternative Spent Fuel Management and Disposal Techniques" was finished. In the framework of this project a repository for spent fuel was planned. In this report we will discuss the safety-related aspects of the mining design and layout of the repository, namely the radiological exposure of the plant personnel and the inhabitants during normal operation and after accidents.

I. INTRODUCTION

Spent fuel reprocessing is the reference concept for the back-end of the fuel cycle in the Federal Republic of Germany. Due to a political decision in 1979, disposal of spent fuel is also being investigated. In the framework of the R&D-program "Alternative Spent Fuel Management and Disposal Techniques" efforts in the repository development sub-program are focused on the adaption of existing plans for the repository at the Gorleben salt dome, which is designed for emplacement of reprocessing wastes, and to the specific requirements of disposal of spent fuel.

In Phase 1 of the R&D-program, which lasted up to mid-1982, several disposal techniques were investigated and one concept was selected based on criteria such as safety-related aspects, technical feasibility and costs. It was recommended that disposal in galleries with self-shielded containers should be the reference concept.

This concept of repository containers consists of three parts: the so-called dry storage bin, the disposal canister and the lost shielding (Figure 1). In a first step, three intact spent fuel elements are enclosed in the austenitic stainless steel dry storage bin which has welded lids and a wall-thickness of 8 mm. This dry storage bin is to guarantee gas tightness of the package during the 50 year operating phase of the repository. Holding

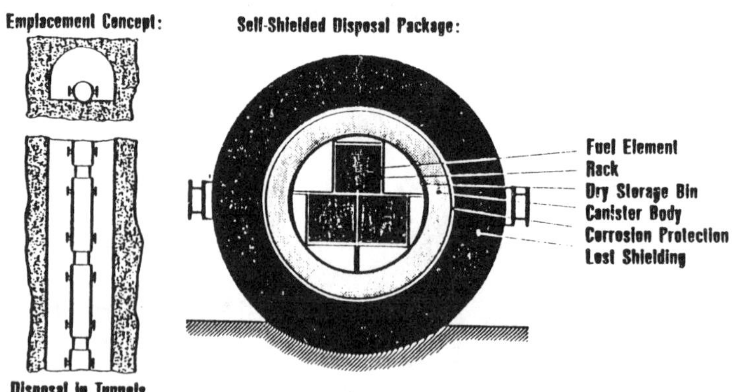

FIGURE 1. Reference Concept for Spent Fuel Disposal

about 1.5 Mg of heavy metal, such a package has a length of about 6.0 m, a diameter of 1.4 m and weights 55 MT. For an annual capacity of 700 MT of spent fuel roughly two disposal packages per day have to be manufactured in the conditioning and encapsulation plant and handled in the repository, totalling 437 packages per year.

This concept considers a mining design with the following layout: The repository features two shafts. Shaft 1 is for material and personnel transport and air inflow. Shaft 2 is designed to accommodate radwaste transportation and the outflow of the exhaust air. The diameter of the shafts is 7.5 m allowing the transport of the disposal packages in the cage in a horizontal position. The distance between the two shafts is 400 m. Two parallel access galleries are driven at a distance of 770 m from each other. Every 200 m, drifts are branching off at right angles. Each of these connection drifts belongs to an actual emplacement field comprising fifty parallel disposal tunnels, 5 m wide and 4 m high, which are separated by pillars 10 m wide. At an areal heat load of 180 kW/ha (about 70 kW/acre) the longitudinal spacing (center to center) between the spent fuel packages in the disposal tunnel is 7.5 m.

The whole repository is subdivided into three sectors: The eastern wing accommodates the spent fuel packages. Secondary waste from the conditioning and encapsulation facility, from nuclear power stations of 26 GW capacity, from industry and other sources which cause only negligible thermal loading are disposed of in the western sector. In between those two wings is the central area which accommodates the two shafts, the infrastructure, etc.

In this report we will discuss the safety-related aspects of the mining design and layout of the repository, namely the radiological exposure of the plant personnel and the inhabitants during normal operation and after accidents.

II. RADIOLOGICAL EXPOSURE IN THE OPERATING PHASE OF THE REPOSITORY

Under the Radiation Protection Ordinance valid in the Federal Republic of Germany, the exposure dose to the personnel operating nuclear facilities and to the public in the environment must be kept as low as reasonably achievable irrespective of any maximum levels defined. This applies both to the normal operation of a plant and to accident conditions.

In order to be able to consider aspects of the radiological safety as design criterion in the planning for a repository, one must examine the exposure pathways of the persons exposed to radiation and quantify the doses encountered. An adequate dosimetric quantity for this purpose is the collective dose commitment, which takes into account all exposure pathways. This quantity was calculated in accordance with the recommendations by ICRP (International Commission on Radiological Protection)[1].

A. Normal Operation

Under normal operating conditions, the operating personnel in a repository will be exposed mainly through the inhalation and direct irradiation pathways. The exposure from inhalation is due to the radioactive substances contained in the air in the mine building as gases and aerosols. These are released from the waste package handling and the waste stored in the repository. Waste packages are not absolutely impervious to the radionuclides they contain.

The magnitude of the radiation dose caused by inhalation is a function of the residence time of the respective persons and of the concentration of radionuclides in the air. In order to minimize the residence time of persons in contaminated air, this aspect was taken into account in planning

- the excavation,
- the air flow patterns, and
- the transport and emplacement processes.

All this resulted in the fact that excavating occurs only under fresh air conditions and that the return air from the emplacement locations is passed to the shaft for exhaust air through a separate floor where no personnel will stay under normal conditions.

In order to minimize the nuclide concentration and, hence, the absolute quantity of the radionuclides dispersed in the environment with the exhaust air, the following criteria were taken into account in designing the mine building:

- Emplacement is carried out only in the retreating mode, i.e., the most remote regions of the mine are filled up first.

- Small rooms and galleries are used for emplacement.

- Early backfilling and closing of the emplacement spaces is planned.
- Filling and gastight closure of entire disposal fields by sealing structures is required.

As a consequence of these measures soon after emplacement, the radionuclides released from the package disposed of in the repository will no longer constitute a radiological burden to the operating personnel and the environment because they will be retained by these additional barriers.

At present, only very few experimental findings and computer models exist to describe the release behavior of radionuclides from waste packages. The radiation exposure of the operating personnel and the radionuclide emission from the repository was determined on the basis of guiding values for planning purposes.[2] With ventilation data and cumulated radionuclide inventory data taken into account, maximum release rates for the waste forms were estimated on this basis. These release rates represent criteria to be met by the waste packages. In conjunction with the meteorological data and the population distribution around the site of the Gorleben salt dome, the emissions from the repository estimated in this way were used to determine the collective dose commitment to the public in the environment at short (50 km) and long distance (radius of 200 km).

The radiation exposure resulting from direct irradiation is a function of the radiation intensity of the waste package, the residence time in the radiation field, and the distance from a radiation source.

Within the framework of our studies we determined the location of the operating personnel as a function of time and space. The complete sequence of emplacement steps was studied in this way; at a few working places, where persons are required to do work in high radiation fields for long periods of time, shielding was planned and taken into account in determining collective doses.

The Table 1 below shows the collective dose commitment for the operating personnel of the repository during normal operation. The letter E stands for the emplacement personnel, S for the radiation protection personnel, T for the technical personnel of the surface facilities, and B for the mining personnel.

As a result, it can be stated that the limits under the German Radiation Protection Ordinance are being observed for all persons working in the repository as well as for the public in the environment.

TABLE 1. Occupational Collective Dose Commitment of the Operation Personnel in a Repository

Depart-ment	Number of workers	Collective dose contributions: Direct irradiation, manSv/year	Collective dose contributions: Inhalation, manSv/year	total collective dose, manSv*/year
E	43	2.4 E-1	3.8 E-5	2.4 E-1**
S	22	8.0 E-3	7.2 E-6	8.0 E-3
T	119	3.7 E-2	6.6 E-6	3.7 E-2
B	152	-	9.7 E-5	9.7 E-5
Total	336	2.8 E-1	1.5 E-4	2.8 E-1

B. Accidents

Based up on an analysis of the planned operation sequences - above ground and underground - potential accident events were listed. The analysis was carried out for casks with spent fuel and also for other types of waste. The planning of the operation steps was adapted to the results of analysis, thereby minimizing the impacts of accidents and/or their frequency of occurrence.

For those accidents whose occurrence cannot be excluded, accident analyses were carried out. They were based on the following boundary conditions:

- The radionuclides released during an accident correspond to the specific radionuclide spectrum of the waste package studied.
- Only those nuclides were considered whose activity was higher than 1 E+6 Becquerel (Bq) per waste package.
- The calculations were performed on the basis of the Accident Guidelines adopted by the German Federal Ministry of the Interior.
- The dose factors were taken from the International Commission for Radiological Protection (ICRP) recommendations.
- The exposure pathways of the operating personnel included β/γ-submersion, inhalation and direct irradiation, those of the public in the vicinity γ-radiation from the ground, β/γ-submersion, inhalation and ingestion.

---

*1 Sv = 100 Rem.
**2.4 E-1 = 2.4×10^{-1}

- All particles released with less than 100 µm equivalent diameter can become airborne; particles of less than 10 µm equivalent diameter are respirable.

The following retention factors in the repository were taken into account:

- 20% retention in the shaft,

- sedimentation of small particles in accordance with the laws governing fluid dynamics, and

- Retention by planned filter systems in accordance with their assumed efficiency.

To determine the source terms acting during these accidents, the mechanical and thermal energies were estimated. The source strengths were determined from evaluations of experiments carried out and from the literature. On this basis, and including planned ventilation and technical operating data, the collective dose commitment resulting from accident conditions was calculated.

The frequencies of occurrence of the individual accidents investigated were determined by evaluating accident records of conventional mining operations and of nuclear handling processes. These frequencies indicate how often such an accident must be assumed to occur per year. The sum total of the multiplication of the collective dose commitment by the frequency of occurrence of each individual accident yields the expected annual accident dose to the operating personnel. It was determined to be 6.6 E-4 man Sv/a. In determining the accidental collective dose commitment, again the specific data applying to the Gorleben repository site were used as a basis. The impacts were calculated for an area with a radius of 50 km around the site.

The results of the accident analysis show that the impacts of accidents, taking into account their frequencies of occurrence, are not very great. Compared with the risk of normal operation the accident risk, i.e., the product of consequence and frequency of occurrence is much lower. This is also demonstrated by Figure 2. The ordinate is a plot of the frequency of occurrence of the events, the abscissa shows the collective dose commitment of the operating personnel. The dashed line in this diagram represents the total accident risk. The square on the right hand side represents the risk of normal operation, the crosses on the left hand side of the curve the risk of the worst accidents studied. It is evident that there is a distance of at least two orders of magnitude between the accident risk and the risk of normal operation.

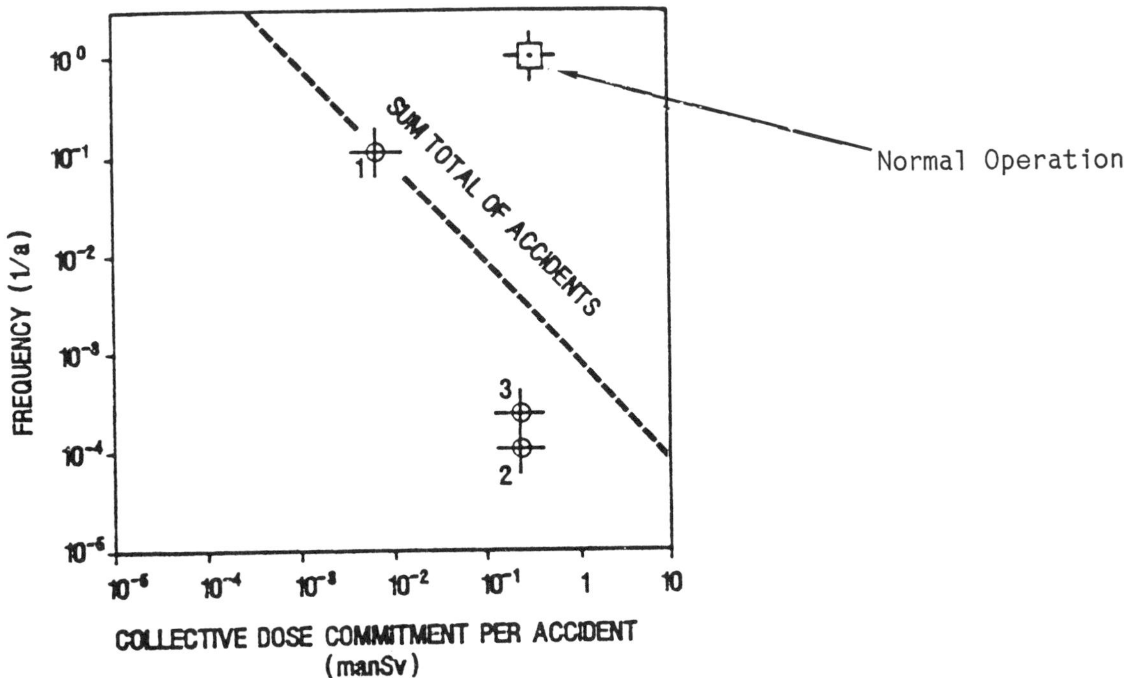

1 - Casks Crashing in Galleries

2 - Collisions of Means of Conveyance Accompanied by Fires at Points (Rail Transport)

3 - Collisions of Means of Conveyance Accompanied by Fires in the Emplacement Gallery (No Rail Transport) Normal Operation

FIGURE 2. Accident Risk as Compared to the Risk of Normal Operation Arising to the Operating Personnel

REFERENCES

1. ICRP-Publication 26, Recommendations of the ICRP Annals of the ICRP, Vol. 1, No. 3 (1977).

2. H. Illi and D. Ehrlich, Strahlenschutz und Radiookologie der Endlagerung Atomenergie-Kerntechnik, Vol. 44, No. 2 (1984).

REPOSITORY DESIGN CALCULATIONS FOR THE ALTERNATIVE STORAGE OF SPENT FUEL

Hans J. Engelmann
Deutsche Gesellschaft zum Bau und Betrieb von Endlagern für Abfallstoffe mbH (DBE)
Woltorfer Strasse 74
D-3150 Peine
Federal Republic of Germany

Dr. Martin Wallner
Bundesanstalt für Gewissenschaften und Rohstoffe (BGR)
Stilleweg 2
D-3000 Hannover 51
Federal Republic of Germany

Dr. Hakon-K. Nipp
Bundesanstalt für Geowissenschaften und Rohstoffe (BGR)
Stilleweg 2
D-3000 Hannover 51
Federal Republic of Germany

ABSTRACT

Parallel to the planning of a final repository in the Gorleben salt dome for radioactive waste from a reprocessing plant the R&D program known as "Alternative Spent Fuel Management and Disposal Techniques" was implemented. In this report we will discuss the safety-related aspects of direct disposal (without reprocessing) of spent fuel supported by rock mechanics calculations. The objective of those computations was to define possible critical effects concerning the thermomechanical impact on the deformation and stresses in the near field and far field as well as to confirm the mining design from a rock mechanics point of view.

I. INTRODUCTION

Parallel to the planning of a final repository in the Gorleben salt dome for radioactive waste from a reprocessing plant, an R&D project on "Alternative Spent Fuel Management and Disposal Techniques" was implemented. Different disposal concepts were developed from which the disposal of spent fuel elements packed in casks and stored in long drifts was considered as a reference concept and examined in a more detailed feasibility study. The specially designed cask (6.0 m length, 1.4 m diameter, 55 MT weight) is able to contain three spent fuel elements. The casks are placed in series in parallel drifts that, after disposal, will be back-filled and sealed. A general layout of the mine is given in Figure 1.

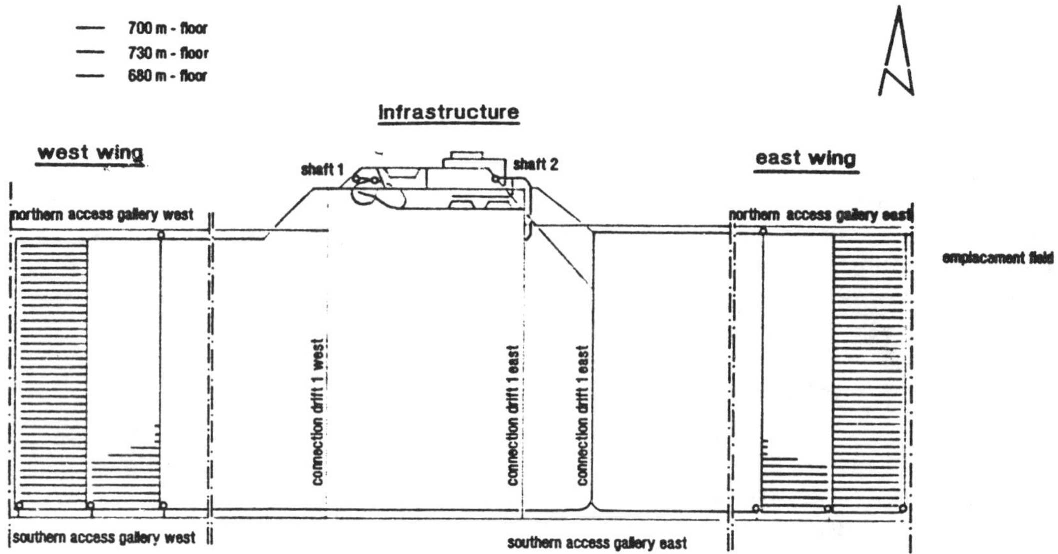

FIGURE 1. Layout of Repository Mine for Alternative Storage of Spent Fuel

Assessment of feasibility and safety-related aspects of direct disposal of spent fuel was supported by rock mechanics calculations. The objective of those computations was to define possible critical effects concerning the thermomechanical impact on the deformation and stresses in the near field and far field as well as to confirm the preliminary mine design from a rock mechanics point of view.

The following calculations were carried out and are discussed in the paper:

- history of temperature distribution in the vicinity of the emplacement drift,
- temperature history in the far-field, especially at the border of the salt dome,
- convergency of open and backfilled drifts,
- stability of pillars between drifts,
- thermally induced stresses and deformations in the far-field, and
- surface uplift.

II. NEAR-FIELD EFFECTS

For the layout of the repository the following parameters, based on mining experience, were set up:

- access tunnels 7.0 m wide and 4.0 m high,
- emplacement tunnels 5.0 m wide and 4.0 high,
- width of pillars 10 m, and
- emplacement level 730 m below ground.

The following boundary conditions in regard to thermomechanical properties were assumed:

- Temperatures must not exceed 200°C at all cost.
- The thermal load caused by 24 canisters was estimated at 57.6 kW.
- For the whole operating time of the repository the thermal load will be 51,250 kW.

The convergency of the access drifts without any backfilling was calculated for the date 5 years after beginning of emplacement. The bottom creep amounted to 13 cm and the roof creep to 48 cm. The stability of pillars between the drifts was also calculated. The results demonstrated that no critical state of stress was developed.

Each specially designed container with self-shielding (6.0 length, 1.4 m diameter and 55 MT weight) is able to contain three spent fuel elements. The container will be placed in series in parallel drifts of the east-wing which after disposal will be backfilled and sealed. Near-field maximum temperatures at the surface of the emplaced spent fuel canisters will increase by 167°C after 30 years. The results are shown in Figure 2.

III. LONG-TERM STABILITY

The long-term stability of the repository depends - among other things - on the local and spatial temperature increase in the salt dome. Therefore, the temperature distribution in the geological formation was analyzed as well as the consequences of the heat load on the stability. The average temperature of the total emplacement area will be 120°C after 65 years. At the boundary of the repository the temperature does not exceed 75°C. At the salt dome top about 250 m below surface the temperature will increase by 15°C after 1,600 years. The long-term thermomechanical effects on the far-field were examined for a time period of 10,000 years, although the calculations were done for a very rough model simplifying the geological structure.

The convergency of backfilled emplacement drifts after 50 years is demonstrated in Figure 3. As a result of the planned heat load of 70 kW/acres, an uplift of the surface above salt dome by 2.7 m after 4,000 years was calculated (Figure 4). The thermomechanical consequences of emplacement of spent fuel elements in the above mentioned layout does not cause any critical stress conditions in the geological barrier salt dome.

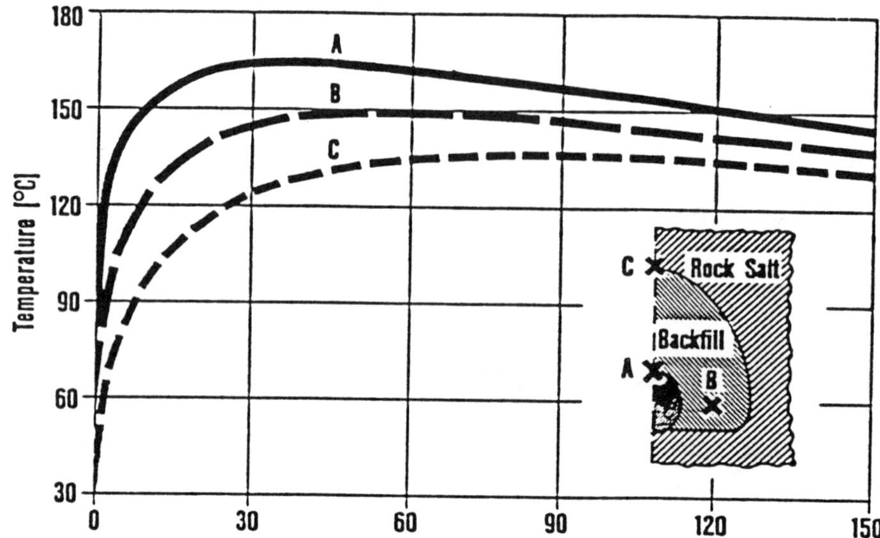

FIGURE 2. Development of Temperature at Characterized Points in a Backfilled Drift

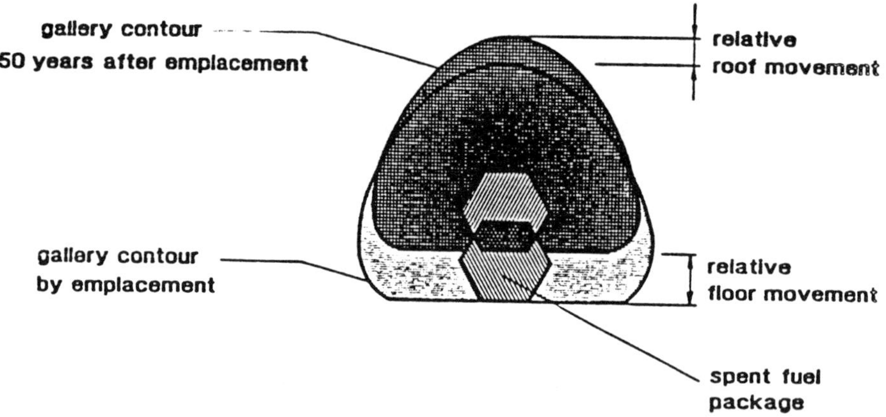

FIGURE 3. Convergency of a Backfilled Drift After 50 Years

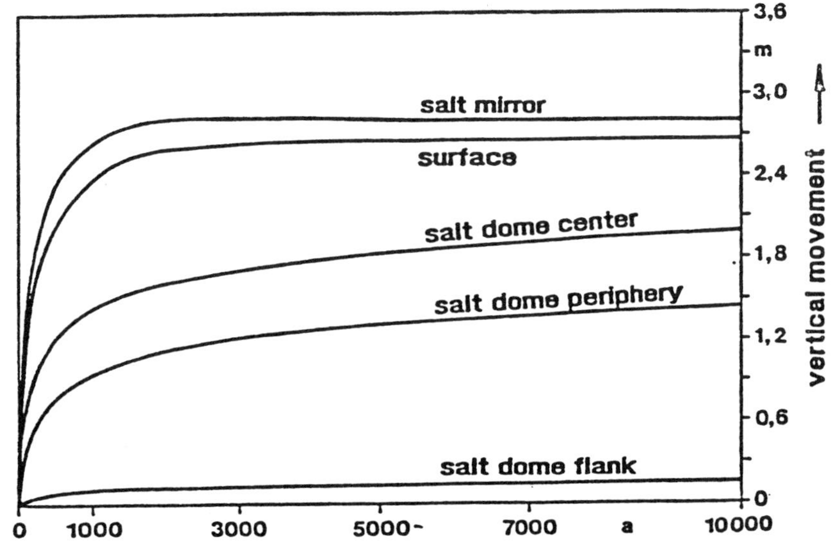

FIGURE 4. Uplift

REPOSITORY WASTE-HANDLING OPERATIONS 1998

L. Connell
A. E. Cottam
Rockwell Hanford Operations
P.O. Box 800
Richland, Washington 99352

ABSTRACT

The Office of Civilian Radioactive Waste Management Mission Plan and the Generic Requirements for a Mined Geologic Disposal System state that beginning in 1998, commercial spent fuel not exceeding 70,000 metric tons of heavy metal (MTHM), or a quantity of solidified high-level radioactive waste resulting from the reprocessing of such a quantity of spent fuel, will be shipped to a deep geologic repository for permanent storage.

The development of a waste handling system that can process 3,000 metric tons of heavy metal annually will require the adoption of a fully automated approach. The safety and minimum exposure of personnel will be the prime goals of the repository waste handling system. A man-out-of-the-loop approach will be used in all operations including the receipt of spent fuel in shipping casks, the inspection and unloading of the spent fuel into automated hot cell facilities, the disassembly of spent fuel assemblies, the consolidation of fuel rods, and the packaging of fuel rods into heavy-walled site-specific containers. These containers are designed to contain the radionuclides for up to 1,000 years.

The ability of a repository to handle more than 6,000 pressurized water reactor spent fuel rods per day on a production basis for approximately a 23-year period will require that a systems approach be adopted that combines space-age technology, robotics, and sophisticated automated computerized equipment. New advanced inspection techniques, maintenance by robots, and safety will be key factors in the design, construction, and licensing of a repository waste handling facility for 1998.

I. INTRODUCTION

Ever since the Manhattan Project was initiated and Enrico Fermi and his collegues successfully demonstrated the world's first atomic reactor at the University of Chicago in late 1942, the United States has been steadily generating quantities of radioactive wastes.

Today, with almost 120 commercial nuclear reactors in operation or under construction in the United States, approximately 150,000 metric tons of heavy metal (MTHM) exist throughout the country in the form of spent nuclear fuel, commercial high-level waste from the processing of spent nuclear fuel, and defense high-level waste from numerous defense programs.

In January 1983, President Reagan signed the Nuclear Waste Policy Act of 1982[1] establishing Federal responsibility and Federal policy for the disposal of high-level radioactive waste and spent nuclear fuel, which would ensure the health, safety, and environment for this and future generations. In addition, the Nuclear Waste Policy Act specified January 1998 as the date when the U.S. Department of Energy would accept high-level waste for permanent disposal at a deep geologic repository.

The Office of Civilian Radioactive Waste Management Mission Plan[2] and the Generic Requirements for a Mined Geologic Disposal System[3] state that the first repository shall be limited to a quantity of spent nuclear fuel or a solidified high-level waste resulting from the processing of spent nuclear fuel not to exceed 70,000 MTHM. In addition, these documents stated that all, or a major portion, of the 70,000 MTHM will be in the form of spent fuel assemblies that will be disassembled and the fuel pins consolidated into waste containers. The waste containers will be designed to last between 300 and 1,000 years after emplacement and be retrievable for up to 50 years after emplacement or until such time as the repository is permanently closed.

Three candidate sites in different geologic media have now been tentatively recommended for Further site characterization. The Basalt Waste Isolation Project is investigating the basalt flows at the Hanford Site in Washington State, the Nevada Nuclear Waste Storage Investigation Project is investigating the tuff rock at Yucca Mountain in Nevada, and the Salt Repository Project is investigating the bedded salt at the Deaf Smith Site in Permian Basin, Texas. Upon approval of the President, exploratory shafts will be drilled and numerous geologic, hydrologic, and geomechanical tests will be conducted at each of the candidate sites before making the selection of the first repository site.

A. Repository Design

In the past, significant emphasis has been placed on the design of the underground facility by the Basalt Waste Isolation Project and other program participants. The Basalt Waste Isolation Project currently is adopting a two-stage, four-quadrant layout (Figure 1) that consists of the two exploratory shafts and the exploratory shaft facilities used to develop stage one of the repository and nine additional shafts that service the total two-stage repository. Each quadrant of the underground facility has the capability to store 9,460 waste containers in the 10 separate waste storage panels (Figure 2). Engineering studies have been completed to determine an optimum underground layout and emplacement configuration and to determine if the repository can perform and isolate the waste from the environment for 10,000 years.

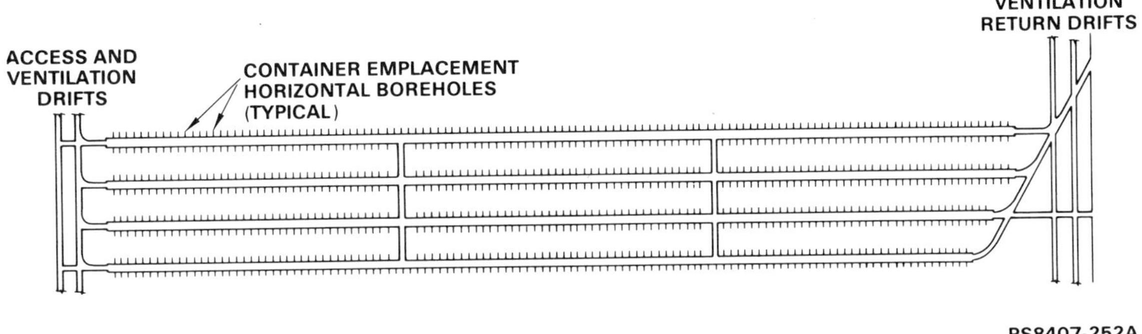

FIGURE 1. The Two-Stage, Four-Quadrant Repository

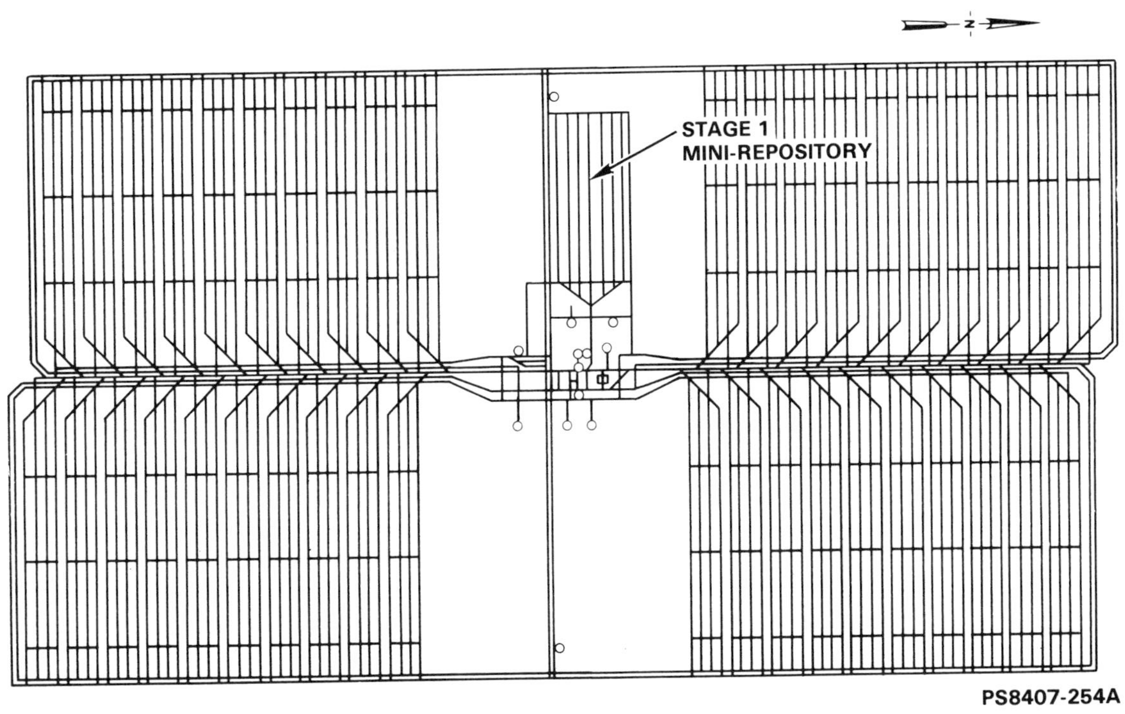

FIGURE 2. Typical Panel Unit

A number of surface facilities have been identified to support the underground activities, and a conceptual arrangement (Figure 3) has been prepared to facilitate the continuous construction and operation of the repository.

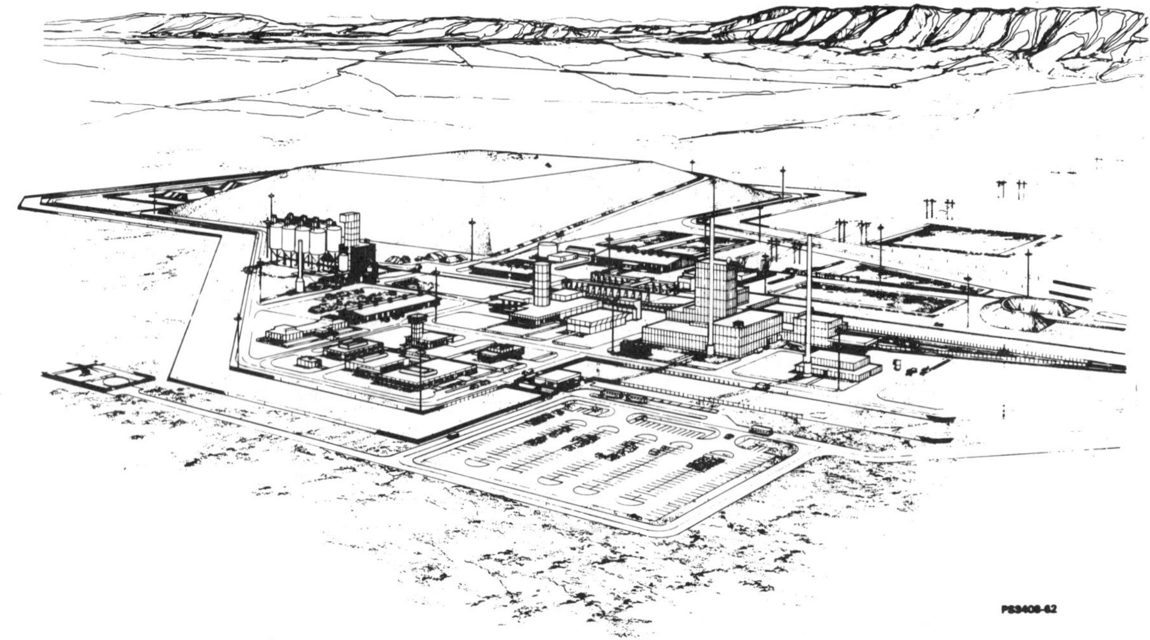

FIGURE 3. Artist's Conceptual View of Surface Facilities

B. Repository Operations

Upon receipt of an operating license from the U.S. Nuclear Regulatory Commission, the repository will operate under conditions similar to those imposed on any other nuclear facility. Safeguards and security, nuclear inventory control, waste handling, administrative, and numerous other functions will be required to maintain the overall operation of the repository. This paper will discuss only the repository waste handling operations that are required to safely dispose of 70,000 MTHM over approximately 23 years of repository operations. This requirement means that the repository waste handling facilities and equipment must handle 3,000 MTHM annually, a throughput of 10 MTHM each day. Repository waste handling operations have been separated in two categories: surface operations and subsurface operations. Each surface and subsurface operation is unique regarding the requirement of each individual operation and the environment in which it is performed. Even though the logistics and throughput are essential to maintain repository productivity, the Basalt Waste Isolation Project considers that safety is the mandatory, controlling factor in repository waste handling operations.

C. Surface Waste Handling Operations

The repository surface waste handling operations play a major role in the overall layout and design of surface facilities. The logistics required to handle large quantities of nuclear waste must be the first operations considered. From some 50 to 90 truck shipments or 8 to 14 rail shipments (2 shipping casks per shipments) will be received weekly at the repository site. These shipments from various nuclear power plants throughout the country will contain up to 156 spent fuel assemblies from pressurized water reactors or up to 396 spent fuel assemblies from boiling water reactors to provide the required daily throughput.

The sequence of surface waste handling activities that are required to support a spent fuel consolidation and packaging program is shown in Figure 4. The sequence required for packaging containers of solidified high-level waste are also included in Figure 4. The major tasks range from initial receipt and inspection of the spent fuel, through the disassembly of the spent fuel assemblies, to the consolidating of the fuel pins into site-specific waste containers. The activities for consolidation and packaging spent fuel encompass 20 major tasks and involve more than 250 complex operations, many of which must be repeated several times. The tasks and operations envisioned for each task based on current receipt and inspection processes now being performed at several facilities are identified in Table 1. The containers are welded and sealed prior to being loaded into a shielded transfer cask for transportation below ground.

Another factor to be considered is that there are several fuel assembly manufacturers and the fuel assembly designs have been modified since-early fabrication. Major manufacturers have fuel assemblies ranging from 7 by 7 arrays (49 fuel pins) to 17 by 17 arrays (289 fuel pins). This means that equipment for unloading and disassembling fuel assemblies and for handling spent fuel pins must be flexible enough to handle the variety of spent fuel.

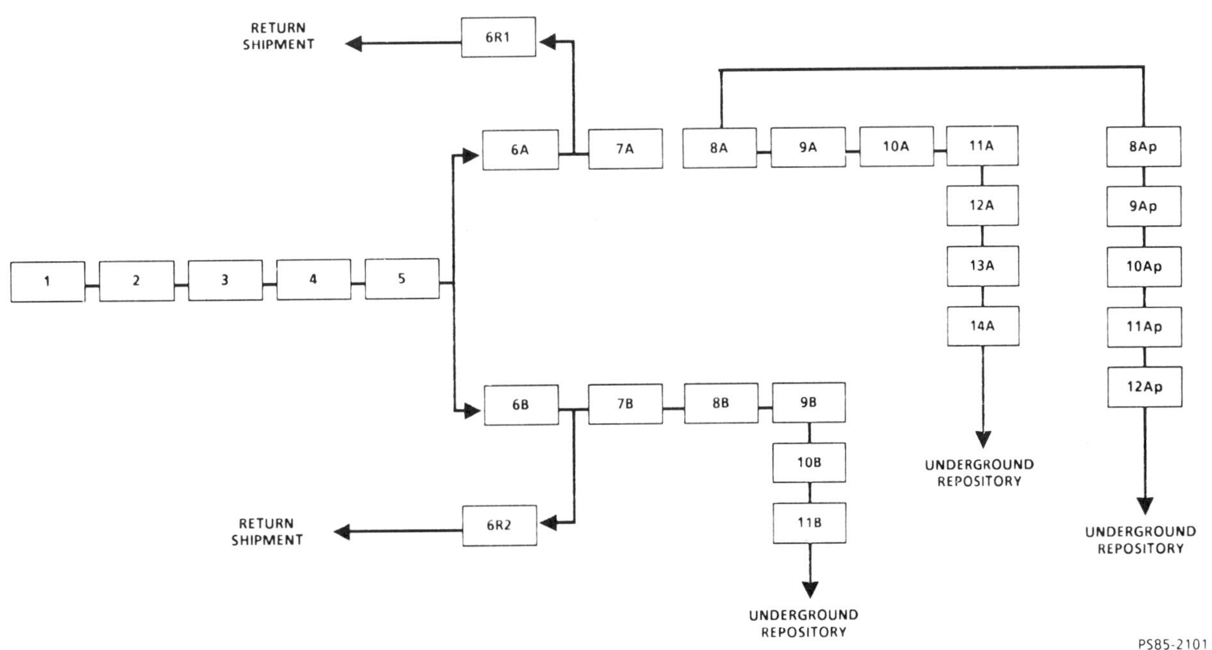

FIGURE 4. Surface Waste Handling Logic

The total logistics involved in completing the operations for processing 10 MTHM per day mandates that a fully automated facility be considered. Handling large quantities of spent fuel pins that have been stored for a long period of time and that are extremely brittle (in some cases, already breached) will create an environment that could be highly contaminated. Equipment maintenance under such conditions could be extremely hazardous; therefore, the prime objective of safety, using the man-out-of-the-loop approach, is an important factor in the facility design. No existing facility has been designed with the capabilities required for the waste handling tasks envisioned for a repository. However, the required facility can be designed and developed using state-of-the-art technology, robotics, and automation.

TABLE 1. Task Description

Number	Task
1	Receive and inspect shipping carrier, truck, or rail
2	Wash down shipping cask and carrier
3	Offload shipping cask from carrier
4	Interconnect shipping cask to hot cell receiving port
5	Unseal shipping cask, purge/assay, inspect shipment
6A	Remove spent fuel assemblies from shipping cask
7A	Inspect/assay and record spent fuel assemblies
8A	Disassemble spent fuel pins from assemblies
9A	Inspect/assay and record spent fuel pins
10A	Consolidate spent fuel pins into bundles,assay, and inspect
11A	Package consolidated fuel pins into waste container
12A	Weld final closure of waste container and inspect
13A	Prepare waste container for loading from hot cell
14A	Load waste container into shielded transfer cask for transportation underground
6B	Remove canister of solidified high-level waste from shipped cask
7B	Inspect/assay and record high-level waste canister
8B	Overpack high-level waste canister into waste container
9B	Weld final closure of waste container and inspect
10B	Prepare waste container for loading from hot cell
11B	Load waste container into shielded transfer cask for transportation underground
8Ap	Process spent fuel assembly hardware
9Ap	Package hardware into waste container
10Ap	Weld final closure of waste container and inspect
11Ap	Prepare waste container for loading from hot cell
12Ap	Load waste container into shielded cask for transportation underground
6R1 and 6R2	Reseal shipping cask, temporarily remove from hot cell port, and transport to decontamination center for return shipment

II. SUBSURFACE REPOSITORY WASTE HANDLING OPERATIONS

The subsurface waste handling operations are different in many ways from the complex number of surface operations that take place in a totally shielded facility. Even though the underground operations appear to be much simpler than those for surface waste handling operations, they must still be approached with the same requirements in mind. The man-out-of-the-loop approach and safety are the prime drivers. If the industry can convince and demonstrate to and convince the public that there is minimum exposure in the operations of a facility, then great strides can be made in gaining public acceptance of the program. There are basically three major tasks involved with underground waste handling activities: transportation, emplacement, and if required, retrieval of waste containers. Transportation consists of transferring the waste container, in a shielded transfer cask, from the surface waste handling building to the underground facilities and then conveying it on an underground transporter to the final emplacement location. This entails 18 specific operations. Waste emplacement is the task of removing the waste container from the shielded transfer cask, placing it into the final location, and then sealing it into place. Twenty-eight operations are involved with this activity. Once the waste container has been sealed into the final emplacement configuration, it will remain there indefinitely unless retrieval is required. Waste container retrieval may take place up to 50 years after emplacement and could be required for such reasons as container inspection, spent fuel reprocessing, verification of an indicated waste container failure, or examination if the repository is not performing as predicted. The number of operations involved with various retrieval scenarios range from a reversal of the emplacement operations (25) to the retrieval of a breached container where many more operations (approximately 44) could be required.

It is believed that most of the operations performed underground could be accomplished automatically with remote equipment or robotics, and, with the introduction of new technology, would greatly enhance underground waste handling operations.

III. AUTOMATION AND ROBOTICS

Since the early 1960s, enormous advances have been made in automated production facility design. This has been necessary to meet the ever-increasing demands in productivity and competition. Automation makes difficult operations simpler, faster, and more economic. It does away with human error and the dozens of safety-related problems associated with manned operations. The advancements in computer application, versatility, and compactness have played a major role in advancing the state of the art in most technical fields. The space program has advanced the state of the art in such areas as material development, microcomputers, and communications. However, in the nuclear industry, only limited progress has been made in automating facilities and the daily operations that are performed remotely. In most nuclear facilities, the operations are still performed manually in glove boxes or with hand-operated master/slave manipulations that were first used in the 1940s and 1950s. Some emphasis has been placed more recently on the development of electrically controlled, master/slave manipulators and servomanipulators, but the nuclear industry in general falls far behind other industries in advanced technology for automation.

Facilities such as the Hot Fuel Examination Facility North in Idaho, the Fuel Material Examination Facility at the Hanford Site, and the Fuel Material Facility in Japan now do a number of operations on a semiautomated basis; however, these facilities were not designed for the high production throughputs that will be required for a repository waste handling facility. Additionally, equipment must be designed with built-in redundancy or self-maintaining capabilities to ensure that productivity can continue uninterrupted even if equipment failure occurs. The Fuel Material Facility, which was based on the design of the Oak Ridge National Laboratory Consolidated Fuel Processing Program, has a capacity of 0.5 to 1 MTHM per day. The process consists of the disassembly of the spent fuel assemblies and the processing of the spent fuel material. However, several problems have been experienced in maintaining equipment in a highly radiated and contaminated environment, and numerous shutdowns have occurred while replacing equipment.

Not only must equipment be developed to automatically perform operations on a production basis in a high radiation environment, but robotics similar to Odex I by Odetics (Figure 5) should be used to perform maintenance operations and to replace equipment in hazardous environments without shutting down the facility.[4] The Oak Ridge National Laboratory Remote Operations and Maintenance Demonstration Facility is continually demonstrating automated robotic types of equipment including the M2 Maintenance System and other remote maintenance equipment.[5]

Artificial intelligence, sensor integration, computer vision, and real-time control vision must all play a major role in the development of equipment that is capable of performing the large variety of complex tasks required to satisfy the repository waste handling operations scheduled to start in 1998.

IV. SAFETY

It was stated earlier that safety is the controlling factor in repository waste handling operations. This should be the case for any production facility and most certainly for any nuclear facility such as a deep geologic repository. Taking the man out of the loop does not guarantee that all safety related problems are overcome; however, it does minimize exposure or contamination to the operators and, consequently, to the public.

Automation of complex operations minimizes the possibility of errors that otherwise could be made if hands-on conditions existed. Automated, remote maintenance by robotic equipment not only minimizes exposure to man but allows production to continue during repair of parts or equipment.

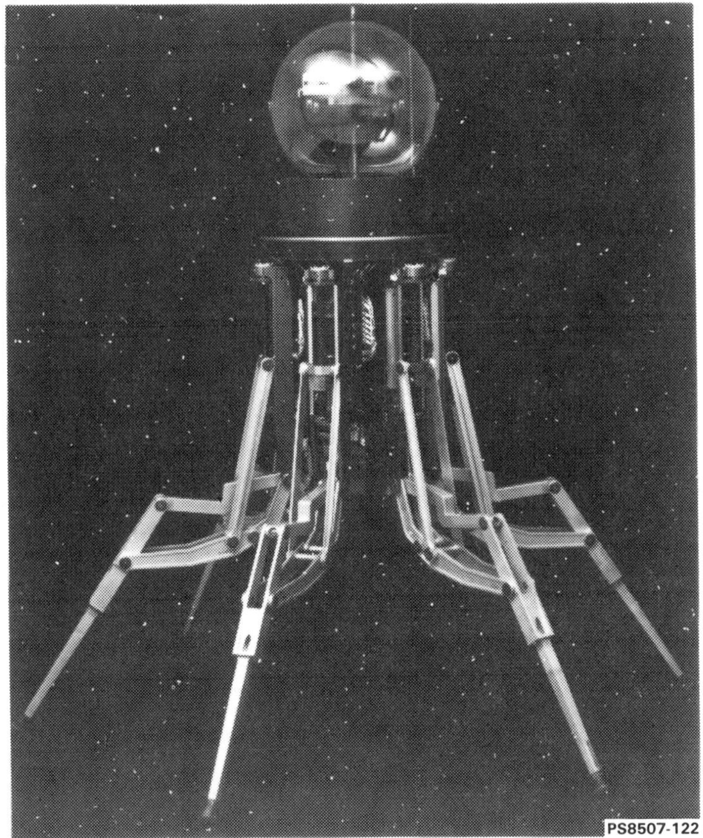

THE FIRST FUNCTIONOID ODEX I

FIGURE 5. Walking Robot, ODEX I by Odetics

V. SUMMARY

The emerging design requirements for a deep geologic repository waste handling operations facility dictate the need for the development of a highly complex, sophisticated system. As a near-term goal, both the U.S. Department of Energy and the nuclear industry must review the development of technology for handling large quantities of nuclear waste. If the technology that is available in other industries is applied effectively, considerable progress can be made in meeting the goals established by the Nuclear Waste Policy Act and the Office of Civilian Radioactive Waste Management. The Basalt Waste Isolation Project recommends that additional design guidelines be established regarding waste handling requirements. These include consolidation requirements, assay and inspection requirements of spent fuel rods, nuclear materials inventory control requirements, material traceability requirements, and quality requirements. These requirements must be initiated immediately to demonstrate that repository operations could commence in January 1998 when an operating license is required.

REFERENCES

1. Nuclear Waste Policy Act of 1982, Public Law 97-425 (1983).

2. Department of Energy (DOE), Mission Plan for the Civilian Radioactive Waste Management Program, DOE/RW-0005 Draft, Washington, D.C. (1985).

3. Department of Energy (DOE), Generic Requirements for a Mined Geologic Disposal System, Washington, D.C. (1984).

4. H. L. Martin and W. R. Hamel, "Joining Teleoperation with Robotics For Remote Manipulation in Hostile Environments," Proceedings of the 1984 National Topical Meeting on Robotics and Remote Handling in Hostile Environments, Gatlinburg, Tennessee (1984).

5. W. R. Hamel and M. J. Feldman, "The Advancement of Remote Systems Technology: Past Perspectives and Future Plans," Proceedings of the 1984 National Topical Meeting on Robotics and Remote Handling in Hostile Environments, Gatlinburg, Tennessee (1984).

TWO-PHASE REPOSITORY CONCEPT

John F. Marron
Rockwell Hanford Operations
P.O. Box 800
Richland, Washington 99352

ABSTRACT

The two-phase repository development is a viable concept to enable limited, early waste emplacement at the Hanford Site and full-scale waste emplacement five years later. This is a primary conclusion of a feasibility study undertaken in 1984 to verify the validity of the two-phase repository development concept. This paper presents a summary of the concepts and conclusions from the study.

I. INTRODUCTION

By the end of this century, the United States plans to begin the operation of the first geologic repository for the permanent disposal of commercial spent nuclear fuel and high-level radioactive waste. The Nuclear Waste Policy act of 1982,[1] specifies the process for selecting a repository site and assigns to the U.S. Department of Energy (DOE) the responsibility for locating, constructing, operating, closing, and decommissioning the repository. The program developed by the DOE to fulfill the requirements of the Nuclear Waste Policy Act is described in the "Mission Plan for the Civilian Radioactive Waste Management Program."[2]

The two-phase repository development concept was devised to provide an interim Phase 1 repository facility capable of receiving high-level radioactive waste at an early date, ensuring compliance with the stated objective of the Mission Plan. Full-scale Phase 2 repository operations would be achieved five years later.

The purpose of the engineering feasibility study[3] undertaken in 1984 was to verify the validity of the proposed two-phase concept in meeting the Mission Plan schedule at the Hanford Site. The study provides 1) a basis for a decision by the DOE on whether to proceed with the two-phase approach, and 2) preconceptual design information to support the Basalt Waste Isolation Project (BWIP) Site Characterization Plan (in preparation) and subsequent repository designs.

A study team comprised of representatives from the DOE, the operating contractor (Rockwell Hanford Operations), the architect/engineer (Raymond Kaiser Engineers, Inc./Parsons Brinckerhoff Quade & Douglas, Inc.), and the construction manager (Morrison-Knudsen Company, Inc.) was assembled at the

Exploratory Shaft site and (with assistance from various home offices) prepared the study. A peer review of the study was conducted by Science Applications Inc.

II. REFERENCE REPOSITORY LOCATION

On the basis of draft environmental assessments[4] of potentially acceptable sites, the DOE has tentatively determined that the reference repository location at the Hanford Site in the State of Washington is one of three sites preferred for site characterization.

The reference repository location is in the west-central part of the DOE-controlled Hanford Site in south-central Washington (Figure 1). The reference repository location lies within the Pasco Basin, a topographic depression in the Columbia River Plateau and, more specifically, in the central part of the Cold Creek syncline. This location was chosen partly because the basalt flows there are nearly flat-lying and should be structurally less disturbed than other areas at the Hanford Site. The terrain at the reference repository location is relatively flat--its features were formed by glacially related floods and more recently developed sand dunes.

The Columbia River Plateau is underlain by a thick sequence of strata (see Figure 1) deposited many millions of years ago in Miocene times. These strata consist entirely of basalt-lava flows in the lower part and of increasing amounts of interbedded sedimentary deposits in the upper part. Semiconsolidated sediments overlie the basalt sequence and attain thicknesses of as much as 370 m (1,200 ft). Approximately 50 basalt flows, with a total thickness of perhaps 5,000 m (16,000 ft), have been identified within the Pasco Basin. Four of these basalt Flows have been identified as candidate host horizons For the repository. Each horizon is continuous through the vicinity of the site. The candidate flows and the range of thickness of each flow in the reference repository location are as follows:

- Rocky Coulee Flow--27 to 47 m (89 to 153 ft),
- Cohassett Flow--73 to 82 m (240 to 270 ft),
- McCoy Canyon flow--34 to 45 m (112 to 148 ft), and
- Umtanum Flow--61 to 71 m (198 to 230 ft).

The Cohassett flow is the reference repository horizon and is used for design purposes and for Exploratory Shaft breakout. This flow is approximately 910 m (3,000 ft) below the ground surface.

The strategy for in situ testing calls for the development of test facilities in phased increments that support progressive assessment of the suitability of a reference location for a repository. One such facility planned for the Hanford Site will be an exploratory shaft facility of a size that provides personnel access to the preferred candidate repository horizon for the purpose of in situ characterization testing and construction authorization application review testing (if required).

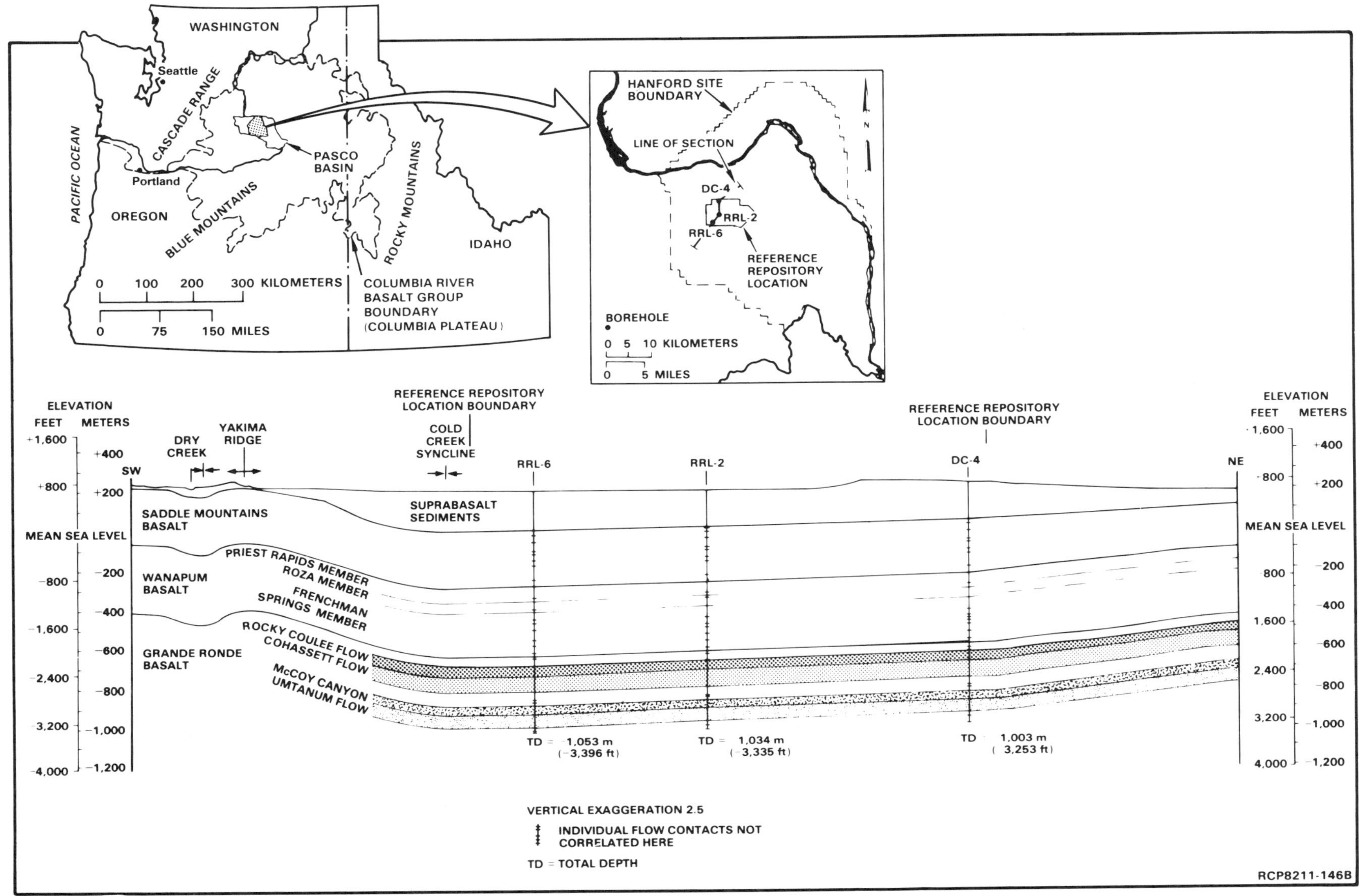

FIGURE 1. Site Location and Stratigraphic Section

III. WASTE PACKAGE CONCEPT

The waste package concept is shown in Figure 2 in its emplacement position in a short horizontal borehole. The waste package consists of the waste, waste container, and packing.

A. Waste

A waste package contains either four pressurized water reactor spent fuel assemblies or nine boiling water reactor spent fuel assemblies. The Phase 1 waste package contains bare (i.e., whole) fuel assemblies. In Phase 2 the spent fuel is disassembled and the package contains consolidated fuel rods. The purpose of consolidation in Phase 2 is to reduce the diameter, length, and weight of the waste package, which will minimize cost and facilitate handling.

For this feasibility study the repository was sized to accept only spent fuel with an age out of reactor of ten years and average burnup conditions. If the spent fuel is younger or older or has different burnup conditions, the number of assemblies per container would be adjusted to not exceed the design-basis container heat load of 2,200 W. The 2,200 W heat load represents the "hottest" waste package anticipated and provides a conservative basis for sizing the repository.

Also, for this feasibility study the repository was not sized to dispose of either commercial, West Valley, or defense high-level wastes. However, with some modifications to handling equipment, the waste handling buildings have the capability to receive and containerize these wastes, and the underground layout can be expanded to store them.

B. Waste Container

The primary function of the container is to ensure waste containment for a minimum period (to be established between 300 to 1,000 yr) as required by 10 CFR 60.[5] A secondary function of the container is to facilitate handling for emplacement and retrieval.

The waste container is a steel cylinder with a welded cap and handling pintle. Reference dimensions for the Phase 1 and Phase 2 waste containers follow:

Container	Phase 1	Phase 2
Wall thickness, cm (in.)	9.9 (3.9)	8.3 (3.3)
Outside diameter, cm (in.)	85.2 (33.5)	50.3 (19.8)
Maximum length, cm (in.)	480 (189)	444 (175)
Quantity, total containers	1,110	37,760

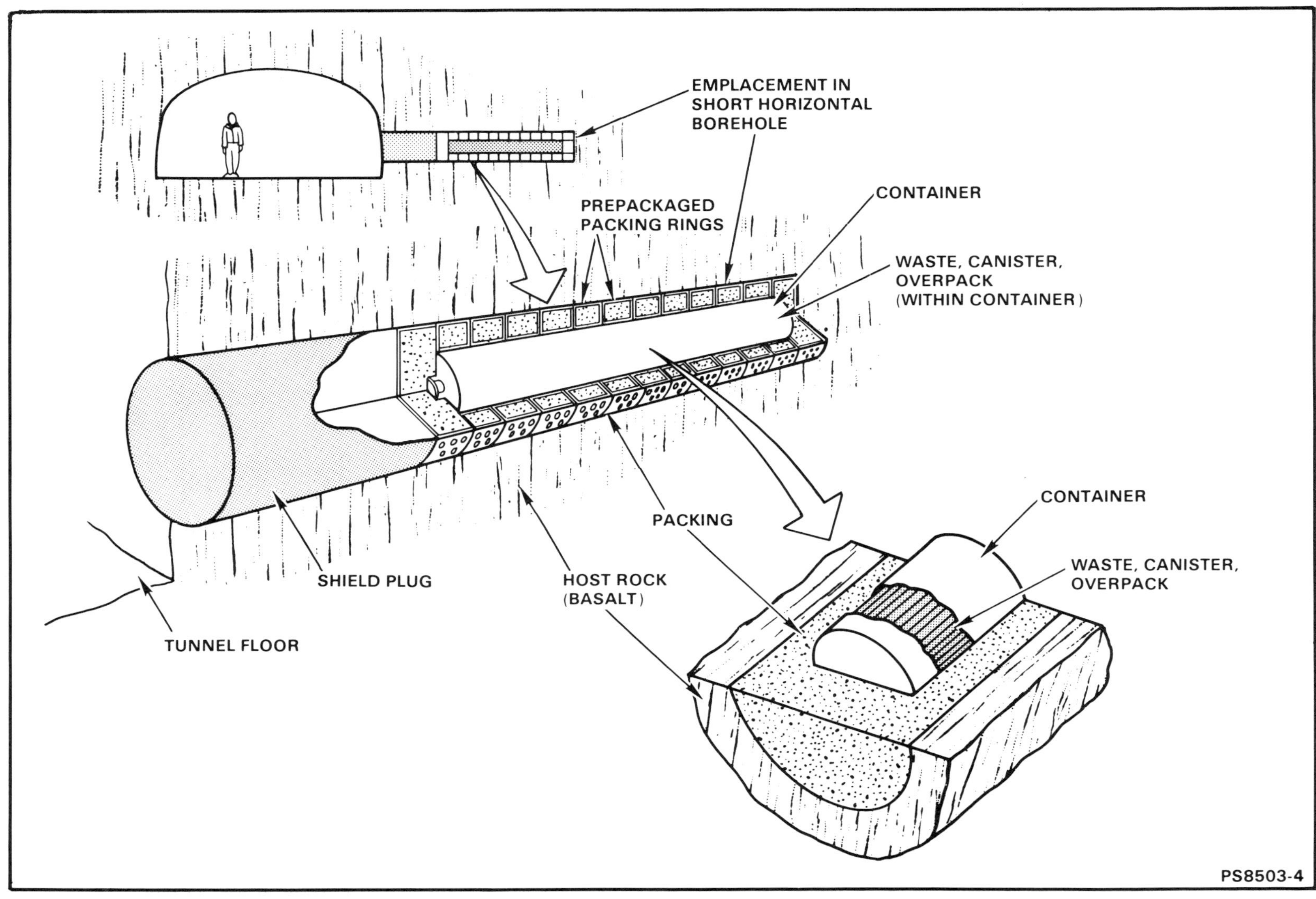

FIGURE 2. Waste Emplacement Package Concept

C. Packing

The function of the packing material is to restrict the migration of radionuclides to the primary natural barrier (the basalt) at the conclusion of the containment period. The reference packing material is by weight 75% crushed basalt and 25% bentonite clay. The packing material will have a minimum thickness of 15.2 cm (6 in.).

IV. SURFACE FACILITIES CONCEPT

The surface facilities are grouped around the central shaft pillar within the reference repository location. A block diagram of the surface facilities is shown in Figure 3. Surface facilities are categorized as Phase 1 process facilities, Phase 2 process facilities, and common support facilities. "Process facilities" are those associated with waste handling and with subsurface activities. The process facilities plus common support occupy an area of 190 ha (470 acres), the approximate center of which is 240 m (800 ft) south of the first exploratory shaft. These surface facilities are surrounded by a double security fence.

A. Phase I Waste Handling Building

Most repository construction schedules show the waste handling building as the major critical path item. Reducing the waste acceptance schedule (to 400 metric tons of heavy metal per year) shrinks the size of the waste handling building and shortens the construction schedule. However, additional shortening of the schedule is necessary to ensure feasibility for the two-phase repository development concept. The following were alternatives for the Phase 1 waste handling building evaluated for the study.

New Conventional Reinforced Concrete Building. A conventional reinforced concrete waste handling building typically requires five to eight years to build and license.

New Earth-Berm Building. This alternative is an innovative concept that uses an earth berm for shielding and a prefabricated steel interior frame for construction acceleration. The earth-berm concept appears promising from a construction cost and schedule viewpoint, but has little licensing precedent.

Existing Waste Handling Facilities. The existing General Services Building at the Washington Public Power Supply System terminated nuclear plant WNP-4 is the preferred waste handling building for the Phase 1 repository and may be usable for Phase 2. Reasons supporting this conclusion are as follows:

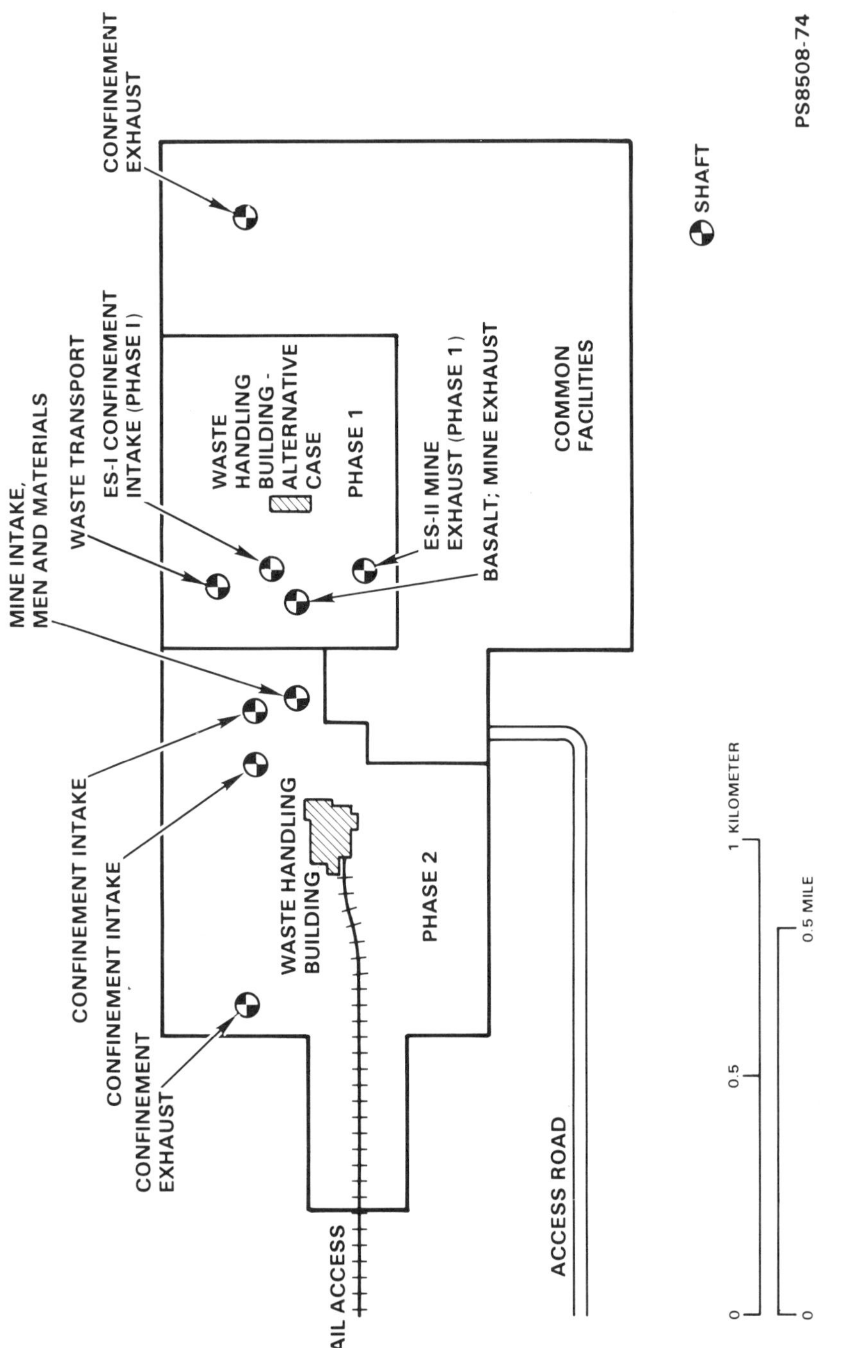

FIGURE 3. Surface Facilities Block Diagram

- meets schedule requirements for Phase 1 repository operation,
- presents minimum risk to program since (a) design concept is well understood, (b) construction and operation are conventional, and (c) civil work is approximately 70% complete,
- cost is less than for construction of a new, conventional waste-handling building, and
- conventional design and construction simplifies licensing.

B. Phase 2 Waste Handling Building

A new multistory, reinforced concrete, 6,300 m^2 (68,124 ft^2) conventional waste handling building maintains confinement ventilation during and after the design-basis earthquake and the design-basis tornado. It contains the process equipment for handling high-level waste, as well as handling and consolidating spent fuel. In addition, it contains administrative and service areas, confinement exhaust facilities, and the treatment area for onsite-generated radioactive waste.

V. UNDERGROUND FACILITIES CONCEPT

The function of the underground facilities is to provide access to a deep basalt flow where high-level radioactive waste will be permanently isolated from the accessible environment. Emplacement in short horizontal boreholes is the reference waste disposal concept for both Phase 1 and Phase 2.

Short horizontal boreholes are drilled into the basalt to a depth allowing emplacement of a single waste package. In general, for mechanical stability of the host rock, the minimum pitch (center-to-center spacing of boreholes) is three borehole diameters. However, when decay heat induces thermal stresses around the borehole and placement room, the pitch is governed by thermal stresses. Stress analyses indicated that a pitch of 7 m (23 ft.) is required for the conditions of the study.

The performed packing assemblies are placed in the borehole prior to emplacement of the waste container. An underground transporter with a shielded transfer cask is then positioned in front of the borehole. The transfer cask is mated to the borehole cover and a waste container is pushed into the annular hole in the packing assemblies. A shield plug is inserted and the transporter is released.

A. Phase I Underground Facilities

The design of the underground facilities is similar to the concept for a single-phase repository[6] using the bow tie design with a central pillar and panels located in the quadrants of the layout. The two-phase repository pre-conceptual layout for the underground facilities is shown in Figure 4. The

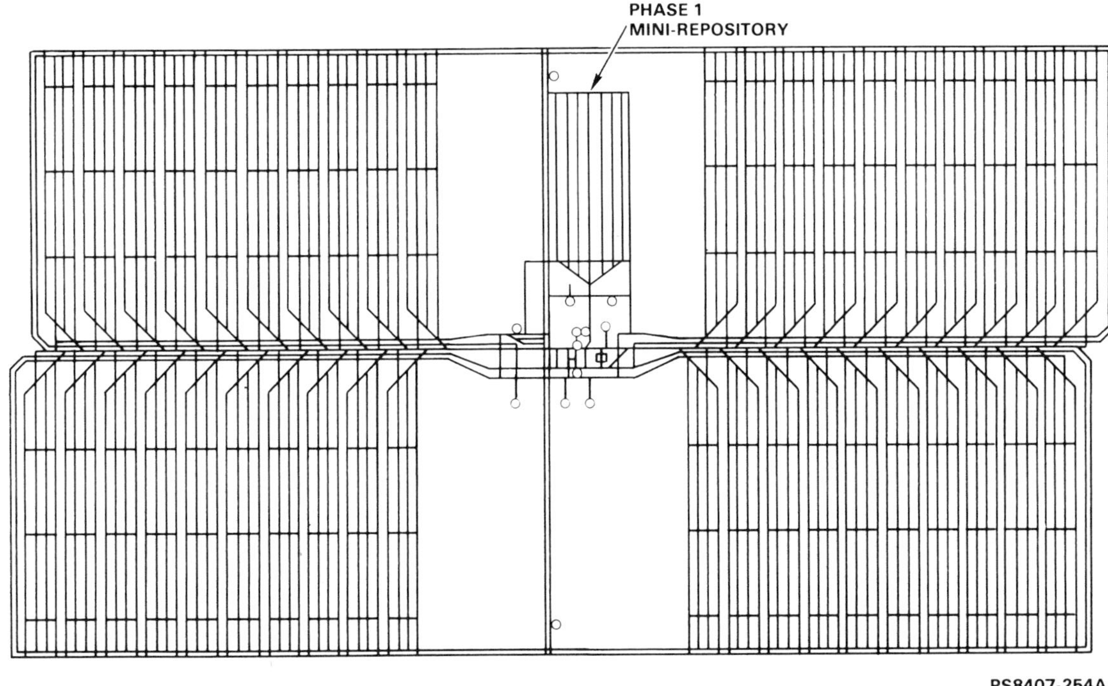

FIGURE 4. The Two-Phase, Four-Quadrant Repository

two-phase concept allows for accelerated exploration and initial development of a section of the underground repository, which is referred to as the Phase 1, or minirepository. The Phase I design incorporates several main objectives:

- Position the two exploratory shafts to provide useful exploration and test data, use these shafts in the initial development, and incorporate in Phase 1.

- Minimize the number of shafts by combining their functions where possible and using the same shafts and shaft pillar area in both phases. Besides the exploratory shafts, two additional shafts will be required for Phase 1.

- Minimize the initial development of the Phase 1 repository by placing the panel area as close as possible to the pillar area.

B. Phase 2 Underground Facilities

The Phase 2 underground layout is included in Figure 4 and involves the development of 5 additional shafts, the remaining pillar area, and a panel area consisting of 40 panels each with 4 placement rooms. The panel areas are located north and south of the shaft pillar area and are divided into 4 quadrants (or compartments) with 10 panels per quadrant. Each panel contains four storage rooms as shown in Figure 5.

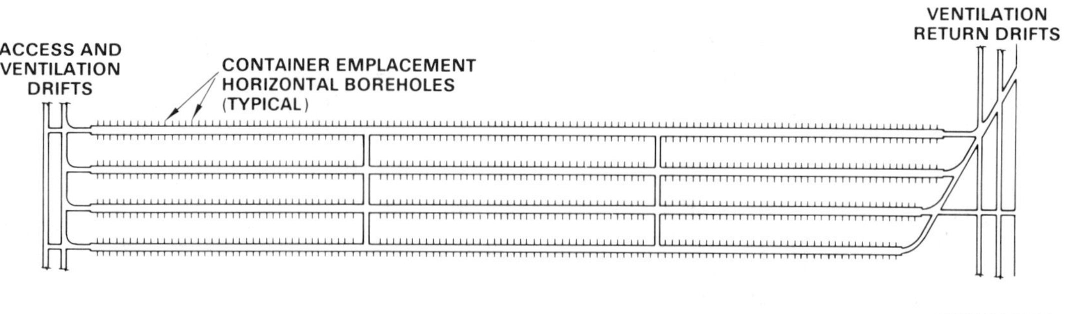

FIGURE 5. Typical Panel Unit

C. Operation Phase

During the operation period, it will be necessary to develop an average of about 1.8 panels or 7 placement rooms to accommodate the 1,666 placement holes needed per year. The annual development of the panels takes into account the simultaneous waste emplacement operation and the development of placement rooms and placement holes.

D. Caretaker Phase

During the caretaker phase, the option to retrieve waste will be maintained. Ventilation stoppings will isolate rooms in which waste placement operations have been completed. A small quantity of air drawn through the occupied placement room will be tested for airborne radioactivity or hazardous material within the room. Instrumentation will also be installed to test the rock stability of the occupied room.

E. Backfill Phase

After the end of the caretaker phase, placement rooms and drifts will be backfilled with a compacted mixture of bentonite and crushed rock. Finally, shafts will be sealed and the repository will be closed.

REFERENCES

1. Nuclear Waste Policy Act of 1982, Public Law 97-425, 42 USC 10101.

2. Department of Energy (DOE), Mission Plan for the Civilian Radioactive Waste Management Program, DOE/RW-0005, Washington D.C. (1985).

3. J. F. Marron et al., Two Phase Repository Feasibility Study, SD-BWI-ES-020, Rockwell Hanford Operations, Richland, Washington (1984).

4. Department of Energy (DOE), Draft Environmental Assessment, DOE/RW-0017, Washington D.C. (1984).

5. Nuclear Regulatory Commission (NRC), "Disposal of High-Level Radioactive Wastes in Geologic Repositories," Title 10, Code of Federal Regulations, Part 60, Washington, D.C. (1983).

6. Michael T. Black et al., "Task V, Engineering Study No. 9 Underground Repository Layout," SD-BWI-ES-023, Rockwell Hanford Operations, Richland, Washington (1985).

HIGH-LEVEL RADIOACTIVE WASTE TEST DISPOSAL ASSE SALT MINE FEDERAL REPUBLIC OF GERMANY

K. Kuhn
T. Rothfuchs
Gesellschaft für Strahlen- und Umweltforschung mbH Munchen
Institut für Tieflagerung
Schachtanlage Asse
3346 Remlingen, Federal Republic of Germany

ABSTRACT

The installation and operation of pilot facilities is considered to be a suitable method for the development and testing of complex technical systems and to define their overall characteristics. From this viewpoint, the High-Level Radioactive Waste (HLW) Test Disposal at the Asse Salt Mine in the Federal Republic of Germany is to be performed by the IfT at the request of the Federal Ministry of Research and Technology of the FRG. Thirty vitrified high-level radioactive waste canisters will be emplaced in six underground boreholes in two test galleries at the 800 m-level below the surface. The duration of test-testing will be approximately five years and all canisters are to be retrieved at the termination of the test due to licensing regulations.

I. INTRODUCTION

This paper describes scopes, objectives, lay out, and the design basis of the planned test disposal of high level radioactive vitrified canisters in the Asse Salt Mine in the Federal Republic of Germany. The underground test field and the equipment to handle the radioactive sources is to be seen as a pilot facility with regard to future final disposal in a national repository in the Federal Republic of Germany and the complete technical system of a high-level waste repository will be tested and proven as far as possible on a one to one scale.

The situation in the field of in situ investigations in salt formations is characterized by a series of in situ heating tests that have been performed without ionizing radiation. The very early test disposal (1965 - 1967) of ETR-fuel elements in a salt formation (U.S. Project Salt Vault, Lyons Mine, Kansas) served for the investigation of thermo-mechanical and radiological conditions. However, the results of Project Salt Vault are of limited validity with regard to the present conditions that are to be taken into account. A significant step with regard to the planned test disposal of high-level radioactive sources is represented by the joint German-American "Brine Migration Test"[1] that is currently conducted at the Asse Salt Mine. However, this test is only of limited validity because the gamma-dose absorbed at the borehole wall is in the range of two orders of magnitude below the dose expected from actual high-level waste.

On the basis of the present German concept, the test disposal in the Asse Salt Mine should have the following characteristics:

- representative heat generation rate by the radioactive sources,
- representative ionizing radiation from the radioactive sources, and
- disposal and thermo-mechanical conditions representative for a real nuclear waste repository.

II. SCOPE, ISSUES

Since 1965 the Institut für Tieflagerung of the Gesellschaft für Strahlen-und Umweltforschung mbH Munchen has developed and tested safe methods for final disposal of radioactive wastes in salt formations. The disposal techniques that were developed for the disposal of low- and medium-level waste at the Asse Salt Mine are part of the program of the Federal Republic of Germany for the national repository to be constructed in the future. In order to test equivalent methods for the disposal of high-level radioactive waste, the Institut für Tieflagerung was asked in October 1981 by the Federal Ministry of Research and Technology to initiate preliminary planning and preparations for a high-level radioactive waste test disposal in the Asse Salt Mine.

Two general aspects have to be considered during planning and operation of the test disposal:

- the conditions of a real repository have to be taken as a planning basis for the test disposal and
- with regard to specific boundary conditions at the Asse Salt Mine the safe retrievability has to be guaranteed over the complete testing period.

Both aspects will have a certain impact on the test but will not really affect the expected results. It is planned to overcome the shielding capability of the necessary lining of the boreholes by the use of adequate radioactive sources (e.g., higher radiation and higher thermal power).

The currently available data that are to be considered for high radioactive waste sources are shown in Table 1. These data are based on the planning data of the Deutsche Gesellschaft zur Wiederaufarbeitung von Kernbrennstoffen (DWK) for the planned WA-350 reprocessing plant.

TABLE 1. Comparison of Data From WA-350 Canisters and PNL-Canisters

	WA-350	PNL
Age of waste after Discharge (a)	7	
Reprocessing after Discharge (a)	7	
Burn up ($MWd \cdot t_u^{-1}$)	40.000	
^{235}U-Concentration (%)	3,6	
Conditioning:		
Density of glass ($kg \cdot m^3$)	no data available	2,8
Waste Concentration (wt %)	14,5	no data
Waste per liter of glass ($t_u \cdot l^{-1}$)	$7,9 \cdot 10^{-3}$	available
Canister O. D. (mm)	430	300
Canister I. D. (mm)	420	287,2
Canister Hight (mm)	1360	1200
Filling Hight (mm)	1038	1000
Glass Volume (l)	150	70
γ-Activity ($Bq \cdot l^{-1}$)	$6,2 \cdot 10^{13}$	$1,4 \cdot 10^{14}$ (^{137}Cs)
Neutron-Activity ($s^{-1} \cdot l^{-1}$)	$1,8 \cdot 10^7$	-
Dose Rate at Canister Surface ($R \cdot h^{-1}$)	$2,5 \cdot 10^5$	$5 \cdot 10^5$
Heat Power ($Watts \cdot l^{-1}$)	$1,7 \cdot 10^1$	$1,8 \cdot 10^1$

III. OBJECTIVES AND EXPERIMENT DESIGN BASIS

The objectives and the design of the HLW experiment at Asse were developed on the basis of data and experience obtained from heater tests that have been performed in the past[2]. For the investigation of important parameters that cannot be varied in the underground test, parallel laboratory and calculational investigations have to be performed. With regard to the interaction of the radioactive material with rock salt and accessory minerals and those water and gas components contained in the salt, the major interest is directed towards the significance of radiolysis (radiation impact on chemical combinations).

A. Objectives

It has been observed in laboratory and in field tests that rock salt contains traces of water and gases that are released into the volume of a borehole containing heat-producing high-level radioactive waste canisters.

The water and also the gases can accelerate corrosion of the waste canister material, thus leaching radionuclides from the waste matrix. Also the released water and gas will build up pressure if the borehole is sealed gastight. Due to creep behaviour of rock salt, the initial gap between the borehole wall and the waste canister will be closed, thereby also increasing the gas pressure inside the borehole, possibly beyond the fracture strength of the salt. The generation of gases by radiolysis will also contribute to the pressure build up.

It is therefore necessary to obtain data from in-situ simulation tests for the final design of borehole plugs that are able to avoid both high pressure build up and the release of radionuclides contained in the borehole atmosphere. Furthermore, the thermomechanical consequences of heat production and radiation are of importance with regard to the deformational behaviour of underground boreholes and openings. Due to the accelerated creep behaviour of rock salt under heat and heat induced stresses, the borehole wall will creep onto the canister surface and will seal the radioactive canister completely. This is considered to be an important advantage, because leaching of radionuclides from the waste matrix is avoided under these conditions. However, creeping of the salt onto the canister surface will also result in a high pressure load to the canister, perhaps under non-hydrostatic conditions. Thereby, the waste canister and also the waste matrix might be mechanically damaged, if its strength is too low. Also rooms and galleries above the nuclear sources will close due to heat and stress induced creeping of salt. This again is considered to be an important advantage of salt, because backfill material that is put into the gallery after complete filling and sealing of the boreholes, will be solidified with time. Thereby, transportation of leached radionuclides from the underground repository to the biosphere may be significantly reduced.

Under consideration of the above mentioned items, the following major objectives were defined for the HLW test disposal at Asse:

- the investigation of rates and amounts of water and gas forms released due to heat production and gamma-radiation of the nuclear sources and the resulting pressure increase inside sealed disposal boreholes,

- the development of a transportation and handling system for high- level radioactive canisters, and

- the investigation of thermally induced stresses and resulting pressure loads to the waste canisters as well as closure and deformation of rooms, galleries and pillars above the nuclear sources.

A fourth but also important objective during in situ testing at Asse is defined by

- Development and testing of suitable methods and measuring techniques to be applied in an underground repository to obtain data on safety aspects related to construction and operation of a repository.

B. Experiment Design Basis

The present German concept foresees the final disposal of HLW canisters in 300-m deep boreholes from a working level approximately 800 m below the surface.

Because a simulation test at Asse cannot be performed in such deep boreholes, a compromise was found that allowed performance of the test under conditions that are acceptable for obtaining useful data. The tests at Asse will be performed in 15-m deep boreholes, and it is expected that this depth is sufficient to gain representative data.

The retrievability of the nuclear sources, that is to be guaranteed over the complete testing period, represents a further aspect which must be taken into account during testing at Asse. For this, the boreholes are to be lined with high-strength tubes to avoid creeping of the salt onto the canister suface.

The design of the experiment is based on the fact, that in the frame of the German/American contract on "Technical Exchange and Cooperation in the Field of Treatment and Disposal of Radioactive Wastes", thirty vitrified glass blocks, containing the radioactive elements Sr-90 and Cs-137, will be delivered by Battelle Pacific Northwest Laboratories (PNL), Richland, Washington.

It has been agreed, to produce differently endowed sources to obtain the following sets of canisters:

Set 1: Ten Canisters spiked with Cs-137 and Sr-90 with a canister gamma-dose rate of 5.0×10^5 R/h at the surface and a heat power of 2065 Watts

Set 2: Ten Canisters spiked with Cs-137 and Sr-90 with a canister gamma-dose rate of 5.0×10^5 R/h at the surface and a heat power of 1680 Watts

Set 3: Ten Canisters spiked with Sr-90 only so that the canisters will have a negligible gamma-dose rate but also a heat power of 1680 Watts.

The test field will consist of a set of eight boreholes. Four of these boreholes will have a maximum salt temperature of 250°C and four of 200°C. Two of the 250°C boreholes will be heated only electrically, and each of the two others will be charged with five radioactive canisters of Set 1. Two of the 200°C boreholes will only be heated by emplacing five Sr-90 sources of Set 3, and each of the remaining two boreholes will be charged with five sources of Set 2. In this manner it is possible to investigate the impact of the gamma radiation and of the heat release at different maximum salt temperatures.

By use of the above mentioned radionuclides only, the most important parameters as heat power and gamma-radiation, that are significant during the first 200 years after HLW disposal, are simulated in a representative way. Therefore, long half-life alpha emitters can be neglected. Because of special source production, neither actinides nor fission products (U-235, Pu-239) are contained in the glass. Each canister will contain 70 liters of glass and will have a height of 1,200 mm and a diameter of 300 mm.

The duration of the test should be as long as necessary to reach steady state conditions. It is expected, that a five-year period is adequate to meet this requirement.

IV. LAYOUT OF THE TEST FIELD

The configuration and dimensions of the underground test field are mainly determined by the number of boreholes that are required to perform the test. It is planned to mine the test field in the form of two parallel test galleries divided by a 10-m thick pillar. In each of the galleries four boreholes or four test sites will be located. Figure 1 shows the potential location and arrangement of the test galleries at the 800-m level at the Asse Salt Mine. Two boreholes in each gallery (A and B) will have the same maximum temperature of 200 and 250°C. Those boreholes containing only electrical heaters, will have a maximum temperature of 250°C and those containing only Sr-90 canisters will have a maximum salt temperature of 200°C. In the direction of the gallery axis, the distance between the borehole will be 15 m and perpendicular through the pillar 20 m. The two test galleries are accessible through two access drifts (necessary for ventilation), one in the western part and one in the eastern part of the test area (see Figure 1).

Those boreholes having the same maximum temperature are differentiated as type A and type B (see Figure 2). In the boreholes of type A the annulus between the liner and the borehole wall will be backfilled with a porous medium to avoid the borehole wall creeping onto the liner and for maintaining a pathway for the released water and gas components. In the boreholes of type B, the annulus will not be backfilled so that the salt is permitted to creep onto the liner. Thereby, the influence of a possibly nearly gastight contact between the rock mass and the waste canisters can be investigated with regard to the release behaviour of water and gases. The layout of both borehole types can be seen in Figure 3. The above mentioned lining of the boreholes is necessary to guarantee the retrievability of the radioactive sources over the complete testing period.

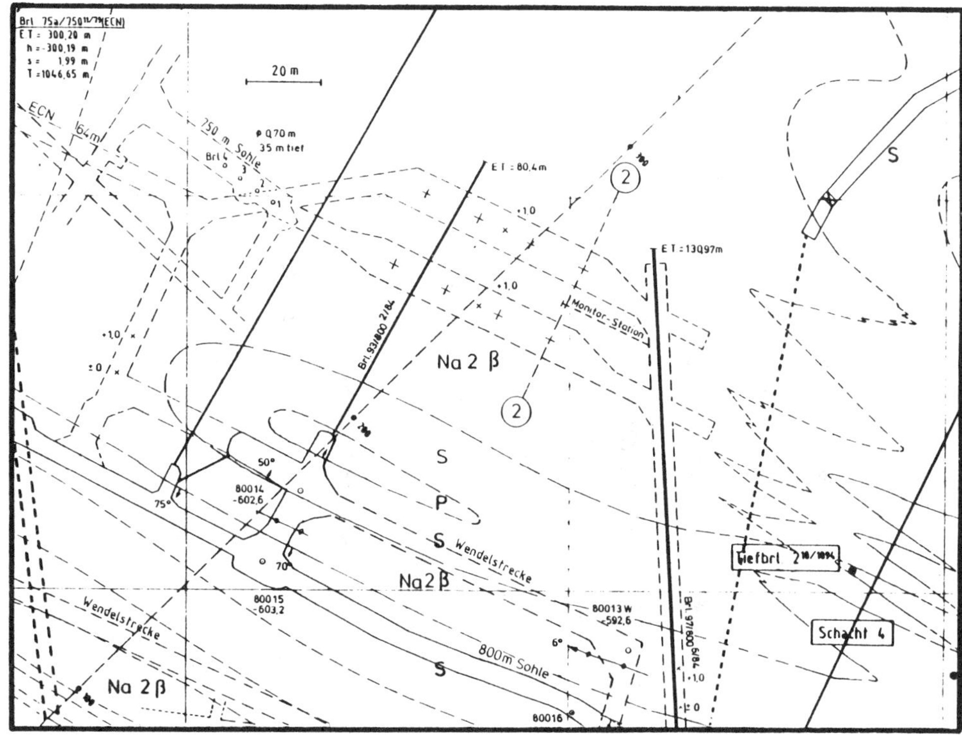

FIGURE 1. HLW Test Field at 800 m-Level of Asse Salt Mine

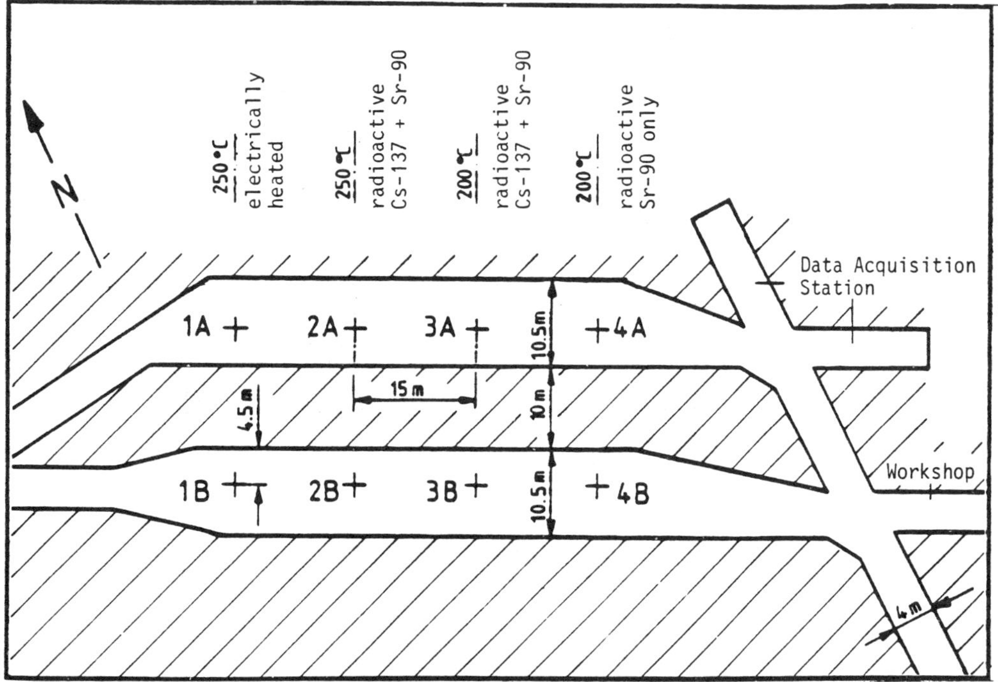

FIGURE 2. Arrangement of Boreholes and Distribution Nuclear Canisters

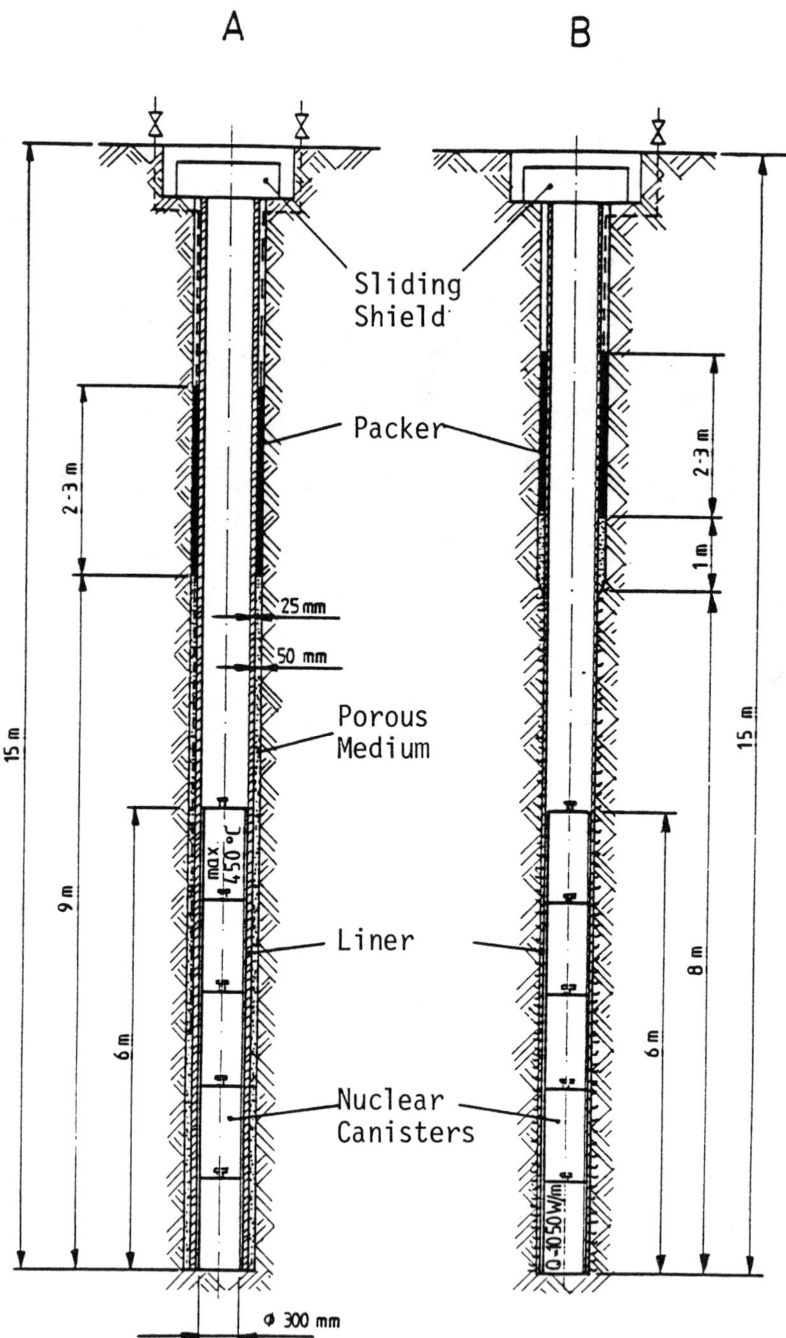

FIGURE 3. Borehole Types

V. SOME EXPECTED RESULTS

Due to heating of the salt formation the salt will expand and creep onto the borehole liner and build up pressure. Preliminary numerical calculations showed that the maximum pressure load at the heated midheight will be about 47 MPa for the 250°C boreholes and about 40 MPa for the 200°C boreholes. The development of this pressure load versus time after emplacement of the radioactive heat sources is shown in Figure 4. The expected water release to the backfilled annulus in the boreholes of type A is shown in Figure 5. In the calculations[3] it was assumed that the water contained in the salt is moving in the form of vapor through the salt porosity and two different models, Darcy-Migration and Knudsen-Migration were considered. It can be seen that the amount of released water in both models is less than 300 g/m of the heated portion of the borehole and the rate of release is continuously decreasing.

VI. OUTLOOK AND TIME SCHEDULE

The high level radioactive waste test disposal at the Asse Salt Mine in the Federal Republic of Germany will probably start in 1987. It is expected that first data will be available in 1989 or 1990. On the basis of this data the preliminary concept for construction and operation of the national nuclear repository will be developed. The mentioned tests at Asse will be terminated in 1992 so that the construction of the underground facilities can be started immediately afterwards.

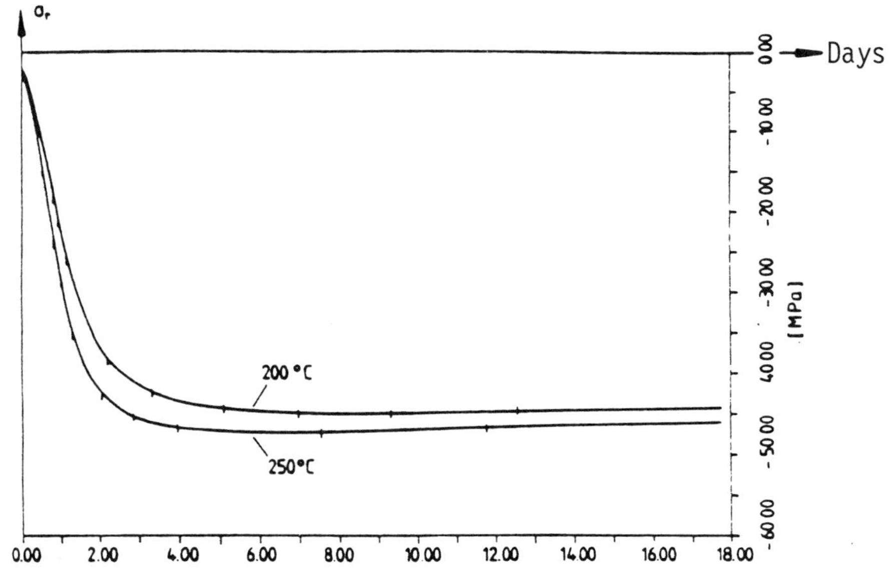

FIGURE 4. Radial Pressure Load on Borehole Liner Versus Time

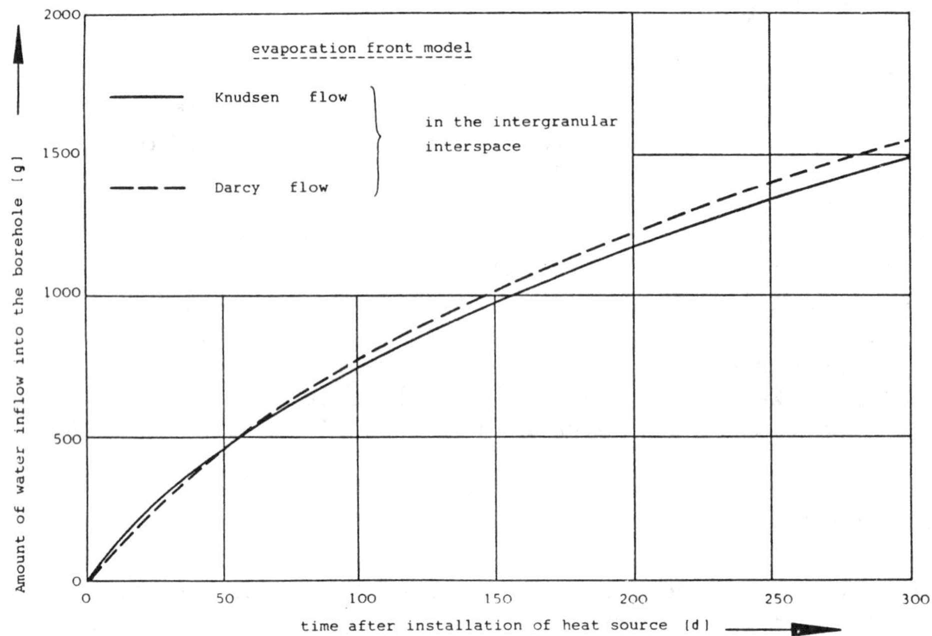

FIGURE 5. Water Release Into a HLW-Borehole

VII. REFERENCES

1. Gesellschaft für Strahlen/Office of Nuclear Waste Isolation, Nuclear Waste Repository Simulation Experiments-Asse Salt Mine, Federal Republic of Germany, Annual Report (1983).

2. T. Rothfuchs et al., "Simulationsversuch im Alteren Steinsalz Na2 im Salzbergwerk Asse - Temperaturversuchsfeld 4 (TVF4), Abschlu bericht zudem Teilvorhaben 1.1, 1.4 und 1.5 im Rahmen der Vertrage 015-76-1 und 058-78-1 WAS D des Indirekten Aktions-programms der Europaischen Gemeinschaften, GSF-IfT (1983).

3. M. Schlich, Simulation der Bewegung von im naturlichen Steinsalz enthaltener Feuchte im Temperaturfeld, Ein Beitrag zur Problematik der Endlagerung hochradioaktiver Abfalle, RWTH Aachen, Dissertation (1985).

THE WIPP RESEARCH AND DEVELOPMENT TEST PROGRAM*

L. D. Tyler
Experimental Programs Division
Sandia National Laboratories**
Albuquerque, New Mexico 87185

ABSTRACT

The WIPP (Waste Isolation Pilot Plant) is a DOE R&D Facility for the purpose of developing the technology needed for the safe disposal of the United States defense-related radioactive waste. The in-situ test program is defined for the thermal-structural interactions, plugging and sealing, and waste package interactions in a salt environment. An integrated series of large-scale underground tests address the issues of both systems design and long-term isolation performance of a repository.

I. INTRODUCTION

The Waste Isolation Pilot Plant (WIPP) has been authorized by the U.S. Congress as a facility with a mission "... for the express purpose of providing a research and development facility to demonstrate the safe disposal of radioactive waste resulting from the defense activities and programs of the United States exempted from regulation by the Nuclear Regulatory Commission."[1] The WIPP is under construction in the salt beds of southeast New Mexico. The facility is to be used to permanently isolate transuranic (TRU) waste generated by the U.S. Defense program, and to provide an "underground laboratory" in which the concepts for safe disposal of defense high-level waste (DHLW) and TRU waste in salt will be validated and demonstrated. The construction at WIPP began in July 1981, and presently more than three miles of drift exist at the facility depth of 655 m. The facility horizon is located in a very thick evaporite sequence of the Salado Salt Formation. This paper will provide an overview of the in-situ test activities of the R&D Program.

II. WIPP R&D PROGRAM

The WIPP R&D Program provides a technical basis for systems design and safety and environmental assessments for the WIPP and future defense waste repositories.[2] The initial phase of the technical development in the WIPP Program involved laboratory testing and small-scale field testing in

*This program supported by the U. S. Department of Energy under Contract DE-AC04-76DP00789.
**U.S. DOE facility.

commercial salt and potash mines. The results of these tests and the results of similar testing done over the past two decades have provided the information to define the phenomena that accompany placement of radioactive waste into salt. The WIPP R&D Program is progressing into a phase in which in-situ experiments on a realistic scale and in an actual environment can be accomplished. This phase of in-situ testing will serve to validate codes, models and designs and to confirm, on a large scale, the results of laboratory experiments. The initial in-situ tests in the WIPP will be conducted in a simulated fashion for both non-heat and heat-producing waste. These simulated experiments will be followed in 1989 by a selected suite of experiments using radioactive waste. The WIPP in-situ simulation experiments are divided into three broad areas: thermal/structural interaction (TSI), plugging and sealing, and waste package interaction experiments. These program areas are described in the following paragraphs, and their locations in the WIPP are shown in the underground layout, Figure 1. Approximately 5000 channels of data will be collected for the in-situ simulated experiments for a "testing" period of at least seven years.

A. Thermal/Structural Interaction (TSI) Experiments

An important issue in repository development concerns the underground openings, and whether the structural response of the drifts, rooms, and shafts can be predicted. Development of a repository in salt should not only provide for stable rooms and entries during the operational period for waste emplacement and possible retrieval, but also sufficient long-term rock-deformation to ultimately encapsulate and isolate the waste.

The prediction of both short-term stability of the rooms and the long-term encapsulation of waste requires reliance on adequate material (constitutive) models, material parameters and numerical computer codes to predict the effects of stress and heat on salt. The TSI in-situ tests are designed to provide the data base needed to validate and to improve the predictive technique which can model the behavior of salt at ambient and elevated temperatures for repository design and long-term isolation assessments.

The TSI underground tests consist of the following six major individual tests:

1. 18-W/m^2 mockup,
2. DHLW overtest,
3. geomechanical evaluation,
4. heated pillar,
5. in-situ stress, and
6. direct shear of clay seams.

The first four tests require excavated rooms and the in situ stress test requires some controlled excavation, Figure 1.

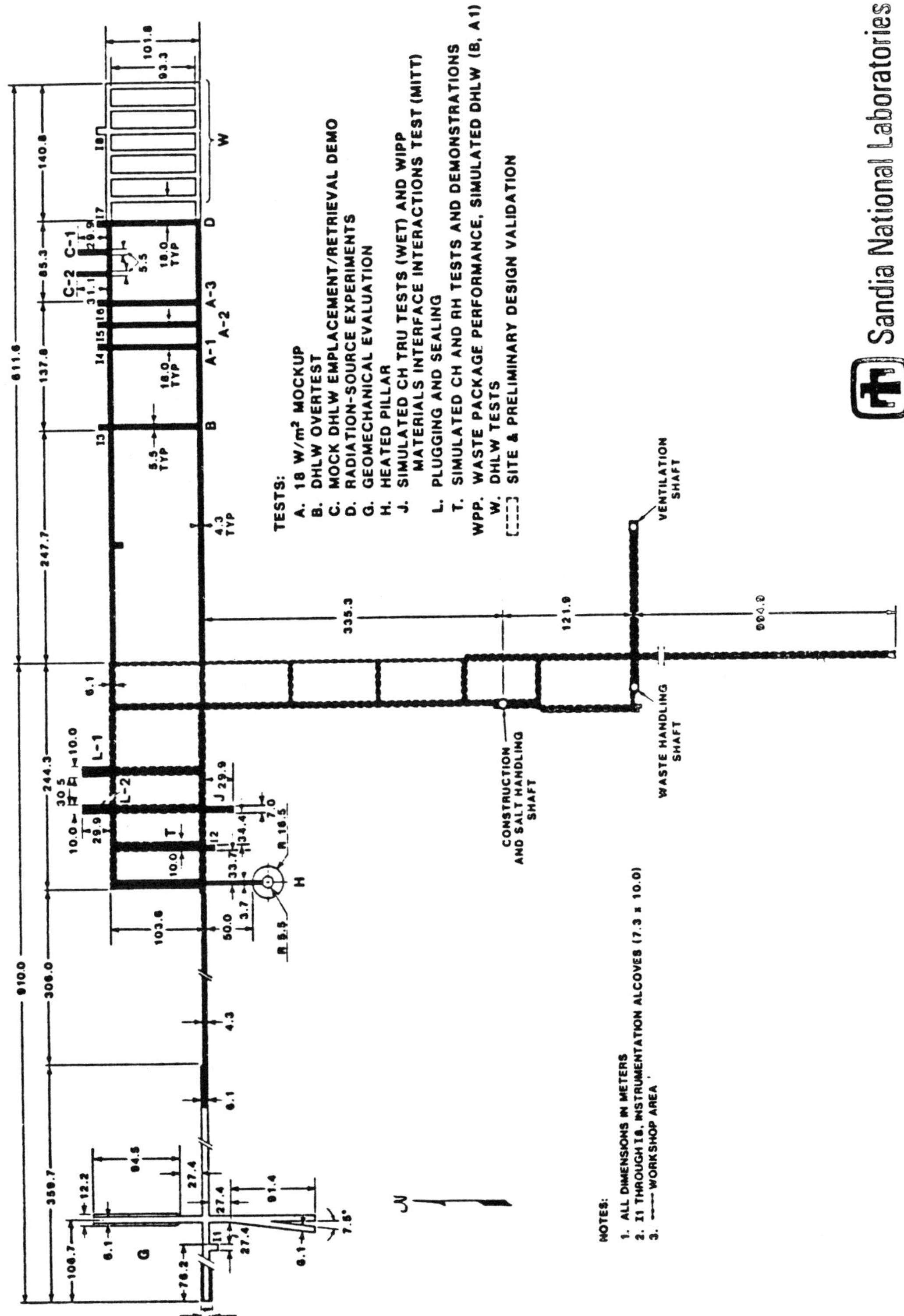

FIGURE 1. Layout of In Situ Tests for WIPP

B. WIPP Waste Package Performance Tests

An extensive body of data on the performance of HLW package material exists from laboratory experiments which simulate most aspects of expected overtest and environments of a salt repository.[4] The in situ testing of simulated DHLW and TRU packages in the WIPP are being done under expected reference and overtest conditions. These tests are a valuable precursor to the actual radioactive waste tests by extending the state-of-the-art technology beyond the laboratory and identifying those aspects which must be resolved with radioactive testing. The present tests are divided into two categories: simulated DHLW package performance tests and TRU waste package performance tests.

The Simulated DHLW Package Performance Tests are to be conducted at a reference repository condition in Room A and at an overtest condition in Room B (Figure 1). Six electrically heated waste canisters (0.47 KW power) will be emplaced in Room A1 to test canister and backfill materials under an expected defense HLW repository environment. All canisters will be 0.6 m in diameter and 3 m long. Three canisters will be made of mild steel and three canisters will be made of mild steel with a 2.2 mm thick TiCode overpack. Two types of backfill materials will be tested: crushed salt and a low density bentonite/ sand backfill. One canister will have brine injected into the emplacement hole to test the canister material.

Twelve waste packages are emplaced in Room B to be tested at an accelerated test condition and a thermal load (1.5 KW/ canister). Eight of the packages contain electrical heaters: the other four packages contain canisters filled with nonradioactive DHLW glass. The same backfill materials as tested in Room A1 also will be tested in the overtest environment. The canister material to be tested includes SS304L canisters with a mild steel overpack and TiCode-12 canisters. The instrumentation for the tests are thermocouples for the canister materials, backfills and nonradioactive glass and pressure gages for the backfill. The canisters will be tested for 3 to 7 years and will undergo extensive post test material testing.

The TRU waste package performance tests will be conducted in the J & T rooms (Figure 1). The in-situ testing of simulated/nonradioactive CH TRU waste drum and RH TRC waste canisters in the WIPP is necessary to evaluate and demonstrate the waste package physical integrity. Testing will be conducted under near reference conditions and severe overtest (accelerated aging) conditions of temperature and brine intrusion. Tailored backfill also will be tested to determine their effectiveness as engineered barriers. The backfills being tested are granular bentonite/crushed salt (30 wt%/70 wt%) and crushed salt. The predominant results of the tests will be in the area of corrosion resistance and backfill applicability for meeting the EPA standards.

C. WIPP Plugging and Sealing Program

The plugging and sealing program provides for the development and demonstration of the technology required to establish plugs and seals in man-made penetrations (boreholes, shafts, drifts, storage rooms) to enhance the long-term isolation of nuclear waste. Basically, the program is designed to test and demonstrate the effectiveness of the various engineered barriers proposed for sealing vertical penetrations into the underground waste disposal facility and for isolating the underground rooms from each other.[5] The test program will quantify the permeability of the salt beds and other layers, especially where altered by the excavation, as a function of time as the walls of the room close upon the backfill. Microcracking due to excavation and stress relief will heal in a salt environment. The rate at which the healing occurs must be established. Both cement grouts and natural (salt) plugs will be tested for use at the WIPP. The current field activities include: (1) in-situ permeability tests of the facility horizon; (2) the plug test matrix for obtaining in situ cured samples for laboratory assessment of geochemical stability; and (3) demonstration and verification of capabilities to install, test and monitor in-situ plugs.

III. SUMMARY

The WIPP R&D Program has entered into the in-situ test phase. This currently addresses nonradioactive tests for thermal structural interaction. plugging and sealing, and waste package interactions. The TSI experiments and simulated DHLW tests are currently emplaced and will be operational from early 1985 for several years. The remaining experiments are scheduled to be operational later in the year. The testing program is designed to provide the technology for developing the design and long-term assessment of waste isolation for the WIPP and future defense waste repositories.

REFERENCES

1. PL9b - 164, December, 1979, U.S. Congressional Authorization B.11.

2. R. B. Matalucci et al., Waste Isolation Pilot Plant (WIPP) Research and Development Programs: In Situ Testing Plan, March 1982, SAND81- 2628, Sandia National Laboratories, Albuquerque, New Mexico (1982).

3. D. E. Munson and R. V. Matalucci, "Planning, Developing, and Fielding of Thermal/Structural Interactions In Situ Tests for the Waste Isolation Pilot Plant (WIPP)," Waste Management '84 Conference, Tucson. Arizona (March 1984).

4. M A. Molecke, J. A. Ruppen and R. B. Diegle, "Materials for High-Level Waste Canisters/Overpacks in Salt Formations," Nuclear Technology, Vol. 53, No. 3 (1983).

5. C. L. Christensen, S. J. Lambert and C. W. Gulick, Sealing Concepts for the Waste Isolation Pilot Plant (WIPP) Site, SAND81- 2195, Sandia National Laboratories, Albuquerque, New Mexico (1982).

BRINE MIGRATION TESTS THAT ARE IN PROGRESS AT ASSE MINE IN THE FEDERAL REPUBLIC OF GERMANY

T. Rothfuchs
Gesellschaft für Strahlen-und Umweltforschung mbH Munchen
Institut fur Tieflagerung
Schachtanlage Asse
3346 Remlingen, Federal Republic of Germany

ABSTRACT

It has been found that when a heat source is placed in salt, traces of water contained in the salt migrate to the heat source. This water can also accumulate around heat producing high-level radioactive waste (HLW) canisters and may corrode the seal, thus releasing radioactive material from the package into the surrounding environment. Since laboratory tests do not truly represent the actual in-situ environment, a large scale brine migration test is being performed in the Asse salt mine to obtain an understanding of the brine migration mechanism at depth. The tests are performed using heat- and Cobalt-60 radiation sources to simulate disposed HLW canisters. Preliminary in-situ measuring data show that the amount of the released water is not of serious concern and also radiolytical decomposition of the released water is of minor significance.

I. INTRODUCTION

The brine migration tests at Asse Mine represent a cooperative effort between the Office of Nuclear Waste Isolation (ONWI), Columbus, Ohio for the USA and the Institut fur Tieflagerung (IfT), Braunschweig, of the Gesellschaft fur Strahlen- und Umweltforschung mbH Munchen (GSF) for the FRG.

In the U.S., the program is part of the Civilian Radioactive Waste Management (CRWM) Program funded by the Department of Energy (DOE) through ONWI and operated by Battelle Memorial Institute (BMI), Project Management Division (BPMD). In the FRG, the program is funded by the Bundesministerium für Forschung und Technologie (BMFT) at the request of the Physikalisch- Technische Bundesanstalt (PTB) and operated by GSF/IfT.

The tests are designed to simulate a nuclear waste repository to measure the effects of heat and gamma radiation on brine migration, radiolytical decomposition of brine, and thermal mechanical behaviour of salt.

II. DESIGN OF THE EXPERIMENTS

To satisfy the objectives of the test plan[1] four tests are being performed at Asse. The tests are conducted at the 800 m-level in areas specially mined for the tests, Figure 1. The main halite (Na_2) present in this area is considered to be a typical repository salt type. The four tests are designed to be performed at a maximum salt temperature of 210°C (410°F) having 3°C/cm (13.72°F/inch) thermal gradient at the borehole wall.

A test setup consists of a central borehole 43.4 cm (17 inches) in diameter. It contains a sleeve 5-m long (16.4 feet). The lower 2 m (6.5 feet) are electrically heated. In two of the tests, to simulate the effect of radiation, Cobalt-60 sources are placed in the lower 2-m long heated portion of the borehole. The central borehole is surrounded by eight guard heater holes that are used to establish the desired maximum temperature and thermal gradient at the central borehole wall.

In the central borehole, the void between the heated sleeve and the borehole wall, which is 5 cm (2 inches) thick, is filled with alumina beads. The brine vapors migrate to the alumina beads and are transported to the collection point in a cold trap by circulating nitrogen. The vapors are condensed in the cold trap and the brine is collected for later analysis. Figures 2 and 3 show a vertical cross section of the test setup.

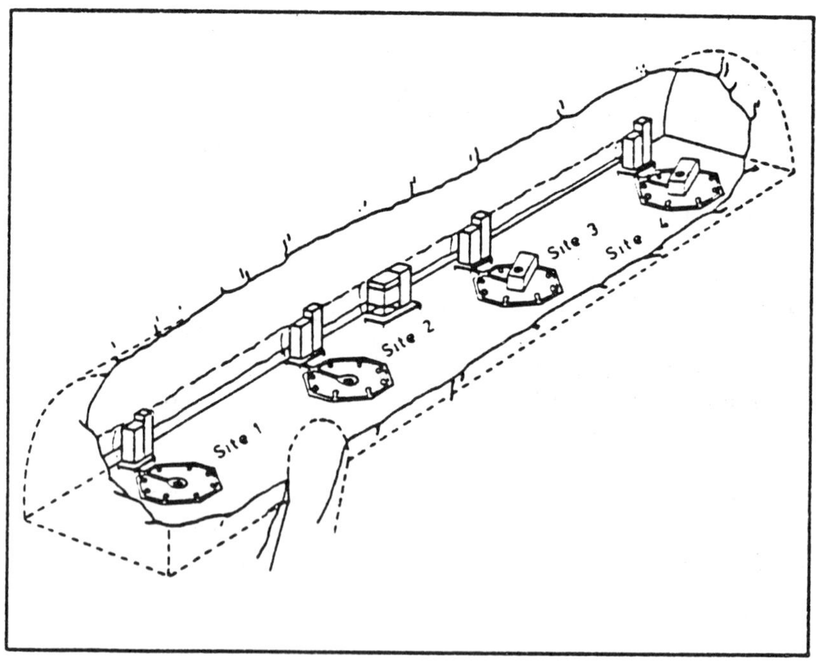

FIGURE 1. Test Gallery

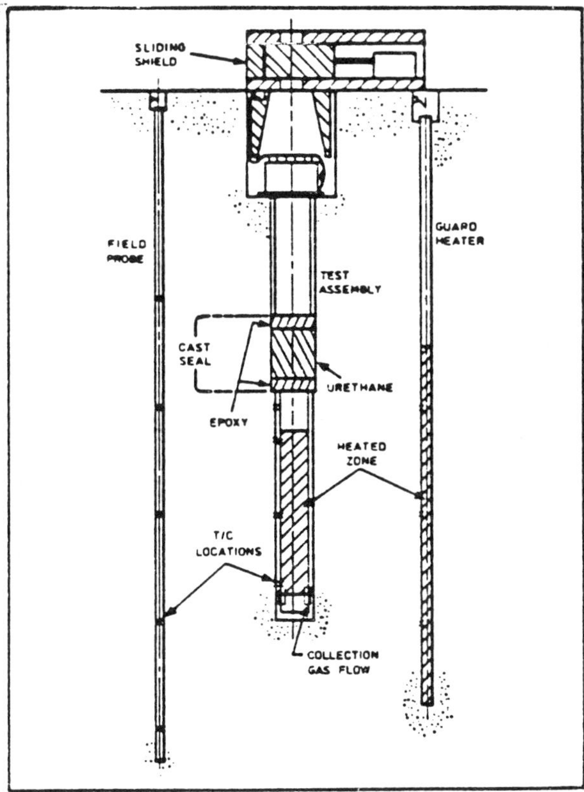

FIGURE 2. Test Site Cross Section

The lower sleeve protects the heaters and the radiation sources from the brine and the lithostatic as well as thermal stresses to assure retrievability at termination of the tests.

The tests are separated by 15 m (45 feet) to avoid any interaction during the testing period.

The basic measurements performed during testing are:

- temperature measurements in the surroundings of each test site,
- brine released to the central borehole,
- other gases produced by thermal release, radiolysis, and corrosion, and
- stresses in the salt and resulting closure of the room along the walls and along the floor and roof.

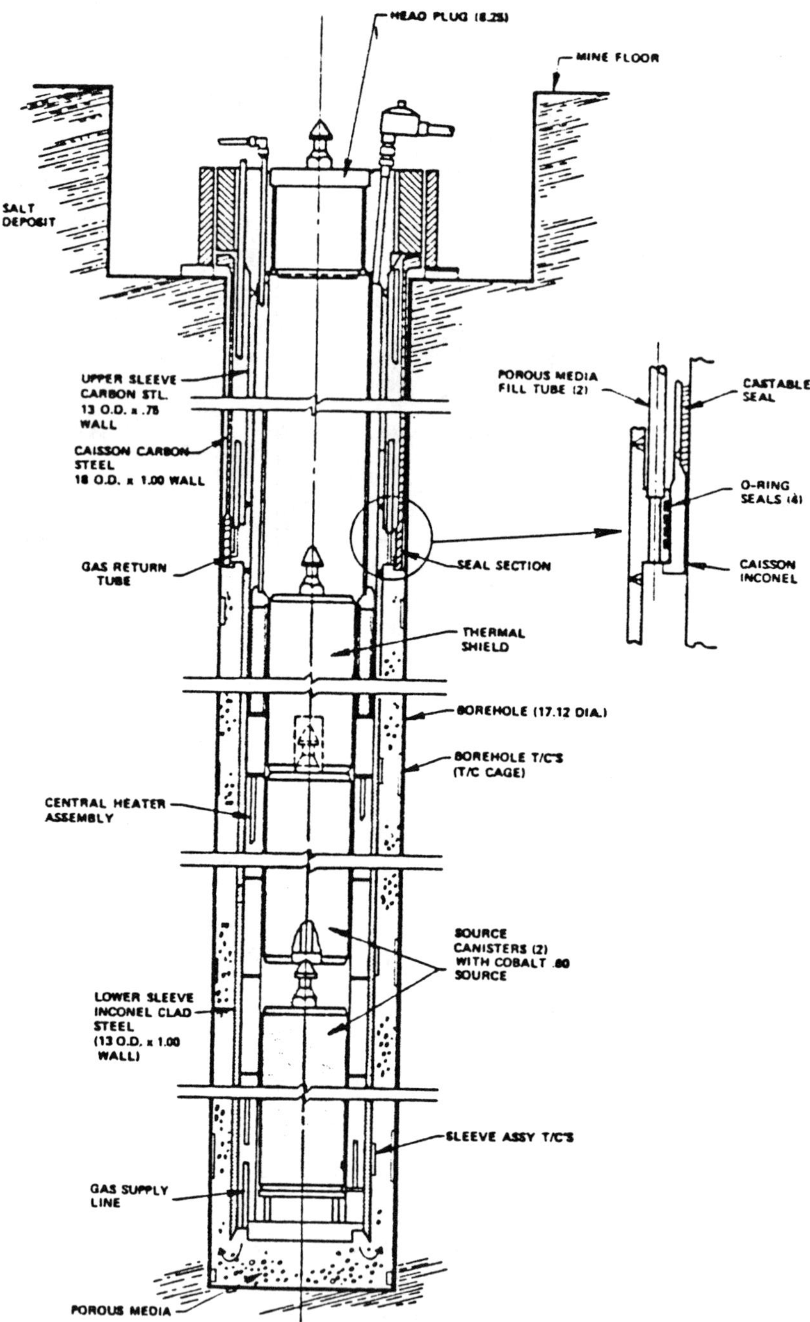

FIGURE 3. Cross Section of Test Assembly

III. THEORIES AND COMPARISON OF MEASURED AND PREDICTED DATA

This section describes all data obtained between May 25, 1983 and the end of March 1985, approximately twenty-two months operation of the nonradioactive test sites 1 and 2 and fifteen months of the radioactive test sites 3 and 4.

A. Temperature

Temperature measurements are taken in the central borehole and the temperature probe boreholes, among others, 4.57 m below the gallery floor at the midheight of the heated section. Figure 4 shows the typical radial temperature distribution at the radioactive test site 3. After start-up of heating on December 15, 1983, the temperature behavior of test site 3 was in good agreement with predictions. The heat power is 2710 Watts for the central heater and 7220 Watts for the eight guard heaters.

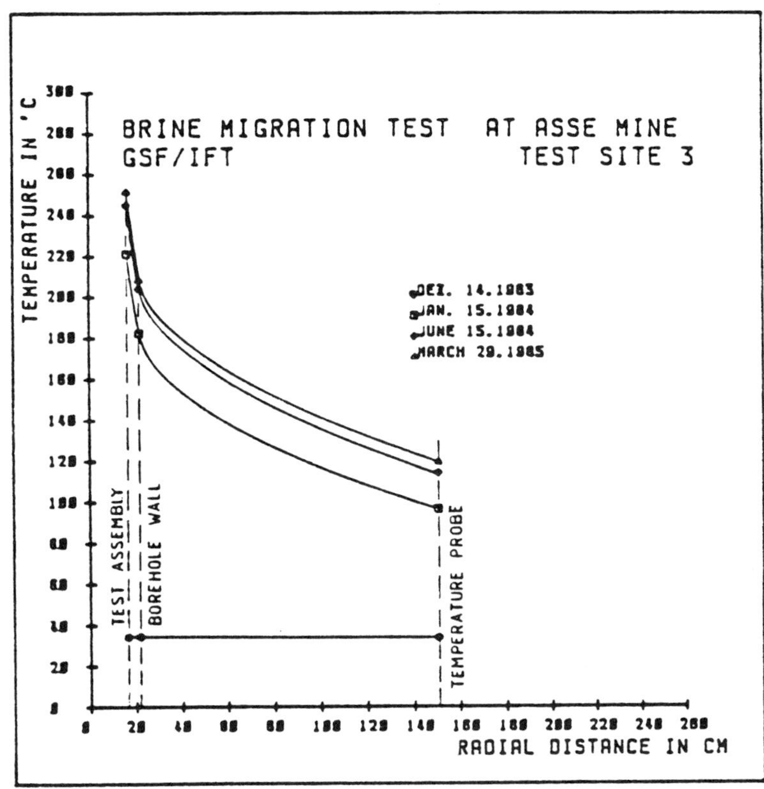

FIGURE 4. Radial Salt Temperature, 4.57 m Below Floor

B. Brine Migration

Two different mechanisms of brine migration must be taken into account. Vapor migration is motion of water vapor which is generated by evaporation of water from brine in the salt porosity or from hydrated minerals. This motion is assumed to be proportional to pressure gradients and to be dependent on the permeability of the salt. Motion of vapor can be described according to Darcy's law

$$V = -\frac{K}{\mu} \nabla Pv \qquad (1)$$

where:

V = velocity of the water vapor (cm/sec)
K = permeability of the salt (cm^2)
μ = viscosity of the water (bar - sec)
Pv = water vapor pressure (bar)

Liquid Inclusion Migration is motion of small inclusions of saturated brine located within the salt crystals or on the grain boundaries. This motion is known to be proportional to the local temperature gradient and increases in velocity at higher temperatures. The following equation[2] for the velocity of liquid inclusions as a function of temperature and temperature gradient describes motion of liquid inclusions

$$V = \nabla Ts \; 10^{(0.00656\, T - 0.6036)} \qquad (2)$$

where:

V = Inclusion Velocity (cm/yr)
Ts = Temperature gradient in the salt (^{o}C/cm)
T = Temperature of the salt (^{o}C)

During the testing period only sites 2 and 4 are operated with continuous brine collection, whereas sites 1 and 3 are closed so that brine and gases will collect in the boreholes and be permitted to build up pressure and react in the boreholes. This is being done to evaluate the impact of the borehole gas pressure on the water release rates, thereby investigating the predominant migration mechanism. Figure 5 shows a comparison of predicted brine collection with measured data of sites 2 and 4. The prediction has been calculated on the basis of a total water content of the salt of 0.05 wt% in the form of liquid inclusions. Mineralogical and chemical investigations of approximately 200 salt samples taken from the test area showed, that the total water content is about 0.23 wt%, but 90% of this water is connected to the hydrated mineral polyhalite and only 10% (0.02 wt% of the salt) is connected to absorbed liquid water on grain boundaries. Since the maximum salt temperature is significantly below the release temperature[3] of the crystalline water of polyhalite it is likely that only the adsorbed water contributes to the water collected in the

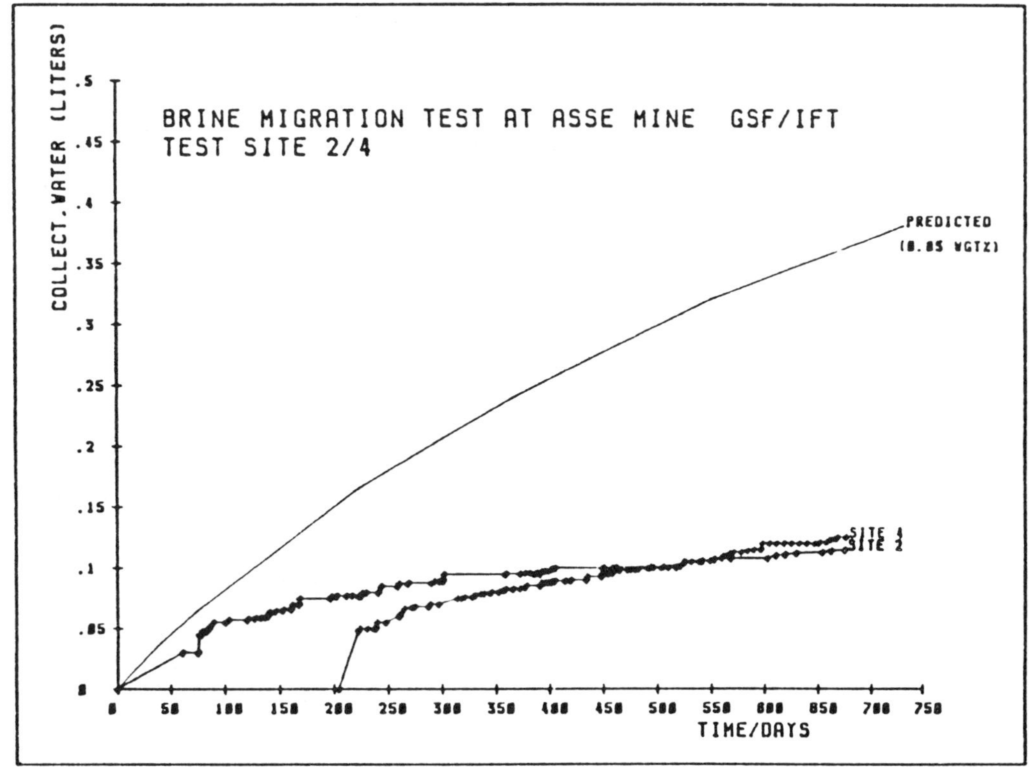

FIGURE 5. Brine Collection Versus Time

boreholes. Therefore, the measured amount of water is smaller than that calculated based on assumed 0.05 wt% liquid water in the salt.

The results of accumulated water up to the end of March 1985 are also presented in Table 1. It appears that a smaller amount of brine is collected in the pressurized boreholes indicating that the gas pressure is retarding the vapor transport mechanism. This phenomena is explainable if vapor transport occurs parallel to liquid inclusion migration. Since the amount of water released to site 1 and 3 is more or less half of that released to sites 2 and 4, it appears that both transport mechanisms have to be taken into account in the unpressurized boreholes, whereas vapor migration is retarded in the pressurized boreholes. A significant difference between radioactive and nonradioactive tests was not observed; in other words, a radiolytical decomposition of the liquid water or of hydrated minerals is of low significance or nonexistent.

TABLE 1. Brine Collection Data

Test Site No.	Days of Operation	Absolute Borehole Gas Pressure (bars)	Average Gas Temperature (K)	Collected Brine (cm^3)
1	673	3.23	480	43 *
2	673	1.05	483	115 **
3	469	1.05	479	80 **
4	469	1.06	480	225 **

* Estimated from water partial pressure
** Cold trap measurement (Test Site 3 was only pressurized up to test day 215 and then connected to the cold trap for continuous brine collection)

C. Gas Release

Gas can be generated by radiolysis as mentioned before, by corrosion of the borehole liner or by release of gas from the salt. The effects of gas are to increase the pressure in the test volume and potentially retard water vapor transport.[4]

Measurements of gas pressure in the test volume that is filled with alumina beads are made at all four test sites. Prior to start-up of the experiments, all boreholes were purged with dry nitrogen gas (N_2) to remove all moisture and other gas components contained in the borehole atmosphere. Figure 6 shows the pressure increase of the closed test volume of the radioactive site 3 until the seal failed on test day 215. Table 2 presents data on the composition of other gases beside water vapor contained in the borehole atmosphere at this time. All gas components contribute to the total gas pressure increase as the temperature increase due to heating. It is of interest that hydrogen was found with nearly the same concentration in the radioactive sites as well as in the nonradioactive sites. It is therefore likely that hydrogen is mostly produced by corrosion reactions and not by radiolysis of the released brine. The production of small amounts of hydrocarbons, carbon dioxide, and carbon monoxide is not surprising since it is known from former investigations that these components usually occur in rock salt.

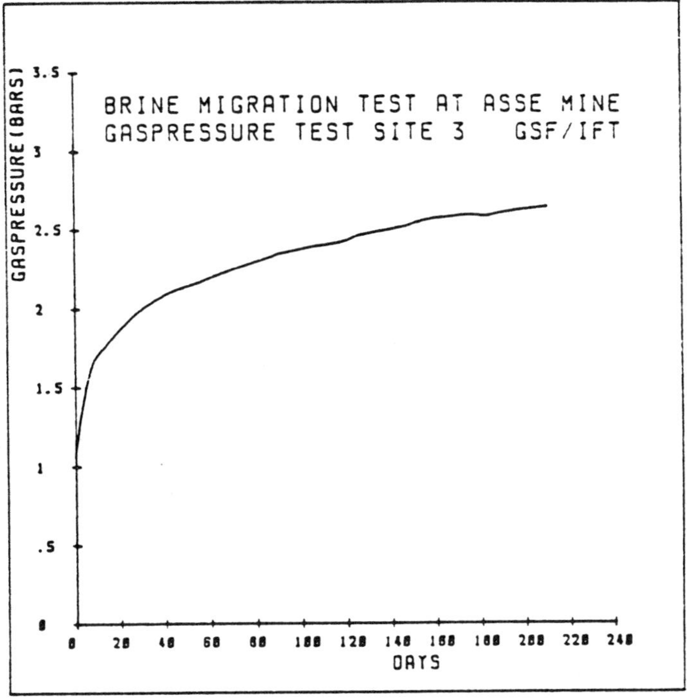

FIGURE 6. Test Volume Gas Pressure

TABLE 2. Composition of the Borehole Atmosphere

Component	H_2	O_2	N_2	CO	CO_2	CH_4	C_2H_6	C_3H_8
Concentration (Vol%)	0.57	0.3	91.3	0.35	6.45	0.23	0.046	0.03

D. Room Closure Observations

Pretest (finite element) calculations[5,6] using the US-code DAPROK and the FRG-code MAUS have been performed to predict the order of magnitude of the thermomechanical rock mass response and to compare them with displacements around the test field. For modeling the induced rock mass response, the following material properties of rock salt and creep law were used:

- Elastic material properties
  Young's Modulus E = 6.9 GPa
  Poisson's Ratio = 0.4

- Creep law adapted for secondary creep rate ε_s
  $\varepsilon_s = A \exp(-Q/RT)^n$

where:

A = structural coefficient (sec^{-1})
Q = activation energy ($KJ \times Mol^{-1}$)
n = stress exponent(*)

(*)For more detailed informations about both finite-element-codes, modeling, and results of numerical pretest calculations, see Reference 5 and 6.

Figure 7 shows the comparison between both pretest calculational results and field measurement data for the horizontal room closure at test site 3. It is obvious that DAPROK produces higher deformations than MAUS in the preheating phase, but produced nearly the same prediction as measured after heating was started. Since both codes are similar, the differences may be due to the fact that DAPROK was applied to a much smaller rock volume than MAUS resulting in higher stress levels and deformations. However, it was known, that the results of these pretest calculations would be of preliminary character, but it can be seen that they are accurate enough for the design of the test program.

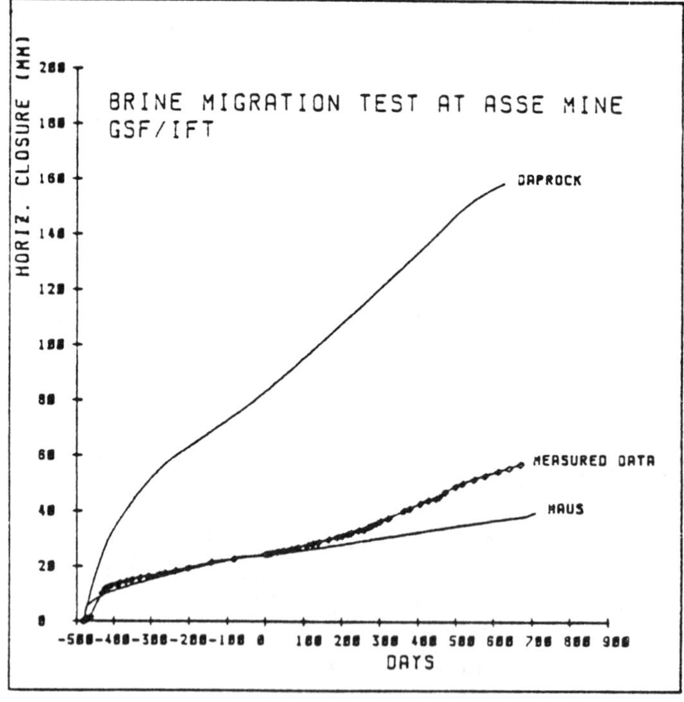

FIGURE 7. Horizonal Room Closure Versus Time Test Site 3

IV. CONCLUSIONS

Preliminary measuring data of the joint US/FRG Brine Migration Test at the Asse Mine show that the heat production of high-level radioactive waste leads to a measurable release of traces of water or brine into disposal boreholes. For the sealed boreholes, the water reacts with metals contained in the borehole, thus producing small amounts of hydrogen. Due to heating other gases are released from the salt but the amounts are of minor significance. Radiolytical decomposition of the water due to gamma radiation was not observed to be significant.

However, the release of gaseous components, such as water vapor and other gases are to be taken into account with regard to the design of underground nuclear repositories but on the basis of in-situ measurements, theoretical considerations and numerical computer code developments it will be possible to estimate the long-time consequences of brine migration and gas generation in a salt repository.

REFERENCES

1. Office of Nuclear Waste Isolation/Gesellschaft für Strahlen und Umweltforschung mbH Munchen, Asse Salt Mine Brine Migration Test, ONWI-245, GSF-T118, (1981).

2. G. H. Jenks, Effects of Temperature, Temperature Gradients, Stress and Irradiation on Migration of Brine Inclusions in a Salt Repository, Oak Ridge National Laboratory, Oak Ridge, Tennessee, ORNL-5526 (1979).

3. N. Jockwer, "Untersuchungen zu Art und Menge des im Steinsalz des Zechsteins enthaltenen Wassers sowie dessen Freisetzung und Migration im Temperaturfeld endgelagerter Abfalle", Gesellschaft fur Strahlen, und Umweltforschung mbH Munchen, GSF-T119 (1981).

4. T. Rothfuchs, "Design and Instrumentation of In Situ Brine Migration Tests at the Asse Salt Mine", CEC/NEA-workshop on the Design and Instrumentation of In Situ Experiments in Underground Laboratories for Radioactive Waste Disposal, Bruxelles, Belgium (1984).

5. Thermomechanical Analysis Brine Migration and Waste/Rock Interaction Field Test Designs, D'Appolonia Report 79-339, Asse Mine, Federal Republic of Germany (1981).

6. G. Albers et al., "Kontrollrechnungen zum gemeinsamen US/FRG Laugenmigrationsversuch", RWTH Aachen (1983).

FIRST TEST OF A DRIFT SEAL UNDER BRINE INFLOW

W. R. Fischle
K. Schwieger
Gesellschaft für Strahlen
und Umweltforschung Munchen mbH
Institut für Tieflagerung
Schachtanlage Asse
D-3346 Remlingen
Federal Republic of Germany

W. H. Stover
Kavernen Bau- und
Betriebsgesellschaft mbH
Rathenaustrasse 13-14
D-3000 Hannover 1
Federal Republic of Germany

ABSTRACT

Safety during the final disposal of radioactive waste in a mine as well as in the case of an accident due to the inflow of water or brine depends on how the propagation of radionuclides into the biosphere is prevented. At termination of the emplacement phase the caverns must therefore be resealed. The flooding of an abandoned mine is enabling the testing of a seal-bulkhead on a 1:1 scale. The aim is to set up a maintenance-free, impermeable, statically stable construction. Besides the testing of measuring devices, data are to be acquired for subsequent safety calculations. The measurements also serve to verify computer programs. After a drying period of more than one year flooding began at the start of 1985 and, therefore, the test of the construction and the measuring devices under the influence of brine.

I. INTRODUCTION

The safety of a final repository for radioactive wastes during water or brine inflow depends on how efficiently the propagation of nuclides into the biosphere is prevented. Besides a favorable rock formation as a natural barrier, suitable sealing measures need to be carried out at the end of the emplacement phase. These are to prevent a transport of nuclides via created cavities and openings such as shafts, drifts, accesses and chambers.

In the Asse salt mine, Federal Republic of Germany, chamber and drift bulkheads have been constructed and monitored by means of measurement techniques for a number of years in rock salt formations. Temperature curves, pressure build-up and stress relief are being measured.

Laboratory investigations led to the development of a salt concrete in which rock salt is used as an additive. Apart from adequate solidity it has an elastic and hydraulic property similar to that of the rock salt and it cannot be destroyed by inflowing saturated brines.

Due to the flooding of the abandoned Hope salt mine near Hannover, it was possible to take in-situ measurements in a mine prior to, during and after the flooding with brine. Besides chemical investigations, physical and geomechanical investigations on a 1:1 scale are being conducted for the first time on a bulkhead that is being constructed and tested under accident conditions 500 m below the surface. The test includes the comparison of the laboratory results to the in-situ conditions as well as the verification of the calculations and prognoses under realistic conditions. No instruments have been employed before for measurements under increased static and hydraulic compression and for transfer of data over large distances. This project calls for the testing of the measuring instruments over a phase of several years.

II. GOALS OF THE TEST

A maintenance-free, impermeable and statically stable construction in a 20 year old drift has to be designed for the post-operational phase of a final repository for radioactive wastes. This is a first opportunity of measuring and estimating the density and consequently the efficiency of a bulkhead in a gallery under a brine pressure of approx. 2.5 MPa, through a later inflow of about 6 MPa. Possible flow paths are to be determined besides the rate of flow. Furthermore, the stress distribution as well as deformations and displacements are to be measured.

III. CONSTRUCTIONAL CONCEPT

The bulkhead consists of two supports with an inner seal and a surface seal on the brine side (Figures 1 and 2). The static pre-calculations are based upon the boundary cases "full restraint" and "free gliding" at the setting in of brine pressure. The construction exhibits a different stress-strain behavior in this instance. Tensile stress occurs in the transitional zone from the brine-faced front to the rock formation. In order to prevent a disintegration, the surface packing was inserted for the load transfer. Contact with the rock formation within the range of the inner seal was obtained by means of a groove. This ensures protection of the construction against brine flow through the rock. Floors and stopes had to be cleared of loose material. The inclined outer surfaces cause an efficient load reduction and reduce the dislocation of individual constructional components. The inclination of the roof is a result of the requirement of a joint-free connection to the rock formation.

A. Building Material: Salt Concrete

Both bulkhead supports were made of salt concrete. Besides the requirement for a minimum compressive strength and a construction material with low permeability, the mixture was to be optimized according to manufacturing and productional techniques. The following mixing ratio was adapted to the capacity of the mixing plant:

$$\text{Cement : Water : Salt} = 1 : 0.5 : 3.0$$

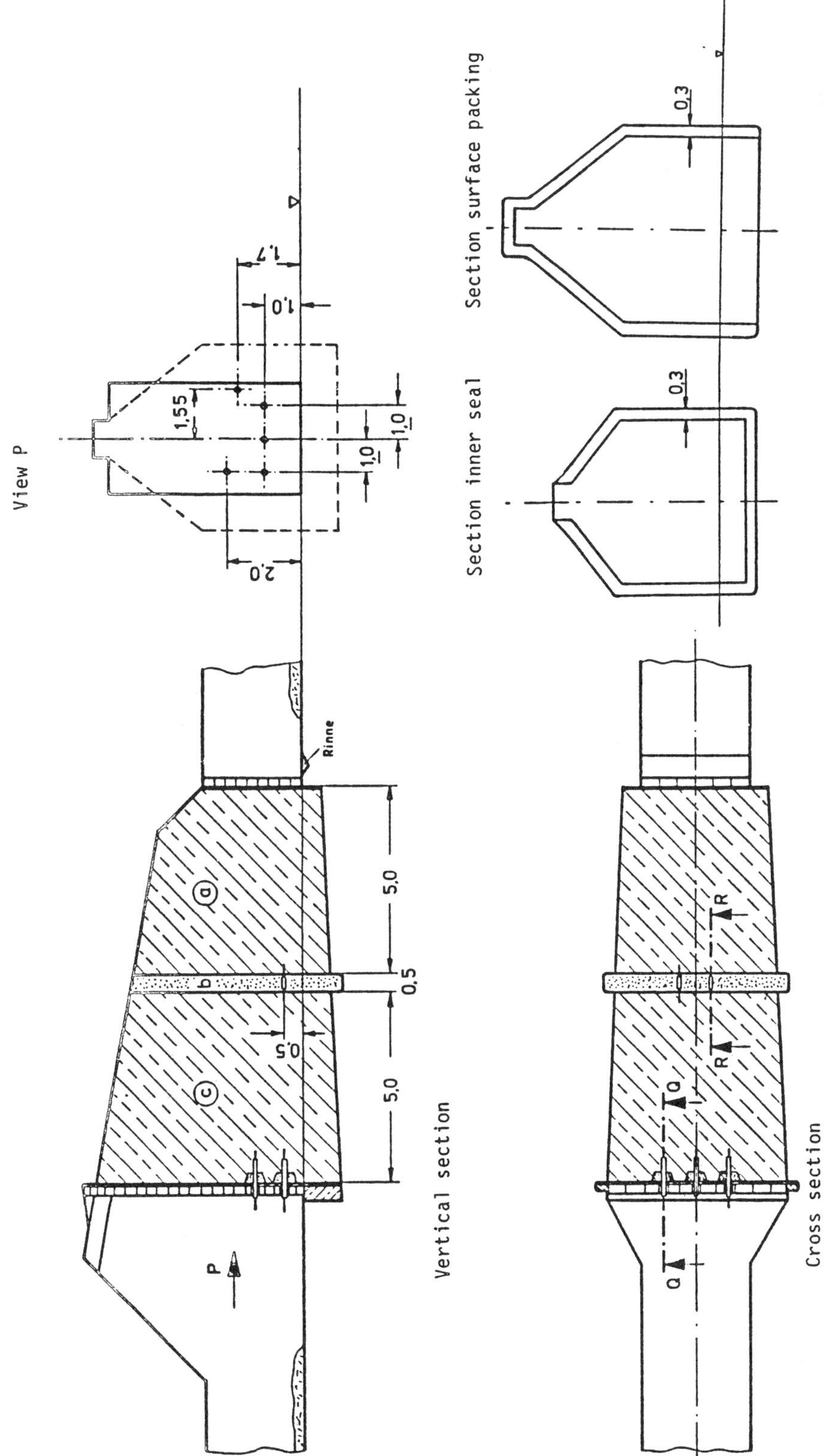

FIGURE 1. Cross- and Vertical Sections of the Bulkhead

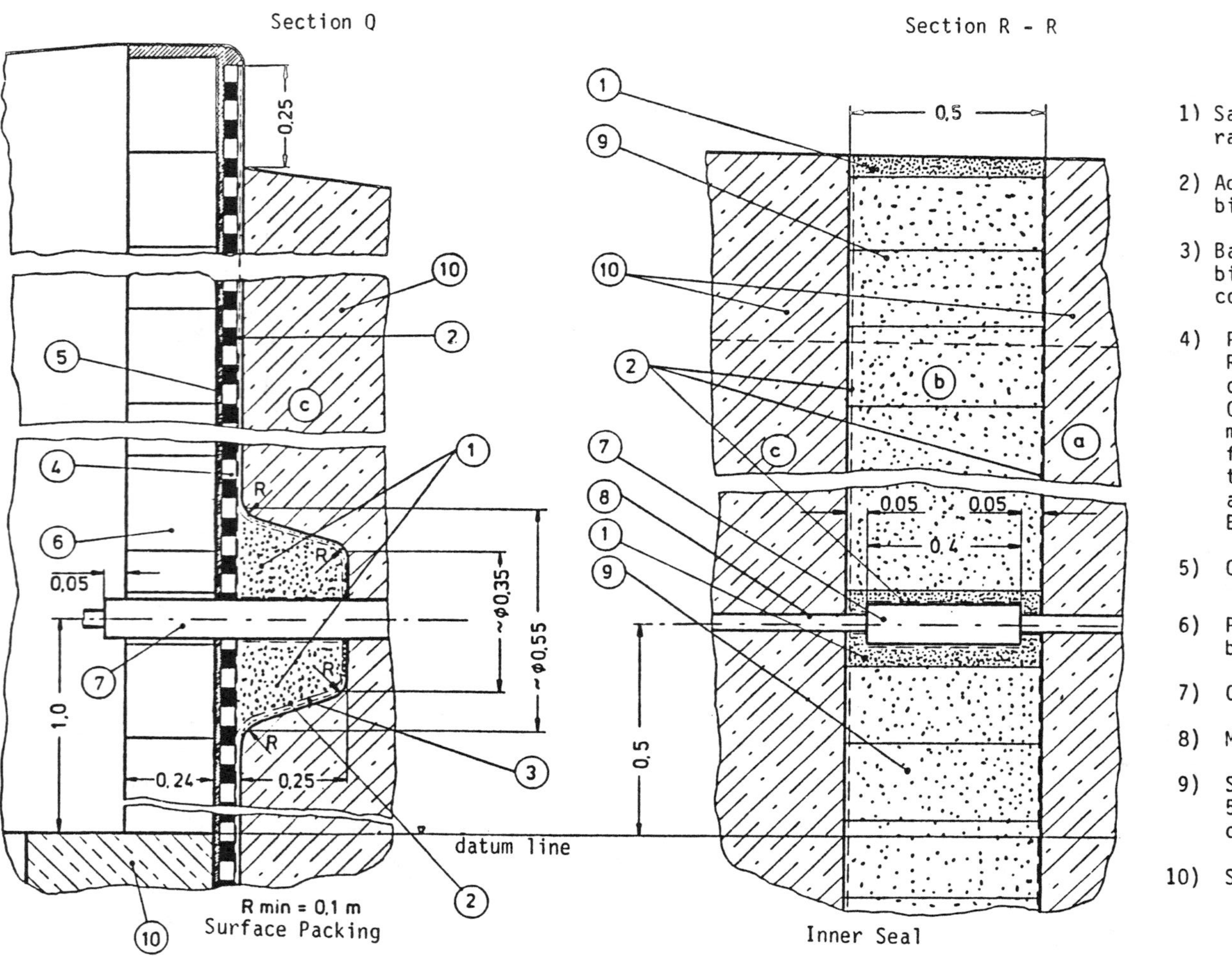

1) Sand asphalt. compacted void ratio < 5 %

2) Adhesive coating of cold bitumen

3) Balancing coating of hot bitumen, possibly two coatings

4) Packer of bitumen layers R 500 N and metallic tape of superior alloy steel 0,05 mm thick, whereby the metallic tapes - as seen from the brine side - have to be installed in the 2nd and 4th layer. Embedding acc. to DIN 18 195

5) Cement jointing 4 cm

6) Protective layer of YTONG-bricks (cemented)

7) Cable passages

8) Measuring circuits

9) Sand asphalt, of bricks 5/ cm x 20 cm x 20 cm cemented in a hot state

10) Salt concrete

FIGURE 2. Details of the Seals

The samples extracted for product control purposes during installation showed a very uniform product with the following properties:

medium bulk density	:	1.94 kg/dm^3
medium dynamic Young's Modulus	:	14.55 GPa
uniaxial compression strength	:	23.71 N/mm^2
tensile fracture strength	:	1.85 N/mm^2
medium permeability with rock salt brine	:	10^{-11} M/s
open porosity	:	4.4%
closed porosity	:	10.1%

B. Sand Asphalt

The following requirements are made upon the inner seal of sand asphalt:

- high impermeability,
- good impermeability after installation,
- speedy closure of pores in the contact zone,
- no migration of bitumen into the rock and into the adjoining structural parts,
- no demixing, resp. sedimentation of the filler,
- adequate natural stability during installation, and
- upper limit for density (cavern content after installation <5%).

A series of pre-investigations resulted in an optimized mixture with the following percentages:

sand (including natural filler)	80 wt%
filler (powdered limestone)	20 wt%
Total mineral content	100 wt%

Bitumen B-80	10 wt%
Filler/bitumen ratio	2/1
Unit density	2.196 kg/dm^3
Cavern content after installation	<5 vol%

During the installation samples were taken and passed on to a laboratory for testing. The compositions investigated lay within the given range. Drilling core investigations showed a pore content of below 5%.

IV. INSTRUMENTATION

The following parameters were measured at the bulkhead:

- pressure distribution within the rock,
- pressure buildup at the supports,
- temperature curves in the bulkhead,
- density of the seal,
- dislocation of the construction, and
- level recording behind the bulkhead.

On the brine and the air sides as well as within the bulkhead 12 absolute pressure receivers are installed in boreholes on three measurement levels (Figure 3). They are to measure the rock pressure and possible changes due to inflowing brine. The load area is 100 x 200 mm. The measuring range between 0 and 200 bar is transformed into a measurement signal between 0 and 5 volts. Two measurement cross-sections within the bulkhead serve to measure the load-bearing capacity from the rock. In the longitudinal axis on the brine and the air sides as well as at the inner seal, strain gages are fitted vertically to the strain axis in order to measure the hydrostatic pressure and its effects.

The curing behavior of the concrete was recorded by temperature sensors. The maximum curing temperature was at approx. 72.4°C. It appeared in the center of building section A two days after termination of the concreting (Figure 4). These also serve to facilitate a temperature compensation for further measurement data. Moisture transducers are to indicate flow paths of the brine in the support on the brine intake side. They are fitted in pairs at different levels. From the measuring point of view the change of resistance of the dry concrete after curing as well as of the moist concrete after inflow of brine is fully utilized.

At the air side of the construction two independent measurement systems are installed in the form of level indicators. The filling levels can be indicated continuously as well as discontinuously. They are to determine the amount of possibly inflowing brine. Two redundantly designed distance gauges are to record possible dislocations of the construction on the brine side. A pressure receiver is to indicate the difference to the brine level (Figure 5).

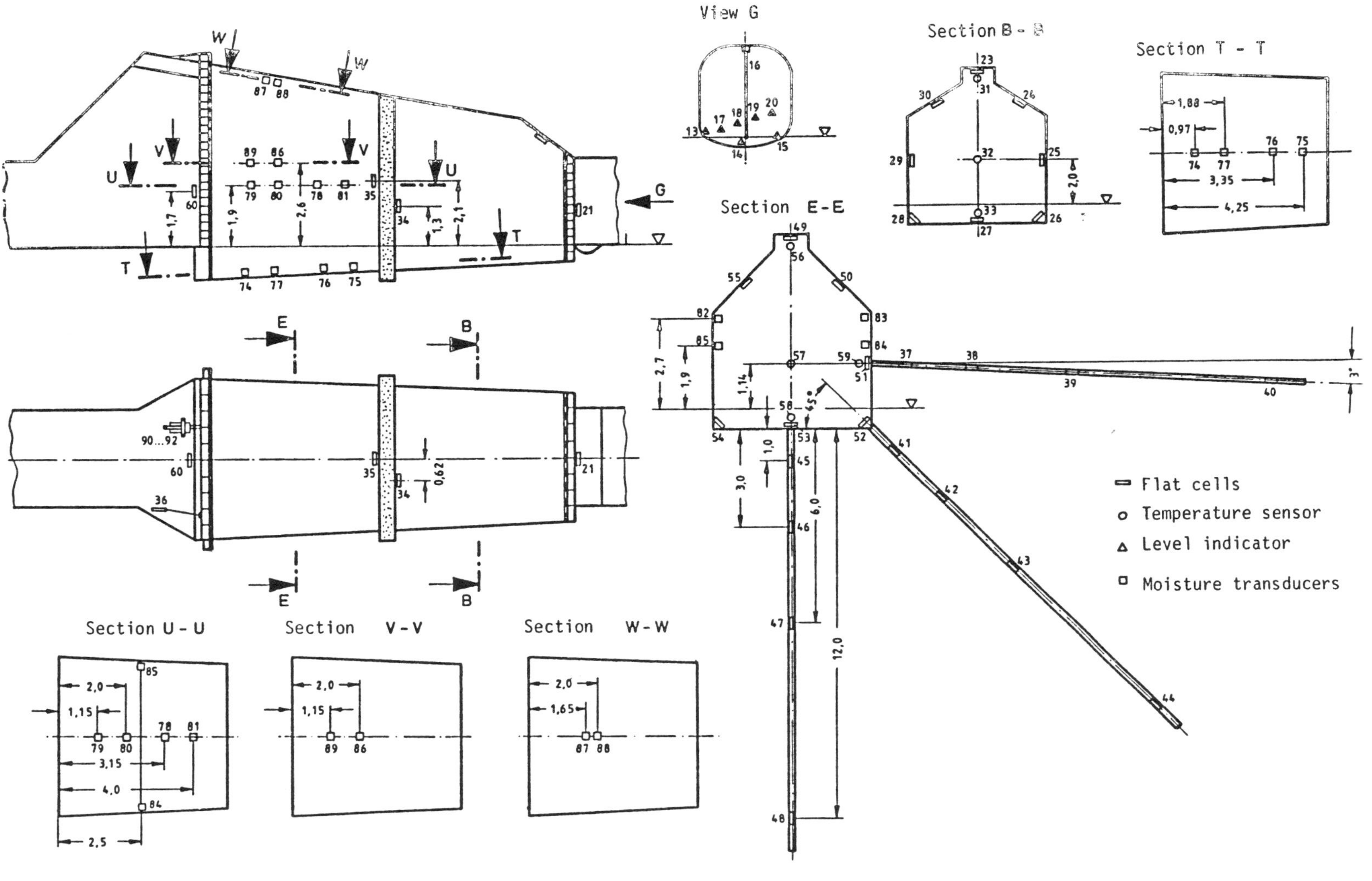

FIGURE 3. Instrumentation

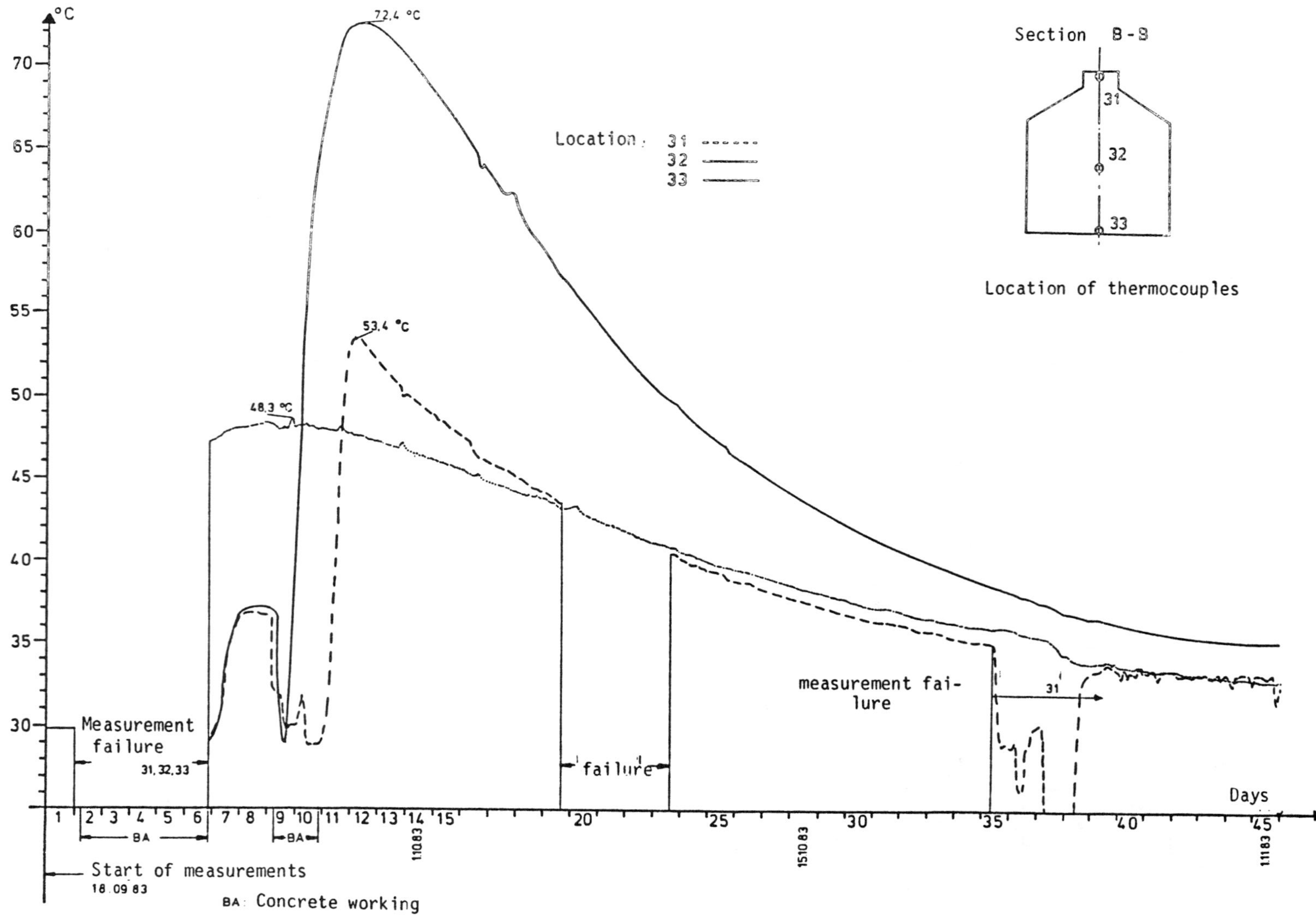

FIGURE 4. Curing Temperatures in Support A

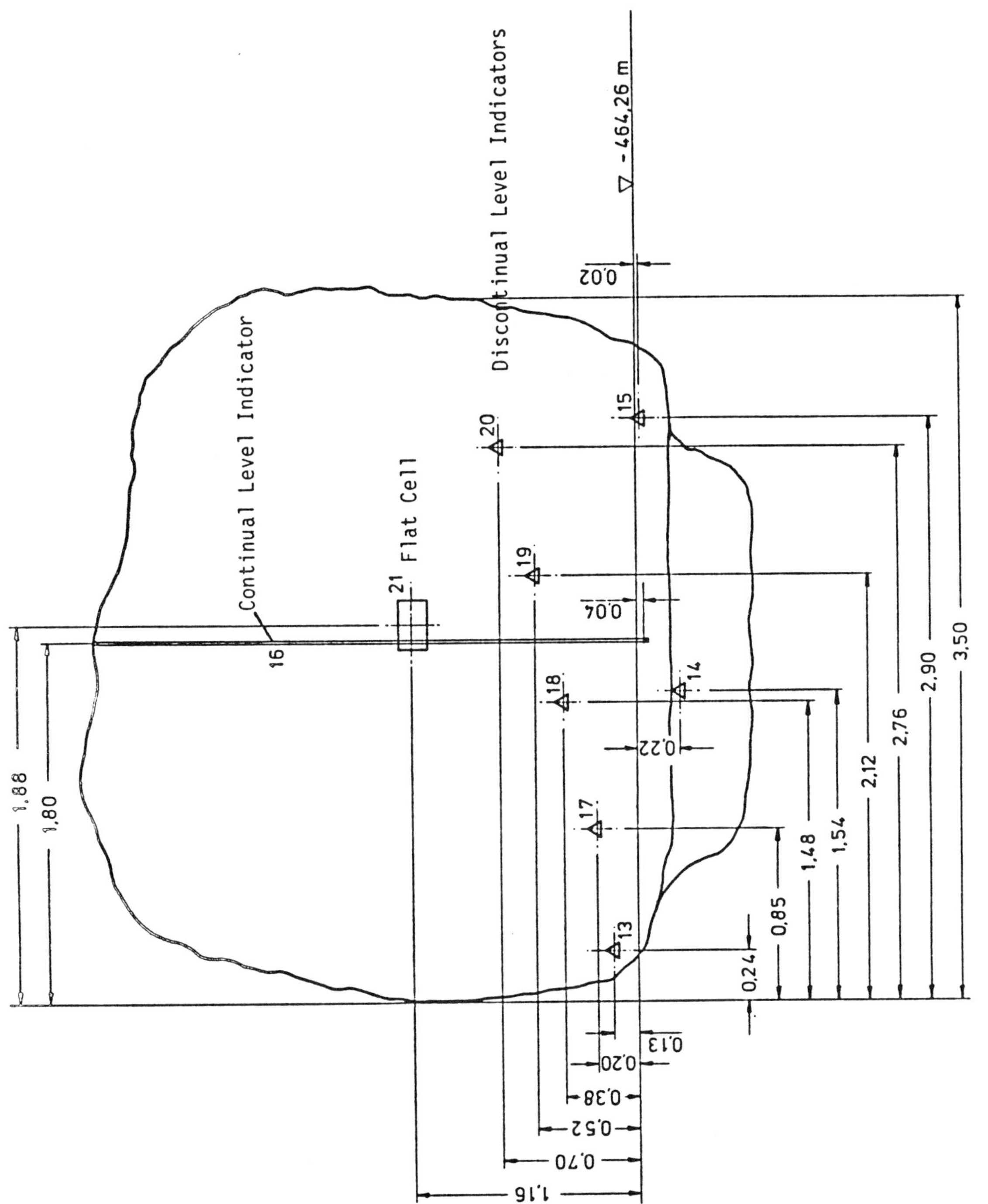

FIGURE 5. Instrumentation of the Air Side of the Bulkhead

The 91 measuring data can be obtained in a cycle between 1 and 24 h. The values are amplified in situ and transmitted for approx. 5 km to a measurement container and then to above ground by means of two independent cable runs. All data are recorded on magnetic tape and a direct control per monitor is given.

V. PRESENT STATUS AND OUTLOOK

After completion of the construction at the end of 1984 the measuring phase began during the day period. Several underground assignments for control surveying and repairs were required. Part of the instrumentation had to be shut down. After a measuring period of 17 months the construction was overflooded by about 30 m in March 1985. The intended brine pressure of 2.5 MPa is expected to be reached in June 1986. The convergences of 1 mm/year measured in the close vicinity showed a significant increase of the pressures on the gages 500 days after finishing contruction. A project-phase until 1989 is aimed at in order to prolong the time of measurement.

If, contrary to all expectations, so many instruments should fail that a definite statement on the density is no longer guaranteed, then the possibility of sinking a borehole in the air cavity behind the bulkhead construction is being considered.

RESEARCH WORK DURING AND AFTER THE FLOODING OF AN ABANDONED POTASH MINE IN NORTHERN GERMANY

Dr. H. J. Herbert
Gesellschaft für Strahlen- und Umweltforschung Munchen mbH
Theodor-Heuss-Strasse 4
3300 Braunschweig
Federal Republic of Germany

W. H. Stover
Kavernen Bau- und Betriebsgesellschaft mbH
Rathenaustrasse 13/14
3000 Hannover 1
Federal Republic of Germany

ABSTRACT

The abandoned potash mine of HOPE in northern Germany is flooded with an NaCl solution. A scientific research program running concurrently is to register and evaluate data on the geochemical, geomechanical and geophysical processes occuring before, during and after the flooding. In addition, a seal bulkhead was built using new materials and seals for testing under a liquid pressure of approximately 2.4 MPa. The work is part of a comprehensive research program investigating the processes occurring in a hypothetical water or brine inflow into a repository for radioactive waste in salt formations.

I. INTRODUCTION AND PROJECT DESCRIPTION

In conjunction with safety considerations for a final repository in salt formations - according to the concept of the Government of the Federal Republic of Germany - investigations into the hypothetical accident scenario "inflow of water or brine into a repository" are of particular significance. Although such a scenario is not very probable, it can, however, not be entirely excluded, especially in the post-operational phase.

The Kali und Salz AG potash salt mine in Hope, abandoned in 1982, was planned to be flooded with NaCl brine resulting from the solution mining of storage caverns in salt formations near Empelde (Figure 1) and thus provided an unique possibility of extending respective research work to the real scale of 1:1, offering the opportunity to compare and/or confirm previous findings, statements and scenarios.

The Hope saltdome (Figure 2) has an almost circular shape and covers an area of about 25 km^2 with a diameter of almost 6 km at its top. Seismic and gravimetric measurements together with deep drilling showed a marked salt overhang of the saltdome in southern, eastern and northern directions. The Zechstein base lies at a depth of 4,000 m. Active uplifting of the saltdome took place between the Triassic and the Muschelkalk. Besides an extensive remineralization (secondary halite, sylvinite, langbeinite, anhydrite, polyhalite), the intensive folding led to oil and gas migration into the salt along microscopic fissures.

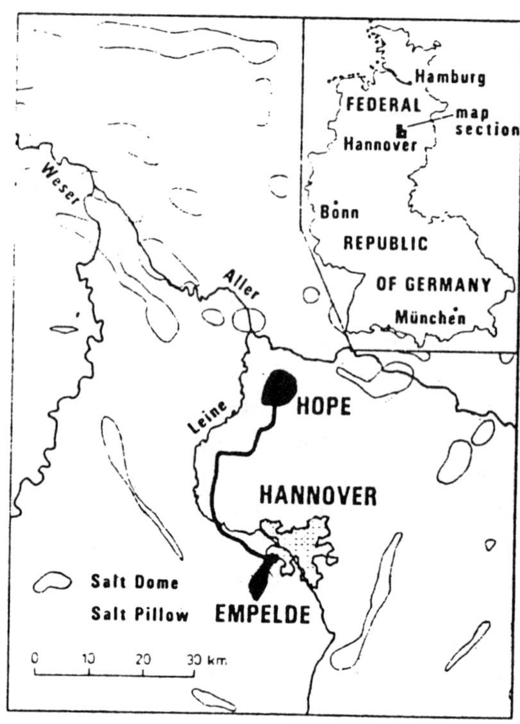

FIGURE 1. Above Ground Brine Pipeline Between Cavern Location at Empelde and Hope Saltdome

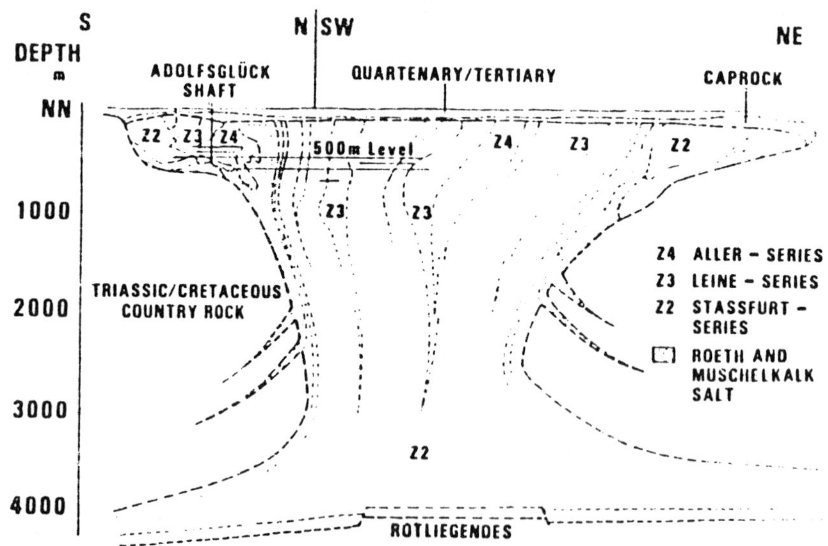

FIGURE 2. Saltdome Hope - Geological Section and Location of Hope Potash Mine

The Hope potash mine dates back to the year 1907, when the sinking of the Adolfsglück shaft began. In 1909 the construction of the second shaft at Hope was started. The mine was in production between 1914 and 1925 and between 1963 and 1981. From 1914 to 1925 about 300,000 t of potash salts with a K_2O content of 20-30% were mined in the southern field at the 400-m and 500-m levels. From 1963 to 1981 a second, steeply inclined deposit between 300 m and 700 m was mined. Production was about 2.9 million t of raw salt with an average K_2O content of 20%. The total mined out volume of the mine corresponds to roughly 1.6 million m^3.

Although the mineralogical and mining aspects of the Hope mine are not directly comparable with a repository for radioactive waste and the investigations concern mainly to the study of general phenomena, they allow a better understanding of the processes involved. The know-how gained at Hope can thus be integrated into the safety analyses of repositories for radioactive wastes in saltdomes.

In early 1983 a research and development program was prepared with the aim of compiling the most important data during and after the flooding of the mine; i.e., geochemical, geophysical, geomechanical and technical findings. The development, testing and implementation of instruments in highly concentrated brines under hydrostatic pressures between approximately 2 and 8 MPa is an additional part of this project, which is sponsored by the Federal Ministry of Research and Technology (BMFT).

A total of 18 measurement stations (Figure 3) were set up on four different levels in the mine (724 m, 621 m, 500 m and 324 m respectively), comprising 5 geochemical, 4 geomechanical and 8 geophysical stations as well as a dam with an elaborate measurement and instrumentation system. The instrumentation work in the Hope Mine and the data compilation station in the survey container aboveground was commenced in August 1983 and concluded successively between November 1983 and February 1984. Flooding started on March 12, 1984 and is scheduled to be terminated at the beginning of 1986.

II. GEOCHEMICAL SURVEY PROGRAM

A. Description

The direct objectives of the geochemical measurement program in Hope may be itemized as follows:

1. Observation of the chemical changes within the introduced NaCl-solution and the occurring dissolution and precipitation processes,

2. Comparison of the theoretical estimations of changes in the solution's composition up to the final equilibrium solution with actual in-situ behaviour,

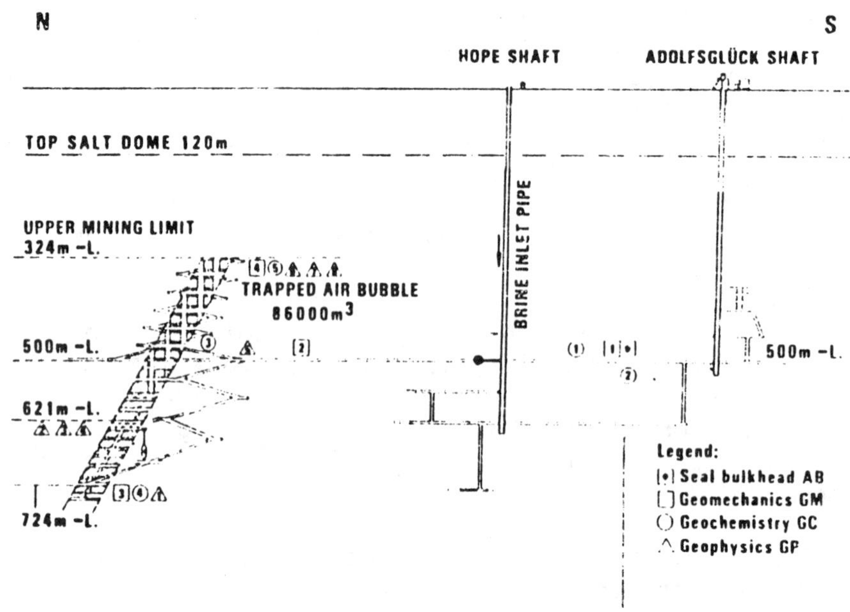

FIGURE 3. Schematic Drawing of Hope Mine and Positions of Measurement Sections

3. Determination of the time required for solutions to achieve saturation with respect to the exposed salt minerals within the mine,

4. Study of the material and heat transport in flooded non-backfilled galleries, chambers and shafts, and

5. Observation of an air bubble formed in the mine during flooding.

Four of the five geochemical survey stations work under brine whereas the fifth records at the highest point of the mine (324 m level) the behaviour of the air bubble. Temperature, pressure, conductivity, sound velocity and flow velocity in the x-, y-, and z-direction are measured in the brine at four locations having different mineralogical composition. The data collected provides both qualitative and semi-quantitative information on dissolution, precipitation, temperature distribution and transport phenomena in the flooded mine. In the air bubble the temperature, pressure, relative humidity, and brine level are measured. In addition to surveys in the mine using fixed probes, surveys are also carried out from above ground using mobile probes in two shafts of the mine. In doing so, continuous profiles down to a depth of 500 m are recorded including density, pH- and Eh-values. Sampling and chemical analyses of the samples taken from various depths form a further part of the program.

B. First Results

Results to date are summarized in Figures 4 and 5. Figure 4 shows the changes of the physical parameters in the NaCl solution introduced into the mine in a sylvinite mining chamber at a depth of 724 m.

The solution temperature shows an initial sharp rise, then gradually stabilizes. This temperature increase generated by the rock heat, results in a simultaneous increase in electrical conductivity of the solution (line 2a). The percentage temperature dependency is shown by curve 2b. The increase in salt content, and the associated change in conductivity, only take place slowly. This is probably related to the solution absorbing KCl at the same rate as NaCl precipitates. The curve 3 plot in Figure 4 shows the change of sound velocity in the solution. This depends on the density, but also on the compressibility and hence on the pressure.

The results of analyses made on various samples are marked in Figure 5. These samples were obtained during the flooding from interesting points along the flow path through the mine.

With increasing distance from the inlet point there is a marked increase of potassium content, particularly in the area of the sylvinite deposit. Together with the rise in concentration of sulphate and magnesium the composition changes correspond on the whole with the processes as predicted. The increase in concentrations corresponds with the known relationships between the different dissolving rates and the respective minerals present.

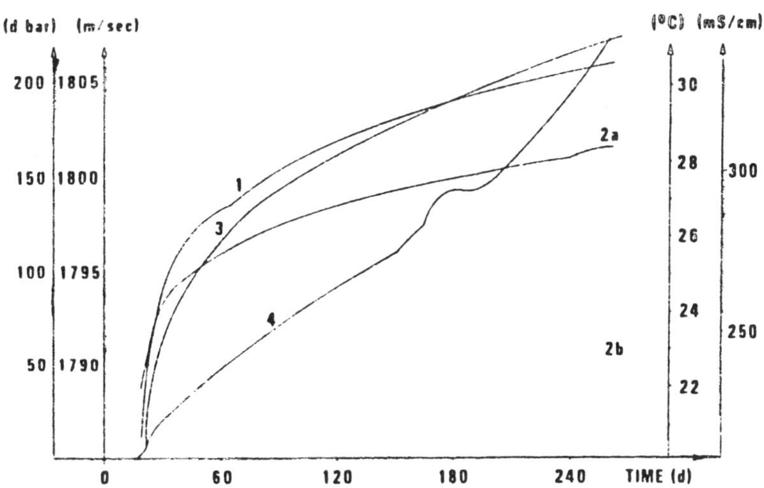

FIGURE 4. Changes in the Physical Parameters of the NaCl Solution in a Sylvinite Area at a Depth of 724 m. 1-temperature, 2a=conductivity, 2b=conductivity without temperature Influence, 3=sound velocity, 4=pressure.

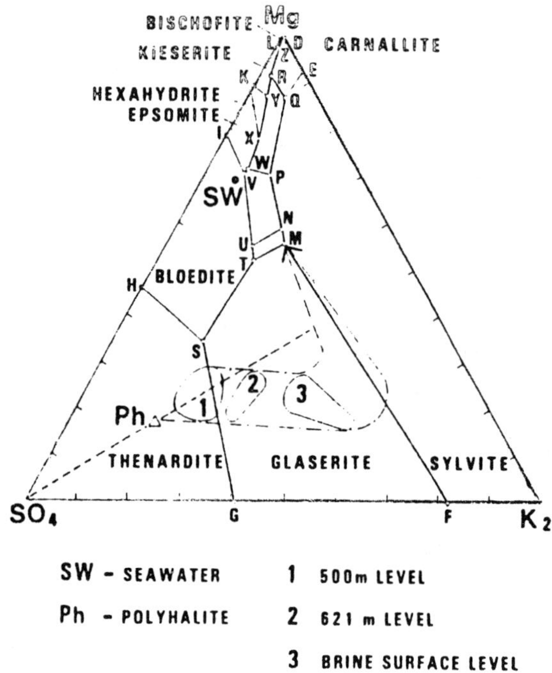

FIGURE 5. Development of the NaCl Solution Along the Flow Paths During the Flooding

The samples taken near the inlet on the 500-m level lie near the point representing polyhalite. This shows that the K_2, Mg and SO_4 contents of the inlet solution were taken up due to incongruent polyhalite decomposition during the leaching process, and further, that in the 500-m level no appreciable dissolution of potassium or magnesium salts is taking place.

As the solution travels further, it comes into contact with the sylvinite deposit where potassium is dissolved faster than Mg and sulphate and so the solution concentrations move towards the point marking sylvite. Upon longer contact with the sylvite the solution will reach the sylvite saturation plane whereby it forms the first constant solution at point M (Figure 5). The metamorphosis of the solution does, however, run in the direction predicted even though the rates of dissolution are smaller than expected.

III. GEOMECHANICAL SURVEY PROGRAM

A. Description

Despite the existence of a number of flooded mines there are few data available concerning the stress/deformation behaviour under such conditions. The geomechanical measurements undertaken in the Hope Mine are thus intended to allow local and general observations with the following direct objectives:

- Test of fixed geotechnical measurement devices in situ in an air bubble to determine the suitability and reliability of standard geotechnical measuring equipment for long-term observations.

- Recording the stress/deformation behaviour in the rock salt at 4 selected cross sections in drifts with increasing pneumatic and hydraulic internal pressure.

- Preparation of in-situ measurement data before, during and after the flooding for the computer models running concurrent to the project.

Four geomechanical survey cross-sections (Figures 3, 6) one on the 724-m level, two on the 500 m level and one in the air bubble on the 324-m level, record the deformation behaviour of the rock salt since November 1983. At hourly intervals surveys are made of the volumetric convergence and inclinometer/extensometer and dilatometer recordings are taken.

B. First Results

Figure 6 shows the basic set up of a geomechanical measurement cross section. Figures 7 and 8 show the first results of the convergence and extensometer measurements before and after the flooding of the cross section. The results obtained so far may, by way of example, be characterized by data collected at measurement cross section GM3 situated at the lowest point of the salt mine (724 m level) and first covered by brine.

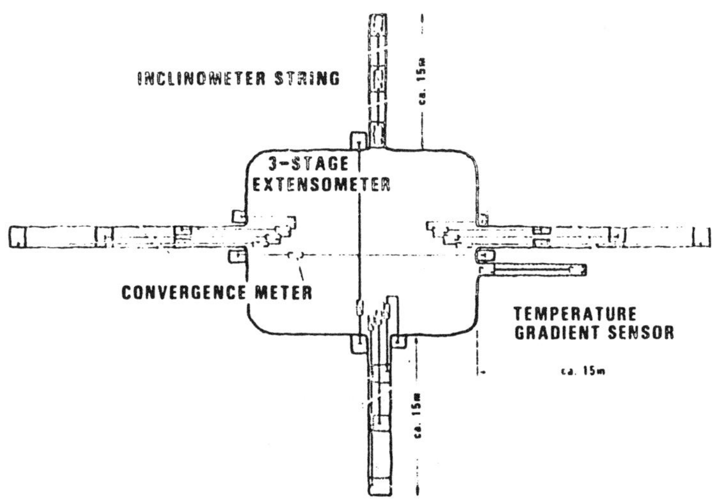

FIGURE 6. Geomechanics - Measurement Cross Section Basic Design

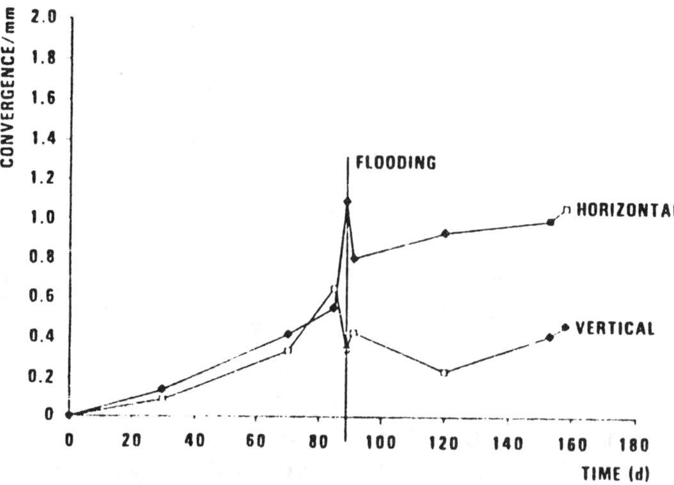

FIGURE 7. Convergence Measurements on the 724-m Level

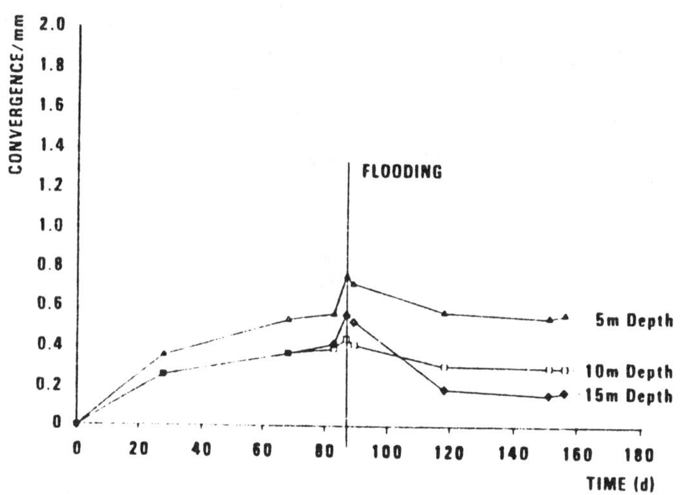

FIGURE 8. Extensometer Measurements on the 724-m Level

Prior to flooding and immersion this measurement cross section supplied data from the dry mine for 85 days (Figure 7). In this period a convergence of roughly 0.8 mm was detected. After start of flooding a convergence decrease in the vertical and an increase in the horizontal can be clearly identified. After only a short flooding time the horizontal salt deformation rate fell to 0.2 mm in 80 days. The extensometers in both the floor and the wall clearly indicated the start of the flooding (Figure 8). They also revealed a reduction in the deformation during flooding. After 160 days in service the measurement cross section GM3 unfortunately failed because of electrical malfunction.

Concurrent to the in-situ surveys, accompanying calculations were performed by the Institute for Statics of the Technical University of Braunschweig. As a preliminary result, the good correspondence in the parameters between the measured and calculated convergence values is worth mentioning.

IV. GEOPHYSICAL SURVEY PROGRAM

A. Description

In the geophysical survey program the microseismic behaviour was recorded before, during and after the flooding using 5 single component geophones on the 724-m, 621-m and 500-m levels as well as a 3-component geophone on the 321 m level (Figures 3, 10).

B. First Results

Valuable measurements have been available since February 1, 1984. Prior to the flooding the number of recorded incidents in Hope was low, but of a similar quantity as in the Asse salt mine. However, this number rose sharply immediately after the start of the flooding. During short interruptions in the flooding operations, a marked decline in these events also occurred, that rose again immediately upon renewed filling with brine (Figure 9). This is a clear indication of the relationship between the flooding and the event frequency.

The majority of events was of short duration (approx. 100 ms) and had frequencies around 200 Hz. The characteristics of the seismograms were not different before or after the flooding. Around mid May 1984 a series of low frequency (approx. 50 Hz) and longer (approx. 30 s) events was recorded. It was possible to identify some as rock falls, but the majority of events is thought to be due to microfracturing induced by the flooding.

Figure 10 shows the depth of the event locations. It is noticeable here that the majority have a depth around 550 m below sea level. This zone was also active prior to flooding. An explanation could be the statics of the mine excavation and the geological structures. These points will have to be clarified during the continuing measurement and evaluation program.

The following explanations for the activities recorded are possible:

- leaching of load-bearing rock sections on the flow paths or in the flooded section of the mine,

- formation of cracks due to volume changes caused by cooling of the rock mass (some of the events located in the roads could be accounted for by such a process),

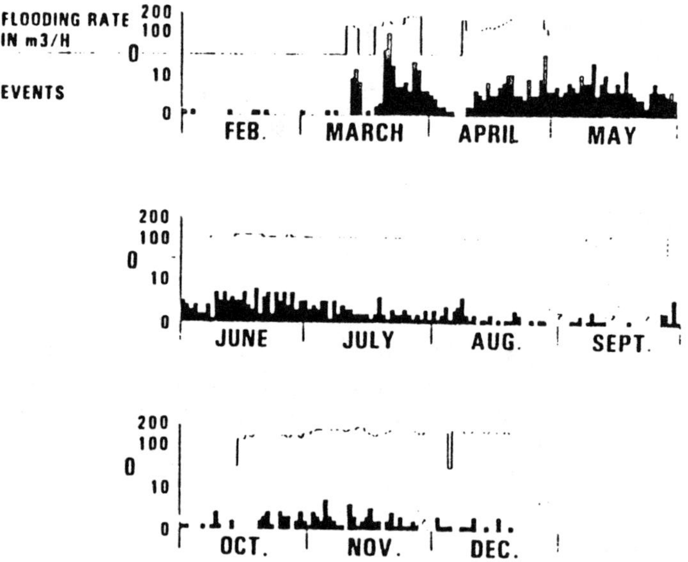

FIGURE 9. Registered Microseismic Events and Brine Fill Rates

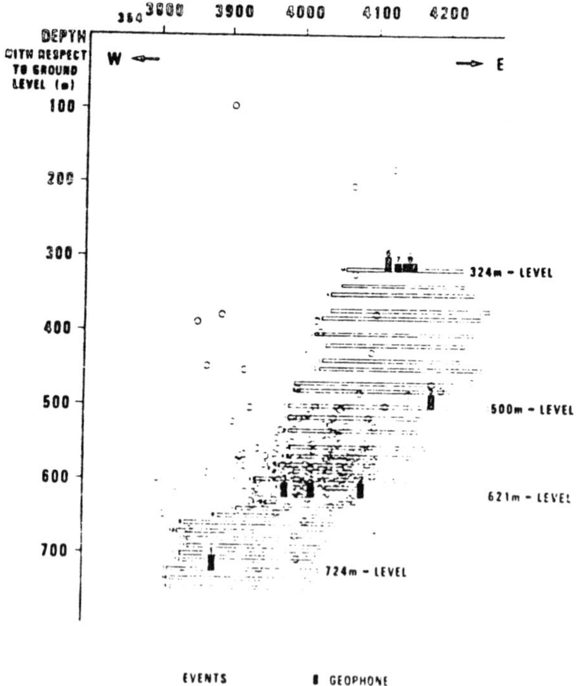

FIGURE 10. Localization of the Microseismic Events

- stress changes due to the mass of the brine introduced (approx. 2,500 to 4,500 tons per day) (this is presently regarded as the most likely cause. The immediate start of activities parallel to start of flooding could be explained thus; exact calculations have, however, still to be carried out), and

- sliding on previously existing cracks, produced by the increased relative humidity (this effect could be classed only as a supporting effect since the relative humidity climbed sharply at start of flooding).

V. CONSTRUCTION AND OBSERVATION OF A BULKHEAD

The seal-bulkhead represents a very costly section of the Hope project as far as its design, performance, control and instrumentation are concerned. Details concerning this part of the project are presented by W. Fischle in a separate paper entitled "First Test of a Drift Seal under Brine Inflow."

VI. CONCLUSION

Although the measuring program of the project Hope is still continuing and the evaluation and assessment of the results is still under way the overall results obtained in the project during the first two years are without question very positive and valuable. From a technical point of view many experiences were gained and documented that could be of great value in the realization of similar projects in the future. From a research point of view some of the objectives set have already been realized and further interesting results are expected. All of which will be of great importance in the assessment of the long-term safety of radwaste repositories in salt formations.

VII. ACKNOWLEDGMENT

Messrs. Dr. Flach, C. Keick, Th. Meyer, W. Sander and M. W. Schmidt were actively involved in the geochemical, geomechanical, and geophysical investigations. Their important contributions to this report are highly acknowledged.

REFERENCE

1. C. Heick and B. Hente, "Accompanying Geophysical Observations During the Flooding of a Salt Mine," at the Second Conference on the Mechanical Behaviour of Salt, September 24-29, Hanover, F.R.G. (1984).

MECHANICAL BEHAVIOR OF GALLERIES IN DEEP CLAY - STUDY OF A CONCRETE EXAMPLE

G. Rousset
L.M.S.
Ecole Polytechnique
91128 Palaiseau Cedex, France

R. Andre Jehan
J.C. Fernique
A.N.D.R.A
33, rue de la Federation
75015 Paris, France

A. Bonne
C.E.N./S.C.K.
Boeretang 200
B-2400 Mol, Belgium

ABSTRACT

At important depths such as those envisaged for the construction of a radioactive waste disposal, clay appears to be a material of rather weak resistance. One of the first problems to be studied is the technical feasibility of a disposal facility. The time-dependent and strain-softening behavior of clay plays a significant role in the long term stability of the lined galleries. The in-situ measurements carried out during the digging of a gallery in Boom clay (Belgium), under a 250 m overburden are presented. The data obtained are analyzed by means of an elastoviscoplastic model including strain softening behavior of the clay.

I. INTRODUCTION

In the case of clay, more than in any other media considered for a disposal (granite, salt), the problem of the ability to construct a facility at several hundred meters depth is a question to be solved with high priority. Even in the case of strongly consolidated clays, it is known that their mechanical resistance is generally weak at such depths. Assuming that well-known techniques could be implemented to build such underground disposal facilities at a reasonable cost, the stability of the structures must also be secured for several decades. It is therefore necessary to adequately line the excavated cavities. An optimal dimensioning of the cavities and of their support can be determined for a given clay by using the long-term mechanical properties of this material, thus providing the data necessary for estimating the cost related to the gallery construction.

II. THE DIMENSIONING OF THE DISPOSAL WORKS

This behavior reflects that of the surrounding rock mass itself and of the interaction between the latter and the lining.

A. Rheology of a Deep Clay

One of the most important characteristics of clays is their time dependent behavior[1]. Undrained triaxial creep tests show that for a given state of stress, samples may undergo delayed strain over a long period. In some cases, this strain stabilizes after a certain time, in other cases it leads to failure. These time-dependent phenomena are also observed in situ. The driving of a cavity is accompanied by a progressive modification of the stresses in the clay which leads to strains within the latter. The strains result in a convergence of the gallery walls, a phenomenon which increases with time. The distinct softening of these materials is a second essential aspect of the time-dependent behavior and has a decisive importance on the long term stability of the galleries.[2] This softening means that the material loses its mechanical resistance when plastic deformation occurs.

During triaxial laboratory tests at an imposed strain rate that a progressive decrease of the axial stress occurs from a maximal value σ_M to a residual value σ_R (Figure 1).

The increase in the strain rate leads to failure at the end of a creep test, where the stresses σ_1 and σ_2 are kept constant (tertiary creep phase shown in Figure 2).

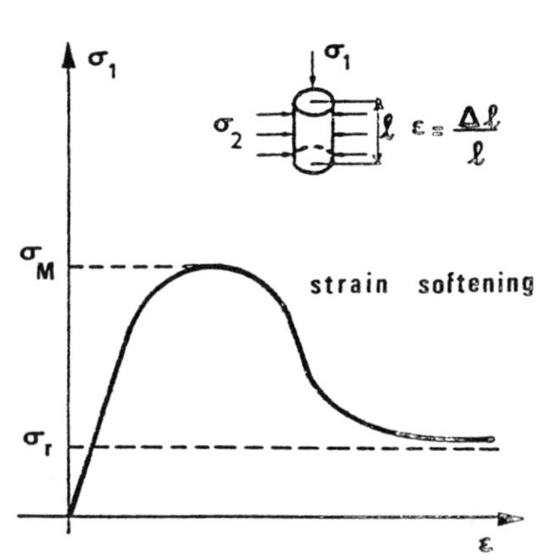

FIGURE 1. Triaxial Creep (σ_2 and ε constant)

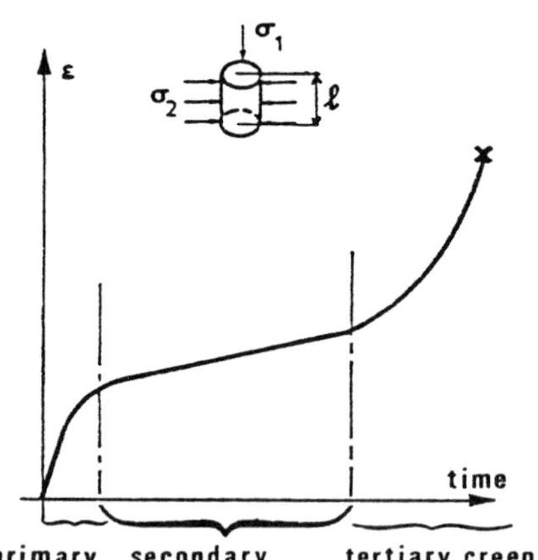

FIGURE 2. Triaxial Creep (σ_1 and σ_2 are constant)

In situ, the softening effect appears in the zone showing the greatest deformation, which is near the cavity. It is characterized by the occurrence of fracture planes in this zone and of spalling of the walls. On a real scale it is reasonable not to particularize each plane of fracturing. Such behavior may be considered as a progressive damage and the material may be considered as homogeneous. The effect of this damage on the mechanical properties of the rock is reflected by a loss of strength during an increasing deformation.

B. Interactions between the Clay Mass and the Lining

Taking into account the delayed response of the clay, which has been mentioned above, the walls of the galleries converge during and after the digging. If a lining is put at the appointed time, contact pressures will gradually buildup and thus limit the convergence of the gallery. After a transition period lasting one or several years, the lining will be submitted to the long-term pressure from the clay mass.

Actions taken to reduce the value of this limit pressure can be based upon the following two parameters:

- First, the mean stiffness of the lining; for a low value of this parameter, the rate of the pressure buildup and the limit pressure will be lower than for a higher value of this parameter. Whatever the nature of the lining may be, there is a compromise to be found between its flexibility and its mechanical resistance.

- The second parameter which can be determined is the time lag necessary for establishing the contact between the lining and the clay mass.

Indeed, if the initial diameter of the excavation is larger than that of the lining or if the zone included between these two diameters is backfilled with a very compressible material, significant plastic strains will develop in the clay mass surrounding the cavity. Then, as soon as the lining is in contact with the clay body, the rate of the confining pressure decreases, and, if the long-term cohesion of the plastified material is adequate, the limit value of this parameter can be much lower than in the case where the lining was put in place immediately after the excavation.

III. MODELLING OF THE BEHAVIOR OF A DEEP CLAY

Our purpose here is to propose a model of the mechanical behavior of a deep clay which accounts for the main properties mentioned above: the time-dependent phenomena, the strain softening and the angle of internal friction.

A. Rheological Model

Figure 3 shows the principle of the proposed elasto-visco-plastic model:[2] if the applied stress F is inferior to the yield point C of the slider (P), only the linear elastic strain is considered ($\varepsilon = F/E$). Upwards from the value C, a viscoplastic strain ε^{vp} superposes on the elastic one ε^{e}.

The softening behavior is accounted for only at the level of the plastic slider. Indeed one may consider that the yield point of the slider is a decreasing function of the strain. More precisely, this yield point is constant, equal to C, in the range $(0, \alpha_1)$ of ε^{vp}, then changes linearly from the value C to the residual value C_0 which is reached when ε^{vp} goes beyond the value α_2 (Figure 4).

The responses of the model to the effects of a given prescribed strain rate or a given stress (creep) are shown on Figure 5. This model does not account for the primary creep phase observed in laboratory, the duration of which is generally much shorter than the duration of the test. However, it must be noted that neglecting these primary delayed effects is still consistent with the long term modeling purpose. The stationary creep phase (secondary phase) comes to an end when the viscoplastic strain goes beyond the α_1 value; the accelerated creep phase (tertiary phase) corresponds to the softening behavior.

B. Generalization to the Three Dimensional Case

Let us maintain the decomposition of the strain tensor in a linear elastic part ε^{e} and a viscoplastic part ε^{vp}:

$$\varepsilon = \varepsilon^{e} + \varepsilon^{vp} \tag{1}$$

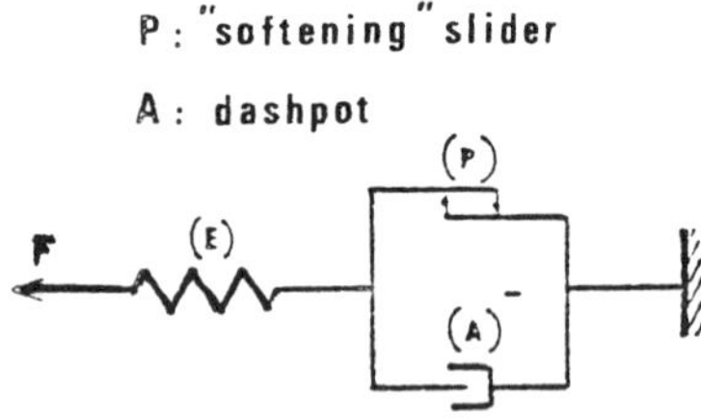

Figure 3. Undimensional Model

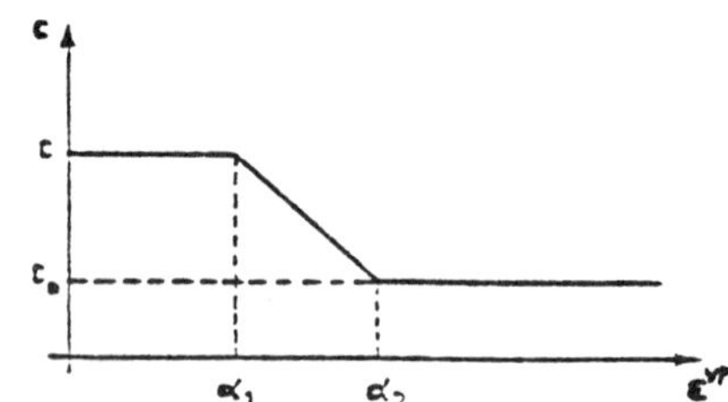

FIGURE 4. Variation of the Yield Point According to the Viscoplastic Strain

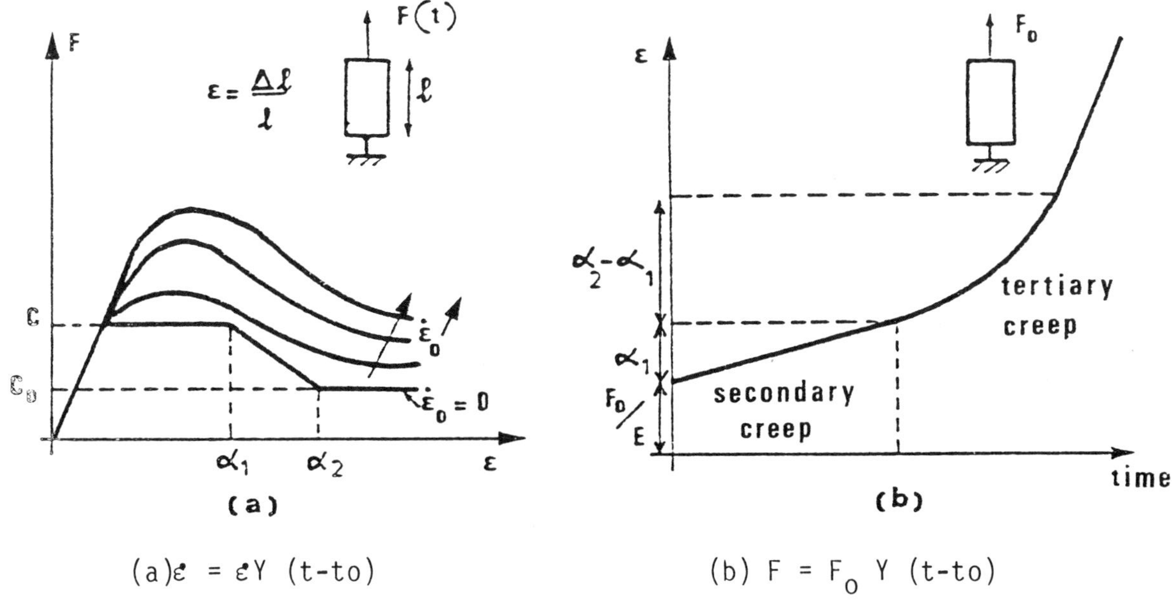

(a) $\dot{\varepsilon} = \dot{\varepsilon} Y\ (t-to)$ (b) $F = F_0\ Y\ (t-to)$

FIGURE 5. Responses of the Model

The viscoplastic behavior law is generalized by writing it as a flow law with a modulus and a direction (this law generalizes the Perzyna one):[4]

$$\frac{\delta \varepsilon^{vp}}{\delta t} = \frac{1}{\eta} < f(\sigma,\alpha) > n \frac{\delta g}{\delta \sigma} (\sigma,\alpha) \tag{2}$$

($<x> = x$ if $x \geq = 0$; $<x> = 0$ if $x < 0$)

α being an internal parameter accounting for the softening.

In the unidimensional case, α equals to the viscoplastic strain modulus. According to the laboratory results, the viscoplastic criterion f is a Coulomb criterion (cohesion C, angle of friction ϕ) which is written:

$$f(\sigma,\alpha) = \sigma_1 - \sigma_3 + (K_p - 1)\ [\sigma_1 - H(\alpha)] \tag{3}$$

The softening is supposed to affect only the cohesion value. As in the unidimensional case, it only depends of the viscoplastic state of strains. The given α parameter is an equivalent viscoplastic strain:

$$\alpha = \int_0^t \sum_{i=1}^{n} |\dot{\varepsilon}|\ dt \tag{4}$$

(σ_i and ε_i are the main values of the tensors σ and ε).

The viscosity of the dashpot A may possibly be non-linear: its response ε to a constant effect σ is such as:

$$\dot{\varepsilon} = \frac{1}{\eta} \sigma^n \qquad (5) \text{ (Norton Law)}$$

Where η is a viscosity parameter.

The non-linearity of the dashpot reflects a phenomenon which is usually observed in laboratory: During the triaxial creep tests, the ratio of strain rate to an imposed stress deviator is indeed an increasing function of the latter.

The potential g depends only on the major and minor principal stresses σ_1 and σ_3 so that:

$$g(\sigma) = \beta\sigma_1 - \sigma_3 \qquad (6)$$

explains a possible dilatancy of the material.

IV. STUDY OF A SPECIFIC CASE

A cooperation between ANDRA of France and CEN-SCK of Belgium enabled the setting up of a joint program on in situ geotechnical measurements in a deep clay formation. In Belgium, an underground experimental facility in clay is operational and the experiments of interest to us were performed at a depth of 250 m below ground level in the so called Boom clay. A schematic view of the facility is given in Figure 6. The construction of the facility has been completed in the framework of a contract between SCK/CEN and CEC. The instrumentation setup aims at acquiring long term data on the behavior of a small lined gallery of 2-m outer diameter and 7-m long (Figure 7).

A. In-Situ Measurements

The in-situ geotechnical experimental program described in reference[5] mainly concerns:

1. The displacements within the clay mass (extensometric measurements)

   The five cells of the extensometer yield very precise and reliable measurements. Figure 8 shows the displacements in millimeters of these 5 points during the drift digging phase. From the recording, an acceleration of the displacements can be observed when the tunnel face gets under the measuring cells. Figure 9 shows these measurements over a year. After a period of slowing down (between 13 and 60 days), these 5 points seem to go on moving but at very low rates.

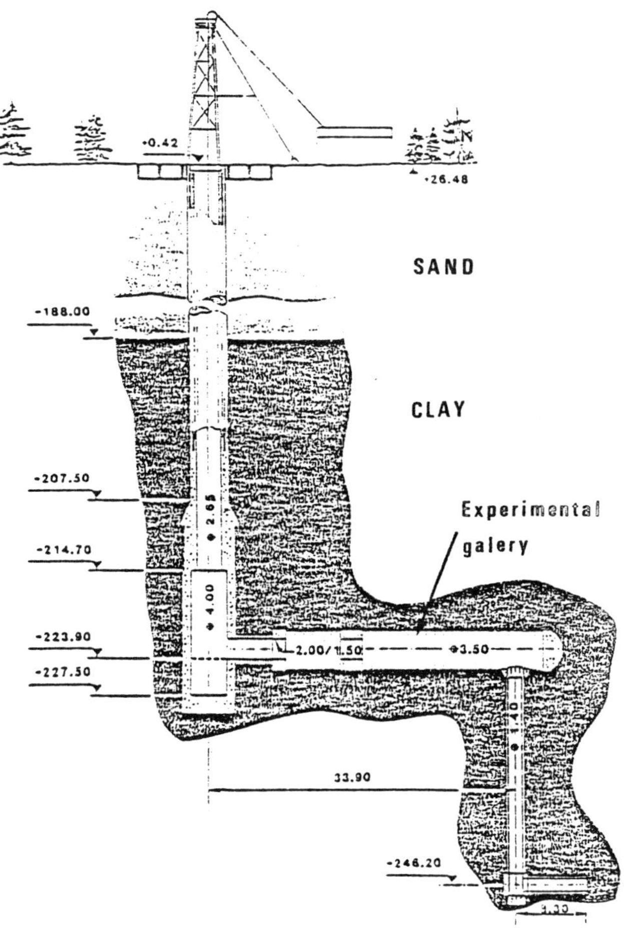

FIGURE 6. Plan of the Mol Facilities

2. The pressure at the extrados of the lining (Figure 10) and that in the lining (Figure 11)

   After an early phase of important variations between 0 and 20 days, the evolution rate of the parameters is perceptibly constant.

   These two series of measurements do not correlate very well, but by favoring force measurements, which are the most direct ones, it can be assessed that the mean pressure exerted by the massif on the lining is inferior to 1 MPa, one year after installation of the work.

3. The convergence of the drift lining

   Each ring of the lining is made of 20 prefabricated concrete blocks (inner diameter 1.4 m). The diametric variations measured on the 10 pairs of reference points are represented in Figure 12. Moreover, the convergence of an unlined cavity was measured by the instrumentation installed in a small diameter (15 cm) borehole (Figure 13).

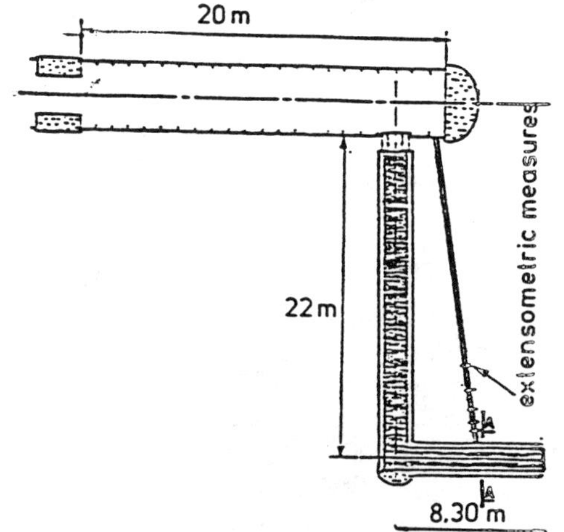

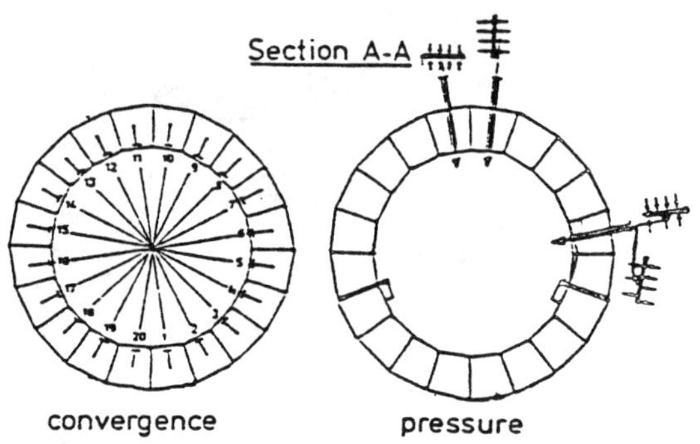

FIGURE 7. Measurements Program

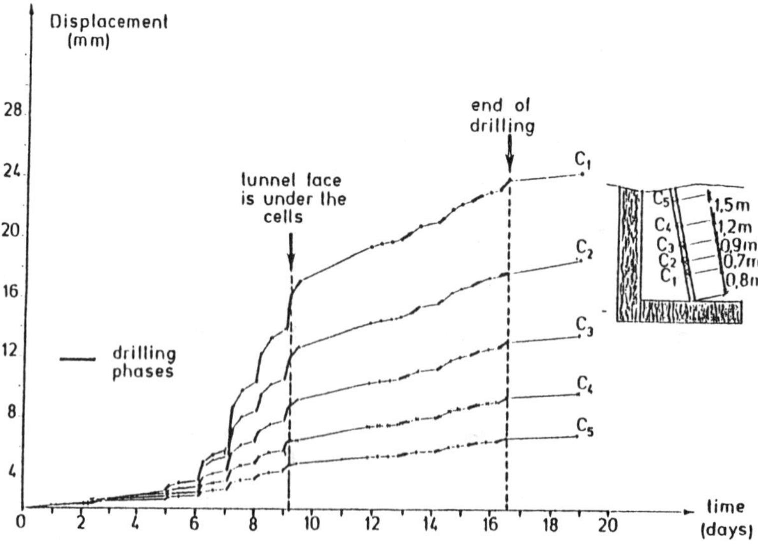

FIGURE 8. Short-Term Displacements

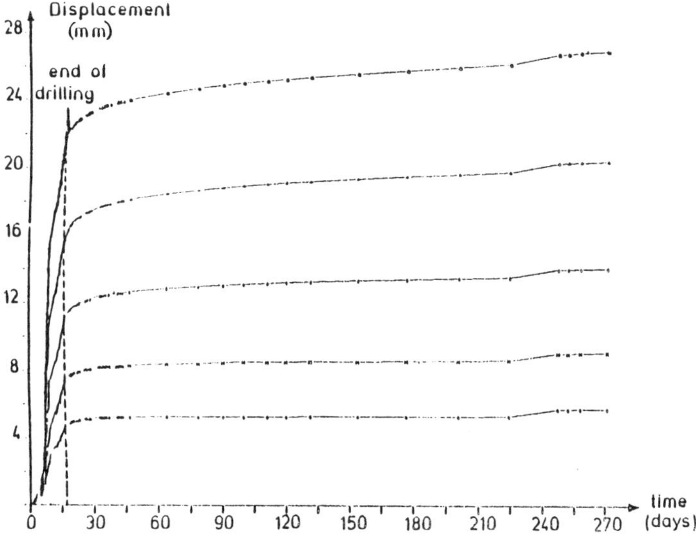

FIGURE 9. Long-Term Displacements

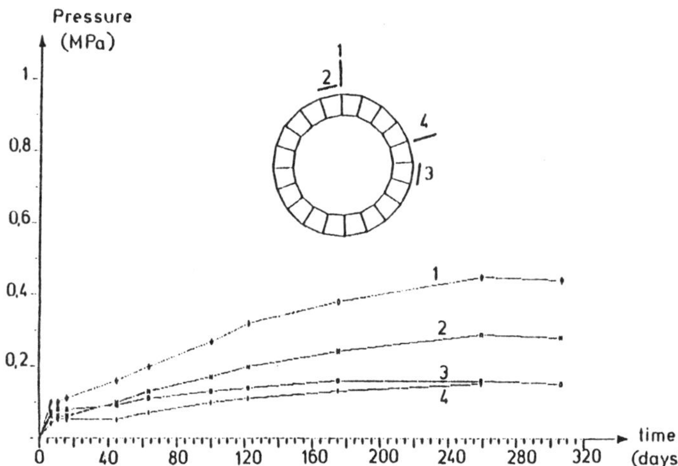

FIGURE 10. Pressure at the Extrados of the Lining

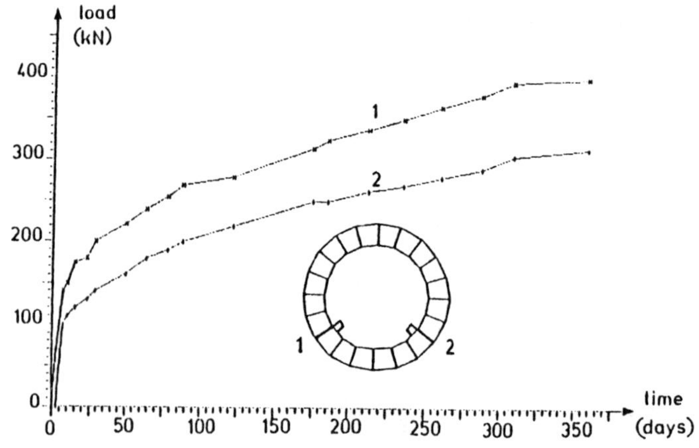

FIGURE 11. Force Between the Blocks of the Lining

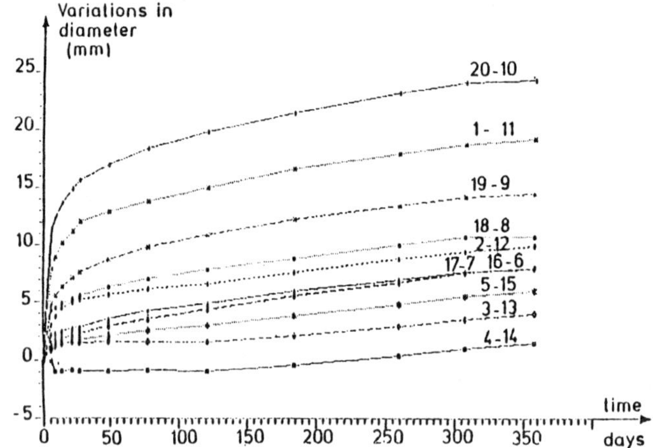

FIGURE 12. Closure of the Gallery

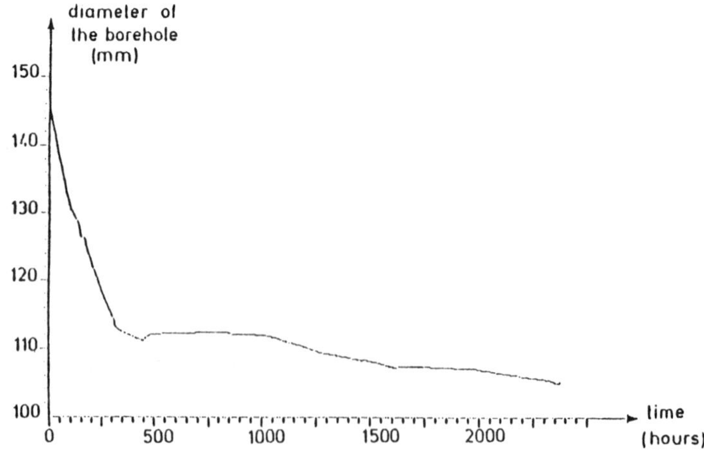

FIGURE 13. Closure of an Unlined Borehole

In short, the following preliminary conclusions can be drawn from the results obtained up to now:

- The general behavior of the gallery is excellent. The total closure of the walls does not exceed 5% whereas, the stress exerted by the clay mass on the lining is inferior to 1 MPa, which is only one-fifth of the lithostatic pressure at this depth.

- The time-dependent effects are very pronounced. Immediately after the digging phase, the measured displacements and stresses substantially increase. During the last ten months, the mean convergence has increased by about 0.4% and the confining pressure by 0.4 MPa.

- The ovalization of the gallery well is very obvious (Figure 12).

- The behavior of the unlined borehole is very different and shows a very rapid and significant closure: more than 30% convergence in less than one month. The process slows down after this phase but the closure should be complete several months after boring.

B. Interpretation

1. The model - In order to simplify, we assume that the natural state of stress in the clay mass is an isotropic state (all the stresses equal P^{∞}). Moreover, according to Panet and Guelleo,[3] it is reasonable to study the problem in plane strain conditions. Therefore the advance of the tunnel face is modelized by a fictitious support pressure, which decreases from a value $P\infty$ to a value 0 when the tunnel face is at sufficient distance from the studied section. The clay is assumed to be homogenous and isotopic and its behavior is described by the above mentioned elastovisco plastic model. With these assumptions, the problem can be treated in cylindrical symmetry. According to the complementary tests program which was carried out at the LMS, the following values have been considered.

$$E = 200 \text{ MPA}, \nu = 0.5$$

The viscoplastic yield point and the softening behavior are described by the 5 following parameters:

$$C = 1 \text{ MPa}, \phi = 5°, C_0 = C/3, \alpha_1 = 4\% \quad \alpha_2 = 6.5\%$$

The considered viscosity parameters are:

$$\eta = 10^7 \text{ and } n = 5$$

2. Results of the calculations with regard to long term behavior - note that on the convergence-confinement curve shown in Figure 14 the two points accounting for the long-term behavior of the

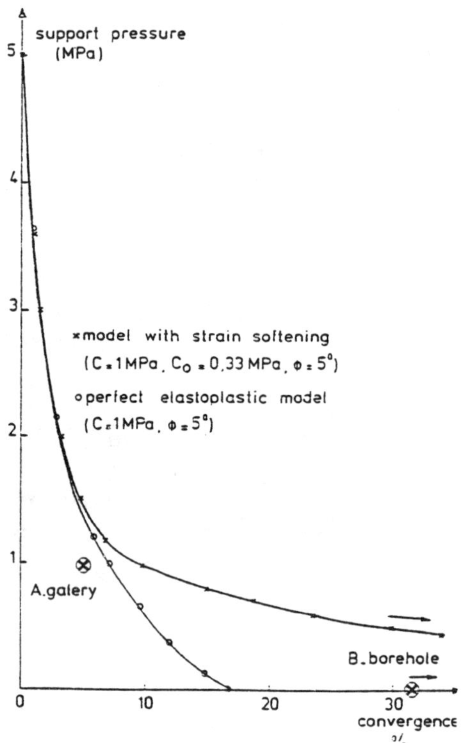

FIGURE 14. Long-Term Convergence-Confinement Curve

gallery (point A: P = 1 MPa and u = 5%) and borehole (point B: P^S = 0 and u^S >30%) as well as the curve resulting from the modelization have been drawn. For comparison, the convergence-confinement curve corresponding to a perfect elastoplastic behavior was drawn on the same figure (constant cohesion equal 1 MPa).

3. Results of the calculations with regard to the time dependent behavior - The viscosity coefficient was chosen in order to match the rapid closure of the borehole during the first days, observed during the in-situ measurements (Figure 15). It is intersting to see on Figure 16, which represents the closure of the gallery, that with this value for the viscosity the 3 above mentioned phases can be found in the case of the experimental gallery. The model anticipates a stabilization of the convergence and of the confining pressure to respectively 5.8% and 1 MPa.

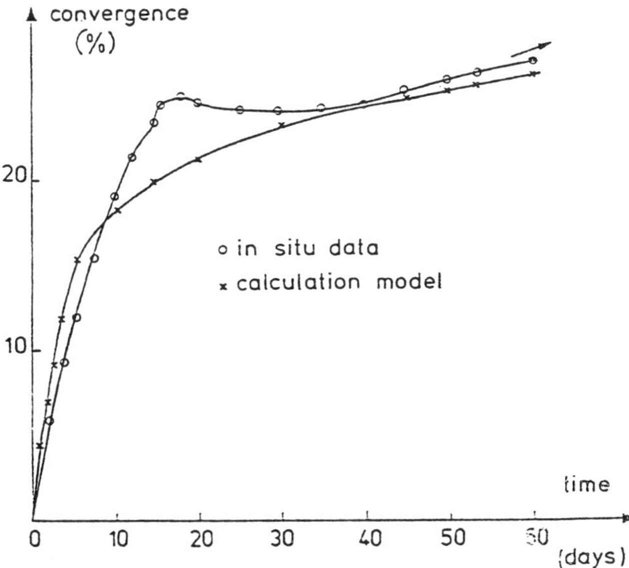

FIGURE 15. Convergence of the Borehole

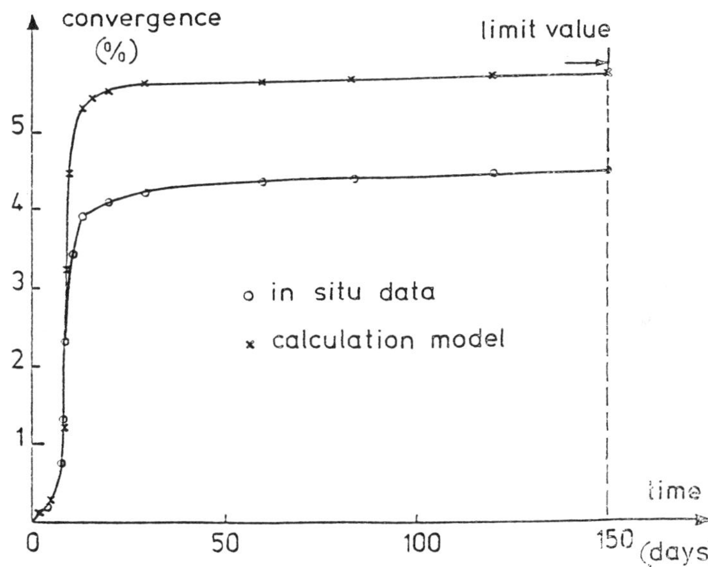

FIGURE 16. Convergence of the Gallery

V. CONCLUSIONS

The knowledge of the long-term behavior of a deep clay is a key element for evaluating the feasibility of constructing an underground disposal facility. This long-term behavior is complex because of the delayed effects which are very significant in the case of clay. An elasto-viscoplastic model, taking into account all important mechanical parameters related to the mechanical behavior of a clay around a gallery, describes fairly well the situation observed experimentally.

However at this stage of the study, the parameters which stand for the secondary effects (anisotropy of the natural state of stress, intricate geometry of the cavities, particular supports . . .) must be simplified. An underground laboratory allows us to have access to real conditions and to set up a comprehensive in-situ experimental program for acquiring the data to understand the long term behavior of clay. It also enables us to characterize various geometries and to define the optimal dimensions of the disposal facility galleries and of their lining.

REFERENCES

1. E. Flavigny, "Comportement Visqueux des Geomateriaux," Ecole d'Hiver des Geomateriaux Sols, Betons, Roches, IMG/CNRS, Aussois, 28 Nov.-5 Dec. (1984).

2. D. Nguyen Minh, "Analyse de la Stabilite des Galeries Creusees dans une Argile Profonde par les Modeles Viscoplastiques," Journee sur les Argiles, E.N.S.M.P., Paris, 13 Dec. (1984).

3. M. Panet, P. Guellec, "Contribution a l'Etude du Soutenement d'un Tunnel a l'Arriere du Front de Taille," 3eme Cong. Int. de Mecanique des Roches, Denvers, Vol. II B (1974).

4. R. Perzyna, Fundamental Problems in Viscoplasticity, Advances in Applied Mechanics, Vol. 9 (1966).

5. G. Rousset, P. Manfroy, B. Neerdael, J.M. Simon, "Participation de l'ANDRA au Programme de Mesures du CEN/SCK (Belgique): Resultats et Interpretations," Journee sur les Argiles, E.N.S.M.P., Paris, 13 Dec. (1984).

Waste Package Design and Testing

THE REFERENCE WASTE PACKAGE DESIGN FOR A NUCLEAR WASTE REPOSITORY IN BASALT

T. B. McCall
J. C. Krogness
Rockwell Hanford Operations
P.O. Box 800
Richland, Washington 99352

ABSTRACT

This paper describes the reference waste package design for the Basalt Waste Isolation Project (BWIP). The waste form, functional requirements, and engineering approach for the reference design are described. A short horizontal borehole emplacement configuration (SHB) containing a single waste package can meet the functional requirements during the operating, containment, and post-containment periods. The waste form will be placed in a thick-walled, low-carbon steel container. Surrounding the container is packing material made from a mixture of crushed basalt and sodium bentonite clay.

I. BACKGROUND AND DESIGN PHASES

The Engineered Barriers Department within the BWIP of Rockwell Hanford Operations (Rockwell) has the ultimate responsibility for the development and design of waste packages suitable for containment of nuclear high-level waste (HLW) in a repository constructed in basalt host rock. These waste packages, according to currently proposed Federal regulations,[1] must successfully provide control of the waste for a period lasting 10,000 yr and isolate the waste from the groundwater environment for a minimum of 300 to 1,000 yr after emplacement. Therefore, successful long-term waste package performance in the expected hydrothermal environment presents a significant engineering design challenge to the BWIP.

The engineering approach for this challenge is a systematic sequence of design phases: conceptual, advanced conceptual, license application, and final procurement and construction design. The present reference waste package design was developed during the advanced conceptual design phase. Presently, the waste package is defined in sufficient detail to allow decisions to be made among design alternatives on the basis of performance, cost, and schedule. Materials testing, development testing, and design analysis will be continued such that a firm basis will be available for license application design that is scheduled to begin in 1988. Final procurement and construction design completes the process begun in license application design.

II. WASTE INTENDED FOR DISPOSAL

The reference waste package design was developed for the requirements of a repository in basalt and specific nuclear waste forms. The spent fuel waste form is either fuel rods from disassembled fuel assemblies or intact fuel assemblies (commercial reactors). The reference waste form for reprocessed wastes is vitrified waste (e.g., West Valley HLW or defense HLW). These waste forms will be enclosed in a thick-walled container to provide containment for handling and emplacement operations and substantially complete containment for at least 300 to 1,000 yr after emplacement. The waste form has a functional requirement to provide resistance to dispersion during the preclosure phase.

III. WASTE PACKAGE FUNCTIONAL REQUIREMENTS

The function of the waste package is to provide short-term containment and long-term control of the release of radionuclides (after breach of containment) in conformance with Federal regulations[1,2] by employing an integrated system of physical and chemical barriers acting in concert with the host repository geology.

The functional requirements for the waste package components can be described best in relationship to three periods in repository history. The first period of the repository history is defined by the conditions experienced by the waste package components prior to permanent closure of the repository. This 84-yr period is designated as the operational/retrieval period. During the operation/retrieval period, the ambient geothermal gradient is perturbed by radionuclide decay heat. The waste package will experience its highest temperature, the ambient pressure will be near atmospheric pressure, and the environment of the waste container will be air and water vapor. The second period of the repository history is defined as the containment period, and is the period of 1,000 yr after permanent closure. During this period, there is a pronounced reduction of heat generation by short-lived radionuclides. The third period occurring after 1,000 yr, is defined as the period of limited release, and is the remaining time of the repository history after assumed waste container breach.

IV. DESCRIPTION OF THE WASTE PACKAGE ADVANCED CONCEPTUAL DESIGN

The reference waste package advanced conceptual design (ACD) is based on the short horizontal borehole (SHB) repository emplacement concept. The reference designs for the three waste forms considered during fiscal year 1985 are similar, varying only in the internal structure of the container and the waste package dimensions to accommodate different waste form sizes and corrosion allowances. For this paper, the discussion will focus on the design for the consolidated spent-fuel waste form.

The waste package ACD for consolidated spent fuel is shown in Figure 1. The packing that is located between the container and a thin-wall, low-carbon steel emplacement shell becomes a low-permeability barrier when saturated. Packing helps to reduce the release rate of radionuclides to the geologic setting once the container barrier is assumed to have breached following the 1,000-yr containment period after repository decommissioning.

Retrieval of the waste package is accomplished by withdrawing the entire waste package (i.e., shell, packing, and container) from the borehole.

The container is designed to hold the rods from four pressurized water reactor (PWR) spent fuel assemblies based on design requirements developed by the BWIP.[3] Once the spent fuel rods are placed into the container, a cover head is installed and is sealed by a welding process. The container wall is sized to resist the external loads expected to be present during the 1,000-yr containment period. The container wall thickness also takes into account corrosion from the surrounding environment and from radiolysis products in the groundwater.

A. Consolidated Spent Fuel Waste Form

The Consolidated Spent Fuel (CSF) waste form, one of several waste forms considered, consists of fuel rods disassembled from four PWR spent fuel assemblies (SFA).

The typical maximum heat generation rate for the CSF waste form is 2,069 W for rods 10 yr out-of-reactor. Younger or higher burnup fuel would require a smaller amount of fuel per container to prevent exceeding the temperature limits on waste package materials.[3] This could be accommodated, without changing the external size of the container, by altering the internal structure of the container, or by adding a nonfissile filler, such as crushed basalt.

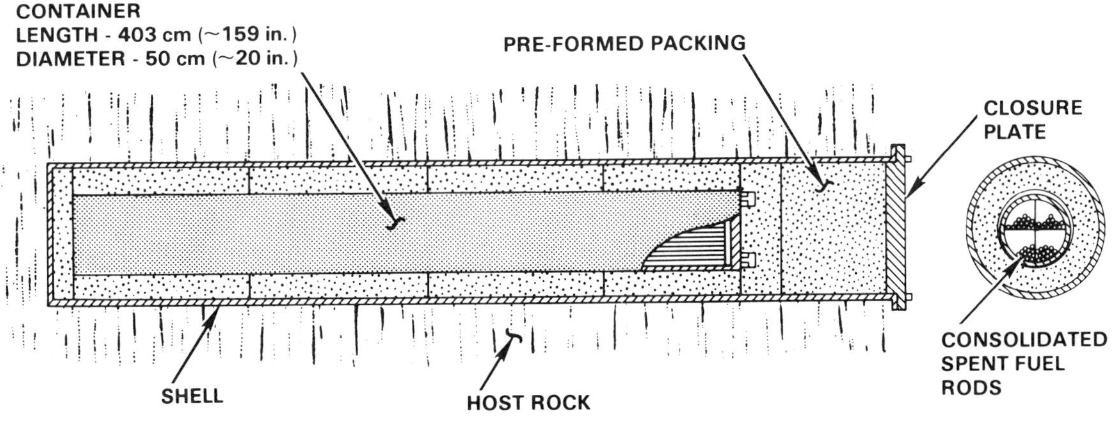

FIGURE 1. Waste Package Advanced Conceptual Design

This paper considers the specific case of a waste form consisting of fuel rods from a Westinghouse PWR 15 by 15 fuel assembly. There are 204 fuel rods per assembly with an outside rod diameter of 10.7 mm (0.422 in.) and length of approximately 3.8 m (152 in.). The rods consist of irradiated UO_2 pellets contained in a Zircaloy-4 tube (cladding).

B. Consolidated Spent Fuel Container

The container is a vessel fabricated from cast steel components. It is designed to resist both corrosion and external pressure, thus providing containment for 1,000 yr. Cast steel was specified by BWIP[3] because of its low cost, good weldability, and resistance to corrosion in the expected basalt geochemical environment. The inside dimensions are designed to be large enough to house 816 PWR rods while providing sufficient clearances to permit remote placement of the rods into the container during rod consolidation operations. A container wall thickness of 50.8 mm (2.0 in.) will resist a uniform hydrostatic pressure of 9.4 MPa (1,363 lbf/in^2) without buckling. A corrosion allowance of 6.3 mm (.25 in.) has been added to the required structural thickness.[3] This is the general corrosion depth calculated for the container. The corrosion depth was calculated using corrosion models developed by BWIP.

The internals of the container are divided into four equal quadrants. Each quadrant is large enough to hold the rods from a PWR spent fuel assembly. The quadrants are formed by two divider plates attached to the interior wall of the container. The divider plate material is carbon steel similar in composition to the container material to minimize the potential for galvanic corrosion.

The container is sealed remotely by making the final head/sidewall weld in a hot cell. Prior to sealing, the container is backfilled with an inert gas (e.g., argon). The purpose of the inert gas is to provide a nonoxidizing atmosphere. Test results have shown that to preserve the fuel pellet UO_2 matrix structure and avoid degradation of the fuel cladding, a nonoxidizing environment is needed when storing fuel dry at temperature in the maximum temperature range specified by BWIP[3] (400°C). The closure weld ensures containment of radioactivity and maintains the nonoxidizing environment.

Lifting pintles designed to interface with repository handling devices are attached to both the upper head and the sidewalls of containers for handling. The pintles, which are machined from low-carbon steel bar stock, are threaded into the head and container body, and welded. The weld prevents thread corrosion that otherwise could destroy the strength of the attachment and prevents the pintles from loosening. The pintle on the head is designed for handling the head in the hot cell to place it on the container. Lifting and handling of the container is done with three pintles located 120° apart on the ends of the sidewalls. During lifting operations,the weight of the container and its waste form are transmitted through the container body and not its upper head. Thus, the closure weld is not subject to the structural loads associated with handling. The container design does not use lifting trunnions on the side of the container body as these would interfere with assembly. The lifting pintles would be used to grapple the container during any retrieval operations.

C. Consolidated Spent Fuel Packing

The application of packing material surrounding the container reflects the current best estimates for the choice of the packing material and its properties. Further experimental data are needed to verify the packing material properties and the resulting packing thickness needed to perform its primary function of aiding the waste form in controlling radionuclide release to the host geology.

The packing material is a mixture of 25% sodium bentonite and 75% crushed basalt (by weight). The ultimate specifications for the materials to be used in waste packages will depend on the results of laboratory testing of the behavior and properties of the material, and the results of process development tests on the manufacture and emplacement of the packing materials. The properties data from testing to date represent a variety of specific particle size distributions for the bentonite and crushed basalt. The nominal value of packing material properties currently specified by BWIP[3] represents the average values obtained from tests of the current reference material. The choice of this material was based on engineering judgment that it will provide acceptable performance properties.

To ensure proper performance as a low permeability barrier, the hydraulic conductivity of the packing material must be sufficiently low so that the mass transport of radionuclides is limited to diffusion processes. The Functional Design Criteria[3] requires an equivalent dry density of 1.7 g/cm^3 (106 lb/ft^3) for the packing material for in-situ emplacement of the mixture.

Designs that use precompressed packing material formed into solid forms will result in void space because of the need to provide assembly clearances. The density for precompressed material was selected by engineering judgment on the basis of ease of fabrication and by the judgment that the assembly tolerances will not result in a void volume greater than 10% of the packing volume. The value selected by BWIP is 2.0 g/cm^3 (125 lb/ft^3).

D. Consolidated Spent Fuel Waste Package Assembly

Waste package components will be fabricated and assembled at various locations including the repository. Below ground emplacement of the containers into the repository boreholes that are prelined with packing is the reference emplacement concept. Described here are assembly and packaging operations necessary for a preferred optional above ground waste-package assembly.

After the container and heads are received at the repository, they will be stored until needed for assembly with the waste form/canisters in a hot cell.

A receipt inspection will be performed and the components will be cleaned as necessary prior to being moved into the hot cell. In addition, the container assembly will be serialized for package identification with the waste form/canister. All subsequent assembly operations are completed remotely in the hot cell because of the radioactivity of the waste form being packaged.

For the CSF waste form, the container must be compatible with the rod consolidation process. The container is remotely loaded with the consolidated fuel rods. After the container is loaded with the waste form, the container head is lifted and set in place. Attachment of the head to the container body is accomplished by making a remote closure weld. Following the welding process, the weld will be inspected both by visual examination via remotely operated TV cameras and by ultrasonic examination. The container is loaded into the thin-wall shell. Between the inside of the shell wall and the outside of the container wall, packing material is emplaced. The thin-wall shell functions to prevent the packing from being exposed to an air-steam environment during the operational period. The shell is also a sacrificial component intended to withstand the abrasion from the borehole host rock when the package is slid into position.

The emplacement scenario described begins with the preparation of the access drifts and boreholes, before the packages are sent underground. Once the boreholes are excavated, the emplacement package is put in place. The waste packages are brought to the emplacement borehole from the assembly hot cell or from the lag storage area in shielded transfer casks. The cask is aligned with the borehole and the container is offloaded into a borehole.

E. Retrieval

Retrieval, if it be necessary, should be performed prior to repository backfilling. Thus, the retrieval scenario will be the reverse of the package emplacement process described above.

V. SUMMARY AND RECOMMENDATIONS

The ACD effort has shown that the SHB reference waste package design concept, for PWR CSF rods, has the potential to meet all requirements, based on available data and the analyses performed to date. These reference waste package designs for a national waste repository in basalt rock will enable the BWIP to further support the basalt "Environmental Assessment" and the "Site Characterization Plan."

The ACD has provided new information regarding the interface between waste package design and repository design and has supported the development of waste package criteria. Further, the ACD has identified the need for future investigations and development testing in specific areas regarding the performance and design of waste package containers and packing.

REFERENCES

1. U.S. Nuclear Regulatory Commission, 10 CFR 60, Disposal of High-Level Radioactive Wastes in Geologic Repositories, Federal Register 46 (130), Proposed Rules (1983).

2. Environmental Protection Agency, 40 CFR 191, Environmental Radiation Protection Standards for Management and Disposal of Spent Nuclear Fuel, High-Level, and Transuranic Radioactive Wastes, Washington, D.C. (1984).

3. W. J. Anderson, Waste Package Subsystem Advanced Conceptual Design Requirements Document, SD-BWI-FDC-009, Rockwell Hanford Operation, Richland, WA (1985).

HIGH-LEVEL WASTE OVERPACK FOR FINAL STORAGE IN THE SWISS GRANITIC BEDROCK: MATERIAL SELECTION, DESIGN AND CHARACTERISTICS

B. Knecht
C. McCombie
National Cooperative for the
Storage of Radioactive Waste (NAGRA)
Parkstrasse 23
CH-5401 Baden, Switzerland

ABSTRACT

Current projects aimed at demonstrating the feasibility of safe final disposal of high-level nuclear waste in Switzerland envisage a repository in the crystalline bedrock of the north of the country. The groundwater is reducing with a mineralisation of typically 10 g/l. The corrosion studies carried out in Switzerland have shown that unalloyed steel is a suitable overpack material under the conditions expected in the repository, the necessary corrosion allowance being 50 mm for a lifetime of 1,000 years. Design work based on the use of a typical cast steel with a tensile strength of 400 MN/m^2 has led to a reference overpack concept for disposal of vitrified HLW. This reference overpack is designed as a self-shielding, self-supporting cylindrical shell with hemispherical ends.

INTRODUCTION

The present concept for final disposal of radioactive waste in Switzerland consists of a repository approximately 1,200 m deep in the crystalline bedrock of Northern Switzerland. In order to delay the return of the radionuclides to the biosphere, and to reduce their concentration there to acceptable levels, reliance is put in the principle of multiple safety barriers. Further to the natural barriers (host rock and overlying sediments), the following engineered barriers are envisaged: the waste form itself (vitrified high-level waste), an overpack, the purpose of which is to ensure isolation of the radionuclides from groundwater for a period of at least 1,000 years and a bentonite backfill within which the overpack is placed, aimed at reducing the transport of water and dissolved species.

The overpacks would be stored horizontally in tunnels and embedded in compacted bentonite. The ambient temperature at the repository depth will be approximately 55°C, but the heat generation from the waste can result in temperatures up to 155°C at the overpack surface for short times.

Because the overpack concept is that of a stressed, thick-walled shell, the radiation levels outside the overpack will be low. The overpack is subjected to an isostatic pressure of up to 30 MPa, being the sum of the hydrostatic pressure and the bentonite swelling pressure. The groundwaters in the crystalline bedrock of Northern Switzerland are reducing with mineralisation levels typically 10 g/l, mainly NaCl with significant amounts of sulphate and carbonates.

The present paper shortly describes the selection procedure used in present feasibility projects in Switzerland that has led to the identification of unalloyed cast steel as the favoured candidate overpack material, the criteria used in the design of a reference overpack, and the most important features of this reference design.

MATERIAL SELECTION AND QUALIFICATION

The approach taken was first to identify candidate materials for the overpack taking into account the required lifetime of 1,000 years or more and the exposure conditions within the repository, i.e. temperatures up to 160°C,[1] radiation intensities expected from vitrified waste at the end of a 40-year intermediate storage period, reducing groundwaters with mineralisation levels up to seawater level and mechanical stresses corresponding to a depth of 1,200 m in crystalline rock. On the basis of a literature survey the following materials were short-listed:[2]

1) Plain carbon steel or nodular cast iron
2) Copper
3) Ti-Code 12
4) Ni-Cr-Mo alloy (Inconel 625 or Hastelloy C)
5) Alumina

In a second stage ease of manufacture and quality control were considered, whereby simple single material designs have an inherent advantage over multiple layer concepts; this led to alumina and Ni-Cr-Mo alloy being given low priority; alumina because manufacture of an object the size of an overpack is not "state-of-the-art" and also because of doubts about the fracture behaviour of this material, particularly over long periods of time; Ni-Cr-Mo alloys because it was judged that only one material needed to be investigated for use in multiple layer concepts and preference in this respect was given to Ti-Code 12. Neither of these materials has, however, been disqualified for Swiss projects; ceramics, particularly alumina, continue to be studied at a low level as a longer term option. Also in this second stage it was decided to select a cast steel and a nodular cast iron with a tensile strength of 400 MN/m^2 as potentially most suitable candidates in order to avoid stress corrosion cracking problems possibly associated with high strength carbon steels.

Critical reviews of the literature [3,4,5] as well as the first experiental results [2,6] under simulated repository conditions indicated that the iron materials were very unlikely to corrode at rates higher than about 20 mm in 1,000 years in the conditions expected in Switzerland. The Swiss overpack programme has thus concentrated on the use of cast steel, although it should be noticed that other standard carbon steels and nodular cast irons are expected to be just as suitable, the final choice for actual repository projects being determined by detailed considerations that are beyond the scope of feasibility projects. It should also be stressed that the evidence obtained in the experimental programme [2,6] supports the conclusion reached in the course of the selection procedure that both Ti-Code 12 and copper are suitable alternatives for the iron materials under the conditions expected in Switzerland. In fact copper has been retained as reference material for the direct disposal of spent fuel with an overpack design essentially the same as one of the options proposed in Sweden (spent fuel bundles cast in lead inside an electron-beam welded copper overpack); again it should be stressed that this choice was made for reasons of convenience, and that carbon steel would be just as suitable for spent fuel as copper in a time horizon of 1,000 years.

The effort in Switzerland at present concentrates on the further qualification of carbon steel as an overpack material, and in particular on further investigation of the general corrosion behaviour as well as of localised corrosion effects.

The investigations of the general corrosion are carried out by immersion corrosion tests in model groundwaters with and without bentonite [2,6,7] as well as by hydrogen evolution measurements.[8,9] The latter method, which allows determinations of the "instantaneous" (or differential) corrosion rate, is well suited for investigations under reducing conditions such as will prevail in the repository once the trapped oxygen has been used up (by corrosion or other processes); no significant amounts of oxidants are generated by radiolysis provided thick-walled overpacks are used. In the absence of oxygen, iron may be oxidised by the hydrogen ion:

$$Fe + 2H^+ \rightarrow Fe^{++} + H_2 \qquad (1)$$

or directly by water

$$Fe + 2H_2O \rightarrow Fe(OH)_2 + H_2 \qquad (2)$$

The reaction with the hydrogen ion is fast but at the pH values in the repository (pH 8-9) the concentration of H^+ is so low (10^{-8} to 10^{-9} mol/l) that the direct reaction with water most probably dominates. In either case one gram atom of iron produces 1 gram molecule of hydrogen. This relationship forms the basis for the measurement of the corrosion rate via the hydrogen evolution rate.

In both cases iron is oxidized to its divalent oxidation state. Thermodynamically Fe_3O_4 is stable under repository conditions; Fe_3O_4 can form directly via the reaction:

$$3Fe + 4H_2O \rightarrow Fe_3O_4 + 4H_2 \qquad (3)$$

or by the conversion of $Fe(OH)_2$:

$$3Fe(OH)_2 \rightarrow Fe_3O_4 + 2H_2O + H_2 \qquad (4)$$

For the purposes of the conversion of the hydrogen evolution rate into a corrosion rate it was assumed that iron was oxidised only to Fe(II). Oxidation to Fe_3O_4 would give lower corrosion rates than those quoted.

The experimental set-up is described in [8,9]. Detection of hydrogen is by gas chromatography; the apparatus at present has a detection limit corresponding to a corrosion rate of about 0.02 µm/a, corrosion rates down to 0.1 µm/a are thus readily measured. Hydrogen absorption in the sample has only a small influence on the initial transient corrosion rate: a 100°C and 0.01 atm H_2-partial pressure the saturation concentration of hydrogen in the sample is less than 10^{-6} wt%, and this will be reached typically within the first 50 hours or so of a given run. Furthermore, as discussed in [9], the determinations of the amount of material corroded by integration of the instantaneous rate and by measurement of the weight loss at the end of the run are in agreement. Figure 1 shows the results obtained in two representative groundwaters on plain carbon steel at 80°C.

It can be seen that after a transient of some 4 days the corrosion rate settles down to a value of about 2.5 µm/a. The duration of the transient (which depends on the geometry and size of the samples) and its magnitude show that integral corrosion rates as measured from weight loss experiments should appear to decrease considerably as the measurement period increases. This has indeed been found to be the case. Once this is taken into account, the results of the measurements shown in Table 1 are considered to be strongly indicative of a general corrosion rate below 10 µm/a.

The overall conclusions concerning the general corrosion behaviour can be summarised as follows:

1) The long-term corrosion rate is most probably well below 10 µm/a.

2) The corrosion rate does not increase monotonically with increasing temperature. The corrosion rate was observed to be lower at 140°C than at 80°C in all the immersion and bentonite experiments. The hydrogen evolution method showed that the corrosion rate reaches a maximum between 50°C and 80°C. It is known that magnetite forms preferentially at higher temperatures; magnetite formation is usually held responsible for the known reduction in corrosion rate above 100°C. A reduction in corrosion rate with temperature indicates a change in mechanism, whether the change between 50°C and 80°C is due to magnetite formation at a lower temperature than usually quoted remains to be investigated.

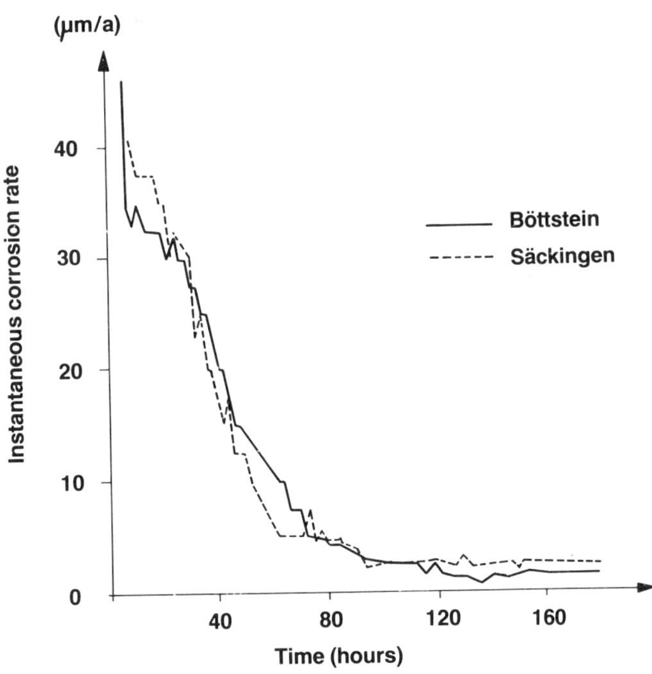

Figure 1. Instantenous corrosion from anearobic corrosion in two representative water compositions with mineralisations of approx. 3 g/l (Säckingen water) and approx. 14 g/l (Böttstein water), as determined from the observed hydrogen evolution rate (10 µm/a corresponds to 4 ml H_2/m^2 h under the assumption that the iron corrosion products are divalent)

3) Bentonite has no large influence on the corrosion kinetics. The corrosion rates in bentonite are comparable to those in the waters at the same temperatures. The more highly compacted bentonite gave lower values, but this may be due to the fact that the amounts of air trapped in the bentonite decrease with increasing compaction.

Also investigated was localised corrosion, which is known [10] and has been confirmed [9] in the course of the investigation programme in Switzerland to be possible in chloride-carbonate solutions. In neither of the two reference waters used in the programme (Säckingen water with total mineralisation about 3 g/l, Böttstein water with total mineralisation about 14 g/l) should, however, such effects be expected, provided the buffer material maintains a pH below 10, as shown by Figure 2. Further work is planned to assess the concentration domains of relevant ions (chloride, carbonate, sulphate) which might cause localised corrosion. Pending the results of these investigations, it has been concluded [11] that a corrosion allowance of 50 mm should be ample for overpack design.

Method	Medium	Corrosion rate (μm/a) at 25°C	50°C	80°C	140°C
Weight loss (2,100 h)	Böttstein water	-	-	14	-
Weight loss (2,160 h)	Compacted bentonite, 30 % Böttstein water, saturated	-	-	29	14
Hydrogen evolution	Böttstein water	1.1	6.5	2.5	-

Table 1. Corrosion rate of carbon steel in oxygen-free waters

Notes: (1) Böttstein water is a synthetic groundwater based on the results of analyses of natural groundwater as found in the bedrock of Northern Switzerland (main components Na^+ 4.8 g/l, Ca^{++} 1.1 g/l, Cl^- 8.1 g/l, SO_4^{--} 1.8 g/l). (2) The higher value obtained in bentonite may only reflect the presence of oxygen trapped in the bentonite at the beginning of the experiment. (3) The integral corrosion rate from weight loss measurments decreases with increasing time, e.g. in Böttstein water the measured rate as deduced from a 6,300 h experiment is 10 μm/a compared with 14 μm/a from a 2,100 h experiment. (4) The corrosion rates deduced from hydrogen evolution measurements correspond to the "steady state" evolution rate and therefore do not include the effect of the initial transient that is always observed (see Figure 1).

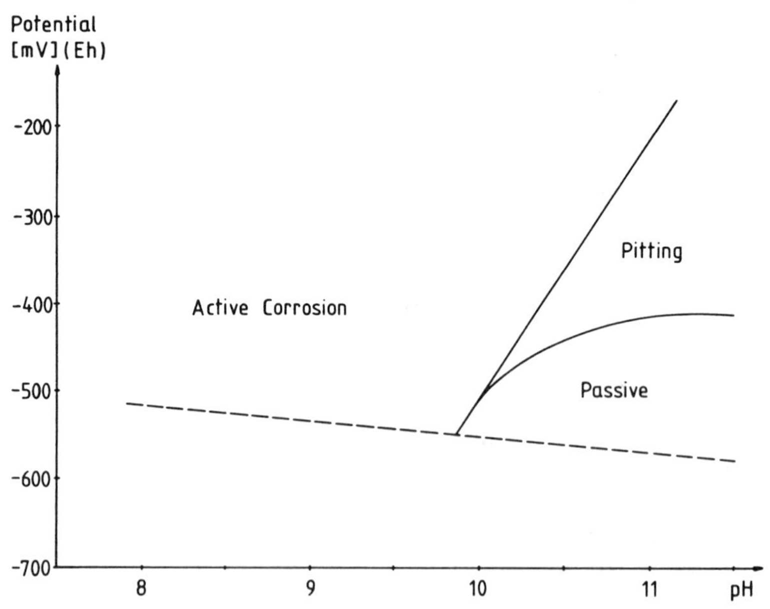

Figure 2. Corrosion behaviour of cast steel in Böttstein water at 80°C (for details see text and notes to Table 1)

REFERENCE OVERPACK: DESIGN CRITERIA AND FEATURES

Based on the results of the corrosion investigations as well as of other programmes in Switzerland, the following criteria were used for the design of a reference overpack for vitrified waste in the framework of current feasibility projects:

- Single layer concept with cast steel GS 40 (essentially equivalent to US Standard ASTM A27-Grade U 60-30) as material

- Stressed-shell overpack (because of the fact that vitrified waste containers from the reprocessing plant contain void space)

- Outer surface dose rate arising from vitrified waste 40 years ex-reactor as low as possible, preferably below 100 mrem/h

- Design external pressure 30 MPa (equal to the sum of the maximum design swelling pressure of the bentonite used as a buffer material and of the hydrostatic head at 1,200 m depth)

- Tensile and bending stresses as low as reasonably achievable

- Leak-tightness achieved through welding. Stresses in the weld as low as reasonably feasible, good inspectability of the weld, possibility to stress-relieve the weld to a large extent despite the limitations imposed on the glass temperature by recrystallisation processes.

The design work carried out [12,13] led to the design shown in Figure 3. The reference overpack consists of a cylindrical body with integrated hemi-spherical bottom and pre-assembled additional shielding; a hemispherical lid, also with pre-assembled additional shielding and with a thread for a gripping device, is pressed onto the body and held in place by means of a conical thread and subsequentely welded.

The total length of the overpack is 2,000 mm and its outside diameter is 940 mm. The bottom inner surface is hemispherical and its outer surface is a spherical segment, its wall thickness being 150 mm at the center increasing to 200 mm at its transition point to the frontal area of the cylindrical container body. The wall thickness of the cylindrical container body is 250 mm. A ring-shaped area at the outer surface serves to position the overpack during the closure operations. At the open side the cask body exhibits a cylindrical extension of larger inner diameter, into which the lid will be positioned. Two cylindrical surfaces, separated by a conical thread, serve to centre the lid. At the outer end of an extension, space for the closing weld is provided. The lid rests on a ring-shaped surface in the container body onto which the external pressure is transmitted by the lid. Thus the external pressure is transmitted directly into the container body without affecting the weld.

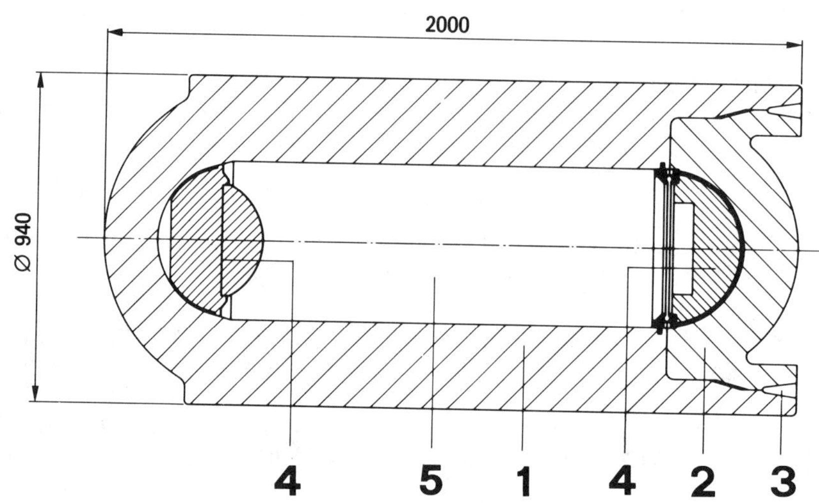

Figure 3. The reference overpack for feasibility projects in Switzerland. All dimensions are in millimetre. 1: Overpack body, 2: Overpack lid, 3: Weld, 4: Additional shielding, 5: Space for accommodating vitrified waste cylinder. For further details see Ref. 12.

The stress analyses carried out in conformity to the ASME-Code, Section VIII, Division 1 for an external pressure of 30 MN/m^2 and for a wall thickness reduced by the amount of the design corrosion allowance, i.e. 50 mm, resulted in maximum stresses of 80 MN/m^2, i.e. a margin of a factor 5 against the tensile strength of 400 MN/m^2.[12] The margin against buckling was found to be almost 4, and therefore well above the margin of 3 incorporated into the appropriate sections of the ASME-Code. Further calculations have been carried out to assess the effects of a one-sided corrosion both on the stress distribution and on the stability behaviour, showing that even such very conservative assumptions do not lead to a significant decrease of the above margins. Finally, it could be shown that even under very conservative assumptions creep buckling can be neglected during the required lifetime of 1,000 years.

Radiation dose calculations have shown that handling by the operating staff should be fairly straightforward, the maximum dose rate at the cylindrical part of the container being less than 100 mrem/h, and at the bottom and the lid less than 30 mrem/h. These low doses also imply that radiolysis effects are insignificant during the lifetime of the overpack of 1,000 years.

CONSEQUENCES OF THE USE OF A CARBON STEEL OVERPACK ON REPOSITORY PERFORMANCE

Work carried out in connection with the use of carbon steel overpacks [14,15] has identified two important effects with regard to repository performance. Firstly, large quantities of hydrogen will be generated by the iron corrosion (typically 50 kg in 1,000 years);[14] unless it can escape from the near field, this hydrogen could lead to a considerable pressure rise. It could, however, be shown that the hydrogen will in fact escape through the bentonite buffer by a percolation process.[16] Secondly, the large quantities of iron corrosion products will form an effective redox buffer even after breach of the overpack;[15] this implies that the release of many nuclides from the near field is limited by their solubility under reducing conditions.

CONCLUSIONS

The corrosion investigations as well as the design and performance assessment studies carried out in the course of the Swiss nuclear waste disposal programme have shown that it is possible to achieve isolation of the waste from groundwater in deep crystalline formations through the use of cast steel overpacks with a lifetime of at least 1,000 years.

ACKNOWLEDGEMENTS

The authors take pleasure in acknowledging the contributions of all the persons and organizations involved in the Swiss overpack programme.

REFERENCES

1. Nagra, Project Gewähr 1985: Endlager für hochaktive Abfälle - Das System der Sicherheitsbarrieren, Report NGB 85-04, Baden/Switzerland, Nagra (1985).

2. J. P. Simpson, Experiments on Container Materials for Swiss High-Level Waste Disposal Projects, Part I, Technical Report NTB 83-05, Baden/Switzerland, Nagra (1983).

3. E. Heitz, zur D. Megede, Korrosionsverhalten von unlegiertem Stahl, Stahlguss und Gusseisen als Endlagerbehälterwerkstoff in wasserführendem Granitgestein, Technical Report NTB 82-08, Baden/Switzerland, Nagra (1982).

4. H. Gräfen, and E. Heitz, Korrosion und Korrosionsschutz von Endlagerbehältern aus Eisenwerkstoffen unter besonderer Berücksichtigung des schweizerischen Endlagerkonzeptes, Technical Report NTB 84-04, Baden/Switzerland, Nagra (1984).

5. R. Grauer, Behältermaterialien für die Endlagerung hochradioaktiver Abfälle: Korrosionschemische Aspekte, Technical Report NTB 84-19, Baden/Switzerland, Nagra (1984).

6. J. P. Simpson, Experiments on Container Materials for Swiss High-Level Waste Disposal Projects, Part II, Technical Report NTB 84-01, Baden/Switzerland, Nagra (1984).

7. J. P. Simpson, and B. Knecht, Corrosion Behaviour of Unalloyed Steel and Cast Iron in Groundwaters of the Bedrock of Northern Switzerland." Presented at and to be published in the proceedings of the International Seminar on Radioactive Waste Products - Their Suitability for Final Disposal. Jülich (Fed. Rep. of Germany), Kernforschungsanlage (1985).

8. R. Schenk, Experimente zur korrosionsbedingten Wasserstoffbildung in Endlagern für mittelaktive Abfälle, Technical Report 83-16, Baden/Switzerland, Nagra (1983).

9. J. P. Simpson, R. Schenk, and B. Knecht, "Corrosion Rate of Unalloyed Steels and Cast Irons in Reducing Granitic Groundwaters and Chloride Solutions." To be presented at and published in the Ninth International Symposium on the Scientific Basis for Nuclear Waste Management, Stockholm, Materials Research Society and SKB (1985).

10. G. P. Marsh et al., "Corrosion Assessment of Metal Overpacks for Radioactive Waste Disposal," European Appl. Res. Rept. - Nucl. Sci. Technol. 5(2):223-252 (1983).

11. The Nagra Working Group on Container Technology, An Assessment of the Corrosion Resistance of the High-Level Waste Containers Proposed by Nagra, Technical Report NTB 84-32, Baden/Switzerland, Nagra (1984).

12. Steag Kernenergie GmbH, Motor-Columbus Ingenieurunternehmung AG, Behülter aus Strahlguss für die Endlagerung hochradioaktiver Abfälle, Technical Report NTB 84-31, Baden/Switzerland, Nagra (1984).

13. H. Bienek, and W. Wick, "Final Storage Container Design for Highly Radioactive Waste." To be presented at and published in the proceedings of the Ninth International Symposium on the Scientific Basis for Nuclear Waste Management, Stockholm, Materials Research Society and SKB (1985).

14. I. Neretnieks, Some Aspects of the Use of Iron Canisters in Deep-Lying Repositories for Nuclear Waste, Technical Report NTB 85-35, Baden/Switzerland, Nagra (1985).

15. I. G. McKinley, The Geochemistry of the Near-Field, Technical Report NTB 84-48, Baden/Switzerland, Nagra (1984).

16. R. L. Pusch, L. Ranhagen, and K. Nilsson, Gas Migration through MX-80 Bentonite, Technical Report NTB 85-36, Baden/Switzerland, Nagra (1985).

A FLEXIBLE APPROACH FOR EARLY PRODUCTION OF HIGH-LEVEL WASTE GLASS: A PROCESS CANISTER

C. C. Chapman
L. R. Eisenstatt
West Valley Nuclear Services Co., Inc.
P.O. Box 191
West Valley, New York 14171-0191

ABSTRACT

A process canister, which has several advantages, is being considered for use at the West Valley Demonstration Project. The plan is to cast an acceptable high-level waste glass in process canisters. These canisters are low-cost molds which are suitable for casting the high-level waste glass in the production plant and for storing on-site. Before shipping to a federal repository, the processing canisters could be packaged in approved repository disposal canisters. This approach would allow the Project to fix the design of the process equipment and facility at an early stage while providing the flexibility of meeting disposal canister acceptance criteria that are not yet final. Due to the unique nature of the West Valley Demonstration Project, the process canisters appear to provide a simple, yet efficient, way for providing a waste package with uncontaminated exterior surfaces for interim storage before shipping to the repository. This factor may be a real cost advantage to the Project. This method also provides flexibility as to the type of disposal canister that is ultimately used.

I. INTRODUCTION

Two of the requirements of the West Valley Demonstration Project Act[1] are to solidify the high-level waste stored at West Valley, New York, and to design canisters suitable for the disposal of the waste. The West Valley Demonstration Project (WVDP) will solidify the waste into borosilicate glass that will be poured into canisters during the vitrification process. Once the glass has been poured, the canisters will have to be handled at the vitrification facility to move them to interim storage. When the time comes to ship the waste, the canisters will have to be handled to place them into a transportation cask, and finally the waste will be handled at the repository.

In addition to being capable of being handled at different facilities, the canistered waste must meet disposal specifications. Although not specifically mentioned as requirements for the canister, 10 CFR 60[2] and 40 CFR 191[3] do discuss requirements on the whole waste package, which may influence the selection of the canister. Ultimately, the repositories will decide how the criteria in the regulations will be implemented and what role the canister will play in meeting the requirements. To assist the WVDP the Department of Energy has developed draft interim waste acceptance specifications for West Valley HLW[4] for use as guidance. This document would require that the disposal canisters be fabricated from 304L stainless steel, have a standard lifting pintle, have exterior dimension of 61-cm diameter by 300-cm length, and have a permanent label. The programs responsible for developing the geologic repositories in the various potential host media are beginning to issue waste acceptance specifications[5-7], but these are in draft and/or conceptual stages. Furthermore, because the high-level waste packages must be transported from West Valley to the repository, the waste packages will have to adhere to applicable transportation regulations (e.g., 10 CFR 71[8], 49 CFR 172[9], and 173[10]). In addition to interfacing with the repository equipment, the canisters will have to interface with a shipping cask.

II. A FLEXIBLE STRATEGY FOR CANISTER SELECTION

As mentioned above, the repositories will decide how to implement the regulations that apply to them and that affect the canister. They have begun to do this, but their requirements are in an early stage of development. The Waste Acceptance Specifications are entitled interim waste acceptance specifications, published as a draft, or are conceptual. There are some undetermined specifications. Because they are subject to change, it is conceivable that a difficult retrofit operation may be needed at West Valley if the present interim specifications were followed and later changed. For example, the present plan of the repositories is to take no containment credit for the waste producer's disposal canister. A basis for existing specifications is to only prevent the canister material from being detrimental to the waste glass or repository container. However, a change in this approach could cause significant changes in the Project's canister design.

The WVDP has considered using a canister design similar to the one developed for use at the Defense Waste Processing Facility (DWPF) at the Savannah River Plant. This canister would have to interface with WVDP equipment and the repository. However, the WVDP cannot wait until the repositories have definitively determined their canister requirements to select a canister for its process, because the WVDP will begin HLW vitrification during 1988, and the earliest repository will not be operating until 1998 at the earliest. Therefore, the WVDP is investigating the use of process canisters for its HLW packaging.

A. Description of the Process Canister Concept

The main component of this approach involves the use of a low-cost "process canister" that will be used as the receptacle of the waste glass during vitrification and for interim storage at West Valley. It will be sized small enough so that it can fit inside a disposal canister acceptable to the repository (see Figure 1). This disposal canister may be similar to the canister currently planned to be used at the DWPF. Alternately, a canister similar to that being developed for remote-handled transuranic waste could be used.[11] Within the constraints of the known requirements, decisions on the process canister geometry, weight and disposal canister interfaces can be deferred until a

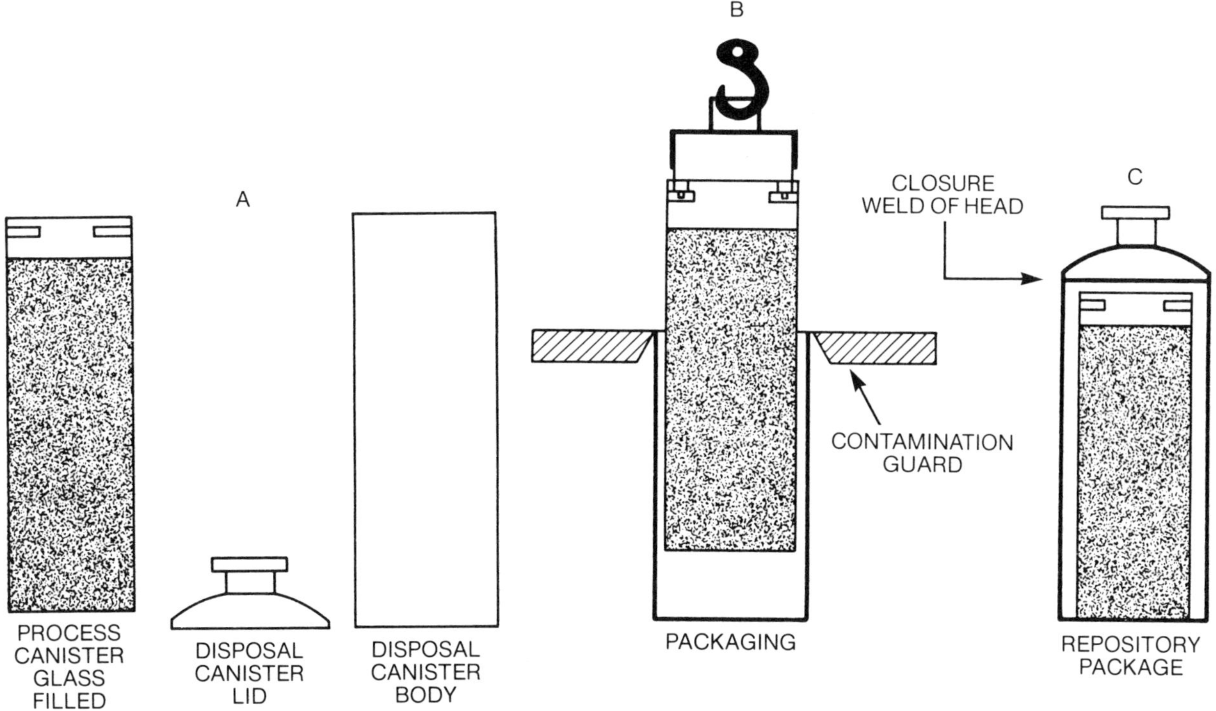

FIGURE 1. Process Can Being Loaded into a Disposal Canister

few months prior to actual production. This approach will provide the opportunity to more completely meet the specifications being discussed at that time at the lowest cost. Conceptually, the flow sheet at West Valley for the process canister will consist of the following steps (see Figure 2).

1. The empty process canister is put into the Vitrification Facility (VF) transfer cart in the Equipment Decontamination Room (EDR).

2. The cart and canisters are moved into the VF, loaded into the turntable, filled, and removed to the decontamination station.

3. A removable canister lid or seal is secured on the processing canister by mechanical means such that water infiltration from Step 4 is precluded.

4. The canister (with cover) is decontaminated of loose contamination by a decontamination solution and water spray and then dried. The spent solutions are recycled to the tank farm or concentrator.

5. The decontaminated process canister is loaded into the transfer cart and transferred to the Chemical Process Cell (CPC) where it is stored until shipped to the federal repository.

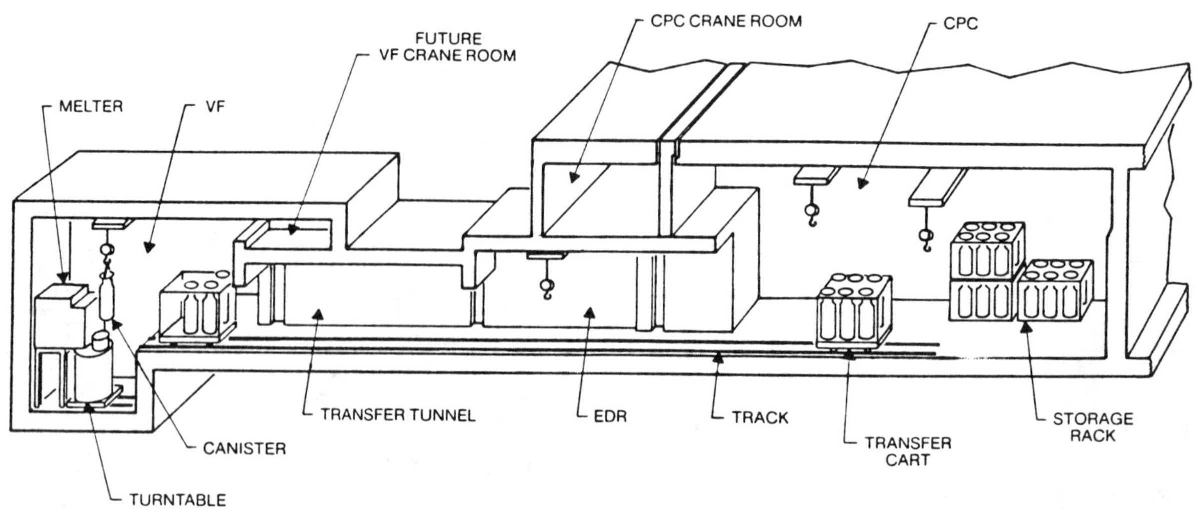

FIGURE 2. Vertical Section of the Vitrification Facility, Equipment Decontamination Room and Chemical Process Cell

At this point, the processing cycle starts over during the vitrification of the West Valley waste. However, when the repository begins to receive waste, handling of the glass canister might continue as follows:

6. Using the canister cart, the process canister is removed from CPC to EDR. The processing canister is loaded into the repository specified disposal canister giving an uncontaminated exterior surface.

7. The disposal canister head is welded closed according to a technique approved by the repository.

8. The disposal canister is loaded into the shipping cask from the VF-EDR tunnel by a five-ton monorail crane which extends over the uprighted shipping cask.

B. Advantages of the Process Canister

The approach presented above allows the WVDP to fix its facility design now while maintaining the ability to meet even the most stringent repository requirements later. This approach reduces risks and defers, and possibly reduces (if the work is done elsewhere), significant costs to the Project in development, design and hardware associated with the disposal canister. The flexibility and options available with this approach are presented in Figure 3. Once the on-site storage period is over, the waste could be shipped without any further packaging at West Valley. The transportation cask would meet applicable transportation criteria, and the repository would package the process can in its container prior to disposal. Otherwise the process can could be packaged on-site in a disposal canister fabricated from carbon steel, stainless steel, or another material. Other aspects of the disposal canister

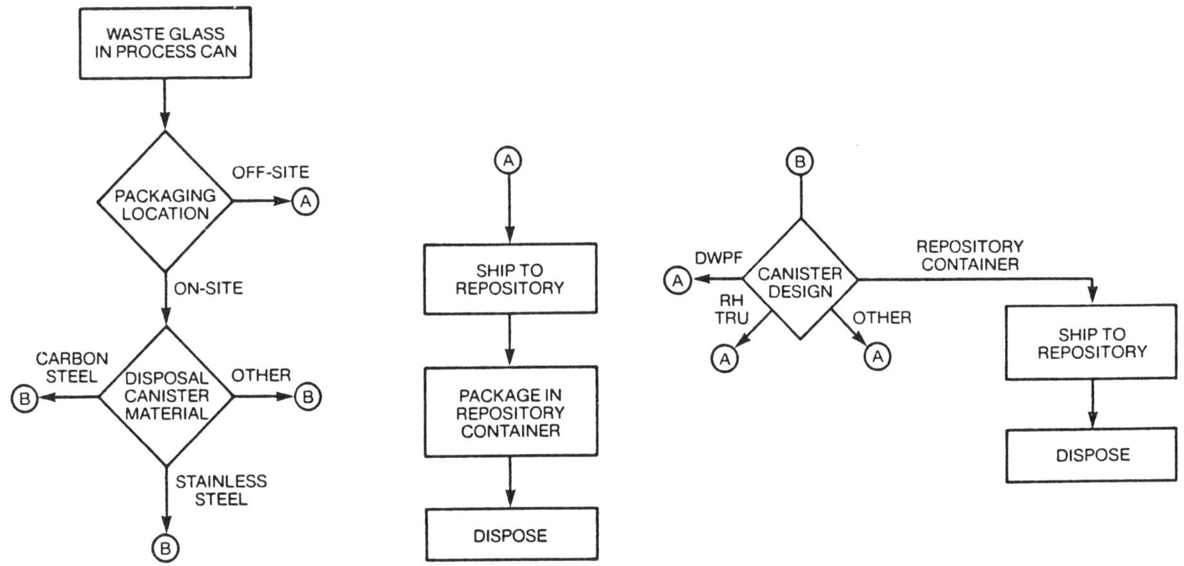

FIGURE 3. Flexibility of the Process Canister Concept

design can be based on the DWPF canister, the remote-handled TRU container, the repository container or another type. It is expected that if the DWPF canister or the TRU container is used, the waste package would have to be placed in a repository container. The less expensive options involve packaging the process cans directly in the repository containers. The least expensive option is to package at the repository. Packaging at the repository would be the least expensive, because one less canister is required and transportation weight is minimized. Other advantages of this approach are discussed below.

1. The disposal canister will not have been exposed to the vitrification process environment and, therefore, will not need a difficult-to-perform decontamination operation to remove tenacious contamination.

   A canister decontamination system for West Valley would require extensive design and modification of any existing systems. At the present time, there are no test results available that confirm that a canister decontamination system will work efficiently at West Valley. The DWPF frit blaster is inappropriate for use at West Valley, because as much as 100 to 150 tons of contaminated frit would be produced at the time of shipment. Similarly, electropolishing may work, but large volumes of contaminated aqueous solutions would need to be treated long after the majority of the Project is over.

   All decontamination systems suffer from the possibility that a "spot" is not successfully decontaminated and not detected until it is received at the repository. The only recourse in this case is to overpack. Any contamination that fell on the disposal canister would be relatively loose and more easily cleaned. Thus, by overpacking the process canisters, the Project assures itself and the accepting repository that the disposal canisters are clean.

2. Canister material properties can be more easily controlled on a disposal canister not exposed to the vitrification process environment.

   There is the possibility that any canister exposed to the vitrification environment will be automatically unqualified and must be overpacked. For example, this may be because the filled canisters may have experienced "unknown" thermal histories and will be suspected of being sensitized. Much information on canister heating will be obtained during cold testing that will be used to explain thermal history during hot operations. However, if the repositories decide that this is insufficient, then the proposed strategy accommodates this possibility automatically.

   The ultimate repository acceptance specifications can be achieved by the proposed strategy. Disposal canister wall thickness, canister material, and canister QA/QC can be made to be acceptable "outside" the radiation area in the future when these requirements are fixed.

3. Without the process canister (with its mechanically attached, removable lid) it is possible that both a canister lid welder and an overpack welder would be required.

   In many unacceptable product and canister scenarios (e.g., improper material or weld, excessive contamination), the practical solution will be overpacking which, with the proposed strategy, is inherent. If the Project selected a canister lid welder, it is likely that a full diameter overpack welder would also be required. By planning to overpack, we avoid cost and effort while satisfying the requirements. In this strategy, the disposal canister welder would be less likely to be contaminated and, therefore, could be contact maintained. This could provide cost savings.

   Finally, since the repositories plan to develop and use welders for containerized canisters, the Project could simply procure a design that is already completed, tested and verified by the U.S. Department of Energy.

4. If the repositories require the high-level waste processors to sample their glass, this would be possible.

   The WVDP does not plan to sample any of its waste glass, but some of the repositories may require it periodically. One draft waste acceptance specification document[7] already discusses waste glass sampling. The proposed approach of using a removable lid on the process canister provides access for a sample any time before shipment. If samples are required from places other than near the top surface, they may be removed from anywhere in the process canister by drilling through it. Patch welding of the disposal canister may not be acceptable for shipping or for disposal. It would likely require overpacking, which again could be accommodated by the proposed approach.

5. The cost of this approach at West Valley appears competitive. The cost disadvantages are the cost of the process canisters and the cost of additional waste packages required by volume and weight limits. However, these costs may be offset by the cost savings of not having to decontaminate the canisters to make them acceptable for transportation and repository acceptance and of not processing decontamination waste streams in the year 2000. Also, there is the development, design, and capital expense of a canister closure system that is capable of achieving its objective. Furthermore, there is the potential cost of overpacking some or all of the waste packages, if an unacceptable product is generated when a process canister is not used.

   The processing canister is limited to being a casting mold for the glass and being transferred within the hot cells at West Valley. Carbon steel or alloy steels with thin walls are conceivable which would minimize costs of the processing canister. Canisters with 0.64-cm (0.25-inch) wall thickness are cast routinely in pilot plant tests. All but one canister produced by Pacific Northwest Laboratories (PNL) for the WVDP have had 0.64-cm (0.25-in.) wall thickness and were made of carbon steel. Much of the glass produced by PNL for WVDP has been cast into 208-liter (55-gallon) carbon steel drums with wall thickness of 0.15 cm (0.06 inches). A full-size (61-cm-diameter by 300-cm-long) 304L stainless steel canister with 0.25-cm (0.1-inch) wall thickness has been filled with glass and successfully handled, lifted, and transferred. Thus, practical experience has demonstrated that an opportunity exists to minimize the cost of the process canister.

C. Future Decisions

Areas yet to be investigated include the selection of the process canister material, the selection of the diameter and length of the process canister, and the transportation interface. Testing at the West Valley Component Test Stand (CTS) and reviewing available data on the properties of waste glass in the presence of various materials will be used to recommend the most economical material for the process canister that can meet handling, thermal distortion, and expected repository requirements. Once the material has been recommended, it will be incorporated into the product verification strategy (e.g., glass durability tests may be performed in the presence of the process and disposal canister materials). The material recommendation will be available for review by the repository projects and the U.S. Nuclear Regulatory Commission before a final selection is made. The diameter and length of the process canister is important, because eventually it must fit inside the container of a size specified by the repository. Similarly, it must be capable of fitting inside a transportation cask. It must also interface with the West Valley vitrification and process equipment.

Recently experiments have been completed at the WVDP to assess the thermal/mechanical responses of potential process canister designs. Two 208-liter carbon steel drums of 1.21-mm wall thickness and three carbon steel canisters have been filled at rates ranging from 45 to over 200 kg/h. The higher rates are about four times the reference rate or are typical of a batch dumping processing strategy. The test process canisters have been about 0.56 mm in diameter and 2.63 m tall with wall thickness of 4.76 mm, 4.16 mm, and 3.4 mm. The maximum diametrical expansion has been less than 1.6 mm. The straightness of the canisters has been maintained within 1.6 mm.

Discussions are being held with the developers of HLW shipping casks to determine the acceptability of this package. Using this approach, the shipping package will consist of waste glass inside a process canister inside a disposal canister inside a shipping cask. If there is a concern about the process canister being able to move inside the disposal canister, this problem may be solved by using inserts or filler material (acceptable to the repository) to fill in the void space. Furthermore, there is the possibility that the process canister could be shipped without the disposal canister.

III. CONCLUSION

The requirements that will have an impact on the selection of a HLW disposal canister are included in 10 CFR 60, 40 CFR 191 (proposed), and various interim repository specifications. These requirements, as well as any other applicable requirements in existence at the time, will form the basis for West Valley HLW disposal canister selection. When this selection is made, an analysis will be made to show that the canister complies with the applicable technical requirements on the disposal of HLW. Because the repositories will decide how to implement the regulations, and because the repository specifications are not fixed yet, the WVDP will continue to develop a strategy in which a process canister will be used to interface with the vitrification and interim storage equipment. Eventually this process canister could be placed inside a repository specified disposal canister which will be developed later. By deferring the use of a disposal canister the WVDP removes the risk of using the wrong type of disposal canister and allows the Project to develop a process canister that may be more suited to its needs. Costs could be reduced if it is decided that the disposal canister can serve the same purpose as the container that some repositories plan to use. Savings would result from not having to procure an additional container and reducing the number of canisters produced by being able to put more glass in each canister. The basis for selection of the process canister will be to meet on-site equipment interface and safety requirements.

REFERENCES

1. Public Law 96-368, October 1, 1980.

2. U.S. Nuclear Regulatory Commission (NRC), "Disposal of High-Level Radioactive Wastes in Geologic Repositories Technical Criteria, " 10 CFR 60, Washington, D.C., June 21, 1983.

3. U.S. Environmental Protection Agency, "Environmental Standards for the Management and Disposal of Spent Nuclear Fuel, High-Level and Transuranic Radioactive Wastes," 40 CFR 191 (draft), Washington, D.C.

4. U.S. Department of Energy, Interim Waste Acceptance Requirements for High-Level Waste Form and Canisters for Disposal of West Valley High-Level Waste (WVHLW) in Geologic Repositories, Draft, Washington, D.C. (1984).

5. Conceptual Waste Package Interim Product Specifications and Data Requirements for Disposal of Borosilicate Glass Defense High-Level Waste Forms in Salt Geologic Repositories, ONWI-464, Office of Nuclear Waste Isolation, Battelle Memorial Institute, Columbus, OH (1983).

6 "Draft Waste Acceptance Requirements for the Basalt Waste Isolation Project," SD-BWI-CR-018, Rockwell Hanford Operations, Hanford, WA (1983).

7. V. M. Oversby, The Nevada Nuclear Waste Storage Investigations Project Interim Acceptance Specifications for Defense Waste Processing Facility and West Valley Demonstration Project Waste Forms and Canisterized Waste, UCID-20165, Lawrence Livermore National Laboratory, Livermore, CA (1984).

8. U.S. Nuclear Regulatory Commission (NRC), "Packing of Radioactive Material for Transport and Transportation of Radioactive Material Under Certain Conditions," 10 CFR 71, Washington, D.C.

9. U.S. Department of Transportation, "Hazardous Materials Tables and Hazardous Materials Communications Regulations," 49 CFR 172, Washington, D.C.

10. U.S. Department of Transportation (DOT), "Shippers - General Requirements for Shipments and Packagings," 49 CFR 173, Washington, D.C.

11. "User's Manual for Remote-Handled Transuranic Waste Container," RHO-RE-MA-7, Rockwell Hanford Operations, Richland, WA (1984).

TMI-2 DEFUELING CANISTERS

P. C. Childress
E. J. McGuinn
D. A. Nitti
W. G. Pettus
J. M. Storton
Babcock & Wilcox, Co.
3315 Old Forest Road
Lynchburg, VA 24506-0935

ABSTRACT

The accident at TMI-2 resulted in a significant portion of the reactor core being reduced to debris ranging in size from very small fines to partial fuel assemblies. Babcock & Wilcox (B&W) has developed designs for three types of waste packages (canisters) which will be used with different defueling techniques to encapsulate the core for transportation to and subsequent storage at INEL. Analysis and testing, including a series of drop tests, were used to ensure characteristics for operational and faulted conditions that comply with criticality and NRC/ASME Code criteria.

I. INTRODUCTION

Babcock & Wilcox has developed, under contract to GPU Nuclear Corporation, three canister designs consistent with the different defueling techniques to be used at TMI-2. Compatible with the storage pool environment, these canisters (Figures 1, 2, and 3) provide an effective containment for the storage of the TMI-2 core debris. The canisters will retain and encapsulate debris ranging in size from very small fines (0.5 micron) to partial-length and full cross-section fuel assemblies. Special features were factored into the canister design to ensure that the canister contents will remain subcritical under all postulated TMI-2 site and transport conditions as well as interim storage conditions at DOE's Idaho facility. Most components are fabricated from AISI 300 Series low-carbon stainless steel.

II. GENERIC CANISTER FEATURES

The canisters are circular pressure vessels that encapsulate one of the three possible internals modules depending on the function of the canister. Except for the upper closure head, the basic pressure vessel (outer shell) is identical for all three canister designs and consists of a 14 inch OD, .250 inch wall pipe, a reversed dished formed lower head (.375 inch wall) and a machined plate upper head (about 3 inches thick).

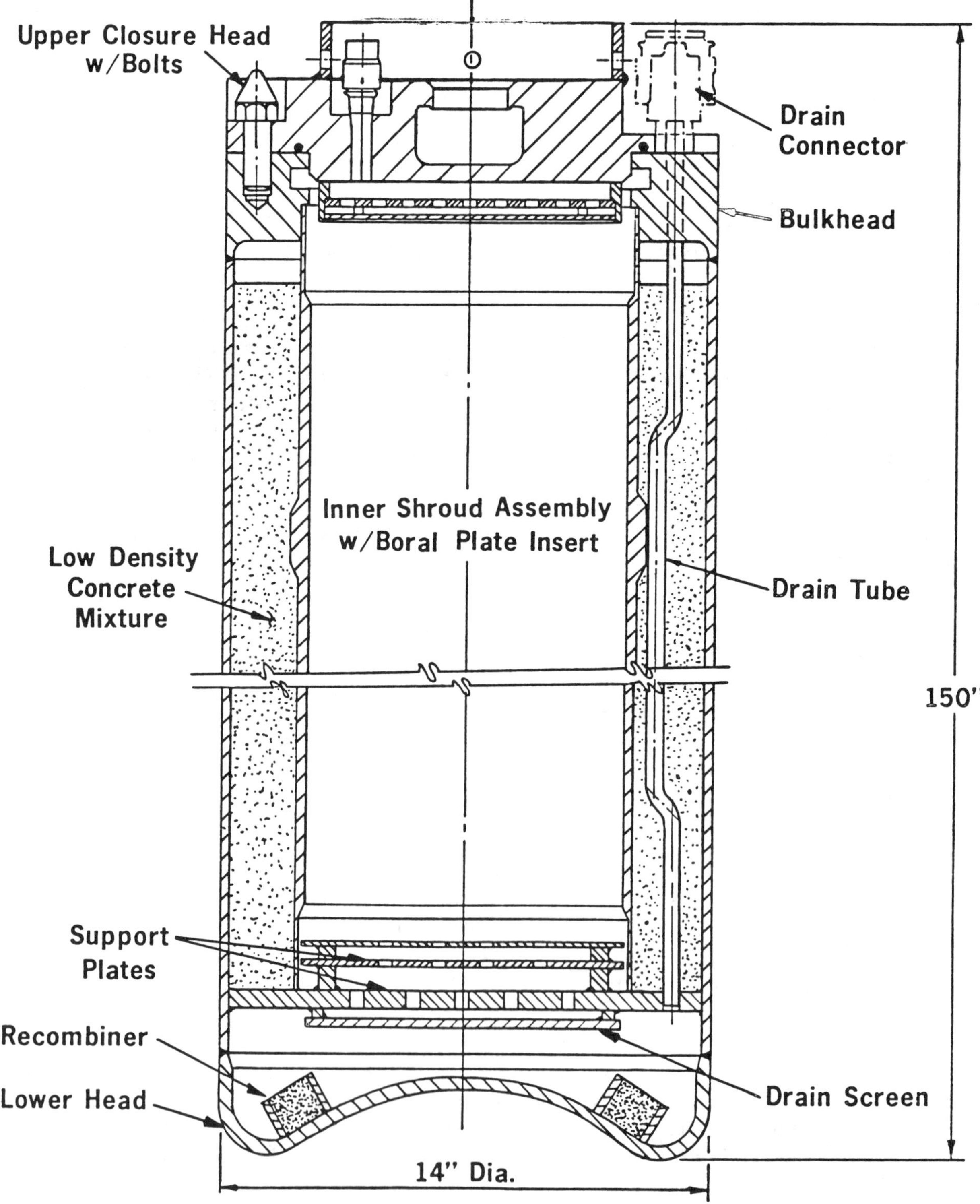

FIGURE 1. Fuel Canister

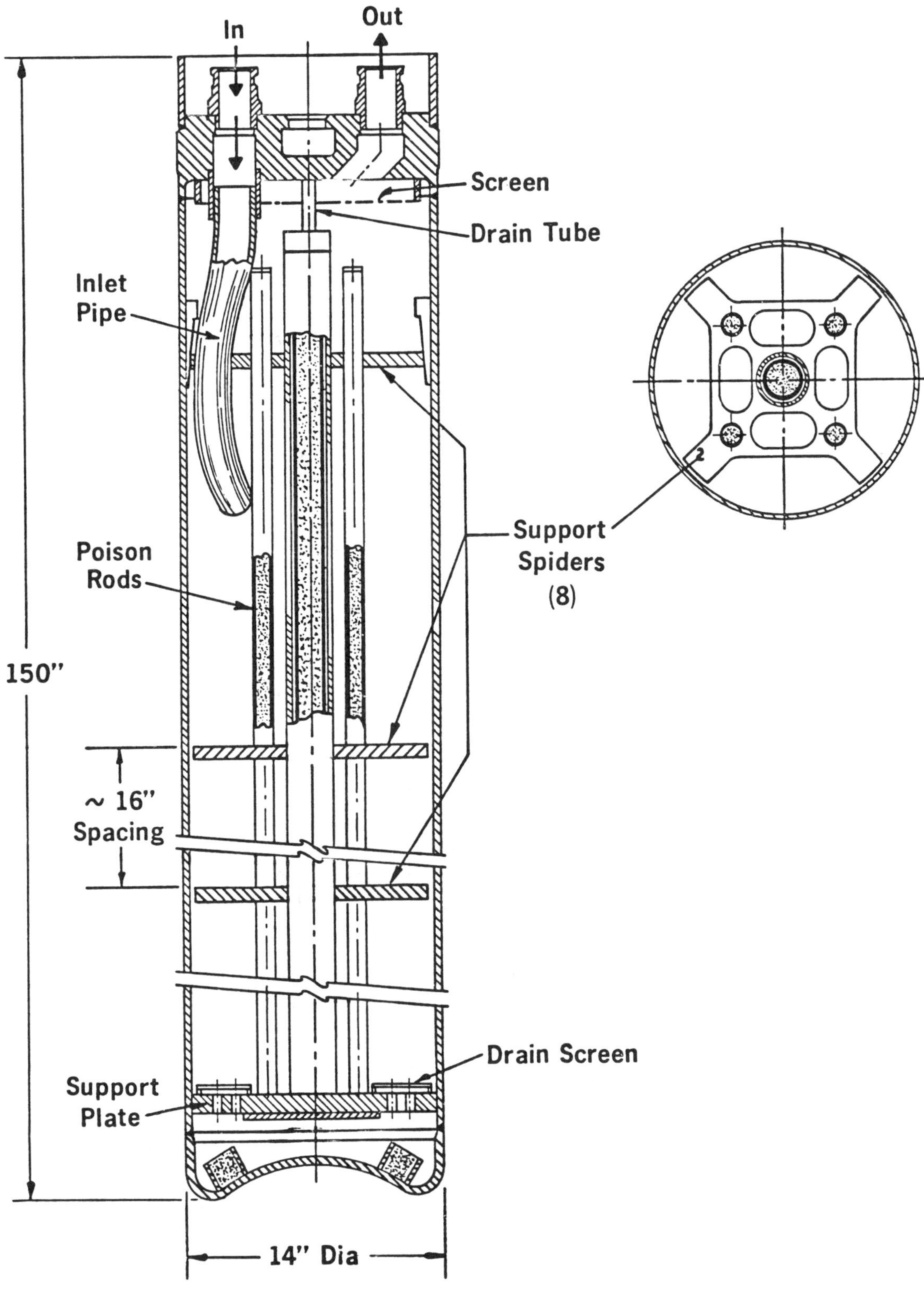

FIGURE 2. Knockout Canister

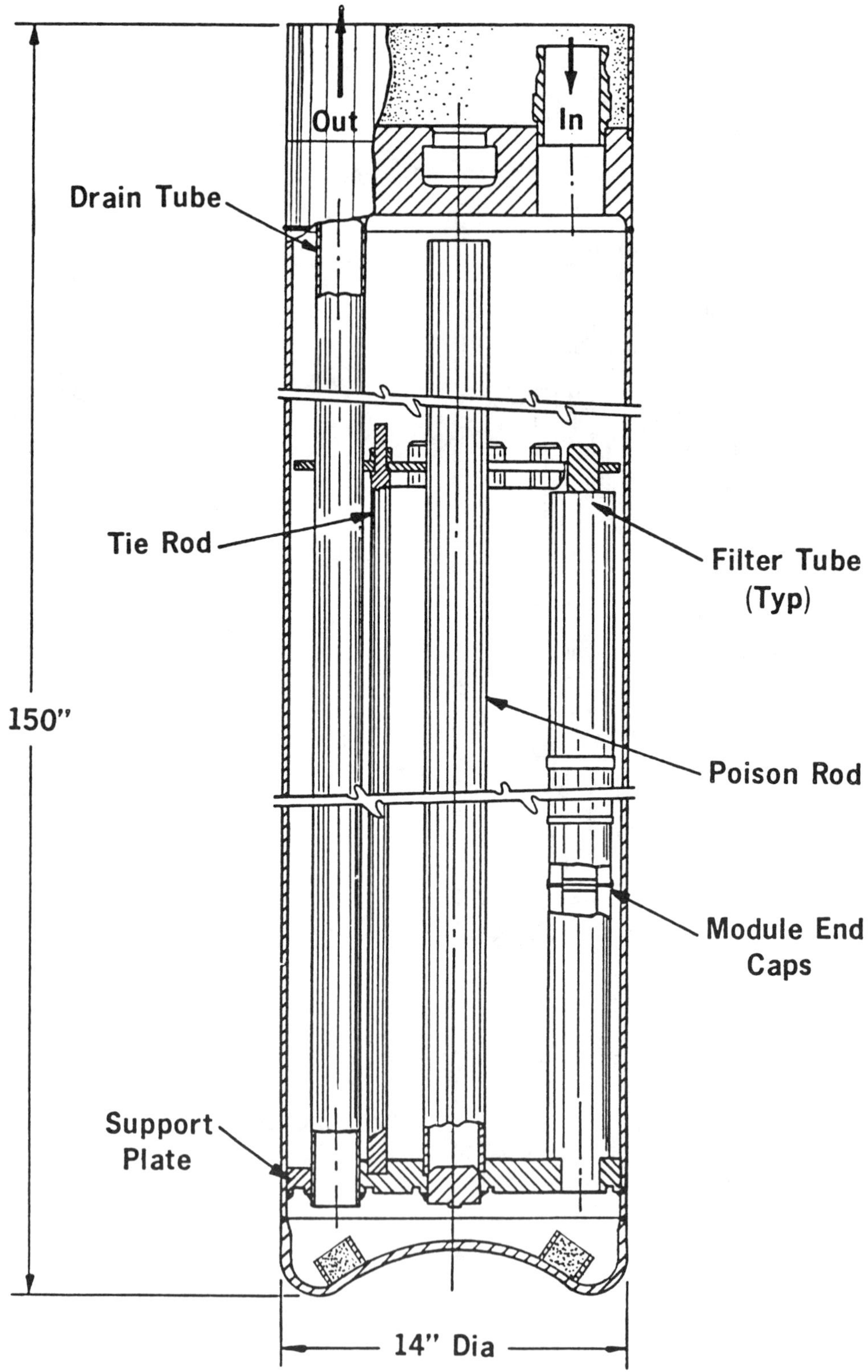

FIGURE 3. Filter Canister

The outer shell protects against leakage of the canister's contents as well as provides structural support for the neutron absorbing materials. It is sized to withstand the stresses associated with normal operating and postulated accident conditions. Material and fabrication requirements for the outer shell are consistent with ASME Code Section VIII[1] criteria. All connection fittings are quick disconnect and are located in the upper head for easy access. A skirt on the upper head protects the fittings from side impacts. All fittings are torqued in place using a sealant capable of maintaining the seal for the design life of the canister. A recessed grapple interface in the center of the upper head provides single pickup with low susceptibility to shipping damage. Water trapped by the skirt and handling recess is drained by a machined hole in the upper head.

The same basic dewatering system is used to remove the slosh water in all three types of canisters prior to shipment. Inert gas introduced into the top of the canister drives the free water to the bottom sump, through the internal drain line and out of the canister to a water processing system. The inert gas is then stabilized at about 30 psia to serve as a cover gas for long term storage. All dewatering fittings are Hansen self-sealing types that can be capped or hold a removeable pressure relief valve. The fitting and cap seals are made of Ethylene Propylene (EP) to provide adequate radiation resistance.

A recombiner catalyst package is incorporated into the upper and lower head of all the canisters. The catalyst recombines the hydrogen and oxygen gases formed by radiolytic decomposition of the residual water entrapped in the wet debris. This reduces the buildup of internal pressure in the canister and keeps the gases below their flammability/explosive limits. The redundant catalyst locations and dewatering operations ensure that an adequate amount of catalyst is available to assure their proper function with the canister in any orientation. The catalysts have been placed in locations consistent with maintaining their function even after a drop accident.

III. SPECIFIC CANISTER FEATURES

The fuel canister, Figure 1, is a receptacle for large pieces of core debris and fuel assemblies that can be mechanically placed in the canister. For this reason, the upper closure head is removable to permit easy access for debris loading. The closure head is attached to the canister body by eight Inconel 625 bolts and is sealed by the use of Inconel X-750 O-rings. An internal square shroud controls the size of the inner cavity and encapsulates the neutron absorbing material, Boral™, used for criticality control. The space between the shroud and the outer shell is filled with LICON™, a B&W-developed low density, high strength concrete that provides lateral structural support during accident conditions.

As part of the core debris vacuum system, the knockout canister, Figure 2, separates the medium size debris from the water by suddenly reducing the flow velocity and reversing the flow direction, thereby allowing the larger particles, from fuel pellet size down to about 140 microns, to settle out. An internal array of four rods around a larger center rod, all containing boron carbide (B_4C) pellets, is included for criticality control. The center poison rod is contained within a thick-walled strongback tube that is welded to lateral supports on a 16-inch axial pitch. The vacuum system fittings are 2-inch camlocks that are sealed with Thaxton plugs after the hoses are disconnected.

To remove very small fines, the filter canister, Figure 3, utilizes permanent internal filter elements fabricated from a sintered stainless steel media. These elements are joined together to form a filter bundle permitting a flow rate up to 125 gpm while filtering out particles as small as 0.5 microns. The elements are sized such that the canister will run continuously until completely filled with fines (400-1000 pounds, depending on particle size.) The performance of the media, a recent development of Pall Trinity Micro Corporation, was tested by B&W to ensure a steady state effluent turbidity of less than 1 Nephelometric Turbidity Unit (NTU) while being challenged by a variety of ZrO_2, Fe_2O_3, stainless steel and UO_2 particles. Loading characteristics, operational performance and backwash capabilities were also investigated using an innovative see-thru housing which allows video taping of the tests. A center rod containing B_4C pellets ensures that the canister contents remain subcritical. Like the knockout canister, the system flow connections are sealed with Thaxton plugs after use.

IV. ANALYSIS/TESTING

The canisters were designed to satisfy the acceptance criteria for normal operation as stated in the ASME Pressure Vessel Code [1] Section VIII, Division 1 (Lethal). Faulted conditions, either from a drop accident during handling or a transportation accident within the shipping cask, were analyzed using the ASME Code Section III or NRC Regulatory Guide 7.6. The most deformed geometries predicted for each canister were then conservatively modeled using the KENOIV[2] computer code to ensure an acceptable margin to the appropriate criticality criteria.

The filter and knockout canisters were analyzed for faulted condition using the ANSYS[3] computer code and classical engineering techniques. The knockout canister, for example, had its internals module simulated using a 299 element model that calculated stresses and deflections for each citicality control rod about every 4 inches down its length. All analyses showed compliance with the appropriate requirements for each regulatory scenario. Additional drop testing of the knockout canister to confirm the analytical results has been scheduled for late 1985.

Because of the complex geometry of the fuel canister B&W felt that a series of drop tests performed using simulated canisters would more reliably confirm its design adequacy than would a computer model. Vertical and horizontal drops from heights ranging up to 30 feet were performed by B&W and confirmed that the resultant deformations of the fuel canister were within the modeling assumptions used in the criticality analyses. Table 1 displays the pertinent parameters of the fuel canister drop testing program.

REFERENCES

1. American Society of Mechanical Engineers (ASME), Boiler and Pressure Vessel Code (1983).

2. Babcock and Wilcox, KENOIV, An Improved Monte Carlo Criticality Program (B&W Version of ORNL Code, KENOIV), NPGD-TM-503, Rev. B.

3. "ANSYS" Revision 4.1, Swanson Analysis System, Inc., Houston, PA (1983).

TABLE 1. Summary of Drop Test Conditions for Fuel Canisters

	VERTICAL DROP TEST		HORIZONTAL DROP TESTS			
TEST NO.	V-1	V-2	H-1	H-2	H-3	H-4
Orientation of Impact	Axial Impact on Std Elliptical Head	Axial Impact on Inverted Dished Head	Lateral Impact on Flat Side of Shroud	Lateral Impact on Edge of Shroud	Lateral Impact on Edge of Shroud	Lateral Impact on Flat Side of Shroud
Shroud Impacts	Support Plate	Support Plate	Lateral Support Tubes	Lateral Support Tubes	Lateral Support Tubes	LICON Cement
Shroud Length (in.)	~159	~159	~79	~ 80	~80	~30
Type of Load	Gravel + Steel Scrap	Gravel + Steel Scrap + Water	Gravel + Steel Scrap	Gravel + Steel Scrap	Gravel + Steel Scrap	Steel Scrap
Load in Shroud (lbs)	~1,800	~1,550	~.575	~575	~600	~370
Total Weight (lbs)	2,700	2,850	1,000	1,000	1,025	600
Drop Height (ft)	30	18	15	15	30	30
Impact Velocity (ft/s)	44	34	31	31	44	44
Impact Energy (ft-lbs)	81,000	51,300	15,000	15,000	30,750	18,000

GENERAL CORROSION STUDIES OF CANDIDATE CONTAINER MATERIALS FOR THE BASALT WASTE ISOLATION PROJECT

J. M. Lutton
W. F. Brehm
H. P. Maffei
C. L. Rivera
Westinghouse Hanford Company
P.O. Box 1970
Richland, Washington 99352

R. P. Anantatmula
Rockwell Hanford Operations
P.O. Box 800
Richland, Washington 99352

ABSTRACT

The general corrosion studies program consists of three types of tests to determine the corrosion behavior of candidate container materials for the Basalt Waste Repository. Air/steam tests simulate conditions in the repository operating period, and static and refreshed pressure vessel tests simulate post-closure conditions. Copper and iron-based alloys and their weldments are being tested. Weight loss and surface analysis data from two short-term tests are reported. In these, the effect of weldments in static groundwater, and that of packing material in an air/steam environment were studied. Early results from long-term testing are also included.

I. INTRODUCTION

The Basalt Waste Isolation Project (BWIP) has the responsibility for the development and design of waste packages suitable for containment of high-level nuclear waste in a repository constructed in basalt. These waste packages, according to Federal regulations, must provide substantially complete containment of the waste for 300 to 1000 years after emplacement. Therefore, successful long-term performance in the environment of the repository is an important criterion in the selection of a material for container fabrication.

In support of the BWIP, Westinghouse Hanford Company (WHC) is performing a series of studies to evaluate the corrosion behavior of candidate container materials in simulated repository environments. The anticipated environment during the repository operating period is an air/steam environment, in which the containers may reach temperatures as high as 300°C. In the post-closure period, the environment will consist of groundwater, which has equilibrated with the basalt surrounding the repository and the packing material surrounding the containers. The maximum container surface temperature expected during this period is 200°C.

II. EXPERIMENTAL ARRANGEMENTS AND CONDITIONS

A. Types of Tests

The general corrosion studies program includes three kinds of tests:

1. air/steam tests,
2. static pressure vessel tests, and
3. refreshed autoclave tests.

The air/steam tests simulate the repository operating period and will provide long-term corrosion data in an oxic, humid atmosphere. In these tests, ambient air is passed through a heated humidifier and then through heated lines to controlled temperature chambers containing the specimens. Each chamber has four sections to accommodate specimen racks, the partitions functioning as flow directing baffles. Exhaust air passes through condensers and flow meters, which monitor the moisture content and flow rate of the gas.

The post-closure environment is simulated in both the static pressure vessel and refreshed autoclave tests. Small, general purpose laboratory pressure vessels, made of titanium, having a volume of about 125 ml, are used in the static tests. The vessels are loaded with about 55 ml of packing material saturated with groundwater, and corrosion specimens are inserted so that they are buried in the slurry. Then 18 more milliliters of groundwater are added to cover the packing. After sealing, the vessels are exposed in controlled temperature ovens. For the refreshed autoclave tests, one and four liter titanium lined autoclaves are used. Test specimens are surrounded by packing material in Hastelloy C capsules which have porous ends. Groundwater is pumped at very low rates into the autoclaves at the test pressure and flows out through regulating valves. The autoclaves operate at controlled temperatures and pressures. In all tests, specimens are electrically isolated from each other and all other metallic materials.

B. Testing Conditions

The work performed in this program is designed to develop base case corrosion data; radiation effects are not included. A companion program is in progress in which similar experiments are being performed to determine the effect of radiation on corrosion.

For the present tests, synthetic groundwater, formulation GR-4, was used. The composition is given in Table 1. Packing material, as currently formulated, consists of 25% bentonite and 75% fine basalt. Fine basalt is defined as:

6 parts -11 to +16 mesh
11 parts -16 to +60 mesh
2 parts -60 to +120 mesh
1 part -120 to +230 mesh.

TABLE 1. Composition of Synthetic GR-4 Groundwater

	Concentrations, mg/liter (ppm)								
Substance:	F^-	Cl^-	$SO_4^=$	Na^+	K^+	Ca^{+2}	Si	Mg^{+2}	C*
As Prepared:+	20	398	4	349	14	2	46	0	20

*Total Carbon
+The initial pH of Synthetic GR-4 groundwater is 9.7 ± 0.2

This material was prepared from natural basalt, collected from outcroppings in the field, then crushed and sieved by the Rockwell Hanford Operations (RHO) Materials Testing Laboratory. Before use, the basalt was blended and split to appropriate sample sizes, using a riffle sample splitter[1]. Individual samples were blended with bentonite.

The air/steam tests are carried out at four temperatures: 150°C, 200°C, 250°C and 300°C at atmospheric pressure. The maximum temperature was selected based on BWIP design studies which project 300°C as the maximum container surface temperature expected for a spent fuel container of reference design. The operating period of the repository is expected to be not more than 90 years. The air flowing through the test chambers has a dew point of 40°C. This condition was selected based on the mean expected temperature of the basalt horizon in the repository. Steam conditions were obtained by passing the air through two humidifiers in series, the first filled with demineralized water and the second with synthetic GR-4 groundwater, both operated at 49°C. The heat-traced tubing carrying the air operates at temperatures in excess of 60°C to prevent condensation. The air flow rate (234 liters/day) was chosen to produce an approximate linear velocity of 0.4 meter/hour along the specimens, and is the same as that to be used in the radiation effects tests.

In all tests simulating the post-closure period, the exposure temperatures are 50°C, 100°C, 150°C and 200°C, based on a predicted maximum container surface temperature of 200°C for this condition and a minimum of 50°C. In the repository, the actual temperature of a container at the time of closure will depend on the heat output from racioactive decay. Thereafter, the temperature will slowly decline since the peak temperature is expected to be attained prior to closure. In the refreshed autoclaves, groundwater flow rates were set by the lowest flows achievable in the smallest autoclaves, 6 ml/hr. For the larger autoclaves, the flows are set to produce the same effective rate per unit volume of the autoclave. These flows, together with the zero flow of the static tests provide an upper and lower bound to the flow conditions expected in the repository period. The specimens for these tests were loaded and handled under anoxic conditions, and all groundwater used was also oxygen free. This condition was selected because the strongly reducing character of the basalt is expected to produce oxygen free groundwater in the natural situation.[2]

Rectangular corrosion specimens, approximately 1/16 inch thick, were used in all types of tests. In the air/steam tests the specimens are 1" by 2", while those for the static pressure vessel tests are 1/2" by 1". Two sizes are used in the refreshed autoclaves, 1" by 3/8", and 1/2" by 3/4". Before testing, the specimens, after being cut to size, were polished through 600 grit and engraved with identification markers. The specimens are then dimensionally measured, cleaned and degreased with trisodium phosphate, dried and weighed. Five specimens are used for each specific test condition. Also, welded specimens were used in some plates, cut to the specimen sizes described above, so as to include the base material, fusion zone, and heat affected zone. Preparation of the welded specimens before testing was as described above.

Post-test analysis of corrosion specimens consists of weight change measurements, characterization of surface corrosion products, and metallography of the cleaned surfaces. The results from these analyses will be applied to characterize the corrosion behavior of the materials and will be used to develop and update container corrosion models. Weight loss data are reported in terms of mills/year of penetration in the present paper.

III. TESTING COMPLETED

The test program includes long-term tests, scheduled to run for at least 42 months, and a variety of short-term tests, aimed at evaluating the effect of specific variables on the corrosion process.

A. Short-Term Tests

Two short-term tests have been completed to date. In both tests, several candidate materials and their weldments were evaluated.

1. Air/Steam Packing Material Effects Test

The first of these tests was aimed at determining whether the presence of packing material influences corrosion behavior in an air/steam environment. Materials studied were cast carbon steel (ASTM A-27 Gr 60-30), Fe9Cr1Mo steel, their weldments, wrought low-carbon steel (AISI 1020), and cupronickel 90-10 (CDA 706.) In addition, specimen sets of Fe9Cr1Mo steel prepared by surface grinding rather than polishing were included to compare the effects of surface condition on the corrosion process. Specimen sets of each material were loaded into packing material in teflon beakers and five corresponding specimens were strung on hangers. Specimens were exposed in air/steam test chambers at 150°C and 250°C for one month.

After completion of the test, all specimens from each set were photographed, dried and weighed. Four specimens were then cleaned with formaldehyde-inhibited hydrochloric acid to remove corrosion product films and reweighed. The fifth specimen was reserved for surface characterization.

Corrosion data, based on weight loss measurements, from this one-month test are presented in Table 2. In general, very little corrosion occurred. (It should be noted that a corrosion rate of 0.01 mils/year corresponds to a weight loss of about 0.5 mg in these tests.) Packing material generally has little or no effect on the corrosion of any of the ferrous materials. Statistical tests indicate that only two of the differences shown in the table are significant and these are in opposite directions, indicating that packing material has no real effect on the corrosion of a particular material. In packing material, the wrought carbon steel corroded at significantly higher rates than the A27 cast steel, though in the absence of packing material the rates were not detectably different. For the cupronickel 90-10 at 250°C, where the corrosion was markedly higher than for any of the ferrous alloys, the specimens in packing material corroded significantly less than those in the air/steam. Data from other tests presented below indicate that these one-month tests are not very reproducible, so the differences in detail noted should be interpreted cautiously.

TABLE 2. Short-Term Air/Steam Packing Material Effects Test

Material:	Loading	Corr. Rates ± Stand. Dev., mils/yr	
Fe9Cr1Mo	Condition	150°C	250°C
Base	Bare	0.040 ± 0.047	0.039 ± 0.015
Base	+Pack. Mat.	0.019 ± 0.047	0.039 ± 0.015
Weldment	Bare	0.028 ± 0.010	0.038 ± 0.016
Weldment	+Pack. Mat.	0.019 ± 0.004	0.027 ± 0.005
Surf. Ground	Bare	0.031 ± 0.023	0.020 ± 0.010
Surf. Ground	+Pack. Mat.	0.014 ± 0.005	0.015 ± 0.001
A27 Cast Steel			
Base	Bare	0.052 ± 0.007	0.110 ± 0.026
Base	+Pack. Mat.	0.039 ± 0.003	0.094 ± 0.010
Weldment	Bare	0.056 ± 0.006	0.080 ± 0.008
Weldment	+Pack. Mat.	0.027 ± 0.009*	0.089 ± 0.009
AISI 1020 Steel			
Base	Bare	0.063 ± 0.012	0.109 ± 0.008
Base	+Pack. Mat.	0.068 ± 0.009	0.156 ± 0.010*
Cupronickel 90-10			
Base	Bare	0.113 ± 0.012	0.643 ± 0.084
Base	+Pack. Mat.	0.141 ± 0.012	0.282 ± 0.024*

* Denotes that the difference in rate between the bare metal and that in packing material is significant at the 95% confidence level.

Corrosion of the weldments was essentially the same as that of the base materials. The different surface preparation treatment given to Fe9Cr1Mo steel specimens had no observable effect on the corrosion within the limits of the error. Cupronickel 90-10 and the carbon steels all show significantly higher corrosion at the higher temperature. The ferritic Fe9Cr1Mo specimens all corroded about the same amount and substantially less than any of the other materials.

2. Weldment Effects, Static Pressure Vessel Test

The purpose of this test was to compare the corrosion behavior of candidate ferrous container materials and their weldments. Welded specimens include base material, fusion zone, and heat affected zone. The tests were carried out in static pressure vessels, with specimens loaded in packing material saturated with synthetic GR-4 groundwater. Materials studied were Fe9Cr1Mo steel and Cast A27 Gr 60-30 steel. Sets of five specimens were loaded in individual pressure vessels for each material. The coupons were placed in a slurry of groundwater and packing material so that they were isolated from each other and the titanium walls of the vessel. Enough groundwater was then added to make the ratio of water to packing solids equal to 2.36. All loading operations were done in an argon glove box and argon-sparged groundwater was used to obtain anoxic conditions.

Sealed pressure vessels were exposed for one month at 100°C and 200°C and then unloaded in the argon glove box. Groundwater samples from the vessels were analyzed. Results from the test are presented in Table 3. The data again show no significant difference between the corrosion rate of the parent materials and their weldments at either temperature. The data indicate that the corrosion rate of Fe9Cr1Mo steel increases with increasing temperature, though it is lower than that for the carbon steel under both conditions in this test. Conversely, the A27 cast steel was attacked much more severely at 100°C than at 200°C. This is in accord with results previously reported[3] and is presumably associated with a more porous and less adherent reaction product film at 100°C. Structural studies of these films are underway.

B. Long-Term Testing

Long-term testing has been started in all three types of test. Five basic materials are included:

Ferritic Steel, Fe9Cr1Mo
Cast Carbon Steel, A27 Gr 60-30
Oxygen-Free Copper, CDA 102 (2 Heats)
Phosphorus Deoxidized Copper, CDA 122 (2 Heats)
Cupronickel 90-10 (CDA 706)

These long-term tests are planned to run at least 42 months, with intermediate exposure and examination periods of 1, 4, 10, 18 and 28 months. Pressure vessel and air/steam tests are operating at all four temperatures. Five

TABLE 3. Weldment Effects in Static Pressure Vessels

Material:	Corr. Rates ± Stand. Dev., mils/yr	
Fe9Cr1Mo Steel	100°C	200°C
Base metal	0.169 ± 0.011	0.329 ± 0.016
Weldment	0.171 ± 0.014	0.354 ± 0.038
A27 Cast Carbon Steel	100°C	200°C
Base metal	2.546 ± 0.141	0.400 ± 0.011
Weldment	2.607 ± 0.100	0.420 ± 0.014

refreshed autoclaves are also operating, two with copper-base alloys (100°C and 200°C) and three with iron-base materials (50°C, 100°C and 200°C). At this time, most of the one-month exposure data has been obtained.

The results from the air/steam tests are shown in Table 4, and from the pressure vessel tests in Table 5. All of these data represent only one-month exposures and are to be viewed with caution with respect to interpretation of any long-term trends. Corrosion rates can be expected to be relatively high this early in testing. Some general observations can be made from the data, though no attempt should be made to extrapolate to long-term conclusions at this point. No significant differences in the corrosion rate of the Fe9Cr1Mo steel and the cast A27 steel were observed in the air/steam test, except at 300°C, where the corrosion rate of the cast A27 steel was considerably higher. The copper-base materials all did well in the oxygen free water of the pressure vessel test, but less well at the higher temperatures in the air/steam environment. At 150°C and 200°C, the high purity coppers behaved similarly in both environments, while the cupronickle 90-10 showed higher rates in the liquid at these common temperatures. Limited data for the pure coppers from one month autoclave tests, given in Table 6, show rates similar to, but lower than that observed in the pressure vessels. The iron-base alloys, on the other hand, perform well in the air/steam atmosphere. Though Fe9Cr1Mo in the pressure vessel test shows slightly higher corrosion rate than the copper alloys, it is superior to the cast steel, particularly at intermediate temperatures. That the rates observed are not representative of longer-term results is suggested by the data shown in Table 7 for four months exposure of the two iron-base alloys in the 200°C autoclave. These rates are considerably lower than the one-month rates observed in the pressure vessel test.

All of the materials showed a general increase in corrosion rate with temperature in air/steam and pressure vessel type tests, (under anoxic aqueous conditins with packing), with the exception of the A-27 cast carbon steel, where the rate peaks at 100°C in the pressure vessel environment. This behavior was noted also in the weld effects test discussed above. Pertinent results from the short-term are included in Tables 3 and 4 so that direct comparison

TABLE 4. Long-Term Air/Steam Tests, One-Month Data

Test Temp:	Corrosion Rate ± Stand. Dev., Mils/yr 150°C	200°C	250°C	300°C
Copper-base				
Cu-Ni 90-10	0.046±0.003	0.096±0.017	0.148±0.037	0.268±0.024
Cu-Ni 90-10*	0.113±0.012		0.643±0.084	
O_2-free Cu-1	0.091±0.007	0.300±0.005	0.844±0.026	1.605±0.243
Ph-deox-Cu-1	0.085±0.004	0.390±0.038	1.318±0.039	1.438±0.096
Iron-base				
Fe9Cr1Mo	0.051±0.005	0.029±0.009	0.041±0.010	0.082±0.044
Fe9Cr1Mo*	0.040±0.047		0.039±0.015	
A-27	0.043±0.005	0.036±0.003	0.054±0.006	0.236±0.015
A-27*	0.052±0.007		0.110±0.026	

*Comparable data from Short Term Test (Table 2)

Table 5. Long-Term Pressure Vessel Test, One-Month Data

Test Temp:	Corrosion Rate ± Stand. Dev., Mils/yr 50°C	100°C	150°C	200°C
Copper-base				
Cu-Ni 90-10	0.033±0.166	0.166±0.003	0.088±0.011	0.304±0.018
O_2-free Cu-1	0.041±0.003	0.074±0.006	0.088±0.005	0.261±0.057
O_2-free Cu-2	0.011±0.003	0.045±0.012	0.092±0.010	0.239±0.045
Ph-deox-Cu-1	0.035±0.007	0.065±0.005	0.119±0.009	0.537±0.041
Ph-deox-Cu-2	0.008±0.003	0.047±0.007	0.092±0.005	0.256±0.037
Iron-base				
Fe9Cr1Mo	0.096±0.012	0.213±0.015	0.404±0.011	0.100±0.066
Fe9Cr1Mo*		0.169±0.011		0.329±0.016
A-27	0.529±0.010	2.660±0.015	1.424±0.049	0.530±0.011
A-27*		2.546±0.140		0.400±0.011

*Comparable data from Short Term Test (Table 3)

TABLE 6. Long-Term Autoclave Tests, One-Month Data

	Corrosion Rates ± Stand. Dev., Mils/yr	
Test Temp:	100°C	200°C
Copper-base (1 Month exp.)		
O_2-free Cu-2	0.033±0.009	0.153±0.073
Ph-deox-Cu-2	0.017±0.003	0.125±0.018

TABLE 7. Long-Term Autoclave Tests, Four-Month Data

	Corrosion Rate ± Stand. Dev., Mils/yr
Test Temp:	200°C
Iron-base (4 Months Exposure)	
Fe9Cr1Mo	0.028±0.004
Fe9Cr1Mo (Weldment)	0.024±0.001
A-27	0.134±0.006
A-27 (Weldment)	0.136±0.009

can be made. The most obvious case of disagreement is that of cupronickel 90-10 specimens from the 250°C air/steam tests. The short-term test specimens, which show a much higher rate, also showed the same behavior relative to those in packing material in the earlier test. Inspection of the exposed coupons showed that the oxide film on the pure copper alloys from the air/steam test was very flaky and non-adherent, particularly on the 250°C and 300°C coupons, while on the cupronickel 90-10, the films were more adherent, but still somewhat flaky at the highest temperatures, a factor which may produce much of the non-reproducibility. On the cast steel specimens, the coupons from the 100°C and 150°C pressure vessels had voluminous and loosely adherent reaction product build-ups.

IV. SUMMARY AND CONCLUSIONS

The work completed to date does not permit definitive conclusions at this time. However, it appears that in the case of the ferrous materials, welding does not affect the general corrosion process. Tentatively, over the short term, it also appears that packing material does not have a marked effect on the corrosion of any of the materials tested in the air/steam environment. These conclusions are subject to confirmation by longer testing. The relative performance of the various candidate materials is also a matter for much longer term evaluations currently in progress.

REFERENCES

1. ANSI/ASME C 702-80, Methods for Reducing Field Samples of Aggregate to Testing Size.

2. D. L. Lane et al., "The Basalt/Water System: Considerations for a Nuclear Waste Repository," Mat. Res. Soc. Symp. Proc. 44, 95 (1985).

3. R. P. Anantatmula, "Effect of Grande Ronde Groundwater Composition and Temperature on the Corrosion of Low-Carbon Steel in the Presence of Basalt-Bentonite Packing," Mat. Res. Soc. Symp. Proc. 44, 113 (1985).

WIPP WASTE PACKAGE PERFORMANCE TESTING ON SIMULATED DHLW: EARLY DATA*

Martin A. Molecke
Sandia National Laboratories
Division 6332
Albuquerque, NM 87185

ABSTRACT

Several series of waste package performance experiments are currently being conducted at the Waste Isolation Pilot Plant (WIPP) facility. These experiments involve 18 full-size, simulated (nonradioactive) defense high-level waste (DHLW) containers of several waste package designs. These DHLW packages are emplaced in vertical boreholes in bedded rock salt, under both (thermal) near-reference and accelerated-aging repository conditions. A major purpose of these experiments is to evaluate the in-situ materials performance (i.e., degradation and alteration) of the various waste package barriers: metallic waste canisters and overpacks, backfill materials, and high-level glass waste form (nonradioactive). The waste package test emplacements are heavily instrumented, providing remote readouts of in-situ temperatures, pressures, moisture inflow, and container tilt. No materials samples have been removed for laboratory analyses to date, but about five months of instrumentation data have been acquired. This paper describes the current technical status of these WIPP tests and discusses the available, early results.

I. INTRODUCTION

A total of 16 full-size, simulated (nonradioactive) DHLW test packages were emplaced in the WIPP facility in 1984[1] for both materials and concept-validation testing. All of these DHLW packages (i.e., containers, backfills and waste form) are located in vertical boreholes in the rock-salt floor of two test rooms in the WIPP. The test rooms are located at a depth of about 650 m below the surface. Six DHLW packages are located in WIPP Room A1 under near-reference repository-thermal conditions. Each of the containers incorporate an internal electric heater with a thermal output of 470 W each; these Room A1 packages are part of a three-room-test unit cell with an overall thermal power density of 18 W/m^2. The other ten DHLW test packages are located in WIPP Room B, under accelerated-aging, overtest conditions. These conditions include a thermal output of 1500 W/container, a thermal flux into the salt about three times greater than in Room A1, the artificial injection of 100 L of brine into 5 of 12 test emplacements (to accelerate interactions), and the introduction of four containers with intentional surface defects. Two of the ten test packages contain nonradioactive DHLW glass from the Savannah River Laboratory (SRL) and

*This work was performed by Sandia National Laboratories and supported by the U.S. Department of Energy under Contract DE-AC04-76DP00789.

experience thermal loading from external sources. Another two glass-filled containers will be added to the existing test matrix in May 1986. The simulated DHLW tests in Room B have been in heated operation since April 1985, those in Room A1 since October 1985.

The primary purpose of these 3- to 7-year duration experiments is to evaluate the in-situ materials performance of all high-level waste package barriers, including their near-field interactions with the host rock-salt environment. The lack of high-intensity radiation fields from fully radioactive waste form is considered the only simulation in these experiments. Basically, the in-situ durability or containment integrity of DHLW packages in salt "repository" emplacement is being evaluated. Also being compared are the overall performance adequacy, engineering feasibility, and cost-effectiveness of two reference-design DHLW containers[2,3] and a developmental canister-overpack. Because these experiments are still in their initial phases, no material samples have yet been obtained for lab evaluations. However, thermal and pressure data obtained from installed, remote reading instrumentation plus current status of these tests will be presented.

II. EXPERIMENTAL DETAILS

A. Test Rooms

The simulated DHLW experiments are being conducted in two separate test rooms (mechanically mined tunnels in the rock salt) in the WIPP facility, Rooms A1 and B. Both rooms are 5.5 x 5.5 x 93.3 m in size and are spaced 80-m apart. In Room A1, the waste packages are emplaced into vertical boreholes in the salt floor in a single-row configuration, 1.9-m apart. Each borehole is either 0.76 m or 0.91 m in diameter and 5.5-m deep. In Room B, the waste packages are in boreholes 0.91 m in diameter and 4.9-m deep, in a double-row configuration. The rows are 2.3-m apart. The simulated DHLW packages in both test rooms are co-located with multiple other (similarly sized) electric heaters being used in separate thermal-structural interactions experiments.[4] As such, the waste packages are part of larger heated-room thermal arrays, as they would be if located in a full-scale operating repository.

B. DHLW Containers

Three basic designs for DHLW containers are used in the experiments. The containers in Room B have been fabricated to be essentially identical to the two current, proposed "reference" designs for DHLW. The first is the TiCode-12 disposal container, as proposed and developed at Sandia National Laboratories[2] for testing in WIPP. TiCode-12 is an alloy of titanium, ASTM grade 12, containing 0.8% Ni and 0.3% Mo. These containers are 3.0-m long, 0.61 m in diameter, and have a 0.64 cm-thick wall. They also have a hemispherical dome top 1.91-cm thick, adequate to resist expected lithostatic pressures, and are illustrated in Figure 1. Four of the TiCode-12 containers have internal 1500-W electric heaters; all internal void volume has been filled with fine-grained silica sand to provide crush-resistance during these tests. Two other

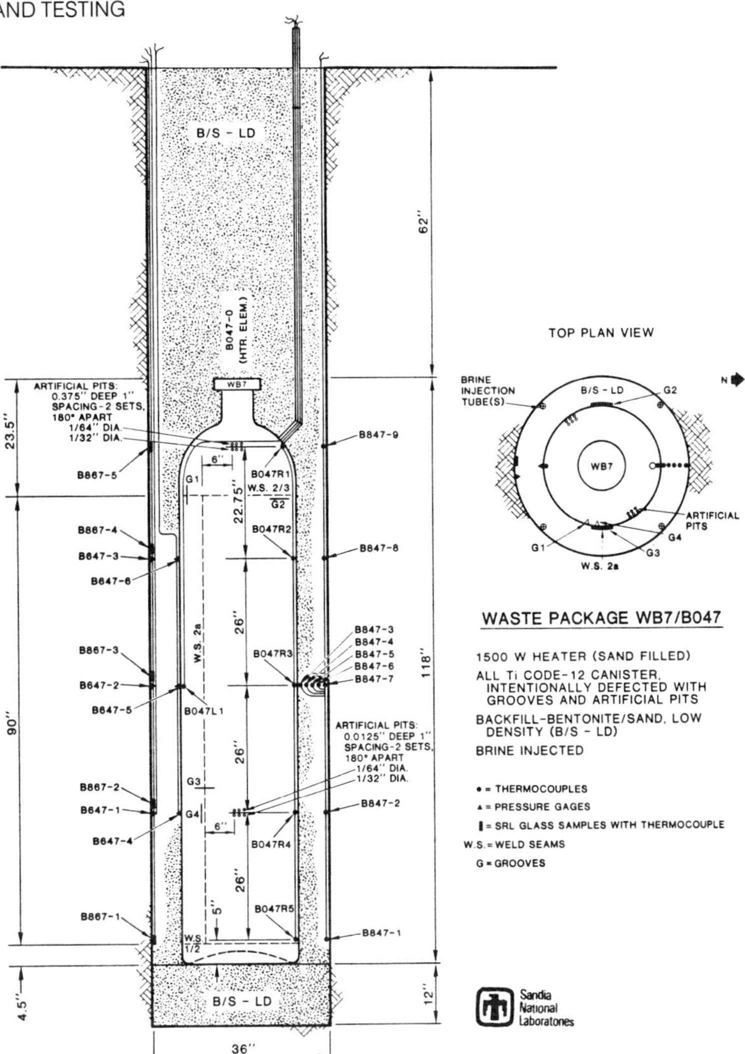

FIGURE 1. Simulated DHLW TiCode-12 Container Emplaced in WIPP

TiCode-12 containers will be filled (to a height of 2.60 m) with nonradioactive glass DHLW (to be described) provided by SRL, and will be added to the Room B test matrix in May 1986.

The second proposed "reference" DHLW containers used in Room B are those designed by the Office of Nuclear Waste Isolation (ONWI) in 1983[3] and fabricated by Sandia National Laboratories. The ONWI-design containers, shown in Figure 2, consist of an inner, stainless steel 304L glass-pour canister (3.0-m long, 0.61 m in diameter, and with a 0.95-cm-thick wall), and a thick, outer overpack of cast mild steel A216 grade WCA (3.35-m long, 0.80 m in diameter, and with an overall 8.6-cm-thick wall). Two of the four test stainless steel canisters contain 1500-W electric heaters, the other two have been filled (to a height of 2.31 m) with nonradioactive DHLW glass at SRL.

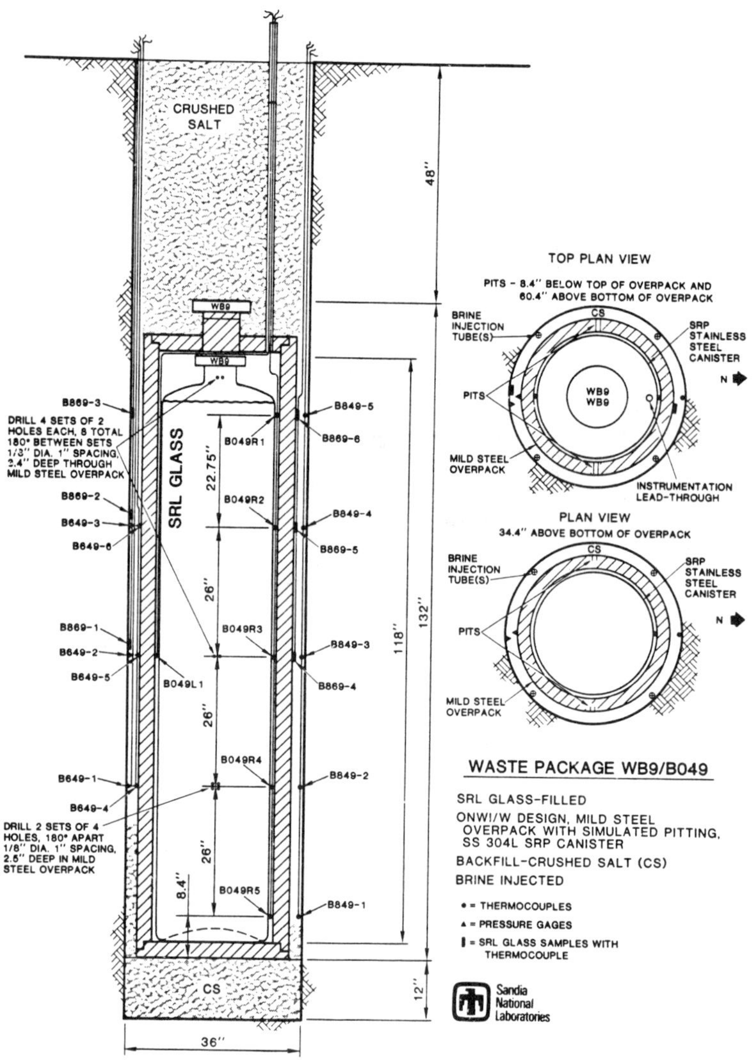

FIGURE 2. Simulated DHLW ONWI-Design Container Emplaced in WIPP

The test containers in Room A1 are basically a developmental canister-overpack design for DHLW. They are fabricated from schedule 60 mild steel pipe, nominally 3.0-m tall, 0.61 m in diameter, and with a 2.5-cm-thick wall. They have flat top and bottom heads and a top handling pintle identical to that used on the other "reference" design containers. Half of these test canister-overpacks have a 2.2-mm-thick TiCode-12 overlay wrapped (with no appreciable air gaps) and welded around the mild steel, as well as TiCode-12 pintles, rather than mild steel ones. The Room A1 test containers, shown in Figure 3, all have internal 470-W electric heaters. These test containers will be further designed and optimized, in order to serve as an overpack for future, actual DHLW canisters. They are currently being tested only for materials behavior, not design feasibility. The TiCode-12 overlay, mild steel canister-overpacks are similar to a previously suggested ONWI design.[5]

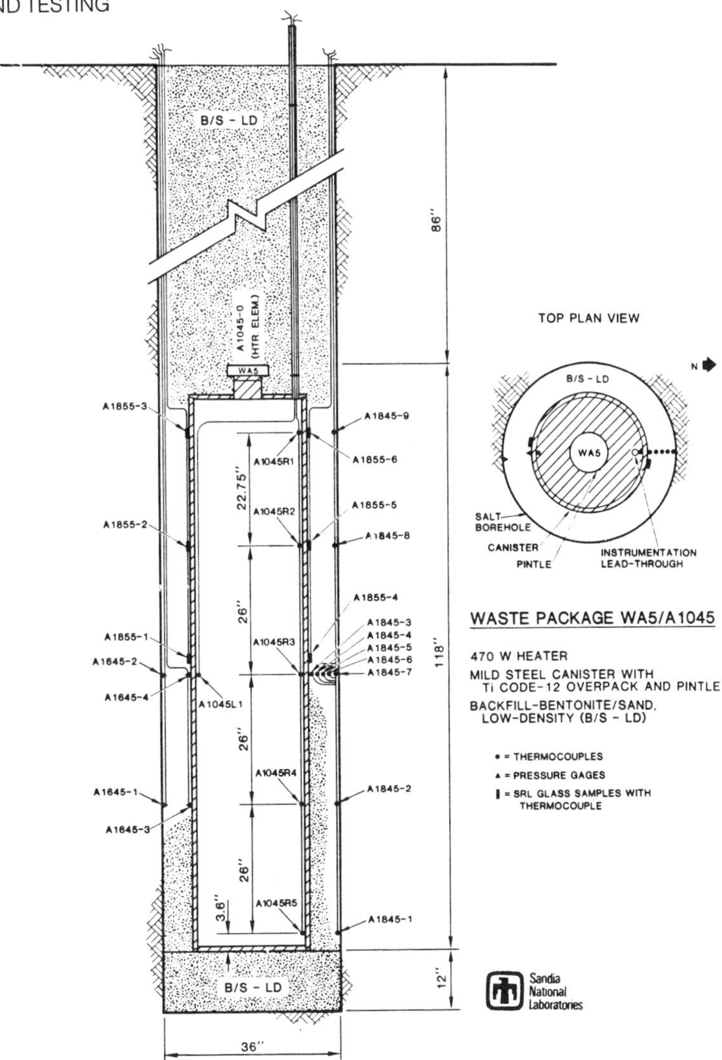

FIGURE 3. Developmental DHLW Canister-Overpack Design, Emplaced in WIPP

C. DHLW Glass Waste Form

Two full-size DHLW containers in Room B were filled (and another two will be) with nonradioactive borosilicate glass (TDS 165 black frit) at the TNX pilot-scale vitrification facility at SRL, with equipment prototypic of the future Defense Waste Processing Facility (DWPF). The glass waste form inside these test containers will not be subjected to in-situ leaching, but will be used to evaluate the effects of glass filling, transportation, and repository handling, emplacement, and retrieval operations on glass physical integrity properties. Alterations in glass mechanical behavior and fracturing, increases in glass respirable particle concentrations, etc., will be quantified. This testing and evaluation is a joint WIPP/Sandia National Laboratories Savannah River Laboratory cooperative effort.

The four glass-filled test packages contain no internal electric heaters. Instead, they are heated externally by adjacent hot salt--as heated by nearby electric heaters. A maximum glass temperature of about 110°C is anticipated. After about three years of heated emplacement in the WIPP, these four containers will be retrieved for posttest analyses.

The effects of the salt repository-environment and multi-year emplacement on glass waste form leaching, surface alterations, corrosion, etc., will also be evaluated in these experiments with the inclusion of small glass samples external to the waste containers. Small, right-circular cylinders (12.7-mm high x 12.7-mm diameter) of SRL TDS 165 glass were core-drilled from a 0.61-m diameter slice of a prototypical DWPF waste canister, and then were polished on both flat ends. A total of 55 of these glass samples were emplaced in the waste package backfill materials, adjacent to multiple waste containers. The samples are attached directly to thermocouples, to both monitor actual in-situ glass temperatures and to assist in sample retrieval. The glass samples will be retrieved on a periodic basis for laboratory analyses (at SRL). The first batch of glass samples is to be retrieved in November 1985 by using sample recovery tube/devices installed at the same time as the samples.

In Room A1, these glass samples are located on the outer surface of several waste containers, at multiple vertical locations, as shown in Figure 3. The samples are intended to be representative of waste glass in a container that is "potentially" breached during the operational phase of a repository. In Room B, the glass samples are located similarly, either adjacent to the surface of externally heated, glass-filled containers (refer to Figure 2), or at the borehole wall-backfill interface in 1500-W heater emplacements, in a comparable temperature region (about 90°-120°C) (refer to Figures 1. Emplacing glass samples at several different locations within multiple test boreholes permits evaluation of waste glass surface alterations and leaching as a function of time, temperature, backfill composition, and moisture content, all within a salt "repository-relevant" environment.

D. Waste Package Backfills

Two separate types of backfill materials have been selected for use in these tests.[6] Both are granular materials and were poured in place under, around, and above the waste containers. The first backfill material is a tailored mixture[7] of 70 wt% bentonite clay (predominantly sodium and calcium montmorillonite, MX-80) and 30 wt% silica sand. This backfill material was emplaced primarily adjacent to TiCode-12 DHLW containers, as shown in Figure 3. The second, nontailored backfill material is crushed salt, as mined in the WIPP, then screened so that no particles are greater than 6 mm in size. Crushed salt backfill was used primarily adjacent to mild steel-overpacked, ONWI-design waste containers, as shown in Figure 2.

Due to the effect of borehole closure from salt creep, both granular backfills will be compacted in time to higher densities. This compaction will also result in an increase in the effective, in-situ thermal conductivities of the backfill materials.[7,8] Effective thermal conductivities will be calculated as a function of time and at multiple locations within each test emplacement, as derived from temperatures measured by remote-reading thermocouples (to be described). Changes in emplaced backfill density, moisture content, and geochemical alterations (if any) will be evaluated by means of retrieved material samples. Backfill samples will be retrieved by coring operations (scheduled to start in April 1986), packaged carefully, and then sent to the laboratory for analyses. Cores will be taken over the entire vertical length of multiple test emplacements, at two radial locations per hole. After each coring operation, the remaining "holes" will be filled with replacement backfill materials.

E. Test Accelerants and Techniques

In order to significantly increase the rate of interactions (e.g., corrosion and leaching, as compared to that occurring under expected, dry backfill conditions) during the planned course of these experiments, significant quantities of concentrated brine are being injected into 5 of the 12 waste package emplacements in Room B. 100 l of Brine A,[9] a concentrated Na-Mg-K-Cl-SO_4 brine is being injected at four longitudinal positions on the borehole-backfill interface of selected test emplacements. The brine injection tubes, shown in Figures 1 and 2, release the brine over the center 50% of the test containers' surface. This injected brine is intended to be representative of inflow expected from the phenomenon of brine migration.[10] However, the test injection volume, 100 L, is about a factor of ten greater than the nominally expected volume per container from brine migration.[10] The brine has been injected into the test emplacements at the rate of 5 L/day, 4 times/week. Each injection takes place into successive, porous injection tubes, spaced at 90-degree intervals around the borehole circumference.

As an additional test accelerant, several of the waste containers in Roon B have been intentionally degraded (before emplacement) with several small drill holes and grooves as shown in Figure 1. The drill holes are 1.6 mm in diameter, at multiple locations on two of the TiCode-12 containers, and penetrate into, but not through the metal thickness. On two of the ONWI design overpacks, the drill holes are 3.2 mm in diameter, at multiple locations shown in Figure 2, and penetrate both 6.35 cm into and through the mild steel. These holes are intended to be representative of potential localized attack-pitting; changes in these "pits" will be quantified in posttest lab analyses. The grooves into two of the TiCode-12 containers are 10-cm long x 2.5-mm deep, and were made by a slitting saw. The grooves are located along the axis of the container in base metal, across a circumferential weld, and parallel to a longitudinal weld in its heat-affected zone. The artifical grooves are intended to be representative of potential, very low probability weld-joint or base metal cracking. Posttest metallurgical properties and observations on container segment samples (surrounding the grooves) will be compared to pretest, representative samples properties, to determine any effects of multi-year repository emplacement on materials properties.

Two test emplacements in Room A1 and two in Room B have been temporarily dedicated (for about a two-year time span) to the monitoring of moisture release.[11] The moisture is due to the near-field phenomenon of brine inclusion migration as well as the migration of free moisture in any host rock clay seams. These four waste packages have no backfill material around and above the containers. There is a vapor barrier plate and borehole plug-packer seal about 10 cm above the top pintle of the test containers. The void space is continually flushed with dry nitrogen and any transient or steady-state moisture released (vaporized water from brine) is trapped in a separate dessicant system for analyses.[11] This technique has been used successfully in other field tests.[12,13]

E. Test Instrumentation

Up to 20 thermocouples have been installed within each simulated DHLW test emplacement. They are on or in containers, on the salt borehole wall, in horizontal arrays through the backfill materials, and connected to each small glass sample, as shown in Figures 1, 2, and 3. The Chromel-Constatan thermocouples are clad in Inconel 600 sheaths, 3.2 mm in diameter, and are accurate to ±0.5°C. These remotely read thermocouples continually monitor in-situ temperatures and also allow the determination of effective backfill thermal conductivities at multiple locations.

Similarly, a total of 4-6 remotely read stress transducer-pressure gages have been installed directly on most waste canisters and salt borehole walls, as shown in Figures 1, 2, and 3. These gages allow the continual monitoring of the pressures exerted on the container and backfill system due to thermally induced salt creep of the borehole walls, thermal expansion of the intact salt mass and waste package materials, and backfill swelling pressures caused by brine sorption.

Manually read container tilt-measurement systems have been installed on three of the waste packages in Room B. The tilt-measurement system consists of a plumb bob and target apparatus attached to a stiff instrument-leadthrough pipe attached to the container top. The apparatus is capable of measuring 1/4 degree of tilt. Relative tilt between containers in a two-row emplacement configuration can be caused by thermally induced creep or floor heaving. Tilt is being measured to assist in the design of container retrieval equipment.

III. RESULTS AND DISCUSSION

These waste package performance technology experiments for simulated DHLW are a subset of a much larger WIPP in-situ experimental program conducted by Sandia National Laboratories for the U.S. Department of Energy, and described elsewhere.[10] The experimental objectives, techniques, materials, package designs, and instruments used in these tests are also described in more detail elsewhere.[6] The design and goals of these in-situ tests are directly based on the research and results of earlier laboratory and field tests conducted by Sandia and other laboratories.[1,6,8,10,12,13]

Fabrication and emplacement activities of the simulated DHLW packages have provided some interesting insights on representative waste package designs. The delivered cost of the TiCode-12 containers, fabricated out of a relatively expensive titanium alloy, was $11,000 each. The fabricated and delivered cost of the ONWI-design cast steel overpack was $23,000 each; this expense must be added to the cost of the stainless steel glass-pour canister (as provided by SRL), about $6,500, yielding total waste container cost of about $29,500. this is a factor of about 2.6 more costly than the single-shell TiCode-12 design. The cost of adding a thin TiCode-12 overlay and pintle to a mild steel "pipe" container (the developmental Room A1 canister overpack) was about $4,000 per container. The relative costs of these designs are more important than the actual values--the cost would be expected to drop somewhat on any design if ordered in great quantities. In terms of relative bulk, repository handling ease, and weight, the TiCode-12 waste container also shows a distinct advantage over the ONWI-design container; the former weighs about 2000 kg, the latter about 7300 kg, both filled with waste glass.

Since no waste package materials samples have yet been retrieved for detailed laboratory analyses, this paper limits discussion to a review of instrumentation data currently obtained. Table 1 gives a sample of measured temperature and pressure data at four months (of heated operation) in Room B. This is basically raw data for comparative purposes only.

The points of measurement above can be discerned on Figures 1 and 2. The container wall temperatures for the ONWI-design containers were measured on the wall of the internal stainless steel canister. All measured temperatures and pressures are still very slowly increasing with time. In all cases, the measured temperatures are greatest at, or very near the center-height of the container, decreasing in value near both (container) ends. All measured temperatures for all emplaced simulated DHLW packages in WIPP are currently being compared to ongoing, thermal-structural modeling calculations at Sandia, for modeling validations. These particular calculations focus on the near-field waste package environment.

TABLE 1. Preliminary WIPP Instrumentation Data

	TiCode Can WB7 (1500 W)	ONWI Container WB5 (1500 W)	ONWI Container WB9 (glass-fill)
	Temperatures, °C (top - center - bottom)		
Container Wall	173-174-134	156-141-133	74- 85- 85
Salt Hole Wall	90-112- 97	96-105- 95	77- 95- 91
Glass, on Salt	108-125- 86	98-109- 99	66- 74- NA
Glass, on Can	- - -	- - -	77- 91- NA
	Pressures, MPa (top - center - bottom)		
On Container Wall	NA - 0 -2.3	NA-24.7-18.5	NA-17.0-16.6
On Salt Hole Wall	3.1-2.4-4.1	NA-11.0- 6.2	5.9-24.7-27.1

The glass-filled package WB9 is not as hot as the two other, 1500-W internally heated containers shown. The temperatures in test emplacement WB9 are, however, appreciably increasing in value; the initial, ambient salt temperature in Room B was 27°C. The temperatures indicated for the small glass samples located outside the waste containers (as measured by attached thermocouples) are within the same general range as expected for waste glass inside of "reference thermal output" (470 W) actual DHLW containers.

The measured temperatures throughout the canister-backfill region(s) are also currently being used to determine the in situ thermal conductivities of the backfill materials as a function of time and location; these values will clarify the effects of backfill composition, compaction density, and moisture content (from brine injection and/or brine migration), on the effective thermal conductivities. Results and comparisons of the measured temperature values, parallel modeling calculations, and calculated thermal conductivities will be presented in the future, as available.

The measured pressure values are a bit more difficult to interpret. However, it does appear evident at this time that the pressures exerted on the ONWI-design DHLW containers are significantly higher than those on the TiCode-12 containers. The basic reason for the pressure differences appears to be due to the difference in the available annular space between the container wall and the salt borehole wall; this is, initially, 15 cm for the TiCode-12 container and only 5.7 cm for the ONWI-design container. With a larger annulus there is basically more available volume for the inwardly creeping salt to move into and more possible consolidation of backfill material. There is also a significant concern over pressure transducer longevity and reliability (in the severe environment in which they have been installed) particularly in regard to how reliability may affect indicated pressures. Another series of thermomechanical modeling calculations being conducted for Sandia are being compared to the measured pressures, for modeling validation. Major purposes of these calculations are to determine the pressures exerted on the test waste packages in WIPP as a function of time, container thermal output (470 and 1500 W), backfill materials (trapped air/none, crushed salt, and granular bentonite/sand), and initial backfill annulus (5, 10, 15, and 30 cm). The results of these modeling calculations and comparisons with actual measured pressures will be presented in the future.

Remotely measured data and future in situ material results will be compared with previous laboratory and field test data for analytical modeling verifications, to help provide long-term predictions for high-level waste isolation in salt and to aid in interpreting the significance of these in-situ results.[1] The initial results and testing experience from this series of non-radioactive experiments are presently being used to refine the test plans[14] for fully radioactive, actual DHLW packages to be emplaced and tested in the WIPP beginning in FY 1990. The results from both the simulated[6] and future, actual[14] DHLW package tests, and parallel operational tests and demonstrations in the WIPP will help provide the technical basis for adequately resolving the performance of high-level waste packages in salt. This resolution will be important in assuring the general public and the technical community that the concept of high-level waste disposal in (a future licensed repository in) salt is both valid and safe.[1]

REFERENCES

1. M. A. Molecke, "WIPP Waste Package Testing On Simulated DHLW: Emplacement," in Scientific Basis for Nuclear Waste Management VIII, ed. J. Stone and C. Jantzen, North-Holland (1985).

2. M. A. Molecke, J. A. Ruppen, and R. B. Diegle, "Material for High-Level Nuclear Waste Canister/Overpacks in Salt," Nuclear Technology, Vol. 63, 476-506 (1983).

3. Westinghouse Electric Corp., "Waste Package Reference Conceptual Designs for a Repository in Salt," WSTD-TME-001 (prepared for the Office of Nuclear Waste Isolation, Columbus, OH) (1983).

4. D. E. Munson, TEST PLAN: Overtest for Simulated DHLW, Thermal-Structural Interactions, Sandia National Laboratories (June 1983).

5. Westinghouse Electric Corp., Engineered Waste Package Conceptual Design: Defense High-Level Waste (Form 1), Commercial High-Level Waste (Form 1), and Spent Fuel (Form 2) Disposal in Salt, ONWI-438, Office of Nuclear Waste Isolation (April 1983). Formerly printed as AESD-TME-3131 (1982).

6. M. A. Molecke, TEST PLAN: Waste Package Technology Experiments for Simulated DHLW, Sandia National Laboratories (1984).

7. M. Moss and M. A. Molecke, "Thermal Conductivity of Bentonite/Quartz High-Level Waste Package Backfills," in Scientific Basis for Nuclear Waste Management VI, ed. D. Brookins, North-Holland (1983).

8. M. A. Molecke and T. M. Torres, "The Waste Package Materials Field Test in Southeast New Mexico Salt," in Scientific Basis for Nuclear Waste Management VII, ed. G. McVay, North-Holland (1984)

9. M. A. Molecke, A Comparison of Brines Relevant to Nuclear Waste Experimentation, SAND83-056, Sandia National Laboratories (1983).

10. R. V. Matalucci, C. L. Christensen, T. O. Hunter, M. A. Molecke, and D. E. Munson, Waste Isolation Pilot Plant Research and Development Program: In Situ Testing Plan, SAND81-2628, Sandia National Laboratories (1982).

11. E. J. Nowak, TEST PLAN: Moisture Release Experiment for Rooms A1 and B, Sandia National Laboratories (1985).

12. R. I. Ewing, Preliminary Moisture Release Experiment in a Potash Mine in Southeastern New Mexico, SAND81-1318, Sandia National Laboratories (1981).

13. R. I. Ewing, Test of a Radiant Heater in the Avery Island Salt Mine, SAND81-1305, Sandia National Laboratories (1981).

14. M. A. Molecke and R. V. Matalucci, Preliminary Requirements for the Emplacement and Retrieval of Defense High-Level Waste Tests in the Waste (October 1984).

WASTE FORM AND WASTE PACKAGE CHARACTERIZATION STUDIES RELATED TO DISPOSAL IN A GEOLOGICAL CLAY FORMATION

R. De Batist
P. Van Iseghem
F. Casteels
Study Center for Nuclear Energy
Boeretang, 200
B-2400 Mol, Belgium

ABSTRACT

Waste form and waste package characterization studies are described which are performed to generate an extended data base for the safety analysis required for the evaluation of a deep clay formation as a candidate high-level waste repository. The following topics are treated:

- in-situ glass and container material corrosion experiments in a surface clay quarry,
- in-situ glass and container material corrosion experiments in an under-ground laboratory (200-m deep), building on the experience gained with the surface in-situ tests; these experiments are still in progress and no detailed results are available yet,
- laboratory corrosion tests with container and structural materials,
- laboratory glass leaching experiments in argillo-aqueous media, and
- compound experiments with waste form, container material and backfill simultaneously exposed to argillo-aqueous media in the presence of a radiation field (experiments in progress).

Some of the more striking results of the clay compatibility tests are discussed in more detail.

I. INTRODUCTION

Nuclear waste management in Belgium is currently based on reprocessing and final disposal in a deep geological formation of the vitrified high-level and alpha-contaminated waste. Inventorization of potentially suitable geological formations showed clay to be the most likely candidate. Following a preliminary seismic survey, it was decided to set up an extensive R&D program aimed at evaluating the suitability of a 100-m thick clay formation underlying the

---

Work supported in part by the Commission of the European Community and by NIRAS/ONDRAF (National Agency for Radioactive Waste and Fuel).

nuclear site at Mol[1]. The program comprises in part, a detailed risk analysis based on geological evolution scenario's in both normal operating conditions and under selected accidental circumstances, engineering studies related to construction and handling techniques in deep clay formations as well as a wide range of laboratory and in-situ experiments for collecting the data base related to the relevant clay properties and to its interaction with structural materials, waste forms and waste package materials.

In this paper we describe the waste form and waste package characterization experiments and we discuss some of the results obtained so far. Preliminary in-situ tests have been carried out in a surface clay quarry; further in-situ experiments are presently underway in the underground laboratory constructed at the Mol nuclear site. More detailed laboratory experiments have been set up for determining the corrosion behavior of various candidate container and structural materials and for studying the release rate of radionuclides from various types of vitrified waste forms in the presence of a clay-water environment. Finally, we give some details about an "integral" type of experiment designed for studying possible synergistic effects occurring when waste form, container material and backfill material are simultaneously exposed to a clay-water environment in the presence of a radiation field.

II. IN-SITU EXPERIMENTS IN A SURFACE CLAY QUARRY

Prior to the construction of the underground laboratory at the S.C.K./C.E.N. site, a number of preliminary experiments were performed in a surface clay quarry at Terhagen, some 50 km to the west of Mol and covering the same clay formation.[2] In addition to a series of experiments for the determination of the thermal properties of the clay,[3] two types of compatibility tests were carried out. In this way, it was possible to obtain the experience required for designing the experiments to be introduced in the 200-m underground laboratory.

The two types of experimental configuration used in the surface quarry are aimed at 1) corrosion studies in direct contact with clay and 2) corrosion studies in a humid clay atmosphere. Experiments were performed over extensive time periods covering in some cases up to 64 months and at temperatures between ambient (13°C) and 150°C. A wide range of metals and alloys (in particular iron, aluminium, copper, nickel and titanium) has been studied, thus allowing a preliminary selection of the more promising materials to be used for the in-situ tests in the underground laboratory.[2]

In addition to metallic samples, simulated vitrified waste samples were also included in some of the experiments. Both borosilicate glasses, as used for the incorporation of high-level wastes, and ferro-alumino-silicates as obtained by high temperature slagging incineration of alpha-contaminated wastes,[4] were used in these experiments. From these screening tests, the following conclusions can be derived for the vitreous materials.[5] Specific weight losses are smaller at 150°C than at 50°C; in some cases, especially at 13°C, weight increases are observed. Infrared reflectance spectroscopy[6] can be used to obtain qualitative information about the intensity of surface deterioration and about changes in surface composition. The results of the metallic corrosion experiments are discussed in Section IV.

III. LABORATORY CORROSION EXPERIMENTS

Using gas corrosion chambers and inert furnaces, corrosion tests have been performed on a very wide range of the materials mentioned above using a series of corrosive environments. The metallic alloys were used in the as-received condition, following thermal treatments to simulate the thermo-mechanical history of the container materials used for the vitrified HLW, with or without anti-corrosive coatings for the candidate structural materials; also, the susceptibility to stress corrosion has been evaluated.[7] The corrosive environments used comprise direct clay contact and a humid atmosphere loaded with corrosive products extracted from the clay; temperatures up to 300°C were used. Further tests were performed at 49°C and 98°C in interstitial clay water and in a ground water composition representative for the aquifer encountered at the Mol site.

IV. SURVEY OF CORROSION RESULTS

In the humid clay-derived atmospheres, the corrosion rate of candidate canister materials decreases with increasing exposure temperature, as a result of the pronounced influence of relative humidity on the corrosion and deposition processes. At 50°C, the losses are between 2 $\mu m/yr^{-1}$ for the 1803 T ferritic steel and 0.2 $\mu m/yr^{-1}$ for the titanium alloys. At 300°C, these values are 0.05 $\mu m/yr^{-1}$ and 0.01 $\mu m/yr^{-1}$, respectively.[7]

In direct contact with clay, the differences between the various materials are much more pronounced, at least at temperatures exceeding 100°C. Whereas, at 300°C, the aluminium alloys are completely deteriorated, thick, non-protective reaction layers are formed on carbon steel and nickel and high weight losses are observed for inconel 600 and incoloy 800 (of the order of mm yr^{-1}), the corrosion rate of the titanium alloys and of hastelloy C remains in the order of $\mu m\ yr^{-1}$.

Corrosion rates remain fairly small in interstitial clay water (25°C, 98°C; e.g., 0.1 $\mu m/yr^{-1}$ for titanium alloys) and in Antwerpian ground water (49°C, between 5 $\mu m/yr^{-1}$ for chromized steel and 0.05 m/yr^{-1} for titanium alloys). Some of the more promising materials were tested in the surface clay quarry.[2] In the humid clay atmosphere obtained at 13°C and at 50°C, the metal loss remained below 1 μm after 3 years of exposure for titanium alloys and for Hastelloy C.

Detailed results of these corrosion experiments, with a discussion of the structure, thickness and composition of the interaction layers and of the occurrence (or not) of localized attack are presented e.g., in Ref. 3.

V. GLASS LEACHING IN CLAY-RELATED ENVIRONMENTS - LABORATORY EXPERIMENTS

To enable the safety assessment of a high-level nuclear waste repository in a deep clay formation to be performed using as extensive a data base as possible, the interaction mechanisms between vitrified waste forms and the geological environment have to be qualitatively and quantitatively described and if possible modelled over a wide time span.

Over the past several years, work has been engaged on a rather extensive research program for determining and characterizing the mechanisms governing the interaction between clay environments and vitrified waste forms and for studying the influence of a number of potentially relevant parameters on this interaction.[8,9] Characterization and quantification of the interaction mechanisms has been attempted by measuring, as a function of the interaction time, the variation in specimen mass, the development and the structural and compositional characteristics of the surface layers and the chemical composition of the leachate. The experimental parameters considered representative for repository conditions are temperature, leachant composition (clay to water ratio), redox conditions and specimen surface area to leachant volume ratio (S:V). Pressure effects are considered to be unimportant (the clay layer at the Mol nuclear site is situated between -160 m and -260 m). Taking into account the very low permeability of clay, water flow rate is also considered not to be a relevant parameter and all experiments reported here were carried out in static conditions. Whereas this approach can be justified in "normal" operating conditions of the repository, it is not necessarily so under "accidental" conditions of rapid water ingress during the operational period of the repository (i.e., before definite closure), or even in the post-closure period. Although the data obtained at various values of S:V can be used to evaluate the consequences of such a high flow rate situation, it may be necessary to perform additional experiments to obtain information on the effect of flow rate on the leaching mechanism. Obviously, results obtained with a range of clay to water ratios may also be used in this context. Redox conditions are of considerable importance since the Boom clay formation underlying the Mol site contains a fair concentration of pyrite, resulting in a reducing redox potential of ≅ -260 mV.[10] Exposure to ambient air, as will occur during the operational period of the repository, rapidly oxidizes the clay resulting in a drastic decrease of the pH. The buffering capacity of the clay is sufficient to restore the initial redox and pH conditions following closure of the repository. Clearly, information about the influence of redox on the corrosion of the glass is required. Temperature has been used as a parameter, not so much because it is expected to cover a wide range under actual repository circumstances (ambient temperature at -200 m is about 15°C, the repository design aims at a maximum temperature at the clay-gallery interface of about 100°C)[11] but to evaluate whether increasing the experimental temperature can be used as an accelerating factor for modelling the long term behavior of vitrified waste forms in contact with the geological formation.

In the following section, the main results and conclusions of this research program are succinctly described. More extensive discussions can be found in the recent annual progress reports.[11,12]

VI. SURVEY OF GLASS LEACHING RESULTS

A. Waste Forms used in Leaching Experiments

Although leach testing has been carried out on borosilicate HLW simulants as well as on a wide range of basalt-like alumino-ferro-silicate glasses representative of the production of a high temperature slagging incinerator for conditioning alpha-contaminated wastes[14], the present discussion will be restricted to the HLW form simulants.

The HLW forms studied in this program have been selected in the frame of the joint research program co-sponsored and coordinated by the CEC and are representative of materials as used in some CEC countries. They comprise a glass-ceramic (proposed by the FRG) and a number of borosilicate glasses designed for reprocessing waste existing in France, the UK and Belgium.[8]

B. Effect of Leachant

To evaluate the influence of the presence of clay on the leaching behavior of waste glasses, experiments were performed using either wet, solid clay or various concentrations of clay mixed with distilled water. Figure 1 illustrates the increase in leach rates observed with increasing clay to water ratio based on the data obtained for one of the waste glasses, using static conditions with T = 90° and S:V = 100 m^{-1}.[15] It is seen that the decrease in leach rate with increasing leach time, which is observed in distilled water, is strongly reduced in a clay environment. It was also observed that in clay environments (at least for the experimental periods covered so far) no crystallized deposits are formed in the leachate or on the glass surface, in striking contrast with the pronounced surface crystallization phenomena observed in distilled water at 150° and 190°C.[5] These observations should also be related with results from leaching experiments in which fresh glass monoliths were leached in pre-concentrated leachates (loaded with elements leached from powders of the same glass type). In these conditions, the observed mass losses are strongly dependent on the time-temperature scale of the experiment: after one month at temperatures of 90°C or 120°C, mass losses were smaller, but at 150°C and 190°C, they were one to two orders of magnitude larger than for continued leaching in initially "pure" distilled water. Furthermore, extensive surface layer formation is found to occur.[16] These observations suggest that a decrease in leach rate results from a combination of an increase in leachate concentration for certain elements and a change in surface layer composition. Both crystallization processes and adsorption on clay particles may lead to the removal of specific elements from the leachate and hence to the continuing of the leaching at a barely reduced rate.

The effect of redox conditions[15] appears to be restricted to clay-water mixtures, where the decrease in leach rate which is observed in oxidizing conditions is absent under reducing conditions. In addition to the effect of redox on the solubility of polyvalent elements,[17] it is conceivable that changes in the surface layer characteristics also play a role in this phenomenon.

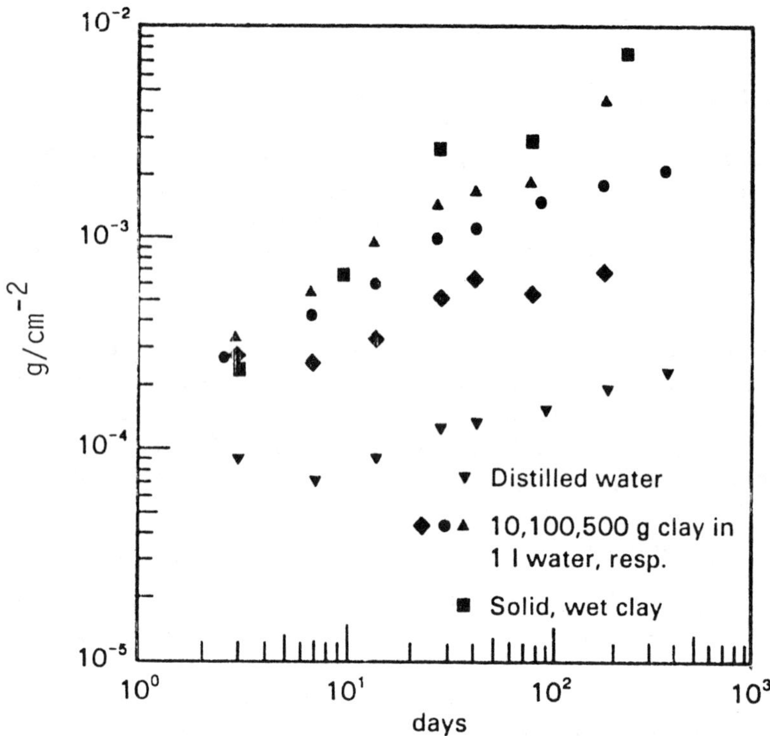

FIGURE 1. Effect of Clay. Water ratio on the specific mass loss of a high alumina borosilicate glass SAN60 (composition, see Ref. 8) in static conditions at 0°C and with S:V=100 m^{-1}

C. Effects of Temperature and of S:V

Some of the effects of the temperature of the leachant have already been hinted at above. In distilled water, where crystallization is found to have a pronounced effect on the leaching process, it is clear that a single thermally activated process cannot explain the corrosion behavior of these glasses in the temperature range 40°C - 200°C. Indeed, the crystal phases which are formed depend on temperature.[16] Similar conclusions are reached with a clay-water mixture[9] as shown in Figure 2 obtained after 7 days leaching (see also ref. 15 for alpha-waste incinerator type glass). Whereas the high-alumina glass (SAN60) can be described by a unique thermally activated (35 kj mol-1) process, for the other glasses the effective activation energy is smaller at the higher temperatures. The results for the glass ceramic (C31) are not conclusive. It should also be pointed out that this picture is leaching-time dependent, so that it is not warranted to use these values for extrapolations to longer times.

A comparison of the compositions of surface layers formed during corrosion in either DW or in clay-water media indicates that, although in a clay environment components of the clay are incorporated in the surface layer, in terms of glass components there is not a great difference between layers formed in

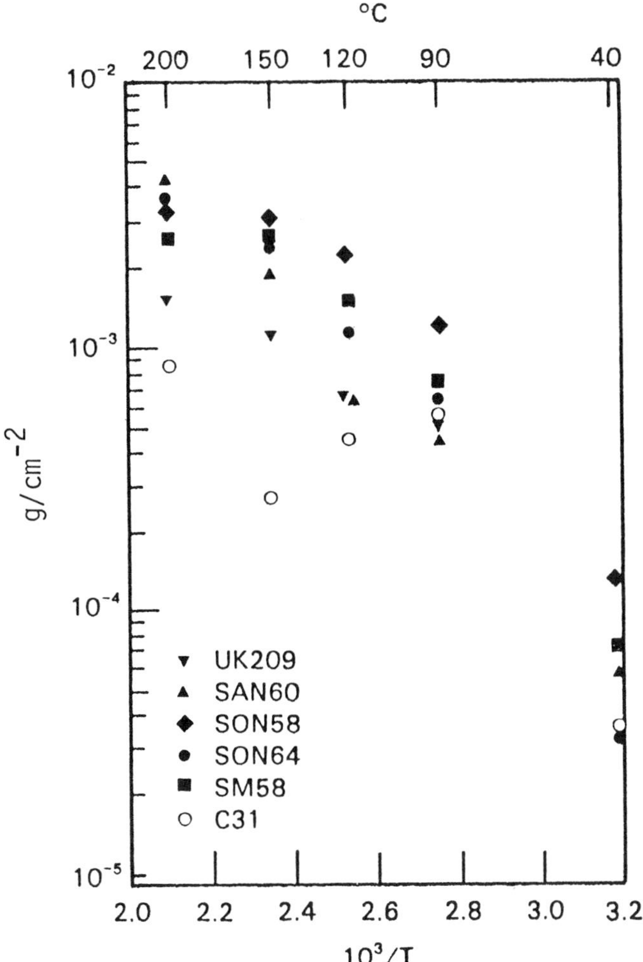

FIGURE 2. Temperature Dependence of the Specific Mass Loss Following 7 Days in a Clay-Water Mixture (100 g/L^{-1}) in Static Conditions with S:V = 100 m^{-1}, for 5 Different Borosilicate Glasses and 1 Glass Ceramic (compositions, see Ref. 8).

the presence or absence of clay.[12] However, the thickness and also the complexity of these surface layers depends strongly on glass composition. An examination of one of the thicker surface layers clearly revealed the presence of at least two sublayers and the incorporation of clay components (such as S and K) in these layers. Although there does not appear to be a simple correlation between overall mass loss and surface layer thickness (similar layer thicknesses are found after corrosion in wet clay and in clay water mixtures, where mass losses differ by an order of magnitude), some similarity is observed when comparing specific mass loss and surface layer characteristics obtained either at 200°C after 28 d or at 90°C after 300 days. This comparison is shown in Table 1.

TABLE 1. Comparison Between Specific Mass Loss and Surface Layer Characteristics

Glass	SM58				SON64	
	WC		CWM		WC	
Corrosion Conditions	200°C 28d	90°C 240d	200°C 28d	90°C 360d	200°C 28d	90°C 240d
ML (kgm^{-2})	0.12	0.13	0.019	0.023	0.2	0.12
Surface layer thickness (m)	30	30	20	50	100	75
Surface layer composition						
enriched in:	Ti,Zr, Nd,Ce S	Zr,Ti, Nd,Ce	Zr,Ti, Nd,S	Zr,Ti, Nd,Al Ca,Si, Ce	Zr,Gd, Al,Ce, Fe,K, Si,S,U	Zr,Gd, Nd,Ce, La,Fe, Al,U,Si
depleted in:	Al,Si, Mg,Na	Al,Mg, Mo,Na, Si	Mg,Si	Mg,Mo		

The effect of S:V, covering the range between 10 m^{-1} and 10^4 m^{-1} has been studied using distilled water at temperatures between 90°C and 190°C. At 90°C, changes in corrosion mechanism were observed in going from 10 m^{-1} to 100 m^{-1}, which were also reflected in the structure and thickness of the surface layers.[16] Thick surface layers, frequently consisting of several sublayers appear to be formed only for S:V = 10 m^{-1} (in times less than 1 year). High alumina concentrations in the glass appear to inhibit the formation of thick surface layers, although they also lead to low elemental concentrations in the leachate.

VII. SYNERGISTIC EXPERIMENTS

The radiation field resulting from the radioactive waste forms in the repository environment, together with the simultaneous presence of the various types of materials constituting the repository structure and loading, might conceivably induce synergisms in the long-term material behavior. To evaluate the importance of such synergistic effects, experiments have been started in which waste forms together with container material are exposed to clay-water mixtures in the presence of an external gamma field (^{60}Co; 10^5 rad h^{-1}). The irradiation device can accept up to 37 sample vessels. Tests will run for 1000 h at 90°C in either oxidizing or reducing conditions. In a first phase, waste forms and container materials will be studied separately; simultaneous exposure of the two types of material is planned for a second phase. The experiment allows monitoring of pH and oxygen at 10 d intervals during the test; at the end of the experiment, leachate Eh and composition and also radiolysis products will be analysed.

Preliminary experiments are proceeding; no results are available presently.

VIII. IN-SITU EXPERIMENTS IN THE UNDERGROUND LABORATORY

Based on the experience gained with the clay quarry experiments at Terhagen, two types of test set up have been designed also for the underground laboratory at the Mol site. Both can accept samples of container or structural materials (metallic alloys or concrete) as well as various kinds of solid waste forms. Furthermore, in-situ characterization and monitoring of the clay environment has also been provided for.

To study interaction with solid clay, pressure tubes are used such as shown schematically in Figure 3. The length of the tube (5.3 m) is sufficient to reach a region in the clay which has not been perturbed by the engineering operations required for the construction of the experimental gallery. The test samples are mounted on the periphery of the test loop so as to make good contact with the clay environment. Each loop can accommodate 29 metallic (ring-shaped) samples and up to 64 waste form samples (rectangular plates, 40 x 15 x 5 mm). A (retractable and hence, if necessary, replaceable) furnace inside the pressure

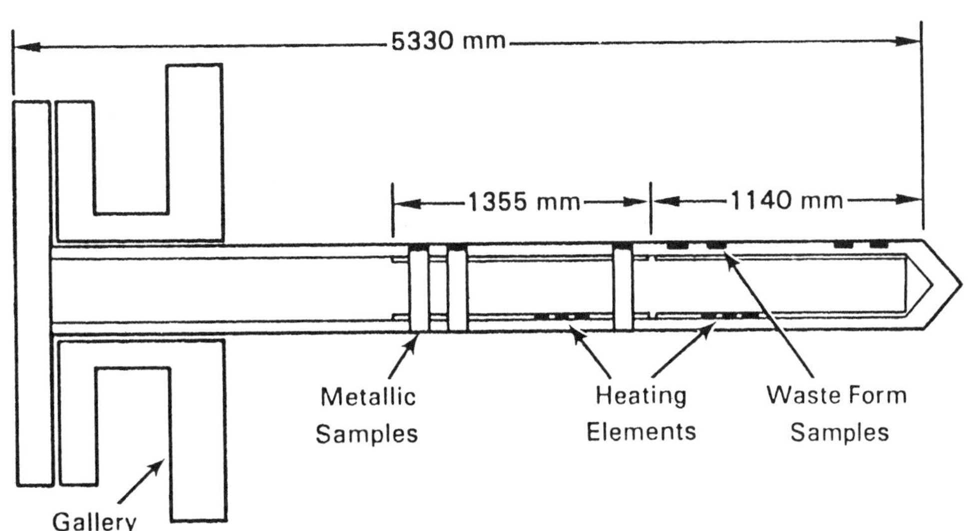

FIGURE 3. Schematic View of the Experimental Loop for In Situ Corrosion Tests in Direct Contact with Clay

tube allows operation not only at ambient temperature (13°C), but also at 90°C and 170°C. Exposure times of up to 50,000 h are planned. At the end of the experiment, an overcoring technique will be used to remove the test loop together with the surrounding clay. This will allow examination of the test samples and of the reaction products at the interface clay-sample as well as evaluation of diffusion (migration) into the clay of corrosion products and relevant radioisotopes from the waste forms.

The second type of experiment is designed to study the corrosiveness of the products emanating from the clay formation. A schematic representation of the loops used for these experiments is shown in Figure 4. A porous, stainless steel plug allows corrosive products perspiring from the clay into the loop to be collected by a carrier gas and circulated over the test samples. Heating elements allow operation at 50°C, 90°C and 170°C, in addition to ambient temperature. Test duration is expected to be up to 50,000 h. Each loop can take up to 36 test samples (30 x 30 x 3 mm).

Monitoring of the corrosion experiments covers not only thermometry, but also redox potential measurements at the interface sample-clay (for the direct clay contact experiment) and analysis of the atmosphere (in the clay atmosphere loops). Furthermore, continuous measurement of the corrosion rates of metallic specimens will be carried out in a number of separate experiments.[18] Analysis of the corrosive gas atmosphere is done using IR absorption for Cl_2, SO_2, NH_3,

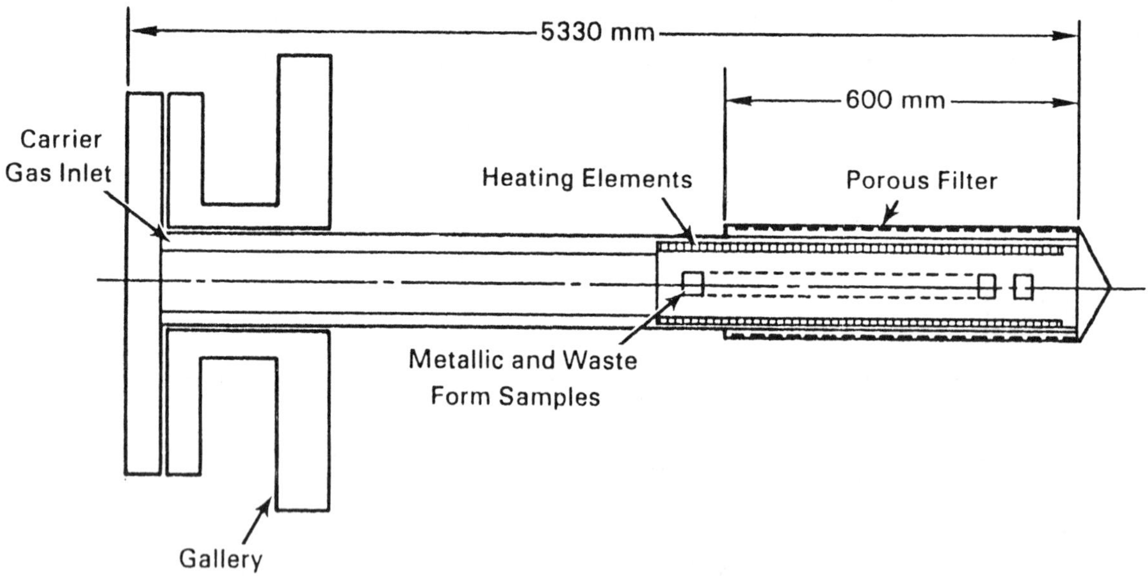

FIGURE 4. Schematic View of the Experimental Loop for In Situ Corrosion Tests in Contact with Clay Derived Atmosphere

F_2, CO, CO_2, NO_x and H_2S. The dew point of the loaded carrier gas is also determined and, following condensation at 8°C and at -80°C, the condensates are analysed (at room temperature) for pH and composition.

Four different systems are to be used for continuous automated monitoring of corrosion rates.[18] The corrosometer deduces changes in specimen cross-section from a measurement of sample resistance; it accumulates corrosion history as a function of corrosion time. The "corrater" set-up is based on a measurement of the linear polarization resistance and yields a value of the instantaneous corrosion rate. The PAIR technique (instantaneous rate of polarization admittance) is based on the same electrochemical relationship and also yields a value of the instantaneous corrosion rate. The last corrosion monitoring technique to be tested in the underground laboratory is based on a measurement of the hydrogen production rate as a measure of the instantaneous corrosion rate.

The first experimental loops were inserted in the underground laboratory in early 1985. No results are available.

REFERENCES

1. R. Heremans, Proc. International Conference on Radioactive Waste Management, Seattle, 1983, IAEA-CN 43/56 (1984).

2. F. Casteels et al., Jül-Conf-42, vol. 2 (1981).

3. R. Heremans, Summary Report 1980-1982, R&D Program on Radioactive Waste Disposal in a Geological Formation, S.C.K./C.E.N. Mol, Belgium.

4. R. De Batist et al., Radioactive Waste Management 1, 171 (1980).

5. R. De Batist et al., Characterization of High Active Waste Forms and Their Interaction With Clay, R 2706, unpublished S.C.K./C.E.N. report (1984).

6. P. Van Iseghem and M. Rotti, Proc. European Workshop on Physical Techniques for Studies of Surfaces and Subsurface Layers of Glasses; F. Lanza and A. Manara, eds., JRC Ispra, 19 (1983).

7. G.P. Marsh et al., "HLW Container Corrosion and Design," 2nd EC Conference on Radioactive Waste Management and Disposal, Luxemburg 1985, to be published.

8. R. De Batist, ed., Testing and Evaluation of Solidified High-Level Waste Forms, Joint Annual Progress Report 1981, EUR 8424 EN (1983).

9. C. Engelmann, ed., Testing and Evaluation of Solidified High-Level Waste Forms Waste Forms, Joint Annual Progress Report 1982, EUR 9268 EN (1984).

10. P. Henrion et al., in Semi-Annual Report n° 15 - R & D Program on Radioactive Waste Disposal in a Geological Formation, S.C.K./C.E.N. Mol, Belgium.

11. P. Manfroy et al., in Semi-Annual Report N° 12 - R&D Program on Radioactive Waste Disposal in a Geological Formation, S.C.K./C.E.N. Mol, Belgium.

12. G. Malow, ed., Testing and Evaluation of Solidified High-Level Waste Forms, Joint Annual Progress Report, 1983, to be published.

13. J.A.C. Marples, ed., Testing and Evaluation of Solidified High-Level Waste Forms, Joint Final Report, to be published.

14. P. Van Iseghem et al., Proc. CEC Seminar on Testing, Evaluation and Shallow Land Burial of Low and Medium Radioactive Waste Forms, Geel (1983).

15. R. De Batist et al., "Behavior of Vitrified Radioactive Waste under Simulated Repository Conditions," 2nd EC Conference on Radioactive Waste Management and Disposal, Luxemburg 1985, to be published.

16. P. Van Iseghem et al., Mat. Res. Soc. Symp. Proc. 44, 55 (1985).

17. B. Skytte Jensen, Riso R-430, 1980.

18. F. Casteels et al., Proc. Joint CEC/NEA Workshop on Experiments in Underground Laboratories, Brussels, 1984.

DEVELOPMENT OF HIGH TEMPERATURE AND PRESSURE ZIRCONIA-BASED pH SENSORS

Mike J. Danielson
Oscar H. Koski
Pacific Northwest Laboratory
P.O. Box 999
Richland, WA 99352

Jonathan Myers
Rockwell Hanford Operations
P.O. Box 800
Richland, WA 99352

ABSTRACT

Yttria-stabilized zirconia pH sensors are suitable for use from 100-300°C. A new Pt internal half cell is discussed which results in a considerable simplification in their calibration. A degradation process takes place after prolonged exposure to 300°C conditions and is manifested by a loss of full Nerstian response at temperature ≤200°C. A hypothesis for the degradation process is discussed.

I. INTRODUCTION

Accurate groundwater Eh and pH data at elevated temperatures are essential for the design of a deep geologic nuclear waste repository. The waste container corrosion behavior; stability of packing (backfill); and radionuclide dissolution, migration, sorption, and precipitation will be effected by Eh and pH conditions existing in and near the repository. This report discusses the research effort to develop sensors in support of the BWIP program at Rockwell Hanford Operations that can be incorporated into autoclaves to continuously monitor the pH conditions existing during nuclear waste/barrier material/host-rock hydrothermal interaction experiments.

Niedrach[1] was the first to report the high temperature (285°C) pH response of yttria-stabilized zirconia closed end tubes. He demonstrated that the pH response was independent of the redox state of the solution. His initial design used an aqueous internal cell, but all later designs featured a solid state Cu/Cu_2O internal half cell. The change to a solid state design was needed because 1) the sensors have an extremely high impedance at lower temperatures so that the complicated electrical/pressure seal could result in a shunting of the internal cell from the external aqueous solution, and 2) the pH of the internal buffer can easily degrade with time resulting in a drift in potential. In later papers,[2,3] his work was extended to lower temperatures and to the geothermal environment. The pH sensors show nearly perfect Nerstian response at 285°C and do not suffer degradation even after weeks at this temperature. However, their response at 95°C has been less ideal, with reports that the probes become sluggish with duration of exposure (at 95°C) and that temperature cycling between 285 and 95°C results in a loss of pH response when evaluated at 95°C. Very little work was carried out at temperatures between 95 and 285°C. One recurring problem with Niedrach's pH measurements is his lack of success in using a high temperature reference electrode. Consequently, his potential data could not be placed on the Standard Hydrogen Electrode (SHE) scale.

Tsuruta and Macdonald[4] examined the pH response of one 17 w% yttria-stabilized zirconia sensor at 100, 150, 200, 225, 250, and 275°C. Design details are sparse on the electrical/mechanical seal, but the sensor contained an aqueous buffer with the Ag/AgCl half cell at ambient temperatures (resulting in a thermal liquid junction potential). They found an excellent Nernstian pH response at temperatures above 200°C, although there was a small deviation on the acid side (pH of 3-5). Large, unexplained deviations from Nerstian response were observed at 100 and 150°C.

Meyers, Ulmer et al.[5] and later, Danielson, Koski, and Myers[6] examined large numbers of 8 w% yttria-stabilized zirconia closed-end sensors and found pH response to be excellent over the temperature range of 100-300°C. Several solid state internal fills were studied (graphite, Au, Ag, and Cu/Cu_2O), and it was discovered that the internal half cell is poised (acting thermodynamically reversible) by oxygen. This observation permits some simplification in the calibration of the sensors. Unfortunately, the sensors require a preliminary autoclaving to 250°C before they become poised to oxygen (when used at lower temperatures), and they lose this property when stored at ambient conditions until reactivated by another autoclaving. Activation energy measurements imply that the primary conduction process in the ceramic involves the oxide ion. Improved pH response at the lower temperatures is thought to be due to the development of an improved electrical/mechanical seal that prevents shunting of the pH response signal. A degradation process, first identified by Niedrach, was verified to be operative. It is observed that when the sensors have been subjected to two or more weeks of exposure to 300°C, they are characterized by a sluggish pH response at lower temperatures. One observation from the study is that it is almost impossible to make a bad pH sensor for use at 250-300°C; even degraded sensors work perfectly.

II. EXPERIMENTAL

All the pH sensors were 8 w% yttria-stabilized zirconia, purchased from Corning (Solen, Ohio) in the standard dimension of 0.64 cm OD (0.25 in.) and 0.48 cm ID (0.19 in.). They were mounted in standard Conax fittings (Buffalo, New York). The latest design with special safety features is shown in Figure 1. It should be noted that the Teflon cap is permeable to air and will allow the internal gas composition to remain fixed by air. Since these are extremely high impedance devices, the electrical termination end is separate from the mechanical seal to minimize the possibility of shunting paths forming. This results in improved pH response at temperatures near 100°C. All autoclave tests were refreshed, and the solutions were sparged with hydrogen since a platinum hydrogen electrode was included to directly measure the pH of the solutions (buffers degrade with time) and to serve as the absolute pH standard. An external Ag/AgCl reference electrode of PNL design was also used in the autoclave tests and all the potentials were corrected for the thermal liquid junction potential so that the potentials could be placed on the Standard Hydrogen Electrode (SHE) scale. All these details were further discussed in an earlier reference.[6] Tests below 100°C were carried out in a constant temperature bath using an Ag/AgCl (4.0 molal KCl) reference electrode.

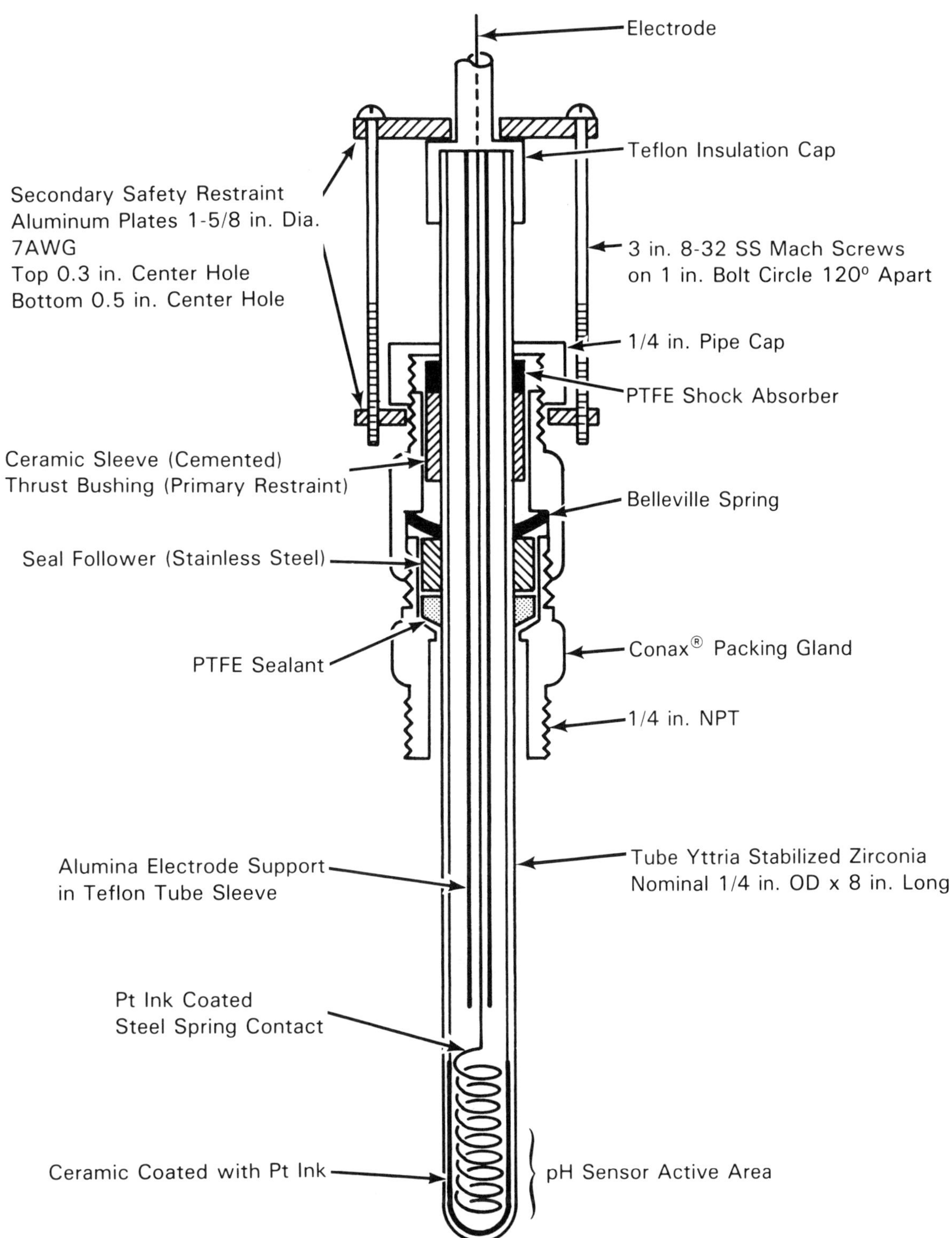

FIGURE 1. pH Electrode

III. PERFORMANCE IMPROVEMENT

Danielson et al.[6] recognized from their potential measurements that the sensors acted reversibly to oxygen (on their internal surface) by the following equilibria:

$$1/2O_2(g) + 2e^- \text{ (metal)} = O^{2-}(ZrO_2)_{\text{inside}} \quad (1)$$

$$O^{2-}(ZrO_2)_{\text{inside}} = O^{2-}(ZrO_2)_{\text{soln. side}} \quad (2)$$

$$O^{2-}(ZrO_2)_{\text{soln. side}} + 2H^+ = H_2O \quad (3)$$

$$1/2O_2(g) + 2e^- + 2H^+ = H_2O \quad (4)$$

This observation has great importance for simplifying the calibration procedure. Previously,[6] all the metallic fills studied (Cu/Cu_2O, Ag, Au) by the authors required a preliminary autoclaving to 250 to 300°C before their potential was that predicted by Eq. (4) and this was very inconvenient for pH studies between 100 to 250°C. Recently, a platinum coat was substituted which results in the sensors acting reversibly to oxygen at 100 to 300°C without the need for initial autoclaving at 250 to 300°C. This offers a great simplification in the calibration and use of these sensors.

The metallic platinum is applied as a thin coating to the inside of the sensor (Figure 1). The Pt-coated sensors will start out at 100°C being poised by oxygen from the air, and it is our experience that sensors which have good pH response at 100°C will operate as well or better at higher temperatures.

The Pt-coated sensors were evaluated in a hydrogen sparged silica-carbonate buffer (F^-, 17.6 mg/L; Cl^-; 335 mg/L; SO_4^{2-}, 2.4 mg/L; Ca^{+2}, 1.3 mg/L; K^+, 10 mg/L; Na^+, 399 mg/L; Si, 43.4 mg/L; CO_3^{2-}, 67 mg/L; HCO_3^-, 32 mg/L; pH = 9.77 @ 25°C) known as GR-4 (ground water) buffer.

Table 1 shows the calculated pH (based on experimental potential data) for the hydrogen electrode and the zirconia sensors. The old zirconia sensors had a varied autoclave history including about 30 days of autoclaving before this test. Sensor 1 is degraded (doesn't give full Nernstian response at 100°C) and is included for comparison.

The pH values were calculated by using the E° (Table 2) of Eq. (4), the measured voltage E_{meas}) of the sensors on the hydrogen scale, and the Nernst equation (taking account that $fO_2 = 0.2$). Equation (5) was used for this calculation:

TABLE 1. pH Measurement Data

Sensor	Description	pH [Calculated from Eq. (5)] 100°C	150°C	200°C
Pt	hydrogen electrode	8.77	8.29	7.99
1	old/degraded	8.39	8.36	7.71
2	old	8.73	8.29	7.91
3	new	8.69	8.36	7.92
4	new	8.97	8.41	8.11

TABLE 2. E° Data[6]

T	E°
100°C	1.67 V
150°C	1.127 V
200°C	1.088 V
250°C	1.050 V
300°C	1.013 V

$$E_{meas} = E^0 - \frac{2.3\ RT}{F}(pH) + \frac{2.3\ RT}{4F}(fO_2) \qquad (5)$$

The hydrogen electrode pH value is considered to be the most reliable measured value. This is an especially severe test of the pH performance of the zirconia sensors since pH sensors are normally standardized in a known buffer and their pH response measured relative to this value. Except for the response of Sensor 1 at 100°C, the measured pH is within 0.2 pH units of the value measured using the hydrogen electrode, and the agreement improves as the temperature increases. Though this data only goes to 200°C, it is our experience that even better performance would be expected at 250 and 300°C.

The following procedure is used to coat the sensors:

1. Coat inside of sensor (depth of 2.5 cm) with Englehard Pt ink (No. A-4338).

2. Dry at 120°C for one hour to harden the coating.

3. Wind piano wire around mandrel to make a spring which is a little larger in diameter than the sensor ID. Coat spring with Pt ink and insert.

4. Fire in air to 400°C for 18 hours.

5. Epoxy ceramic collar to upper portion of sensor after cooldown.

This development demonstrates a significant simplification in the calibration process [using Eq. (5) and the E° data of Table 2) which should assist in the practical application of the sensors for temperatures of 100 to 300°C. More accuracy can be obtained by using a buffer of known pH to standardize the sensors as is currently done with glass pH sensors at ambient conditions.

IV. DEGRADATION MECHANISM

Somewhere between 3 and 11 days exposure at 300°C, the pH sensors undergo a deterioration in pH response. This deterioration is not evident above 200°C (since all sensors work well $\geq$ 200°C, but only appears when the sensors are used at lower temperatures (85-200°C). In one test, 8 w% sensors from Corning and McDanel were subjected to a 300°C autoclaving for 72 hours. No degradation was observed when evaluated at 85°C. However, after a total of 11 days at 300°C, the sensors gave 77% of their original response. The process continued to develop with increased exposure time at 300°C until (in the range of 30 days exposure) all pH response was lost when evaluated at 85°C.[6] Presumably, all the degradation process continues at a slower rate at temperatures below 300°C. Sensors maintained at 100°C have shown no degradation in pH response in tests of a month duration.

Figure 2 shows the time rate of response of a degraded sensor at 84°C to sudden changes in the pH buffer (9.4 to 7 and 7 to 4). The voltage for perfect Nernstian response is bracketed by horizontal lines. Full pH response was not observed even after 5000 minutes (3.5 days), and the response rate was extremely sluggish. This behavior can be contrasted with a new sensor which will give full Nernstian response in two minutes. Some physical alteration has taken place in the sensor surface with exposure to 300°C.

Under the hypothesis that the pH response would be restored by exposing a fresh surface, about one-third of the wall thickness of a degraded sensor (0.9 mm before grinding) was removed using a diamond wheel. However, no improvement in pH response was noted when evaluated at 85°C. This indicates that the loss in pH response was not due to the selective leaching of a pH sensitive component out of the ceramic surface and that the degradation process involves a significant penetration of the wall.

In another test, a degraded sensor (zero pH response at 85°C) was turned inside-out by placing the metal half cell on the outside of the sensor (formerly the degraded pH sensing surface) and using the internal surface as the new pH sensing surface (formerly the metallic half cell). A pH response of 89% was determined, indicating that the degradation mechanism was not due to a micro-crack developing through the tube wall.

Figure 3 is a cross-section of a degraded sensor. The upper picture, which is observed with both new and degraded sensors, shows a large number of voids. Indeed, the slip-cast Y_2O_3-ZrO_2 sensors do not have theoretical density. The bottom photo has only been observed with a degraded sensor to

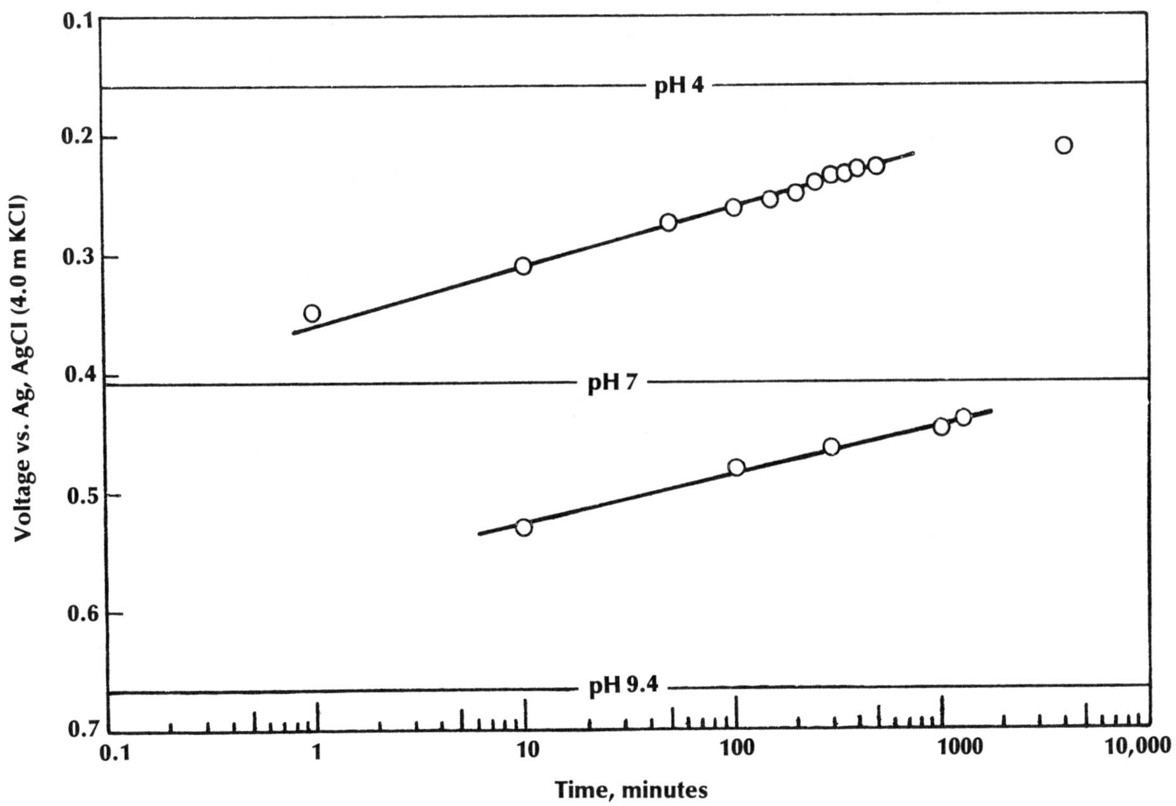

FIGURE 2. pH Response of Degraded sensor (84°C)

date, and it shows a grain boundary connecting with a void. It is hypothesized that the 300°C autoclaving results in a dissolution attack at the grain boundaries and a partial unsuturing of the sintered Y_2O_3-ZrO_2 particles. The hypothesis is more clearly shown in Figure 4 in which a short micro-crack (along a grain boundary) connects with a void. If the pH sensing surfaces on the outside and at the bottom of the crevice can be visualized as a voltage generator, then the total pH response will be a function of the resistance connecting the two (or more) voltage generators to the inside surface. since the wall is thinner at the crevice, the resistance R_2 will be lower than R_1.

Degraded sensors are observed to have a lower impedance, in agreement with this hypothesis. Grinding away the surface would not restore the performance if the micro-cracks were deeper than the depth of material removed. the rate of pH response would also be expected to be slow because mass transport of the buffer down a narrow crack into a much larger void volume would be a slow process. However, pH response would be expected to improve as the temperature increased because mass transport processes are faster. One would expect the depth of attack to increase with exposure time until $R_2 \ll R_1$. At that time, the sensor would completely lose its pH response. Exposure beyond this would result in a leak or fracture of the tube due to a microcrack(s) propogating completely through the tube wall. One way to test this hypothesis would be to

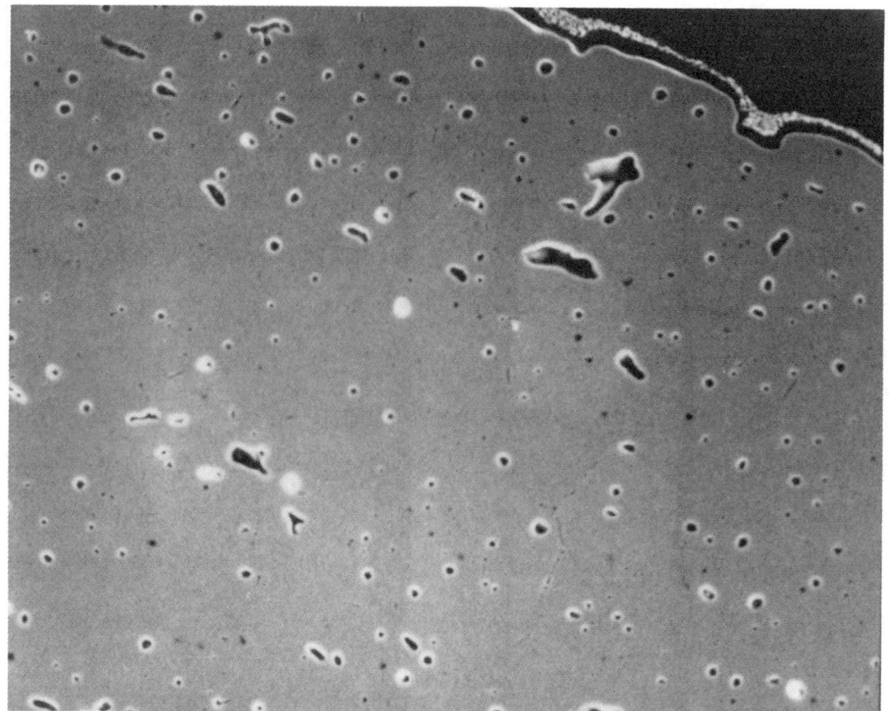

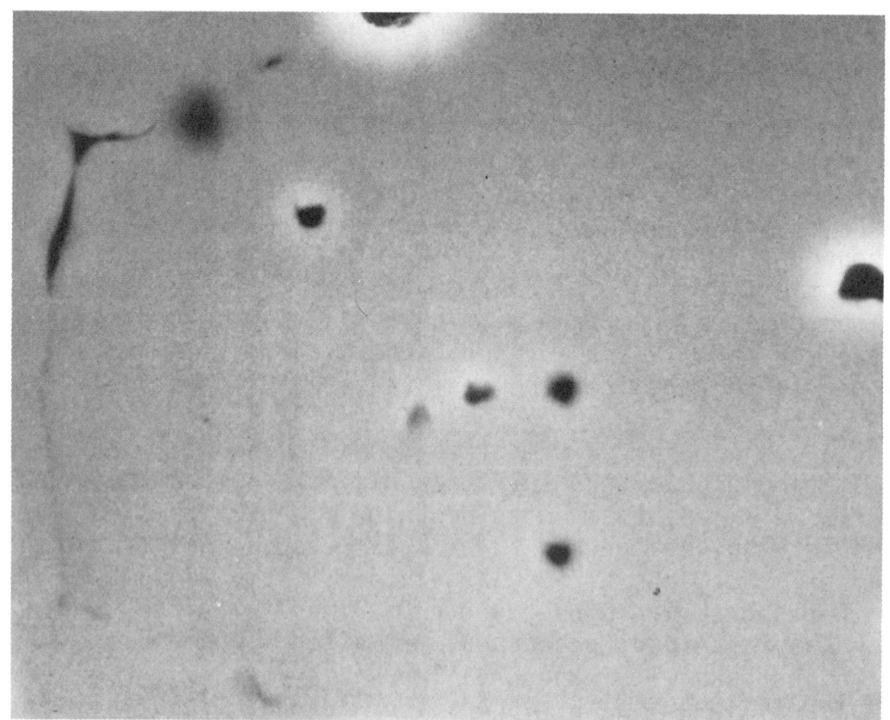

FIGURE 3. X-Section of Degraded Sensor

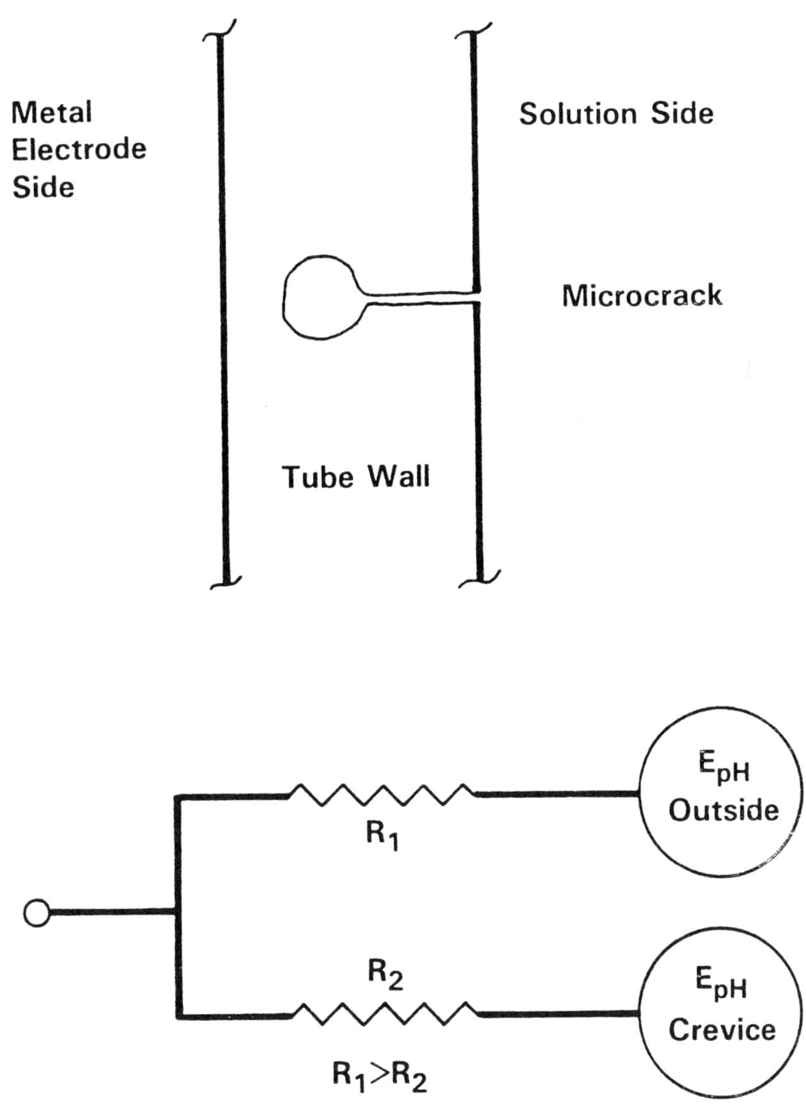

FIGURE 4. Model of Degraded sensor

subject single crystal Y_2O_3-ZrO_2 to a 300°C autoclaving, and we plan to carry this out in the near future. Niedrach et al.[7] has proposed a similar degradation mechanism in which he suggests that the solution is slowly penetrating into grain boundaries and that local functional groups on the occluded surfaces are being titrated by the diffusing solution.

At the present time, we have no way of preventing the degradation process. Sensors for use at or below 200°C should not be subjected to higher temperatures unless absolutely necessary. In a practical sense, the problem can be worked around by categorizing the sensors into age groups with the youngest ones saved for the lower temperature work. The only troublesome case is when the pH sensors are used for weeks at 300°C and then pH data is needed during a cooldown or a temperature ≤200°C.

V. ACKNOWLEDGMENT

This work is supported by the Department of Energy under the Basalt Waste Isolation Project of Rockwell Hanford Operations, Richland, Washington.

VI. REFERENCES

1. L. W. Niedrach, "Oxygen Ion-Conducting Ceramics: A New Application in High Temperature - High Pressure pH Sensors," Science, 207, 1200 (1980).

2. L. W. Niedrach, "Hydrogen Ion Sensor Having a Membrane Sheath of an Oxygen Ion Conducting Ceramic," U.S. Patent 4,264,424, April 28, 1981.

3. L. W. Niedrach and W. H. Stoddard, The Development of a High Temperature pH Electrode for Geothermal Fluids - Task 1, prepared for the Pacific Northwest Laboratory operated by Battelle Memorial Institute for the U.S. Department of Energy, PNL-3857, March 1981.

4. T. Tsuruta and D. D. Macdonald, "Stabilized Ceramic Membrane Electrode for Measurement of pH at Elevated Temperatures," Electrochem. Soc., 129, 1221 (1982).

5. Jonathan Myers et al., "Recent Developments in the Monitoring/Control of Eh-pH Conditions in Hydrothermal Experiments," Geochemical Behavior of Disposed Radioactive Waste. American Chemical Society Symposium Series No. 246, Washington, DC (1984).

6. Mike J. Danielson et al., "Recent Development with High Temperature Stabilized - Zirconia pH Sensors," Electrochem. Soc., 132, 296 (1985).

7. L. W. Niedrach and W. H. Stoddard, "Zirconia Membrane pH Sensors," Industrial Engineer chemical Product Research Development 22, 594-599 (1983).

Disposal System Performance

LISA, A PERFORMANCE ASSESSMENT CODE FOR GEOLOGICAL REPOSITORIES OF RADIOACTIVE WASTE

G. Bertozzi
A. Saltelli
Commission of the European Communities
Joint Research Centre - Ispra Establishment
21020 Ispra (Varese)
Italy

ABSTRACT

LISA, developed at JRC-Ispra, is a statistical code, which calculates the radiation exposures and risks associated with radionuclide releases from geological repositories of nuclear waste. The assessment methodology is described briefly. It requires that a number of probabilistic components be quantified and introduced in the analysis; the results are thus expressed in terms of risk. The subjective judgment of experts may be necessary to quantify the probabilities of occurrence of rare geological events. Because of large uncertainties in input data, a statistical treatment of the Monte Carlo type is utilized for the analysis; thus, the output from LISA is obtained in the form of distributions. A few results of an application to a probabilistic scenario for a repository mined in a clay bed are illustrated.

I. INTRODUCTION

The implementation of nuclear energy at an industrial level requires that the associated radioactive waste be kept in segregation for long time periods. Repositories mined in carefully chosen geological formations can offer an adequate solution for this problem. Such a conviction is based on the geologists' statement that it is possible to select formations which are likely to remain unaltered for geological periods of time, thus assuring that the enclosed waste will be kept adequately segregated.

Qualitative judgments of experts cannot, however, be considered as an acceptable demonstration of the safety of any specific choice. It is necessary to demonstrate scientifically that final storage of nuclear waste can be effected in a given site in a manner consistent with safety requirements. Predictive modelling of the repository evolution and of the radionuclide behaviour in the specific environment can provide the needed answer to the demand for such a demonstration.

Substantial efforts are being made in Europe by individual countries and by the CEC in order to create the basis for the performance analysis of nuclear waste repositories in geologic disposal systems. To this aim, the CEC coordinates a concerted action with the view to establish a common background for the performance assessment of such systems; the exercise is called PAGIS

In this framework, the code LISA (Long-Term Isolation Safety Assessment) has been developed at the JRC, Ispra.[3] It is essentially a statistical code, which calculates, through Monte Carlo simulations, the radiation exposures associated with the release of radionuclides from waste repositories mined in geological formations, and converts the doses into the corresponding risk. It is especially designed to quantify the uncertainty in the model results.

The purpose of this paper is to illustrate the methodological approach which is being implemented at the JRC-Ispra; the various probabilistic aspects of the analysis will be especially emphasized. A few examples of the results obtained with the code LISA in accordance with such an approach are shown.

II. METHODOLOGY AND CODE

The assessment methodology described here consists of a number of steps:[4]

1. identification and description of the different scenarios involving radionuclide release from the repository,
2. assessment of the probabilities of occurrence of the various events capable of triggering or perturbing the release scenarios,
3. modelling of the radionuclide release and transport through the various system components, following the scenario considered; modelling of the radionuclide dispersion in the environment, assessment of their concentrations in food chains and related intakes to man,
4. assessment of the radiological consequences to individuals, in terms of radiation exposure,
5. conversion of doses to risk, and
6. investigation of the relative importance of the various input parameters in governing the output.

Different probabilistic aspects appear in some of these points; an additional one arises implicitly from the uncertainties in the input data, which demand that a statistical treatment be applied over the whole analysis. Steps 1 and 2 require the identification of all the events and processes which could either initiate release of radionuclides from the waste, and cause their transport through the geosphere and the biosphere to man, or could influence their release and transport rates. For any given type of geologic formation, there will be some processes which are certain to occur: they constitute the "Normal Evolution Scenario." The assumptions used in analysing this scenario are based on extrapolations into the future of present geological and climatic trends. Probabilistic events and processes, having the capability to perturb the normal situation without generating a completely different scenario, lead to the definition of the "Altered Evolution Scenarios," characterized by sets of parameter values which are outside the normal range. The probabilities that the phenomena will occur are taken into account in the analysis.

A third type of scenario is that in which the radionuclide release is triggered by probabilistic events and occurs in a completely different way, being thus described by a different model. Again, the probabilities of the triggering events have to be considered, when assessing the risk for such scenarios. Events having the potential to cause abrupt, direct releases of radionuclides into the biosphere are grouped into a fourth class, "Disruptive Scenarios." They are very unlikely and - in general - can be ruled out on the mere basis of their probabilities.

Probability evaluation in a geological context is an especially debated point, as most of the events considered are rare, and their probabilities of occurrence can hardly be inferred from a body of statistically significant data. A codified procedure for the assessment of probabilities of geological events and processes does not exist. In the past, we have described a few applications of the Fault Tree Analysis for this aim, with satisfactory results;[5] however, we feel that the binary logic of this tool imposes a too strict working scheme, which is not especially suited for such a multiform context. Thus, for many geological processes the evaluation of the corresponding probabilities is often derived from expert's subjective judgments, on the basis of the current state of knowledge. On the other hand, following the views of the subjectivistic school, probabilities having a large degree of subjective appreciation should not be considered "worse" than those based on relative frequencies.[6]

Steps 3 and 4 require that numerical values be assigned to the parameters used in the analysis. For parameters which cannot be univocally determined, either because of their intrinsic variability in space and time, or because of poor experimental knowledge, probability density functions are assigned to cover their entire range of variability. These distributions have, as a rule, a familiar form: uniform, normal or log-normal distributions are the most common. They are fed into the computer code, which uses them to generate the sets of values for the Monte Carlo simulations, and produces model outputs (e.g., annual doses) which are again in the form of probability distributions (histograms).

The methodology for sampling values of the input variables assumes a particular relevance when the complexity of the model and, therefore, the running time of the code limit the overall number of runs which can be reasonably performed. Some of the submodels utilized in the code involve numerical integration of differential equations, which is a time-consuming procedure; therefore, in LISA, a rationalized sampling technique like the Latin Hypercube[7] is normally preferred to the simple Random Sampling. This latter, however, may be optionally chosen, when more careful estimates of the extreme values are needed.

The input data matrix of the hypercube is elaborated in LISA with a procedure which purges undesired spurious correlations among the input variables; alternatively, it is possible to impose required correlations without affecting the original distributions of the single variables.[8]

Coming to step 5, the way of expressing the results is a very critical issue, which is still the subject of some debate, both in national and international contexts. In the past, only radiation doses have been considered as a suitable analysis output, no consideration being paid to the various stochastic terms. The present trend, on the contrary, is very much towards a more comprehensive approach, in which risks are calculated as the product of the probability of occurrence of the scenario, the dose, and the dose-to-risk conversion factor.[9,10] This latter approach has been adopted for LISA, where the following relationship is used:

$$R = H \times r \times P \tag{1}$$

where

- R = risk to an individual (1/yr)
- H = committed annual dose equivalent (rem/yr)
- r = dose to risk conversion factor (1/rem)
- P = probability that the annual dose, H, will be received by an individual.

Thus, the output from LISA is a distribution of the risk. It is built, first, by generating a large number of dose rate curves as a function of time (Figure 1), each curve being associated with a particular set of input data; then, the peak doses are gathered into a histogram. This latter is finally transformed into a histogram of the risk, by multiplying each column by the risk factor (10^{-4} rem^{-1}).

For probabilistic scenarios, whose probability of occurrence is small, the peak doses are also multiplied by the probability of the scenario itself, which is assumed to identify with the probability P (Eq. 1) that individuals will receive exposures. Thus, a distribution of the risk linked to the given accidental scenario is generated (Figure 2). This latter histogram can be further elaborated, by summing up the various columns, each one being weighted by its frequency ϕ_i:

$$R_T = \sum_i R_i \times \phi_i \tag{2}$$

where R_T is the total risk (1/yr) associated with the given scenario. This figure of risk may then be compared with any pre-established risk limit.

In this approach, peak doses are gathered together, regardless of their time of occurrence. Actually, as time-dependent discount parameters are not considered in the analysis, the relevance of radiological consequences is independent of their time of occurrence, provided that they fall within the limits established for the analysis. Such an approach, however, might be questioned, as doses occurring at different times are combined together to

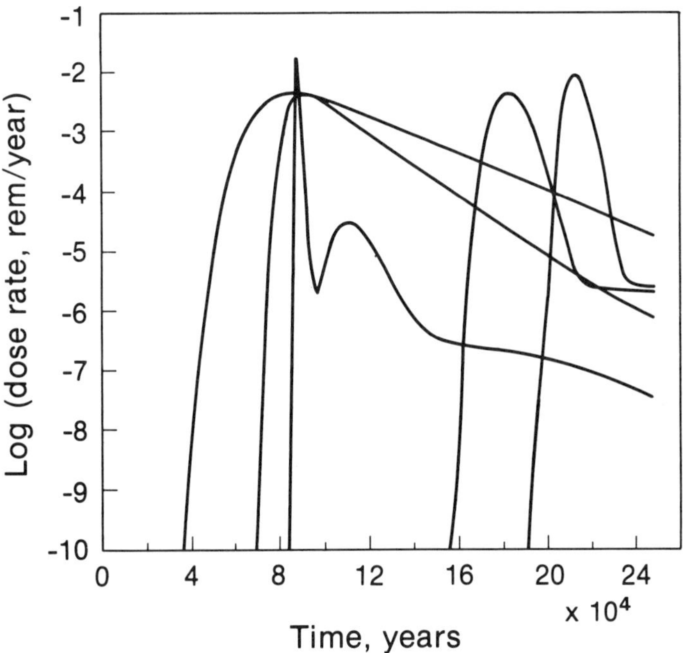

FIGURE 1. Annual Doses as a Function of Time, Generated by Five Different Simulations

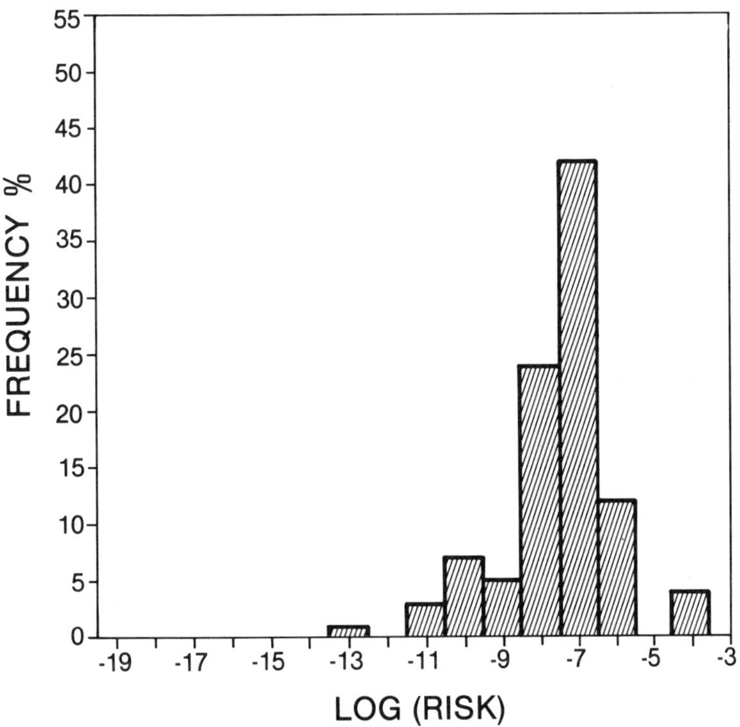

FIGURE 2. Distribution of the Risk for the Scenario Analyses

calculate the risk for an individual. An alternative approach, which permits avoiding such a drawback, consists in generating a set of distributions for a set of time points, each histogram showing the frequency distribution of doses at a given time. From such histograms, the curve of expected annual dose as a function of time can be drawn; this latter, finally, is easily converted to a risk versus time curve (Eq. 1). This second approach permits taking account of the time dependence of the quantity assessed.[10] The above procedure is in agreement with the proposal contained in a recent NEA Experts Report.[9]

Finally, (step 6) the statistical analysis of the results permits the investigation of the relative importance of the various uncertain parameters, thus pointing out those parameters and research areas which deserve further investigations.

III. TEST CASE

A hypothetical radioactive waste repository is supposed to exist in a specific clay formation and to contain known quantities of vitrified high-level waste.[11] A release of radionuclides is assumed to occur as a consequence of the accidental creation of a permeable fracture in the clay bed, through the repository, leading to interconnection of aquifers.[12] Groundwater may thus come into contact with a fraction of the waste; the contaminated leachant moves then into and along the adjacent aquifer to a water extraction point.

The probability of occurrence of such a scenario has been previously assessed, on the basis of site specific investigations;[5] the cumulative value of 10^{-2} over the time period considered for the analysis (250,000 years) has been adopted. Such a value is not small enough to permit consequences to be neglected. They were, therefore, analysed with the code LISA; some significant results are illustrated herewith.

The histogram of Figure 3 shows the frequency distribution of the peak dose rate values, regardless of their time of occurrence. The bimodal shape of the histogram is not casual; it is due to the existence of a high-dose range, centered at about 10^{-3} rem/yr, controlled by the Np-237, and a low-dose range, around 10^{-6} rem/yr, in which peak doses are due to a number of different nuclides, such as Tc-99, Th-229, and Ra-226. The special role of Np-237 is due to a particularly unfavourable concurrence in its properties: long life, large mobility, and high radiotoxicity.

The possibility of identifying in the output histograms the various radionuclides which are responsible for the peak doses enables to address further investigations to the parameters specific for those nuclides, neglecting those which are of minor importance.

The average and the 95th percentile dose rate curves are reported in Figure 4, as a function of time; it can be seen that, after an initial transient of no dose at all, the average curve starts to increase, due to those runs

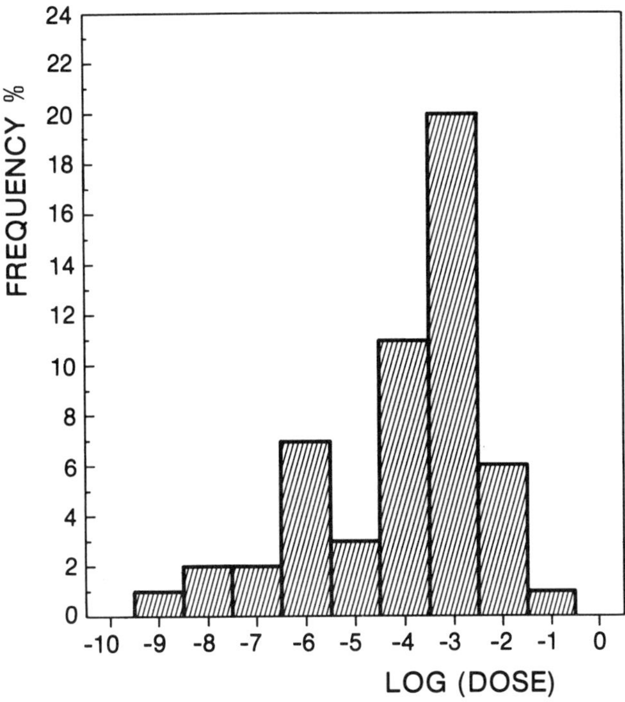

FIGURE 3. Frequency Distribution of Peak Doses

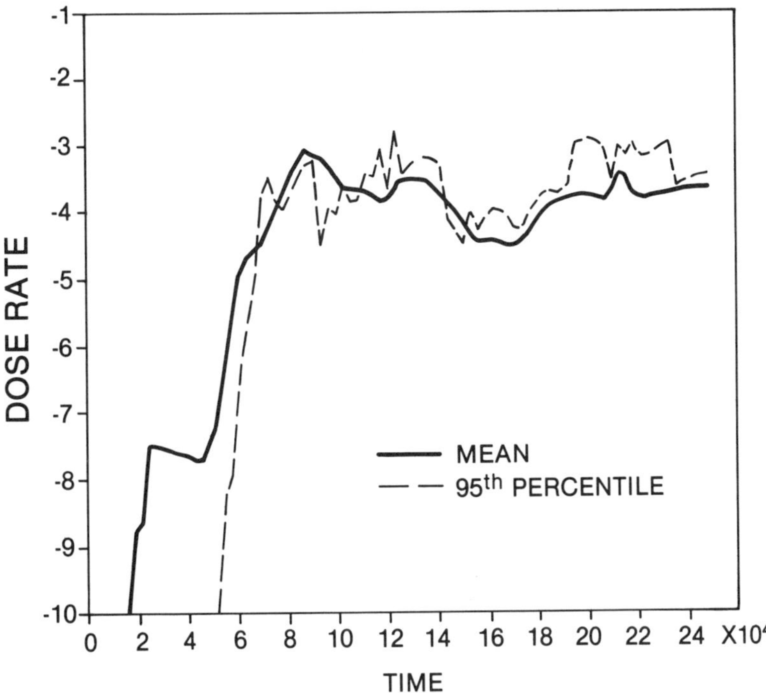

FIGURE 4. Average and 95th Percentile Doses as a Function of Time

for which the time of occurrence of the fault is sampled to fall immediately after the repository closure. The average is dominated by these few runs, while the percentile curve starts to increase later, when a more consistent fraction of runs produces non-negligible doses. After about 80,000 years, the two curves reach and fluctuate around a plateau at about 10^{-4} rem/yr, the 95th percentile curve staying about one order of magnitude higher than the average.

From the above results, it appears that the annual doses possibly received by the most exposed individual of a hypothetical critical group would fall in the range 10^{-9} to 10^{-1} rem/yr, the largest doses occurring with a frequency of about 1%. It is apparent that such dose levels are acceptably low, and largely comply with the ICRP recommendations concerning life-long exposures,[13] independently of the probability that such a scenario will really occur, and that any individual will be exposed to such radiation levels, in the future.

However, our approach requires that the model results be based on risk rather than on mere exposures, taking account of the various probability components. These latter identify with the release event probability (10^{-2}) and with the probability that a health effect will occur to any individual if the dose has been received. On this basis, the risk associated with the highest doses (about 0.1 rem/yr) sampled with a frequency of 1% may be easily calculated as $2 \cdot 10^{-9}$/yr, which is quite an acceptable level.

Actually, the whole dose rate histogram should be elaborated in this way, by multiplying each value on the abscissa by the corresponding probability terms. However, because of the logarithmic scale, the risk value practically coincides with that calculated above. The comparison between the input data values associated with the highest doses and the probability distributions assigned to the corresponding parameters permits the identification of the parameters which are more critical in governing the model results, following a procedure already utilized for the Canadian code SYVAC.[14]

IV. CONCLUSIONS

The approach and the computer code utilized at JRC-Ispra to analyse the performance of any radioactive waste repository in geological formations have been described. The probabilistic components of the analysis are particularly emphasized. They may be identified with:

a. the probability of occurrence of different triggering or perturbing events, as causes of release scenarios,

b. the probability that any individual, having been exposed to a given irradiation level as a consequence of the defined scenario, will undergo a health effect, and

c. the probability of a given exposure level, as due to a particular combination of parameter values.

Such an approach leads one to produce model results in terms of risk rather than mere radiation exposures.

REFERENCES

1. PAGIS, Performance Assessment of Geological Isolation Systems, Summary Report of Phase 1, Report EUR-9220EN (1984).

2. N. Cadelli, F. Girardi and S. Orlowski, "PAGIS, Common European Methodology for Repository Performance Analysis," 2nd Europ. Comm. Conf. on Rad. Waste Manag. and Disposal, Luxembourg (1985).

3. A. Saltelli, G. Bertozzi and D. A. Stanners, LISA, A Code for Safety Assessment in Nuclear Waste Disposal, Program Description and User Guide, Report EUR- 9306EN (1984).

4. G. Bertozzi, M. D. Hill, J. Lewi and R. Storck, "Long-Term Risk Assessment of Geological Disposal: Methodology and Computer Codes," 2nd Europ. Comm. Conf. on Rad. Waste Manag. and Disposal, Luxembourg (1985).

5. M. D'Alessandro and A. Bonne, "Rad. Waste Disposal into a Plastic Clay Formation: a Site Specific Exercise of Probabilistic Assessment of Geological Containment," Rad. Waste Manag., Vol. 2, Harwood Acad. Publ., Paris (1981).

6. G. Apostolakis, Nuclear Safety, 19, 305 (1978).

7. M. D. McKay, R. J. Beckman and W. J. Conover, Technometrics (C), 21, 2, 239-245 (1979).

8. R. L. Iman and W. J. Conover, Communications in Statistics, 11, 311-334 (1982).

9. Long-term Radiation Protection Objectives for Rad. Waste Disposal, NEA Experts Report, OCDE/NEA, Paris (1984).

10. PAGIS Workshop on Methods for Estimating Risks from High-level Waste Disposal and Presenting Results, Ispra, Italy, unpublished (1985).

11. A. Bonne et al., "Safety Analysis of a HLW-Repository in a Clay Formation," in these proceedings (1986).

12. G. Bertozzi, M. D'Alessandro and A. Saltelli, Waste Disposal into a Plastic Clay Formation: Risk Analysis for Probabilistic Scenarios, Report EUR-9641EN (1984).

13. ICRP Publication 26, Ann. ICRP 1 (1977).

14. T. Andres, Oral Comnunication, 16th Canad. Waste Manag. Infor. Meeting, Winnipeg (Canada), 26/27 September 1983.

PREDICTING THE SAFETY OF A HLW REPOSITORY

Charles McCombie
National Cooperative for the Storage of Radioactive Waste (NAGRA)
Parkstrasse 23
5401 Baden, Switzerland

I. INTRODUCTION

In Switzerland, legal and regulatory requirements for projects demonstrating the feasibility and long-term safety of disposal of all categories of nuclear wastes were formulated in 1979. These led to the submission to the appropriate governmental bodies at the beginning of 1985 of the so-called Project Gewaehr (Guarantee) of which an overview is given in another paper to this conference.[1] The part of the project treating the repository for high-level wastes is described in more detail in other papers covering the geological database[2] and the engineering concept.[3] The present document covers the quantitative safety analyses performed in order to predict the long-term performance of the repository for comparison with the the relevant safety guidelines. A fuller account is contained in the corresponding volumes of the Project Gewaehr Reports.[4,5]

II. SUMMARY OF SWISS HLW DISPOSAL CONCEPT

The main waste management option in Switzerland involves the reprocesssing of spent fuel abroad with subsequent return of vitrified high-level wastes and also other processed wastes. Reprocessing fuel from 240 GW(e)/yr of reactor operation yields approximately 6,000 cylinders (approx. 1,100 m^3) of borosilicate glass with 12% fission and activation products and actinides. This glass is assumed to be stored for 40 yr before encapsulation in a massive carbon steel overpack and disposal into a mined repository.

The repository consists mainly of drilled tunnels of 3.7 m diameter. The glass cylinders are emplaced horizontally in the tunnel axis surrounded by blocks of highly compacted sodium bentonite. The host rock is deep-lying granite covered by a few hundred metres of sedimentary rock; repository depth is 1,200 m. The repository can also contain other actinide wastes disposed in a different configuration, but these will not be discussed further in this paper.

Table 1 gives an overview of the safety barriers and the principal functions which they are intended to fulfill. Data on the performance of technical barriers was gathered by complementing international work with specific Swiss

TABLE 1. Safety Barriers in a HLW Repository System

Barrier	Functions Accounted For In Safety Analyses	Main Data Sources
Glass matrix (0.42 m diam.)	- Limits radionuclide release rates through diffusion or corrosion	- Reprocessors' specifications - Extensive foreign programs - Joint Japanese/Swedish/Swiss program
Steel canister (0.94 m diam.)	- Prevents water access for more than 1,000 years - Reduces radiation fields - Ensures reducing conditions in near field	- Nagra studies and experimental programs on mechanical properties and corrosion
Bentonite buffer (3.7 m diam.)	- Ensures diffusive rather than convective nuclide transport - Filters any colloids formed - Delays break-through of radionuclides to host rock	- KBS studies - Stripa Project data - Complementary Swiss laboratory work
Deep granitic host rock	- Ensures low water flows in near field - Provides long-term tectonic stability - Ensures retardation under reducing conditions	- Regional geological program[2] - Extensive measurements in 6 boreholes (1,300 m - 2,500 m)
Overlying granite and sediments	- Carry more water and provide large dilution - Reduce probability of direct access to repository	- Regional geological program[2] - Extensive measurements in boreholes

laboratory investigations, including studies on an interactions dataset representative of a potential siting area in north-central Switzerland.[2] The purpose of the safety analyses described below is to quantify the behavior of the combined barrier system in such a way as to produce predictions of safety performance; these should be as realistic as possible while remaining conservative, i.e., overpredicting any negative consequences.

III. THE APPROACH TO PREDICTING SAFETY

In Project Gewaehr repository operational safety is not treated in detail, the implicit assumption being that current technology can provide the acceptable levels of safety for such an underground nuclear facility. Emphasis was placed on post-operational safety because more challenging technological issues are involved in predicting system behavior over very long times up to about one million years. It is recognized[6] that this task must be based upon system models allowing long extrapolations. Conceptual models reflecting our understanding of system behavior must be translated into mathematical models producing quantitative results. Appropriate data are required at all stages, both for input and for validation of the models. The choice of models is governed by the objectives of the safety assessment, the criteria to be fulfilled, the available methodology and data and, of course, the timescales in which the analyses must be performed.

A. Repository Safety Criteria in Switzerland

The relevant Swiss safety criteria were formulated by the nuclear safety authorities in 1980. In guidelines[7] published then, two important objectives were defined. A geologic repository should be capable of being sealed off so that it provides adequate safety without requiring any further attention or control. The maximum dose which any individual might receive at any time in the future is 0.1 mSv/yr (10 mrem/yr).

Important features of the guidelines are as follows:

- They are global system requirements which leave the choice of subsystems open to the disposer. No explicit criteria are placed on subsystem performance.
- They are basically deterministic, although the actual wording is framed so as to allow varied treatment and assessment of release scenarios with very low probabilities.
- They are individual dose criteria, i.e., population doses or risks are not required.

Since publication of the guidelines further documents have given more indications on how to interpret the requirements contained[8] therein.

The nature of the guidelines, along with an assessment of the status of probabilistic safety analyses for geologic repositories, led at the outset of Project Gewaehr to the decision that a largely deterministic approach would be taken. This is described below in sections explaining the choice of release scenarios treated and the consequence models employed.

B. Scenario Selection

In a first step, the importance of all processes and events possibly leading to or influencing nuclide release was evaluated for Swiss conditions and for various timescales. As usual for crystalline rock disposal, the normal, most probable release scenarios are via groundwater transport. Accordingly, the basic modeling tools are aimed at the best possible handling of this scenario. Other less probable scenarios can be then grouped into a large set which can be studied by variation of input parameters within the standard groundwater chain, and scenarios requiring different models (e.g. direct drilling). Finally, some of these scenarios which would require special modeling can be excluded from quantitative consequence analysis because of their extremely low probabilities or dominating nonnuclear effects (e.g. large meteor impact). This selection process, which is summarized in Figure 1, is similar to approaches in other countries.[9]

C. Basic Modeling Chain

The chain of models used for treating the normal groundwater release scenario and its derivatives is displayed in Figure 2. The original Project Gewaehr documentation[4,5] contains references to detailed reports on all models. Only a few special features can be mentioned here.

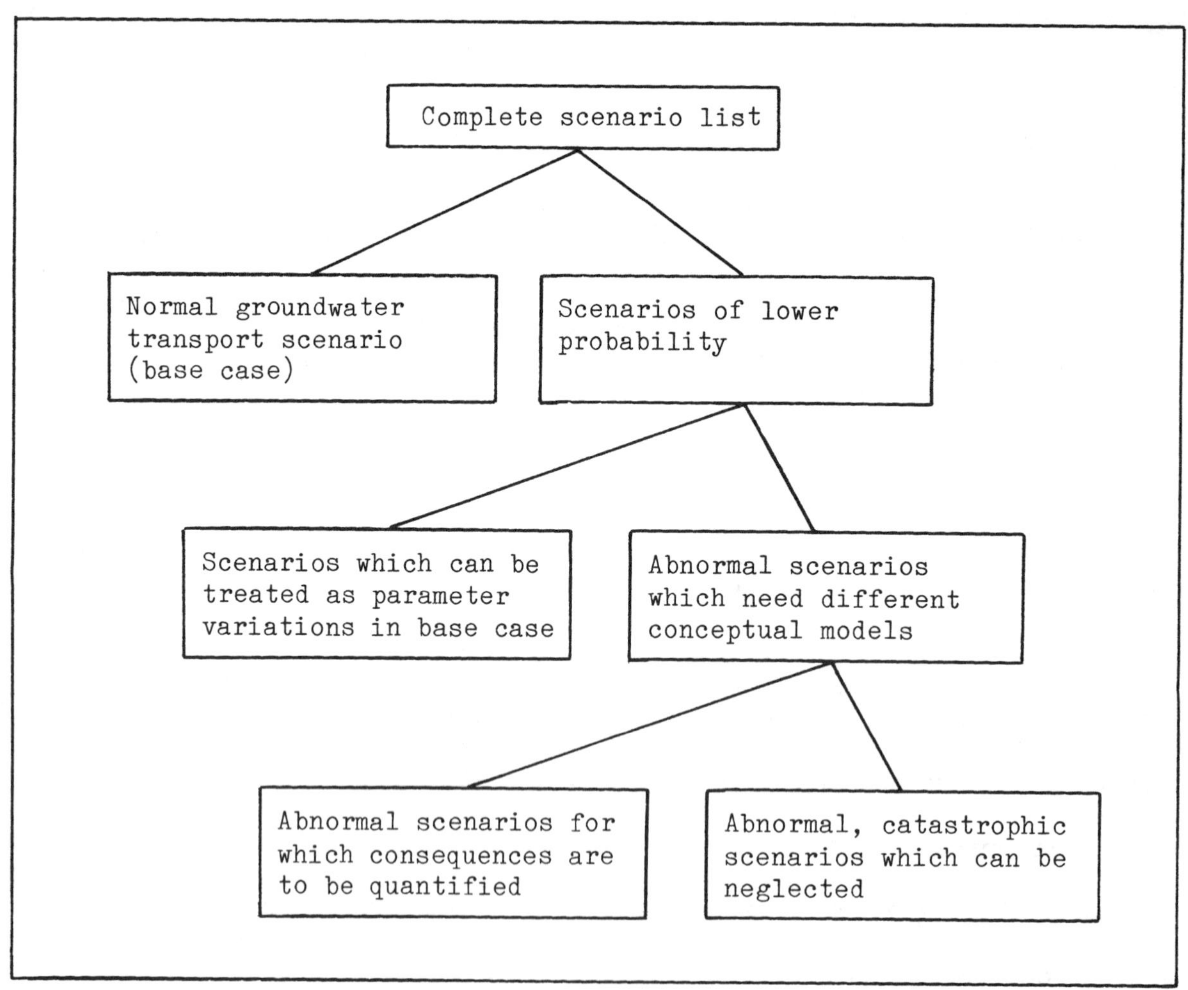

FIGURE 1. Classification of Possible Release Scenarios

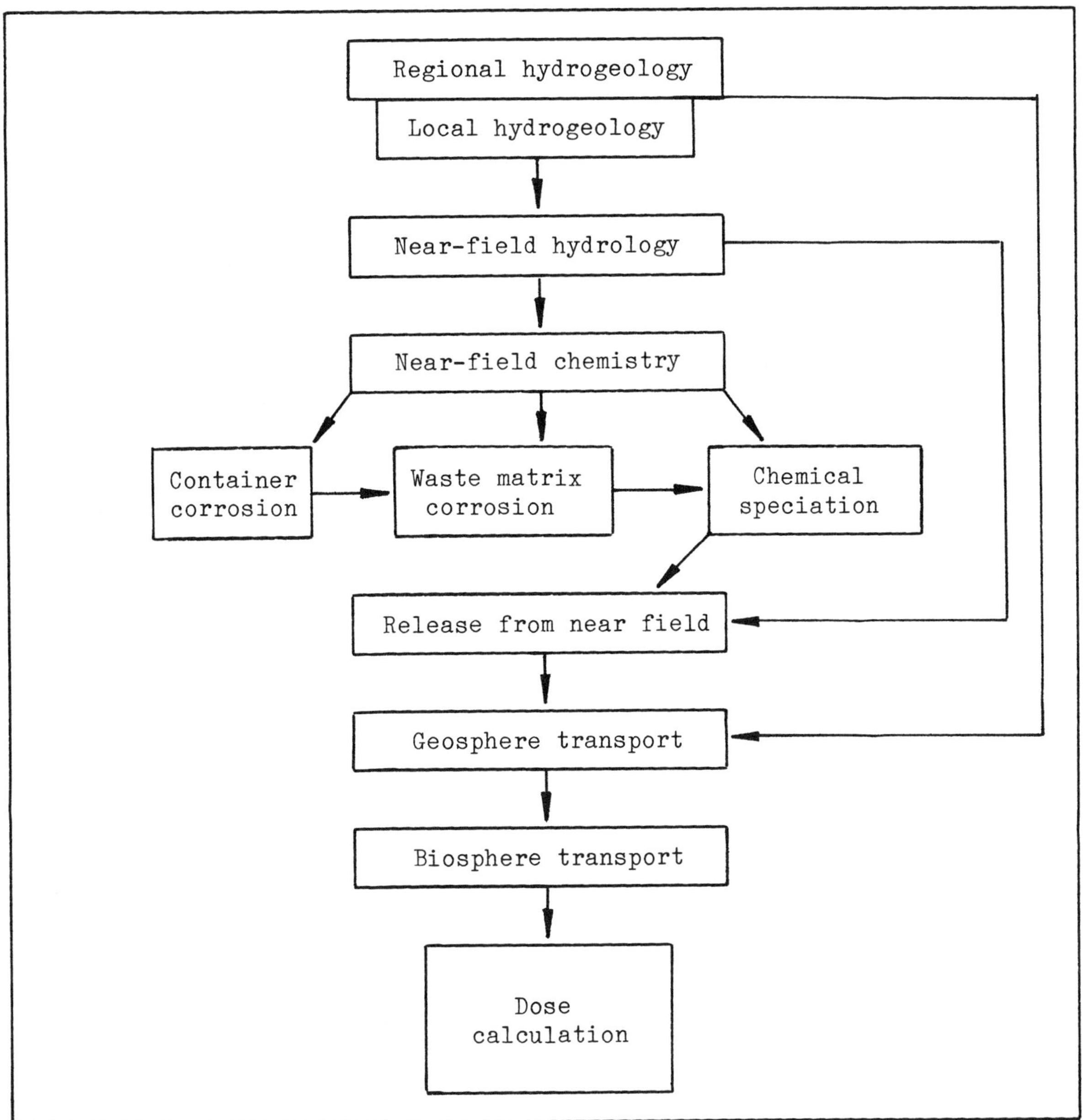

FIGURE 2. Modeling Steps

The hydrogeology on regional and local scales is complex because of the large areas of the flow systems to be modeled (23,000 km^2 and 900 km^2, respectively). Current uncertainties in the necessary input data meant that no precise output data on flow directions or velocities could be used with confidence in the safety analysis. Low predicted hydraulic gradients and volumetric flows in the deep crystalline rocks were used along with conservative choices of flow paths. Strenuous efforts have been, and are being, made to validate the hydrologic model to the maximum extent.[10]

In the near field, various interacting processes will occur and these have been quantitatively analyzed using models of different complexity. Container corrosion was studied in detail; the results are directly used to predict the duration of complete containment and to give input to near-field chemistry modeling. A chemical thermodynamic model is used to study groundwater evolution and to predict elemental solubilities. Diffusive transport through the buffer allowing for retardation and decay can be modeled with a variety of differing conceptual descriptions of the system geometry.

The radioactive nuclides released from the near field must be transported through the geosphere and this can be modeled only when the detailed flow paths are characterized. In deep granites of the type studied, water flow occurs in discrete channels within disturbed shear zones (kakirites) of increased porosity, or else in fractures in intrusive dykes (aplite or pegmatite). Conceptual models for both flow systems were developed and the mathematical analyses performed with versions of the RANCH radionuclide transport code.[11]

Using a standard compartment biosphere transport model, radionuclide uptakes and radiation doses were calculated for a variety of scenarios describing the entry into the biosphere and the state of the biosphere at entry (i.e. geographic, climatic and demographic variations were allowed for).

D. Choice of Data

An important boundary condition for Project Gewaehr was that data be based as far as possible on laboratory and field measurements and observations. In order to avoid the deterministic analyses giving a misleading, unrealistic picture of possible consequences of waste disposal, a base case of the normal groundwater scenarios was defined in which the data were chosen to be realistic but conservative. Another set of parameters was defined choosing values which represented pessimistic or worst-case estimates. Finally wider variations were performed for important parameters which cannot be well defined with knowledge available at present (e.g., extent of weathered zones around water-bearing features). The partly subjective judgment entailed in selection of data in this way is unavoidable for the type of deterministic analysis performed but can perhaps be circumvented by developments in methodology as indicated in Section V.

IV. PREDICTED PERFORMANCE OF THE REPOSITORY SYSTEM

A. Near-Field

As a first step the base-case release rate of radionuclides from the engineered repository system into the geosphere is calculated, taking into account the expected behavior of the technical barriers. The steel canisters are designed to have a minimum lifetime of 1,000 years so that there is zero release up to this time. When canister failure has occurred, release rates into the rock can, in principle, be limited by 3 effects. These are:

1. the leach or corrosion rate of the vitrified waste form (base case 10^{-7} g/cm^2.day),

2. the small amounts of poorly-soluble radionuclides which can be transported away in the restricted water flow at the repository (base case flow 0.7 l/canister/yr), and

3. the maximum rate of diffusion through the compacted bentonite of the buffer region.

For the most important radionuclides in vitrified HLW, the second effect is dominant. Accordingly, much effort has gone into quantifying the water chemistry near the repository and the corresponding maximum elemental solubilities. For elements of which various isotopes are present, this solubility is shared between the isotopes according to their relative abundance at anytime. An example of a significant solubility limited nuclide is Np-237; in the reducing conditions at depth in the crystalline host rock only 2×10^{-9} mol/l can go into solution, which leads to a total predicted release time of over 10^7 years.

In the base-case analyses in Project Gewaehr, no credit was explicitly taken for the diffusion resistance of the clay buffer. This is because the calculated effects depend strongly upon the conceptual models describing the transfer of radionuclides from the clay outer surface into the host rock and the corresponding concentrations in the groundwater. However, even very conservative conceptual models (for example with zero concentration in the rock and hence maximum concentration gradient for diffusion) clearly show that the buffer can delay nuclide transport long enough for many isotopes even with relatively long half-lives to decay very significantly. Realistic calculations indicate, for example, that Cm-245 and Am-241 do not emerge in significant concentrations through the buffer. Very long-lived nuclides like Np-237 can eventually reach saturation concentration in the low water flows outside the buffer; but if flow rates were to become higher, the diffusion resistance would ultimately limit the long-term release from the near-field even for such nuclides.

B. Releases through the Far-Field

As mentioned in Section IIIC the water-carrying zones in granite are discrete and they are also relatively far apart. Based on geologic observations, a conceptual model of the host rock in which the repository lies was developed. Tight granite ($k = 10^{-12}$ m/s) with predominantly in-filled fractures includes, every 75 m, weaker kakirite zones (k = approximately 10^{-9} m/s) which carry the moving groundwater. Although many canisters would then lie distant from a kakirite, the conservative base case neglected this and assumed all nuclides leaving the tunnel had immediate access to a flowing zone. The extremely heterogeneous granite is assumed to exist only at depth; the uppermost 500 m of granite ($k = 10^{-7}$ m/s) is modeled as homogeneous and permeable because of the increased frequency and conductivity of flowing zones observed there. This upper granite thus provides only dilution and no further retardation of radionuclides. The same is assumed true for the hundreds of meters of sediments (including aquifers) which lie above the granite.

The barrier effect of the geosphere is therefore provided by retention processes in the kakirites or in fissured dykes within the lower crystalline rock. Of great importance is the extent to which radionuclides can diffuse out of flowing zones into stagnant water in adjacent porous zones. It was conservatively decided that credit would be taken for diffusion only in weathered zones of increased porosity (higher than 2 to 3%). These certainly exist around the flow paths; the extent of the zones is, however, difficult to define. For the fissures in dykes only 1 mm was assumed; observations on kakirites led to a base-case diffusion depth of 50 cm around the water-carrying veins. Variations of this parameter lead to large variations in nuclide concentrations at the geosphere outlet.[11]

The released radionuclides from the host rock in the deep granite are diluted in overlying layers and then disperse through the biosphere. This dispersion is calculated using appropriate uptake and transfer factors and can lead eventually to doses to human populations. Figure 3 shows the results of one of the base-case variations calculated. Doses are predicted only at far-future times and at totally insignificant levels.

C. Conclusions on Repository Safety

Project Gewaehr was initiated specifically to prepare within a defined timescale an overall synthesis of waste disposal planning in Switzerland in order to allow a global assessment of the achievable safety. This synthesis has been made possible by using a combination of refined models (for well-understood parts of the system) and simplified, conservative models (in areas still needing development) and by using a combination of measured data and pessimistic estimates.

The conclusion reached is that, under all reasonably assumable circumstances, individual doses resulting as a consequence of a deep repository of the type described will always be very far below the Swiss regulatory guidelines of 10 mrem/yr. To provide this level of safety, both technical and geological barriers are needed. Given that the geological environment ensures a stable repository setting in low groundwater flows, the near field provides excellent containment of almost all nuclides for very long times. Thereafter, the geosphere can efficiently attenuate any releases of long-lived radionuclides.

Not all questions have been fully answered and work is still required before realization of a repository project. In the current project, uncertainties which could have a negative effect on predicted safety (e.g., lack of data on colloid transport) must be weighed against large conservatisms

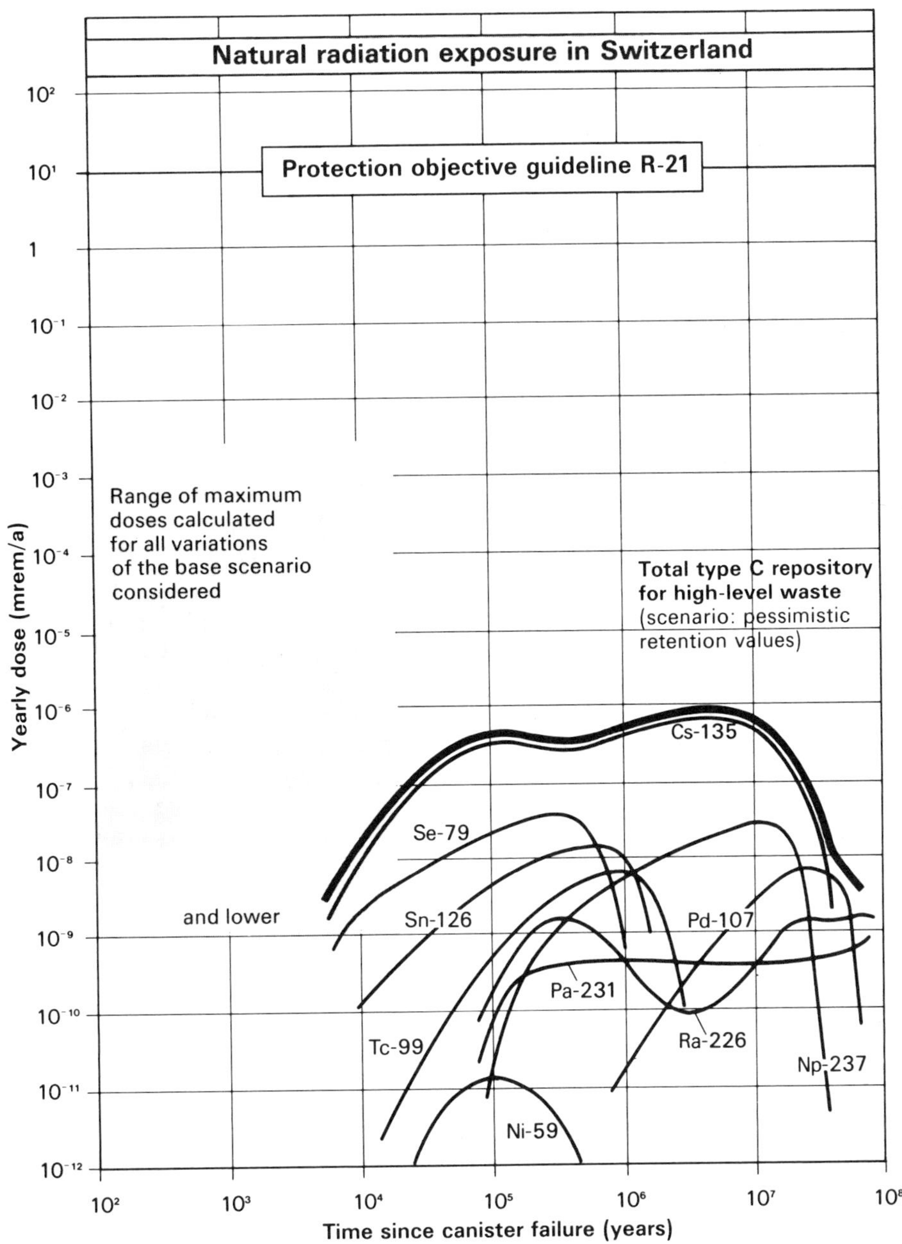

FIGURE 3. Calculated Radiation Doses

in modeling (e.g., neglect of near-field diffusion) or in data (e.g., choice of solubility and sorption values). This approach suffices for the broad assessment needed in Project Gewaehr; in preparation for the ultimate realization of optimized projects, improvements will continue to be made.

V. DIRECTIONS FOR FUTURE WORK

As indicated above, the present modeling methods will continue to be improved in key areas. These include hydrologic modeling in fractured media, thermodynamic modeling in complex chemical systems, and radionuclide transport modeling with better handling of chemical effects. Closer integration of models treating coupled physical and chemical effects is also needed. Perhaps even more important is the need to extend and refine the database with respect to geological parameters in the field and to selected parameters from the laboratory (e.g., thermodynamic constants). Finally, continuing validation of the conceptual models currently employed is an obvious major topic for future work. Here, studies on natural analogue systems are beginning to provide useful input to the validation process.

It is also planned that the technical approach to performance assessment will be amended to become less deterministic. The Swiss guidelines and timescales for Project Gewaehr made a deterministic approach appropriate, but regulatory bodies are moving towards risk criteria for disposal, as for other technological undertakings. A full risk assessment entails difficult quantification of human, technical and natural scenarios over long times, and exact treatment will never be possible. More immediately applicable is the use of probabilistic methods to quantify the likelihood of a particular result within a given scenario described by parameters, each of which can be better specified by a probability distribution than by a fixed value. Developments in this direction in other lands[12,13] are being closely followed. Current Swiss performance safety assessment efforts are, however, mainly concentrated on improving further our scientific understanding of the deterministic process occurring in a repository system.

REFERENCES

1. R. Rometsch and H. Issler, "The Swiss Nuclear Waste Disposal Program - Repository Projects and Feasibility Studies," in these proceedings (1986).

2. M. Thury, "The Swiss Site Selection Program for High-Level Nuclear Waste Disposal," in these proceedings (1986).

3. A. L. Nold, "The High-Level Waste Repository Concept for the 1985 Gewaehr Project in Switzerland," in these proceedings (1986).

4. Nagra, Projekt Gewaehr 1985 - Endlager fur hochaktive Abfalle: Das System der Sicherheitsbarrieren, Nagra NGB 85-04, Baden/Switzerland (1985).

5. Nagra, Projekt Gewaehr 1985 - Endlager fur hochaktive Abfalle: Sicherheitzsbericht, Nagra NGB 85-05, Baden/Switzerland (1985).

6. NEA, Long-Term Radiation Protection Objectives for Radioactive Waste Disposal, Paris (1984).

7. HSK, Schutzziel für die Endlagerung radioaktiver Abfalle, Richtlinien für Kernanlagen R-21, Eidgenossische Kommission für die Sicherheit der Atomanlagen KSA und Hauptabteilung für die Sicherheit der Kernanlagen HSK (ASK), Wurenlingen (1980).

8. HSK, Endlager für hochaktive Abfalle im kristallinen Grundgebirge - Modellstudie, HSK-E9, Hauptabteilung fur die Sicherheit der Kernanlagen HSK, Wurenlingen (1985).

9. PAGIS, Performance Assessment of Geological Isolation Systems - Summary Report of Phase 1: A Common Methodology Approach Based on European Data and Models, EUR 9220 EN, Commission of the European Communities, Luxembourg (1984).

10. R. W. Andrews et al., "Validation of Hydrogeologic Models to Describe Groundwater Flow in the Crystalline of Northern Switzerland," Contribution to the 9th International Symposium on the Scientific Basis of Waste Management, Stockholm, Sweden (1984).

11. H. Hadermann et al., "Transport of Radionuclides in Inhomogeneous Crystalline Rocks," Contribution to the 9th International Symposium on the Scientific Basis of Waste Management, Stockholm, Sweden (1985).

12. K. W. Dormuth and G. R. Sherman, SYVAC - A Computer Program for Assessment of Nuclear Fuel Waste Management Systems, Incorporating Parameter Variability, AECL-6814, Atomic Energy of Canada Ltd., Chalk River (Ontario) (1985).

13. A. Saltelli et al., LISA - A Code for Safety Assessment in Nuclear Waste Disposal, EUR 9306 EN, Commission of the European Communities, Luxembourg (1984).

SAFETY ANALYSIS OF A HLW REPOSITORY IN A CLAY FORMATION

A. Bonne
J. Marivoet
Study Center for Nuclear Energy
Boeretang 200
2400 Mol, Belgium

ABSTRACT

The Boom clay layer at Mol (Belgium) is a geological formation which can be considered for the construction of a high-level waste repository. A methodology developed at JRC ISPRA has been applied to perform site specific (Mol) risk assessment studies. The method consists in the identifications of possible scenarios for the release of radionuclides, assessment of corresponding occurence probabilities and consequences, and calculation of the risk. For the normal scenario, which considers the natural degradation of the repository, preliminary calculations have been performed and show that low maximum individual dose rates may be expected. The clay layer of 50-m thickness is a very efficient barrier for the disposed radionuclides. Some risk assessment exercises have been carried out already for disruptive scenarios such as a large tectonic displacement and glacial erosion. Their results indicate that for these cases, also, low risks may be expected.

I. INTRODUCTION

In 1974 studies on the possibilities for disposing of vitrified high-level waste and conditioned alpha-waste were started in Belgium. An inventory of potential host formations for geological disposal[1] indicated that only clays and shales could be considered in our country. One of the seven potential areas that were identified is situated in the north-eastern part of Belgium where the Belgian Nuclear Research Establishment (SCK/CEN) and some industrial nuclear facilities are located. It was decided in 1975 to focus the Belgian R&D programs concerning high-level waste disposal on the specific case of the oligocene Boom clay formation at Mol. The R&D program of SCK/CEN is performed within the scope of contracts with the Commission of the European Communities, a research agreement with the Belgian institute for the management of nuclear waste NIRAS/ONDRAF and several cooperation agreements with other research institutions. SCK/CEN is participating within the CEC's PAGIS action (performance assessment of geological isolation systems). This action coordinates the studies concerning continental clay formations. Granite, salt and also some sub-seabed areas are covered by other contractors. An important part of the safety analysis studies for the Belgian clay repository is performed in close collaboration with the Joint Research Centre of the European Communities at ISPRA which developed its own methodology for those analyses.[3,4]

The safety assessment exercises which have been performed or started are based on the repository concept and on release scenarios specific for the Boom clay at the Mol site. The geoscientific data to be used in the safety analysis are specific for the geological context of the area. The disposal facility concept is based on a design[2] to accomodate the vitrified high level waste and alpha waste resulting from the operation of nuclear power plants with a total hypothetical capacity of 10 GW(e) during a 30-year period.

II. METHODOLOGY

Based on the JRC Ispra methodology, the long-term assessment for the proposed repository in clay, will be performed in three steps:

1. scenario analysis, aiming at identification of the various phenomena and events that can lead to a release of radionuclides to the biosphere,

2. consequence analysis, aiming at the estimation of the radiation doses to man resulting from the selected scenarios, and

3. risk analysis, where the probability of the scenario and its consequences are combined into an estimated risk, which can be compared with relevant acceptability criteria.

III. SCENARIO ANALYSIS

Three types of scenarios can be distinguished:

1. a normal-evolution scenario, describing the natural degradation of the repository and the corresponding release of radionuclides,

2. altered-evolution scenarios, where the natural evolution is perturbed by events of a probabilistic nature modifying the conditions of the normal evolution, and

3. disruptive scenarios, describing events of probabilistic nature which have the potential to cause direct and abrupt release of radionuclides into the biosphere.

A. Normal Evolution Scenario

It is assumed that after the sealing of the repository its various components will eventually be contacted by the interstitial water in clay. This interstitial water will act as a medium which facilitates interactions among the different repository components (including the adjacent host rock). Natural degradation of the repository will thus occur resulting in the corrosion of canisters and glass blocks with release of radionuclides. These conditions form the source term for migration of the radionuclides through the clay formation. Under normal evolution conditions the characteristics of the host rock will be altered by the presence of the waste package because of its heat dissipation, irradiation effects and the release of corrosion products. The quantitative effects of these phenomena form the subject of continuing research on the properties of the host rock. However, in the absence of significant advective transport, the only way for the radionuclides to reach the adjacent aquifers is by migration through the clay formation. Once the aquifers are reached, the radionuclides will be conveyed by the ground water flow system. The normal scenario assumes the existence of a water supplying well in the migration plume of the radionuclides within the aquifiers. Presently such a well is considered as the main entrance to the biosphere, wherein different biological pathways and exposure modes are further considered. This scenario is schematically shown in Figure 1. The probability of occurrence of this scenario is assumed to be equal to one.

B. Altered Evolution Scenarios

An identification of the failure of the geological containment was performed by applying the fault-tree analysis technique.[5] This technique attempts to identify the possible failure modes and to estimate the probability of their occurrence. By this analysis one identified as well, disruptive scenarios such as altered evolution scenarios resulting from natural causes and the events resulting from human activities, inspired by the need for natural resources. This study identified three possible release receptors: ground water, land surface and atmosphere. For each receptor, failure probability ranges were estimated for different time spans ranging from 2000 years to 250,000 years. The analysis revealed that the most important release scenarios, which can be classified as altered evolution scenarios are human intrusion for the short time span, smaller tectonic displacement for the long time span (more than 25,000 years). Climatic changes, which on their own do not cause a direct

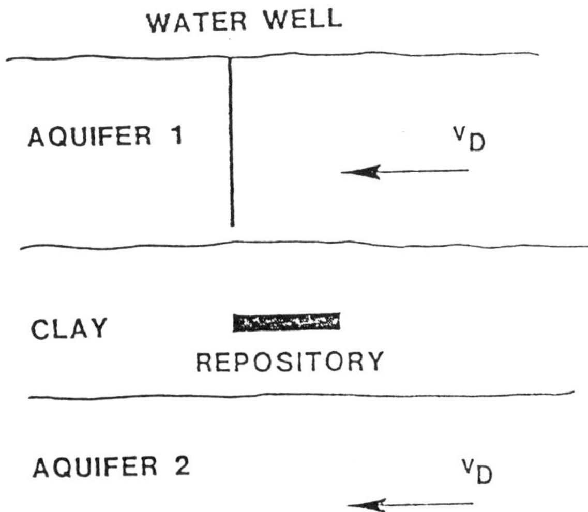

FIGURE 1. Configuration of the Modelled Near-Field, Far-Field and Biosphere System

release of radionuclides but may alter considerably the normal evolution scenario, and which, in the long term (more than 25,000 years) are generally accepted as very important. The estimated probabilities of occurrence of the altered evolution scenarios are given in Table 1.

TABLE 1. Probabilities of Occurrence for the Altered Evolution Scenarios

Scenario	Time Period	Probability of Occurrence Events/Year
human intrusion	short-term	10^{-5} to 10^{-6}
tectonic displacement	long-term	10^{-6}
climatic changes	long-term	10^{-3} to 10^{-5}

C. Disruptive Scenarios

A few disruptive scenarios have been identified for the Mol site: meteorite impact, magmatic activity, large tectonic displacements and glacial erosion. Probabilities of occurrence of these disruptive events have been estimated[5] and are given in Table 2 for a 250,000-year time span.

TABLE 2. Probabilities of Occurrence for Some Disruptive Scenarios for a 250,000-Year Time Span

Scenario	Probability
meteorite impact	$8.5\ 10^{-7}$
extrusive magmatic activity	$2.5\ 10^{-6}$
explosive magmatic activity	$6\ 10^{-7}$
tectonic displacement	10^{-5}
glacial erosion	$5\ 10^{-5}$

IV. CONSEQUENCE ANALYSIS

The aim of consequence analysis is to estimate the radiation doses to man resulting from the disposal of radioactive waste.[6] Both the yearly dose rates to a maximum exposed individual as well as the collective dose rates to the population are considered.

Because of the uncertainties associated with the model parameters, two sets of calculations are performed. The first set is designed to produce best estimates of dose rates; the calculations are based on the best estimate or available values for the model parameters, and carried out with the most accurate models available to describe the behaviour of the repository, the surrounding geology and the biosphere; they also incorporate an extended inventory of radionuclides. The other set is designed to account for the uncertainties on the values of the model parameters. These calculations are based on distributions of parameter values, using Monte Carlo techniques to compute the corresponding distribution of dose rates. In this case a reduced inventory of radionuclides and, as far as possible, simplified transport models are used.

A. Normal-Evolution Scenario

1. Best-estimate calculation--The computer code developed has a modular structure and is built up with submodels describing the radionuclide inventory in the waste form, the migration in the clay formation and in the aquifers, and the transport to man in the biosphere. In the near-field an extremely simple model is used because it has been shown by a sensitivity study with respect to the model assumptions that the near-field phenomena have a very small influence on the calculated fluxes of radionuclides leaving the clay layer. Migration in the bulk of the clay layer is based on the NUCDIS model[7] which uses a numerical solution of the one-dimensional advection-dispersion equation. A three-dimensional analytical aquifer model MICA[8] has been developed to calculate the concentrations of radionuclides in the water pumped from a water well located in the aquifers overlying or underlying the clay formation. A simple biosphere model[9] is used to calculate dose rates to the maximum exposed individual, accounting for both ingestion and inhalation pathways. The ingestion pathway includes doses due to the ingestion of contaminated drinking water, vegetables, cow milk and beef. The inhalation pathway estimates dose rates due to the inhalation of resuspended dust from a contaminated soil. Collective dose rates will be calculated by making use of the comprehensive biosphere model BIOS[10].

   Preliminary calculations of the yearly dose rates to a maximum exposed individual show that Tc-99, Cs-135, Np-237, U-233, Th-229 and U-236 may give rise to dose rates above 10^{-10} Sv/yr. The maximum dose rates for the ingestion and inhalation pathways and their time of occurence are given in Table 3. For most of the considered radionuclides the most important pathway in the biosphere model is the drinking water intake, which contributes more than 90% of the total dose rate.

2. Calculations with sampled parameter values--This type of calculation uses a modified version of the LISA code.[11] The far field module of the original LISA code has been replaced by two computer codes: the analytical code[12] MICOF for the description of the migration of radionuclides in the clay layer and the code MICA already mentioned which describes the transport in the aquifer. Calculations with this modified LISA code are planned after detailed evaluation of the best estimate results. These calculations will be complemented by sensitivity and uncertainty analyses to obtain an estimation of the uncertainties on the calculated dose rates. The sensitivity analysis will also provide a basis for assigning priorities to R&D work on model parameters.

TABLE 3. Peak Dose Rates and Times of Occurrence for the Important Radionuclides in the Case of the Normal Evolution Scenario

Radionuclide	Time (million years)	Ingestion Dose Rate (Sv/yr)	Inhalation Dose Rate (Sv/yr)
Tc-99	1.6	$9.2\ 10^{-10}$	$2.5\ 10^{-15}$
Cs-135	7.4	$3.5\ 10^{-8}$	$5.2\ 10^{-12}$
Th-229	12	$3.5\ 10^{-9}$	$7.4\ 10^{-9}$
U-233	12	$6.0\ 10^{-10}$	$1.1\ 10^{-12}$
Np-237	12	$2.3\ 10^{-7}$	$9.7\ 10^{-11}$
U-236	50	$1.6\ 10^{-10}$	$3.0\ 10^{-11}$

B. Altered-Evolution Scenarios

In this part, a first exercise of consequence analysis for the tectonic displacement scenario[13] with subsequent migration within the aquifers over- and underlying the Boom clay has been performed. The occurrence of a fault with a displacement greater than 10 m is assumed. This type of fault accounts for about 6% of the total faults observed in the Mol area.[5] The considered pathways to man are the pumping of ground water in a contaminated water plume for direct consumption as drinking water and for irrigation purposes. Within these assumptions, a maximum expectation value of about 10^{-5} Sv/yr is obtained for the dose rate to the maximum exposed individual.

Because of the very long time spans that have to be considered in safety studies, climatic changes have a high probability of occurrence. They may influence the hydrological system of the region, but will not affect the very strong retardation capacity of the clay layer and have a negligible effect on the flux of radionuclides entering the aquifers. For this scenario individual dose rates of the same order of magnitude as for the normal scenario may be expected. Consequence analysis for the human-intrusion scenario will be started in the next months.

C. Disruptive Scenarios

The highest probability of occurrence in this group is attributed to the glacial erosion scenario and a first attempt of a consequence analysis[13] has been performed. The scenario assumes the destruction of the geological containment by the digging action of an overlying ice cap. After the glaciation period the waste will be dispersed as small fragments in the remnants of the glacier ridge. Possible pathways to man are the pumping of well water for domestic and agricultural purposes and the utilization of river water. The maximum expectation value of the dose rate is estimated to be about 0.02 Sv/yr to the maximum exposed individual.

V. RISK ASSESSMENT

The contribution of each scenario to the total risk at a given time is defined by the product of its probability of occurrence with its calculated risk. This risk is defined as the product of the calculated individual dose rate and a risk conversion factor. This risk factor is 0.0125 1/Sv for effective dose rates below 1 Sv/yr, which is the case in HLW disposal, where only stochastic health effects (fatal cancer) need to be considered.

Applying this method yields a maximum risk of 3×10^{-9} per year as an estimate for the normal scenario. In the case of the tectonic displacement scenario the probability of occurrence is 0.06 for a 10^6 years time span. The contribution to the total risk for this scenario is then 10^{-8} per year. In the case of the disruptive glacial scenario, the contribution to the total risk is 5×10^{-9} per year.

These values for such a catastrophic event are still much lower than the maximum individual risk objective of 10^{-5} per year which is a value recommended by the NEA Expert Group;[14] this value corresponds to the health risk due to natural background radiation at sea level.

VI. CONCLUSIONS

Although the calculations of the maximum individual dose rates have still a preliminary character, the results obtained show that the risks associated with a HLW repository in the Boom clay at Mol may be expected to be much lower than the criteria presently recommended at an international level for nuclear waste disposal. The results have to be completed with probabilistic calculations, with the calculations of the collective dose rates and with sensitivity and uncertainty analyses. Risk assessment studies have to be performed for other scenarios particularly for scenarios where breaching of the clay layer is considered. Preliminary versions of these complementary studies form part of the PAGIS program and they will be completed by the end of 1986. Detailed studies will be carried out in the framework of a Belgian Safety Analysis Report for the designed HLW disposal facility; this report is expected to be finished in 1988.

REFERENCES

1. Commission of the European Community, "Confinement Geologique des Dechets Radioactifs dans la Communauté Européenne", EUR 6891 FR (1980).

2. P. Manfroy et al., "Conception d'une Installation pour l'Enfouissement dans l'Argile de Dechets Radioactifs Conditionnés," Proc. IAEA Symp. on Underground Disposal of Radioactive Waste, Otaniemi 1979, vol. II, p. 59 (1980).

3. G. Bertozzi and M. d'Alessandro, "A Probabilistic Approach to the Assessment of the Long-Term Risk Linked to the Disposal of Radioactive Waste in Geological Repositories," Rad. Waste Management and the Nucl. Fuel Cycle, Vol. 3(2), p. 117 (1982).

4. B. Bertozzi and A. Saltelli, "On the Risk Analysis of Radioactive Waste Repositories," in publication (1985).

5. M. d'Alessandro and A. Bonne, Radioactive Waste Disposal into a Plastic Clay Formation, Harwood Academic Publ., London (1981).

6. I.A.E.A., "Safety Assessment for the Underground Disposal of Radioactive Wastes," Safety Series No. 56, I.A.E.A., Vienna (1981).

7. J. Marivoet, "NUCDIS. Numerical Solution of the Advection-Dispersion Equation Applicable to the Migration of Radionuclides in Clay Layers," in publication (1985).

8. C. Van Bosstraeten, "MICA. Computer Program for the Calculation of the Migration of Radionuclides from a Repository in a Clay Layer to the Biosphere," in publication (1985).

9. A. Saltelli, "Extensions and Improvements to the Code LISA," technical note, JRC Ispra (1985).

10. G. Lawson and G. M. Smith, BIOS. A Model to Predict Radionuclide Transfer and Doses to Man Following Releases from Geological Repositories for Radioactive Wastes, Report R 169, NRPB, Chilton (1985).

11. A. Saltelli, G. Bertozzi and D. A. Stanners, LISA. A Code for Safety Assessment in Nuclear Waste Disposals; Program Description and User Guide, Report EUR 9306 EN, JRC Ispra (1984).

12. M. Put, "MICOF. A Uni-Directional Analytical Model for the Calculation of the Migration of Radionuclides in a Porous Geological Medium," in publ. in Rev. Waste Management (1985).

13. G. Bertozzi, M. d'Alessandro and A. Saltelli, Waste Disposal into a Plastic Clay Formation: Risk Analysis for Probabilistic Scenarios, Report EUR 9641 EN, JRC Ispra (1985).

14. NEA, Long Term Radiation Protection Objectives in Radioactive Waste Disposal, Report of the joint CRPPH/RWMC Expert Group, NEA (OECD), Paris (1984).

DETERMINING PROBABILITIES OF GEOLOGIC EVENTS AND PROCESSES[1]

R. L. Hunter
R. M. Cranwell
Waste Management Systems Division
Sandia National Laboratories
Albuquerque, New Mexico 87111

C. John Mann
Department of Geology
University of Illinois
Urbana, Illinois 61801

ABSTRACT

The Environmental Protection Agency has recently published a probabilistic standard for releases of high-level radioactive waste from a mined geologic repository. The standard sets limits for contaminant releases with more than one chance in 100 of occurring within 10,000 years, and less strict limits for releases of smaller probability. The standard offers no methods for determining probabilities of geologic events and processes, and no consensus exists in the waste-management community on how to do this. Sandia National Laboratories is developing a general method for determining probabilities of a given set of geologic events and processes. In addition, we will develop a repeatable method for dealing with events and processes whose probability cannot be determined.

I. INTRODUCTION

The U.S. Environmental Protection Agency (EPA) has recently published a proposed environmental standard for the management and disposal of high-level radioactive waste.1 The standard is probabilistic. It sets certain limits for contaminant releases that are estimated to have more than one chance in 100 of occurring within 10,000 years, and a less strict set of limits for releases estimated to have between one chance in 100 and one chance in 10,000 of occurring within 10,000 years. No limits are set for releases of lesser probability. The proposed standard requires performance assessments that estimate probabilities of events and processes that might lead to releases of radioactive waste, but it does not offer any guidelines for methods of determining probabilities.

No consensus exists in the waste-management community as to how probabilities of geologic events and processes should be determined. Methods used in the past have differed widely. In some cases, events and processes have

---

[1]This work was supported by the U. S. Nuclear Regulatory Commission, Office of Nuclear Material Safety and Safeguards and performed at Sandia National Laboratories, which is operated by the U. S. Department of Energy under contract number DE-AC04-76DP00789.

been chosen for preliminary performance assessments without any published consideration of probabilities.[2] Some workers have assumed probabilities.[3] Some probabilities have been calculated based on fairly sophisticated mathematical techniques, but using limited or uncertain data[4,5,6]. Other studies have used a combination of these methods, depending on the availability of data.[7] The general geological literature also contains examples of probabilistic analysis of geological events and processes that are based on a variety of techniques.[8,9,10,11,12,13]

Sandia National Laboratories in Albuquerque, under the sponsorship of the U.S. Nuclear Regulatory Commission, (NRC) Office of Nuclear Material Safeguards and Safety, is developing a general method that can be used to determine probabilities for a given set of geological events and processes considered to be potentially hazardous. Our approach is to model probability determinations in a completely general fashion for all possible geologic conditions, emphasizing events and processes that are thought to be likely at sites the Department of Energy (DOE) is currently considering. The work described here represents only a preliminary approach to probability determination, not our final product. Neither does this paper necessarily represent the opinion of the NRC or its staff. Wherever possible, the final method will be compatible with existing NRC procedures for probabilistic risk assessment of nuclear reactors.[14]

II. PROPOSED METHOD

Most geologic events and processes can be sorted into one of three groups when attempting to assign probabilities of future occurrence. These groups are those for which probabilities can be determined with near certainty, fairly accurately, or only with limited confidence. In the past, however, workers have often assumed some probability, without reference to available data or techniques, for events or processes whose probability could have been fairly accurately determined. A flow chart (Figure 1) offers a systematic, repeatable method of determining whether the probability of a given event or process can be estimated with confidence.

First, the timing and nature of some events or processes can be predicted with near certainty, because our knowledge of them and the conditions under which they occur is excellent. Many of these events or processes are virtually deterministic, both because the area of concern is well understood and because quantitative development of the subject is well established on either theoretical or empirical grounds. For example, heat production and inventory change in a radioactive source can be predicted with great accuracy, given data on the initial inventory. Ground-water flow within some aquifers and basins can also be predicted, although the accuracy of the prediction depends on data that may not be readily obtained. The physics of fluid flow through porous media is well established theoretically;[15,16] numerous quantitative models have been developed and applied to real geologic situations to predict ground-water movement; and hydrologists generally agree that several established methods and

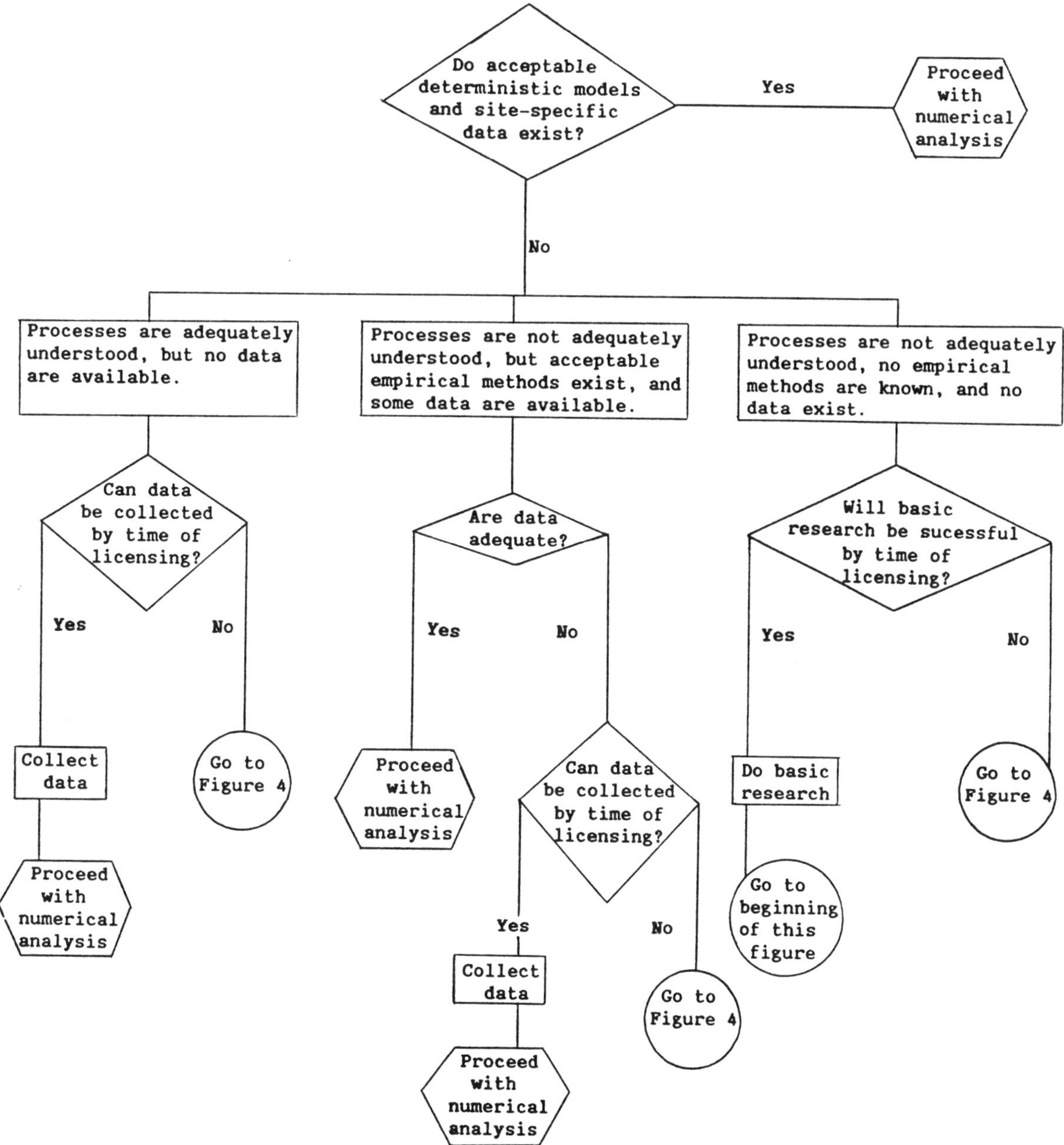

FIGURE 1. Tentative Flowchart for Sorting Geologic Events and Processes During the Application of the Method for Determining Probabilities

techniques provide acceptably similar results.[17] Generally speaking, however, ground-water data from sites currently being considered by DOE are sparse. Inadequacies in the data make uncertainty and sensitivity analysis of the predicted flow necessary. If we assume that a DOE site is in a basin in which porous flow dominates, then an appropriate path through the flow chart (Figure 2) implies that ground-water flow will be well understood after appropriate data are collected and ground-water flow is modeled.

Second, other phenomena can be predicted reasonably accurately today because adequate data bases exist or may be obtained easily to estimate the probability of certain geologic events and processes. Some phenomena can be predicted based on general geologic knowledge to be nearly certain to occur in a given area during a long enough period of time. For example, Quaternary volcanism is entirely restricted to the West (western United States), where an upper lithospheric plate is above a subducting plate margin. Here active volcanoes exist today.[18] Their frequencies of eruption during the Quaternary Period are generally well known (NOAA Data Center) from geologic data. Some sites being considered, such as the basalt, tuff, and Paradox Basin sites, are well inside this area of Quaternary volcanism. Because young volcanoes tend to be easily identified in the arid West, existing data are usually adequate. An appropriate path for determining probabilities of volcanism would likely involve more probabilistic analyses than did ground-water flow (Figure 3). Even though it is not possible to predict deterministically when or where a volcano will erupt next, the process is well enough understood that it can be

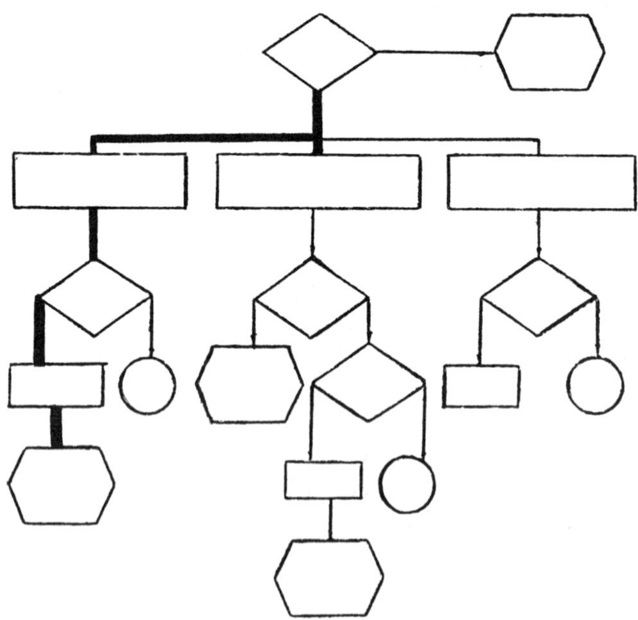

FIGURE 2. A Path in Figure 1 Likely to be Followed in Assessing Risk from Ground-Water Flow in a Drainage Basin with Few Existing Hydrologic Data

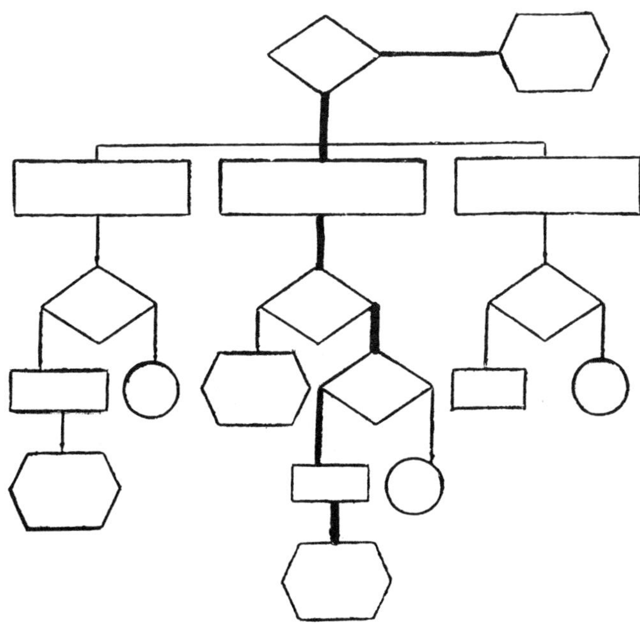

FIGURE 3. A Path in Figure 1 Likely to be Followed in Assessing Risk from Volcanism in an Area of Inadequate Existing Data on Volcanic History

stated that some will occur in the West. Moreover, enough data exist to determine the past rate of new eruptions in the West, and therefore (assuming that the rate continues unchanged), the probability of new eruptions. In fact, just such a calculation has been made for the tuff site in southern Nevada[6]; the probability of new eruptive vents was estimated from the known number of volcanoes and their frequency of eruption. Because Gulf Coast salt domes being considered for nuclear waste disposal are well outside the area of Quaternary volcanism, volcanism probably need not be considered in salt dome performance assessments. The validity of assumptions about rates and changes in rates will be one topic of investigation during this project.

Extreme values of some geologic parameters may be difficult to predict with great accuracy even when the parameters are well understood and available data are excellent. For example, extreme uncertainty exists that future seismic events will not exceed some stated level of energy release. Seismologists have used Gumbel's theory of extreme values,[19] Markovian models,[20] and maximum entropy,[21] as well as assuming various probability distributions to estimate maxima that may be expected for future extreme values. Many such extreme values may have small probabilities, however, and Appendix B of the most recent draft of the EPA standard[22] suggests that performance assessments need not consider events and processes that are estimated to have less than one chance in 10,000 of occurring over 10,000 years. For extremely small probability events, a rough estimate of probability may be adequate to demonstrate compliance with the standard.

Phenomena that fall into a third and final category, however, present major difficulty because they can be predicted only poorly and with limited confidence, either because the event or process is inadequately understood or because data are inadequate to make accurate predictions. Geologic events and processes that are included in this third category of poorly predictable phenomena offer the greatest uncertainties to risk evaluation in nuclear-waste disposal. A second flow chart (Figure 4) offers a preliminary method for dealing with these events and processes. The flow chart is arranged more-or-less in order of increasing uncertainty and greater undetermined risk. By following the first branch, both uncertainty and risk of a given event or process may be eliminated. For example, the unknown risks associated with raising large volumes of rock to high temperatures might be eliminated by reducing the thermal loading of the repository. Thus a design change may eliminate or reduce the probability of a risk or decrease the uncertainty associated with it.

Following the second branch (Figure 4) may result in calculation of such a small upper bound on probability that the event ceases to be of concern. The probability of new faulting that would affect the WIPP site in southeastern New Mexico has been calculated to be less than 4×10^{-11} per year.[4] Even if uncertainty is a few orders of magnitude, such a small probability removes that event from regulatory concern. As discussed above, the most recent draft of the EPA standard[22] suggests that performance assessments need not consider events that are estimated to have less than one chance in 10,000 of occurring over 10,000 years. Nevertheless, these estimates of probabilities, even though acceptably small, retain great uncertainty and must be used cautiously.

If risk cannot be eliminated by changing the repository design, and no data that place a small bound on the probability of the event or process exist, then it is appropriate to model the consequences of the event or process of concern. Some events and processes, even if assumed to have a probability of 1, have only a negligible risk because consequences are extremely small compared with other possible releases. Appendix B of the most recent draft of the EPA standard[22] suggests that performance assessments need not include events and processes with releases that would not significantly change the estimate of cumulative release.

If consequences are not acceptable, then it becomes important to obtain some estimate of the probability. Although it has been determined that data neither exist nor can be obtained in an reasonable amount of time that can be used in calculating or bounding a probability, expert opinion might nevertheless be useful in estimating the probability. Expert opinion should be incorporated in assessments in some probabilistic form, presumably by using Bayesian or Delphi methods. If the probability is judged very small by experts (such as meteorite impact or volcanism in the Gulf Coast), then moderate or large consequences are of small concern, although at this stage in the process, confidence in the predicted results may not be great. If, on the other hand, the probability is judged moderate or great, then risk can be calculated using the consequences determined previously and expert judgment of the probability.

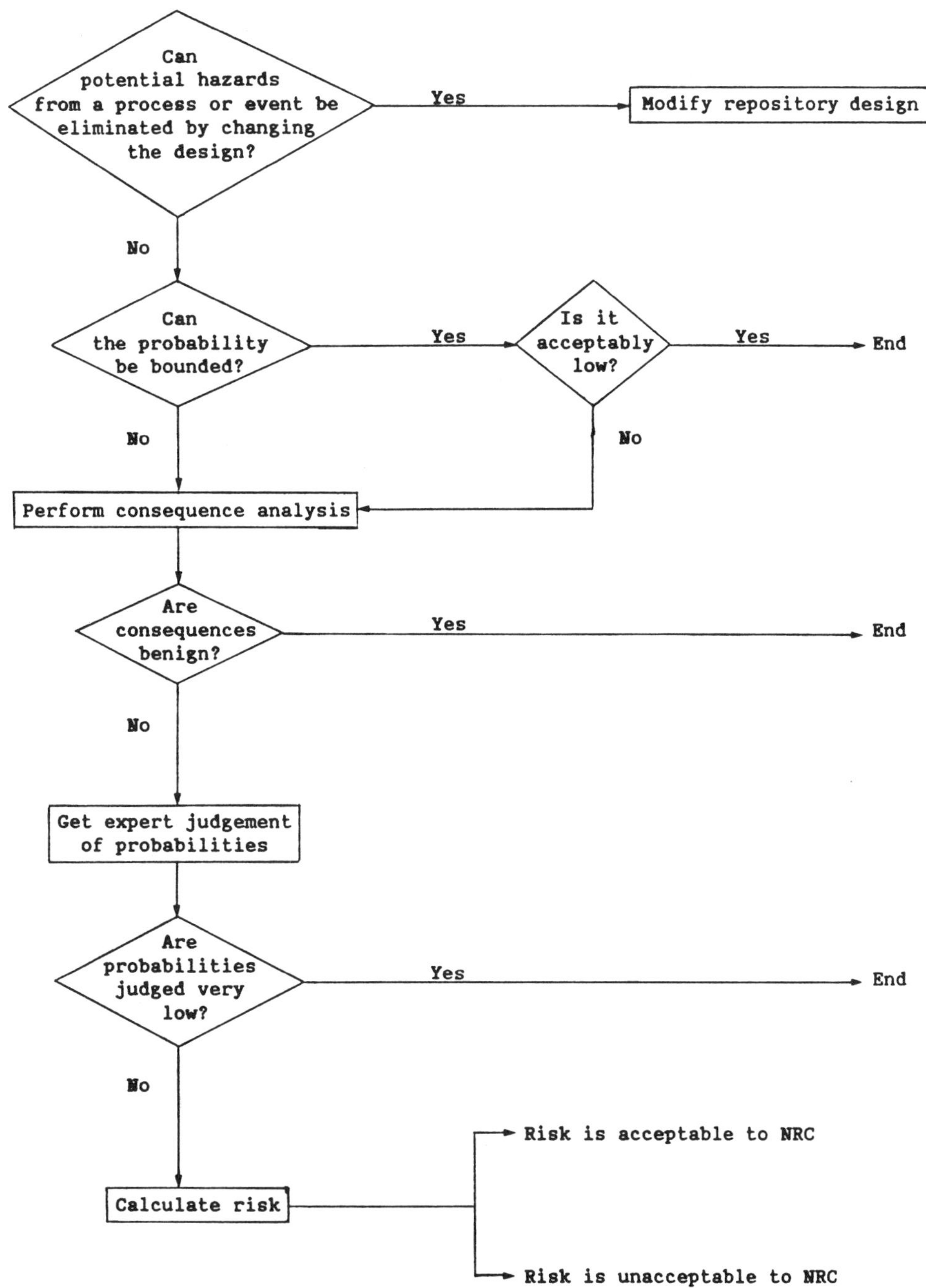

FIGURE 4. A Flowchart for Dealing with Events and Processes Whose Probabilities Fall into Category C Can Be Determined with Only Limited Confidence

If DOE does not wish to withdraw the site from consideration, then the NRC must use its regulatory discretion to determine whether or not the calculated risk is acceptable, bearing in mind the uncertainty that may be present in the expert judgment of probability. The appropriateness of the use of expert judgment in predictions and qualitative judgement in determining compliance is explicitly recognized by the most recent draft of the final EPA standard.[22]

III. ANALYSIS OF INACCURACIES

Regardless of which category geologic events and processes fall into, all may contain some inaccuracy that has entered into the calculations at each stage of the process. Uncertainty in the probabilities may arise from two sources. First, the physical model may be inaccurate. For example, volcanism might be modeled as a spatially random process, even though it is known to be controlled by deep-seated crustal weaknesses. If the location of crustal weaknesses is unknown, spatial randomness may be the best model available, but nonetheless it is not the correct model. Second, uncertainty may exist about the values of parameters used to calculate the probabilities. For example, the probability of volcanism in a given area might be calculated based in part on the ages of existing volcanoes. The radiometric ages used always carry some uncertainty. Performance assessments for nuclear-waste repositories will estimate releases that could affect large numbers of people. Some estimate of uncertainity in probabilities is necessary to technically defend decisions about the risks for proposed repositories.

IV. SUMMARY

Many natural events and processes that will be considered in a performance assessment can be treated deterministically or probabilistically using a systematic and repeatable method such as that outlined tentatively here. It would be useful to the waste-management community to reach some agreement on the steps to be included in establishing probabilities for risk assessment methods. In the final performance assessments, however, some expert judgement and irreducible uncertainty may be present, necessitating a final decision from the NRC on the acceptability of the residual uncertainty and risk presented by the repository.

REFERENCES

1. U.S. Environmental Protection Agency, "Environmental Standards for the Management and Disposal of Spent Nuclear Fuel, High-Level and Transuranic Radioactive Wastes," Federal Register, v. 47, p. 58195-58206 (1982).

2. R. C. Arnett et al., Preliminary Hydrologic Release Scenarios for a Candidate Repository Site in the Columbia River Basalt, RHO-BWI-ST-12, Rockwell Hanford Operations, Richland, WA (1980).

3. R. E. Pepping, M. S. Y. Chu, and M. D. Siegel, Technical Assistance for Regulatory Development: A Simplified Repository Analysis in a Hypothetical Bedded Salt Formation, SAND82-0996, Sandia Natl. Labs., Albuquerque, NM (1983).

4. H. C. Claiborne and F. Gera, Potential Containment Failure Mechanisms and Their Consequences at a Radioactive Waste Repository in Bedded Salt in New Mexico, ORNL-TM-4639, Oak Ridge Natl. Lab., Oak Ridge, TN (1974).

5. R. M. Cranwell et al., Risk Methodology for Geologic Disposal of Radioactive Waste: Scenario Selection Procedure, NUREG/CR-1667, SAND80-1429, Sandia Natl. Labs., Albuquerque, NM (1982).

6. B. M. Crowe and W. J. Carr, Preliminary Assessment of the Risk of Volcanism at a Proposed Nuclear Waste Repository in the Southern Great Basin, U.S. Geological Survey Open-file Report 80-357 (1980).

7. R. L. Hunter, G. E. Barr, and F. W. Bingham, Scenarios for Consequence Assessments of Radioactive-Waste Repositories at Yucca Mountain, Nevada Test Site, SAND82-1227, Sandia Natl. Labs., Albuquerque, NM (1983).

8. T. R. Carr, "Log-linear Models, Markov Chains, and Cyclic Sedimentation," Jour. of Sedimentary Petrology, v. 52, p. 905-912 (1982).

9. H. E. Clifton, "Storms, Floods and the Coastal Sedimentary Record," (abst.), Amer. Assoc. of Petroleum Geologists Research Colloquium, Oklahoma City (1978).

10. P. E. Gretner, "Significance of the Rare Event in Geology:" Amer. Assoc. of Petroleum Geologists Bulletin, v. 60, p. 2197-2206 (1967).

11. S. E. Rantz, "Surface Water Hydrology of Coastal Basins of Northern California," U.S. Geological Survey Water Supply Paper 1758 (1964).

12. P. M. Sadler, "Sediment Accumulation Rates and the Completeness of Stratigraphic Sections," Jour. of Geology, v. 89, p. 569-584 (1981).

13. G. G. Simpson, "Probabilities of Dispersal Through Time," Amer. Museum of Natural History Bulletin, v. 99, p. 163-176 (1952).

14. American Nuclear Society and The Institute of Electrical and Electronics Engineers, PRA Procedures Guide, A Guide to the Performance of Probabilistic Risk Assessments for Nuclear Power Plants, Final Report, Volumes 1 and 2, NUREG/CR-2300, U.S. Nuclear Regulatory Commission (1983).

15. R. J. M. DeWiest, ed., Flow Through Porous Media, Academic Press, NY, 516 pp. (1969).

16. C. M. Marle, Multiphase Flow in Porous Media, Gulf Publ. Co., Houston, TX, 254 pp. (1981).

17. S. N. Davis and R. J. M. DeWiest, Hydrogeology, John Wiley and Sons, NY, 447 pp. (1966).

18. H. Williams and A. R. McBirney, Volcanology, Freeman, Cooper, and Co., San Francisco, CA, 400 p. (1979).

19. A. F. Shakal and D. E. Willis, "Estimated Earthquake Probabilities in the North Circum-Pacific Area," Seismological Soc. of Amer. Bull., v. 62, p. 1397-1410 (1972).

20. A. S. Patwardhan, R. B. Kulkarni, and D. Tocher, "A Semi-Markov Model for Characteristic Recurrence of Great Earthquakes," Seismological Soc. of Amer. Bull., v. 70. p. 323-347 (1980).

21. J. B. Berrill and R. D. Davis, "Maximum Entropy and the Magnitude Distribution," Seismological Soc. of Amer. Bull., v. 70. p. 1823-1831 (1980).

22. U.S. Environmental Protection Agency, Working Draft 5--Final Environmental Standards for the Management and Disposal of Spent Nuclear Fuel, High-Level and Transuranic Radioactive Wastes (1985).

THE VAULT SUBMODEL FOR THE INTERIM ASSESSMENT OF THE CANADIAN CONCEPT FOR NUCLEAR FUEL WASTE DISPOSAL

D. M. LeNeveu
J. R. Walker
Atomic Energy of Canada, Ltd
Whiteshell Nuclear Research Establishment
Pinawa, Manitoba ROE 1LO, Canada

ABSTRACT

The Canadian concept for nuclear fuel waste disposal involves immobilization of the waste and emplacement in plutonic rock. Consequences well into the future are predicted with the aid of the computer code, SYVAC. SYVAC accepts a description of the disposal system in the form of simplified submodels, and samples input data from distributions that reflect the uncertainty and variability of the data. At present, there are submodels to represent the disposal vault, the geosphere and the biosphere.

The vault submodel is described in detail in this paper and some analysis of the performance of the vault is presented. The uncertainty in the analyses due to uncertainty about submodel assumptions and in specifying probability distributions for significant input parameters is investigated.

I. INTRODUCTION

In the Canadian concept for nuclear fuel waste disposal, the waste is immobilized in a matrix material, and emplaced deep underground in a vault located in stable plutonic rock.[1] Immobilization technology is being developed with two options in mind: 1) disposal of intact used CANDU™ fuel and 2) disposal of the wastes that would result from reprocessing and recycling this fuel.

The post-closure assessment of this concept covers the period of time following closure of the vault, and is performed using the computer code SYVAC.[2-4] SYVAC contains a set of submodels representing the major components of the disposal system: the vault, the geosphere and the biosphere. Parameters that define the behaviour of the system are entered as probability distributions rather than as single values. Parameter values are sampled from these distributions to characterize a possible state of the system, i.e., to define a scenario. For each scenario, the transport of radionuclides from the disposal vault to the biosphere is simulated deterministically and a consequence is calculated. Repeated sampling of scenarios and calculation of consequences result in a histogram of estimates of consequence versus frequency.

An initial version of SYVAC, called SYVAC1, was used in an interim assessment of used-fuel disposal in plutonic rock[5] and of recycle-waste disposal under the seabed.[6] Subsequent development resulted in a revised version,[7-10] SYVAC2, which was used in the second interim assessment of the Canadian concept for nuclear fuel waste disposal.[11]

This paper describes the vault submodel used in SYVAC, and the predicted performance of the vault in the disposal system. The uncertainty in the analysis due to uncertainty about submodel assumptions and in specifying probability distributions for significant input parameters is investigated.

II. THE VAULT SUBMODEL

The vault submodel is based on a disposal vault consisting of excavated rooms with cylindrical holes, called boreholes, drilled into the floor of the rooms. Thin-walled titanium containers encapsulating the wastes would be placed in the boreholes and surrounded by a compacted buffer material, and the remaining excavation would be backfilled. The buffer would be composed of a mixture of bentonite and sand. The backfill would be composed of bentonite and crushed rock.[12]

Five major processes are considered in the vault submodel: the failure of the waste containers, the release of radionuclides from the waste form, the coupling between the mass transfer in the vault and the mass transfer in the geosphere, the transport and sorption of radionuclides in the vault, and the precipitation of radionuclides in the vault.

A. The Failure of the Waste Containers

To describe the rate of failure of the waste containers, a container failure function (CFF) is used, which gives the fraction of containers failing as a function of time. A detailed study has been carried out to characterize the CFF.[8] The CFF takes into account the spatial and temporal variations of temperature in the vault and uses experimental data for the rates of uniform corrosion of a thin-walled titanium container. The study suggested that the CFF can be described by a truncated normal probability distribution. The uncertainty in the mean lifetime of the container, due to uncertainties in the estimated temperature profile and the rates of corrosion, is taken into account by treating the mean lifetime as a parameter selected from a normal distribution with a mean value of 1.7×10^4 yr and a standard deviation of 3.2×10^3 yr. The mean lifetime is also used to define an early truncation time, before which all containers are assumed to remain intact.[8]

B. The Release of Radionuclides From the Waste Forms

The characteristics of the waste forms will depend on the disposal option used. For the disposal of intact used fuel, the primary waste form will be the used UO_2 fuel. If used fuel is recycled to remove valuable components (such as plutonium), a variety of special waste forms will be produced. Both options include several waste forms, and those considered in the second assessment are listed in Table 1.

TABLE 1. Waste Forms

Disposal Option	Waste Type	Waste Form
Used fuel	Zircaloy activation products	Zircaloy sheaths
	Fission products and actinides	Used UO_2 fuel
	Container infilling material	Lead-5% antimony*
Recycle fuel waste	Fission products and actinides (except for U, Pu)	Sodium-calcium aluminosilicate glass or borosilicate glass
	Iodine	Bismuth oxyiodide or barium iodate
	Carbon	Barium carbonate
	Zircaloy activation products	Compacted Zircaloy hulls

* The lead-5% antimony is assessed for its chemical toxicity.

The rate of release of radionuclides from the waste forms is represented by three source boundary conditions: a short-term, or, "instant", release boundary condition and two long-term congruent release boundary conditions.[8] In one of the congruent release boundary conditions, the radionuclide release is controlled by a constant concentration of the dissolved matrix material. In the other, the radionuclide release is controlled by the constant dissolution rate of the matrix material. The source boundary conditions for each waste-form component are summarized in Table 2.

C. Vault-Geosphere Coupling

The radionuclide flow across the vault-geosphere boundary is assumed to be proportional to the radionuclide concentration at the boundary. The constant of proportionality is called the mass transfer coefficient. Expressions for the mass transfer coefficient as a function of assumed groundwater flow direction have been derived for steady-state conditions, idealized geometry, a uniform velocity field, and diffusional transport in the buffer.[8] Through these expressions, the radionuclide flow from the vault has been coupled to the groundwater flow field in the geosphere. For example, an increase in the groundwater flow in the rock around the vault would be represented by an increase in the mass transfer coefficient and would result in an increase in the rate of mass transfer of radionuclides from the vault to the geosphere.

TABLE 2. Source Boundary Conditions

Source Boundary Condition	Waste Form or Waste-Form Component
Instant release: characterized by an impulse input	Inventory of the radionuclides in the gaps and grain boundaries of the used fuel, readily leachable fraction of the borosilicate sodium-calcium aluminosilicate glass.
Solubility-controlled congruent release: determined from a separate solution to the convecton-diffusion equation[13] with a constant concentration boundary condition	Used-fuel matrix, sodium-calcium aluminosilicate glass matrix, Zircaloy sheaths, compacted Zircaloy hulls, lead-5% antimony infilling material, barium iodate matrix, bismuth oxyiodide matrix and barium carbonate matrix
Leach-rate controlled congruent release: characterized by a constant matrix dissolution rate	Borosilicate glass matrix

D. Mass Transport and Sorption in the Vault

The mass transport and sorption of radionuclides in the buffer are represented by the set of one-dimensional convection-diffusion equations for a decay chain.[13] The effect of sorption of radionuclides in the buffer is characterized by an element-specific retardation factor, whose value is a function of the distribution coefficient of the element.[14] The retardation factor is defined as, $1 + \rho k_d/\varepsilon$, where ρ is the buffer density, k_d is the distribution coefficient, and ε is the buffer porosity. The distribution coefficient is defined as the ratio of the concentration of a radionuclide in the buffer to that dissolved in the groundwater.

An analytical solution to the set of equations has been obtained by Laplace transform methods, for an impulse input flow and an outlet flow characterized by the mass transfer coefficient. The solution has been separated into impulse response functions, which give the contribution from each chain member. To obtain the time-dependent radionuclide flow into the geosphere, two convolutions are done. The radionuclide release from the waste form, determined from the source boundary conditions listed in Table 2, is convoluted with the CFF to obtain the total radionuclide release into the vault. The total radionuclide release into the vault is convoluted with the impulse response functions to obtain the radionuclide flow into the geosphere.

E. Precipitation of Radionuclides

Mass transport in the buffer could be affected by the solubility of the radionuclides in groundwater. A full treatment of this effect would be complicated because several isotopes could contribute to the dissolved fraction, and precipitation could occur at any place in the buffer because of buildup of a parent radionuclide or spatial variability in chemical conditions. Solubility constraints can be applied, however, to provide an upper limit to the radionuclide flow from the buffer. This upper limit can be defined to be the product of the mass transfer coefficient, the effective vault area for flow, and the radionuclide solubility.

III. PERFORMANCE OF THE VAULT IN THE DISPOSAL SYSTEM

The maximum radionuclide flow from the vault and the maximum dose have been used as measures of the performance of the vault. For the purposes of the SYVAC assessment, maximum dose is defined to be the maximum annual effective dose equivalent to a member of the most exposed group, occurring up to a specified time.[11] The analysis described in this report is for the borosilicate glass waste form. Initially, 1000 simulations with SYVAC2 were performed to estimate the uncertainty in the vault performance. In all of these but one, ^{99}Tc was the major contributor to maximum dose and the only contributor in the high-dose region (greater than 10^{-3} mSv).[15] To simplify the analysis, the only radionuclide used in subsequent investigations of vault performance was ^{99}Tc.

In the vault submodel, borosilicate glass was assumed to dissolve at a constant rate, Z, whose value was sampled from a loguniform distribution with a lower limit of 2.71×10^{-3} and an upper limit of 1.34×10^{1} $kg \cdot m^{-2} \cdot yr^{-1}$. Solubility constraints were placed on the release of ^{99}Tc. For these conditions, the maximum flow of congruently released ^{99}Tc from the vault, F_t, can be approximated by

$$F_t = \text{minimum of } ZAI_t/I_g \text{ and } k_v C_t A \qquad (1)$$

where

I_t is the inventory of ^{99}Tc (mol)

I_g is the inventory of glass (kg)

Z is the glass dissolution rate ($kg \cdot m^{-2} \cdot yr^{-1}$)

A is the effective vault area for flow (m^2)

k_v is the mass transfer coefficient ($m \cdot yr^{-1}$)

C_t is the solubility of ^{99}Tc ($mol \cdot m^{-3}$).

Of the input parameters in Equation (1), k_v, Z, and C_t have the largest variability and, thus, would be expected to be the most significant in determining the variability in F_t. This was supported by a study of the linear correlation between the maximum flow of ^{99}Tc from the vault, F_t, and the vault input parameters. Assumptions made to determine k_v are given elsewhere.[8] The uncertainty in the dissolution rate of borosilicate glass, Z, and the uncertainty in specifying a probability distribution for the solubility of ^{99}Tc, C_t, are examined in more detail below.

There is evidence to suggest that, instead of dissolving at a constant rate, the dissolution of borosilicate glass might be controlled by the solubility of silica in groundwater.[16,17] The solubility of silica depends mainly on pH and temperature. Based on a temperature range in the vault from 15 to 100°C and a pH from 5 to 10, and using data from Stumm and Morgan[18] and Okamato et al.,[19] the solubility of silica can be represented by a lognormal distribution with an average of 1.00, a median of 0.464, and a mode of 0.100 kg.m^{-3}. Figure 1 illustrates the flow of congruently released ^{99}Tc into the value for the two glass-dissolution mechanisms, using median values of the vault input parameters.

A lognormal solubility distribution for ^{99}Tc, with an average of 0.100, a median of 0.0464, and a mode of 0.0100 kg.m^{-3}, was used in SYVAC2. However,

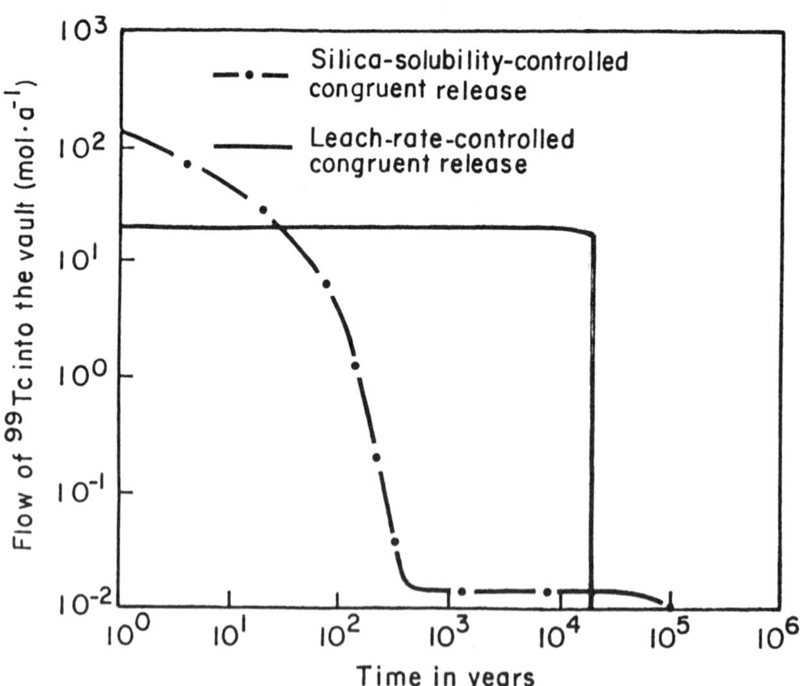

FIGURE 1. The Flow into the Vault of Congruently Released ^{99}Tc for Median Values of the Vault Input Parameters

Paquette and Lawrence[20] have suggested that the solubility of ^{99}Tc in deep groundwaters would be controlled by the carbonate concentration and accordingly, a loguniform solubility distribution with a lower value of 10^{-5} and an upper value of 10^{-2} mol.m^{-3} might be more appropriate. These two distributions are illustrated in Figure 2.

To investigate the influence of these parameters, four cases of 10000 simulations each were run:

Case 1: Leach-rate-controlled congruent release with lognormal solubility for ^{99}Tc.

Case 2: Silica-solubility-controlled congruent release with lognormal solubility for ^{99}Tc.

Case 3: Leach-rate-controlled congruent release with loguniform solubility for ^{99}Tc.

Case 4: Silica-solubility-controlled congruent release with loguniform solubility for ^{99}Tc.

In all four cases, the sampling was done so the values of all sampled parameters, except those listed above, were identical. Downward cumulative distributions for the maximum flow of ^{99}Tc from the vault and for the maximum dose for the four cases are shown in Figures 3 and 4, respectively.

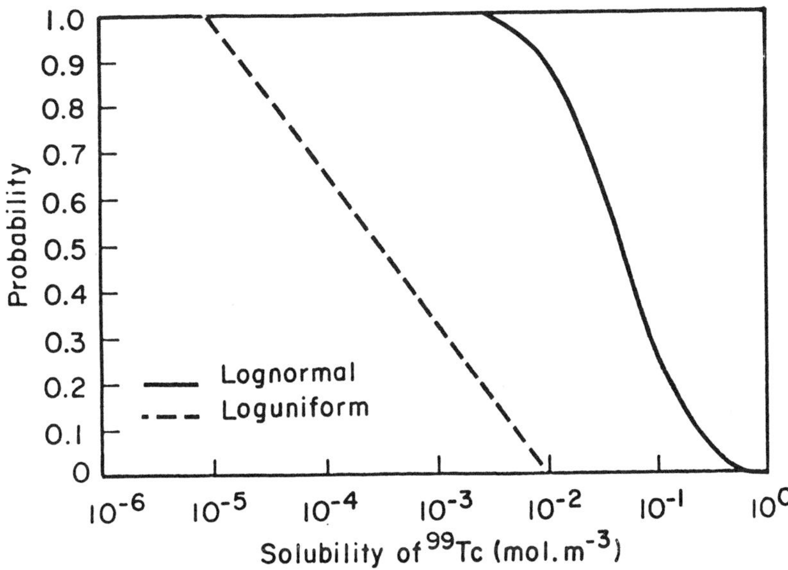

FIGURE 2. Downward Cumulative Probability Distributions for the Solubility of ^{99}Tc

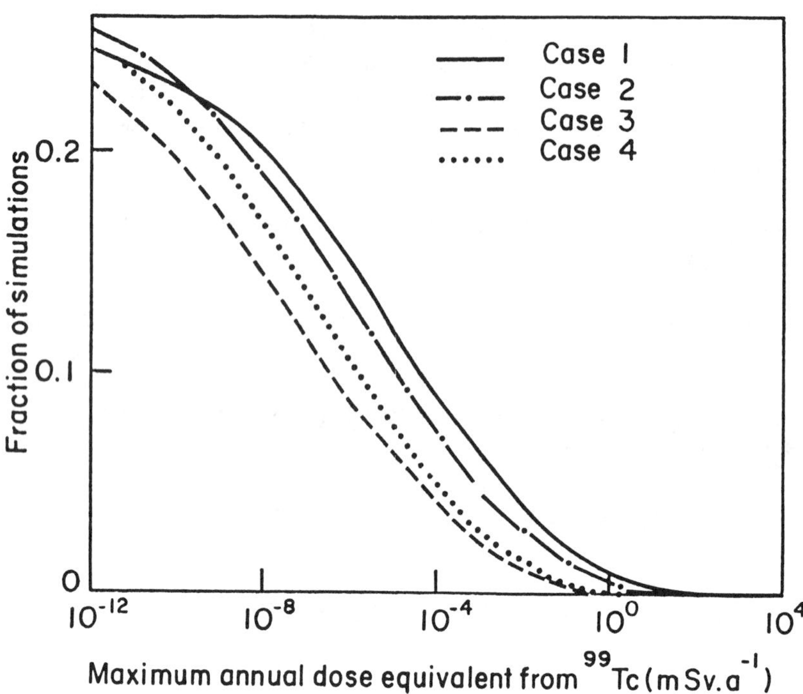

FIGURE 3. Downward Cumulative Distributions for the Maximum Flow of ^{99}Tc From the Vault

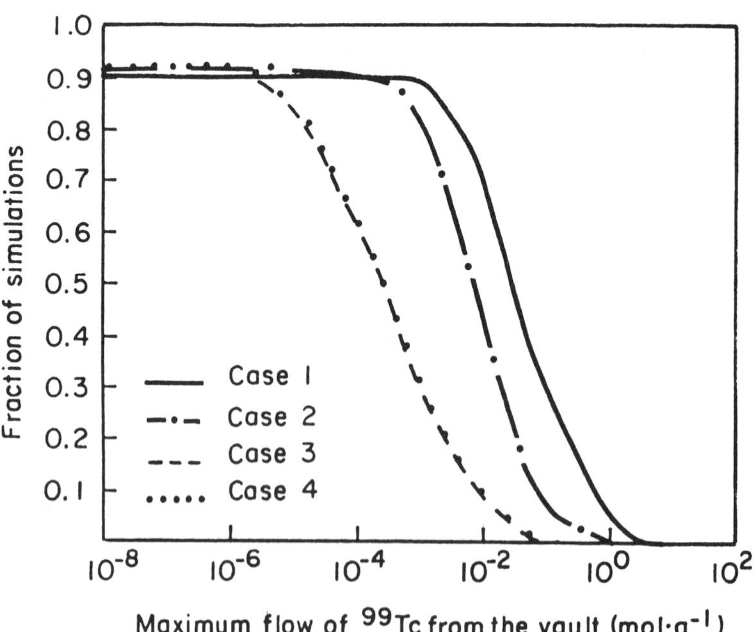

FIGURE 4. Downward Cumulative Distributions for the Maximum Dose from ^{99}Tc

The effect of changing from an assumption of a constant glass-dissolution rate to a silica-solubility-controlled glass-dissolution rate when the solubility of ^{99}Tc is sampled from a lognormal distribution, can be seen by comparing the results of Case 1 and Case 2. In these cases, the assumption of silica-solubility-controlled glass dissolution leads to lower maximum flows and lower maximum doses except in the low-flow and low-dose region. In Figure 3, the maximum flows of ^{99}Tc from the vault are concurrent for Case 3 and Case 4 above 10^{-5} mol·yr^{-1}, illustrating that, when the loguniform distribution is used, the maximum flow of ^{99}Tc is determined by the solubility of ^{99}Tc independent of the glass dissolution mechanism. The difference between the downward cumulative plots for Case 3 and Case 4 of Figure 4 can be explained by the fact that the geosphere submodel responds differently to input flows of different shapes, even though the maximum flows may be the same.

IV. CONCLUSIONS

The major contributor to maximum dose for the disposal of the borosilicate glass waste form is ^{99}Tc. The mass transfer coefficient, the solubility of ^{99}Tc and the glass-dissolution mechanism have the most influence on the vault performance. The uncertainty in the vault performance depends not only on the uncertainty in the vault input parameters, as defined by their input probability distributions, but also on the uncertainty in the glass-dissolution mechanism and in the probability distribution used for the solubility of ^{99}Tc.

REFERENCES

1. T. E. Rummery and E. L. J. Rosinger, "The Canadian Nuclear Fuel Waste Management Program," Proc. of the ANS International Topical Meeting on Fuel Reprocessing and Waste Management, held in Jackson Hole, Wyoming, August 26-29, 1984, 1, 14 (1984). Also AECL-8374, Atomic Energy of Canada Limited (1984).

2. K. W. Dormuth and R. D. Quick, "Accounting for Parameter Variability in Risk Assessment for a Canadian Nuclear Fuel Waste Disposal Vault," Int. J. Energy Systems 1, 125 (1981).

3. K. W. Dormuth and G. R. Sherman, SYVAC - A Computer Program for Assessment of Nuclear Fuel Waste Management Systems, Incorporating Parameter Variability, AECL-6814, Atomic Energy of Canada Limited (1981).

4. G. R. Sherman et al., SYVAC2 - A Systems Variability Analysis Code for Assessment of Nuclear Fuel Waste Disposal Concepts, TR-317*, Atomic Energy of Canada Limited, in preparation (1985).

*Unrestricted, unpublished report available from SDDO, Atomic Energy of Canada Limited Research Company, Chalk River, Ontario KOJ 1J0.

5. D. M. Wuschke, et al., Environmental and Safety Assessment Studies for Nuclear Fuel Waste Management. Volume 3: Post-closure Assessment, TR-127-3*, Atomic Energy of Canada Limited (1981).

6. D. M. Wuschke et al., Environmental Assessment of Subseabed Disposal of Nuclear Wastes: A Demonstration Probabilistic Systems Analysis, TR-206*, Atomic Energy of Canada Limited (1983).

7. G. R. Sherman et al., "The Program SYVAC, for Stochastic Assessment of Nuclear Fuel Waste Disposal," presented at Waste Management '85 Symposium, Tucson, Arizona, USA (1985).

8. D. M. LeNeveu, Vault Submodel for the Second Interim Concept Assessment of Nuclear Fuel Waste Disposal: Post-closure Phase, AECL-8383, Atomic Energy of Canada Limited, in preparation (1985).

9. W. F. Heinrich, Geosphere Submodel for the Second Interim Assessment of Nuclear Fuel Waste Disposal: Post-closure Phase, TR-286*, Atomic Energy of Canada Limited (1984).

10. K. K. Mehta, Biosphere Submodel for the Second Interim Concept Assessment of Nuclear Fuel Waste Disposal: Post-closure Phase, TR-298*, Atomic Energy of Canada Limited, in preparation (1985).

11. D. M. Wuschke et al., Second Interim Assessment of the Canadian Concept for Nuclear Fuel Waste Disposal, Volume 4: Post-Closure Assessment, AECL-8374-4, Atomic Energy of Canada Limited, in preparation (1985).

12. W. L. Wardrop and Associates, "Buffer and Backfilling Systems for a Nuclear Fuel Waste Disposal Vault," TR-341*, Atomic Energy of Canada Limited (1984).

13. D. H. Lester et al., "Migration of Radionuclide Chains Through an Nuclear Fuel Waste Disposal Vault," TR-341*, Atomic Energy of Canada Limited (1984).

14. L. H. Baetsle, "Computational Methods for the Prediction of Underground Movement of Radionuclides," Nuclear Safety, 8, 576 (1967).

15. B. W. Goodwin, "SYVAC Approach for Long-Term Environmental Assessment." presented at Symposium on Groundwater Flow and Transport Modeling for Performance Assessment of Deep Geologic Disposal of Radioactive Waste a Critical Evaluation of the State of the Art, Albuquerque, New Mexico, 1985, sponsored by U.S. Department of Energy and the U.S. Nuclear Regulatory Commission (1985).

---

*Unrestricted, unpublished report available from SDDO, Atomic Energy of Canada Limited Research Company, Chalk River, Ontario KOJ 1JO.

16. A. E. Hughes et al., The Significance of Leach Rates in Determining the Release of Radioactivity from Vitrified Nuclear Waste, AERE-R10190, AERE Harwell (1981).

17. D. Savage, The Geochemical Interactions of Simulated Borosilicate Glass, Granite and Water at 100°-350°C and 50 MPa, British Geological Survey Natural Environment Research Council Report No. FLPU84-3, Fluid Processes Research Group, British Geological Survey (1984).

18. W. Stumm and J. J. Morgan, Aquatic Chemistry, John Wiley and Sons, New York (1981).

19. G. Okamoto et al., "Properties of Silica in Water," Geochim Cosmochim. Acta, 12, 123 (1957).

20. J. Paquette and W. E. Lawrence, A Spectrochemical Study of the Technetium (IV)/Technetium (III) Couple in Bicarbonate Solutions, AECL-8529, Atomic Energy of Canada Limited (1985).

---

*Unrestricted, unpublished report available from SDDO, Atomic Energy of Canada Limited Research Company, Chalk River, Ontario KOJ 1J0.

BIOSPHERE MODEL FOR THE INTERIM ASSESSMENT OF THE CANADIAN CONCEPT FOR NUCLEAR FUEL WASTE DISPOSAL

J. W. Barnard
Atomic Energy of Canada, Ltd
Whiteshell Nuclear Research Establishment
Pinawa, Manitoba ROE 1LO, Canada

ABSTRACT

The Canadian Nuclear Fuel Waste Management Program has been established to assess the concept of geological disposal of nuclear fuel waste in plutonic rock. The Systems Variability Analysis Code (SYVAC) has been developed to estimate the range and probabilities of effects on man and the environment that could result from the disposal of nuclear fuel waste. SYVAC is composed of three submodels that represent the vault, the geosphere and the biosphere. The biosphere submodel calculates the transport of radionuclides through the environment following release from the geosphere and estimates the resulting radiation dose to man. This paper describes the biosphere submodel.

I. INTRODUCTION

The System Variability Analysis Code (SYVAC)[1] was developed to assess the Canadian concept for geological disposal of nuclear fuel waste. It consists of three submodels that simulate the vault, the geosphere and the biosphere. The code models the release of radionuclides from the vault and their transport through the geosphere to the biosphere over long time periods. Uncertainties in the parameter values are handled by Monte Carlo sampling from distributions of parameter values.

II. GENERAL DESCRIPTION OF THE BIOSPHERE SUBMODEL

In SYVAC, the biosphere includes the unconsolidated overburden, shallow groundwater region, surface waters, soil and atmosphere.[2] The biosphere submodel simulates the transport of radionuclides through various environmental pathways and calculates dose to man.

The radionuclide transport component of the submodel is depicted in the flow diagram of Figure 1. Radionuclides entering the biosphere from the geosphere are distributed into six main compartments. In four of the compartments (air, lake, sediment and well) radionuclides are assumed to be mixed instantly so that uniform concentration throughout each compartment is maintained. It is assumed that transfers from these four compartments are governed by first-order processes; that is, transfer rates are directly proportional to the

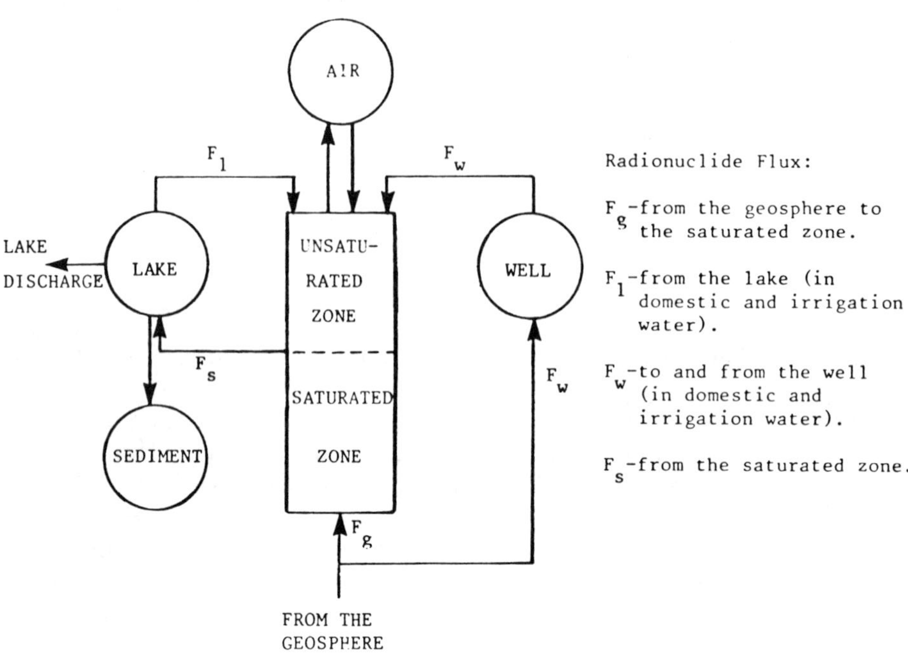

FIGURE 1. Radionuclide Transport in the Biosphere Model

compartment inventory. Radionuclide concentrations in the remaining two compartments (the unsaturated and saturated soil zones) are not uniform. Radionuclide transport in the groundwater passing through these compartments is modeled explicitly to obtain input and output concentrations and fluxes. The radionuclide flux is influenced by the dispersive properties of the medium and chemical interaction with soil particles, as well as radioactive decay.

The component of the biosphere model that calculates dose is shown in Figure 2. Dose is calculated for a reference group consisting of those people living in the area of the discharge of radionuclides from the geosphere. This group could receive external dose by direct exposure to air, standing on the ground, or bathing in the local water. The group could receive internal radiation dose by inhalation of air, drinking local water, and ingesting locally produced foods.

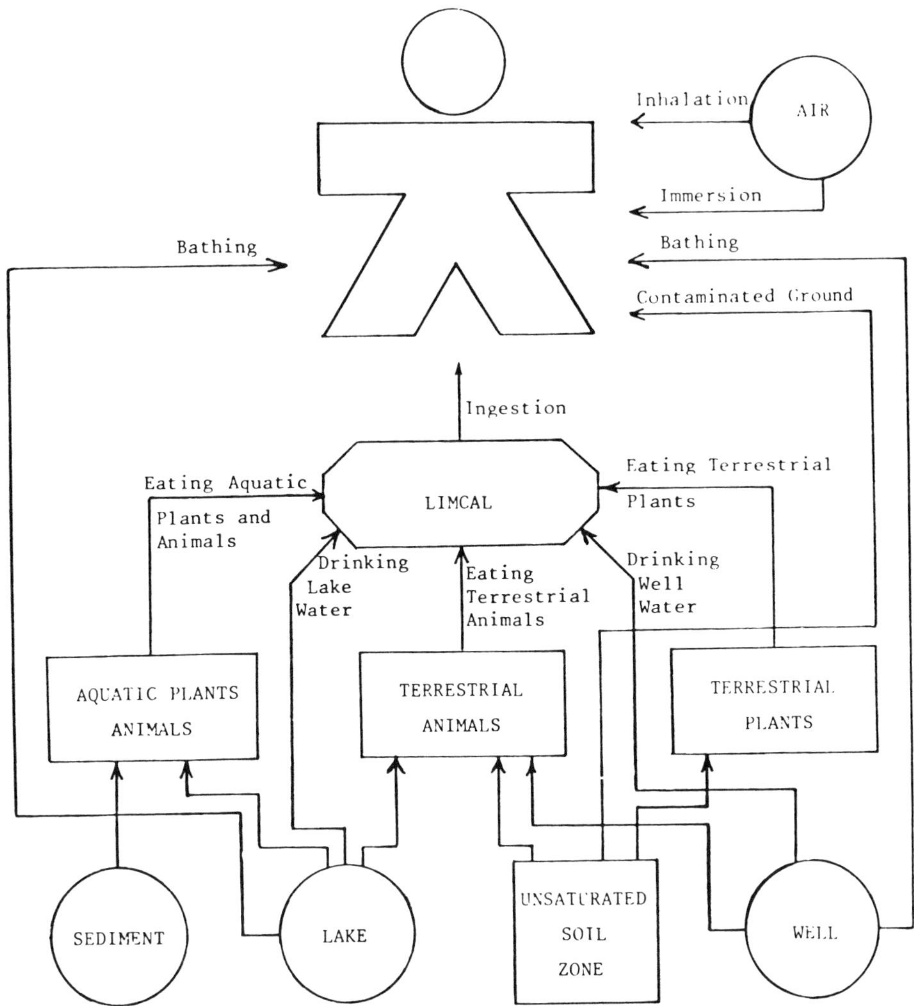

FIGURE 2. Dosimetry Model for SYVAC

III. DETAILS OF THE SUBMODEL

A. The Biosphere Response

It is assumed that the time span of radionuclide input to the biosphere compartments can be subdivided into a series of discrete time steps or intervals and that the continuous input of radionuclides can be replaced by a series of impulses at the beginning of each time step equal in magnitude to the total input of radionuclide i for that interval. The quantity of radionuclide passing through the compartment as a function of time (t) is then given by

$$Q_i(t) = \sum_{t'} R_i(t,t') \cdot I_i(t') \cdot t' \qquad (1)$$

where

$Q_i(t)$ = amount of the "ith" radionuclide in the compartment at time t.

$I_i(t')$ = amount entering the compartment at the time interval commencing at time t'.

$R_i(t,t')$ = amount of radionuclide i in the compartment at time t due to a unit impulse (Dirac delta function) input at time t'. $R_i(t,t')$ includes the effect of radioactive decay.

At present, the spacing of the time between pulses in SYVAC is 5×10^4 yr.

B. Geosphere-Biosphere Interface

The concentration of radionuclide i at the geosphere-biosphere interface is given by

$$C_{g,i}(t)_i = X\ (t)/H \cdot G \cdot A. \tag{2}$$

when the well-water demand (see Section C) is less than H·G·A. Otherwise:

$$C_{g,i}(t) = 0.$$

where

$C_{g,i}(t)$ = concentration of radionuclides in groundwater discharged from the geosphere (moles·m^{-3})

$X_{g,i}(t)$ = flux of radionuclide i from the geosphere (mole·yr^{-1})

H = hydraulic conductivity of the upper geosphere layer (m·yr^{-1})

G = hydraulic gradient in the upper geosphere layer (m·yr^{-1})

A = groundwater discharge area (m^2)

All parameters on the right hand side of Equation 2 are passed from the geosphere model[3] except for the area of the discharge, A. Mehta[2] has suggested that a multi-dimensional computer code modeling the regional flow system could be used to estimate the discharge area for a specific site. Until more information is available on discharge areas, the parameter distribution for A recommended by Mehta[2] will be used. This is a lognormal distribution whose logarithm has a mean value (base ten) μ = 6.6 and standard deviation σ = 0.2. This distribution is truncated at the $\pm 3\sigma$ values.

C. Well Compartment

The well used by the members of the reference group is assumed to be drilled into bedrock. The concentration of radionuclide i in the well $C_{w,i}$

is assumed to be equal to $C_{g,i}(t)$ (Equation 2). However, if the well-water demand exceeds the discharge rate from the geosphere within the area A, then the total radionuclide flux from the geosphere is assumed to be intercepted by the well. In this case the concentration of radionuclide i in well water is equal to the geosphere flux of radionuclide i divided by the annual well-water demand. The annual well-water demand is calculated from statistics on well use and household size compiled by Beals.[5]

Geosphere modeling by Thunvik[6] and Carlsson et al.[7] produced estimates of dilution factor for bedrock wells in Sweden. In the Swedish assessment of high-level waste disposal[8], the dilution factor was the ratio of the annual water flux through the vault to the amount of water withdrawn annually from the well. The comparable ratio in SYVAC would be

$$R = A_V/H \cdot G \cdot A \quad \text{for } H \cdot G \cdot A \geq Q_D \qquad (3)$$

$$R = Q_V/Q_D \quad \text{for } H \cdot G \cdot A < Q_D$$

where

R = the SYVAC "dilution factor"

Q_V = annual water flux through the vault ($m^3 \cdot yr^{-1}$).

Q_D = annual well-water demand ($m^3 \cdot yr^{-1}$)

Although dilution factors are not used to calculate radionuclide concentrations in the well, it is interesting to compare the distribution of the ratio R with the range of dilution factors from the Swedish modeling.

Figure 3 shows that the distribution of R in SYVAC ranges from approximately 10^{-7} (the smallest value for dilution factor estimated in the Swedish studies) to 10^{-1} (one order of magnitude greater than the highest value for the Swedish study). The median value for the distribution of R centres around 10^{-4}, which is the value of the dilution factor adopted for use in the deterministic Swedish assessment.

D. Overburden

The overburden includes the two compartments labeled saturated soil zone and unsaturated soil zone in Figure 1. Figure 4 shows a profile of the overburden modeled in SYVAC. It is assumed that radionuclides discharged from the geosphere migrate in groundwater directly to lakes and that only a small fraction contacts the top of the water table and contaminates the unsaturated zone. Hydrogeological modeling for an area near Atikokan, Ontario, indicated that deep groundwater discharge is, in general, to lakes[9].

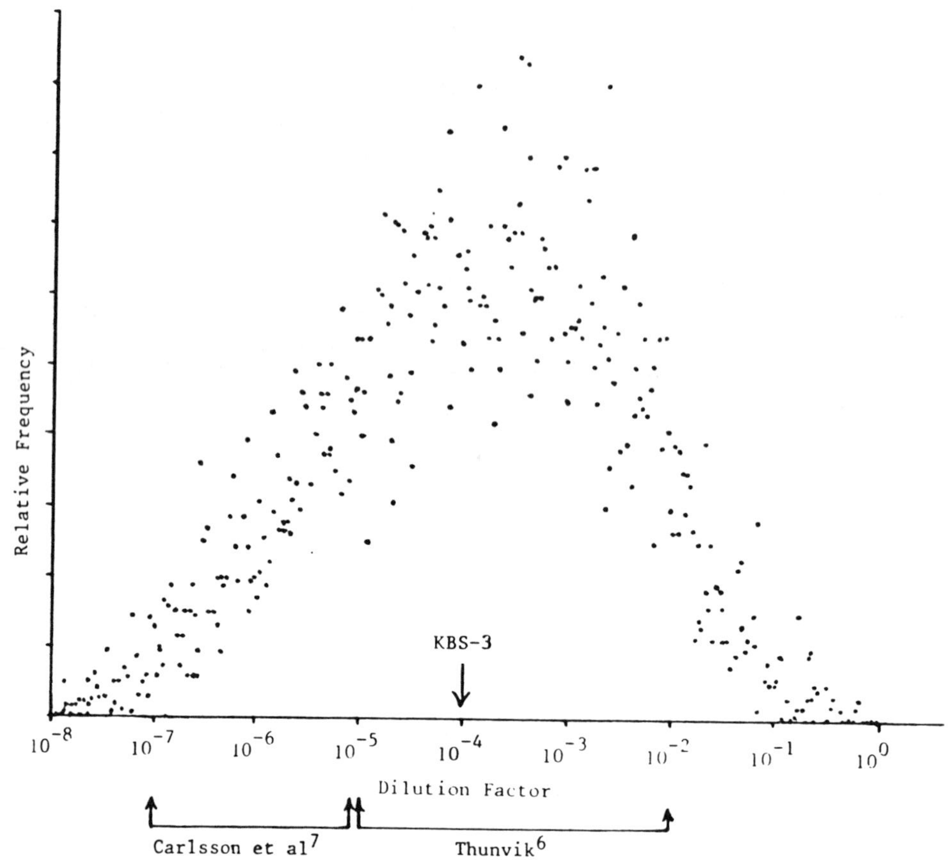

FIGURE 3. Distribution of Dilution Factor, R, in Present Model. Also shown are the range of values from the Swedish well modeling studies and the value adopted in KBS-3[8].

It is further assumed that the concentration of radionuclides bound to soil is linearly related to their concentration in groundwater. The proportionality constant is k_d ($m^3 \cdot kg^{-1}$). Solutions to the one-dimensional convection-diffusion equation subject to a unit pulse input at time t=0 and zero concentration at infinity are the response functions used to model transport of radionuclides through the saturated zone.[10]

The concentration at the bottom of the saturated zone is equal to $C_{g,i}(t)$ and is given by Equation 2. The annual input to the saturated zone, $I_{s,i}(t)$ ($moles \cdot yr^{-1}$) from the geosphere is

$$I_{s,i}(t) = X_{g,i}(t) - C_{w,i} \cdot Q_d \qquad H \cdot G \cdot A \geq Q_D \tag{4}$$

$$I_{s,i}(t) = 0 \qquad H \cdot G \cdot A < Q_D$$

The response functions are used to convert these inputs into output radionuclides fluxes and concentrations at the top of the water table. The annual flux from the saturated zone is assumed to discharge directly to the lake.

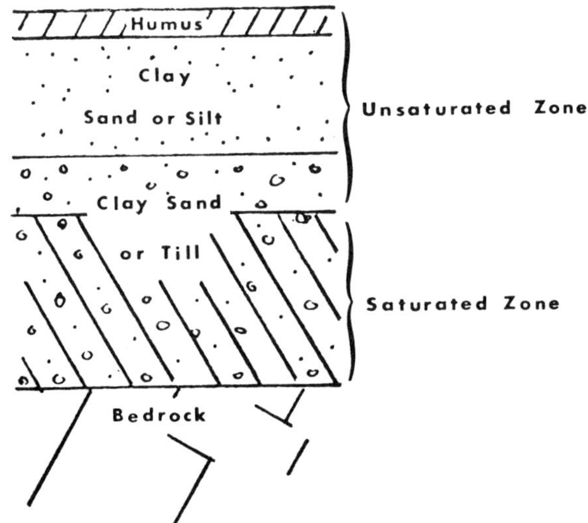

FIGURE 4. Soil Profile Modeled in SYVAC

During the summer months on the Canadian Shield, water rises upwards through the soil, driven by evaporation. In late fall and early spring, water moves downward because of permeation of precipitation from the surface. The annual net flow is generally downward, but a fraction of the trace elements dissolved in groundwater at the top of the water table is transferred to the surface by molecular diffusion. This exchange with the surface reaches steady state quickly in comparison to one SYVAC time step.

Sheppard[1] has adapted the soil model SCEMR to model radionuclide transport through the unsaturated soil zone of the Canadian Shield. Although evapotranspiration depends on several factors, Sheppard[12] has demonstrated that evapotranspiration is strongly correlated with annual precipitation. A set of tables has been prepared using SCEMR, which give average radionuclide concentration in the root zones (the two top layers in Figure 4) per unit concentration in saturated zone groundwater as a function of k_d, annual precipitation and unsaturated zone thickness. These data are used by SYVAC to convert saturated zone concentrations to root-zone concentrations.

E. Lake and Sediments

It is assumed that radionuclides deposited on the soil surface are transferred through the unsaturated zone and the upper water table to the lake in a time that is short in comparison to one SYVAC time step. Therefore irrigation and domestic water discharge from the well add to the radionuclide content of the lake but water withdrawn from the lake for these purposes has no effect.

A constant lake volume has been assumed so that the flow rate from the lake is the product of the net precipitation, P (precipitation less evaporation in $m \cdot yr^{-1}$), and the catchment area A_d(in m^2). The first order rate constant for transfer of radionuclide i from the lake due to flow out of the lake, $\alpha_i (yr^{-1})$, is

$$\alpha_i = \frac{P \cdot A_d}{V_1} \tag{5}$$

where V_1 = lake volume (m^3).

Transfer of radionuclides from lake water to sediment is governed by a residence time in the water column τ_i. This is the time required for the concentration of radionuclide i in the water column to be reduced to 1/e of its initial value by sedimentation alone. Residence times for radionuclides in a variety of lakes have been determined by Cornett and Ophel.[13] The top layer of lake sediment is the portion of the sediment compartment that is accessible as a nutrient source for marine plants and animals. This is always the most recently deposited layer. The concentration in this layer is given by

$$C_{sed,i}(t) = \frac{X_{1,i}(t)}{A_1 \cdot \tau_i \cdot R_s} \tag{6}$$

where

$C_{sed,i}(t)$ = concentration of radionuclides in the sediments as a function of time ($moles \cdot m^{-3}$)

$X_{1,i}(t)$ = quantity of radionuclide i in the lake at time t moles.

A_1 = lake area (m^2)

R_s = lake sedimentation rate ($m \cdot yr^{-1}$).

F. Air Compartment

The concentration of suspended soil particles in air is computed using an atmospheric dust load recommended by Zach.[14] The dust load is extremely variable and has been represented in SYVAC by a lognormal distribution whose logarithm has a mean value $\mu = -7.0$ (i.e., mean value of dust load = 10^{-7} $kg \cdot m^{-3}$) and standard deviation $\sigma = 0.4$. The concentration of radionuclides in suspended particles is assumed to be the same as in soil.

The only gas taken into consideration is radon evolved from soil. The concentration of radon in air, $C_{a,Rn}$ ($Bq \cdot m^{-3}$) is correlated to the radon emission rate from soil X_{Rn} ($Bq \cdot m^{-2}$) by formula recommended by Davis.[15]

$$C_{a,Rn}/X_{Rn} = -4.95 + 27.1\ A^{1/8} \tag{7}$$

where A = discharge area (m^2).

The radon emission rate is related directly to the concentration of the precursor ^{226}Ra in soil by a normalized radon emission rate, q (q = 2.7 × 10^9 $Bq \cdot m^{-2}$ per mole of ^{226}Ra per kg of soil).

G. Food Chain and Dose Calculations

The section of the biosphere submodel that establishes man's diet and calculates radionuclide ingestion by man is a special version of the food chain model LIMCAL called LIMCAL-S.[14] In LIMCAL, it is assumed that man's energy requirements remain constant throughout life. Man may consume a number of food types; the sum of fractional caloric intake from each food type must equal unity. Man's consumption of water is correlated to caloric intake.

The internal annual effective dose equivalent due to radionuclide i, H_i^{int}, is calculated as

$$H_i^{int} = I_i^{int} \cdot DCF_i \tag{8}$$

where I_i^{int} = the total intake of radionuclide i ($Bq \cdot yr^{-1}$).

DCF is the 50-year committed effective dose equivalent conversion factor ($Sv \cdot Bq^{-1}$). Johnson[15] has shown that the use of these DCF_i factors rather than an annual dose conversion factor results in conservative and realistic estimates of the annual dose equivalent. External effective dose equivalents for immersion in air and water and standing on contaminated ground are calculated using dose-rate conversion factors from Barnard and D'Arcy.[16] The annual dose is the sum of the annual internal and external components of dose.

IV. WORK IN PROGRESS

Additional pathways for radionuclide transport in the biosphere have been examined for inclusion in the SYVAC biosphere submodel. These include the irrigation pathway,[17] animals' inhalation pathway,[18] and animals' soil ingestion pathway.[19] These all appeared to be significant pathways according to the criteria used by each investigator. Their importance must be judged in relation to pathways, that are presently modeled within SYVAC. Evaluation of their significance within SYVAC showed that animals' soil ingestion and irrigation are important, but animals' inhalation is not.

Sheppard[17] has shown that root uptake of radionuclides deposited by irrigation in soil is more important than consumption of radionuclides deposited on leaves. He has recommended that irrigation rates between 0 and 500 $mm \cdot yr^{-1}$ be used in long-term assessment studies but has given no recommendation for the probability of irrigation aside from noting that irrigation occurs now on a limited scale through the watering of gardens.

Work is in progress to use SCEMR to calculate factors relating concentration of radionuclides in the root zone to areal deposition rates of radionuclides on the soil surface from irrigation water.

The possibility of reselecting weather parameters, particularly precipitation, in each time step to represent long-term shifts in climate is being studied by Heinrich.[20]

V. SUMMARY

The large quantity of information available about the environment of the Canadian Shield allows us to model the biosphere in more detail than is possible in other submodels (the vault and the geosphere). In addition, there is more possibility of change with time in the biosphere parameters, so that extra detail is warranted. This need for realistic and detailed modeling in the biosphere are constrained by the amount of computation required, so there is an incentive to reduce the number of pathways by eliminating those that are insignificant contributors to dose.

LIMCAL provided a major improvement in dose calculation for the second assessment[4] of the Canadian concept for nuclear fuel waste disposal, compared to the method used in the first assessment[12]. More modifications are being made to the radionuclide transport component of the submodel for use in a third assessment. The well model has been altered so that well-water concentrations are more directly related to geosphere parameters. The present model for radionuclide transport in overburden represents a considerable advance over earlier models, especially since the SCEMR code is supported by extensive validation obtained using soils representative of the Canadian Shield.[22]

REFERENCES

1. K.W. Dormuth and R. D. Quick, "Accounting for Parameter Variability in Risk Assessment for a Canadian Nuclear Fuel Waste Disposal Vault," Int. J. of Energy Systems 1, 125.

2. K.K. Mehta, "Biosphere Submodel for the Second Interim Assessment of the Canadian Concept for Nuclear Fuel Waste Disposal - Post-Closure Phase," Atomic Energy of Canada Limited Technical Record, TR-298* (1985).

3. W. F. Heinrich, "Geosphere Submodel for the Second Interim Concept Assessment," Atomic Energy of Canada Limited Technical Record, TR-286* (1984).

4. D. M. Wuschke et al., Second Interim Assessment of the Canadian Concept for Nuclear Fuel Waste Disposal-Volume 4: Post Closure Assessment, Atomic Energy of Canada Limited Report, AECL-8373-4, Pinawa, Manitoba (1985).

---

*Unrestricted, unpublished report available from SDDO, Atomic Energy of Canada Limited Research Company, Chalk River, Ontario KOJ 1JO.

5. D. Beals, "Hydrologic and Hydrogeologic Parameters for Post-Closure Biosphere Assessment of Nuclear Fuel Waste Disposal," Atomic Energy of Canada Technical Record, TR-287*, in preparation (1985).

6. R. Thunvik, Calculation of Fluxes through a Repository Caused by a Local Well, A KBS Technical Report, TR 83-50 (1983).

7. L. Carlsson, A. Winberg and B. Grundfelt, Mode Calculation of the Groundwater Flow at Finnsjon, Fjallveden, Gidea and Kanlunge, KBS Technical Report, KBS-83-45 (1983).

8. KBS, Final Storage of Spent Nuclear Fuel, KBS Report, KBS-3 (1983).

9. INTERA Environmental Consultants Inc., "A Regional Hydrology Simulation of the Atikokan Site (Research Area 4)," Atomic Energy of Canada Limited Technical Record, TR-156* (1981).

10. G. L. Moltyaner, Migration of Radionuclides in Unconsolidated Materials: Equilibrium Transport Models, Atomic Energy of Canada Report, AECL-8253 Chalk River, Ontario (1983).

11. M. I. Sheppard, "SCEMR: A Model for Soil Chemical Exchange and Migration of Radionuclides in Unsaturated Soil Part 1: A Users' Manual," Atomic Energy of Canada Limited Technical Record, TR-175* (1981).

12. M. I. Sheppard, Atomic Energy of Canada Limited, Pinawa, Manitoba, Private Communication (1985).

13. R. J. Cornett and J. L. Ophel, "Sedimentation of Radiocobalt in a Small Shield Lake," Accepted for Pub. Can. J. Fisheries and Aquatic Science (1985).

14. R. Zach, "Preliminary Probability Density Distributions for the Parameters of the Food Chain Model LIMCAL," Atomic Energy of Canada Limited Technical Record, TR-205* (1982).

15. J. R. Johnson, "Dose Conversion Factors Used in the Current Canadian High Level Waste Disposal Assessment Study," Radiation Protection Dosimetry 3 47-50 (1982).

16. J. W. Barnard and D. D'Arcy, "EDEFIS. Code for Calculating Effective Dose Equivalent For Immersion," Atomic Energy of Canada Limited Technical Record, TR-244*, in preparation (1985).

17. S. C. Sheppard, Use of the Foodchain Model FOOD III and the Soil Model SCEMR to Assess Irrigation, Atomic Energy of Canada Limited Report, AECL-8380 (1985).

*Unrestricted, unpublished report available from SDDO, Atomic Energy of Canada Limited Research Company, Chalk River, Ontario KOJ 1JO.

18. R. Zach, "Contribution of Inhalation by Food Animals to Man's Ingestion Dose," Submitted for Publication in Health Physics (1985).

19. R. Zach and K. Mayoh, "Soil Ingestion by Cattle: A Neglected Pathway," Health Physics 46, 431 (1984).

20. W. H. Heinrich, Atomic Energy of Canada Limited, Pinawa Manitoba, Private Communication (1985).

21. D. M. Wuschke et al., "Environmental and Safety Assessment Studies for Nuclear Fuel Waste Management, Volume III: Post-Closure Assessment," Atomic Energy of Canada Limited Technical Record, TR-127-3* (1981).

22. M. I. Sheppard and S. C. Sheppard, "A Solute Transport Model Evaluated on Two Experimental Systems with Different Nuclides Soil, Nuclide Placement and Aeration States," in preparation (1985).

---

*Unrestricted, unpublished report available from SDDO, Atomic Energy of Canada Limited Research Company, Chalk River, Ontario K0J 1J0.

MECHANICAL EFFECTS ASSOCIATED WITH SURFACE LOADING OF DRY ROCK DUE TO GLACIATION[1]

Krishan K. Wahi
SAIC, Suite 1200
505 Marquette Ave. NW
Albuquerque, New Mexico 87102

Regina L. Hunter
Waste Management Systems Division
Sandia National Laboratories
Albuquerque, New Mexico 87185

ABSTRACT

Many scenarios of interest for a repository in the Pasco Basin begin with glaciation. Loading and unloading of joints and fractures due to the weight of ice sheets could affect the hydrologic properties of the host rock and surrounding units. Scoping calculations performed using two-dimensional numerical models with simplifying assumptions predict stress changes and uplift or subsidence caused by an advancing glacier. The magnitudes of surface uplift and subsidence predicted by the study agree well with previous independent predictions. Peak stress unloading near the repository horizon is a small fraction of the ambient stress. Any resultant aperture increase is likewise small. Based on the results of this study, mechanical loading caused by a glacier is expected to have a minimal effect on rock permeability, assuming that the excess compressive loads do not crush the rock. Possible rock failure due to shear or compression was not analyzed because the focus was on fracture aperture change due to change in normal stress across a joint or fracture.

I. INTRODUCTION

The Hanford Site in the Pasco Basin of the Columbia Plateau is under consideration as a possible repository location for high-level nuclear waste. Sandia National Laboratories, Albuquerque, (SNLA) has been assisting the U.S. Nuclear Regulatory Commission (NRC) in developing a performance assessment methodology for use in evaluating high-level waste (HLW) disposal sites. One part of this methodology consists of screening scenarios for subsequent detailed analysis. A preliminary scenario screening exercise for the Hanford Site indicated that many release scenarios could begin with glaciation.[1] The object of this paper is to assess potential changes in the rock stresses and, subsequently, in rock permeability as a result of the mechanical loads imposed by an advancing glacier. The calculations presented here have been performed in support of the performance-assessment methodology demonstration in basalt.[2]

---

[1]This work supported by the United States Nuclear Regulatory Commission and performed at Sandia National Laboratories, which is operated for the U.S. Department of Energy under contract number DE-AC04-76DP00789.

From a modeling perspective, the problem is to determine the stress-strain response of rock mass to a variable external load. The conceptual model consists of an advancing glacier that approaches the repository regionfrom the north at a prescribed rate, comes within roughly 150 km, then retreats at a prescribed rate different from the advance rate. Two numerical simulations were designed to calculate the stress-strain response to glacial loading. These scoping calculations are expected to help determine whether further consideration of glaciation scenarios is warranted in the performance assessment analyses. If the effects of loading and unloading joints and fractures due to the weight of the glacier are minimal, perhaps certain scenarios can be dismissed on the basis of low consequence. The effects of heat generated by the high-level waste were neglected, because thermal effects are expected to have subsided before any glacier approaches the region.

Changes in recharge or regional ground-water flow patterns due to glacial activity are beyond the scope of this paper. Moreover, some effects or processes that could influence the results were ignored in this analysis for various reasons. Simplifying assumptions were made for one or more of the following reasons: (1) this is a preliminary scoping type of analysis; (2) appropriate modeling capabilities do not exist; (3) the data base for meaningful inclusion of certain mechanisms or processes is not available. The effects that were not considered in this analysis are pore-pressure, penetration of permafrost, lubrication of fractures, faulting, and alteration of the geothermal gradient.

II. BACKGROUND

As an ice sheet advances, its weight depresses the crust below it. In addition, the crust is depressed to a lesser extent in front of the ice margin. Several investigators have postulated that still farther in front of the advancing ice margin, the crust is raised into a peripheral bulge. These effects have been indicated both by computer modeling of the crustal and mantle response to loading[3,4,5] and by examination of field data.[4] Walcott's calculations[4] suggest a maximum surface heave of 17 meters (m) at roughly 500 kilometers (km) from the ice edge for an ice sheet comparable in size to the Laurentide sheet. The crust-mantle relationship in eastern Washington may be anomalous because of the proximity of the subduction zone of the East Pacific crustal plate; however, isostatic submergence and emergence of western Washington in response to loading by the Puget lobe of the Cordilleran ice sheet has been well documented.[6] Apparently there are no field data on a peripheral bulge. Possible crustal warping in eastern Washington, to our knowledge, has not been investigated in the field since 1937.[7] However, for this study it has been assumed that both the depression and peripheral bulge will occur at typical distances from the margin of the advancing ice. This assumption is based on the observation that crustal warping is wide-spread and well-documented along ice margins.

Calculations by Brotchie and Silvester[3] suggest a much lower maximum heave of 3.86 m at 150 km from the ice edge. No field data are available for crustal deflection beyond the edge of the Cordilleran ice sheet.

Virtually all ground-water flow in the basalt flows is through fractures and joints; therefore, hydrologic properties are expected to be strongly dependent on the state of stress. The loading and unloading of joints and fractures due to the weight of ice sheets may significantly affect the hydrologic properties of the host rock and surrounding units. This could in turn alter ground-water travel times and the associated transport of radionuclides. Earlier heuristic modeling efforts[8] indicated that these glacial effects could dominate all other changes in hydrologic properties of the rock mass.

III. MODELING APPROACH

A two-dimensional finite-difference computer code, STEALTH 2D,[9] was used to perform the numerical simulations. Linear elastic behavior was assumed for each rock type. Both a regional and a repository-scale (local) model were formed. The regional model extends 500 km horizontally and 155 km vertically. The model dimensions, stratigraphy, and mechanical boundary conditions for the regional model are shown in Figure 1. The repository may be envisioned as being 250 to 300 km from the left boundary and at mid-height of the top layer ("Basalt 2" in Figure 1). Assumed material properties for the various layers are summarized in Table 1. Values of P and S wave velocities reported in Holmes' textbook[10] were used to derive the moduli for the mantle and the soft layer. The boundary conditions represent idealizations, with the surface boundary (i.e., top boundary) loading based on advance and retreat rates predicted by Foley and others.[8] The imposed stress, over that portion of the surface to which the glacier has advanced, corresponds to a column of ice 730 m tall. The model calculations commence at 5,000 years from the present. It was assumed that the glacier advances at a constant rate of 0.057 km/year between 5,000 and 10,000 years from present and retreats at 0.077 km/year after 10,000 years. Roller boundary conditions (i.e., zero normal displacement) were prescribed on the side and bottom boundaries. The numerical mesh for the regional model consisted of 51 vertical lines and 21 horizontal lines, with a total of 1,071 computational nodes.

Selection of boundary conditions is always a difficult task in a numerical modeling effort. It is even more difficult when the spatial extent of the model is limited and there are no data to guide the selection of boundary values. For the repository-scale model, information from the literature was used to establish a displacement history along the bottom boundary. The top boundary was a free surface, and the left and right boundaries had a constant stress equal to twice the overburden.[11] The model dimensions are 11 km wide and 1.7 km deep. The numerical mesh for the local model had 51 vertical and 33 horizontal lines, giving 1,683 computational nodes. Both two-dimensional models have a translational symmetry; that is, the dimension perpendicular to

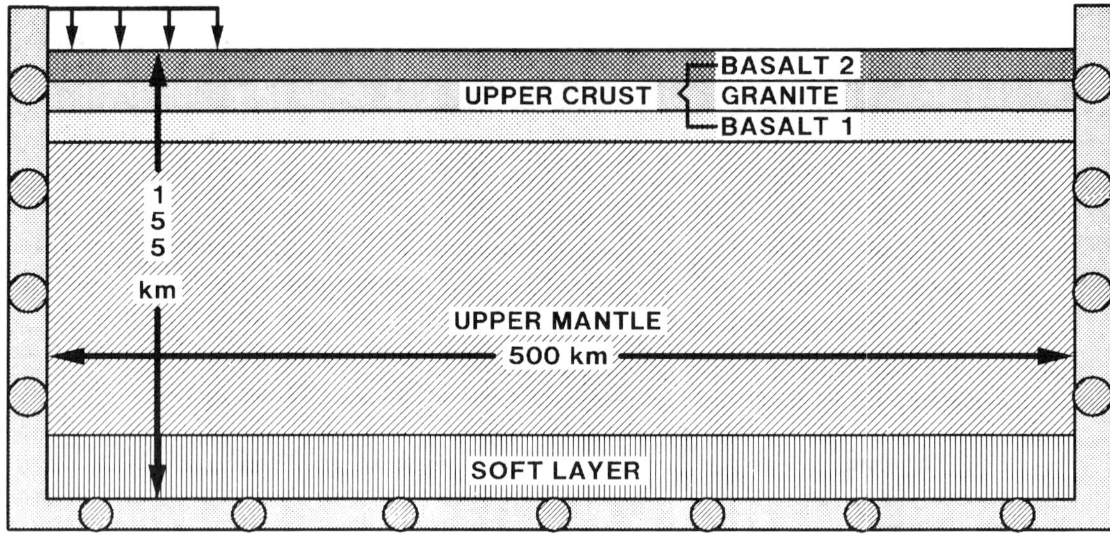

FIGURE 1. Regional Model for Surface Loading Due to Glacier

TABLE 1. Mechanical Properties of Rock Types

Rock Type	Density (kg/m^3)	Bulk Modulus (Pa)	Shear Modulus (Pa)
Soft Layers	3200	1.523×10^{11}	5.92×10^{10}
Mantle	3600	1.585×10^{11}	7.62×10^{10}
Basalt 1	2900	7.40×10^{10}	5.10×10^{10}
Granite	2500	4.50×10^{10}	3.38×10^{10}
Basalt 2	2800	4.71×10^{10}	2.69×10^{10}

Note: 1 psi = 6,895 Pa

the plane of the cross-section shown in Figure 1 is infinite. The direction of glacier advance is from left to right. The calculations reported here did not include the possibility that glacial loading may reactivate one or more of the faults in the Pasco Basin.

IV. ANALYSIS OF RESULTS

Two quantities of primary interest result from this study. They are the displacements along the surface and stress changes in the rock mass at repository depths. The results of the repository-scale model are discussed first, followed by those of the regional model.

The peak surface displacement for the repository-scale model is predicted to be roughly 3.4 m, and the maximum stress reduction (i.e., unloading of fractures or joints) at repository depth is 110 MPa (16,000 psi). The displacements calculated for the repository-scale model seem realistic and in good agreement with Brotchie and Silvester's calculation[3] of 3.86 m at a distance of 150 km from the glacier tip. However, the predicted stress changes are too large to be physically reasonable because the weight per unit area of the glacier that presumably causes the surface and at-depth displacements ahead of the glacier is too small. A likely explanation for the anomalous stress change is that the imposed bottom boundary displacements are not representative of the actual subsurface motion at that depth. To a lesser degree, the assumption of constant horizontal stress on the left and right boundaries is also suspect.

For the regional model, the displacements as well as stress changes appear realistic. Clearly, the remoteness of the boundaries from the area of interest and the type of boundary conditions assumed result in minimal to moderate effects on the model response. The largest subsidence, not surprisingly, is directly under the left edge of the model and equals 6 m. This value is the incremental (not total) subsidence as a result of an additional 140 km advance of the glacier. Similarly, the maximum (incremental) surface uplift associated with a 140 km advance is predicted to be 0.27 m. The surface profile is shown in Figure 2 at a simulation time of 2,500 years, when the glacier advances to its maximum extent. Note that the vertical scale in Figure 2 has been grossly exaggerated relative to the horizontal scale. The maximum stress reduction (i.e., decrease in the ambient compressive stress) near the repository depth is approximately 0.5 MPa (~70 psi). A time history of stress changes near the repository is plotted in Figure 3. The next step in the analysis was to relate the predicted stress changes to possible changes in the rock permeability.

Qualitatively, a decrease in the normal compressive stress across a joint or fracture should result in a larger aperture and, hence, increased permeability. Witherspoon and others[12] have found that the cubic law is valid for flow in a deformable fracture. They further believe, based on previous work,[13] that permeability is uniquely defined by fracture aperture and is independent of stress history.

Iwai[13] performed experiments on granite, basalt, and marble samples at room temperatures to study the flow response in closed fractures where the aperture is changed by changing stress. Plots of "$Q/\Delta h$" versus axial stress (σ_e) were obtained by Iwai that provide a relationship between stress and permeability; Q is the volumetric flow rate and Δh is the head difference.

Assuming a reference stress of 13.8 MPa (2000 psi) at repository depth, which is comparable to the initial stress near repository horizon in the model, and using Iwai's data for the unloadng portion of the first cycle for a fracture in granite, $Q/\Delta h$ values of 5.63×10^{-10} m^2/s and 5.49×10^{-10} m^2/s are obtained at σ_e values of 13.3 and 13.8 MPa, respectively. This implies

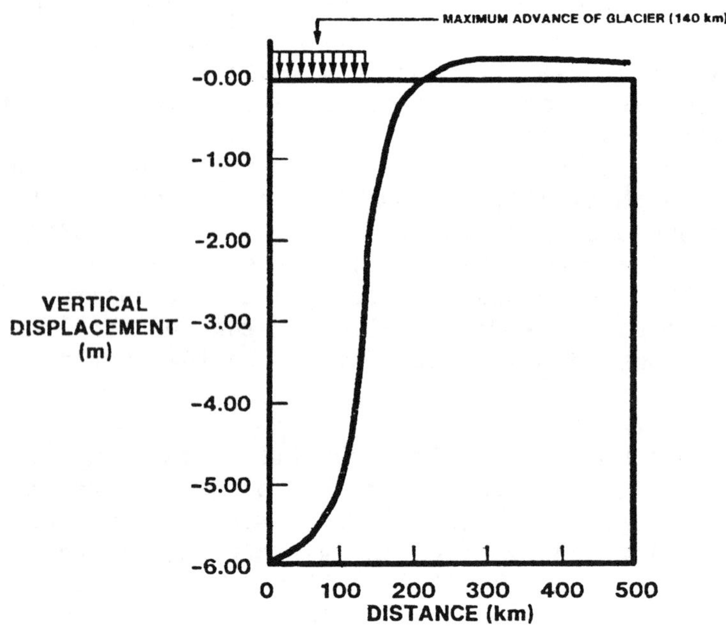

FIGURE 2. Surface Deformation Profile at a Simulation Time of 2500 Years

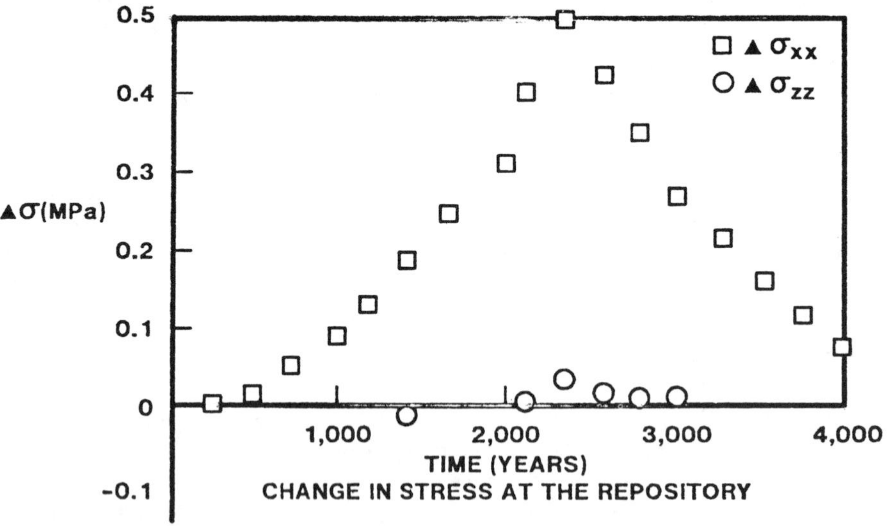

FIGURE 2. Surface Deformation Profile at a Simulation Time of 2500 Years

a permeability change of only a few percent (~3%). Higher-percentage changes, however, can be expected at shallower depths, because of a highly non-linear relationship between $Q/\Delta h$ and σ_e. For example, at a reference σ_e value of 1.38 MPa (200 psi), the increase in permeability corresponding to a stress unloading of 0.5 MPa (70 psi) is about 35 percent.

A simple exercise qualitatively supports the above findings. Assume that a joint or fracture has a certain stiffness. This stiffness is lower than the rock mass modulus, which in turn is lower than the intact rock modulus. Based on engineering judgment, it reasonable to assume that the joint stiffness is one-hundredth of the intact modulus and represents a width of 1 mm across a joint. Using a typical basalt intact-modulus value of 7×10^{10} Pa (10^7 psi), the strain across the joint corresponding to a stress change of 0.5 MPa is 7×10^{-4}. An aperture change of 1 μm would result from that strain across the fracture opening. If the initial aperture were 50 μm, the permeability change would be of the order of $(51/50)^3$, or 6 percent.

V. CONCLUSIONS AND RECOMMENDATIONS

The relatively simple linear elastic regional model of this study suggests that the mechanical effects of an approaching glacier do not significantly alter the fracture apertures at repository depths. The associated permeability changes are also small. At shallower depths, the effects are more pronounced, but not of a magnitude that would be of concern given the uncertainties and inhomogeneities of hydrologic parameters. The anomalous stress response for the repository-scale model needs further investigation before it can be rejected completely. The findings of this study cannot be extrapolated to possible changes in recharge, discharge, and regional flow patterns caused by glacial activity.

REFERENCES

1. R. L. Hunter, "Preliminary Scenarios for the Release of Radioactive Waste From a Hypothetical Repository in Basalt of the Columbia Plateau," NUREG/CR-3353, SAND83-1342, Sandia National Laboratories, Albuquerque, NM (1983).

2. E. J. Bonano et al., "Development of Performance Assessment Methodology for Nuclear Waste Isolation in Geologic Media," presented at Symposium on Groundwater Flow and Transport Modeling for Performance Assessment of Deep Geologic Disposal of Radioactive Waste, Albuquerque, NM (1985).

3. J. F. Brotchie and R. Silvester, "On Crustal Flexure," Jour. Geophys. Research, v. 74, p. 5240-5252 (1969).

4. R. I. Walcott, "Past Sea Levels, Eustasy and Deformation of the Earth," Quaternary Research, v. 2, p. 1-14 (1972).

5. W. R. Peltier and J. T. Andrews, "Glacial-Isostatic Adjustment--I. The Forward Problem," Geophys. J. R. Astr. Soc., v. 46, p. 605-646 (1976).

6. R. M. Thorson, Isostatic Effects of the Last Glaciation in the Puget Lowland, Washington, OF-81-0370, U.S. Geological Survey (1981).

7. R. F. Flint, "Pleistocene Drift Border in Eastern Washington," Geol. Soc. Amer. Bull., v. 48, p. 203-232 (1937).

8. M. G. Foley et al., Geologic Simulation Model for a Hypothetical Site in the Columbia Plateau: Volume 2: Results," PNL-3542-2, Battelle Pacific Northwest Laboratory, Richland, WA (1982).

9. R. Hofmann, STEALTH--A Lagrange Explicit Finite-difference Code for Solids, Structural, and Thermohydraulic Analysis, Volume 1 A and B: User's Manual," NP-2080-CCM, prepared for Electric Power Research Institute, Palo Alto, CA, by Science Applications, Inc., San Leandro, CA (1981).

10. A. Holmes, Principles of Physical Geology, second edition, The Ronald Press Co., New York (1964).

11. U.S. Department of Energy, Draft Environmental Assessment: Reference Repository Location, Hanford Site, WA, DOE/RW-0017 (1984).

12. P. A. Witherspoon et al., "Validity of Cubic Law for Fluid Flow in a Deformable Rock Fracture," Water Resources Research, v. 16, p. 1016-1024 (1980).

13. K. Iwai, Fundamental Studies of Fluid Flow through a Single Fracture, Ph.D. Thesis, 208 pp., University of California, Berkeley, CA (1976).

EVALUATION OF TEMPERATURES FOR A NUCLEAR WASTE REPOSITORY IN SALT

E. Gregory McNulty
Leslie A. Scott
Office of Nuclear Waste Isolation
Battelle Project Management Division
505 King Avenue
Columbus, Ohio 43201

ABSTRACT

Differences among thermal models of TEMP, HEATING6, THAC-SIP-3D, STEALTH, and SPECTROM-41 may affect the calculation of temperatures for a nuclear waste repository in salt. These codes agree closely in the very near field and approximately in the near- and far-field regions of a nuclear waste repository. Uncertainty in the thermal conductivity of salt affects calculated temperatures much more than any uncertainties in the heat transfer models. In particular, nonsalt stratigraphy appears to affect temperatures much less than the uncertainty in the thermal conductivity of salt. Differences in geometrical representations of CHLW heat sources and sequence of emplacement have little effect on calculated temperatures.

I. INTRODUCTION

Effective designs for the isolation of nuclear waste in salt require data on the temperatures within the repository. Temperatures affect virtually all processes within the repository region including fluid movement, waste package corrosion, mechanical response of the repository, and radionuclide transport. This paper compares calculation of temperatures for vertically emplaced waste packages of commercial high-level waste (CHLW) by five computer codes: TEMP, a semianalytical heat transfer code for finite length line sources in three dimensions; the HEATING6[1], THAC-SIP-3D[2], and the STEALTH[3], three-dimensional finite difference codes; and SPECTROM-41[4,5], a two-dimensional finite element code. Table 1 briefly describes each of these codes. McNulty[6] describes the basic solution technique for finite length line sources for an earlier version of the TEMP code. In addition, McNulty[8] gives principal input data and calculational details for these benchmark comparisons. This paper discusses how differences among the codes given in Table 1 may affect the calculation of temperatures in the very near field (canister region), near field (room region), and far field (repository region) of a nuclear waste repository in salt.

TABLE 1. Brief Code Descriptions

Code	Models	Principal Applications	Comments
TEMP	Semianalytical Solution Temperature Dependent Isotropic Conductivity Heat Transfer by Conduction Only	Single Source, Infinite Array, Finite Array with Sequential Emplacement 3-Dimensional Problems	Infinite Medium Model No Stratigraphy Vertical/Horizontal Emplacement
HEATING6	Explicit/Implicit Finite Difference Temperature Dependent Anisotropic Conductivity	Single Source, Infinite Array 1, 2, or 3-Dimensional Problems	Can include details of waste package and stratigraphy
THAC-SIP-3D	Strongly Implicit Finite Difference Temperature Dependent Anisotropic Thermal Conduct-ivity	Single Source, Infinite Array 3-Dimensional Problems	Rectangular parallele-piped heat source
STEALTH	Lagrange Explicit Finite Difference Temperature Dependent Anisotropic Thermal Conductivity Radiation at Boundaries	Single Source, Infinite Array 1, 2, or 3-Dimensional Problems	
SPECTROM-41	Finite Element Temperature Dependent Anisotropic Thermal Conductivity	Single Source, Infinite Array 2-Dimensional or Axisymmetric Problems	Can model actual CHLW package materials and geometry in axisymmetric model

II. VERY-NEAR-FIELD

Thermal calculations for the very-near-field region around a waste package provide temperatures needed in the calculation of thermally induced brine movement and waste package corrosion. Calculations were made for a single CHLW source, an infinite array of CHLW sources, and sequential emplacement of a finite array of CHLW sources. For the simplest case of a single CHLW package in an infinite salt medium, Figure 1 shows excellent agreement among HEATING6 and TEMP temperatures calculated at the surface of waste package.

Typically, however, several waste packages will be placed within a short period of time so that they can be modeled as an infinite array. In addition, the salt formation may or may not include nonsalt interbeds having lower thermal conductivities than those of salt. For the case of no nonsalt interbeds, Figure 2 shows good agreement among temperatures calculated by THAC-SIP-3D, TEMP, and SPECTROM-41 at the surface of the waste package. Figure 2 also suggests that heat source geometries such as a rectangular parallelepiped (THAC-SIP-3D), a finite-line (TEMP), and a cylinder (SPECTROM-41) have little effect on calculated temperatures. In addition, Figure 2 indicates that temperatures from an axisymmetric model such as the one used by SPECTROM-41 can agree closely with three-dimensional representations used by THAC-SIP-3D and TEMP.

For the case of nonsalt interbeds, Figure 3 shows a temperature profile at five years from TEMP that agrees closely with that from SPECTROM-41[5]. SPECTROM-41 can account for nonsalt interbeds while TEMP cannot. However, the close agreement shown in Figure 3 suggests that nonsalt interbeds may have little effect on temperatures near the waste package. Similarly, Loken et al[5] varied the thicknesses and position of nonsalt interbeds in the Permian Basin. Loken et al[5] then calculated temperatures on the surface of the waste package with SPECTROM-41 for each case of stratigraphy. Loken et al[5] found that these stratigraphic permutations would cause a probable change in surface temperatures of only 8°C and a maximum possible change of 30°C. However, the greatest uncertainty in the calculation of surface temperatures on the waste package depends on the accuracy of the thermal conductivity of salt, not the modeling of nonsalt interbeds. McNulty[6] shows that estimated uncertainties in thermal conductivity due to testing technique and sampling disturbance could raise peak surface temperatures by 120°C for CHLW. In addition, TEMP agrees closely with SPECTROM-41 even though TEMP uses a finite line heat source and SPECTROM-41 modeled the material and geometric details of the CHLW package.

Finally, sequential emplacement of different waste regions of a repository represented by a finite array of waste packages may give different temperatures than those calculated for an infinite array. When an emplacement schedule such as the one proposed by Stearns-Catalytic[7] is used. Figures 4 and 5 suggest there will be little difference between thermal calculations for an infinite array and a finite array with sequential emplacement. Figure 5 shows that backshifting all emplacement times to zero makes the infinite array and sequential emplacement temperatures almost coincide. Therefore, a point in the middle of a finite array of waste packages behaves essentially as one in an infinite array. As an aside, note that TEMP calculated slightly higher

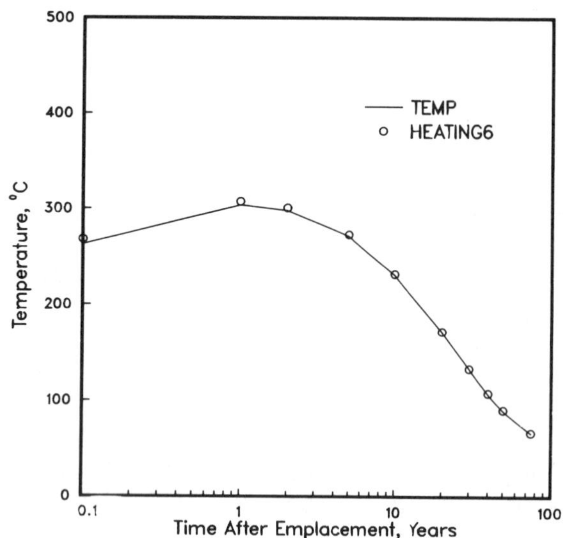

Figure 1. Comparison of Temperatures from TEMP and HEATING6 with Time at Salt-Overpack Interface of a Single CHLW Heat Source in Salt with Temperature Dependent Thermal Conductivity

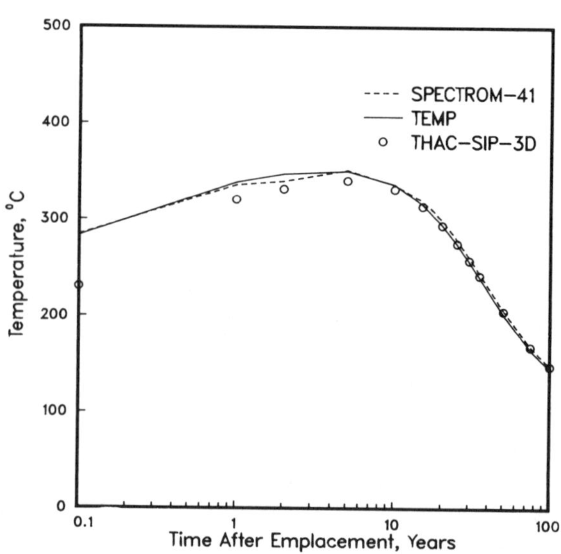

Figure 2. Comparison of Temperatures from TEMP, THAC-SIP-3D, and SPECTROM-41 with Time at Salt-Overpack Interface for an Infinite Array of Heat Sources in Salt with Temperature Dependent Thermal Conductivity

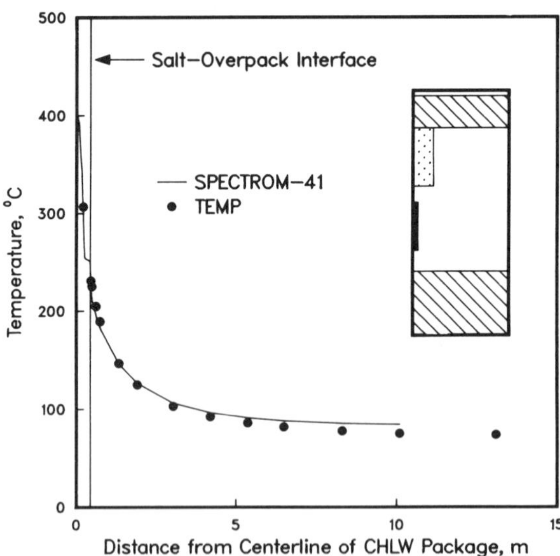

FIGURE 3. Comparison of Temperatures from TEMP and SPECTROM-41 with Distance from Centerline and Mid-Height of CHLW Package at Five Years

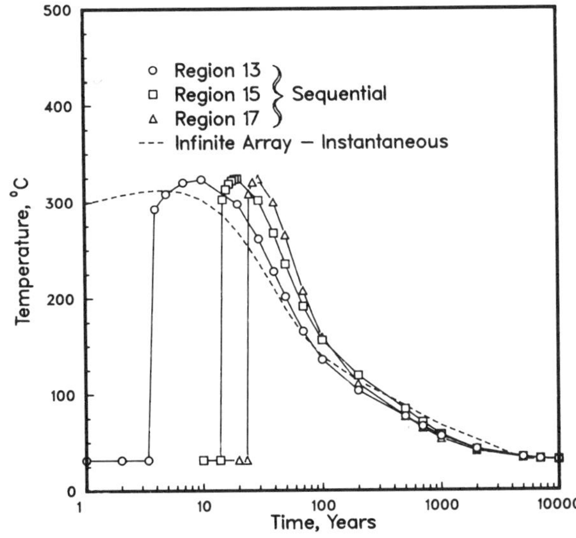

FIGURE 4. Comparison of TEMP Temperatures for CHLW at Salt-Overpack Interface for Sequential Emplacement and Infinite Array Solutions

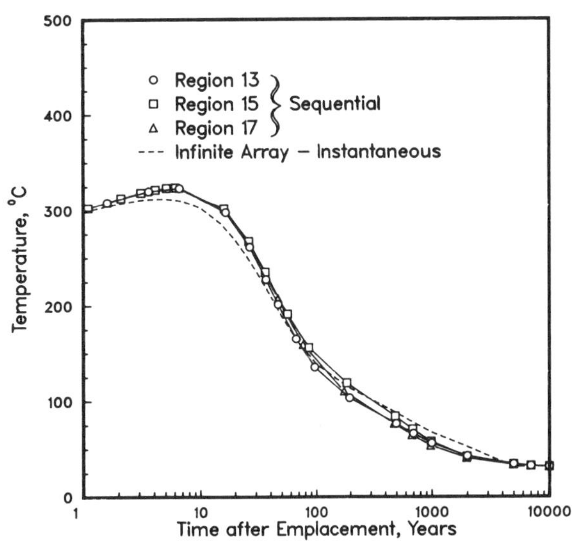

FIGURE 5. Comparison of TEMP Temperatures for CHLW at Salt-Overpack Interface for Sequential Emplacement After Backshift to Time Zero and Infinite Array Solutions

temperatures for sequential emplacement as opposed to the infinite array. The infinite array calculation used the package and tunnel spacings cited by Stearns-Catalytic. In contrast, TEMP currently restricts tunnels spacings to a fixed value for the finite array option. So, to compare the infinite array and sequential emplacement calculations, we calculated an "equivalent" package spacing for the finite array to keep the same areal thermal loading. However, this "equivalent" package and tunnel spacing for the finite array reduced the radial distance to the nearest waste package. Consequently, for the same areal thermal loading, this reduced radial distance causes calculation of slightly higher temperatures for the finite array.

III. NEAR-FIELD

Thermal calculations for the near field region provide temperatures needed in the calculation of the mechanical behavior of a repository room. Near-field calculations were made with TEMP, STEALTH,[8] and SPECTROM-41[5] on an excavated repository room for CHLW having nonsalt stratigraphy. Figure 6 shows temperatures with time for three points within the near-field region of a CHLW room. The temperatures calculated by these three codes agree only approximately because of three reasons. First, STEALTH and SPECTROM-41 can model the lower thermal conductivity of nonsalt stratigraphy while TEMP cannot. Partly because TEMP cannot account for the lower nonsalt thermal conductivity, TEMP underpredicts temperatures at Points A and C in Figure 6. STEALTH and SPECTROM-41

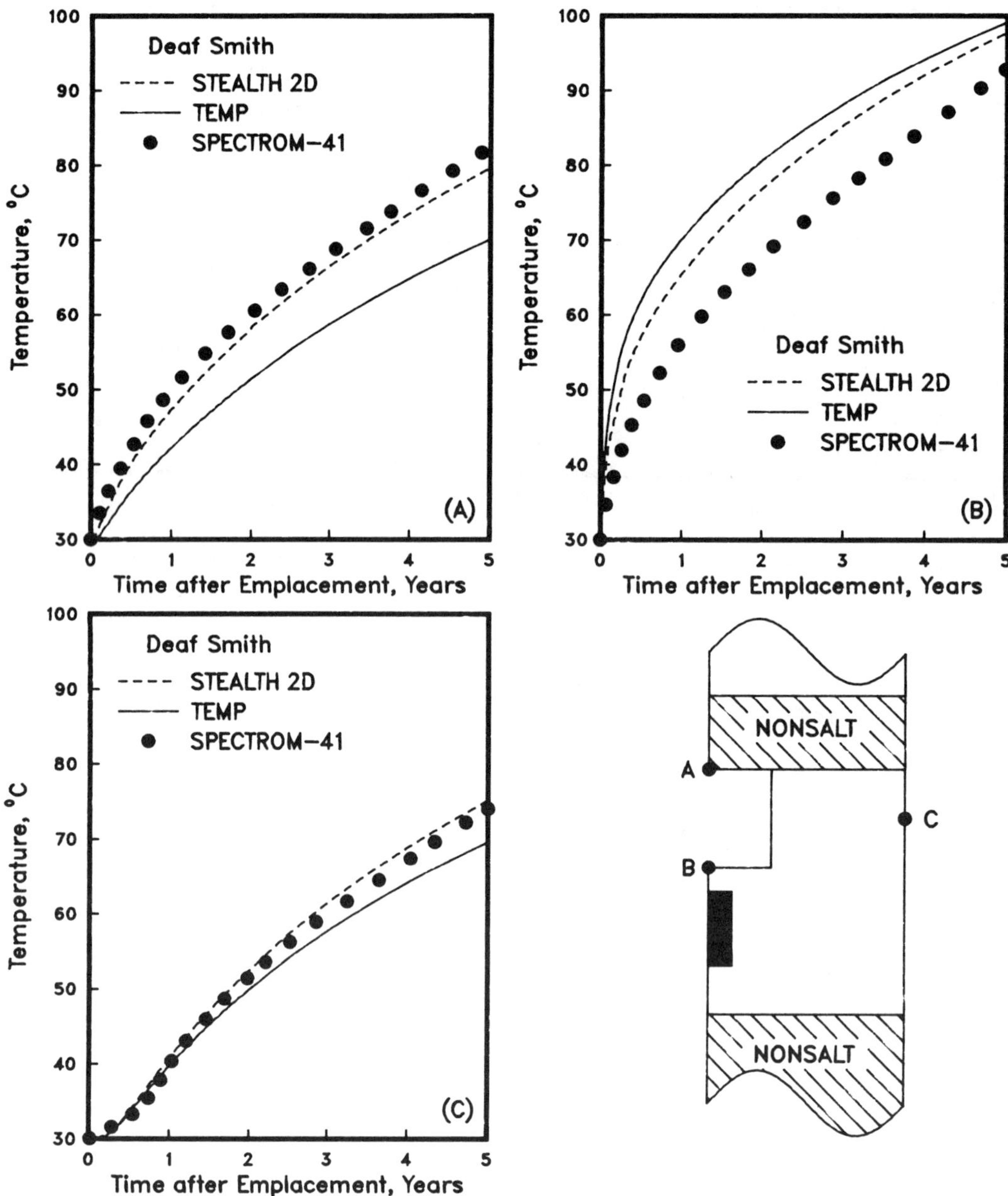

FIGURE 6. Temperatures from TEMP, STEALTH 2D, and SPECTROM-41 at Three Separate Points in CHLW (30 W/m^2) Room Region for Deaf Smith, No Backfill

agree rather closely in part because these codes account for the nonsalt thermal conductivity. Next, TEMP does not account for heat transfer by radiation as does STEALTH or by equivalent conduction through the air as does SPECTROM-41. Consequently, TEMP partly underpredicts temperatures at Point A and overpredicts temperatures at Point B. In comparison to SPECTROM-41, STEALTH overpredicts temperatures (by about 20%) at Point B because STEALTH must extrapolate temperatures from the center of a finite difference zone in a region of high thermal gradients. Finally, STEALTH and SPECTROM-41 simulated the CHLW near-field region in 2 dimensions while TEMP used three dimensions. Therefore, STEALTH and SPECTROM-41 modeled the CHLW heat generation as distributed over the volume of a long slab as opposed to being concentrated near the finite line heat source modeled by TEMP. Consequently, temperatures calculated by TEMP directly above the finite-line source should be higher than those calculated by STEALTH and SPECTROM-41 as shown in Figure 6 at Point B.

IV. FAR-FIELD

Thermal calculations for the far field provide temperatures needed in the calculation of uplift and subsidence and far-field changes in permeability and hydrology that could accelerate radionuclide transport. Figure 7 compares temperatures from TEMP and SPECTROM-41[5] above the centerline of the CHLW region shown in Figure 8. Figure 7 suggests that the far-field stratigraphic differences given in Figure 8 may not have a major effect on calculated near-field or far-field temperatures. TEMP predicts slightly different temperatures as compared to SPECTROM-41 for 3 reasons. First, TEMP cannot account for stratigraphic differences in thermal conductivity as can SPECTROM-41. Therefore, TEMP temperatures should not match those of SPECTROM-41. However, Figure 7 shows rather close agreement among temperatures calculated by SPECTROM-41 and TEMP above the 560 meter depth. Second, the two-dimensional SPECTROM-41 code must reduce the areal thermal loading to account for the impact of the non-heat generating haulageways and corridors on far-field temperatures. This reduction

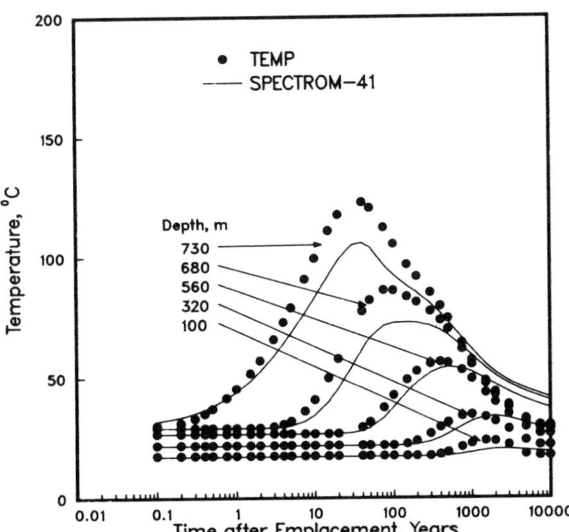

FIGURE 7. Comparison of CHLW Temperatures from TEMP and SPECTROM-41 for Deaf Smith Repository Region

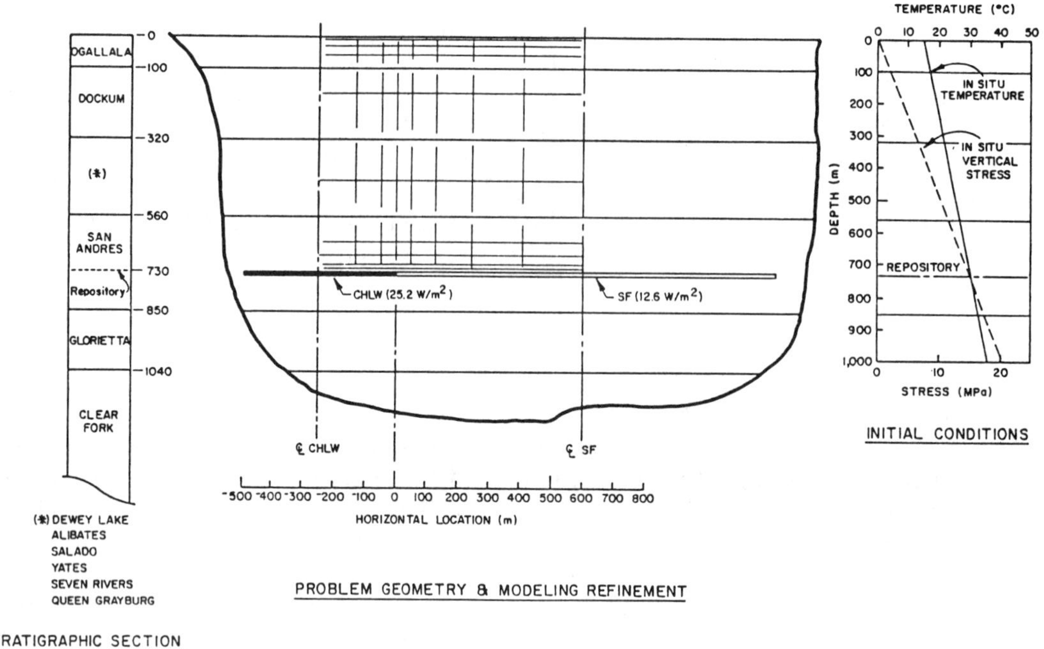

FIGURE 8. Repository Region Used in Comparing TEMP and SPECTROM-41 Near-Field and Far-Field Temperatures (Adapted from Reference 5)

of areal thermal loading for far-field analyses causes SPECTROM-41 to underpredict near-field (730 m) temperatures as shown in Figure 7. Third, the current version of TEMP cannot include a nearby spent fuel (SF) region while calculating temperatures above the centerline of the CHLW region. Consequently, Figure 7 shows that TEMP underpredicts the SPECTROM-41 temperatures at long times.

V. SUMMARY AND CONCLUSIONS

In the very-near-field, all the codes considered (HEATING6, THAC-SIP-3D, TEMP, and SPECTROM-41) agree closely with or without interbeds of nonsalt stratigraphy. Whether a code can model nonsalt stratigraphy affects temperatures on the surface of the waste package less than the uncertainty in the accuracy of thermal conductivity. In addition, the representation of the CHLW heat source as a finite line (TEMP), rectangular parallelepiped (THAC-SIP-3D), or as a detailed waste package (SPECTROM-41) seems to cause few differences in predicted surface temperatures of a CHLW package. Finally, sequential emplacement appears to have little effect on canister surface temperatures.

For the near-field, two-dimensional calculations by STEALTH and SPECTROM-41 give accurate enough temperatures if calculations are not made too close to the heat source. Even though TEMP gave temperatures within 15°C of STEALTH and SPECTROM-41 in the room region, nonsalt stratigraphy and the excavated room did reduce the accuracy of TEMP.

For the far-field region, TEMP agreed rather closely with SPECTROM-41 in spite of stratigraphic variations. Also, TEMP gave more accurate temperature predictions in the near-field in this far-field problem because the two-dimensional SPECTROM-41 had to reduce the areal thermal loading for calculation of far-field temperatures. However, SPECTROM-41 still gave more accurate far-field temperatures at long times because TEMP did not account for heat contributed by nearby SF region.

VI. ACKNOWLEDGMENTS

The authors gratefully acknowledge that Darrell Svalstad and Marc Loken of RE/SPEC, Inc. provided the SPECTROM-41 temperatures, Barry Dial, Don Maxwell, add W. Young of Science Applications International Corporation provided the STEALTH temperatures, and Matt Feldman, George Llewellyn, and Moshe-Siman-Tou of Oak Ridge National Laboratory provided the HEATING6, and THAC-SIP-3D temperatures.

REFERENCES

1. W. O. Turner, O. C. Elrod, and G. E. Giles "HEATING6: A Multidimensional Heat Conduction Analysis Program with a Finite Difference Formulation," Sect. F10 in SCALE: A Modular Code System for Performing Standardized Computer Analysis for Licensing Evaluation, ORNL/NUREG/CSD-2 (1981).

2. W. O. Turner, THAC-SIP-3D - A Three-Dimensional, Transient Heat Analysis Code Using the Strongly Implicit Procedure, K/CSD/TM-24 (1978).

3. R. Hofmann, STEALTH - A Langrange Explicit Finite Difference Code for Solids, Structural, and Thermohydraulic Analysis Introduction and Guide, EPRI NP-2080-CCM-SY, Electric Power Research Institute, Palo Alto, CA (1981).

4. O. K. Svalstad, User's Manual for SPECTROM-41: A Finite-Element Heat Transfer Program, ONWI-326, Office of Nuclear Waste Isolation, Battelle Memorial Institute. Columbus, OH (1983).

5. M. C. Loken et al., Thermomechanical Analyses of Conceptual Repository Designs for the Paradox and Permian Basins, Office of Nuclear Waste Isolation, Battelle Memorial Institute, Columbus, OH, Draft (1985).*

6. E. G. McNulty, Expected Near-Field Thermal Performance for Nuclear Waste Repositories at Potential Salt Sites, Office of Nuclear Waste Isolation, Battelle Memorial Institute, Columbus, OH, Draft (1985).*

7. Stearns-Catalytic, Basic Repository EA Design Basis - Permian Basin - Deaf Smith County Site, Office of Nuclear Waste Isolation, Battelle Memorial Institute, Columbus, OH, Draft (1984).*

8. B. W. Dial, D. E. Maxwell, and W. Yeung, Room and Canister Scale Thermal-Mechanical Benchmarks Between STEALTH 2D and SPECTROM-41/SPECTROM-21, Office of Nuclear Waste Isolation, Battelle Memorial Institute, Draft (1985).

*Draft references available upon request from:

John Bellay
(614) 424-4961
Controlled Technical Document Center
Room 15-1-127
Battelle Project Management Division
505 King Avenue
Columbus, OH 43201-2693

ANALYTICAL AND SEMIANALYTICAL MODELS FOR ESTIMATING TEMPERATURE, BRINE MIGRATION, AND RADIONUCLIDE TRANSPORT IN A SALT REPOSITORY

Sanford G. Bloom
Gilbert E. Raines
Office of Nuclear Waste Isolation
Battelle Project Management Division
505 King Avenue
Columbus, Ohio 43201

ABSTRACT

Many types of mathematical models are used to estimate the performance of a nuclear waste repository in salt. Most of these models are based on solutions to partial differential equations which simulate the various mechanisms that affect performance. Analytical and semianalytical solutions have been developed for the partial differential equations that simulate temperature, brine migration, and radionuclide transport in salt. Models based on these analytical and semianalytical solutions have been used in preliminary performance assessments and will be extended for use in more detailed assessments and in sensitivity analyses. The derivation of these models and their limitations are presented in this paper.

I. INTRODUCTION

Many types of mathematical models are used to estimate the performance of a nuclear waste repository in salt.[1] Among these are models for temperature, brine migration, and radionuclide transport. These models are based on solutions to partial differential equations which simulate the various mechanisms. Finite-difference and finite-element methods have been developed for the general solutions of these differential equations but these solutions are generally time-consuming and subject to errors due to the selection of time and distance intervals. Analytical and semianalytical methods have also been developed for special solutions to these differential equations in order to circumvent some of the above problems. The derivations of the models based on the analytical and semianalytical methods are presented here.

II. TEMPERATURE MODEL

The temperature model approximates the heat generation of the waste packages as a system of finite line sources and the repository as an infinite, isotropic medium. Heat transfer is by conduction only. The heat generation rate of the sources is continuously decreasing with time and the temperature at a specific location is obtained by superpositioning contributions from the sources.

The differential equation plus the boundary and initial conditions for each source are

$$\sum_{i=1}^{3} \partial(k\partial T/\partial x_i)/\partial x_i = \rho c \partial T/\partial t \qquad (1)$$

$$T = T_I \text{ at } t = 0 \text{ and } |x_i| > 0 \text{ for } i = 1,2,3 \qquad (2)$$

$$T \to T_I \text{ as } |x_i| \to \infty \text{ and } t > 0 \text{ for } i = 1,2,3 \qquad (3)$$

$$\lim_{r \to 0} [2\pi r k \partial T/\partial r] = -q(t)/(2L) \text{ for } t>0 \text{ and } -L < x_3 < L \qquad (4)$$

where: T = temperature (°C)
x_i = the three principal coordinate directions (m)
t = time (s)
k = thermal conductivity (J/m-°C-s)
ρ = density (kg/m^3)
c = heat capacity (J/kg-°C)
r = $(x_1^2 + x_2^2)^{1/2}$
T_I = the initial temperature of the infinite medium (°C)
q = the heat release rate for the entire line source (J/s)
L = the half-length of the line source whose midpoint is at $x_1 = x_2 = x_3 = 0$ (m).

If k depends only on temperature, Equation (1) can be put into a quasi-linear form by a simple transformation,[2] which is based on defining a new function:

$$\phi = \int_{T_I}^{T} k \, dT \qquad (5)$$

where: ϕ = the new function (J/m-s).

In terms of ϕ, Equations (1) through (4) become:

$$\alpha \sum_{i=1}^{3} \partial^2\phi/\partial x_i^2 = \partial\phi/\partial t \qquad (6)$$

$$\phi = 0 \text{ at } t = 0 \text{ and } |x_i| > 0 \text{ for } i = 1,2,3 \qquad (7)$$

$$\phi \rightarrow 0 \text{ as } |x_i| \rightarrow \infty \text{ and } t>0 \text{ for } i = 1,2,3 \tag{8}$$

$$\lim_{r \rightarrow 0} [2\pi r \partial\phi/\partial r] = -q(t)/2L \text{ for } t>0 \text{ and } -L < x_3 < L \tag{9}$$

where: $\alpha = k/\rho c$ is the thermal diffusivity (m^2/s).

If α were constant, Equation (6) would be a linear differential equation and the solution to Equations (6) through (9) could be derived using methods given in Carlslaw and Jaeger.[2] The appropriate solution is

$$\phi = (0.5\alpha/L) \int_{-L}^{L} \int_{0}^{t} q(t-t') \exp\left[-(r^2 + (x_3 - x_3')^2)/(4\alpha t')\right] [4\pi\alpha t']^{-3/2} dt' dx_3' \tag{10}$$

Also, the ϕ contributions from each source could be superimposed at a given location and transformed by inverting Equation (5) to yield the temperature at the location. In general, α will be approximately proportional to k. The effect of variable α is incorporated by the following scheme: 1) a temperature, T, is calculated for a given location and time using α evaluated at T_I (i.e., α_I), (2) α is evaluated at T (i.e., α_T), and 3) a final temperature for the given location and time is calculated using the average (i.e., $(\alpha_I + \alpha_T)/2$). This scheme is used in the TEMP model[3] and benchmark calculations indicate the error in using this scheme is small.

A closed form of the integral of Equation (10) for the general case is not known. One procedure for evaluating the integral, used in the FLLSSM model,[4] is to first integrate analytically with respect to length and then numerically integrate the result with respect to time. However, the method used in the FLLSSM model gives poor results for long time spans. An alternative method is used in the TEMP model.[3] The heat rate is empirically fit to a sum of exponential terms such that

$$q(t) = q_o \sum_{j=1}^{J} a_j \exp(-b_j t) \tag{11}$$

where: q_o, a_j, b_j = empirically-fitted constants
J = the number of terms in the sum.

Equation (10) can then be integrated first with respect to time to give

$$\phi = q_o \, [8\pi L]^{-1} \int_{-L}^{L} [r^2 + (x_3 - x_3')^2]^{-1/2} \tag{12}$$

$$\sum_{j=1}^{J} a_j F \, ([r^2 + (x_3 - x_3')^2]^{1/2} [4\alpha t]^{-1/2}, \, (b_j t)^{1/2}) dx_3'$$

where: $F(u,v) = \exp(-u^2) R \, [\exp((u+iv)^2) \operatorname{erfc}(u+iv)]$
R refers to the real part of the function.

The F function can be easily evaluated by a procedure based on formulas presented by Salzer.[5] Equation (12) is then integrated numerically over the length. The numerical integration is very simple and is easily performed using Simpson's rule. The method is fast and gives good results for all time intervals.

III. LIMITING FORMS OF THE TEMPERATURE MODEL

There are limiting forms of Equation (12) that apply to certain ranges of distance and time. These limiting forms are used to further increase the speed of solution. For early times with slowly varying q, the average value of q can be used in Equation (10) to give

$$\phi = q_A \, [8\pi L]^{-1} \int_{-L}^{L} [r^2 + (x_3 - x_3')^2]^{-1/2} \tag{13}$$

$$\operatorname{erfc} \, ([r^2 + (x_3 - x_3')^2]^{1/2} \, [4\alpha t]^{-1/2}) dx_3'$$

where: q_A = the average value of q (J/s)

$$q_A = t^{-1} \int_0^t q(t') \, dt'. \tag{14}$$

Equation (13) is equivalent to Equation (12) for $q_o = q_A$, $J=1$, $a_1=1$, and $b_1=0$.

Another limiting expression can be developed for later times with slowly varying q. Another average q value is defined as

$$q_B = (t_m)^{-1} \int_{t-t_m}^{t} q(t')\,dt' = (t_m)^{-1} \int_0^{t_m} q(t-t')dt'. \quad (15)$$

where: $t_m = 11\ (r^2 + x_3^2 + L^2/3)/\alpha$
t_m = a measure of the time necessary to approach 99.9% of the steady-state temperature change due to the introduction of one or more constant, finite line sources.

With q_B, the time integral in Equation (10) can be divided into an integral from 0 to t_m, and an integral from t_m to t over which the exponential term has approached 1.0. The result is

$$\phi = q_B\ (8\pi L)^{-1} \Big(\ln\{[r^2+(x_3+L)^2]^{1/2} + x_3 + L\}$$

$$-\ln\{[r^2+(x_3-L)^2]^{1/2}+x_3-L\}\Big)$$

$$+ q_o\ (64\pi^3\alpha)^{-1/2} \sum_{j=1}^{J} a_j \int_{t_m}^{t} \exp\ (-b_j(t-t'))\ (t')^{-3/2}dt'. \quad (16)$$

The integral in Equation (16) can be expressed in terms of Dawson's integral which, in turn, can be evaluated by procedures based on Salzer's formulas.[5] A major advantage of Equation (16) is that the integral term is usually small and can often be neglected.

A final limiting expression applies to any time for regions far from the source. For values of $r^2 + x_3^2$ much greater than L^2, the integrand in Equation (12) does not vary significantly over the range of integration and the integral can be approximated as the product of 2L and the integrand evaluated at $x_3 = 0$. The result is the solution for a point source and is

$$\phi = q_o\ [4\pi]^{-1}\ [r^2+x_3^2]^{-1/2} \sum_{j=1}^{J} a_j\ F([r^2+x_3^2]^{1/2}[4\alpha t]^{-1/2},\ (b_j t)^{1/2}). \quad (17)$$

Equation (17) is equivalent to the point source solution given by Beyerlein and Claiborne.[6]

IV. BRINE MIGRATION MODEL

Brine migration is approximated as convective transport in an infinite isotropic medium which has a uniform initial concentration of brine inclusions. Vertical and azimuthal uniformity is assumed so that only radial variation is considered. It is also assumed that the inclusions move toward the waste package at a velocity determined by temperature and temperature gradient which are functions of radial position and time. The separation between waste packages is assumed to be large enough to prevent significant interaction on the brine migration.

The differential equation plus the boundary and initial conditions near each waste package are

$$r\partial C/\partial t + \partial(rCv)/\partial r = 0 \tag{18}$$

$$C = C_0 \text{ at } t = 0 \text{ and } r > 0 \tag{19}$$

$$C = C_0 \text{ at } t > 0 \text{ and } r \to \infty \tag{20}$$

where: C = brine concentration (volume fraction)
C_0 = initial brine concentration (volume fraction)
v = brine migration velocity (m/s).

The brine flow per unit length of waste package is given by

$$R = 2\pi(rCv)_I \tag{21}$$

where: R = brine flow per unit length of waste package (m^2/s)
I = subscript to indicate conditions at the waste package borehole.

If v is not a function of C but only depends on r and t, Equation (18) can be transformed by substituting

$$s = r^{-1} \int_0^t v(t')\, dt' \tag{22}$$

for r and t. In order to be compatible with Equation (20), s must approach 0 for large r. If v can be expressed as a product of a function of t only and r raised to some power, such that

$$v(t,r) = v_0(t)\, r^p, \tag{23}$$

the solution to Equations (18) through (20) is

$$\ln(C/C_0) = [(1 + p)/(1 - p)] \ln(1 + ps - s). \quad (24)$$

Equation (20) requires that $p < 1$.

In general, the velocity function will not be in the form of Equation (23). However, the transformation can still be used to develop an approximate solution to Equations (18) through (21). The appproximate solution is

$$g_1(t) = \int_0^t (\partial v/\partial r - v/r)_I \, dt' \quad (25)$$

$$g_2(t) = \int_0^t [\partial v/\partial r + v/r]_I \, [1 + g_1(t')]_I^{-1} \, dt' \quad (26)$$

$$C_I = C_0 \exp(-g_2). \quad (27)$$

The integrations in Equations (25) and (26) are easily done numerically.

The velocity function that has been used with Equations (25) and (26) is based on the function assumed by Jenks[7] to represent the maximum velocity of brine inclusions in bedded salt. The function takes the form

$$v = (\partial T/\partial r) \exp(aT + b) \quad (28)$$

where: a,b = empirical constants.

Application of Equation (28) to Equations (25) and (26) requires estimates of T, $\partial T/\partial r$, and $\partial^2 T/\partial r^2$. All of these estimates can be obtained from the temperature model presented earlier. However, because of the need for superpositioning, it is more convenient to calculate only T and to approximate the derivatives from a simpler model based on a single, infinitely long line source at steady state. The errors in these approximate derivatives tend to offset each other since the actual sources have finite lengths and the derivative contributions of neighboring sources are neglected. The net result of the errors usually leads to overestimates of the magnitudes of the derivatives since the derivative contributions of neighboring sources decreases with increasing distance from the source, much faster than the corresponding temperature contribution.

It should be noted that, for constant k, the temperatures due to a single, infinitely long line source at steady state, along with Equation (28), would produce a velocity function equivalent to Equation (23) and lead to an exact solution of Equations (18) through (21). Consequently, Equations (25) through (27) are very good approximations at the borehole radius since the temperature behavior close to the middle of a single source is close to that of an infinite line source. The equations are probably less accurate for a more arbitrary velocity function. If v were also a function of C, a two-step approximation, similar to that used for temperature, may be adequate.

V. RADIONUCLIDE TRANSPORT MODEL

There are two models to simulate radionuclide transport corresponding to two widely different scenarios. Yet, the mathematical form of the resulting expressions are very similar. Both models assume the radionuclide source can be approximated as an infinite plane in an infinite, isotropic medium and the transport is in the direction perpendicular to this plane. Both models also assume that the released radionuclides must first dissolve in brine that accumulates around the waste packages and, after dissolution, the release occurs instantaneously.

In the first model, a small amount of brine accumulates due to the brine migration model presented earlier. After the temperature decline, there is no longer a driving force for brine migration toward the waste package and it is postulated that the excess brine will tend to migrate back into the salt formation. This back migration is assumed to follow a diffusion-like mechanism. Dissolved radionuclides will migrate with the brine and, if there were no radioactive decay, maintain a constant concentration with respect to the brine. Because the brine is dispersing and radioactive decay is occurring, the radionuclide concentration in the formation decreases with time and distance from the source. The equations for the first model are

$$D\partial^2C/\partial y^2 = \partial C/\partial t \tag{29}$$

$$C = 0 \text{ at } t = 0 \text{ and } |y| > 0 \tag{30}$$

$$C \to 0 \text{ at } t > 0 \text{ and } |y| \to \infty \tag{31}$$

$$\int_{-\infty}^{\infty} C dy = B \text{ at } t > 0 \tag{32}$$

$$C_R = C_{RO}\, C \exp(-\lambda t) \tag{33}$$

where: y = distance from the source (m)
D = "diffusivity" for the diffusion-like migration of the brine (m^2/s)
B = volume of brine accumulated around the waste packages per unit area (m)
C_R = radionuclide concentration in the formation (Ci/m^3)
C_{RO} = initial radionuclide concentration in the brine (Ci/m^3)
λ = radioactive decay rate (s^{-1}).

In the second model, a larger amount of brine is assumed to contact the waste packages due to some intrusion which connects an aquifer to the repository. The connection will eventually close due to the creeping of salt and precipitation of salts from the brine. For convenience, the same symbol, B, is used for the accumulated brine. After dissolving radionuclides, the brine is instantaneously released into a flowing aquifer. The equations for this second model are based on equations derived by Harada et al.[8] and are

$$D_L \partial^2 C_G/\partial y^2 - V \partial C_G/\partial y = K \partial C_G/\partial t + \lambda K C_G \tag{34}$$

$$C_G = 0 \text{ at } t = 0 \text{ and } |y| > 0 \tag{35}$$

$$C_G = 0 \text{ at } t > 0 \text{ and } |y| \to \infty \tag{36}$$

$$\varepsilon K \int_{-\infty}^{\infty} C_G \, dy = B\, C_{RO} \exp(-\lambda t) \text{ at } t > 0 \tag{37}$$

where: D_L = dispersion coefficient (m^2/s)
V = interstitial ground-water velocity (m/s)
K = radionuclide retardation factor
ε = porosity of the aquifer
C_G = concentration of radionuclide in groundwater.

The dependence of the initial radionuclide concentration on solubility takes the same form for both models. The expressions are

$$C_{RO} = I_R/B \text{ if } I/B < S \tag{38}$$

$$C_{RO} = S\, I_R/I \text{ if } I/B > S \tag{39}$$

where: I_R = inventory of the radionuclide per unit area (Ci/m^2)
I = inventory per unit area of the radionuclide and all other isotopes of the element (kg/m^2)
S = solubility of the element in the brine (kg/m^3).

If D represents either D or D_L, the solution to both models is

$$C_G = B\ C_{RO}\ (\varepsilon K)^{-1}\ (4\pi Dt/K)^{-1/2}\ \exp\ [-\lambda t - (y-Vt/K)^2\ (4Dt/K)^{-1}]. \tag{40}$$

The concentration of the radionuclides in the formation is related to the concentration in the groundwater by

$$C_R = \varepsilon K C_G. \tag{41}$$

For the first model, K and ε are 1.0 while $V = 0$.

A quantity that must be estimated for regulatory requirements is the cumulative radionuclide flow past some specified distance from the repository. The cumulative flow is defined by

$$Q = \frac{1}{K}\int_0^t [VC_R - D\partial C_R/\partial y]\ dt' \tag{42}$$

where: Q = cumulative radionuclide flow per unit area (Ci/m^2).

Upon substituting Equation (40) into Equation (42), the result for $y > 0$ is

$$Q = 0.25\ BC_{RO}\ [(1+\nu/\gamma)\ \exp\ (0.5\ (\nu-\gamma)y/\alpha)\ \mathrm{erfc}\ ((y-\gamma t)\ (4\alpha t)^{-1/2})$$

$$+\ (1-\nu/\gamma)\ \exp\ (0.5(\gamma+\nu)y/\alpha)\ \mathrm{erfc}\ ((y+\gamma t)\ (4\alpha t)^{-1/2})] \tag{43}$$

where: $\nu = V/K$ (44)

$\alpha = D/K$ (45)

$\gamma = (\nu^2 + 4\ \alpha\lambda)^{1/2}.$ (46)

For very long times, the first erfc in Equation (43) aproaches 2.0 and the second approaches zero. Therefore, for very long times

$$Q = 0.5\ BC_{RO}\ (1 + \nu/\gamma)\ \exp\ (0.5\ (\nu - \gamma)\ y/\alpha). \tag{47}$$

Since $\gamma > \nu$ [see Equation (46)], the cumulative radionuclide flow for large times decreases exponentially with distance from the source.

VI. CONCLUSIONS

Analytical and semianalytical models have been developed to simulate special cases of temperature, brine migration, and radionuclide transport in a salt repository. These models are relatively simple to use, the calculation procedures are very fast, and the results provide some insight into the behavior of the mechanisms. They have been very useful in preliminary performance assessments and will be extended for use in more detailed assessments, benchmark exercises with numerical codes, and in sensitivity analyses.

REFERENCES

1. Office of Nuclear Waste Isolation. Performance Assessment Plans and Methods for the Salt Repository Project, ONWI-545, Battelle Memorial Institute, Columbus, Ohio (1984).

2. H. S. Carslaw and J. C. Jaeger, Conduction of Heat in Solids, 2nd ed., p. 10, p. 256, Oxford University Press, New York (1959).

3. E. G. McNulty. Expected Near-Field Thermal Performance for Nuclear Waste Repositories at Potential Salt Sites, Office of Nuclear Waste Isolation, Battelle Memorial Institute, Columbus, Ohio, Draft (1985)*

4. Kaiser Engineers, Inc., Finite-Length Line Source Superposition Model (FLLSSM), ONWI-94, prepared for Office of Nuclear Waste Isolation, Battelle Memorial Institute, Columbus, Ohio (1980).

---

*Draft reference available upon request from:

John Bellay
(614) 424-4961
Controlled Technical Document Center
Room 15-1-127
Battelle Project Management Division
505 King Avenue
Columbus, OH 43201-2693

5. H. E. Salzer, "Formulas for Calculating the Error Function of a Complex Variable," Math Tables Aids Comp. 5, 67 (1951).

6. S. W. Beyerlein and H. C. Claiborne, The Possibility of Multiple Temperature Maxima in Geologic Repositories for Spent Fuel from Nuclear Reactors, ORNL/TM-7024, Oak Ridge National Laboratory (1980).

7. G. H. Jenks, Effects of Temperature, Temperature Gradients, Stress, and Irradiation on Migration of Brine Inclusions in a Salt Repository, ORNL-5526, Oak Ridge National Laboratory (1979).

8. M. Harada et al., Migration of Radionuclides Through Sorbing Media, Analytical Solutions - I, LBL-10500, Lawrence Berkeley Laboratory (1980).

EFFECTS OF SORPTION HYSTERESIS ON RADIONUCLIDE RELEASES FROM WASTE PACKAGES

G. S. Barney
D. T. Reed
Rockwell Hanford Operations
P.O. Box 800
Richland, Washington 99352

ABSTRACT

A one-dimensional, numerical transport model was used to calculate radionuclide releases from waste packages emplaced in a nuclear waste repository in basalt. The model incorporates both sorption and desorption isotherm parameters measured previously for sorption of key radionuclides on the packing material component of the waste package. Sorption hysteresis as described by these isotherms lowered releases of some radionuclides by as much as two orders of magnitude. Radionuclides that have low molar inventories (relative to uranium), high solubility, and are strongly sorbed, are most affected by sorption hysteresis. In these cases, almost the entire radionuclide inventory is sorbed on the packing material. The model can be used to help optimize the thickness of the packing material layer by comparing release rate versus packing material thickness curves with Nuclear Regulatory Commission (NRC) and Environmental Protection Agency (EPA) release limits.

I. INTRODUCTION

Radionuclide releases from waste packages emplaced in deep geologic repositories must be estimated to evaluate repository performance. Release predictions must be calculated over the 10,000 year period that the disposal system is required (by federal regulations) to isolate radioactive waste. The most credible means of radionuclide release in the deep basalts of the Columbia River formation is via groundwater transport. Several barriers must be penetrated to allow radionuclides to reach the host basalt rock. These barriers include the waste form (spent fuel or vitrified waste), the metal container, and the packing material. Release calculations must include quantification of the processes involved in penetration of these barriers by groundwater and waste radionuclides. The estimated releases must be consistant with performance requirements set by the NRC[1] and the EPA[2].

The objective of this study was to determine how sorption hysteresis affects (1) radionuclide transport through the packing material and (2) releases from the waste package. Laboratory measurements of radionuclide sorption and desorption in packing material-groundwater systems[3] have demonstrated that Freundlich sorption and desorption isotherms for most radionuclides are non-singular and show a strong tendency for sorption hysteresis. This phenomenon is expected to have a significant effect on radionuclide releases when desorption of these radionuclides occurs on the packing material.

A one-dimensional, numerical model was developed to calculate radionuclide releases from the waste form to the host basalt rock through a packing material. This model incorporates both sorption and desorption parameters (Freundlich constants) for key radionuclides using techniques developed by van Genuchten and Wierenga,[4,5] to predict the effects of sorption hysteresis on transport of pesticides through saturated soil.

The release calculations described here do not include the effects of waste package barriers other than packing material. It is expected that the waste form and container will significantly reduce releases from the waste package.

The approach to modeling waste package release was similar to that used by Relyea and Wood[6]. The model describes the behavior of a single waste package. The calculated release rates are therefore greater than those expected for the whole repository since probabilistically distributed failures of waste containers are not accounted for. Solubility limits are used to define radionuclide concentrations at the waste form. Diffusion of dissolved radionuclides is the major transport mechanism described by the model. Radioactive decay of the radionuclides is not accounted for in the calculations.

II. PHYSICAL AND CHEMICAL CONDITIONS

The current waste package design consists of the waste form (spent fuel or processed waste) sealed inside a metal container (steel or copper) which is surrounded by a layer of packing material (currently a mixture of 75% crushed basalt and 25% bentonite clay by weight). The waste package will be placed inside holes bored into the dense interior of a thick basalt flow. The container is designed to prevent releases for a minimum period of 1000 years. After the containment period, the packing material will minimize releases to the host rock by promoting precipitation and sorption of the radionuclides and by slowing advective transport.

Groundwater containing radionuclides from dissolution of the waste must travel through container corrosion products, packing material, and into fissures in the dense basalt. The intact basalt has almost no effective porosity. The fissures are filled with secondary minerals (mainly smectite clays) that are expected to have porosities and sorption properties similar to packing material. The one-dimensional model presented in this paper represents the waste package as a column of packing material connected to a smaller diameter column of basalt secondary minerals. The dimensions of the packing column are approximated by the thickness and cross-sectional area of the packing in the waste package. The basalt secondary mineral column is long enough to approximate an infinite length and has a cross-sectional area corresponding to to the estimated area of basalt fissures at the packing material-host rock interface.[11] Actual dimensions are given later in this paper.

III. DESCRIPTION OF THE MODEL

A. One Dimensional Radionuclide Transport.

The one-dimensional transport of dissolved radionuclides by groundwater through porous packing material is assumed to obey the following differential equation:[5]

$$\frac{\rho}{\theta}\frac{\partial S}{\partial t} + \frac{\partial C}{\partial t} = D\frac{\partial^2 C}{\partial x^2} - V\frac{\partial C}{\partial x} \tag{1}$$

where C is the concentration of radionuclide in solution, S is the amount of radionuclide sorbed per gram of packing material, D is the dispersion coefficient, ρ is the bulk density of the packing, θ is the porosity, v is the average pore velocity, x is the distance, and t is the time. Assuming that both sorption and desorption can be described by the Freundlich equation,

$$S = KC^N \tag{2}$$

where K and N are empirical constants, the transport equation can be expressed in terms of one dependent variable, C, by differentiating equation (2) and substituting the results into equation (1) as described by van Genuchten and Wierenga[4], which yields:

$$\frac{\partial C}{\partial t} = \frac{1}{R_f}\left(D\frac{\partial^2 C}{\partial x^2} - V\frac{\partial C}{\partial x}\right) \tag{3}$$

The retardation factor, R_f, is

$$R_f = 1 + \frac{\rho}{\theta} KNC^{(N-1)} \tag{4}$$

The values of Freundlich coefficients, K and N, will depend on whether sorption or desorption is occurring. The value of K_{des} (K when desorption occurs) is not constant during transport, but depends on the amount sorbed before desorption occurs. The following equation can be used to calculate K_{des}:

$$K_{des} = K_{ads}^{(N_{des}/N_{ads})} S_{max}^{(1-N_{des}/N_{ads})} \tag{5}$$

where S_{max} is the amount of radionuclide sorbed when desorption begins. The ratio, N_{ads}/N_{des}, is a measure of the extent of sorption hysteresis and is obtained from measured sorption and desorption isotherms.[3] When either N_{ads} or N_{des} are not equal to 1, no analytical solution to Equation 3 exists and numerical techniques must be used. The finite difference approximation of Equation 3 is:

$$\frac{C_i^{t+\Delta t} - C_i^t}{\Delta t} = \frac{1}{R_f}\left[D^* \frac{C_{i+1}^t - 2C_i^t + C_{i-1}^t}{\Delta x^2} - v\,\frac{C_i^t - C_{i-1}^t}{\Delta x}\right] \tag{6}$$

where i is the index for the layer or compartment number. The porous media (in this case packing material) is divided into a number of layers (NL) of a given thickness (Δx). The finite time step is Δt. The numerical dispersion introduced in this difference approximation is corrected for by using a corrected dispersion coefficient:[4]

$$D^* = D - \tfrac{1}{2} v \left(\Delta x - \frac{v}{R_f}\Delta t\right) \tag{7}$$

Since R_f will change with radionuclide concentration when $N \neq 1$ (Equation 4), it must be evaluated at some point between t and $t + \Delta t$. An average concentration ($\overline{C}_i$) was used and appears to be adequate (results are comparable to those obtained from analytical calculations), where:

$$\overline{C}_i = \tfrac{1}{2}(C_i^t + C_i^{t+\Delta t}) \tag{8}$$

Substituting $C_i^{t+\Delta t}$ from Equation (6) into Equation (8) and then substituting the resulting expression for C_i into Equation (4) gives:

$$R_f = 1 + \frac{\rho K N}{\theta}\left\{C_i^t + \tfrac{1}{2}\frac{\Delta t}{R_f}\left[\left(D - \tfrac{1}{2}v\,\Delta x + \frac{\tfrac{1}{2}v^2\,\Delta t}{R_f}\right)\left(\frac{C_{i+1}^t - 2C_i^t + C_{i-1}^t}{\Delta x^2}\right) - \frac{v\,(C_i^t - C_{i-1}^t)}{\Delta x}\right]\right\}^{(N-1)} \tag{9}$$

This equation can be solved for R_f at each layer for each time step by using the Newton-Raphson method. The $C_i^{t+\Delta t}$ is calculated using these R_f values and Equation (6). The quantity sorbed at $t+\Delta t$, $S_i^{t+\Delta t}$, can then be calculated from the Freundlich equation.

The model considers transport of radionuclides through two media, the packing material and the adjacent host rock. The boundary conditions at the waste container-packing material interface are:

(1) the concentration of radionuclide at this interface, C_o = the solubility of the radionuclide in moles/liter as long as the inventory of the radionuclide in the waste has not been depleted or,

(2) $C_o = C_1$ when the radionuclide inventory is depleted (C_1 is calculated from the previous time step)

The second condition prevents diffusion back into the waste form that occurs if C_o is allowed to go to zero after depletion. The radionuclide concentrations outside the basalt column are conservatively set at zero and the length of the basalt column is set long enough to approximate an infinite length. Effects of the basalt column length are presented later in this paper. Initial concentrations of the radionuclide in the packing and basalt columns were set at zero.

Radionuclide release rates and cumulative releases were calculated at the boundary between the last layer of packing (NL) and the first layer of basalt (NL+1). The fractional release rate at this boundary for any given time is

$$q = \frac{A\theta}{I}\left[vC_{NL} + D^* \frac{(C_{NL} - C_{NL+1})}{\Delta x} \right] \qquad (10)$$

where q is the fractional release rate, A is the cross-sectional area of the basalt column, θ is the porosity of the secondary minerals, v is the groundwater pore velocity in the basalt column, and I is the initial inventory of the radionuclide in the waste. The cumulative release is

$$Q = \sum_{j=1}^{n} q_j \Delta t_j \qquad (11)$$

where j is the index of the time steps for n time steps. A computer program (TRANSCOL) was written in BASIC for performing this analysis.

B. Transport Parameters Selected for the Model.

Parameters that must be evaluated for use in the model include chemical parameters for the individual radionuclides, physical dimensions of the waste package, and physical parameters of the packing material and host rock. Chemical parameters were estimated from prior measurements of sorption, desorption, and solubility of the radionuclides under expected repository conditions. The sorption and desorption data for the packing material were reported earlier[3] and the solubility data were taken from a report by Salter and Jacobs[7]. Sorption and desorption parameters for the secondary mineral deposits in basalt fissures are assumed to be the same as those for the packing material. This assumption is based on measured sorption values reported for a sample of secondary minerals from a large basalt vug which indicate that radionuclide distribution in secondary minerals and packing material are similar.[8,9] Ranges of values for the chemical parameters are given in Table 1.

The diffusion coefficient for radionuclides in groundwater saturated packing material was estimated from measurements of tritium and chloride diffusion in this material reported by Relyea et al.[10] Their reported values for 60°-90°C (~ 1×10^{-9} m^2/sec or ~300 cm^2/year) was used in this model. The average groundwater pore velocity in the basalt was calculated

TABLE 1. Chemical Parameters Used in the Model

Nuclide	Inventory, Ci/MTHM[a]	Solubility, moles/liter Expected	Solubility, moles/liter Conservative	K_{ads}, cm^3/g[b] Expected	K_{ads}, cm^3/g[b] Conservative	$\frac{N_{ads}}{N_{des}}$
^{238}U	0.32	10^{-6}	10^{-4}	1000	400	70
^{237}Np	0.95	10^{-7}	10^{-5}	4000	500	61
^{226}Ra	0.001	10^{-10}	10^{-6}	300	50	17
^{99}Tc	13	10^{-6}	10^{-4}	1300	100	65
^{79}Se	0.35	10^{-6}	10^{-3}	50	10	84

[a]Spent fuel at 1000 years.
[b]A linear adsorption isotherm is assumed ($S=K_{ads}C^{N_{ads}}$, where N_{ads} = 1).

to be 10^{-8} cm/sec (0.3 cm/year), assuming a hydraulic conductivity of 3 x 10^{-6} and a head gradient of 10^{-3}. This corresponds to a velocity in the packing material of 0.003 cm/year because of the relationship:

$$A_1v_1 = A_2v_2 \tag{12}$$

where the subscripts 1 and 2 correspond to packing material and host rock, respectively. Since the model considers transport through the packing material and then into basalt fissures filled with secondary minerals, the cross-sectional area of the packing material and basalt columns will be proportional to their bulk porosity. The bulk porosity of basalt is about 100 times less than the packing material and essentially all of this porosity is in the fissures. Therefore, A_1/A_2 = 100 and v_1/v_2 = 0.01. The value of A_1 is the approximate calculated surface area of the packing material-host rock interface for a cylindrical waste package 80 cm in diameter and 400 cm long. The porosity and bulk density of the packing material are measured values[6]. The secondary minerals in basalt fissures are assumed to have the same values for ρ and θ. The values of all the physical parameters used in the model are summarized in Table 2.

IV. RESULTS AND CONCLUSIONS

Sorption hysteresis for a given radionuclide in packing material will only occur after the inventory of that radionuclide in the waste form has been depleted. For radionuclides having large inventories, (e.g., uranium-238), depletion over the time period of interest (10,000 years) is not likely. For these elements, sorption and desorption behavior is of secondary importance and the most significant chemical parameter is radionuclide solubility[6]. In this case release rates and cumulative releases are proportional to solubilities. For many key radionuclides, (neptunium-237, technetium-99, selenium-79, radium-226, iodine-129, and carbon-14), however, inventories are low enough and/or solubilities are high enough to deplete the initial waste form inventory over the time

TABLE 2. Physical Parameters Used in Model

Parameter	Value Used for Packing Material	Value Used for Basalt	Description
D	300 cm^2/yr	300 cm^2/yr	Diffusion coefficient for a radionuclide in packing or basalt infilling
v	0.003 cm/yr	0.3 cm/yr	Average pore velocity of the groundwater
ρ	1.8 g/cm^3	1.8 g/cm^3	Bulk density of the packing or basalt infilling
θ	0.3 cm^3/cm^3	0.3 cm^3/cm^3	Porosity of the packing or basalt infilling
X	20 cm	100 cm +	Thickness of the packing or basalt
A	100,000 cm^2	1,000 cm^2	Cross-sectional area of the packing or basalt infilling

period when the present model is applied. If the radionuclides sorb appreciably, most of the inventory is retained on the packing material. This feature is illustrated in Figure 1 for technetium transport. The inventory of technetium is depleted at about 1000 years into the release period. After this time, sorption hysteresis lowers the release rate significantly and increases the amount of technetium sorbed in packing material near the canister-packing material interface. Effects of sorption hysteresis on release rates and cumulative releases for technetium and other radionuclides are summarized in Table 3. Sorption hysteresis has the most impact on releases of radionuclides having combinations of high solubilities, high K_{ads} values, and low inventories. Each of these conditions increases the probability that the entire inventory of a radionuclide will be sorbed on the packing material.

The results of analyses using combinations of expected and conservative values for solubility C_0 and K_{ads} are shown in Table 4 for the radionuclides. Since uranium has the highest molar inventory of any radionuclide in spent fuel, desorption does not occur from packing material over the ten thousand year release period. The maximum release rates and cumulative releases for uranium are proportional to the solubility and are relatively insensitive to K_{ads}, as suggested by Relyea and Wood[6] for an unlimited radionuclide inventory. For the other radionuclides, desorption occurs for at least some combinations of C_0 and

TABLE 3. Effects of Sorption Hysteresis on Release Rates and Cumulative Releases

Nuclide	C_o, M	K_{ads}	$\frac{N_{ads}}{N_{des}}$	Maximum Fractional Release Rate, Yr-1	Cumulative Release[a] Ci/MTHM	CR[b]/EPA Limit
^{237}Np	10^{-5}	4000	1	4.42×10^{-7}	2.94×10^{-3}	2.94×10^{-2}
			61	6.00×10^{-8}	4.14×10^{-4}	4.14×10^{-3}
^{99}Tc	10^{-4}	1300	1	1.60×10^{-6}	1.24×10^{-1}	1.24×10^{-2}
			65	9.64×10^{-8}	6.94×10^{-3}	6.94×10^{-4}
^{226}Ra	10^{-6}	300	1	1.98×10^{-5}	6.62×10^{-5}	6.62×10^{-4}
			17	2.69×10^{-6}	8.88×10^{-6}	8.88×10^{-5}
^{79}Se	10^{-3}	50	1	4.39×10^{-5}	3.44×10^{-2}	3.44×10^{-2}
			84	1.62×10^{-6}	7.75×10^{-4}	7.75×10^{-4}

[a]Over 10,000 years
[b]CR = Cumulative release over 10,000 years

K_{ads}. When desorption occurs, it reduces both the maximum release rates and cumulative releases. For example, neptunium and technetium releases are lower than expected by several orders of magnitude when desorption occurs. As shown in Table 4, this model predicts that none of these radionuclides will exceed NRC or EPA release limits, even though the effects of other engineered barriers (waste form and container) are not considered in the analysis. The two maximum fractional release rates calculated for selenium-79 (at $C_o=10^{-6}$) are greater than the NRC limit of 10^{-5}. This is because the isotopic fraction of selenium-79 (approximately 0.1) was ignored in the calculations. If this isotopic fraction is taken into account, the release rates are lowered by about a factor of 10.

Effects of packing material thickness on radionuclide release were also examined. Figure 2(A) shows maximum neptunium release rates for packing thicknesses of 1-50 cm. As expected, the slope of the curve is steep for small thicknesses and then begins to level off at about 20 cm. The shape of this curve is obviously dependent on values of C_o, K_{ads}, and inventory for a given radionuclide. However, similar curves could be used to specify the required thickness of backfill in waste package designs.

The basalt column used in this model is an approximation of a semi-infinite column. The longer the basalt column used in the model, the closer it approximates a semi-infinite column. This must be balanced against long computer run times for longer columns. Several runs were made to determine the minimum necessary basalt column length for the model. Figure 2(B) shows that no significant difference in the results for neptunium were obtained with basalt columns longer than about 10 cm.

TABLE 4. Effects of Radionuclide Solubilities C_o and K_{ads} on Release Rates and Cumulative Releases

Nuclide	C_o, M	K_{ads}	Desorption	Maximum Fractional Release Rate[a], Yr-1	Cumulative Release[a] Ci/MTHM	Cumulative Release[a] CR[b]/EPA Limit
^{238}U	10^{-6}	1000	no	4.64×10^{-10}	1.19×10^{-6}	1.19×10^{-5}
		400	no	4.62×10^{-10}	1.05×10^{-6}	1.05×10^{-5}
	10^{-4}	1000	no	4.64×10^{-8}	1.19×10^{-4}	1.19×10^{-3}
		400	no	4.61×10^{-8}	1.05×10^{-4}	1.05×10^{-3}
^{237}Np	10^{-7}	4000	no	2.61×10^{-8}	1.14×10^{-4}	1.14×10^{-3}
		500	no	3.32×10^{-8}	2.37×10^{-4}	2.37×10^{-3}
	10^{-5}	4000	yes	6.00×10^{-8}	4.14×10^{-4}	4.14×10^{-3}
		500	yes	2.89×10^{-6}	4.16×10^{-3}	4.16×10^{-2}
^{226}Ra	10^{-10}	300	yes	1.45×10^{-6}	4.34×10^{-6}	4.34×10^{-5}
		50	yes	2.91×10^{-5}	3.33×10^{-5}	3.33×10^{-4}
	10^{-6}	300	yes	2.69×10^{-6}	8.88×10^{-6}	8.88×10^{-5}
		50	yes	1.53×10^{-5}	3.43×10^{-5}	3.43×10^{-4}
^{99}Tc	10^{-6}	1300	no	2.81×10^{-7}	2.88×10^{-2}	2.88×10^{-3}
		100	no	2.77×10^{-7}	2.16×10^{-2}	2.16×10^{-3}
	10^{-4}	1300	yes	9.64×10^{-8}	6.94×10^{-3}	6.94×10^{-4}
		100	yes	9.47×10^{-6}	6.73×10^{-2}	6.73×10^{-3}
^{79}Se	10^{-6}	50	yes	2.74×10^{-5}	4.61×10^{-3}	4.61×10^{-3}
		10	no	3.52×10^{-5}	5.51×10^{-2}	5.51×10^{-2}
	10^{-3}	50	yes	1.62×10^{-6}	7.75×10^{-4}	7.75×10^{-4}
		10	yes	8.97×10^{-6}	1.25×10^{-3}	1.25×10^{-3}

[a] Over 10,000 years
[b] CR=cumulative release rate over 10,000 years

In summary, it has been shown that sorption hysteresis in packing material can have a significant effect on releases of several key radionuclides. In some cases, release rates and cumulative releases are reduced by several orders of magnitude. Sensitivity measurements show that sorption hysteresis is most important when the entire inventory of a radionuclide can be sorbed in the packing material. The thickness of the packing material in waste package designs may be specified by estimating the releases of key radionuclides. Finally, even though the waste form and container were not considered in these calculations, releases estimated by the model for the radionuclides studied were within NRC and EPA release limits.

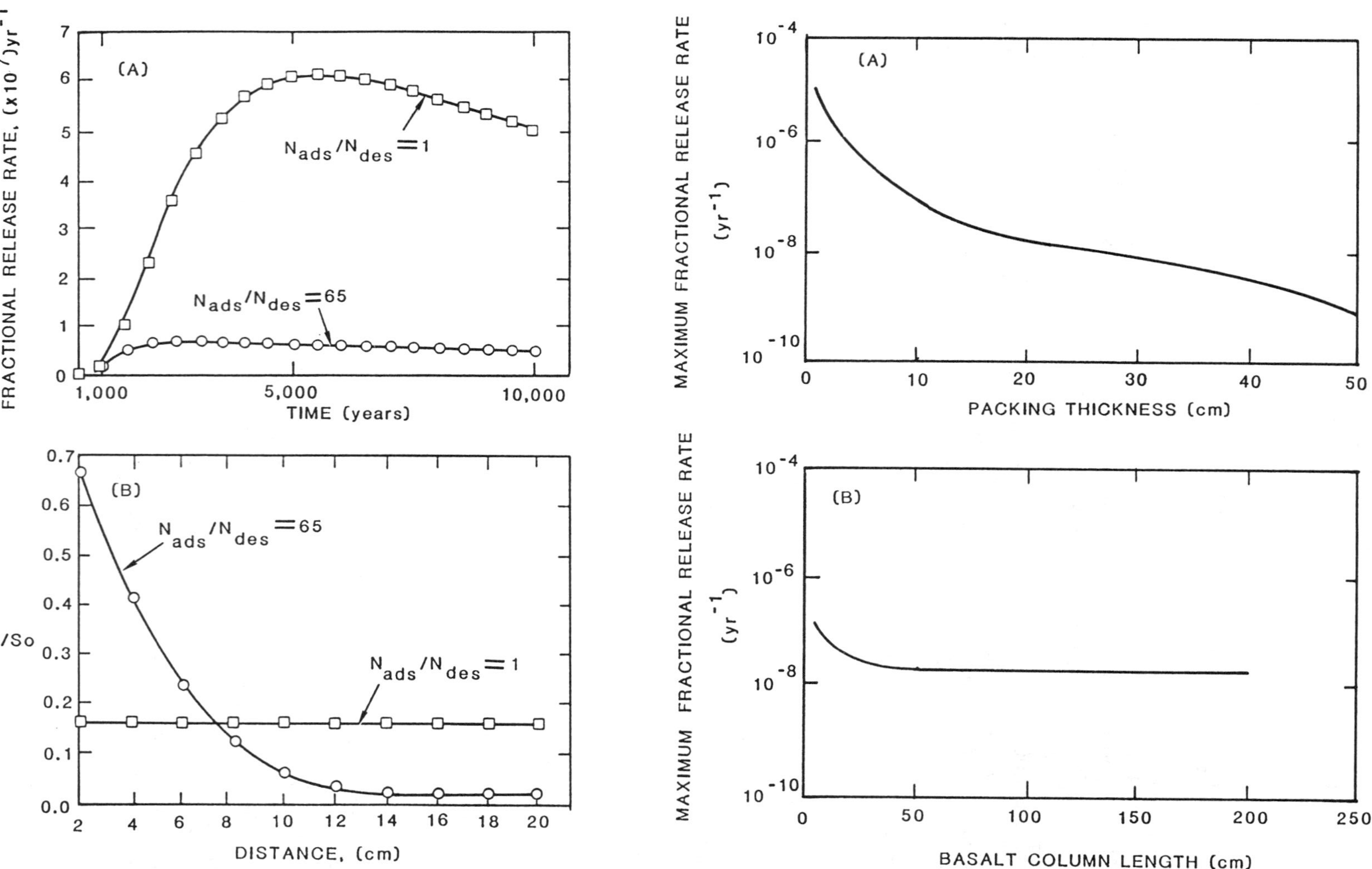

Figure 1. Effects of Sorption Hysteresis on Fractional Release of Technetium Over 10,000 Years(A) and Amount of Technetium After 10,000 Years As a Function of Distance Into the Packing Material(B) (S_o is the amount sorbed at $C=C_o$; $C_o=10^{-4}\underline{M}$, $K_{ads}=1300$)

Figure 2. Effects of Packing Material Thickness(A) and Basalt Column Length(B) on Maximum Fractional Release Rates of Neptunium ($Co=10^{-5}\underline{M}$, $K_{ads}=4000$)

REFERENCES

1. NRC, "Disposal of High-Level Radioactive Wastes in Geologic Repositories: Technical Criteria," 10 CFR 60, Code of Federal Regulations, U.S. Nuclear Regulatory Commission, Washington, D.C.

2. EPA, "Environmental Radiation Protection Standards for Management and Disposal of Spent Nuclear Fuel, High-Level and Transuranic Radioactive Waste," Draft 5, 40 CFR 191, Code of Federal Regulations, U.S. Environmental Protection Agency, Washington, D.C.

3. G. S. Barney et al., Sorption and Desorption Reactions of Radionuclides with a Crushed Basalt-Bentonite Packing Material, RHO-BW-SA-416 P, Rockwell Hanford Operations, Richland, Washington (1985).

4. M. Th. van Genuchten and P. J. Wierenga, "Simulation of One-Dimensional Solute Transfer in Porous Media," Bulletin 628, New Mexico State University Agricultural Experiment Station, Las Cruses, New Mexico (1974).

5. M. Th. van Genuchten, J. M. Davidson, and P. J. Wierenga, "An Evaluation of Kinetic Equilibrium Equations for the Prediction of Pesticide Movement Through Porous Media," Soil Sci. Soc. Amer. Proc., 38, 29 (1974).

6. J. F. Relyea and M. I. Woods, An Analytical One-Dimensional Model for Predicting Waste Package Performance, SD-BWI-TI-232, Rockwell Hanford Operations, Richland, Washington (1984).

7. P. F. Salter and G. K. Jacobs, BWIP Data Package for Reference Solubility and K_d Values, SD-BWI-DP-001, Rockwell Hanford Operations, Richland, Washington (1983).

8. G. S. Barney, Radionuclide Reactions With Groundwater and Basalts from Columbia River Basalt Formations, RHO-SA-217, Rockwell Hanford Operations, Richland, Washington (1981).

9. P. F. Salter, L. L. Ames, and J. E. McGarrah, Sorption of Selected Radionuclides on Secondary Minerals Associated with the Columbia River Basalts, RHO-LD-43, Rockwell Hanford Operations, Richland, Washington (1981).

10. J. F. Relyea et al., Diffusion of Tritium and Chloride in Basalt-Bentonite Mixtures, RHO-BW-SA-431 P, Rockwell Hanford Operations, Richland, Washington (1985).

11. W. W. Loo and R. C. Arnett, Effective Porosities of Basalt: A Technical Basis for Values and Probability Distributions Used in Preliminary Performance Assessments, SD-BWI-TI-254, Rockwell Hanford Operations, Richland, Washington.

AUTOMATED SENSITIVITY ANALYSIS OF THE RADIONUCLIDE MIGRATION CODE UCB-NE-10.2

Francois G. Pin
Brian A. Worley
Edward M. Oblow
Richard Q. Wright
Oak Ridge National Laboratory
P.O. Box X
Oak Ridge, Tennessee 37831

William V. Harper
Office of Nuclear Waste Isolation
Battelle Project Management Division
505 King Avenue
Columbus, Ohio 43201

ABSTRACT

The Salt Repository Project (SRP) of the U.S. Department of Energy is performing ongoing performance assessment analyses for the eventual licensing of an underground high-level nuclear waste repository in salt. As part of these studies, sensitivity and uncertainty analyses play a major role in the identification of important parameters, and in the identification of specific data needs for site characterization. Oak Ridge National Laboratory has supported the SRP in this effort resulting in the development of an automated procedure for performing large scale sensitivity analysis using computer calculus. GRESS, GRadient Enhanced Software System, is a pre-compiler that can process FORTRAN computer codes and add derivative taking capabilities to the normal calculated results. The GRESS code is described and applied to the code UCB-NE-10.2 which simulates the migration through a sorption medium of the radionuclide members of a decay chain. Conclusions are drawn relative to the applicability of GRESS for more general large-scale modeling sensitivity studies, and the role of such techniques in the overall SRP sensitivity/uncertainty program is detailed.

I. INTRODUCTION

Since nuclear waste repositories must by necessity be designed and analyzed for safety by analytic methods, an assessment program is essential for establishing the level of confidence that can be placed in such a design. In a classical engineering design, verification is usually supplied by prototypic experiments, the results of which can also be used to test analytic methods. Since it is not possible to verify the processes occuring in a repository experimentally, considering the time span needed, other approaches must be developed. Sensitivity and uncertainty analysis is one of the possible alternatives to experimental risk assessment. The purpose of a sensitivity analysis is to relate the uncertainty in "key" design parameters to an uncertainty in the responses that measure design performance.

Analytical methods based on standard statistical techniques and direct numerical methods based on perturbation techniques are available to treat sensitivity analysis problems. When applied to large sensitivity studies, however, these methods have major drawbacks. The statistical methods are generally unable to handle all parameters constituting a model, and their specificity to a given model or code requires considerable analytical development to analyze a reduced problem. As a consequence, direct numerical methods have traditionally been used to generate sensitivity results in studies involving large computer codes. In these methods, sensitivities are estimated by observing the changes in the results of interest resulting from perturbation of selected parameters, hence requiring a considerable number of model runs. Sensitivity analyses with these methods are, therefore, limited solely by the size of the data base whose sensitivities are of interest.

An automated procedure based on the use of computer calculus was developed to perform large-scale sensitivity analyses. The procedure was embodied in a FORTRAN pre-compiler called GRESS, which automatically processes computer codes adding derivative-taking capabilities to the normal calculational scheme. All sensitivities of interest can be obtained in a single run from the caculated derivatives. The procedure allows derivatives of any real variable used in the code to be calculated with respect to another variable or an input parameter. In this paper, we describe the GRESS code and apply it to UCB-NE-10.2, a code which simulates the migration through a sorption medium of the radionuclide members of a decay chain. Sensitivities are calculated and analyzed for a sample problem and the GRESS procedure is validated using analytical and perturbation analysis results.

II. THE GRESS PRE-COMPILER

GRESS[1] is a FORTRAN pre-compiler that enhances conventional FORTRAN programs with analytic differentiation of arithmetic statements. Basically, GRESS reads the FORTRAN source code text, redefines the variable storage locations, searches for arithmetic statements, enhances these for gradient calculation, and then generates a new FORTRAN source program that can produce derivatives for any real variables if desired. GRESS thus allows any standard FORTRAN code to be enhanced to allow the calculation of any required derivatives, whether they be for use internally (e.g., for iteration acceleration) or externally (e.g., for sensitivity studies).

The independent variables (referred to as parameters), with respect to which GRESS will generate partial derivatives in the enhanced FORTRAN source code, may be selected by the user. The only restriction to this choice is that the sequence of computations in the source code must be continuous and in differentiable form for the derivatives generated to have any practical meaning. In the enhanced code the derivatives are calculated according to the chain rule of differential calculus and are propagated from operation to operation analytically by implicit differentiation rules. Inherently, therefore, no numerical difference scheme is used and the derivatives are as accurate as the analytical computations in the reference code. In processing the original code with GRESS, the analytic formulas representing the derivatives are neither generated explicitly nor saved. At any stage during enhanced program execution, therefore, the only additional information made available to the code is the numerical value of the derivatives of interest. In this regard, GRESS automatically manages the storage requirements of these enhanced computations so that memory is allocated as dynamically and efficiently as possible.

From the viewpoint of GRESS, there are three classes of variables: independents (i.e., parameters), dependents, and constants. The parameters must be explicitly declared (although this declaration is automated if desired), and may be dynamically altered during the course of execution. Conversely, a variable whose value is computed from either another dependent variable or a parameter becomes a dependent variable and acquires a derivative. If it already was a dependent variable its derivative is modified accordingly. When it is recalculated solely from parameters and constants, it is redefined and loses its prior derivatives. All real input variables are defined to be parameters if the automatic parameter definition option is chosen.

To make use of GRESS results in a practical manner, the dependent variables used in the source code being enhanced should be differentiable. This is the only practical limitation in GRESS use which requires some words of caution. The requirement for differentiability is a simple one but its violation is not always easily detected. In essence, the program should be written so that an unbroken chain of real (floating point) algebraic calculations link the dependent and independent variables. If this is so, then in principle, if not in practice, it should be possible to condense the source program into a series of enormous equations, one for each dependent variable, in which one or more of the independent variables are used. The main limitation which this description conceals is that integer-mode variables cannot propagate derivatives (i.e., by their very nature they are nondifferentiable). They may be freely used in the calculations, but it must be understood that they act merely as constants (i.e., as if they were replaced by their integer values in the equations). On the other hand, the continuity requirement is really rather modest since derivatives are evaluated only at the discrete values attained by the variables. It is therefore only necessary for the derivatives to exist at these points.

The use of GRESS is illustrated in Figure 1. In the preliminary processing step the input program is separated into two subsets. The first, which contains all calculation routines, is hand modified for submission to the GRESS precompiler. The amount and nature of these modifications depends upon the particular application at hand and the limitations of the current version of GRESS. The second subset (possibly null), is composed of subroutines whose only communication with the first subset is through the arguments in their calling sequence. These subroutines are usually associated with input, output, and peripheral program analysis functions. They do not require GRESS compilation and may usually be submitted unchanged to the FORTRAN compiler.

The GRESS precompilation step is the one in which the additional code translation necessary to compute derivatives is performed using automated computer calculus. A rearrangement of the program data structure and a

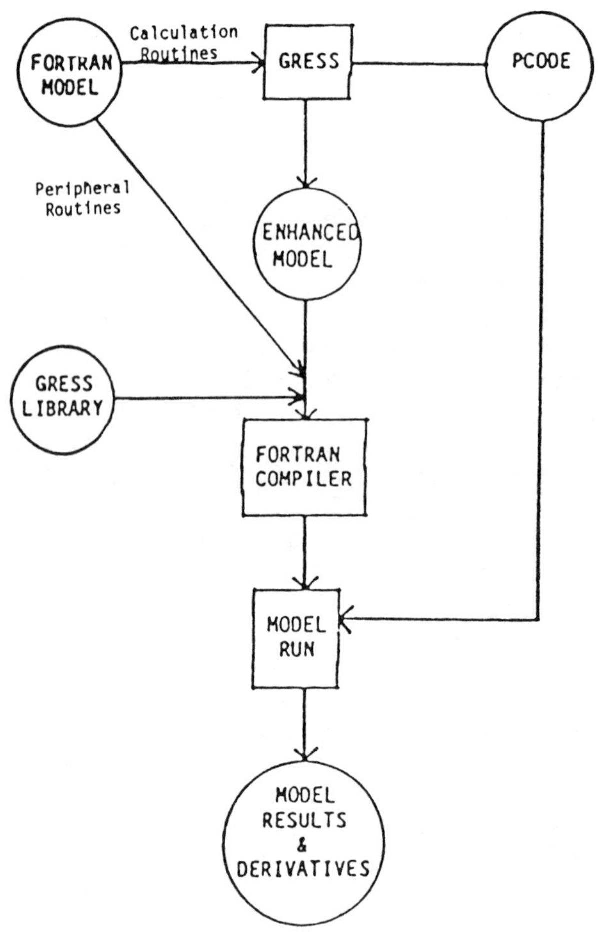

FIGURE 1. Flowchart of Procedures for the Use of GRESS

substitution of calls to the GRESS interpretive software are performed automatically by GRESS for all arithmetic lines of coding. The modified program is then compiled into pseudo-machine code (the GRESS P-CODE) for use during program execution. The two output files of this step are therefore, the enhanced FORTRAN code and the binary P-CODE file.

In the final stage of the GRESS procedure, the transformed FORTRAN subroutines are combined with the untranslated subset of subroutines and both are submitted to the normal FORTRAN compiler. The relocatable object modules which result from actual compilation are then input to the system link-edit loader, which combines them with appropriate portions of the GRESS interpretive library and the P-CODE file to form the complete executable program. The library contains the P-CODE interpreter, a series of support routines, and a set of utility subroutines which may be directly referenced by the translated program to display or manipulate derivatives. The calls to these utility routines are added, as needed, to the original source code before the precompilation step. The resulting translated program will execute identically to the original source program with the option of also calculating derivatives.

III. THE UCB-NE-10.2 CODE

The UCB-NE-10.2 code is one of the codes that SRP plans to use for evaluating deep waste repository systems. It was developed to simulate the migration in groundwater of the radionuclide members of a decay chain. The code is based on the analytical solutions to the one-dimensional transport equation including longitudinal dispersion and adsorption. These solutions for band (i.e., uniform for all nuclides), exponential (for all nuclides), or preferential (i.e., nuclide specific) release modes and various boundary conditions are derived and discussed in References 2 and 3. In the present version of the code, the radionuclide source can be characterized in two different ways using a plane source boundary condition or a concentration boundary condition. The source boundary condition results from specifying a time-dependent dissolution rate of the nuclides at the waste location, assumed to be embedded within an infinite medium. The concentration boundary condition results from specifying the time-dependent concentration of the nuclides at the waste location, on the surface of an infinite half-space in which nuclides are assumed to migrate.

The analytical solutions to the one-dimensional transport equation are given by linear combination of exponential and complementary error functions.[2,3] UCB-NE- 10.2 calculates these solutions by rational approximation[4] or infinite series expansion for a fixed distance from the source, z, or a fixed time, t. Thus the output of the code consists of the time-dependent normalized nuclide concentrations at the fixed location, z, or the spatial distribution of the normalized concentrations at the fixed time, t, depending on the option selected by the user. In the code, the normalized concentration of a nuclide i is defined as the ratio $C_i(z,t)/C_1(0,0)$ of the groundwater concentration of nuclide i to the initial leaching concentration of nuclide 1, the parent nuclide in the chain.

The input data of the code includes parameters characterizing the type of problem to be treated (number of nuclides, ranges of time and distance, fixed z or t, source type, and release mode), parameters describing the geohydrology of the medium (dispersion coefficient, flow velocity and flow rate), and parameters defining the nuclides characteristics (name, half-life, retardation, i.e., sorption coefficient in the medium, initial concentration, and time of leaching). This set of parameters represents the data base whose sensitivities are of interest in a problem to be solved by the code.

A sample problem involving a decay chain of three radionuclides was used for this study. In this problem, the radionuclides, ^{234}U, ^{230}Th, and ^{226}Ra are initially present in the waste with respective concentrations of 1.0, 0.01, and 0.004 C_i/m^3, and have respective half-lives (λ_i) of 2.445 × 10^5, 7.7 × 10^4, and 1.6 × 10^3 yr, and respective retardation factors (R_i) of

3.0×10^2, 2.0×10^4 and 1.0×10^4. A band release mode, with leaching occurring between 0 and 10^5 yr, and a concentration boundary condition are used to characterize the source. The groundwater velocity (v) and flow rate (g), and the dispersion coefficient (D) in the medium are 1.0 m/yr, 1.0 m^3/yr and 50.0 m^2/yr, respectively. The fixed location option with z = 500 m is selected.

The code results for this sample problem are given, in graphical form, in Figure 2. In the figure, the dimensionless concentrations C_i (z = 500 m, t)/C_1(0,0) are plotted versus time for each nuclide. These results are consistent with the analyses presented in References 2 and 3. Due to its relatively low retardation factor the ^{234}U concentration peak is the first to reach the fixed location point. The travel time of the peak is 1.5×10^5 yr which corresponds to the time required for the peak to travel 500 m at the ^{234}U migration speed ($v_i = v/R_i$) of 3.33×10^{-3} m/yr. Due to its long half life, ^{234}U can migrate a relatively long distance without significant decay, therefore showing a high ratio of concentration at the peak 500 m from the source. The migration speed of ^{226}Ra is greater than that of its precursor ^{230}Th. However, in spite of its greater migration speed, ^{226}Ra cannot survive long in the absence of its parent because of its short half-life. Its concentration pattern is, therefore, strongly controlled by the concentration pattern of its parent, and the arrival of its concentration peak closely follows the arrival of the ^{230}Th peak. In fact, with migration speeds of 5.0×10^{-5} and 1.0×10^{-4} m/yr respectively, the ^{230}Th and ^{226}Ra that have leached from the source should arrive at the z = 500 m location at times of 1.0×10^7 and 5.0×10^6 yr. At these times, ^{230}Th and ^{226}Ra have gone through 130 and 3125 half-lives, respectively, and have almost entirely decayed. The peaks that are observed at the z = 500 m location therefore correspond exclusively to ^{230}Th and ^{226}Ra that result from decay of ^{234}U during its migration. This means that the decay chain is controlled only by its parent nuclide and is expected to migrate over long distances because of the long half-life and the relatively large migration speed of the parent nuclide.

The dashed lines in Figure 2 show the solutions for ^{234}U in the case of no dispersion and no decay (square function) and in the case of no decay. As can be seen from comparing the curves, the strong dispersion considered in the sample problem (D = 50 m^2/yr) has considerably smoothed the concentration profile. The ^{234}U in the frontal part of the profile has gone through only 1/5 or 1/4 of a half-life when it reaches the z = 500 m location and, therefore, does not show significant effects of decay. Conversely, in the trailing edge of the profile, ^{234}U has gone through two or four half lives when reaching z = 500 m and the effect of decay is increasingly larger. The results of the code were also verified for conservation of mass and showed only small errors well within the range of the truncation errors from the numerical scheme. Before being submitted to the GRESS procedure and in order to insure compatibility with the GRESS pre-compiler, the data statements in the code were regrouped within a data block and the input and output files were renamed.

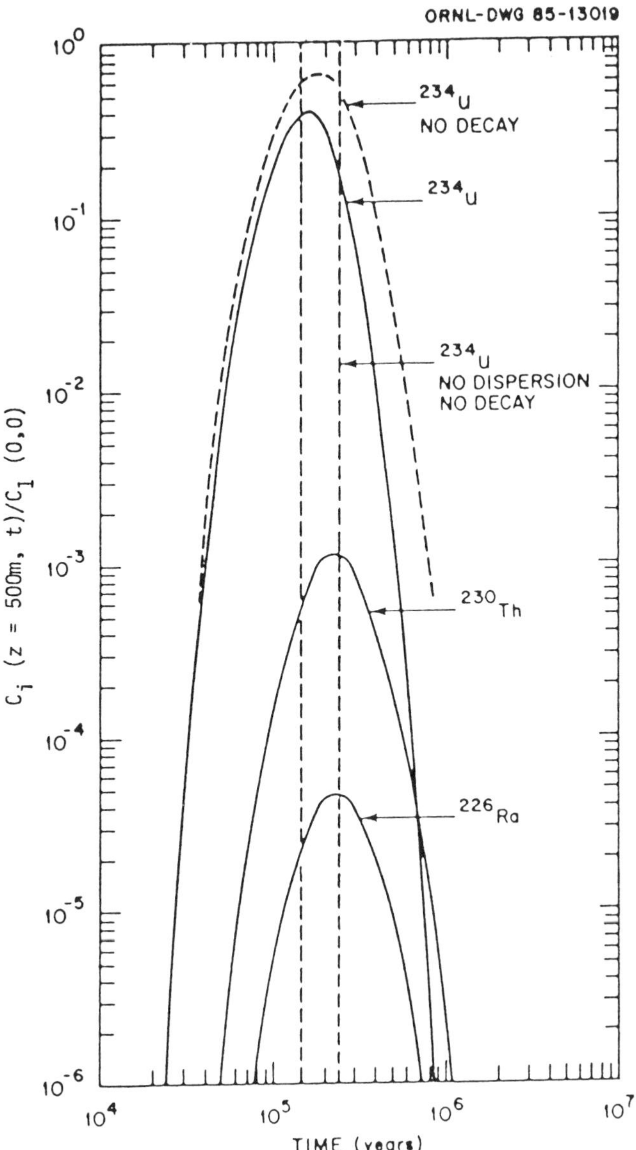

FIGURE 2. ^{234}U, ^{230}Th, and ^{226}Ra Dimensionless Concentrations Versus Time at the z = 500 m Location

IV. GRESS SENSITIVITY CALCULATIONS

The sensitivity of a calculated variable y to an input parameter or code datum x can be expressed by a normalized sensitivity coefficient defined by the ratio of the percent change in y per percent change in x. Examination of sensitivity coefficients reveals valuable information about the influence of data, input parameters, and even physical assumptions upon the calculated variables. In addition, sensitivities are a necessary tool to evaluate the uncertainties associated with the results of a given problem as a function of the uncertainties in the data. Ranking the sensitivities reveals the most crucial data, and ranking the products of the sensitivity coefficients times the respective data uncertainties identifies the potential leading sources of uncertainty in the calculated variables.

As mentioned earlier, the GRESS precompiler automatically calculates partial first derivatives for all data, input parameters, and dependent variables. Based on these derivatives, the "localized" sensitivity coefficients of interest are determined. By localized, it is meant that first derivatives and sensitivity coefficients are calculated at a given solution point and therefore, the information provided relates to the importance of the data and to the behavior of the solution only in a neighborhood of the given solution point.

The application of GRESS to the UCB-NE-10.2 computer code was tested for the sample problem described in the preceding section. This problem calculates the time-dependent nuclide concentrations of ^{234}U, ^{230}Th, and ^{226}Ra at a distance of 500 m from the waste source, relative to an initial concentration of ^{234}U at the source location. Thus, the sensitivity of the three nuclide ratios $^{234}U(z,t)/^{234}U(0,0)$, $^{230}Th(z,t)/^{234}U(0,0)$, and $^{226}Ra(z,t)/^{234}U(0,0)$ to the input parameters used to describe the physical properties of the sorping medium and the transporting fluid, the radionuclide half-lives, and the leaching mode were selected for the sensitivity calculation and verification study.

In the case considered here, the application of GRESS to the UCB-NE-10.2 code allowed normal results and sensitivities of 60 performance measures with respect to 18 parameters to be calculated in a single run using 8 times the CPU time required for running the original UCB-NE-10.2 code. Note that the calculation of sensitivities using reruns would have required at least 18 runs in this case. Moreover, the run-timing results indicate that the additional CPU time is almost entirely spent in GRESS library routines setting up and interpreting the information contained in the P-code. Since the P-code represents, in logical form, the arithmetic equations in the original model, the major part of the additional CPU time represents the overhead necessary to set up the machinery needed to compute derivatives for each arithmetic operation. Once the overhead is spent, the CPU time required to perform the added load of differentiation is only weakly linear with respect to the number of parameters defined.

The efficiency of GRESS, therefore greatly increases with the number of parameters considered for the study. The calculation of derivatives also requires additional core memory to store the P-code and the calculated gradients. Although the additional core requirement is problem- and code-dependent (in this case a factor of two was needed) experience[5-6] has shown that the maximum storage overhead needed by GRESS amounts to only a factor of two to three times the storage requirements of the original code.

This study was performed as part of the validation study of the GRESS procedure and included visual verification of the GRESS enhancement of the

UCB-NE-10.2 code, and verification of the GRESS sensitivity calculations. The GRESS-calculated sensitivities were verified by comparisons with reruns. Note that the first derivatives at a point can only be approximated by rerunning, and then only for small perturbations in the input parameters because of the possible influence of nonlinear effects in going from one solution point to another. By progressively decreasing the magnitude of the change in a particular input parameter, we were able to verify the accuracy of the GRESS-calculated sensitivities to better than 0.01% for all performance measures and input parameters defined previously.

V. RESULTS AND DISCUSSIONS

Tables 1-3 list selected results of sensitivity calculations at t = 10^4, 10^5, and 10^6 years respectively. In these tables the "performance measures" are the three nuclide concentration ratios. The normalized sensitivity coefficients are the percent changes in the three nuclides ratios per percent change in the input parameters. For example, in Table 1, a one percent change in the ^{234}U retardation factor is estimated to cause a -37.8% change in the concentration ratio of ^{234}U at z = 500 m and t = 10^4 years. This sensitivity coefficient is negative since an increase in the retardation factor would delay the arrival of the front at the z = 500 m location and result in lower concentrations at time t = 10^4 yr. The coefficient is large since a slight delay in the arrival of the front would correspond to a large difference in concentration due to the steep slope in the frontal part of the concentration pattern (see Figure 2). The sensitivity coefficients of the three nuclide concentrations to the ^{234}U retardation factor are found nearly equal since, in the frontal part of the migration pattern (t = 10^4 yr), ^{230}Th and ^{226}Ra are created exclusively from local ^{234}U decay. Also note that, because radioactive decay is an irreversible process, the retardation factor of a nuclide of the chain influences only the daughter nuclides and has no influence on the parent nuclides (indicated by sensitivity coefficients of zero in Tables 1, 2, and 3). This is also observed for the sensitivity coefficients to the nuclide half-lives at all times (Tables 1, 2 and 3).

At t = 10^4 yr, only the very frontal part of the ^{234}U concentration profile has reached the z = 500 m location. That part of the profile results exclusively from dispersion of the original band profile (Figure 2). For this reason, an increase in the dispersion coefficient would disperse the contaminant further downstream and increase the concentration at the z = 500 m location. This is indicated by the large positive sensitivity coefficient of the ^{234}U concentration ratio to the dispersion coefficient. Similarly, the large positive sensitivity coefficient of the ^{234}U concentration ratio to the flow velocity indicates that an increase in the flow velocity at t = 10^4 yr would advect the high values of concentration further downstream and increase the concentration at the z = 500 m location. Sensitivity coefficients for ^{230}Th and ^{226}Ra to the dispersion coefficient and the flow velocity are nearly equal to those for ^{234}U since, in the frontal part of the concentration profile, ^{230}Th and ^{226}Ra are created exclusively from ^{234}U decay and their concentration profiles are controlled by the ^{234}U profile. In Tables 1 and 2, the

TABLE 1. Normalized Sensitivity Coefficients at z = 500 m, t = 1.0 x 10^4 Years*

Nuclide Ratio	$\frac{^{234}U\ (z,t)}{^{234}U\ (0,0)}$	$\frac{^{230}Th\ (z,t)}{^{234}U\ (0,0)}$	$\frac{^{226}Ra\ (z,t)}{^{234}U\ (0,0)}$
Unperturbed Value	5.77×10^{-16}	6.26×10^{-21}	2.53×10^{-23}
INPUT PARAMETERS			
Half-Life			
^{234}U	0.028	-0.972	-0.973
^{230}Th	0.0	0.002	-0.998
^{226}Ra	0.0	0.0	0.094
Retardation Factor			
^{234}U	-37.829	-37.760	-38.557
^{230}Th	0.0	-1.015	-0.015
^{226}Ra	0.0	0.0	-1.028
Dispersion	33.154	34.093	34.912
Velocity	4.675	4.683	4.689
Leach Time	0.0	0.0	0.0

*Percent change in the nuclide ratios per percent change in selected input parameters.

TABLE 2. Normalized Sensitivity Coefficients at z = 500 m, t = 1.0 x 10^5 Years*

Nuclide Ratio	$\frac{^{234}U\ (z,t)}{^{234}U\ (0,0)}$	$\frac{^{230}Th\ (z,t)}{^{234}U\ (0,0)}$	$\frac{^{226}Ra\ (z,t)}{^{234}U\ (0,0)}$
Unperturbed Value	1.78×10^{-1}	1.41×10^{-4}	5.31×10^{-6}
INPUT PARAMETERS			
Half-Life			
^{234}U	0.283	-0.757	-0.762
^{230}Th	0.0	0.130	-0.874
^{226}Ra	0.0	0.0	0.901
Retardation Factor			
^{234}U	-3.054	-2.877	-2.968
^{230}Th	0.0	-1.014	-0.014
^{226}Ra	0.0	0.0	-1.003
Dispersion	0.714	1.239	1.299
Velocity	2.340	2.652	2.686
Leach Time	0.0	0.0	0.0

*Percent change in the nuclide ratios per percent change in selected input parameters.

TABLE 3. Normalized Sensitivity Coefficients at z = 500 m, t = 1.0 x 10^6 Years*

Nuclide Ratio	$\frac{^{234}U\ (z,t)}{^{234}U\ (0,0)}$	$\frac{^{230}Th\ (z,t)}{^{234}U\ (0,0)}$	$\frac{^{226}Ra\ (z,t)}{^{234}U\ (0,0)}$
Unperturbed Value	3.32×10^{-8}	1.92×10^{-6}	8.15×10^{-8}
INPUT PARAMETERS			
Half-Life			
^{234}U	2.835	-0.338	-0.338
^{230}Th	0.0	6.900	5.879
^{226}Ra	0.0	0.0	1.021
Retardation Factor			
^{234}U	15.697	2.131	0.057
^{230}Th	0.0	-0.943	-1.000
^{226}Ra	0.0	0.0	0.139
Dispersion	10.499	13.856	0.139
Velocity	-26.196	-1.326	-1.326
Leach Time	2.145	1.345	1.345

*Percent change in the nuclide ratios per percent change in selected input parameters.

sensitivity coefficients to the total leach time T are both found to be zero. This is consistent with the analytical solutions to the transport problem (see References 2 and 3) and the input value of T = 10^5 y. Before leaching ceases, T does not appear explicitly in the solution equations since the nuclide concentration ratios depend on prior leaching only and not on future events such as the end of the leaching period. After leaching ceases (i.e., $\lambda t > T$), the Bateman coefficients, B_{ij}, are adjusted by the factor $e^{-\lambda_j T}$ (see Equations 2.1.18 and 2.1.19 in Reference 3) and partial derivatives with respect to T propagate through the gradient calculations. At t = 10^6 years (Table 3), the sensitivity coefficients to the leach time are found to be positive since an increase in the leach time would extend the high concentration ratios to the trailing edge of the concentration profiles.

The sensitivity coefficients at times $t = 10^5$ and $t = 10^6$ yr, shown in Tables 2 and 3 are also consistent with the physical combined phenomena of advection, diffusion, adsorption and decay. For example, the sensitivity coefficients to the dispersion coefficient reach a minimum at the concentration peak where the effects of dispersion are the least, and increase away from the peak in the frontal zone and trailing edge zone of the concentration profiles where the concentration ratios are increasingly dependent on dispersion. The change in sign and magnitude of the sensitivity coefficients to the flow velocity and the ^{234}U retardation factor from 10^4 to 10^6 yr can be interpreted in similar ways since they both effect the major advection parameter, i.e., the migration speed ($v_i = v/R_i$) of the parent nuclide which controls the chain. An increase in flow velocity or a decrease in sorption will result in faster downstream migration of the high concentrations. If this faster migration occurs before arrival of the peak (e.g., 10^4 years), it would result in bringing higher concentrations at the observation point. If it occurs after the passage of the peak (e.g., 10^6 years), it would tend to advect high concentrations away from the observation point. Finally, since decay is expressed as an exponential function of the half-life, the normalized sensitivity coefficients of a nuclide concentration with respect to the nuclide half-life is observed to increase steadily with time. In fact, for the parent nuclide ^{234}U, the normalized sensitivity coefficient S = /C dC/d can be easily calculated as S = (t ln 2)/ = $2.835 \times 10^{-6} \times t$ with which the results of Tables 1 through 3 agree perfectly.

In addition to this physical insight into the problem solution, the sensitivity coefficients provide a means of quantifying uncertainties in the solution given data uncertainties. Consider the values shown in Table 1. At time $t = 10^4$ yr, a one percent uncertainty in the retardation factor of ^{234}U will result in about a 38% uncertainty in the nuclide concentrations. Unfortunately, the retardation factor is probably only known to within an order of magnitude. Thus, the product of the sensitivity coefficient times the uncertainty in the value of the ^{234}U retardation factor gives an uncertainty in the nuclide concentrations of 38 x 1000% = 38,000%. The uncertainty may well be less in reality, but even if the half-lives were only known to an order of magnitude, the uncertainty in the solution at $t = 10^4$ yr would still be swamped by the assumed ^{234}U retardation factor. The product of sensitivity coefficients times the uncertainties makes clear where additional research is most beneficial for the purposes of reducing the overall uncertainty in the calculated nuclide concentrations. Considering the half-lives are known to within a few percent, and fluid velocity and dispersion coefficient to within a factor of about five, the major sources of uncertainty in this sample problem arise from the uncertainty in the ^{234}U retardation factor and from the uncertainty in the dispersion coefficient. These parameters would therefore be the most important in planning or performing the site characterization activities to gather data and in interpreting the results of the code for a site performance assessment.

VI. SUMMARY

The GRESS automated sensitivity analysis procedure has been presented. The procedure makes use of the GRESS pre-compiler and computer calculus to automatically add derivative taking capabilities to FORTRAN computer codes. The procedure has been applied to the UCB-NE-10.2 code and the sensitivity analysis results have been verified using analytical and perturbation analysis methods. The calculated sensitivity coefficients are directly applicable to uncertainty analysis for site characterization planning or performance assessment. This study, together with similar studies using other source codes,[5,6] provide a validation of the GRESS procedure and demonstrate its efficiency and applicability to a wide range of FORTRAN computer codes. Future work will include addition of a machine independence capability to the GRESS pre-compiler and restructuring of the core/storage allotment for automatic optimization and direct adjoint generation.

REFERENCES

1. E. M. Oblow, An Automated Procedure for Sensitivity Analysis Using Computer Calculus, ORNL/TM-8776, Oak Ridge National Laboratory, Oak Ridge, Tennessee 37831 (1983).

2. M. Harada et al., Migration of Radionuclides Through Sorbing Media Analytical Solutions - I, ONWI-359, LBL-10500, UC-70 (1980).

3. T. H. Pigford et al., Migration of Radionuclides Through Sorbing Media Analytical Solutions - II, ONWI-360, LBL-11616, UC-70 (1980).

4. SAND77-1441, Sandia Mathematical Program Library (1978).

5. F. G. Pin et al., Automated Sensitivity Analysis Using the GRESS Language, ORNL-6148 (1985).

6. F. G. Pin et al., GRESS Translation of the ORIGEN2 Code, ORNL/TM-9694 (1985).

THE INFLUENCE OF THE DAMAGED ZONE, INTERFACE, AND VARIOUS SEALING COMPONENTS ON SHAFT SEAL PERFORMANCE FOR A REPOSITORY IN BASALT

R. A. Lundstrom
J. B. Case
P. C. Kelsall
S. R. Cullinan
IT Corporation
2340 Alamo S. E., Suite 306
Albuquerque, New Mexico 87106

ABSTRACT

A stochastic finite element groundwater model was developed to study the performance of a seal system for a repository shaft in basalt. The model examined the influence of various seal components on flow through the plug, interface, and damaged zone. Performance was measured using total flow rate, minimum travel time, and a contamination breakthrough curve, which shows the cumulative flow of contaminants through the model at times following introduction of the contaminant at the repository level. Damaged zone flow was found to be important to seal system performance. The breakthrough curve was a particularly useful performance measure.

I. INTRODUCTION

Performance requirements for nuclear waste repositories are presently expressed as the allowable cumulative release of radionuclides to the accessible environment for a 10,000-year period after placement. Depending on factors such as the locations of shafts relative to the waste and to hydraulic gradients, the degree of damage resulting from shaft excavation, and the performance of seals placed in the shafts, flow through the shafts could represent a relatively high proportion of the 10,000-year cumulative release from the repository, especially if travel times through the undamaged host rock approach or exceed 10,000 years. This study examines flow through potential shaft seal systems for a repository in basalt to determine some of the implications the shafts may have for repository performance. The study was supported by the Basalt Waste Isolation Project (BWIP).

The paper has four specific objectives: (1) to examine the relative influence of the three components of the total flow through a shaft seal system--plug flow, interface flow, and damaged zone flow; (2) to examine various measures by which the performance of a seal system might be evaluated; (3) to show how various sealing components improve seal system performance; and (4) to study the uncertainty of system performance given uncertainties in fundamental hydrologic parameters.

It should be emphasized that the purpose of the study was to examine the relative importance of various flow zones and the relative effectiveness of various seal components. Because the purpose was not a rigorous absolute performance assessment it was possible to incorporate a number of assumptions or simplifications:

- steady-state Darcy flow was assumed,
- dispersion was not considered,
- sorption was not considered,
- degradation of seal material properties over time was not considered, and
- engineering judgment was used to determine input parameters.

II. PROBLEM DESCRIPTION

A. Components of Flow

The performance of a shaft seal system is determined by plug flow, interface flow, and damaged zone flow. Plug flow is defined here as flow through engineered materials placed in the shaft opening. Interface flow is flow along the interface between the plug and the adjacent rock. Conceptually, the interface could act as a cylindrical fracture surrounding the plug with a microscopic aperture providing a small flow area. Damaged zone flow is flow through a zone of rock surrounding the shaft opening which has been damaged during excavation. Since flow in basalt occurs primarily through fractures, rock damage could increase conductivity by increasing the number of fractures (perhaps as a result of blasting) or by widening existing fractures by relieving stress across them.[1,2]

B. Performance Measures

Three aspects of seal-system performance may be influenced by changing seal designs: (1) the total flow rate through the seal system, (2) the minimum travel time from the repository through the seal system to the environment, and (3) the discharge rate in flow paths which have relatively short travel times. These aspects are illustrated by the conceptual contamination breakthrough curve in Figure 1. This curve is a plot against time of contaminated discharge at a specified point. The curve assumes that contaminants placed at the base of the shaft at time zero take various paths through the seal system. They travel at the seepage velocity which will vary among flow paths. Flow paths contributing to the contaminated discharge are those in which contaminants have reached the specified end point by a given time. The remaining paths represent the uncontaminated discharge. At the minimum travel time (point 1), contaminated discharge increases from zero to a value equal to the discharge through the fastest travel path. Contaminated discharge continues to increase as contaminants in other, slower paths reach the specified point at later times. The curve levels off when contaminants have reached the specified end point in all travel paths (Point 2).

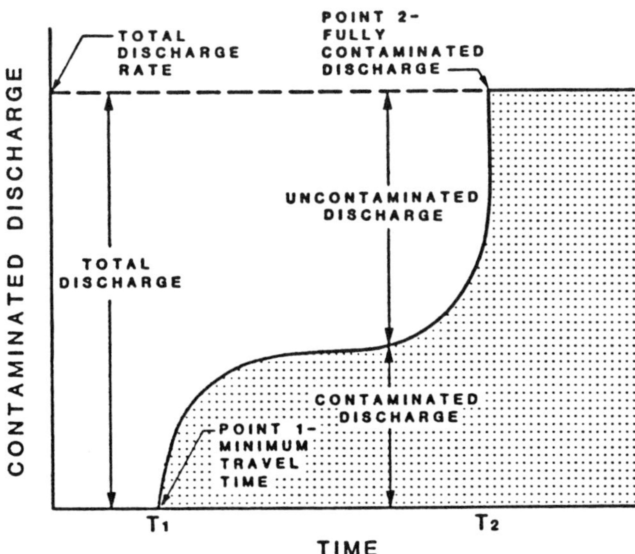

FIGURE 1. Conceptual Contamination Breakthrough Curve

C. Seal System Components

Several different types of sealing components were studied for possible use to reduce flow through the plug, interface, or damaged zone. Backfills having a low conductivity (within an order of magnitude of the rock conductivity) were studied as a means to control plug flow. Earthen backfills also have a high effective porosity relative to basalt resulting in low seepage velocities. Concrete bulkheads were considered, with keyways excavated carefully through the zone of blast damage near the shaft wall, to reduce flow through the shaft and the blast damaged zone. While shaft diameter enlargement in this manner may extend the zone of stress relief damage into the rock, the compensating advantage of cutting off flow through the highly conductive blast damage zone may improve overall seal performance. The study also considered bulkheads with drilled cutoffs[2] excavated through the zone of stress relief. Conceptually, drilled cutoffs might be constructed by drilling overlapping holes into the rock perpendicular to the shaft. Each hole would be filled with concrete and the concrete allowed to cure before drilling the adjacent holes. In this way, only a small volume of rock would be exposed to stress relief at any one time. Drilled cutoffs should reduce flow through the stress-relieved zone and lengthen the interface flow path.

D. Effect of Uncertainty

Performance requirements for a repository presently include a probability of conformance. This requires evaluation of the influence of the uncertainty in basic parametric values on the level of confidence in predicted results. For this paper, only uncertainties in geological parameters were considered in order to develop some understanding of their effect on seal system performance. One deterministic model was selected for stochastic analysis using the Monte Carlo method.

III. COMPUTER CODES

A system of computer codes, referred to as the SHAFTFLOW[3] modeling system, was developed to facilitate input preparation, perform the analysis, and interpret the output. Required input to the system includes the problem geometry, material conductivities and porosities, distributions for any parameters which are to be varied stochastically, and guidance on where to project flow paths. A preprocessor randomly generates multiple sets of material parameters according to the specified probability distributions. A finite element groundwater analysis code determines potentials and flow velocities. A postprocessor traces travel paths through the model and calculates travel times, flow rates, and the contamination breakthrough curve. The system includes 4-node continuum elements along with 2-node "line elements" which can be used to model discrete fractures or material interfaces.

IV. DETERMINISTIC MODELS

Several axisymmetric models were developed to include the following features: horizontal stratigraphy based on the stratigraphy at the Reference Repository Location (RRL)[4] at the Hanford site; a shaft excavation; an interface between the shaft and rock; a damaged zone within the rock which included a blast-damaged zone and a stress relief zone; one of several alternative shaft seal systems; and an assumed constant hydraulic gradient.

A. Stratigraphy

The stratigraphy at the RRL consists of a series of flatlying basalt flows with occasional sedimentary interbeds. The model included four basalt flows from the candidate repository horizon to the base of the Vantage Interbed, which was the designated point at which groundwater travel times were measured. Travel time between this point and the accessible environment was not considered. Each basalt flow was idealized as having two distinct zones; a low conductivity flow interior overlain by a more highly conductive and porous flow top. Both zones were modeled as equivalent porous media with defined hydraulic conductivity and effective porosity based upon fracture characteristics. Conductivity values of 10^{-12} and 2.3×10^{-7} m/s were used for the flow interiors and flow tops respectively. These are similar to values reported in the Draft Environmental Assessment.[4] Corresponding effective porosity values were assumed to be 10^{-4} and 5×10^{-3} respectively based on guidance by BWIP staff.

For purposes of the stochastic analysis, each hydraulic conductivity and porosity was assumed to be log-normally distributed. The stochastic analysis assumes that the effective hydraulic porosity is functionally dependent on the independent hydraulic conductivity according to the cubic law. This analysis assumed that fracture frequency and fluid kinematic viscosity are known constants. Standard deviation values (on the log of conductivity or porosity) were assumed, based on guidance by BWIP staff. The values which were used were 1.0 and .34 for flow interior conductivity and porosity, respectively, and 1.85 and .62 for flow top conductivity and porosity.

B. Shaft and Interface

A 4.9-m diameter shaft was considered and assumed to be backfilled throughout. An interface zone was included at the boundary between the rock and the backfill or seal component. The flow characteristics of the interface zone were obtained from an interpretation by the authors of a bench-scale laboratory test on a cement-grout borehole plug emplaced in basalt.[5] Evaluation of this test indicated that flow through the interface predominated over flow through the plug itself, and that the variation of the interface aperture under changing effective stress was similar to that of a single fracture in basalt. The interface was therefore modeled as a deformable fracture in which flow is given by the cubic law relation.[6] A theoretical aperture of 4.4 m was determined from the laboratory test using the cubic law.

C. Damaged Zone

The high-conductivity damaged zone in the rock was considered to result from two mechanisms: stress relief resulting from excavation, and blast damage. Stress relief damage across existing fractures was modeled at 1000-m depth assuming inelastic deformation properties. The reduction of stress across existing fractures was assumed to increase fracture aperture and, consequently, the conductivity. The increase in conductivity was estimated from the results of laboratory tests of fracture conductivity at different confining stresses. Blast damage was assumed to increase the fracture frequency by a factor of ten within a one meter distance from the shaft wall. These two mechanisms were combined to determine a step function defining several zones of increased conductivity in the rock as shown in Figure 2. The conductivity of the damaged zones was expressed as a multiplier to be applied to the undamaged rock conductivity in both the flow interiors and flow tops. Effective porosity was calculated by assuming it is proportional to the cube root of the hydraulic conductivity based on the cubic law.[6] The constant of proportionality was evaluated using the undamaged rock properties. Details of the derivation of damaged zone properties are discussed elsewhere.[1,2]

D. Alternative Seal Designs

The seal systems studied included backfills, bulkheads with keyways, and bulkheads with keyways and drilled cutoffs, as specified on Table 1. Backfills were modeled as porous media with conductivity and porosity values

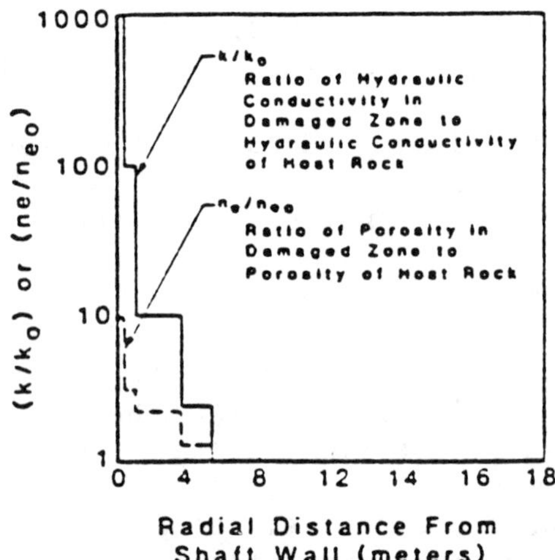

FIGURE 2. Damaged Zone Model

TABLE 1. Summary of Deterministic Results

Run No.	Description	Normalized Flow Rate	Min. Travel Time (yr) Assuming i = .03
0	Flow through an equal area of intact rock	0.3	130,000
1	Backfilled shaft ($K = 10^{-10}$ m/s)	1.0	900
2	Shaft with four bulkheads	0.8	2,200
3	Shaft with four bulkheads with cutoffs	0.8	9,300
4	Backfilled shaft ($K = 10^{8}$ m/s	12.6	50
5	Shaft with two backfills ($K = 10^{-10}$ m/s and 3×10^{-9} m/s)	1.0	6,000

of 10^{-10} m/s and .38 (corresponding to a densely-compacted 25%/75% bentonite/crushed-rock mixture), except where different values are noted. These values are based on preliminary BWIP testing. Bulkheads with keyways were given conductivity values for a concrete with a basalt aggregate (10^{-11} m/s with a porosity of .19).[7] The keyway was assumed to be excavated through the one-meter wide zone of blast damage and backfilled with concrete. Drilled cutoffs, constructed of 0.5 meter diameter overlapping holes, extend radially from a bulkhead through the zone of stress relief to a radius of 10 m. The location of bulkheads within the model for Cases 2 and 3 is shown in Figure 3A. One run was performed to study the effect of placing a zone of increased conductivity within a low conductivity backfill in a shaft. The shaft backfill zoning for this run is shown in Figure 3B.

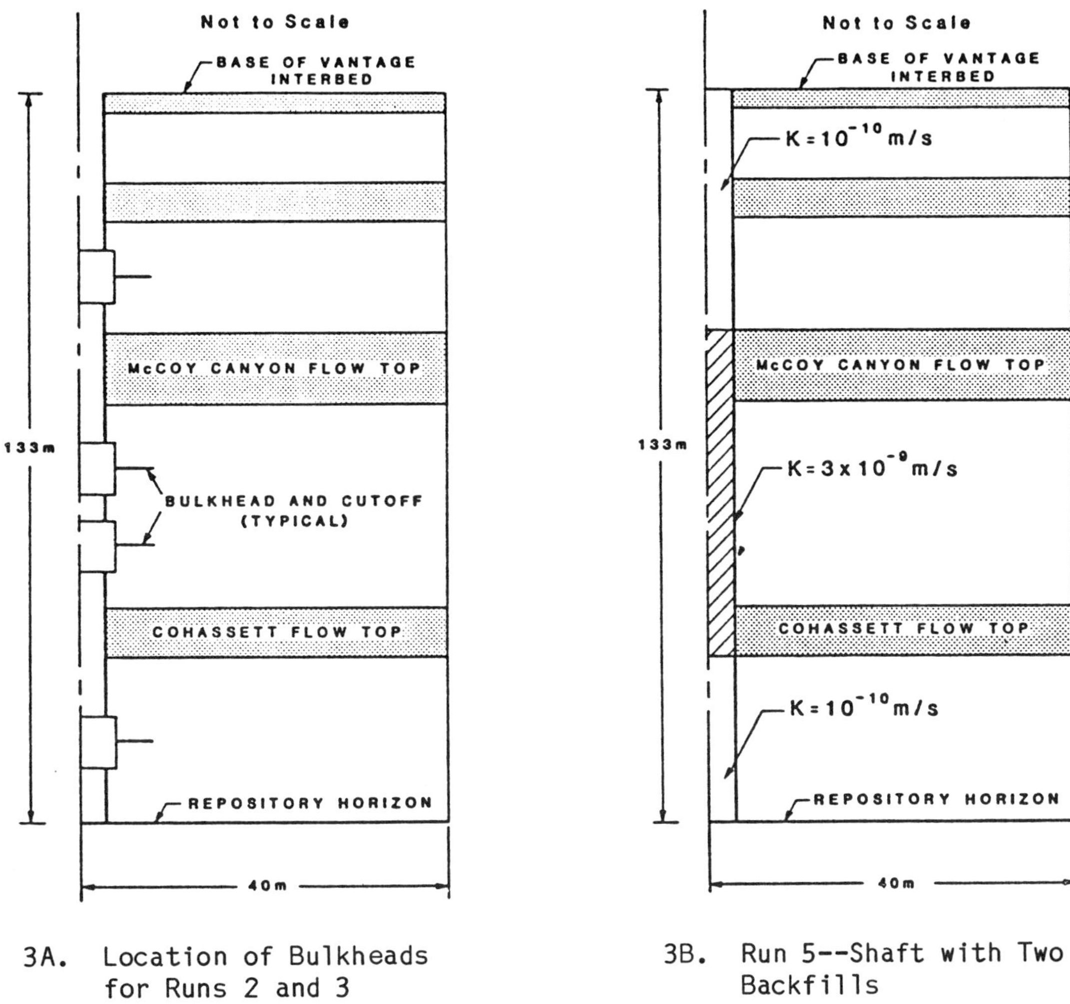

3A. Location of Bulkheads for Runs 2 and 3

3B. Run 5--Shaft with Two Backfills

FIGURE 3. Geometry of Deterministic Models

E. Hydraulic Gradient

The results were obtained for a nominal hydraulic gradient of 0.03, which corresponds approximately to the peak upward gradient due to buoyancy occurring during the thermal cycle of the repository.[4] Constant head conditions were set at the top and bottom of the model to achieve this average gradient and a no-flow condition was specified at the lateral boundaries.

V. STOCHASTIC MODEL

Case 2 (Table 1) was selected for stochastic analysis using the Monte Carlo method. Seal system parameters were held constant and only the characteristics of the basalt were varied stochastically. A total of 61 simulation runs were performed.

VI. DISCUSSION OF RESULTS

Table 1 presents the deterministic results for flow rate and travel time for an assumed hydraulic gradient of 0.03. Figure 4 presents the contaminated discharge curves for selected cases. Discharge in both Table 1 and Figure 4 has been normalized to the total flow for Case 1. The study reached several conclusions regarding the initial objectives.

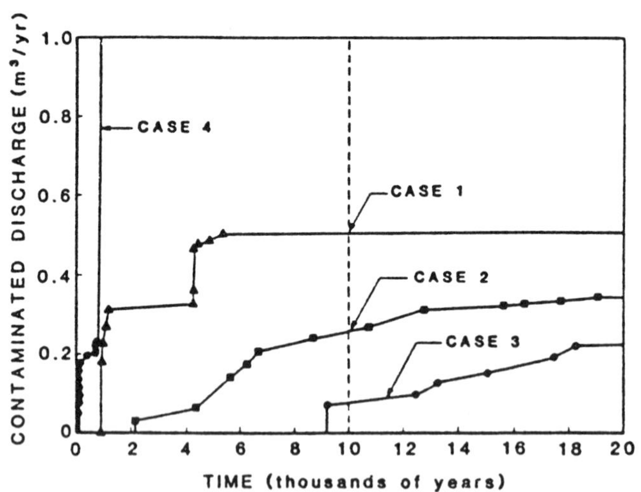

FIGURE 4. Contamination Breakthrough Curves

A. Comparison of the Three Flow Components

A perspective for comparing the various flow components is found by examining the results of Case 0 (flow through undamaged rock). The normalized flow rate was 0.3 while travel time for a gradient of 0.03 was 130,000 years.

The damaged zone flow appeared to be the most important component of flow through the seal system. This is evident from an examination of the results of Case 1 (backfilled shaft) in which the damaged zone flow was 0.25 m^3/year as opposed to 0.04 m^3/year for the backfill and 0.03 m^3/year for the interface. The damaged zone also contained the fastest travel path which was through the zone of blast damage adjacent to the shaft wall. Travel time along this path was two orders of magnitude faster than through the undamaged rock (see Case 0). In addition, the damaged zone conducts virtually all of the contaminated flow arriving at the Vantage interbed at early time. Shaft flow and flow through the undamaged rock both arrive at much later times. The interface flow rate is relatively insignificant. Damaged zone flow may therefore control radionuclide release to the environment in the first 10,000 years.

The damaged zone is the most important flow component even in Cases 2 and 3 which include seal components intended to treat it. Damaged zone flow, while reduced, still exceeded plug flow and interface flow. The critical travel path for Case 2 (Figure 5) follows the blast-damaged zone leaving it only when necessary to circumvent a low conductivity bulkhead. A similar trend was found for travel paths for Case 3.

Plug flow could become an important factor only if low conductivity materials are not used to backfill the shaft. This can be seen in the results of Case 4 in which the backfill conductivity was two orders of magnitude higher than for Case 1. Shaft flow dominated total flow for this case, and total flow was increased significantly.

B. Evaluation of the Three Performance Measures

While total flow rate may serve as an index of seal-system performance it is of little help in evaluating the radionuclide release within 10,000 years since actual contaminated flow may be lower.

Travel time is a fairly sensitive index of seal system performance and the seal system may represent the critical pathway with respect to travel time. This factor alone, however, cannot determine radionuclide release within 10,000 years unless the minimum travel time exceeds 10,000 years assuring no release.

The contamination breakthrough curves give a complete history of the contaminated flow rate which should facilitate evaluation of the radionuclide release rate at any time. The area under the curves is equal to the total volume of contaminated water which has been released through the seal system.

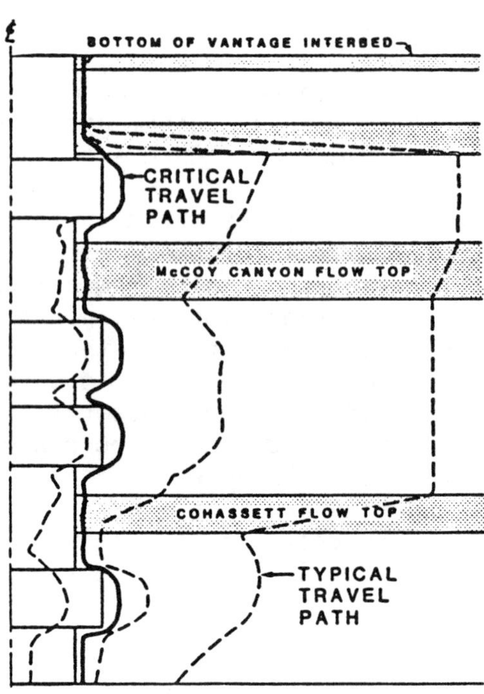

FIGURE 5. Typical Travel Paths - Case 2

This volume should influence the total cumulative radionuclide release. These curves imply that the radionuclide release within 10,000 years can be reduced by either an increase of the travel time or a reduction of the flow rate in the fastest travel paths. Most of the seal components made both types of improvements.

It should be noted that travel times and the breakthrough curves would be changed by an analysis considering dispersion or sorption phenomena. Sorption would increase travel times while dispersion would reduce them. The authors expect that the same general comparative trends would be evident, however.

Regardless of what performance measures are used, flow through the shafts may be quite small compared with total flow through the repository. For the purpose of this comparison, flow through the repository roof was calculated assuming vertical flow through the stratigraphy (Section IV.A) using the same rock hydraulic conductivity values and the same gradient as was applied to the shafts. Assuming a repository area of 4×10^6 m^2, flow through five 4.9m-diameter shafts, with damaged zones and backfilled with low conductivity backfill (10^{-10} m/s), would be only 2 percent of the total flow from the repository. Flow through five similar shafts backfilled with high conductivity backfill (10^{-8} m/s) would be 25 percent of the total.

C. Evaluation of Various Sealing Components

The analyses of various sealing components indicate that each one has the potential to improve seal system performance. This can be seen by comparing total flow rates and travel times in Table 1 and especially by comparing the breakthrough curves in Figure 4. As would be expected, the use of low conductivity backfill resulted in considerable improvement over the use of higher conductivity backfill (compare Case 1 with Case 4). Low conductivity bulkheads (Case 2), especially with cutoffs (Case 3), improved performance further by forcing flow into the lesser damaged zones of the flow interiors, which in turn increased travel times and reduced flow rates in the fastest travel paths. The use of an increased conductivity backfill to interrupt the low conductivity backfill (Case 5) was also effective. This served to draw flow from the damaged zone into the high porosity shaft backfill, where seepage velocities are low, resulting in increased travel time.

D. Evaluation of the Effect of Uncertainty

A cumulative distribution for total flow rate is shown in Figure 6A. It depicts the probability that flow will be less than a given value--or probability of conformance to a given flow rate requirement. The cumulative distribution for travel time presented in Figure 6B depicts the probability that the travel time will be less than a given value--or probability of nonconformance to a given travel time requirement. The distribution does not appear to follow any known standard distribution and no attempt to fit the results to particular distributions was made. Total flow and travel time vary widely (over approximately three and four orders of magnitude respectively) even though the analysis accounted for uncertainties in only the rock characteristics. The travel time value which has a 90% probability of conformance (or 10% probability of nonconformance) is approximately 25 years which is

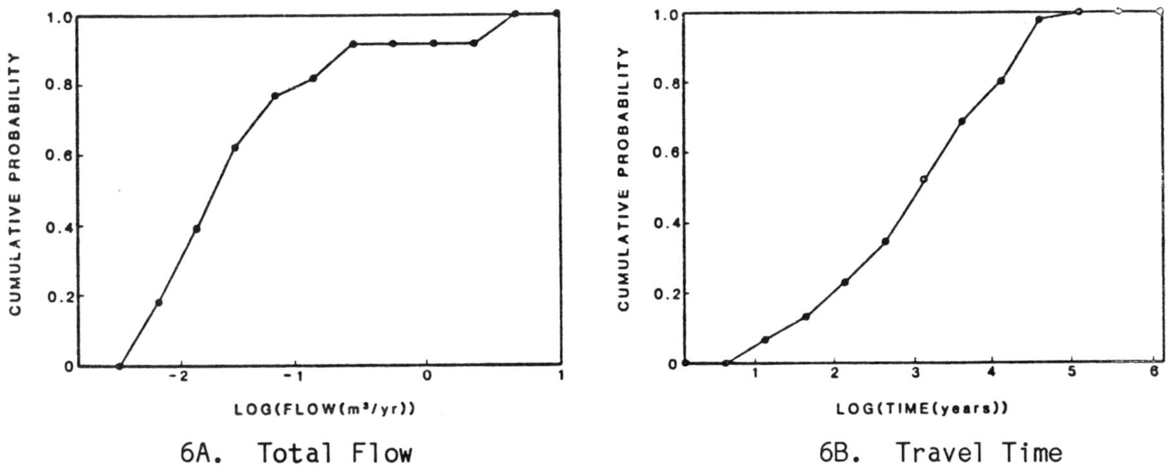

6A. Total Flow

6B. Travel Time

FIGURE 6. Probability Distributions

one to two orders of magnitude lower than the deterministic value. These results must be considered preliminary since only 61 simulation runs were performed. More runs were not considered warranted given the various simplifications and assumptions involved in the analyses.

VII. SUMMARY

The SHAFTFLOW modeling system is a useful tool for optimizing seal designs and evaluating their performance. Flow through the sealed shafts, while small compared to the total flow from the repository, may have a significant influence on the release of radionuclides during the first 10,000 years. Simple seal components such as low permeability backfills can improve shaft performance significantly. It is noted that the damaged zone may have an important effect on seal system performance and that these conclusions are based on analyses which used a preliminary damaged zone model. More site-specific study is needed before final conclusions can be reached.

REFERENCES

1. Kelsall, P. C., J. B. Case, and C. R. Chabannes, A Preliminary Evaluation of the Rock Mass Disturbance Resulting from Shaft, Borehole or Tunnel Excavation, ONWI-411, Office of Nuclear Waste Isolation, Columbus, OH (1982).

2. Kelsall, P. C., J. B. Case, and C. R. Chabannes, "Evaluation of Excavation-Induced Changes in Rock Permeability," International Journal of Rock Mechanics and Mining Sciences, Vol. 21, No. 3, pp. 123-135 (1984).

3. Lundstrom, R. A., J. B. Case, and S. R. Cullinan, User's Manual, SHAFTFLOW, Groundwater Flow Computer Program for Repository Seal Systems, prepared for Rockwell Hanford Operations by IT Corporation, Albuquerque, NM (1985).

4. U.S. Department of Energy, Draft Environmental Assessment--Reference Repository Location--Hanford Site, Washington, DOE/RW-0017 (1984).

5. Lingle, D. R., and D. D. Bush, Full-Scale Borehole Sealing Test in Basalt Under Simulated Downhole Conditions, report prepared for the Office of Nuclear Waste Isolation by Terra Tek, Salt Lake City, UT (1982).

6. Witherspoon, P. A., J. S. Wang, and J. E. Gale, "Validity of the Cubic Law for Fluid Flow in a Deformable Rock Fracture," Water Resources Research, Vol. 16, No. 6, pp. 1016-1024 (1980).

7. Goto, S. and D. M. Roy, "The Effects of W/C Ratio and Curing Temperative on the Permeability of Hardened Cement Paste," Cement and Concrete Research, Volume II, pp. 575-579 (1981).

BASALT WASTE ISOLATION PROJECT PRECLOSURE PERFORMANCE ASSESSMENT: METHODS AND OBJECTIVES

A. Lynn Franklin
Pacific Northwest Laboratory
P.O. Box 999
Richland, Washington 99352

ABSTRACT

A subsurface nuclear waste repository in basalt (NWRB) is being considered at the Hanford site in Richland, Washington. Based on a conceptual design of the NWRB, the facility would accept commercial high level waste (CHLW), spent fuel (SF), and low level transuranic wastes (LL-TRUW) for a period of twenty years. Each canister will be configured for disposal, emplaced in a placement hole and monitored for a period of 50 years in a state of potential retrievability. When the monitoring period is complete, backfill will be placed in the annulus around the waste canister and the surface facilities will be decommissioned.

As part of the Basalt Waste Isolation Project (BWIP), which provides support for the design and licensing of the NWRB, a preclosure performance safety assessment was performed to analyze the:

- initial construction of repository surface facilities,
- mining activities required for construction of shafts, pillars, shaft pillar, and waste panels,
- backfilling of subsurface waste panels, and
- decommissioning of the repository.

The assessment consisted of a system characterization, a preliminary hazards analysis (PHA), an accident scenario analysis, a consequence characterization, and a development of suggestions for accident prevention and consequence mitigation.

I. ANALYSIS PROCESS

Safety analysis activities can interact with facility design activities at several points within the design operations. Figure 1 presents a simplified diagram of several of these review/design interactions. Valid safety review conclusions can be drawn as early in the design process as the preconceptual design stage. Reviews occurring at this time are limited by the detail of descriptive information and by the degree of commitment to particular design

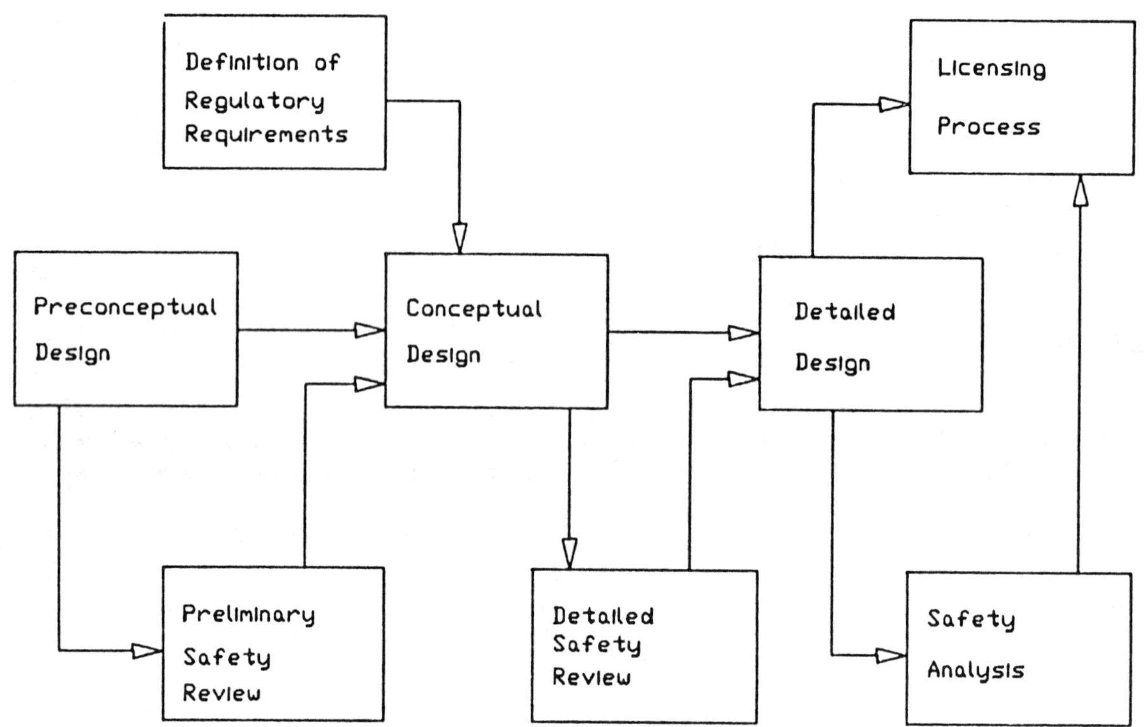

FIGURE 1. Design and Review Process

features. Preliminary reviews are generally directed at prioritizing design activities and providing insight into the compliance with regulatory requirements.

As the design matures, an increased amount of facility information becomes available and a stronger commitment to particular design features begins to form. Because of the increasing commitment to particular design features, the emphasis of the safety review shifts from guiding the design to verifying particular safety system performances. This type of analysis result requires both increased design detail and more elaborate and accurate analysis capabilities. By nature, this analysis becomes more time consuming and expensive. Impacts of the safety review upon the facility design become more subtle as the design is refined.

In preparation for licensing activities, a final safety analysis activity is initiated. This analysis is directed at providing specific quantitative performance information for use in required licensing documentation. In many cases the safety analysis effort will include obtaining and reporting results from specific system performance and verification tests. At this point, the facility design is generally considered fixed and safety analysis becomes a design support activity rather than a design influencing activity.

For each of these levels of safety review and analysis, a similar approach will be used. Figure 2 presents one approach for proceeding with the standard safety analysis. As the design progresses from preconceptual to detailed design, the level of effort in each analysis area will change and the particular tools selected for the analysis will vary. The preliminary safety review will tend toward qualitative analysis tools that provide comparative results that can be used to aid in design tradeoff decisions. As the design becomes more specific, the analysis tools become more quantitative and more rigorous. For analysis in support of the design during licensing activities, the analysis tools must adhere to specific quality controls and their application must be documented with particular detail.

A. Study Objectives

The purpose of the Pacific Northwest Laboratory (PNL) study of the BWIP facility was to provide guidance on where significant opportunities might exist for improving the safety of the facility through focused engineering considerations. This guidance was to take the form of a sorting of facility hazards into ranges which indicated their relative importance to the facility safety. Within the framework of Figure 1, this corresponds to the preliminary safety analysis.

The overall objective of the analysis was to provide comparative risk information for normal and abnormal facility operations to guide later design activities in considering possible safety implications. To provide the proper perspective on potential design considerations and to focus future safety analyses, the PNL study provided an intentionally broad overview of all hazard types including radiation exposure, radioactive material release, and common industrial accidents. The advantage of this approach is the identification of the relative importance of various hazards, independent of their entry into public and design awareness. The disadvantage of this approach is that the majority of the identified hazards turn out to be relatively well understood and readily accepted industrial hazards which can sometimes obscure the importance of addressing all hazards in an independent fashion when attempting to improve the facility safety. Taking into account the status of the BWIP at the time of this study and potential for influencing the overall safety of the facility, the PNL study was structured to preserve the indications of relative safety that would result from a first-pass broad analysis of the BWIP facility.

B. Methods Used

For the BWIP study, the analysis was directed at determining a preliminary ranking of the relative hazards of various proposed facility systems. The purpose of this ranking was to direct later more detailed studies into areas of the facility where the opportunities for reducing hazards appeared to be the most significant. The format of the analysis was selected to be consistent with the knowledge available and the state of the design of the proposed BWIP facility. The nature of the BWIP design suggested a combination of predictive and historical analysis techniques would be most appropriate. Of the principle

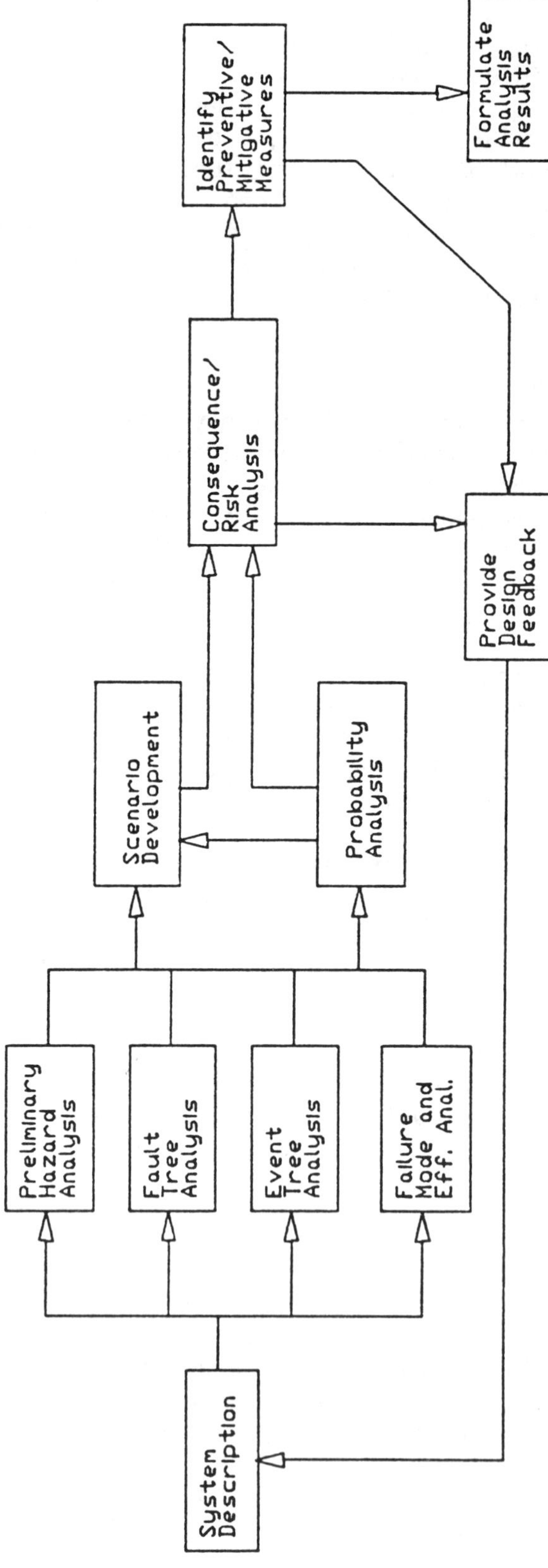

FIGURE 2. Safety Analysis Process

predictive analysis methods available, a preliminary hazard analysis (PHA) appeared to be the most compatible with the information available and the anticipated results. Fault tree and event tree analyses required a commitment to detailed design information that was not appropriate during the study of a conceptual facility. Historical information was used to indicate the relative probabilities and to corroborate the list of identified hazards.

Qualitative estimations of hazard consequence rather than quantitative estimates were chosen as the appropriate approach because of the tentative nature of the reference facility description and the certainty that the facility design would be changed substantially in the near future. Qualitative representations of consequence provided an adequate means of sorting accident scenarios into relative risk categories for directing the search for possible preventive/mitigative measures.

C. Additional Analysis Opportunities

The analysis tools selected for use in this initial BWIP analysis reflected the nature of the analysis information available and the status of the facility design. As additional design detail becomes available, the preclosure performance assessment should reconsider the analysis tools being used. Additional predictive techniques such as fault tree and event tree analysis become more viable as the descriptive information becomes more reliable. Furthermore, as the facility safety analysis matures, enhanced detail and accuracy will become necessary in those areas that indicate, by the PHA, substantial potential for safety improvement.

II. ANALYSIS RESULTS

A. Hazard Ranking

The analysis yielded a qualitative understanding of a variety of potential hazards at the repository. The vast majority of these hazards are common to standard industrial practices and do not reflect unusual hazards associated with repository construction and operation. Their inclusion in the analysis was justified as a means of placing those hazards that are unique to the construction and operation of the repository in proper perspective relative to their overall impact on facility safety. To provide an indication of the importance of each hazard to the overall system, a relative measure of probability and consequence, Table 1, was assigned to each accident scenario. The qualitative descriptions of probability and consequence were combined into three catagories of risk as shown in Table 2. The scenario value assignments are presented in Table 3 according to general risk categories for a portion of the accident scenarios considered in the study.

TABLE 1. Measures of Probability and Consequence

Probability

H - High probability of occurring during the preclosure period
M - Moderate probability of occurring during the preclosure period
L - Low probability of occurring during the preclosure period
U - Unlikely to occur during the preclosure period

Consequence

H - Life threatening injuries, fatality, or potential public exposure
M - Serious injury, fractures, or exposure greater than standards
L - Minor injuries or exposure to personnel

TABLE 2. Risk Category Assignments

	Consequence		
Probability	L	M	H
H	2	3	3
M	1	2	3
L	1	1	2
U	1	1	1

B. Preventive/Mitigative Measures

The information in Table 3 is intended to highlight specific safety issues which appear to provide significant impact on the safety of the conceptual repository design. Issues which were grouped into risk categories 2 and 3 were further examined to determine whether immediately apparent preventive/ mitigative measures could be identified to address the hazard. Risks from falling objects associated with the shafts appear to be adequately addressed and no additional actions were identified. A partial listing of the recommendations developed from this analysis includes:

- prepare subsurface areas with several inches of washed crushed rock to improve moisture drainage and general working conditions,

- place electrical, communication, monitoring, and moisture control systems in the prepared gravel floor to improve tolerance of normal traffic and survival during rock fall,

- adopt roof and wall finishing techniques that preserve necessary ventilation characteristic while maintaining access and visibility to ground support systems,

- provide fluorescent egress system markings to identify emergency routes during emergency situations,

TABLE 3. Accident Scenarios by General Risk Category

Accident Scenario	Prob	Conseq	Risk
Impacts with Moving Vehicles			
- surface equipment	L	M	1
- subsurface equipment	M	M	2
- shaft transport	L	H	2
Rock Falls			
- rock from shaft walls	M	M	2
- rock from tunnel roof	M	H	3
- rock bursts	L	H	2
Falling Objects			
- hoist system failure with personnel	L	H	2
- hoist system failure with waste	L	M	1
- hoist system failure with material	L	L	1
- tools or rock from overhead operations	L	H	2
Machinery-Related Accidents			
- caught in moving/rotating machinery	L	M	1
Fires and Explosions			
- methane pocket explosion	M	M	2
- methane explosion in borehole	H	M	3
- fossil fuel fire	L	M	1
Contaminated Transport Vehicles			
- radiation monitor failure	L	L	1
- operator error	L	L	1
Exposure to Unsealed Hot Cell			
- seal verification indicator failure	L	L	1
- operator error	L	L	1
Unshielded Canisters			
- borehole shield not installed	L	H	2
- uncovered loaded transport vehicle	L	H	2
- personnel in hot cell with unshielded canister	L	H	2
Power Failures			
- localized failure due to faulty distribution system	L	L	1
- general off-site failure	M	L	1
- common cause primary and emergency power failure	L	M	1

TABLE 3. Accident Scenarios by General Risk Category (cont'd)

Surface Ventilation Failures			
- loss of pressure gradients	L	L	1
- plugged or failed filters	L	H	2
Subsurface Ventilation Failures			
- loss of mine heading ventilation	M	M	2
- loss of tunnel flushing ventilation	L	M	1
- cross-connect of mining and confinement ventilation circuits	M	M	2
Fission Gas Release			
- rupture of fuel rod in hot cell	M	L	1
- canister failure during subsurface handling operations	L	M	1

- provide mobile electrical and communications distribution cart at active mine headings,

- prevent automatic startup of electrically powered equipment following temporary power outages,

- install permanent voice communications stations at strategic locations throughout the facility,

- provide a hot cell access port interlock to prevent removing the cask until proper seals are confirmed,

- provide borehole shield plug interlock to prevent removing the shield plug or moving the waste transporter until proper positioning is confirmed, and

- apply luminescent coating to areas with potential to exposure to unshielded waste canisters.

C. Recommended Studies

In addition to identifying specific mitigative measures, the study identified areas where additional analyses would be needed to determine the significance of potential safety hazards. The suggested study areas included:

- perform a comparative analysis to support the selection of conventional shaft and tunnel construction techniques or shaft and tunnel boring machines,

- examine the need for and impact of prompt reversal of air flows to accommodate required methane considerations, and

- investigate the potential advantages, disadvantages, and licensing considerations of a sealed hot cell with a batch ventilation approach.

III. SUMMARY

Several fundamental conclusions can be drawn from consideration of the PNL study. It appears that there may be design opportunities associated with the underground construction and operation of repositories, that are not commonly available to traditional mining activities, which can reduce the hazard potential of the facility. Because underground nuclear facilities are not constrained to following ore seams or minimizing rock removal, the fabrication of the underground facilities can be planned to improve the working conditions and safety related services. These include better design for moisture control, ventilation, communication, and equipment movement.

Additional studies are needed to better characterize the design impact and safety considerations from alternative shaft and tunnel construction techniques. Information from existing mine safety studies needs to be incorporated into the selection of shaft and tunnel construction techniques.

IV. DEVELOPING ANALYSIS METHODOLOGIES

The potential for the presence of methane can impose ventilation requirements on a repository facility that are highly unusual for nuclear facility design. The need for prompt reversal of airflow in gassy mines may dictate the need for an elaborate ventilation/filtration system for confinement of potential radioactive material releases.

In the time since the PNL study was made, additional study tools have been developed or are being developed in related programs. The Monitored Retrievable Storage (MRS) Program has developed an exposure comparison (EXCOMP) analysis tool for providing enhanced as-low-as-reasonably-achievable (ALARA) exposure consideration in the conceptual and definitive MRS design. The structure of EXCOMP makes it adaptable to any arbitrary facility and provides a comparative analysis of the anticipated occupational exposure from general facility activities or specific task related activities. With some reduction in the detail of the results, EXCOMP may be adaptable to examining variations in public exposure due to hypothetical facility modifications.

Additional analysis capabilities are also being developed in the area of artificial intelligence (expert systems). Extremely adaptable modeling capabilities and improved event-tree investigations are currently being developed using expert system techniques. The enhancement to modeling capabilities stems from including qualitative system performance descriptions along with traditional quantitative descriptions in facility models. These qualitative descriptions will also provide the improved opportunity for applying event-tree analysis to facility safety considerations during both the design and operations stages.

MATHEMATICAL MODELING OF WASTE REPOSITORY PERFORMANCE: A PEER REVIEW*

Joseph A. Lieberman
OTHA, Inc.
P.O. Box 686
Glen Echo, Maryland 20872

William W.-L. Lee
R. F. Weston, Inc.
2301 Research Blvd.
Rockville, Maryland 20850

ABSTRACT

This paper describes the rationale, membership, operation and major observations of the Performance Assessment National Review Group. The Group was assembled by Weston at the request of the U. S. Department of Energy Office of Civilian Radioactive Waste Management to review performance assessment work in the U.S. basalt, salt and tuff repository projects. The purposes were to evaluate the adequacy of the current methods, identify deficiencies, and suggest potential improvements on repository performance assessment.

To perform the review, Weston retained a group of distinguished consultants who have had extensive experience in disciplines pertinent to management of radioactive wastes including mathematical modeling of fluid transport. The Group developed a list of topics in performance assessment that it deemed important, and requested briefings and discussions from the Projects on these topics. Topics reviewed included flow and transport, source term and uncertainty analysis. While the emphasis was on methodologies, the Projects were specifically requested to show currently available results so that the way they utilized familiar methodologies could be evaluated.

Two meetings were held in which the Projects, the Environmental Protection Agency (EPA) and the Nuclear Regulatory Commission (NRC) made presentations to and engaged in discussions with the Group. While there was some subsequent exchanges, most of the Group's review was based on the presentations and discussions. The Group's final report was transmitted to Weston and the U.S. Department of Energy in March 1985. All parties concerned reviewed the draft report and the Group considered the comments received in producing the final report.

This paper will highlight some of the technical observations of the Group as well as some managerial and institutional issues.

---

*Work supported in part by U. S. Department of Energy contract DE-AC01-83-NE44301. Opinions expressed herein are those of the authors.

INTRODUCTION

This paper presents the rationale, membership, procedures and the major findings of the Performance Assessment National Review Group. The Group was assembled in 1984 to provide a peer review of the performance assessment activities in the basalt, salt and tuff repository projects in U.S. Department of Energy's civilian radioactive waste management program. The Group carried out its activities under the direct sponsorship of R.F. Weston, Inc., the technical support contractor to US DOE's Office of Civilian Radioactive Waste Management (OCRWM). The first author of this paper was the chairman of the Group and the second author was the executive secretary of the Group. The final report of the Group is available from the authors[1].

The fundamental function of a high-level nuclear waste management system is to contain and isolate the radioactive materials so they cause no undue harm to man or his environment, either now or in the future. Because this requirement extends over many millenia, it is obvious that assurance of fulfillment of the requirement cannot be demonstrated by operational experience. It can best be appraised by a systematic, quantitative evaluation using all relevant field and laboratory data combined with the most up-to-date analytical methodologies and expert scientific judgment. The term used to describe all the steps involved in predicting the potential radiologic impact of a nuclear waste disposal system, taking into account all engineered and natural components of the system, is "performance assessment." Performance Assessment includes the analysis and evaluation of predicted system and component performance to determine compliance with regulatory performance criteria. Performance assessment is the major available tool to accomplish informed technical decisions regarding siting, design, construction, operation and licensing of high-level nuclear waste repositories. More specifically, it was recognized that the application of performance assessment methodologies has five major purposes:

1. evaluation and selection of repository sites;
2. guiding, and setting priorities for supporting programs of research, development, and testing;
3. development and evaluation of repository designs;
4. design and development of engineered barriers; and
5. qualification and licensing of the waste management system.

RATIONALE

The establishment of the Performance Assessment National Review Group had two major origins. First, the Waste Isolations Systems Panel (WISP) of the National Research Council/National Academy of Sciences/National Academy of Engineering published a report entitled A Study of the Isolation System for Geologic Disposal of Radioactive Wastes[2]. The WISP concluded that the selection and licensing of sites for geologic repositories for disposal of high-level wastes would depend heavily on predicted performance of the waste isolation system. The WISP recommended a continuing and systematic technical review of the program with emphasis on relating program efforts to the specific goal of developing a satisfactory method of predicting repository performance.

Second, the US DOE, recognizing the importance of performance assessment in its repository program, especially in eventual licensing, needed an interim evaluation of the status of performance assessment work in the repository projects. Weston, the technical support contractor to US DOE's OCRWM, suggested that such a review could be most effectively accomplished by a group of internationally recognized experts - the Performance Assessment National Review Group. The objectives of the PANRG were to: 1) provide an independent review of the performance assessment work currently being conducted by the basalt, salt and tuff projects, and 2) recommend to Weston and the US DOE any improvements in performance assessment which may significantly contribute to the demonstration of compliance with relevant regulatory standards, and the technical defensibility of the work.

MEMBERSHIP

In order to carry out this review, Weston retained under subcontracts a number of experts, who were:

- knowledgeable in the scientific basis for nuclear waste management,
- knowledgeable in mathematical modeling of radionuclide migration,
- knowledgeable in the licensing process, and
- without significant conflict of interest.

The members of the Review Group were:

Dr. Joseph A. Lieberman, Chair
OTHA, Inc.
P.O. Box 686
Glen Echo, MD 20812

Dr. Stanley N. Davis, Professor
Department of Hydrology and Water Resources
University of Arizona
Tucson, AZ 85721

Dr. Donald R. F. Harleman
Ford Professor of Engineering
Massachusetts Institute of Technology
Cambridge, MA 02139

Dr. Ralph L. Keeney
Professor of Systems Science
University of Southern California
Los Angeles, CA 90089

Dr. David C. Kocher
Health and Safety Research Division
Oak Ridge National Laboratory
P.O. Box X
Oak Ridge, TN 37831

Dr. Donald Langmuir, Professor
Department of Chemistry and Geochemistry
Colorado School of Mines
Golden, CO 80401

Mr. Robert B. Lyon
Acting Director, Nuclear Fuel Waste Management
Atomic Energy of Canada, Ltd.
Pinawa, Manitoba ROE 1L0
Canada

Mr. William W. Owens
Consultant
6269 S. Knoxville
Tulsa, OK 74136

Dr. Thomas H. Pigford, Professor and Chair
Department of Nuclear Engineering
University of California
Berkeley, CA 94720

Dr. William W.-L. Lee (Executive Secretary)
R. F. Weston, Inc.
2301 Research Blvd
Rockville, MD 20850

PROCEDURES

The review was conducted in the following way. The Review Group decided beforehand on the topics they would like to review and requested briefings on these topics by the Projects and relevant DOE Field Offices. While the focus of the review was upon methodology, the Projects were specifically requested to present available data and results. The reason for this was that many approaches to performance assessment are reasonably well known, but in order to evaluate the quality of work by the Projects, it was necessary to examine how they were using well-established techniques and the results that were being obtained. At the first meeting, a general orientation on performance assessment activities and plans was presented by the Projects, and staff of both the NRC and EPA described their related regulatory and standard setting activities. At the second meeting the Projects gave technical presentations in response to an itemized request for briefings. The topics reviewed included the definition and development of the source term, the modeling of flow and transport of radionuclides, and uncertainty/sensitivity analysis of input data. The Group indicated that the response to the briefings request could be oral, written or both.

It must be recognized that the comments in the Group's report were based for the most part, but not entirely, on the oral presentations by the Projects and the discussions during and following these presentations. Accordingly the results of the review must be recognized as a snapshot in time of the performance assessment work in the Projects. Although most presentations included hand-outs of viewgraphs used in the presentations, there were few documented results of performance assessment work provided to the PANRG. This is not necessarily a criticism, but it states the basis for the report. A draft of the Group's report was provided to DOE-HQ, the Field Offices, the Projects, EPA and NRC for comments. The Group considered the comments received in preparing its final report.

MAJOR FINDINGS

The overall conclusion of the PANRG was as follows. Current performance assessment methodologies are still in the developmental stage. Only the simplest of bounding calculations have produced quantitative predictions of radionuclide releases. The methodologies require considerable extension and validation before they can provide answers suitable for major project decisions and licensing. There seems to be preoccupation with method development, without sufficient testing of the models against other predictive techniques and without much use of the models to develop quantitative data for project direction. Nevertheless, the PANRG noted its confidence that greater recognition of the importance of performance assessment as a project management tool, plus improved management of performance assessment activities by the Projects and by OCRWM Headquarters, can provide satisfactory results when needed.

The Review Group paid special emphasis to the work of the Projects in showing compliance with relevant performance requirements of the Environmental Protection Agency and Nuclear Regulatory Commission. A summary of such evaluations is in Table I.

Observations and recommendations of the PANRG fall into two categories - (1) those related to management/institutional factors and (2) those of a technical nature. The PANRG report contains ten observations in the former category and thirteen in the latter. These are highlighted below.

Management/Institutional Factors

- Because the license application is several years in the future and few performance assessment results are available at this time, the Group sought to determine whether, in its judgment, capable staffs are available to do the job. PANRG observed that competent, professional people appear available to carry out the necessary performance assessment work. While the level of experience of individuals is variable, the

TABLE I Summary of PANRG Evaluations

	REQUIREMENTS				
	CONTAINMENT IN WASTE PACKAGE	RELEASE RATES	TRAVEL TIME	CUMULATIVE RELEASE	UNCERTAINTY ANALYSIS
AUTHORITY	10 CFR 60.113(a)(1)(ii)(A)	10 CFR 60.113(a)(1)(ii)(B)	10 CFR 60.113(a)(2)	40 CFR 191	
ALL	Assuming only uniform corrosion a probable licensing issue. Need defensible basis for extrapolation to geologic time scale.	Need to obtain site-specific solubilities of chemical species of relevant radioelements.	Need definition of compliance with this requirement. Need to emphasize that it is mass flux that will transport radioactive elements, pore velocity may not be significant.	Need to identify the nature of release scenarios and their associated probabilities.	Should consider sharing approaches and methodologies.
SALT	Assuming only uniform corrosion and extrapolating to geologic time scale, probable licensing issue.	Good bounding calculations based on physical nature of salt, and several postulated mechanisms of transport. Need site-specific data and a defensible theory for brine migration.	Current estimate of brine diffusion coefficient from a single observation may be questioned in licensing.	Assumes complete containment of radionuclides. Bounding analyses indicate within EPA limits. Need to consider modeling of seals and plugs.	Combination of adjoint and random sampling potentially powerful technique. Need to see practical application to performance assessments.
BASALT	Presentation to PANRG included only a model for uniform corrosion of canister/overpack. Based on this information, the Group considers the approach probably inadequate for licensing.	Bounding calculations initiated. Presentation to PANRG included a one-dimensional mass transfer model through the packing material for which more complete versions are available.	Good models in search of concepts and actual hydrologic data.	Bounding calculations indicate no problem except hydrologic transport system not well defined. Need to consider modeling of seals.	Second-order plus random sampling techniques potentially powerful. Look forward to applica- in practice.
TUFF	Location in a favorable environment helps. Dripping of water from unsaturated rock not a credible corrosion scenario.	Incomplete current understanding of complex physical system precludes accurate predictions.	Flow in partially saturated, fractured rock is difficult problem. Appears to have a good technical approach. Can benefit from some work in other areas.	Bounding calculations indicate no problem if estimate of infiltration rate is correct.	Presentation indicates no established methodologies.

adverse effects of this variability can be offset by providing contact with off-project specialists in academia and industry as well as providing for inter-project dialogue. Three kinds of experience appear to be lacking at present:

(1) experience in making reliable and defensible predictions of long-term performance of engineered and natural systems;

(2) experience in addressing specific issues related to repository performance and bringing them to closure, especially in a regulatory environment;

(3) experience in evaluating and documenting the use of professional judgments throughout the performance assessments.

Obviously, the lack of experience will be overcome as the work progresses and working relationships with the regulatory agencies develop more fully. The Review Group arrived at this observation on the basis of presentations made by the Projects and discussions with Project staffs resulting from these presentations.

- Resources allocated to the conduct of performance assessment activities appear to be inadequate. This problem may be due to inadequate planning, the impact of difficult documentation deadlines apparently imposed by management, arbitrary priority or budgetary constraints, or perhaps to other factors. Determination of more realistic resource needs for performance assessment requires more systematic, rigorous planning and a greater emphasis on performance assessment activities.

- Basic to these managerial/institutional observations and recommendations was the evident need for an integrated performance assessment plan. Such a plan should provide bases for defining both human resources and budgetary requirements and should identify communication mechanisms needed for better management of the overall performance assessment programs.

- The need for more coherent management of the performance assessment program was reflected in the apparently less-than-optimal level of communication between the Projects and Headquarters.

- The need for improved communications at all levels was evident. Within the Projects, closer working relationships between those making performance predictions and those gathering and analyzing data in the laboratory and the field are needed. Similarly, better communications at the project level will promote the exchange of ideas between Projects and the development of more effective solutions to problems of common concern.

- Of particular significance was the need for more effective working relationships between the Projects and the Nuclear Regulatory Commission in order to systematically identify and resolve licensing issues on a continuing, active basis.

- The PANRG recognized that, of necessity, the conduct of performance assessment is based in a fundamental way on the exercise of professional, technical judgment. These judgments should, therefore, be documented carefully, including their supporting rationale and logic.

- Although the proposed EPA standard and NRC regulations do not specifically require the calculation of radiation dose or risk, the PANRG believes that such calculational capability should be available in the Projects and that such calculations should be performed for time periods beyond 10,000 years.

- Continuing independent review of performance assessment activities in the Projects was deemed desirable, if not essential.

Technical Factors

In general, from a technical standpoint, the current situation in performance assessment can perhaps be summarized as follows:

- Much of the necessary site-specific and component-specific data are not yet available, but many of the necessary experiments and tests to acquire such data have been defined and can be carried out.

- Models may well require some refinement or modification based on the data to be acquired. Assurance of model adequacy is largely dependent on in-situ or site-specific tests or investigations yet to be carried out but whose scope and nature are identifiable.

- Techniques and methodologies for handling uncertainties in data and performance sensitivities to specific factors appear reasonably well developed but few results are available and therefore their usefulness and validity cannot be evaluated.

- The Projects' work to date in performance assessment does not reflect sufficient technical evaluation by the Projects themselves of the adequacy of predictive techniques, of the relevance and adequacy of experiments and tests to verify and validate those predictive techniques, and of the adequacy of input data for making such predictions.

- The probabilistic aspect of several facets of performance assessment has yet to be incorporated appropriately into the models and assessment methodologies.

- Sound, acceptable bases for time extrapolation of phenomenological data (e.g., corrosion and dissolution) have yet to be developed.

The Performance Assessment National Review Group's technical observations and recommendations address both the phenomenology and the methodology involved in performance assessment.

Examples of the phenomenological observations include:

- The current lack of adequate data on the characteristics of spent fuel, obviously the major waste form for the first repository, relevant to its behavior under potential repository conditions impairs the ability to predict the source term.

- The need for sufficient understanding of possible corrosion modes of waste canisters other than uniform corrosion is apparent.

- The importance of geochemistry in understanding the nature and behavior of the hydrogeologic system and in assisting in the prediction of radionuclide mobilization and migration was generally recognized by the Projects, but the need was apparent for more comprehensive and detailed investigations and better integration of the results of the investigations into systems analysis and evaluation, and

- There is a need to consider development of appropriate models to characterize the importance and role of the interactions and impacts of coupled phenomena (thermal, chemical, hydrologic, mechanical, and radiological).

From a methodological standpoint, examples include:

- There is urgent need for developing defensible bases for the extrapolation of results from short-term experiments and tests to the long time periods required for postclosure performance assessment. For example, one alternative is to use a verifiable theory for predicting long term behavior that does not require extrapolation, such as mass transfer analysis.

- The desirability of more coordination in the handling of data uncertainties, including the incorporation of uncertainty analysis techniques into the various computer codes, and the application of techniques for sensitivity analysis in the guidance, direction, and setting of priorities for development and testing is apparent.

- The importance of probabilistic approaches to many aspects of performance prediction was generally recognized, but specific applications of probabilistic or stochastic techniques were not yet well defined nor were results of applications available from the Projects.

- Progress in the area of model verification and validation was indicated, but it was also evident that a more definitive, operational definition of what constitutes validation is needed.

CONCLUDING REMARKS

What have we learned from this exercise? We would like to suggest three lessons.

The first lesson is that this peer review appears to have been a constructive effort. This is obviously not unexpected considering the expertise and interests of the Group. The usefulness of the results of the review is evidenced by several things. First, the DOE Projects were given an opportunity to review the Group's draft report. Their response was overwhelmingly favorable. They used terms such as "an excellent report," "particularly insightful" to characterize the report. Second, several of the recommendations of the PANRG have been already implemented. For example, DOE Headquarters staff reacted quite positively to the report and has made the preparation of Performance Assessment Plans a Headquarters-controlled milestone during FY1985.

The second lesson is that the PANRG report highlights several aspects of performance assessment that have not received adequate attention. Although there is sufficient recognition of the importance of performance assessment in licensing, the PANRG highlighted several essential aspects which needed attention. These include the role of professional judgment in performance assessment; the difficulty of extrapolating short term data to geologic time scales; and the concept of predictive reliability.

The third lesson relates to the role of performance assessment in the repository programme. It was the Group's observation that performance assessment has not received sufficient attention from management, nor from other colleagues such as earth scientists and designers. The PANRG, through its access to DOE management, we believe has contributed to upgrading the level of consciousness regarding the overall importance of performance assessment in the repository programme. It is hoped that this is just the beginning of a process that will improve both the effectiveness of performance assessment *per se*, and the application of its results to efficient and effective implementation of the repository programme.

REFERENCES

1. J.A. Lieberman, S.N. Davis, D.R.F. Harleman, R.L. Keeney, D. Langmuir, R.B. Lyon, W.W. Owens, T.H. Pigford, and W.W.-L. Lee. *Performance Assessment National Preview Group*. Weston Report RFW-CRWM-85-01, (1985).

2. T.H. Pigford, J.O. Blomeke, T.L. Brekke, G.A. Cowan, W.E. Falconer, N.J. Grant, J.R. Johnson, J.M. Matusek, R.R. Parizek, R.L. Pigford, and D.E. White. *A Study of the Isolation System for Geologic Disposal of Radioactive Wastes*. National Academy Press, Washington, D.C. (1983).

Disposal and Storage System Cost

CURRENT ESTIMATES OF MRS SYSTEM COSTS

R. D. Izatt
U.S. Department of Energy
Richland Operations Office
Richland, Washington 99352

In order to understand monitored retrievable storage (MRS) system costs, it is necessary to understand what MRS is, why the DOE is evaluating it, what the evaluation status is and how the Federal waste management system with MRS would function. This necessary background is presented first followed by details on the costs.

I. INTRODUCTION

The Nuclear Waste Policy Act of 1982 (NWPA) tasked the U.S. Department of Energy to prepare a proposal for construction of a monitored retrievable storage facility for commercial spent fuel and high-level waste and to submit it to Congress for consideration. The NWPA was, however, very specific in stating that "...disposal of high-level radioactive waste and spent nuclear fuel in a repository developed under this NWPA should proceed regardless of any construction of a monitored retrievable storage facility..."

Specifically with regard to MRS, the NWPA directs the DOE to "...complete a detailed study of the need for and feasibility of, and ...submit to the Congress a proposal for, the construction of one or more monitored retrievable storage facilities for high-level radioactive waste and spent nuclear fuel."

The proposal is to include:

"(A) the establishment of a Federal program for the siting, development, construction, and operation of facilities capable of safely storing high-level radioactive waste and spent nuclear fuel, which facilities are to be licensed by the (Nuclear Regulatory) Commission;

(B) a plan for the funding of the construction and operation of such facilities, which plan shall provide that the costs of such activities shall be borne by the generators and owners of the high-level radioactive waste and spent nuclear fuel to be stored in such facilities;

(C) site-specific designs, specifications, and cost estimates sufficient to (i) solicit bids for the construction of the first such facility; (ii) support congressional authorization of the construction of such facility; and (iii) enable completion and operation of such facility as soon as practicable following congressional authorization of such facility; and

(D) a plan for integrating facilities constructed pursuant to this section with other storage and disposal facilities authorized in this Act."

The NWPA also states "the proposal shall include, for the first such facility, at least 3 alternative sites and at least 5 alternative combinations of such proposed sites and facility designs consistent with the criteria for paragraph (b)(1). The Secretary (of energy) shall recommend the combination among the alternatives that the Secretary deems preferable."

The NWPA directs submission of the MRS proposal and its accompanying documentation by June 1, 1985. Investigations of the need for and feasibility of MRS and of the integration of MRS with other waste management facilities resulted in the recognition that an MRS facility deployed as an integral component of the Federal nuclear waste management system, rather than a backup storage component, offered substantial benefits to the system. Development of this concept for the proposal, however, required substantial additional effort and precluded meeting the June 1985 completion date. The DOE has made substantial progress toward completion of the proposal, the MRS design and the supporting documentation, and plans to submit the proposal package to the Congress on or before January 15, 1986. The current status of the MRS Program is described in another paper presented at this symposium.

II. ROLE AND FEATURES OF AN MRS FACILITY

An MRS facility would serve as a centralized spent fuel and nuclear waste consolidation and packaging facility with storage capacity. Spent nuclear fuel from the nation's commercial nuclear power plants would be shipped to an MRS facility. In the facility, the spent fuel would be prepared for final disposal and then shipped by dedicated train to a geologic repository. If necessary, storage of some spent fuel could be provided at the MRS facility. The waste management system with an MRS facility would be an "improved performance" system. It would provide additional system flexibility, earlier operation, higher initial operation rates and reduced transportation impacts compared to the currently planned system.

Conceptual designs for MRS facilities using two storage concepts, sealed storage casks (preferred) and field drywell (alternate) are nearing completion. Facilities using cask storage concept are being designed for each of three different potential MRS sites in Tennessee: the former Clinch River Breeder Reactor site (preferred), the DOE Reservation at Oak Ridge, and the Tennessee Valley Authority's former Hartsville nuclear plant site near Nashville.

An MRS facility would have two main components--a receiving and handling (R&H) building and a storage area. Support services (administration, maintenance, security, water and sewage treatment, etc.) would also be provided for the facility. An artist's rendition of the facility layout is shown in Figure 1.

A. Receiving and Handling

The R&H building would be the primary preparation area at the MRS facility. Spent fuel would arrive by either rail or truck in heavily-shielded transportation casks and be unloaded inside this building. The nuclear materials would then be consolidated and packaged, ready for shipment to a repository for disposal or movement to the storage area.

The MRS facility would receive primarily spent fuel assemblies. The fuel rods would be removed from the assemblies and close packed before being weld sealed inside a cylindrical steel canister. This disassembly and consolidation procedure would cut almost in half the space required for either storage or disposal of the spent fuel rods. Space will also be required for storage of disposal of the fuel hardware but there still will be a significant overall volume reduction. Because of this reduction in volume, the number of shipments of fuel components to a repository would also be reduced.

The consolidation and packaging operations in the R&H building would be performed remotely inside hot cells to protect workers from direct contact with radioactive materials. The R&H building would be designed to contain radioactive materials being handled within the facility, keeping the radiation exposure of the public and workers well below Federal safety limits.

Following packaging, the spent fuel would normally be loaded into a shipping cask and shipped directly to a repository. The planned throughput rate of the facility is 3000 MTU/year. The design throughput rate is 3600 MTU/year.

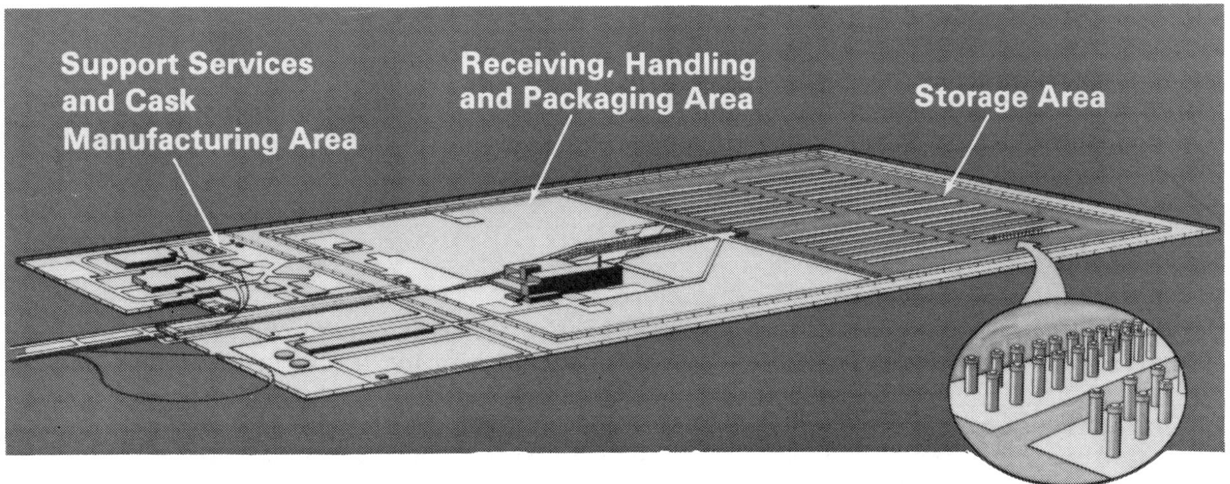

FIGURE 1. Artist's Conception of an MRS Facility

If necessary, spent fuel could be stored temporarily at the MRS facility before its shipment to a repository. As stated previously, DOE has selected sealed storage casks as the primary storage method for an MRS facility; open field drywells were chosen as the alternate method. The MRS facility will have storage capacity for 15,000 MTU of spent fuel.

B. Sealed Storage Cask

A sealed storage cask is a large, steel-lined reinforced concrete cylinder that holds steel canisters of spent fuel and is sealed with a thick concrete shield plug and welded steel lid. For storage of consolidated spent fuel, the casks are about 22 feet tall, measure 12 feet in diameter and weight about 220 tons when loaded. The concrete wall thickness is about 3 feet.

Canisters of spent fuel would be loaded into the casks inside the R&H building, and the casks would then be moved to the storage area and placed upright on a concrete pad. An artist's concept of a sealed storage cask is shown in Figure 2.

C. Field Drywell

The field drywell storage method uses in-ground, dry, sealed, metal containers for storing the spent fuel canisters or drums. Each drywell would hold one canister of spent fuel. The drywells would extend to about 20 feet below the ground surface. An artist's concept of a drywell storage area is shown in Figure 3. Inside the R&H building, spent fuel canisters would be loaded into a shielded transporter vehicle that carries them to a drywell. After the canister and shield plug were lowered into the drywell, the drywells would be welded shut.

D. Schedule and Operation

The schedule and operation of the MRS facility is shown in Figure 4. This schedule assumes congressional approval of the MRS proposal by July 15, 1986. The cost of the MRS program through congressional deliberation on the proposal is expected to be approximately $37 million.

IV. ESTIMATED COSTS*

If Congress approves the construction of the proposed MRS facility, the funding authority required to finance preconstruction activities (site data

*The cost estimates included are very preliminary. The design is nearing completion but not yet final, the repository savings with MRS are still preliminary estimates, and the total system life cycle cost are still rough approximations.

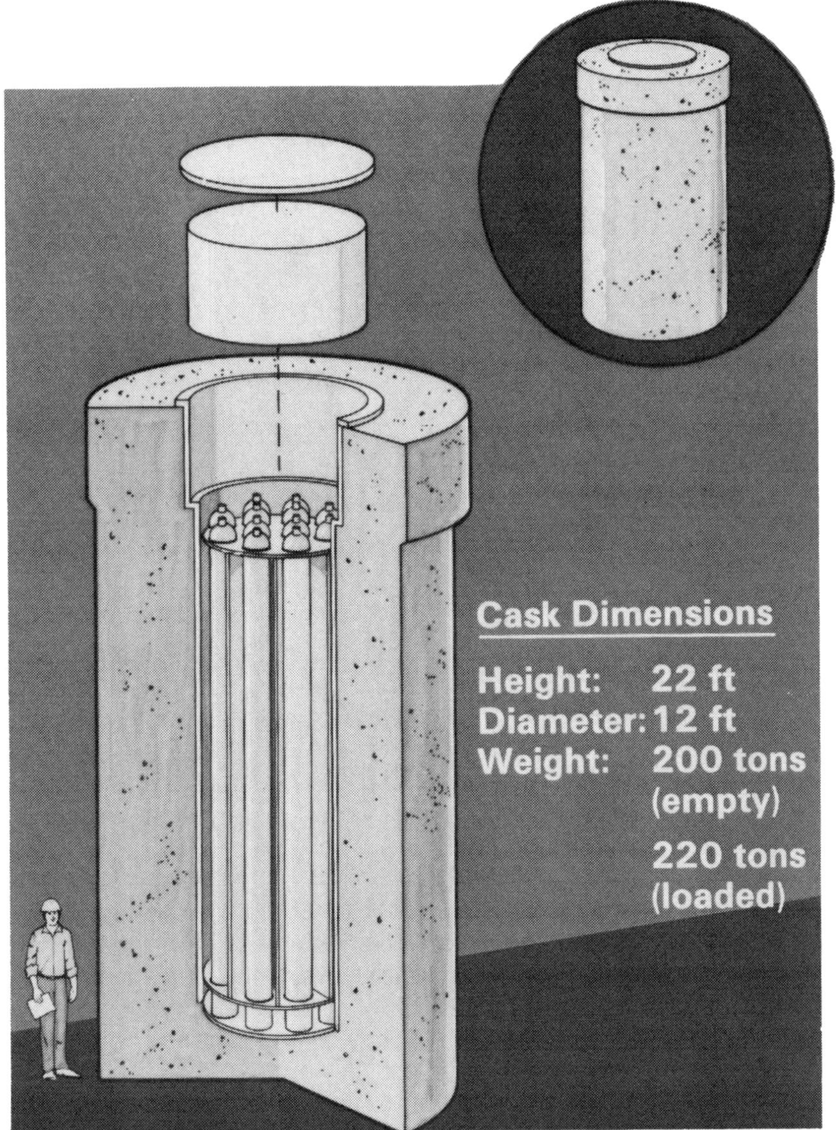

FIGURE 2. Artist's Concept of a Sealed Storage Cask

confirmation, licensing, EIS preparation, design, design verification tests, etc.) is estimated to be $275 million (constant 1985 dollars). The funding authority required for the construction period (assumed to be completed at the end of the pilot scale tests) is estimated to be $700-750 million (in constant 1985 dollars).

The MRS facility life cycle cost estimates shown in Table 1 should not be regarded as the net incremental cost to the Federal waste management system. With an integral MRS facility, some of the functions that would have been performed at the repository, such as spent fuel consolidation, canistering and packaging associated with the operation of the R&H building, are transferred

FIGURE 3. Artist's Conception of Drywell Storage

to the MRS. Therefore, to consider the incremental system costs, it would be necessary to separate out the part of the MRS facility costs that would need to be incurred by the repository, if there were no integral MRS facility.

In approximating the additions to total system costs attributed to the MRS facility, these shared costs need to be properly allocated. Examples of incremental costs due to the inclusion of an MRS facility in the Federal waste management system are shown in Table 2. The incremental costs for an integral MRS facility result in enhancements to the waste management system, including assured timely acceptance of fuel from utilities, improved logistics of shipments to the repository, and smoothing of temporary disruptions which may occur in the system.

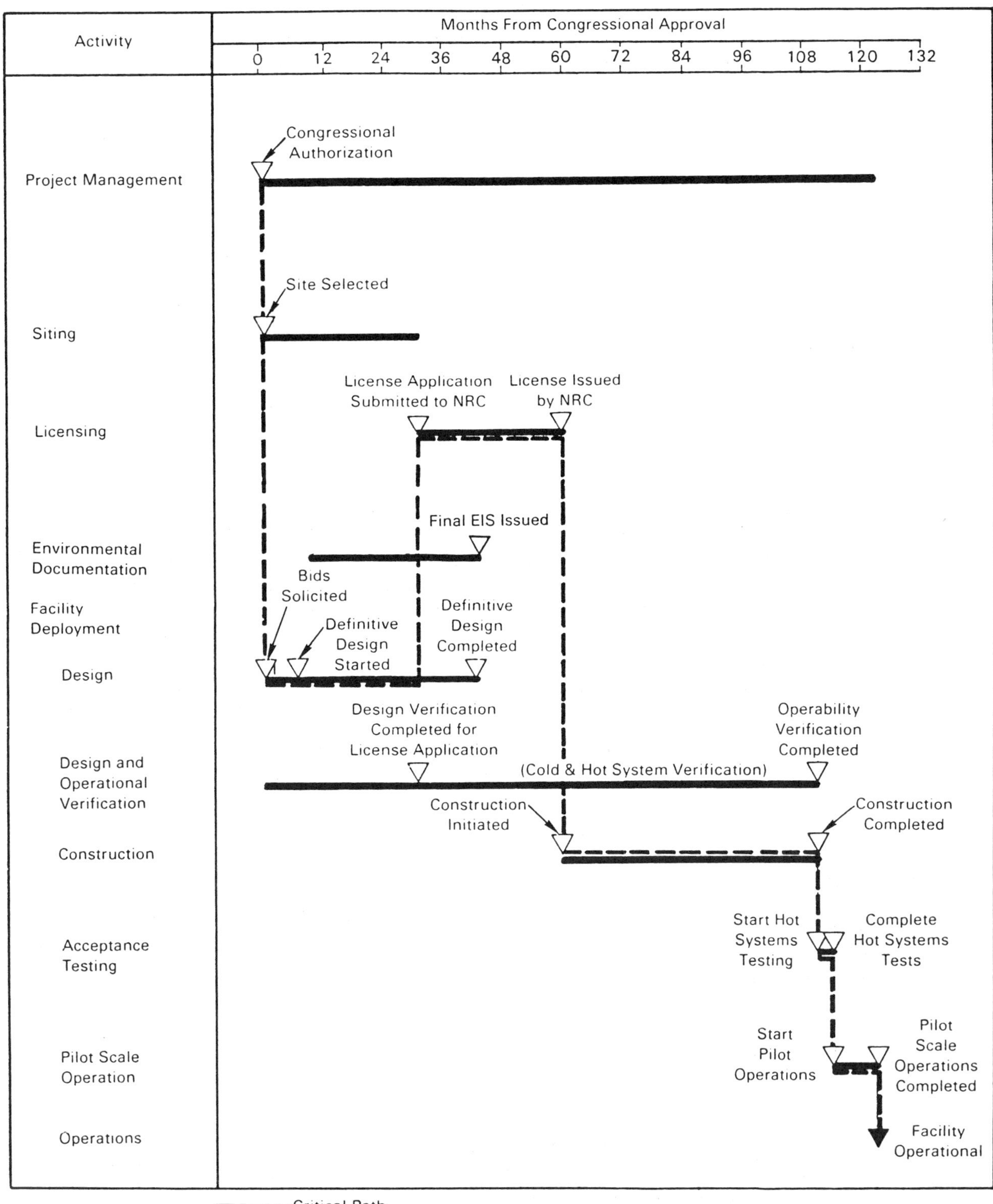

FIGURE 4. Proposed Schedule for an MRS Facility

TABLE 1. Life Cycle Cost Estimates for the Proposed MRS Facility at the Clinch River Tennessee Site for the Sealed Storage Cask Concept, for a Basalt Repository

Item	Constant 1985 Dollars (in Millions)
Preconstruction	275
Construction	700-750
Facility Operation	1850-2000
Decommissioning	75
TOTAL MRS FACILITY[a]	2.9-3.1 Billion

(a) Costs or benefits not included in these estimates are 1) aid for mitigating adverse impacts of MRS facilty construction and operation, 2) grants equal to taxes, 3) DOE headquarters and field office costs related to MRS program planning and oversight, 4) licensing and permitting costs associated with other federal entities, 5) consultation and cooperation costs, and 6) transportation of waste to and from MRS.
(b) Depending on assumptions.

TABLE 2. Incremental System Costs of Integrated Waste Management System with MRS[a]

(Constant 1985 dollars in billions)

Cost Components	Incremental Cost/(Savings)
First Repository	(1.3) - (1.5)[b]
MRS	2.9 - 3.1
Transportation	0.1
TOTAL SYSTEM	1.5 - 1.9

(a) Case represents the basalt repository.
(b) Range based upon further possible cost savings depending on ultimate system configuration and additional value engineering.

REFERENCES

1. Department of Energy (DOE), Selection of Concepts for Monitored Retrievable Storage of Spent Nuclear Fuel and High-Level Radioactive Waste, DOE/RL-84-2, U.S. Department of Energy, Richland Operations Office, Richland, Washington (1984).

2. Department of Energy (DOE), The Monitored Retrievable Storage Proposal Research and Development Report, DOE/S-0021. U.S. Department of Energy, Washington, D.C. (1983).

3. Department of Energy (DOE), The Need for and Feasibility of Monitored Retrievable Storage--A Preliminary Analysis, DOE/RW-0022, U.S. Department of Energy, Office of Civilian Radioactive Waste Management, Washington, D.C. (1985)

4. Department of Energy (DOE), Screening an Identification of Sites for a Proposed Monitored Retrievable Storage Facility, DOE/RW-0023, U.S. Department of Energy, Office of Civilian Radioactive Waste Management, Washington, D.C. (1985).

DESIGN, COST, SCHEDULE AND LICENSING IMPLICATIONS OF AN INTEGRATED WASTE MANAGEMENT SYSTEM

Carl W. Conner
Edwin L. Wilmot
Office of Civilian Radioactive
Waste Management
U. S. Department of Energy
Washington, D.C. 20585

I. INTRODUCTION

The inclusion of the Monitored Retrievable Storage (MRS) Facility as an integral part of the waste management systems will result in improved performance of the whole waste management system.[1] This improved performance does not come without a price and that price is dependent upon the allocation of functions to various system facilities and the impacts of those functions on facility design, cost, licensing, and schedule. For example, the MRS would perform most, if not all, of the spent fuel packaging operations in the repository. The final overpacking could be done at either the MRS or the repository. In April 1985, a task force was established to further evaluate these possibilities. Specific questions addressed were as follows:

1. What would the surface facilities at the repository look like and cost, with an MRS as an integral part of the waste system?

2. How does this compare to a system without an MRS?

3. How is repository licensing affected by adding the MRS to the waste system?

4. How is the repository design and construction schedule affected by adding the MRS to the waste system?

5. Are there clear preferences for the location of the overpacking function and for the location of the packaging operations for spent fuel from western reactors?

Preliminary judgements on these matters need to be confirmed and documented to support the MRS proposal that will be submitted to Congress in January 1986. This paper reports on the work of the MRS/Repository Task Force that was assembled to gather data for use in answering the above mentioned questions. This effort was begun in April 1985, and the final report is now in preparation.

II. APPROACH

To accomplish this task, teams were assembled to address each issue, i.e., design, cost and schedule, and licensing. Knowledgeable individuals on each of these subjects were assembled from DOE, the National Laboratories, the MRS and repository architect-engineers, and the DOE support contractor, Weston.

Various system configuration scenarios were developed in order to define the functional allocations. These scenarios are indicated in Figure 1. It should be noted that these scenarios were developed only for the purpose of gathering design and cost data for various configurations and boundary conditions. They should not be interpreted as being equal, viable alternatives. For example, Scenario 2B was included to determine the minimum possible surface facility at the repository. DOE is not planning to overpack defense waste at the MRS.

Figure 2 summarizes the differences in functional allocations between facilities for the various scenarios. Scenario 1 is the "no MRS" case where all packaging functions are performed at the repository. As you move from left to right on the chart, the MRS functions are increasing, and the repository functions are decreasing. Scenario 2 is the other extreme where all packaging functions are performed at the MRS. Scenarios 4A, 3A and 2A are between these extremes.

In order to facilitate comparisons of designs and costs, a common design basis was selected. The general assumptions that were made are summarized in Figure 3. Facility designs were then prepared for the MRS and the appropriate repository surface facilities corresponding to the assumed functional allocations and the general assumptions. Examples of these designs are given in Figures 4 through 9. Note that underground facilities and operations at the repositories are not considered in this study.

For transportation, an analysis was performed where all shipments were assumed to be by rail. Although this may not be possible, since about 25% of the reactor sites do not currently have rail access, the cost estimates are considered to be very close to what a 70%/30% rail/truck analysis would produce. Figure 10 is a summary of the transportation assumptions.

As designs were developed, corresponding cost estimates were also developed. Then the effect on licensing and schedule were evaluated. A summary of the costing methodology is indicated in Figure 11.

III. FINDINGS

A. Design

Some of the designs that resulted from this exercise are illustrated in Figures 4 through 9. Although there are some slight variations between repository facility designs due to specific site considerations, the designs are adequate for comparing costs.

In the case of the minimum repository, Scenario 2B, it is difficult to determine what capabilities should still be provided for at the repository. For this study it was assumed that the repository would still provide capabilities to: 1) repair any damaged containers received from the MRS; 2) perform confirmation testing required by 10 CFR 60; and 3) inspect and spot check packages received to assure confidence in package certification.

If all of these functions were removed from the repository and performed at the MRS, the decrease in capital cost at the repository would be about $50 million.

B. Cost

Cost estimates were generated based upon the above mentioned designs. These cost estimates should be considered as preliminary based upon the pre-conceptual designs developed for this exercise. Estimates included engineering, construction, operations and decommissioning. A summary of the costs is given in Figure 12. (Refer back to Figure 2 for a description of the scenarios).

In comparing costs, it should be noted that the inclusion of the MRS in the system increases the system's capabilities. These additional benefits are difficult to quantify but are real and should be considered in any cost analysis. These benefits include increased reliability and system flexibility, increased waste acceptance rates in the early years, and increased confidence in meeting schedules.[2]

It should be pointed out again that Scenario 2B shows the overpacking of defense high-level wastes at the MRS. Current plans for the MRS do not provide for this function. It is more likely that defense wastes will be overpacked at the repository or at the defense facility. If Scenario 2B is not considered, the differences in system costs are $1.8 billion to $2.5 billion. Scenarios 3A or 4A are closer to the current planning for the MRS. These scenarios have a system cost differential of $1.5 billion to $2.3 billion or about 5%-9% of the total life-cycle program costs of $25 billion to $30 billion. Notice that the total life-cycle scale costs for the MRS for these scenarios are $2.7 billion to $2.9 billion. Corresponding repository surface facility costs are reduced by $0.7 billion to $1.7 billion. Notice that the costs calculated for the MRS vary from $2.7 billion to $4.1 billion, depending on the functions performed. The repository surface facility costs are reduced by $0.7 billion to $2.7 billion. The differences in costs of the system, including transportation costs are $2.1 billion for salt, $2.3 billion for basalt, and $2.9 billion for tuff. These cost estimates are total life cycle costs including design, construction and operations over a 30 year period.

Also, it should be noted that the cost difference between Scenarios 4A, 3A, and 2A is $0.3 billion or less. Therefore, cost is not a major driver in the allocation of functions between the MRS and the repository. Further discussion of transportation costs will be provided later in this paper.

Figure 13 shows a breakdown of costs between engineering/construction and operation/decommissioning. This chart indicates that the cost differences previously discussed are driven by operation/decommissioning costs. There are very small differences in engineering/construction costs. Therefore, capabilities to perform functions, thereby adding flexibility to the system, could be provided at essentially no additional costs.

Figure 14 is a summary of staffing requirements for the various scenarios. The total staff changes between scenarios are influenced largely by the fixed support staff needed to provide routine operations.

C. Licensing

It does not appear that the 27-month licensing time for the repository can be significantly reduced when an MRS is added to the waste system and various functions are allocated to the MRS verses the repository. It appears that the repository licensing schedule will be controlled by procedural requirements. There are, however, advantages that should be mentioned. For example, the movement of surface facility operations from the repository will reduce the repository licensing workload accordingly since neither the DOE nor the NRC will have to spend time and effort on those operations during the repository licensing process. Also, the licensing of the MRS prior to the repository, will provide useful experience that should help in the licensing of repository surface facilities.

D. Schedule

Overall, it does not appear that the current repository design and construction schedule will be reduced in time when an MRS is added to the waste system. However, by removing functions from the repository, the confidence level of meeting the repository schedule is enhanced.

E. Transportation

The results of the transportation analysis are driven by waste package design, cask capacity and distance. Figure 15 is a summary of the transportation cost analysis. Scenario 2B is again the higher cost scenario because of the shipment of defense waste to the MRS for overpacking. As discussed previously, this is not currently planned and is unlikely to happen.

Scenario 2A is the next highest in cost because of the shipment of overpacked waste packages from the MRS to the repository. These larger packages require more casks to ship the waste. The differences in cost between Scenarios 1, 4A and 3A are very small.

No effort was made to optimize the transportation system in this exercise, such as by the use of large (150 ton) casks. These casks have the potential for reducing costs significantly. Other areas of optimization are also possible.

F. Allocation of System Functions

One of the questions asked at the beginning of this exercise and mentioned at the beginning of this paper was "are there clear preferences for the division of system functions between the MRS and the repository? Specifically, where should the overpacking function be performed, and where should "western fuel" be packaged?

There are many factors that must be taken into consideration in answering this question. Some, but not all, of these factors were addressed in this study, i.e., design, cost, licensing and schedule. Other programmatic and institutional factors were not considered in this study. At this time, it is not clear where the overpacking should be done and where western fuel will be packaged. Further study is underway to resolve these issues. However, these decisions are not critical to determining the need and feasibility of the MRS and therefore can be deferred. These relate more to the optimization of the system and can be addressed in the design phase of the project.

IV. CONCLUSION

This paper reports on a study that has produced data needed for the MRS Needs and Feasibility Report currently in preparation for submittal to Congress. The facility designs are now compatible in terms of functions, operations, technology and costs. The cost estimates are also compatible being based upon compatible designs, operating procedures and unit costs. The licensing and schedule impacts have been systematically examined.

This study does not provide all data needed to answer all systems questions, nor was it intended to do so. It will, however, enhance the Needs and Feasibility Report and enhance the integration of the waste management system.

REFERENCES

1. Department of Energy (DOE), Office of Civilian Radioactive Waste Management Mission Plan, Washington, D.C. (1985).

2. Department of Energy (DOE), The Need for and Feasibility of Monitored Retrievable Storage - A Preliminary Analysis, DOE/RW-0022, Office of Radioactive Waste Mangement, Washington, D.C. (1985).

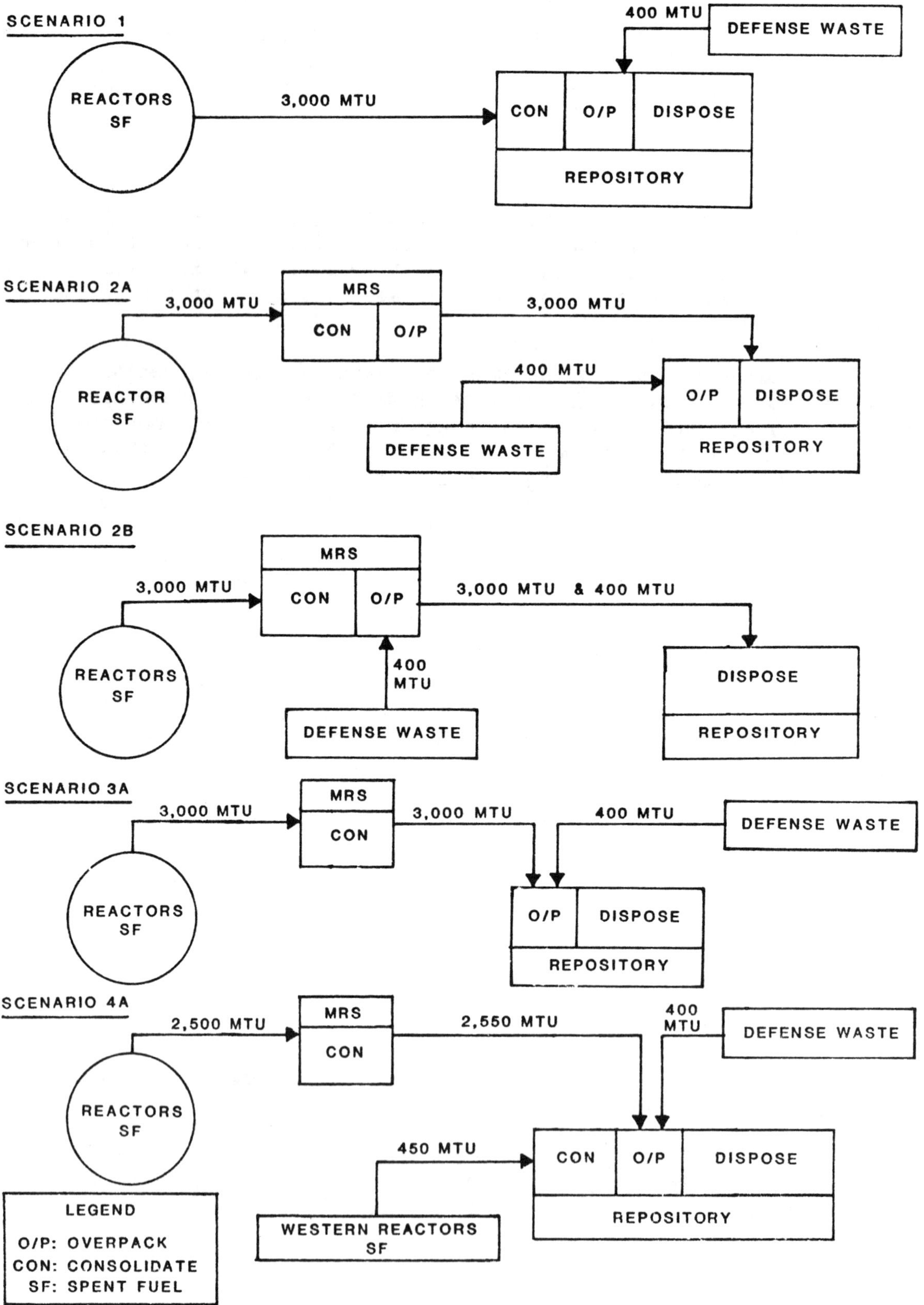

FIGURE 1. System Configuration Scenario

SCENARIO / FACILITY	1	4A	3A	2A	2B
FIRST REPOSITORY	DEFENSE WASTE OVERPACK WESTERN FUEL CANISTER CONSOLIDATE	DEFENSE WASTE OVERPACK WESTERN FUEL	DEFENSE WASTE OVERPACK	DEFENSE WASTE	NONE*
MRS	NONE**	CONSOLIDATE CANISTER	CONSOLIDATE CANISTER WESTERN FUEL	CONSOLIDATE CANISTER WESTERN FUEL OVERPACK	CONSOLIDATE CANISTER WESTERN FUEL OVERPACK DEFENSE WASTE

* "NONE" MEANS REPOSITORY ONLY RECEIVES AND INSPECTS THE WASTE ARRIVING FROM THE MRS

** "NONE MEANS THERE IS NO MRS IN SCENARIO 1

FIGURE 2. Summary of Scenarios

1. STUDY LIMITED TO SURFACE FACILITIES

2. PRIMARY SPENT FUEL DISPOSAL IN THE FORM OF CONSOLIDATED RODS

3. STUDY LIMITED TO ONE-REPOSITORY SCENARIOS

4. BOTH SPENT FUEL AND DHLW CONSIDERED AT THE ACCEPTANCE RATES STATED IN THE MISSION PLAN

5. WASTE FORMS BASED ON THE FOLLOWING:

   -SPENT FUEL MODAL SPLIT 60% PWR, 40% BWR

   -DEFENSE WASTE (DHLW) ASSUMED AT 0.5 MTU EQUIVALENT PER PACKAGE

   -WEST VALLEY (CHLW) PACKAGED IN SAME MANNER AS DHLW

6. GENERAL TRANSPORTATION ASSUMPTIONS AS FOLLOWS:

   -TRANSPORTATION CASKS BASED ON NEW TECHNOLOGY

   -MODAL SPLIT FOR ALL SPENT FUEL FROM REACTORS WAS 70% RAIL, 30% TRUCK (NOTE THAT COST PRESENTED LATER ARE BASED ON THE 100% RAIL

   -TRANSPORTATION FROM MRS TO REPOSITORY EXCLUSIVELY BY RAIL IN DEDICATED TRAINS

7. SPENT FUEL TO BE HANDLED IS 10 YEARS OUT OF REACTOR WITH THE CAPABILITY OF HANDLING UP TO 5% AT 5 YEARS OUT OF REACTOR

8. UNIFORM TECHNOLOGY BASED ON THE CURRENT MRS DESIGN WAS USED FOR ALL SCENARIOS

FIGURE 3. Study General Assumptions

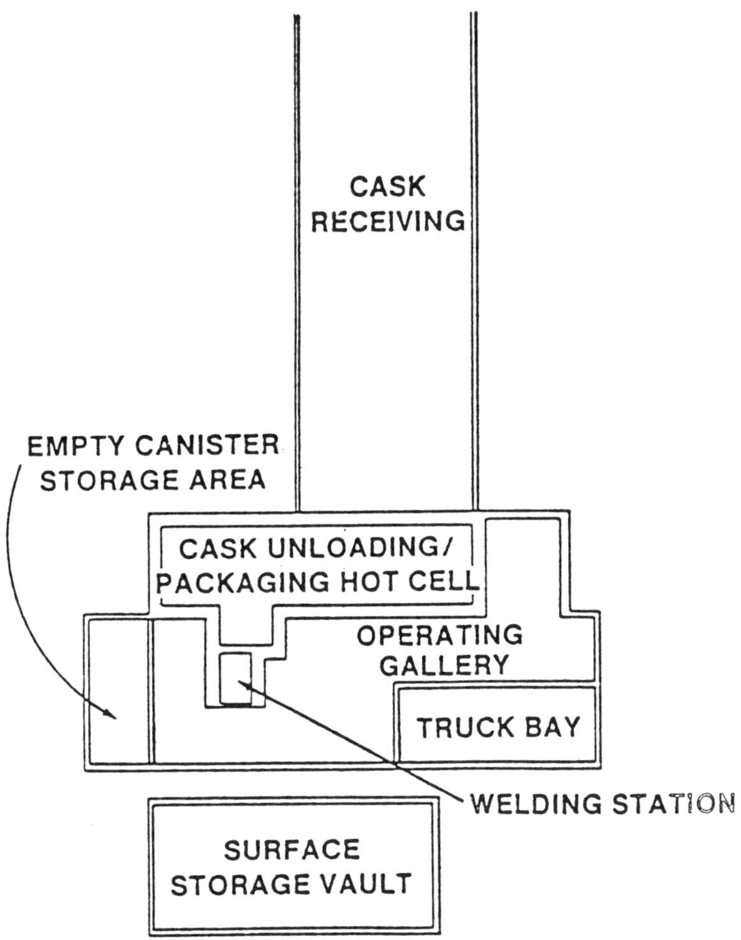

FIGURE 4. Scenario 1 - W.H.B.-1 Two-Phase Repository

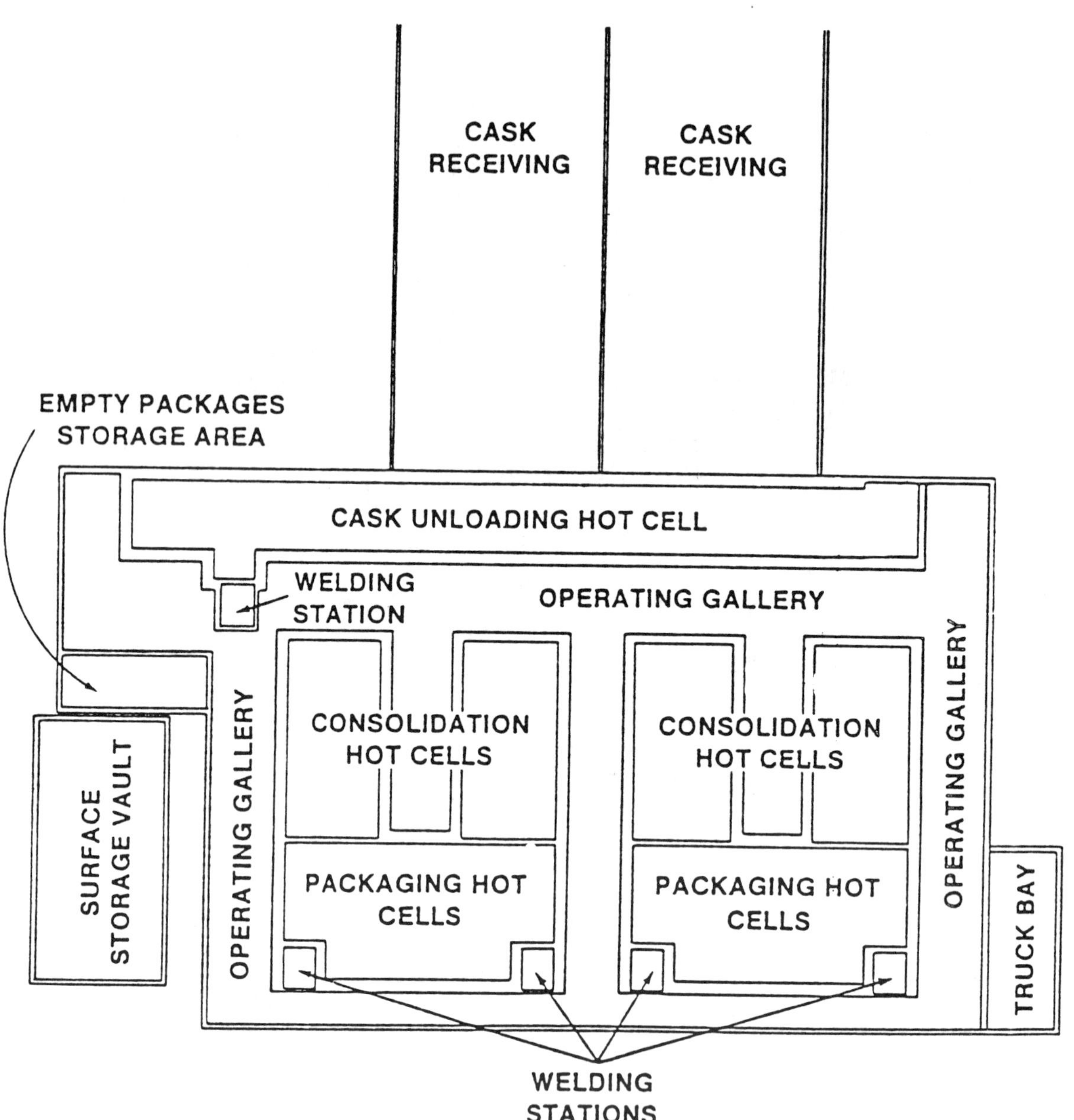

FIGURE 5. Scenario 1 - W.H.B.-2 Two-Phase Repository

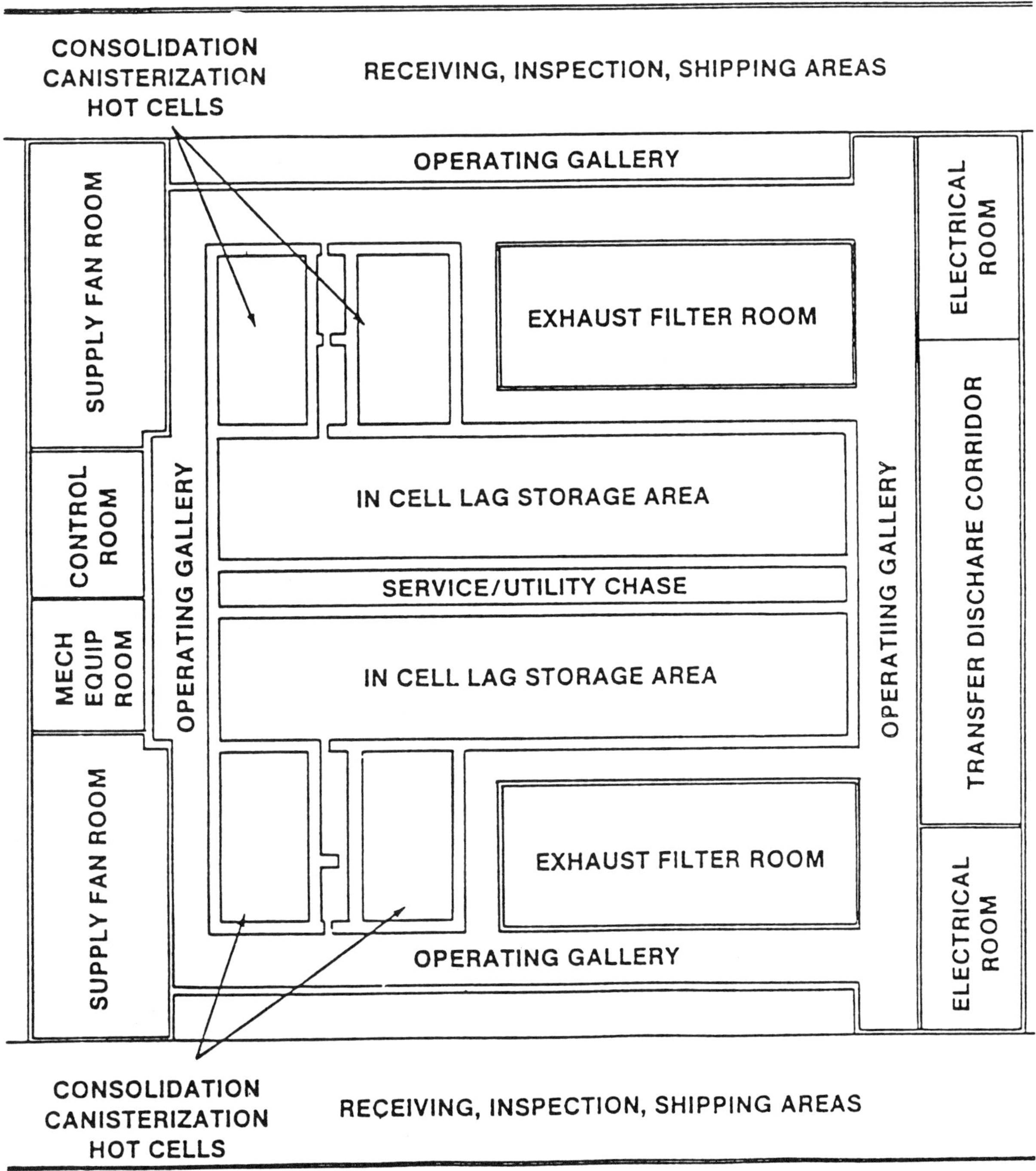

FIGURE 6. Scenario 3A - MRS

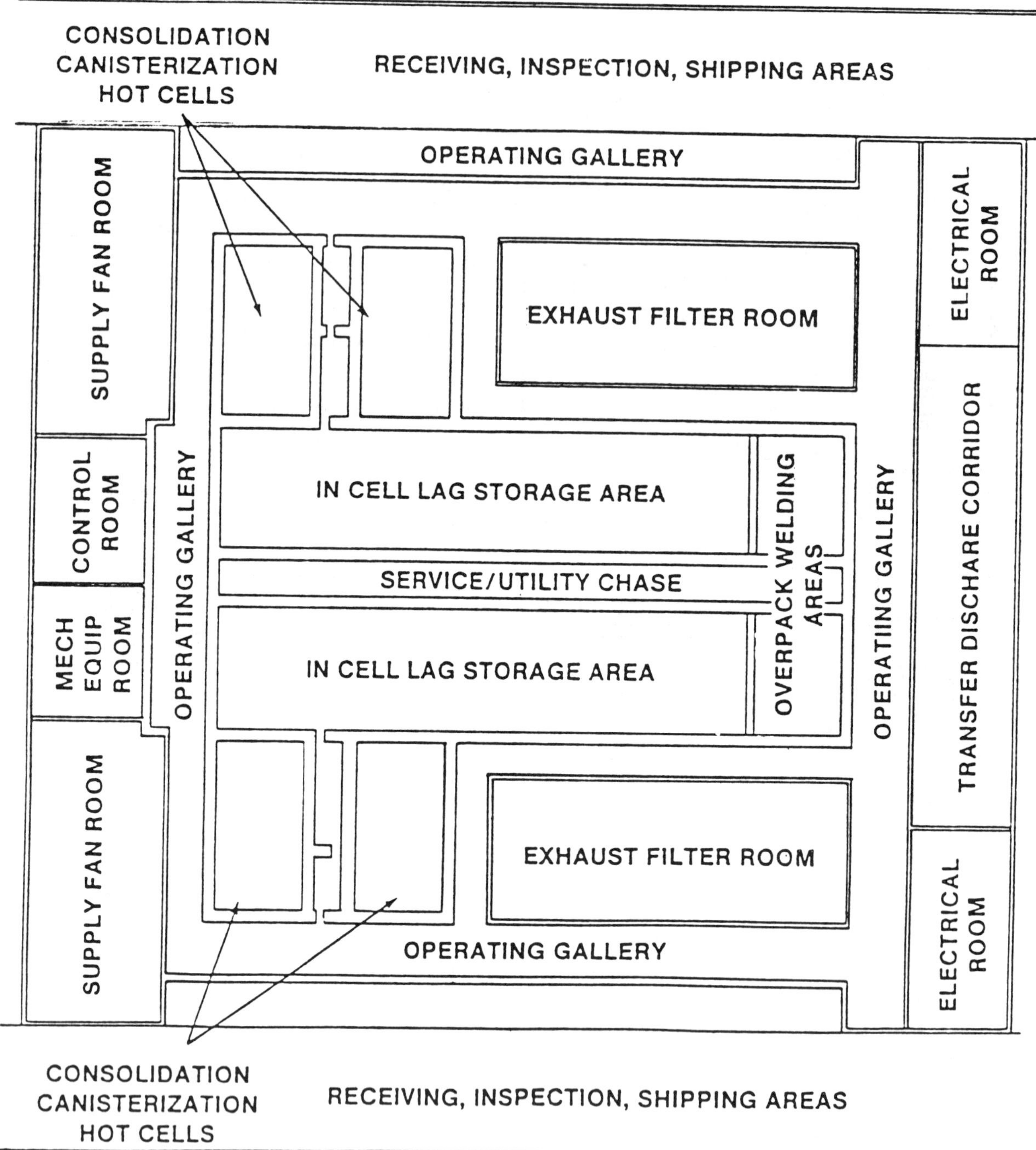

FIGURE 7. Scenario 2B - MRS

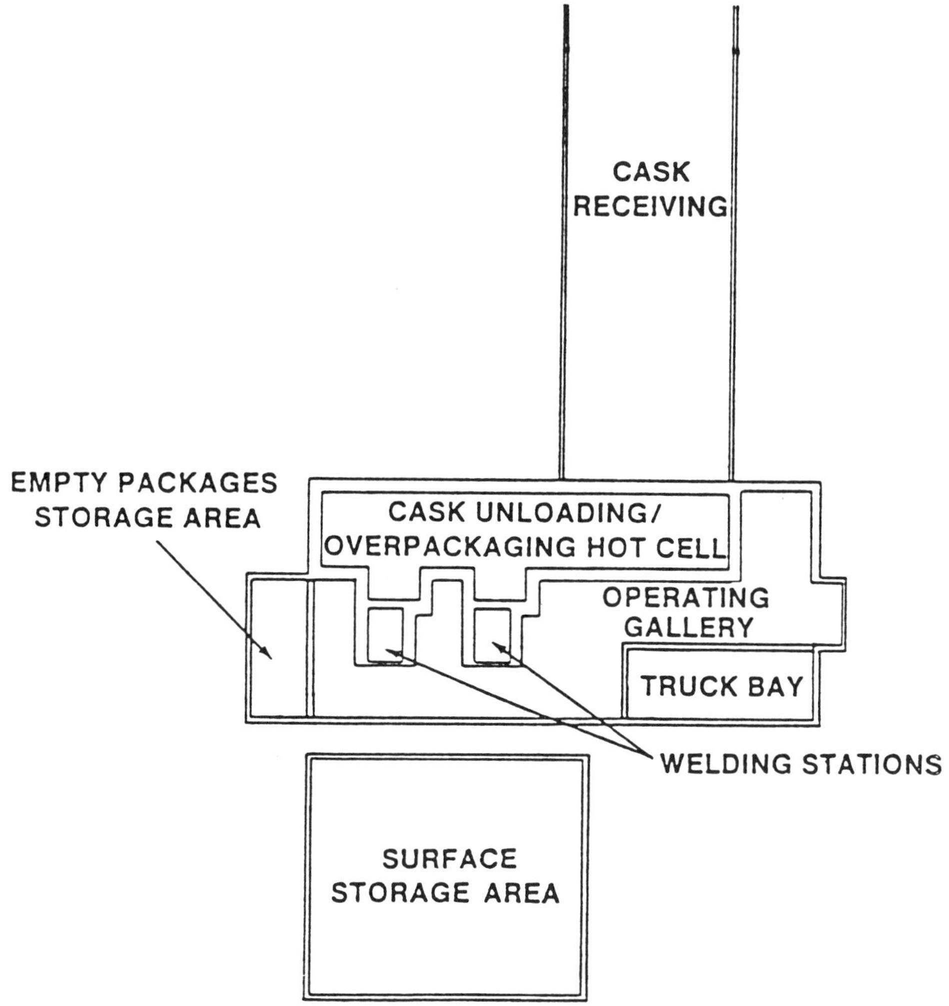

FIGURE 8. Scenario 3A - Single-Phase Repository

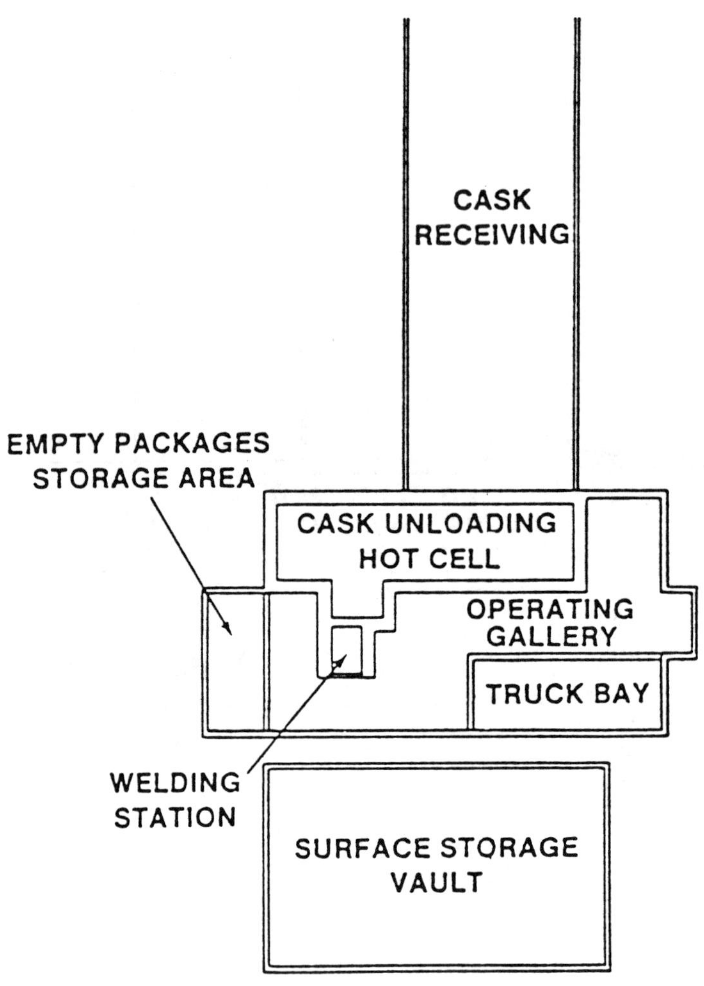

FIGURE 9. Scenario 2B - Single-Phase Repository

CASK TYPES	CAPACITY (PWR/BWR)	
	OVERPACKED CANISTER	WITHOUT OVERPACK
FROM-REACTOR	NA	14/36
FROM-MRS		
To-BWIP	24/25	48/99
To-NNWSI	24/30	36/90
To-Deaf Smith	18/42	24/56
DHLW		
From-Savannah River	3 Canisters	5 Canisters
From-Hanford	3 Canisters	5 Canisters
From-Idaho	3 Canisters	5 Canisters
From-West Valley	3 Canisters	7 Canisters
NON-FUEL COMPONENTS		
Hardware/High Activity	NA	20 Drums
Contact-Handled Transuranic Wastes	NA	72 Drums

FIGURE 10. Transportation Assumptions

CONSTRUCTION COST:

Labor Hours x Unit Rate	= Direct Labor
Direct Labor x 80%	= Indirect Cost
Material Quantity x Material Unit Cost	= Material Cost (MC)
Capital Equipment Costs	= Capital Equipment Costs (CEC)
	Subtotal Construction Costs (SCC)
Sales Tax Rate x (MC + CEC)	= Sales Tax (STX)
SCC x 10%	= Construction Management (CM)
	SCC + STX + CM
(SCC + STX + CM) x (10%-40%)	= Contingency
SCC + STX + CM + Contingency	= TOTAL CONSTRUCTION COST (TCC)

ENGINEERING:

SCC x (16%-27%)[a]	= Engineering Costs (EC)
EC x (10%-40%)[a]	= Engineering Contingency
EC + Contingency	= TOTAL ENGINEERING COSTS (TEC)

OPERATING COSTS[b]:

Average Staffing (FTE's) x Operating Period (Yrs) x Labor Unit Rate	= Operating Labor
Oper. Labor x (7.9%-8.4%)[a]	= Operating Material
SCC x (5.5%-5.6%)[a] x Oper. Period (Yrs)	= Maintenance Materials
Electric Quantity x Operating Period x Unit Rate	= Electric Power
Other Utilities	= Other Utilities
Labor Man-Years x $1266 Per Man-Year	= Employee Transportation
	Subtotal Operations[b] (SO)
SO x 30%	= Operating Contingency
SO + Contingency	= TOTAL OPERATING COST[b]

[a] Percentage Varies By Building
[b] For Emplacement Phase of Operation Only
[c] Transportation Subsidy, (NNWSI Only)

FIGURE 11. Summary of REPCOST Methodology

($ BILLIONS)*

	SCENARIO				
	1	4A	3A	2A	2B
FIRST REPOSITORY					
BASALT	4.6	3.5	3.2	2.2	1.9
SALT	4.3	3.3	3.0	2.1	1.7
TUFF	3.5	2.8	2.5	2.0	1.9
MRS (SERVING):					
BASALT	---	2.7	2.9	3.8	4.1
SALT	---	2.7	2.9	3.7	4.0
TUFF	---	2.7	2.9	3.4	3.5
TRANSPORTATION (SERVING):					
BASALT	1.4	1.3	1.4	1.8	2.3
SALT	1.1	1.3	1.3	1.6	1.8
TUFF	1.3	1.6	1.7	1.9	2.3
TOTALS:					
BASALT	6.0	7.5	7.5	7.8	8.3
SALT	5.4	7.3	7.2	7.4	7.5
TUFF	4.8	7.1	7.1	7.3	7.7

*Costs are life-cycle costs for the surface facilities and transportation costs. Costs for underground operations are not included. These cost estimates are preliminary based on a pre-conceptual design.

FIGURE 12. Preliminary Cost Summary

	1	4A	3A	2A	2B
Basalt					
Eng./Constr.	Ø.9	Ø.6	Ø.5	Ø.5	Ø.5
Op./Dec.	3.7	2.9	2.7	1.7	1.4
TOTAL	4.6	3.5	3.2	2.2	1.9
Salt					
Eng./Constr.	1.2	Ø.7	Ø.7	Ø.7	Ø.7
Op./Dec.	3.1	2.6	2.3	1.4	1.Ø
TOTAL	4.3	3.3	3.Ø	2.1	1.7
Tuff					
Eng./Constr.	Ø.8	Ø.5	Ø.5	Ø.4	Ø.5
Op./Dec.	2.7	2.3	2.Ø	1.6	1.4
TOTAL	3.5	2.8	2.5	2.Ø	1.9
MRS					
Eng./Constr.	Ø.Ø	1.Ø	1.2	1.2	1.2
(T/B/S) Op./Dec.	Ø.Ø	1.7/1.7/1.7	1.7/1.7/1.7	2.2/2.6/2.5	2.3/2.9/2.8
(T/B/s) TOTAL	Ø.Ø	2.7/2.7/2.7	2.9/2.9/2.9	3.4/3.8/3.7	3.5/4.1/4.Ø

FIGURE 13. Preliminary Cost Estimate Breakdown

(Full-Time Equivalents)

Facility	Scenario 1	4A	3A	2A	2B
Basalt					
Waste Buildings	475	288	245	216	203
Support Buildings	469	337	318	292	282
Total Surface	944	625	563	508	485
Salt					
Waste Buildings	353	189	165	114	92
Support Buildings	396	395	309	308	308
Total Surface	749	584	474	422	400
Tuff					
Waste Buildings	276	172	147	135	131
Support Buildings	437	421	400	384	384
Total Surface	713	593	547	519	515
MRS					
Waste Buildings	0	351	315	327	327
Support Buildings	0	301	286	286	286
Storage	0	6	6	6	6
Total Surface	0	658	607	619	619

FIGURE 14. Repository/MRS Surface Staffing Estimates (Emplacement Phase)

	1	4A	3A	2A	2B
BWIP					
Spent Fuel	1.1	Ø.8	Ø.9	1.3	1.3
Non-Fuel		Ø.2	Ø.2	Ø.2	Ø.2
Defense HLW	Ø.3	Ø.3	Ø.3	Ø.3	Ø.8
TOTAL	1.4	1.3	1.4	1.8	2.3
DEAF SMITH					
Spent Fuel	Ø.8	Ø.9	Ø.9	1.2	1.2
Non-Fuel		Ø.1	Ø.1	Ø.1	Ø.1
Defense HLW	Ø.3	Ø.3	Ø.3	Ø.3	Ø.5
TOTAL	1.1	1.3	1.3	1.6	1.8
NNWSI					
Spent Fuel	1.Ø	1.1	1.2	1.4	1.4
Non-Fuel		Ø.2	Ø.2	Ø.2	Ø.2
Defense HLW	Ø.3	Ø.3	Ø.3	Ø.3	Ø.7
TOTAL	1.3	1.6	1.7	1.9	2.3

FIGURE 15. Results of Preliminary Transportation Analysis by Scenario ($ Billion)

LIFE-CYCLE COSTS FOR THE CIVILIAN RADIOACTIVE WASTE MANAGEMENT PROGRAM

Ron Milner
Garet Bornstein
Office of Civilian Radioactive
Waste Management
U.S. Department of Energy
Washington, D.C. 20585

Frank Haines
Andy Leiter
Roy F. Weston, Inc.
2301 Research Blvd
Rockville, Maryland 20850

ABSTRACT

The total system life-cycle cost (TSLCC) analysis for the Department of Energy's (DOE) Civilian Radioactive Waste Management Program is an ongoing activity that helps determine whether the revenue-producing mechanism established by the Nuclear Waste Policy Act of 1982--a fee levied on the producers of nuclear power--is sufficient to cover the cost of the program. This paper summarizes the program cost estimates in the third fee adequacy evaluation conducted in January 1985. Costs for the reference waste-management program were estimated to be 24 to 30 billion (1984) dollars. Costs for the sensitivity cases studied ranged from 21 to 35 billion dollars. Factors such as repository location, quantity of waste generated, transportation cask technology, and repository startup dates were shown to exert substantial impacts on total-system costs.

Costs for the reference program described in this paper show a substantial rise over previous estimates. Two-thirds of this rise stems from increased repository construction and operation costs caused by changing design concepts, assumptions about the effort required to perform the necessary activities, and changes in the source data on which the earlier analyses were based.

TSLCC estimates will continue to change as the waste management program strategy evolves and the work definition becomes clearer. Already, since this analysis was done several major program changes have taken place, i.e., the President's decision, to include defense waste in the civilian repositories and the DOE's serious consideration to recommend a Monitored Retrievable Storage Facility as an integral part of the system. The paper concludes with a preview of the scope of next year's (January 1986) TSLCC analysis.

I. INTRODUCTION

Section 302 of the Nuclear Waste Policy Act (NWPA) established a waste-disposal fee to be collected from utilities that produce electricity from nuclear power plants in order to pay for the cost of disposing of commercially generated radioactive waste. Revenues thus collected are placed into a Nuclear Waste Fund, which is managed by the DOE. The NWPA requires the DOE to annually evaluate and report to Congress whether this fee provides sufficient revenues to cover all the costs of waste disposal. Part of this annual evaluation is the determination of the current total system life cycle costs (TSLCC) of the program.

Thus far, the adequacy of the 1.0 mil/kwh fee, the fee initially determined in drafting the NWPA, has been evaluated and found sufficient in July 1983, July 1984, and January 1985.[1,2,3] However, these evaluations emphasize the substantial amount of uncertainty in both program cost estimates and revenue projections. This paper presents a summary of the TSLCC costs used in the January 1985 Fee Adequacy evaluation. Details of this cost analysis are given in Reference 4.

II. APPROACH

A. Overview of Analysis

To account for the substantial uncertainties in the waste-management program, this TSLCC analysis examines the planned system as of January 1985 plus a limited number of reasonable alternatives; it does not attempt to cover all possible alternatives. This range covers a "success-oriented," on-schedule reference case as well as three sensitivity cases. The reference case has seven variations that differ only by the location of the first and second repository sites. The three sets of sensitivity cases differ from the reference case by: 1) waste generation rate, which is assumed to be lower; 2) transportation-cask technology, which is assumed to be improved; and 3) repository startup dates, which are assumed to be delayed. Ten such sensitivity cases were analyzed (see Figure 1).

Total-system life-cycle costs for these reference cases were calculated by summing the annual costs (expressed in constant 1984 dollars) estimated for the following major cost categories; development and evaluation (D&E), transportation, and repositories (see Figure 2). Before this can be done, it is necessary to establish an estimation method. The first step in this procedure is to define a reference case by determining the components of the waste-management system and the path of waste material flows. The next step is to develop assumptions that characterize the facilities and processes in the system in sufficient detail for the derivation of engineering cost estimates. These assumptions also establish the scope of the system by specifying the quantity and the schedule of waste acceptance. The assumptions for the waste-generation rate, the minimum age at which spent fuel will be accepted by the DOE, and the waste acceptance rate can be used to estimate the annual flows of waste that will occur once the system is operational. The waste flows determine both transportation and repository costs because they determine how much and when waste has to be transported and how long the repository

Case	Repository Host-Rock Type First	Repository Host-Rock Type Second	Cumulative Spent-Fuel Generation (MTU)	Repository Startup Date First	Repository Startup Date Second	Additional Storage Facilities
Reference cases						
	Salt	Crystalline	130,300	1998	2006	None
	Tuff	Crystalline	130,300	1998	2006	None
	Basalt	Crystalline	130,300	1998	2006	None
	Salt	Salt	130,300	1998	2006	None
	Tuff	Salt	130,300	1998	2006	None
	Basalt	Salt	130,300	1998	2006	None
	Basalt	Tuff	130,300	1998	2006	None
Sensitivity cases						
Low waste generation	Basalt	Salt	97,700	1998	2006	None
	Basalt	Tuff	97,700	1998	2006	None
	Salt	Crystalline	97,700	1998	2006	None
	Tuff	Crystalline	97,700	1998	2006	None
Improved transportation cask technology	Basalt	Tuff	130,300	1998	2006	None
	Salt	Crystalline	130,300	1998	2006	None
Repository delay 5-Year	Basalt	Tuff	130,300	2003	2011	MRS
	Salt	Crystalline	130,300	2003	2011	MRS
10-Year[A]	Basalt	Tuff	130,300	2008	2016	MRS
	Salt	Crystalline	130,300	2008	2016	MRS

[A]The waste-receipt rate pattern for the two-stage repository is 400 MTU/yr for the first 3 years of operation, 900 MTU in the fourth year, 1800 MTU in the fifth year, and 3000 MTU/yr thereafter. The pattern for the single-stage repository is 1800 MTU/yr for the first 5 years of operation and 3000 MTU/yr thereafter.

FIGURE 1. Case Structure for the TSLCC Analysis[A]

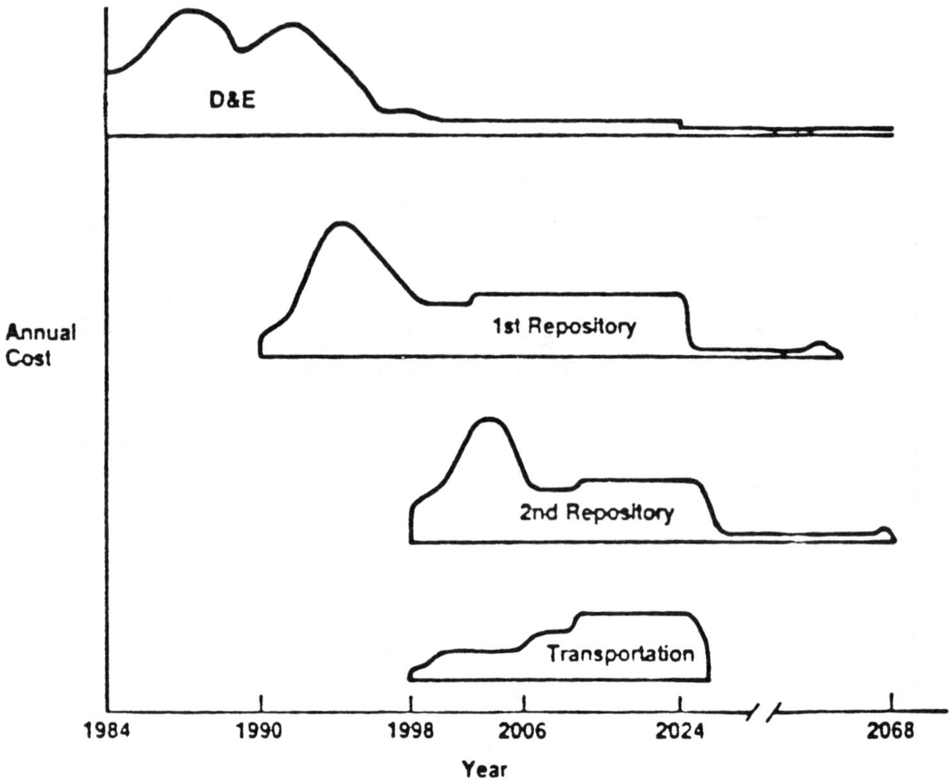

FIGURE 2. Annual Program Expenditures by Major Category

must operate before it is filled to capacity. By using a series of estimating techniques, the total system cost for the single reference case is then calculated.

B. Reference Case Description

The reference case assumes that, after spent fuel has cooled sufficiently for DOE acceptance, it will be transported directly to the repository. Therefore, the only facilities in the reference system are the repository and the required transportation network. The key assumptions for the reference case are:

- Waste type — Spent fuel, low-level waste generated at the repository during spent-fuel consolidation and handling, and commercial high-level waste from West Valley, New York (defense waste is not included)

- Waste quantity — 130,300 MTU, based on the November 1984 midcase projection by the Energy Information Administration (EIA) of the cumulative waste generation through the year 2020[5]

- Minimum waste age for acceptance	Five years
- Number of repositories	Two
- Design capacity	70,000 MTU for the first repository and 60,300 MTU for the second repository.
- Repository design	First repository: 400 MTU receipt rate per year for the first three years, 900 MTU for the fourth year, 1800 MTU for the fifth year, and 3000 MTU per year thereafter to closure. Second repository: 1800 MTU per year for the first 5 years and 3000 MTU per year thereafter to closure.
- Repository startup	1998 for the first repository dates and 2006 for the second repository
- Host-rock type	Basalt, crystalline rock, salt, and tuff
- Transportation cask	Currently licensed technology transportation casks
- Cost basis	Constant 1984 dollars.

Using a selection logic described in detail in reference 4 led to the production of seven reference case pairs based on repository host rock and location. Each of these cases accomodates the same amount of fuel, assumes the same repository design concepts (two stages for the first repository and one stage for the second repository), startup dates, and uses the same type of transportation casks.

C. Sensitivity Case Description

Three categories of sensitivity cases were chosen for analysis: 1) low waste generation, 2) improved transportation-cask technology; and 3) repository delay. To represent a low waste generation case, the EIA "No New Orders" projection of November 1984 was assumed.[5] The cumulative amount of spent fuel generated in this case is 97,700 MTU. The repository waste-acceptance rate and startup schedule remain the same as in the reference case. Four combinations of first-and second-repository locations are assumed: 1) basalt and tuff, the reference-case combination with the highest total system cost; 2) salt and crystalline rock, the reference-case combination with the lowest total system cost; 3) basalt and salt; and 4) tuff and crystalline rock.

These combinations provided an adequate assessment of the cost impacts associated with less waste in the system.

Currently licensed spent-fuel transportation casks assumed for the reference case are inefficient in comparison with the cask designs that are expected to be developed. Current casks are designed to transport spent fuel that is aged a minimum of 120 days from the time of discharge from the reactor. A greater cask capacity can be realized because less shielding will be required by the older and less/radioactive fuel which will be at least five-years old. An alternative, more efficient set of transportation casks was therefore included in this analysis as a sensitivity case to determine the effect on the total system cost. The cost estimates were calculated for two repository-location combinations that represent the greatest (basalt/tuff) and shortest distances traveled (salt/crystalline rock) from among the seven reference-case combinations.

Although the DOE is committed to start accepting spent fuel from utilities by January 1998, two delay cases of five and ten years in the opening of both the first and the second repository were examined. Costs for each of the repository-delay cases were estimated for the reference case repository-location combinations that yielded the highest (basalt/tuff) and lowest (salt/crystalline rock) total system cost.

D. Development and Evaluation Category Costs (D&E)

The D&E cost category covers all the siting, design development, testing, regulatory, and institutional activities associated with the repositories, and the required transportation network; it also covers the D&E activities associated with monitored retrievable storage (MRS). Most of the D&E activities take place before the construction of waste-receiving facilities and the fabrication of waste packages and transportation casks, but some efforts, such as regulatory activities, continue during the facility construction period. Included under D&E is the mitigation of socioeconomic impacts, which is assumed to occur throughout repository construction. Also included in this category is the cost of Federal Government administration of the entire waste-disposal program. As so defined, D&E encompasses all program expenditures both at the present time and for the next several years.

The starting point for D&E cost estimation is a schedule of overall program milestones. Having established a milestone schedule, the activities currently under way to accomplish these milestones must be determined along with the current costs for these activities. Next, activities in the future that must be either continued or initiated to accomplish the entire schedule of milestones are determined, and the time periods over which these future activities must take place are estimated. Finally, the costs of performing each of these activities are judgmentally assumed, drawing on the cost activity relationships of the current and near term activities and independent cost estimates of future activities, where available.

The primary data source for the cost of current program activities is the budget developed by the DOE for the Nuclear Waste Fund. For all of the D&E cost categories, the FY 1986 budget submitted by the DOE to Congress in January 1985 served as the numerical basis through the year 1990.

This reference-case D&E cost estimate pertains to all the TSLCC cases in which the repositories are assumed to start operating according to the program milestone schedule (see Figure 3) and is essentially the same across all host rocks and spent-fuel generation scenarios. In addition to the reference case estimates, the five- and ten-year repository delay sensitivity cases require alternative D&E cost estimates.

Program Milestone	First Repository	Second Repository
Identify potentially acceptable sites	Completed	5/86
Complete draft environmental assessments	12/84	12/90
Complete final environmental assessments	6/85	6/91
Nominate sites	6/85	6/91
Recommend candidate sites	7/85	7/91
Presidential approval of candidate sites	9/85	9/91
Start preparation of exploratory-shaft site	12/85-10/86[A]	9/91
Start exploratory-shaft construction	3/86-02/87[B]	2/92
Start preliminary waste-package design	6/87	6/93
Complete exploratory-shaft construction	3/88-09/88[C]	8/94
Start Title I repository design	2/88	1/94
Complete exploratory-shaft testing for DEIS and recommendation	12/89	12/95
Issue draft environmental impact statement (DEIS)	6/90	6/96
Issue final environmental impact statement (FEIS)	12/90	12/96
Recommendation to President	1/91	1/97
Complete exploratory-shaft testing for construction authorization application	11/90	11/96
Complete Title I repository design	5/90	1/96
Site designation effective, submit construction authorization application	5/91	5/97
NRC grants license, start operations	1/98	2/2006

[A] Earlier date is for basalt and tuff while the later date is for salt.

[B] Earlier date is for basalt, and the later date is for salt. The date for tuff is between these two dates.

[C] Earlier date is for tuff, and the later date is for salt. The date for basalt is between these two dates.

FIGURE 3. Schedule of Program Milestones (as of January 1985)

E. Transportation Costs

The NWPA requires the DOE to take title to the spent fuel at the reactors and to arrange for transportation to storage or disposal facilities. The method for estimating transportation costs is based on an analysis prepared by Pacific Northwest Laboratories, with substantial refinements.[6] This method derives a unit charge for transportation-cask utilization, shipping, and security for each potential transportation pathway; this unit charge is applied to the annual waste material flows to arrive at the total transportation cost. The pathways considered include transportation from the reactors to each repository location, from reactors to an MRS facility (which is assumed to be used only in the repository delay cases), and from the MRS facility to each repository. The total unit transportation cost is the sum of these three unit costs. Each of these estimations is performed for rail and truck transportation. A split between the two modes is assumed in order to calculate the total cost.

Before these unit costs can be derived, the transportation distance for each pathway must be estimated. The distances were estimated in two steps. First, average distances were estimated from the reactors to the repository and the MRS location and from the MRS location to the repository location. These distances are pertinent only when one repository is in operation. Second, a reduction in these distances was calculated to account for the potential savings of optimized routing for the years when two repositories are in operation. Unit costs were estimated for each set of distances and are appropriately applied to the annual waste material flows, depending on whether in a specific year one or two repositories are in operation.

F. Repository Construction, Operation, and Decommissioning Costs

Costs for the two repositories represent the largest component of the TSLCC. By virtue of their relative importance, the repository costs have undergone a substantial review and re-estimation process since the previous set of costs was calculated. The cornerstone of this process was a program strategy shift whereby the design for the first repository was changed from a one stage facility to a two stage facility. This enabled the first repository to accept limited quantities of spent fuel before construction of the complete facility is finished. The feasibility study that was performed for the two stage concept included a cost analysis of the new facility design. The repository cost estimates of the January 1985 TSLCC analysis rely very heavily on the costs emerging from this study.

The method used in estimating the repository costs follows a three-step procedure. First, costs were developed for standard size facilities (70,000 MTU) for each of the alternative host rocks. Second, a scaling technique was applied to the costs of a standard-size facility to derive costs for the facility capacities that are required by the waste material flows for each case in the analysis. Third, the total construction, operation, and decommissioning costs were annualized for inclusion with the other TSLCC components.

In all cases but the low-waste generation case, the first-repository capacity is identical with that of the standard-size facility, or 70,000 MTU. Therefore, no adjustment in the costs of the standard-size costs is required. The second repository capacity, though, is 63,300 MTU, so that the costs of the standard size facility must be scaled downward to estimate the second-repository costs. In the low-waste generation case, both the first and the second repositories are sized below design capacity, and the scaling technique is applied to each.

G. Storage Costs

The April 1984 draft Mission Plan states that if there is a "substantial" repository delay, the DOE will store waste in a monitored retrievable storage (MRS) facility. The five- and ten-year delay cases are assumed to be "substantial" delays, so that storage is provided by an MRS facility in each case. These interpretations of the Mission Plan strategy do not represent DOE policy; they are simply analytical assumptions made for the sole purpose of performing the cost estimation.

Costs were estimated after first determining the amount of the fuel that needs to be stored in each delay case. This was accomplished by assuming that even with a delay in the start of repository operation, the DOE would accept fuel at the reference-case rate. Therefore, the storage requirements were determined by calculating the difference between the reference-case acceptance schedule and the repository acceptance schedule in each delay case. Having made this determination and after deciding on the preferred engineered concept for providing the required storage at an MRS facility, cost estimates were developed by using the best information available on the costs of constructing and operating the required storage facilities. For the 5-year case, a storage capacity of 30,000 MTU is required, while in the 10-year case a capacity of nearly 55,000 MTU is needed. Total costs for each case are $2.0 and $2.4 billion, respectively.

III. RESULTS/COMPARISONS/NEXT YEAR'S ANALYSIS

Repository costs represent 51 to 59 percent of the total-system costs. Development and evaluation costs are the next largest component, accounting for 26 to 33 percent of the total; and, transportation represents 14 to 18 percent of the TSLCC. The relative importance of each of these components is decreased in the repository-delay sensitivity cases because of the inclusion of storage costs. Storage costs may be as high as 8 percent of the total in the 10-year-delay case. Figure 4 summarizes the results for the 7 reference cases and 10 sensitivity cases.

The principal findings of this analysis are as follows:

Case (repository location)	D&E	Transportation	Repository First	Repository Second	Repository Subtotal	Storage	Grand Total[A]
Reference cases							
Tuff/crystalline rock	7.8	3.8	7.0	6.1	13.1	-	24.7
Basalt/crystalline rock	7.8	3.9	10.8	6.1	16.9	-	28.5
Salt/crystalline	7.8	3.3	6.7	6.1	12.8	-	23.8
Tuff/salt	7.8	4.4	7.0	5.8	12.8	-	25.0
Basalt/salt	7.8	4.4	10.8	5.8	16.6	-	28.8
Salt/salt	7.8	4.0	6.7	5.8	12.5	-	24.2
Basalt/tuff	7.8	5.1	10.8	6.1	16.9	-	29.7
Sensitivity cases							
Low waste generation							
Basalt/salt	7.7	3.5	10.4	4.2	10.6	-	25.9
Tuff/crystalline rock	7.8	3.1	6.8	3.9	10.7	-	21.6
Basalt/tuff	7.8	3.8	10.4	4.0	14.4	-	26.1
Salt/crystalline rock	7.8	2.6	6.6	3.9	10.5	-	20.9
Improved transportation-cask technology							
Basalt/tuff	7.8	2.6	10.8	6.1	16.9	-	27.2
Salt/crystalline rock	7.8	1.6	6.7	6.1	12.8	-	22.2
Repository delay							
5-year Basalt/tuff	8.4	5.9	10.8	6.1	16.9	2.0	33.2
Salt/crystalline rock	8.4	4.0	6.7	6.1	12.8	2.0	27.1
10-year Basalt/tuff	9.0	6.6	11.1	6.3	17.4	2.4	35.3
Salt/crystalline rock	9.0	4.6	6.7	6.3	13.0	2.4	28.9

[A] Cost categories may not add to total because of independent rounding.

FIGURE 4. Summary of Total System Life Cycle Cost Estimates (billions of 1984 dollars)

- Reference case TSLCC costs, expressed in constant 1984 dollars, range from $23.8 to $29.7 billion, depending on the location of the two repositories. For the sensitivity cases the TSLCC may be as high as $35.3 billion or as low as $20.9 billion.

- Repository location has a significant effect on the TSLCC; the TSLCC for the repository combination with the highest reference-case cost (basalt/tuff) is 25 percent higher than the TSLCC for the repository combination with the lowest reference-case cost (salt/crystalline rock).

- Waste-generation rate also has a significant effect. The TSLCC for the low-waste-generation case is $2.9 to $3.6 billion less than that for the reference case, depending on repository location. However, on an average unit-cost basis, the low-waste-generation case is 17 percent more expensive than the reference case because of economies of scale.

- Improved transportation-cask technology reduces transportation costs by 50 percent and decreases the TSLCC for the reference case by $1.7 to $2.5 billion, or 7 to 8 percent, depending on repository location.

- Delays of 5 or 10 years in the opening of the repositories increase the TSLCC by $3.3 to $3.5 billion and $5.1 to $5.6 billion, respectively, depending on the repository location combination. These additional costs are due to extended development and evaluation activities, storage facilities that would otherwise not be required, and increased transportation requirements for shipments to and from a monitored retrievable storage facility. (In the 10-year-delay case, repository construction costs also increase because, by definition, the delay is partially due to problems experienced in construction). Both repositories receive 100 percent of the design receipt rate throughout their operating period in the reference case; thus delaying the start of repository operation does not affect repository operation costs because under this delay scenario the system also operates at maximum efficiency.

- Because of the sizable cost impacts of repository location, waste generation rate, transportation cask technology, and repository startup dates, the effects of one parameter may be partially or wholly offset by the effects of another parameter. For example, the TSLCC for the 5-year delay case with repositories in salt and crystalline rock is less than the reference-case cost for repositories in basalt and tuff.

- Cost estimates have risen substantially in each successive TSLCC analysis as is indicated by the data in Figure 5. For example, total estimated program costs for the basalt/tuff repository combination rose 23% and 22%, respectively. More than two-thirds of the latest increase comes from higher repository construction and generation costs. Differences in the TSLCC calculations are due to many factors

Major Cost Category	July 1983 Report (1982 Dollars)	April 1984 TSLCC Analysis (1983 Dollars)	January 1985 TSLCC Analysis (1984 Dollars)
D&E	4.7	7.6	7.8
Transportation	3.9	2.5-3.9	3.3-5.1
Repository	10.7-11.2	10.5-12.9	12.5-16.9
Total[b]	19.3-19.8	20.9-24.4	23.8-29.7

Note: The range in total costs may not equal the sum of maximum/minimum costs for each category because (1) the ranges of each category may not be based on the same case, and (2) independent rounding of the costs for each category was made.

FIGURE 5. Comparison of Total-System Cost Estimates for the Reference Program (Billions of Dollars)

including changes in estimation methods, assumptions, program content, and schedule. Certainly, however, confidence in the January 1985 analysis is substantively greater than in earlier analyses.

Current plans for the upcoming January 1986 TSLCC analysis include consideration of:

- emplacing defense waste in the repositories,
- adding an "Integral MRS System," (this is not to be confused with the MRS included in the January 1985 sensitivity analysis which functions as a storage facility until the "delayed" repository is available to receive shipments; rather, DOE is currently drafting a proposal to Congress to consider including an MRS as an integral part of the system; here the MRS may take on several functions other than storage; this DOE proposal is due to Congress by mid-January 1986 and DOE must receive approval before proceeding with the development of this facility),
- modifying the waste generation rate projections to include the assumption of 2.5% per year extended burnup until 1995, and
- alternative and more detailed calculation procedures for the exploratory shaft portion of the D&E costs, and for the total system transportation costs.

Assumptions, program content, schedule and other influencing elements have changed and may continue to change over the next several years leading to corresponding fluctuations in TSLCC estimates. However, it is expected that these variances will become less frequent and thus enable more precise program cost estimates to be made. In the meantime, data input and cost analysis techniques are being considerably enhanced in preparation of more rigorously defined program assumptions, content, and schedule.

REFERENCES

1. U. S. Department of Energy, Nuclear Waste Policy Act Project Office, Report on Financing the Disposal of Commercial Spent Nuclear Fuel and Processed High Level Radioactive Waste, DOE/S-0020/1 (July 1983).

2. U. S. Department of Energy, Office of Civilian Radioactive Waste Management, Nuclear Waste Fund Fee Adequacy: An Assessment (July 1984).

3. U. S. Department of Energy, Nuclear Waste Fund Fee Adequacy: An Assessment, DOE/RW-0020 (February 1985).

4. U. S. Department of Energy, Analysis of the Total System Life Cycle Cost for the Civilian Radioactive Waste Management Program, DOE/RW-0024 (April 1985).

5. U. S. Department of Energy, Energy Information Administration, Commercial Nuclear Power 1984: Prospects for the United States and the World, DOE/EIA-0438(84) (November 1984).

6. Battelle Pacific Northwest Laboratories, Fiscal Implications of a 1 Mill/kWh Waste Management Fee, PNL-4513, UC-85, Richland, WA (December 1982).

EFFECTS OF WASTE CONTENT OF GLASS WASTE FORMS ON SAVANNAH RIVER HIGH-LEVEL WASTE DISPOSAL COSTS

W. R. McDonell
C. M. Jantzen
E. I. du Pont de Nemours and Co., Inc.
Savannah River Laboratory
Aiken, South Carolina 29808

ABSTRACT

Effects of the waste content of glass waste forms on Savannah River high-level waste disposal costs are evaluated by their impact on the number of waste canisters produced. Changes in waste content affect onsite Defense Waste Processing Facility (DWPF) costs as well as offsite shipping and repository emplacement charges. A nominal 1% increase over the 28 wt% waste loading of DWPF glass would reduce disposal costs by about $50 million for Savannah River wastes generated to the year 2000.

Waste form modifications under current study include adjustments of glass frit content to compensate for added salt decontamination residues and increased sludge loadings in the DWPF glass. Projected cost reductions demonstrate significant incentives for continued optimization of the glass waste loadings.

I. INTRODUCTION

In the Savannah River plan for high-level waste disposal, aqueous radioactive wastes are processed into solid form for final disposal.[1,2] Sludge components of the waste and radionuclides separated from waste salts[3] are converted to glass by melting in the Defense Waste Processing Facility (DWPF),[2,4] and the decontaminated salt solutions are fixed in concrete for onsite burial.[5] After a period of interim storage, the waste glass contained in stainless steel canisters will be shipped to a federal geologic repository for final disposal.[6]

Optimization of the glass waste forms to be produced in the DWPF is being supported by economic evaluations of the impact of the forms on waste disposal costs. Glass compositions are specified for acceptable melt processing and durability characteristics, with economic effects tracked by the number of waste canisters produced. This paper presents an evaluation of effects of variations in waste content of the glass waste forms on the overall cost of disposal, including offsite shipment and repository emplacement, of the Savannah River high-level wastes. The status of recent investigations undertaken to optimize compositions of the sludge-salt glasses are outlined.

II. BACKGROUND

High-level radioactive wastes at the Savannah River Plant, now contained as aqueous sludges and salts in large underground tanks, will be processed to solid form for final disposal.[1] The principal activities involved in the disposal operations are represented in Figure 1. Sludge components, which contain most of the Sr-90 and long-lived actinides, will be washed and treated with sodium hydroxide in tank-farm operations to reduce aluminum hydroxide content, mixed with glass frit and fed as slurry to the DWPF melter.[2] The product of the DWPF melting operation is a borosilicate glass containing nominally 28 wt% waste oxides in stainless steel canisters about 60-cm diameter by 300-cm length.[4] For a representative glass composition, the DWPF output at 75% attainment is 410 canisters per year. Initial DWPF processing will be limited to wastes providing canister heat ratings less than 460 watts; after 5 years operation, wastes providing up to 690 watts/canister may be processed.

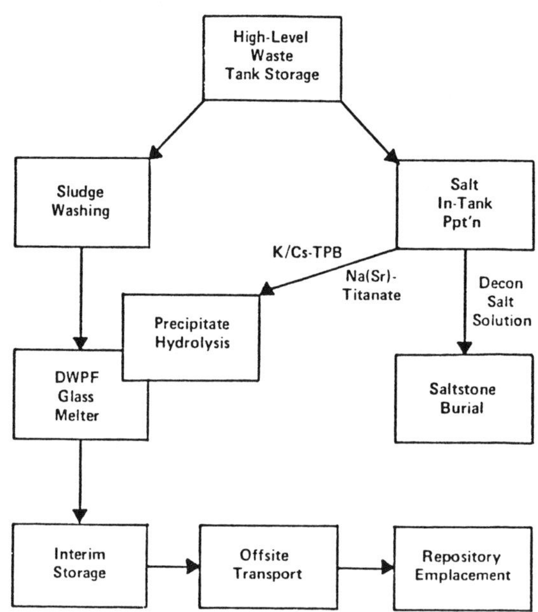

FIGURE 1. Reference High-Level Waste Disposal Activities

Soluble salt components of the wastes will be decontaminated in the waste tanks prior to solidification in concrete for onsite burial. In the salt decontamination operation, radiocesium (principally Cs-137) will be separated along with potassium in the wastes by precipitation as tetraphepylborate (TPB) salts, and residual Sr-90 and actinides will be removed by adsorption on sodium titanate.[3] The precipitated solids, concentrated by filtration and accumulated in the tank, will be treated using a formic acid hydrolysis process (termed "precipitate hydrolysis") to decompose the organic salt, and the product fed along with sludge components of the waste to the glass melter. After filling with glass, the DWPF canisters will be decontaminated, sealed, and stored onsite in a convection air-cooled vault, pending availability of a federal repository for permanent waste disposal.[4]

The decontaminated salt solution will be converted to solid form (termed "saltstone") by mixing with a blended cement for onsite burial as low-level waste.[5] Salt solution will be processed at a rate of about 6-million gallons per year, including recycle washes. Saltstone facilities limit the age of salt solution processed to a minimum of 15 years following reactor discharge.

Following interim storage, the canisters in special casks are assumed to be shipped by truck or rail to the repository site at a rate equal to or somewhat in excess of the rate of DWPF production.[6] At the repository, the canisters will be packaged in appropriate overpack containers and emplaced for permanent disposal in underground facilities. In a representative case, the DWPF canisters would be buried in salt, tuff, or basalt geologic formations in conjunction with commercial nuclear wastes, using a repository design specially augmented to accommodate the defense wastes.[7]

III. SRP WASTE DISPOSAL COSTS

A. Cost Model

The cost model developed for SRP high-level waste disposal defines basic input parameters for the specific processing activities in the waste disposal system.[8] The input parameters include fixed and variable cost components for each activity, with variable cost components dependent on the quantity of waste (for example, number of waste canisters) processed annually. Since economic effects of the waste content of the glass waste forms arise primarily from changes in the number of waste canisters processed, only the variable costs associated with DWPF processing (including interim storage), offsite transport, and repository emplacement of the canisters are evaluated in this study. Variable costs associated with sludge washing, salt precipitation, and saltstone processing activities are not affected by changes in the number of canisters processed, and fixed costs representing capital expenditures for all the waste processing activities are considered constant.

B. Reference DWPF Operation

Derivation of incremental costs from input parameters of the cost model requires projection of reference operating scenarios for onsite and offsite waste processing facilities. For the onsite (DWPF) facilities, it is assumed that SRP reactors will operate at approximately the current level of waste generation to the year 2000. This is an arbitrary assumption of the cost model, since actual SRP operation will depend on demand for nuclear materials. For the year 2000 operation as represented in Figure 2, current and future inventories of waste will utilize full DWPF operating capacity until completion of sludge processing about the year 2005, following which the DWPF is assumed to be maintained in standby condition to allow batching of a residual inventory of 15-year aged salt for workoff in a final campaign. Typically 7500 canisters would be produced over the approximately 25 years of DWPF operation (including standby) required for processing waste sludge and salt generated to year 2000.

Incremental canisters resulting from adjustments of glass compositions impact the reference DWPF operating scenario by increasing (or decreasing) the time of full-capacity DWPF processing before shutdown (standby) on completion of sludge processing operations. For a nominal 3.5% decrease in the number of canisters produced (equivalent to 1% increase in canister waste content), the time of DWPF operation at full capacity is decreased by somewhat less than one year, as shown in Figure 2. Incremental costs assigned correspond to the difference in costs of full-capacity and standby DWPF operation over this period. The incremental annual costs of DWPF operation, summarized in Table 1, are about $46.4 million corresponding to $113,000 per canister.

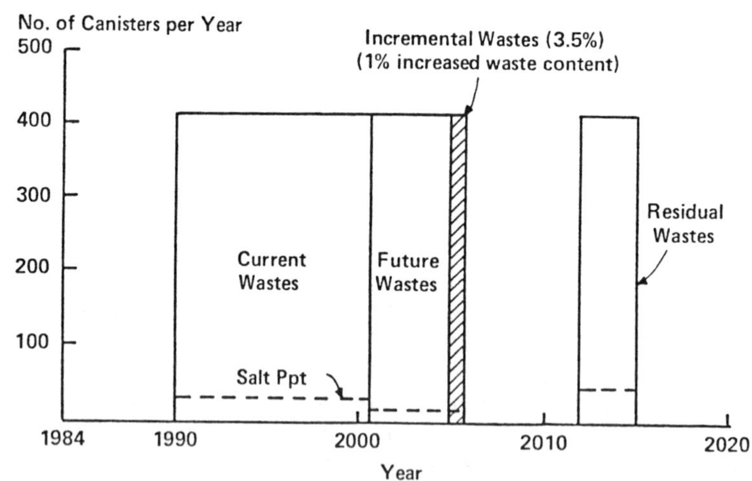

FIGURE 2. Schedule for Processing Year 2000 High-Level Waste Inventory DWPF Canisters

TABLE 1. Incremental Costs of SRP High-Level Waste Disposal Activities

Activity	Incremental Costs, 10^3 \$(1985) Annual	Unit*
DWPF Operation	46,400	113.2
Offsite Shipment	--	11.0
Repository Emplacement	--	55.0
Total		179.2

*Unit costs assume 410 DWPF canisters per year.

C. Offsite Transport

Transport of the SRP waste canisters to a federal repository following interim onsite storage is projected in accord with previous studies.[8-10] Shipment of the SRP canisters contained in a single canister truck cask of appropriate design to a repository located in the Northwest is assumed at a rate of 500 canisters per year beginning about year 2000. The incremental canisters would be shipped and emplaced in the repository near the end of the SRP waste processing campaign in year 2015. Incremental costs estimated using updated tariff rates are projected in 1985 dollars at about $11,000 per canister.

D. Repository Packaging and Emplacement

Repository costs of SRP high-level waste disposal are projected on the assumption that the wastes will be emplaced in a commercial repository augmented to accommodate the defense waste canisters. As previously detailed,[7] emplacement of 7500-DWPF canisters in a salt, tuff, or basalt repository of standard design would require an underground area typically about 5% of that committed to the commercial high-level wastes. The incremental costs incurred under these circumstances are principally variable costs proportional to the number of DWPF canisters handled.

Prior to emplacement, the waste canisters would be packaged in appropriate overpack containers as required to meet regulatory requirements for the specific repository geology involved.[11] The overpack containers consist generally of carbon steel or stainless steel assemblies of sufficient thickness to withstand hydrostatic or lithostatic pressures of the repository, sometimes encased in a titanium alloy sheath for corrosion resistance.

The waste packages are assumed to be emplaced in boreholes excavated either vertically or horizontally in underground rooms or tunnels of the repository, with spacing dependent on heat output of the packages and thermomechanical properties of the geologic medium.[7] Placement room areas for defense wastes in vertical boreholes are typically 30 m^2 per package (15 to 23 W/m^2 for 460 to 690 watt DWPF canisters) compared to commercial (spent fuel) wastes ranging from 85 to 250 m^2 per package (15 W/m^2) depending on heat output. The lower placement room areas compared to commercial wastes are a consequence of the lower heat loading (lower radioactivity content) of the defense waste packages. The defense waste packages are generally not emplaced at heat load limits for the repository because of mechanical constraints on borehole spacing, so that nominally increased waste loadings can be tolerated with the same placement room areas, especially for lower heat wastes processed in early DWPF operations.

A repository fee for defense waste disposal analogous to that charged utilities for commercial waste disposal is under current consideration in the Department of Energy. The fee established for commercial waste disposal, equal to 1.0 mill per kWh electricity generated using nuclear facilities, depends primarily on the quantity of radioactivity in the commercial wastes, and not on the volume of waste or number of waste packages emplaced. An analogous assignment of repository fee for defense wastes would thus provide no dependence of repository costs on the number of waste canisters emplaced. Over the long term, however, the repository fee for defense waste must cover the real costs of disposal, so that reduced numbers of canisters emplaced would be reflected in lower repository fee assessments.

The real costs of repository emplacement of defense waste canisters are projected in the Savannah River model in two general categories, 1) direct costs representing incremental capital and operating expenditures and 2) indirect costs representing prorated charges assessed for use of facilities in common with commercial wastes.[7,8] Since incremental expenditures are of primary interest for evaluation of waste content effects, only the direct costs are considered in this study. Variable components of the direct costs dependent on the number of canisters emplaced include incremental expenditures for waste packaging components and operations, for mining of placement rooms and boreholes, for emplacement and monitoring operations, and for backfill operations. Such costs depend on repository geology, waste packaging requirements, and underground emplacement patterns. Representative values for SRP waste disposal are derived as average costs for canisters in steel overpacks emplaced in characteristic patterns (vertical or horizontal boreholes) in salt, tuff, and basalt repositories. The variable costs are projected in 1985 dollars at about \$55,000/canister, including packaging costs of about \$31,000/canister and mining and emplacement costs of \$24,000/canister.

E. Incremental Costs of Increased Glass Waste Content

The cost impacts of increasing the waste content of DWPF glass, summarized in Table 2, are established by the foregoing projections. As shown in Table 1, costs dependent on the number of canisters processed, including costs of DWPF processing, offsite transport, and repository emplacement, total \$179,000 per canister. For a total of 7500 waste canisters produced by

TABLE 2. Cost Impact of Increased Waste Content in DWPF Glass

Waste Content Increase	Incremental Canisters	Cost Reduction, 10^6 \$(1985)* Nominal	Present Value 3%	Present Value 10%
Unit (1 wt%)	268	48	24	5.5
Frit Adjustment (8 wt%)	2200	394	198	45
Increased Sludge (7 wt%)	1500	269	135	31

*SRP wastes generated to year 2000.

processing SRP waste generated to the year 2000, a nominal 1% increase in waste loading represents a decrement of 268 canisters, equivalent to $48 million in 1985 dollars in reduced waste disposal costs. Assuming DWPF processing of the incremental canisters about the year 2005 in accord with Figure 2 and transport and repository emplacement of the incremental canisters about the year 2015, present values of the incremental waste loading costs are $24 million and $5.5 million (1985 dollars) discounted at 3% and 10%, respectively.

Two modifications of waste loading are under current study at Savannah River. In a first modification now being optimized as the reference glass composition, precipitate hydrolysis product is substituted for frit rather than added as a waste component of the glass.[8] This reduces the frit content from 72 to 64% and effectively increases the waste content by about 8%. The accompanying decrease of 2200 waste canisters required for SRP waste generated to year 2000 results in a cost reduction of $394 million in 1985 dollars before discounting, with present values of $198 million and $45 million after discounting at 3% and 10%, respectively. Changes in frit composition at this waste loading needed to optimize properties of the glass do not additionally affect the cost reduction produced by the frit adjustment.

In a second modification, sludge-salt glasses with increased sludge content are being investigated. There are no present plans for increasing sludge loadings in the DWPF glass, but the cost effects indicate this may be a fruitful area for development after DWPF startup. An increase of the sludge content from 28 to 35% further decreases the frit content from 64% to about 57%, providing an additional 7% increase in waste content. The accompanying 1500 decrease in number of waste canisters produced for SRP reactor operation to the year 2000 results in a cost reduction of $269 million in 1985 dollars, with present values of $135 million and $31 million at discount rates of 3% and 10%, respectively.

The status of property measurements undertaken to optimize the compositions of the modified waste glasses is indicated in the following section.

IV. Optimization of Glass Waste Compositions

A. Frit-Adjusted Compositions

Optimization of the composition of the frit-adjusted sludge-salt glasses is necessary to provide acceptable processability and durability properties. Additions of precipitate hydrolysis product from the salt decontamination process introduces excess alkali (Na, K, and Cs) and boron oxides, tending to decrease melt viscosities during DWPF processing and increase leachability of the glass. Properties of the frit-adjusted sludge-salt glasses are being determined by measurement of viscosity and leachability of compositions with frit compositions selected to compensate for the precipitate hydrolysis product. Representative frit compositions, developed by modification of compositions previously optimized for sludge-only wastes,[12] are of three general types, illustrated in Table 3.

TABLE 3. Frit Compositions for Sludge-Salt Glasses*

Component	Type 1 (Frit 165)	Type 2 (Frit 165 Mod)	Type 3 (Frit 165 Alk)
SiO_2	68	76	77
Na_2O	13	10	6
Li_2O	7	6	7
B_2O_3	10	7	8
MgO	1		
ZrO	1	1	2

*Reference Frit 165 for sludge-only glasses modified for sludge-salt glasses: Type 1 - Frit content decreased with unchanged composition; Type 2 - Frit composition decreased in alkali and boron oxides to compensate for Na_2O, Cs_2O, and B_2O_3 in the precipitate hydrolysis product; Type 3 - Frit composition additionally decreased in alkali and boron oxides to compensate for K_2O in the precipitate hydrolysis product.

Results of both leach resistance (durability) and melt viscosity tests correlate with the molar ratio parameter $(SiO_2)/\Sigma(\text{alkali} + B_2O_3)$ of the glasses. As shown in Figure 3, the best leach resistance in deionized water is exhibited by sludge-salt glasses containing Types 2 and 3 frits with reduced alkali-B_2O_3 content. In accord with the molar ratio parameter, the durability of the Type 3 frit glasses (Table 3) is generally comparable to that of the current reference Frit 165 sludge-only glass. All the candidate sludge-salt glasses fall within calculated molar ratio parameter limits correlating empirically with acceptable melt viscosity values.

B. Increased Sludge Compositions

Increased sludge loadings up to 35 wt% produce leachabilities about the same as the reference 28 wt% waste loadings without deleteriously affecting viscosity in sludge-only glasses.[13] A similarly increased sludge loading may therefore be acceptable for the sludge-salt glasses. Durability tests of 35 wt% sludge glasses with adjusted frit compositions analogous to those of

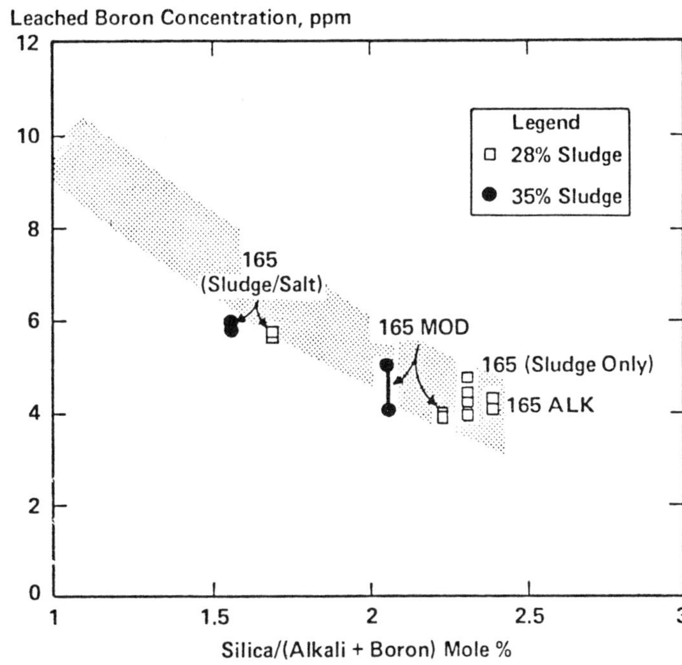

FIGURE 3. Durability of Sludge-Salt Glasses in Deionized Water (MCC 28-day Static Test). Leached boron concentrations are shown as function of the molar ratio parameter for glasses containing modified frits given in Table 3. Background shading indicates general dependence on leachability on the molar ratio parameter of SRP glasses.

Table 3 are in progress, and other property measurements including melt viscosity are projected. Preliminary results included in Figure 3 show leachabilities for sludge-salt glasses with 35 wt% sludge content comparable to those with 28 wt% sludge content. Minimum leachability values with modified frit compositions are equivalent to those for the optimized Frit 165 sludge-only glass. Such results suggest that acceptable sludge-salt glasses with increased sludge content may be feasible for future DWPF operations.

ACKNOWLEDGMENT

The information contained in this article was developed during the course of work under Contract No. DE-AC09-76SR00001 with the U.S. Department of Energy.

REFERENCES

1. R. Maher, L. F. Shafranek, J. A. Kelley, and R. W. Zeyfang, "Solidification of Savannah River Plant High Level Waste," Trans. Am. Nucl. Soc., 39, 228 (1981).

2. M. D. Boersma, "Process Technology for Vitrification of Defense Waste at Savannah River," Proceedings - Fuel Reprocessing and Waste Management, Jackson, Wyoming, August 1984, Vol. 1, p. 131, American Nuclear Society.

3. L. M. Lee and L. L. Kilpatrick, "Precipitation Process for Decontamination of Water Soluble SRP Radioactive Waste," Trans. Am. Nucl. Soc., 43, 124 (1982); H. D. Martin, M. A. Schmitz, M. A. Ebra, D. J. Walker, L. E. Lee and L. L. Kilpatrick, "In-Tank Precipitation Process for Decontamination of Water-Soluble SRP Radioactive Waste," Waste Management 84, Tucson, Arizona, Vol. 1, p. 291, Univ. of Arizona (1984).

4. R. G. Baxter et al., "The Defense Waste Processing Facility at the Savannah River Plant," Waste Management 84, Tucson, Arizona, Vol. 1, p. 275, Univ. of Arizona (1984).

5. C. A. Langton, M. D. Dukes and R. V. Simons, "Saltstone: Cement-Based Waste Form for Disposal of SRP Low-Level Waste," Waste Management 84, Tucson, Arizona, Vol. 1, p. 297, Univ. of Arizona (1984).

6. Department of Energy (DOE), The Defense Waste Management Plan, DOE/DP-0015 (June 1984).

7. W. R. McDonell, "Costs for Disposal of Savannah River High-Level Waste in a Commercial Repository," Waste Management 84, Tucson, Arizona, Vol. 1, p. 553, Univ. of Arizona (1984).

8. W. R. McDonell and C. B. Goodlett, "Systems Costs for Disposal of Savannah River High-Level Waste Sludge and Salt," Proceedings - Fuel Reprocessing and Waste Management, Jackson, Wyoming, August 1984, Vol. 1, p. 414, American Nuclear Society.

9. General Atomic Company, High Level Waste Transportation System Development Program, Final Report for Fiscal Year 1980, GA-A-16124 (1981).

10. W. B. Andrews et al., Defense Waste Transportation Cost and Logistics Studies, Battelle Memorial Institute, Pacific Northwest Laboratory, PNL-3721 (1982).

11. Advanced Energy Systems Division, Westinghouse Electric Corp., Engineered Waste Package Conceptual Design: Defense High-Level Waste (Form 1), Commercial High-Level Waste (Form 1), and Spent Fuel (Form 2) Disposal in Salt, ONWI-438 (1983); Waste Package Conceptual Designs for a Nuclear Repository in Basalt, RHO-BW-CR-136P/AESD-TME-3142 (1982); Conceptual Waste Package Designs for Disposal of Nuclear Waste in Tuff, ONWI-439 (1983).

12. P. D. Soper, D. D. Walker, M. J. Plodinec, G. J. Roberts and L. F. Lightner, "Optimization of Glass Composition for the Vitrification of Nuclear Waste at the Savannah River Plant," Am. Ceram. Soc. Bulletin, 62(9), 1013 (1983).

13. W. D. Rankin and G. G. Wicks, "Chemical Durability of Savannah River Plant Waste Glass as a Function of Waste Loading," J. Am. Ceram. Soc., 66, 389, 1983.

ECONOMIC CONSIDERATIONS/COMPARISONS FOR THE DISPOSAL OF DEFENSE HIGH-LEVEL WASTE

D. B. Leclaire
E. G. Lazur
Office of Defense Waste and By-Products Management
U. S. Department of Energy
Washington, D.C. 20545

ABSTRACT

This paper provides a summary, in a generic sense, of the economic considerations and comparisons of permanent isolation of defense high-level waste (DHLW) in a licensed geologic repository.

INTRODUCTION

Economic considerations addressed in the paper include costs for:

1. development & evaluation (D&E),
2. system acquisition including:
   - land and improvements
   - environmental impact assistance (EIA)/other
   - design and construction,
3. operations,
4. decontamination and decommissioning (D&D), and
5. transportation.

Costs associated with the following activities specifically excluded from this paper are:

1. interim storage (DOE's current method of storing DHLW),

2. preparation of DHLW into final waste form for disposal (i.e., vitrification, etc.), and

3. temporary storage of prepared waste pending availability of licensed geologic repository.

DISCUSSION

Comparisons will consist of illustrating the cost of permanent isolation of defense high-level waste through the utilization of a defense-only waste repository versus the use of a combined commercial and defense waste repository. While only one combined repository is used for comparison purposes it is recognized that defense waste could be placed in one or more combined repositories. No attempt is made nor should any assumptions be concluded regarding what costs should be paid by the Federal Government (to the special account) for the option of utilizing a combined repository for the disposal of DHLW. Table 1 is used to illustrate the economic consideration and comparisons.

TABLE 1. Cost Estimates for the Disposal of DHLW (in millions of constant 1984 dollars)

Cost Element or Option	A DHLW Only	B Commercial Only	C A & B	D Combined	E Delta (D vs C)
Land/Improvements	61-83	61-83	122-166	61-83	61-83
EIA/Other	17-240	17-240	34-480	17-240	17-240
Design & Constructions	728	1175-1294	1903-2202	1279-1427	624-775
Operations	1244	3975-4551	5219-5795	4609-5459	336-610
D&D	173	238-239	411-412	259-260	152
Totals	2223-2468	5466-6407	7689-8875	6225-7469	1406-1464

The delta column represents the cost effectiveness of using a combined repository vs. the construction of separate repositories for defense and commercial HLW.

A generic case is developed by providing the range of cost estimates for salt and hard rock repository options as they appear in the "Evaluation of Commercial Repository Capacity for the Disposal of DHLW" (DOE/DP-0020).

The costs associated with overpacks (if they are required for DHLW) are not shown as they are basically the same for any alternative. These costs would add $400-700 million to each alternative.

D&E costs are also not incorporated in the above table illustrating economic considerations and comparisons for DHLW. These costs including such activities as technology development, socioeconomic studies, site identification/characterization/approval, construction authorization and consultation and cooperation are estimated to be approximately $4 billion in 1984 dollars for the commercial repository. Any additional D&E costs for a combined repository are considered insignificant; therefore, D&E costs for a combined repository are assumed to be the same as those for a commercial repository without defense waste. The D&E costs for a defense-only repository would be less than for a commercial repository because some Nuclear Waste Policy Act provisions do not apply; i.e., site nomination procedures. However, costs would be incurred for site characterization and state and Indian tribe participation and consultation activities which apply for development of any repository.

Land and improvement costs consist of land and land rights and onsite improvements including clearing, fencing, roads, parking, etc.

EIA/other costs are assumed to include offsite development and improvements including offsite rail and road systems and improvements, offsite water and power systems, State and community assistance, etc.

Design and construction costs include architectural/engineering services and the costs of all facilities and equipment.

Operation costs include all labor, materials, equipment replacement, and utilities to operate the repository(s) for 25 years. D&D costs are costs associated with decontamination, final backfilling, sealing, etc.

The cost of transportation of defense high-level waste to a repository site (also not shown in the table above) depends on both the quantity of the waste and the distance that waste must travel. This paper assumes that a total of 20,000 canisters will be transported from three DOE sites. The total transportation costs for defense high-level waste include the capital and maintenance costs for the casks and carrier transportation charges. The capital and maintenance costs depend on the number of casks required which, in turn, depends on the transportation distance, travel time, cask turnaround time at the repository, the number of canisters transported per trip, and the annual rate of waste transport to the repository. Carrier transportation charges depend on the distance traveled, the weight of the cargo, the mode of transport, and any expenses associated with handling hazardous cargo. The range estimate of $110-260 million represents the high and low costs associated with alternatives which include different repository regions, rail and truck transport modes, and different numbers of casks. The numbers would be the same whether DHLW is shipped to a defense-only or combined repository.

CONCLUSIONS

In conclusion, since a combined repository costs on the order of $1.5 billion less than would separate repositories for DHLW and commercial HLW, and if no other factors are considered or warranted, then the economic considerations and comparisons and other factors considered suggest this country proceed with the combined repository option.

S.C.U.F.F. -
A COMPUTER CODE FOR SIMULATING AND COSTING USED FUEL MANAGEMENT STRATEGY

J. M. Cipolla
Ontario Hydro
700 University Avenue
Toronto, Ontario, M5G 1X6
Canada

ABSTRACT

By 1992, Ontario Hydro's nuclear generating program will consist of 20 units producing over 14,000MW(e). Over 1,700,000 used fuel bundles will have been generated by the year 2000. A computerized code called SCUFF (System Costing of Used Fuel Facilities) has been developed to provide a tool to model the numerous used fuel management strategies and parameters that are possible. SCUFF allows variations to the nuclear generation program and to the storage, transport, and disposal phases. Major output from SCUFF includes predictions of:

1. used fuel production,
2. storage inventories and used fuel flows,
3. storage, transport, and disposal schedules and facility requirements, and
4. detailed cost breakdown for each phase.

This information can be advantageously used to evaluate and optimize used fuel management strategy, compare alternate storage, transport and disposal concepts, and can provide a rapid means of predicting costs and conducting sensitivity analysis. Additional capabilities are being incorporated into the code to increase its usefulness, including the addition of alternate fuel management concepts, inclusion of automatic optimization subroutines, and modifications to allow operation on an IBM-PC.

I. BACKGROUND

Ontario Hydro is a publicly owned electrical utility serving over 700,000 municipal, rural, and direct customers in the province of Ontario. It has an installed generating capacity of over 25,000MW(e) with energy production of over 110 billion kWh in 1983, 35% of which was produced by nuclear generating units. At present two nuclear stations are in full operation and an additional three stations are in various stages of construction, commissioning, and partial operation. Each station consists of four CANDU reactor units and it is expected that a total of 20 units will be in operation by the year 1992, generating over 14,000MW(e).

Under the terms of the Canada/Ontario Nuclear Fuel Waste Management Program, Ontario Hydro was given responsibility for the storage and transportation of used fuel and through its Technical Assistance Program, Ontario Hydro assists Atomic Energy of Canada Limited (AECL) in their Fuel Waste Disposal Program. Although various studies have been conducted to assess and attempt to optimize components within each phase of the used fuel management cycle, few programs have studied the interaction between each phase or the optimization of the entire cycle. Because of the highly complex nature of used fuel management, the SCUFF (System Costing of Used Fuel Facilities) computer code[1,2] has been developed to model the storage, transportation, and immobilization/disposal phases and to evaluate logistics and costs from an overall systems approach.

II. REFERENCE USED FUEL MANAGEMENT SYSTEM

The current version of SCUFF models Ontario Hydro's reference used fuel management strategy although work is progressing on the incorporation of capabilities to model alternate concepts for storage, transportation, and disposal. As of the end of 1984, over 300,000 used fuel bundles have been discharged and stored at the Pickering and Bruce nuclear sites. Projections indicate that over 1.7 million bundles will require storage by the year 2000 and considerably more fuel will be generated prior to the disposal inservice year.

At present, used fuel is stored in onsite water pool facilities dedicated to each station. Initial storage is provided in primary fuel bays and as these become filled, fuel is transferred either by underwater conveyor system or by an above ground cask and transporter to secondary fuel bays. At some future date, existing storage bays will become full to capacity requiring construction of additional facilities at these sites. The reference strategy is to build additional water bays although other dry storage options such as dry vault and canister storage concepts are under evaluation for potential implementation.

The reference concept for off-site transportation is to employ dry shipment of used fuel in a 192-bundle capacity road cask. A prototype cask is expected to be in operation by 1987. Alternate concepts being considered for large-scale operation include rail and water (ship/barge) transport.

The current disposal concept is a conventional room and pillar design 500- to 1000-m deep in granitic or plutonic rock of the Canadian Shield. The used fuel is directly immobilized in a corrosion resistant container and placed in the vault surrounded by a bentonite buffer material. The vault is backfilled with a clay/sand mix and the vault shafts and boreholes are sealed with bentonite plugs.

III. THE SCUFF MODEL

A. Rationale and Structure

Because of the uncertainty and complexity of predicting future used fuel management strategy, it was recognized that a method was necessary for assessing the numerous options and parameters that are possible. As a result, SCUFF was developed to provide the following major capabilities:

1. a tool to allow modelling and economic evaluation of the entire used fuel management program (although each phase can be analyzed individually, optimization of the entire fuel management cycle can only be conducted if all three phases are modelled simutaneously),

2. a consistent method for conducting comparisons of alternative storage, transport, and disposal options (meaningful comparisons can only be conducted if their assessment is based on consistent assumptions and terms of reference), and

3. a quick method for conducting sensitivity analysis.

The SCUFF code employs a user-friendly, menu-driven screen display with the code design employing three main features:

1. flexibility (a variety of storage, transport, and disposal options and parameters can be selected for modelling),

2. simplicity of operation (the code contains all required reference data internally, freeing the user from having to provide large amounts of input data; the code prompts the user for data which is to differ from the onboard reference values), and

3. modularity (SCUFF consists of many separate subroutines each simulating individual fuel management options; this structure allows later expansion of the code to be easily accommodated).

Figure 1 illustrates a simplified flowchart of the SCUFF structure. The code consists of various functional blocks made up of subroutines called into operation by the main executive program. The first block contains subroutines which allow user selection of output requirements. Based on this selection, these subroutines also determine what portions of SCUFF must be executed to produce the desired output. The second block contains input data subroutines. These subroutines store all reference data required for running the code and prompt the user for alternative data should any nonreference values or scenarios be desired.

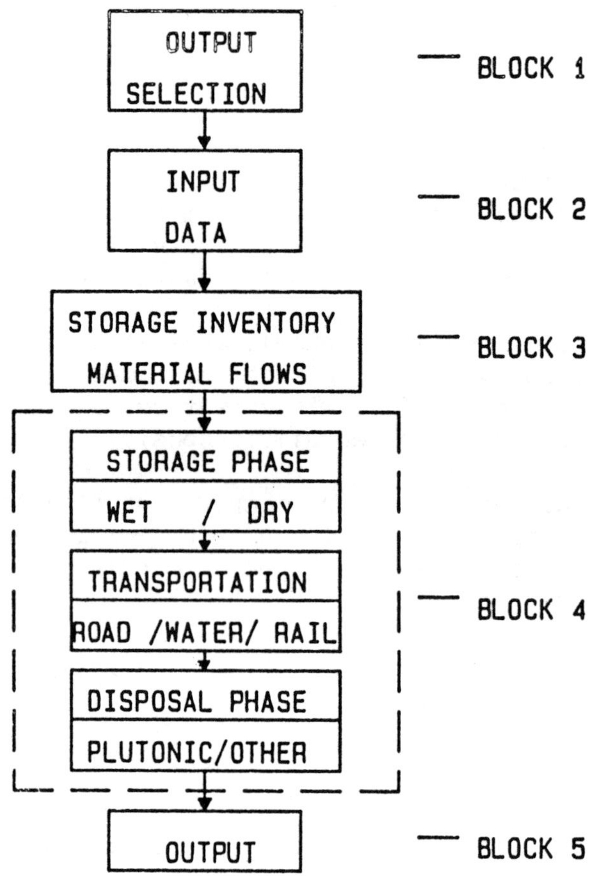

FIGURE 1. SCUFF Structure

Two levels of variables are modelled by SCUFF. The first-level variables provide a general selection of the overall nuclear program and fuel management strategy to be simulated, including options for storage, transport and disposal of used fuel. The second-level variables provide detailed data corresponding to the overall strategy selected. A list of the major selections is provided in Table 1.

The third block contains subroutines which calculate yearly used fuel bundle storage inventories, off-site fuel bundle transfers, and quantities of used fuel immobilized/disposed. A series of prioritized requirements and constraints govern the storage, transport, and disposal strategy. This logic takes into account the addition of new stations to the nuclear program as well as retirement of old stations which have operated for their design life. The simulation proceeds on a yearly basis accumulating fuel at each station until the disposal start date is reached at which point transportation of fuel to the disposal facility is simulated. This material flow data forms the basis for subsequent cost calculations.

TABLE 1. Options and Parameters

	Major Options	Variable Parameters
Generation Program	- Growth in Nuclear Program	- Annual Growth Rate - Station Details, e.g., Fuel Discharge rates, Design Life, In-service dates
Storage Phase	- Water Pools - Dry Vaults - Modular Canister - Centralized Storage	- Storage Capacities - Capital Costs - Unit Operating Cost Components
Transportation Phase	- Road Transport - Rail Transport - Barge/Ship	- Transport Distances to Disposal - Cask Size - Equipment Design Life - Unit Operating Cost Components - Capital Costs
Disposal Phase	- In Plutonic Rock - Under Sedimentary Shale	- Disposal Start Date - Disposal Fuel Age Constraint - Maximum Disposal Rate - Immobilization Costs - Capital Costs - Unit Operating Costs
Economics	- Current Dollars - Present Worth Dollars	- Costing Year - Present Worth Discount Rate

SCUFF then works sequentially through the storage, transport, and disposal phase sections of the fourth subroutine block. In this block, facility and operating requirements, in-service schedules, and detailed capital and operating costs are determined for each phase of the used fuel management program. The final block contains output subroutines. The output data provided corresponds to that requested by the user at the beginning of the run.

B. Cost Calculations

Cost calculations for each phase of the used fuel management program include estimates of both capital and operating components. Capital costs for the storage facilities include direct costs for buildings and equipment, construction indirects, engineering, overheads, and contingencies. Costs for existing storage bay facilities are input directly as data whereas capital requirements for additional storage facilities (of varying capacities as determined by SCUFF) are estimated within the code by interpolation from a function relating storage capacity to capital costs. Operating costs for the storage phase are applied on an annual basis and consist of five components:

1. maintenance and material costs,
2. safeguards,
3. storage container costs,
4. seismic stacking frames, and
5. operator costs for transfer into storage facility.

Components 1 and 2 are charged on a per facility basis. An annual cost is allotted for each facility in operation. Components 3 to 5 are charged on a per bundle basis with annual charges applied depending on the number of used fuel bundles discharged to storage.

Capital costs for the transportation phase are based on supplier quotations for casks, trucks, and trailers. Operating costs are applied annually and are based on cost estimates produced by an external haulage company. The following items are included:

1. maintenance and materials (e.g., gas, etc),
2. driver salaries and overheads,
3. cask and truck loading/unloading costs, and
4. miscellaneous costs (e.g., escort requirements, etc).

Capital and operating costs for the immobilization/disposal phase were derived from estimates provided in a conceptual design study conducted for AECL3. Capital costs include direct charges for surface facilities and equipment, construction indirects, engineering, overheads, taxes, and contingencies. Operating costs include the following components for the surface facilities:

1. container and immobilization materials,
2. operator salaries and overheads, and
3. maintenance and materials.

Costs for operation of the disposal vault itself include estimates for panel development, backfilling, room and vault sealing, and maintenance and monitoring. Proportionality constants were developed for both capital and operating cost components to estimate charges for facilities of varying capacities as determined by the SCUFF code.

From the above cost calculations, SCUFF provides an annual estimate of costs for the entire used fuel management program. However, economic comparisons and ranking of various options can only be conducted if future expenditures can be expressed in terms of equivalent dollars. The present worth concept is used to discount costs. All annual costs are discounted to any base year (selected by the user) by the application of a discount factor which allows for cost escalation. The value of this factor is determined by the net difference between escalation (e) and interest (i) rates and is of the form

$$1/(1 + f)^n$$

where f = i - e
n = number of years from time zero

The major output data provided by SCUFF is discussed in the following section.

C. Output Data

Output from the SCUFF code is provided in tabular form. A detailed description of the simulation results as well as summary information is available. The user can request any or all of the available outputs which include:

1. annual and cumulative fuel bundles discharged to storage,
2. inventory of fuel bundles stored at each station,
3. additional storage facility requirements,
4. annual storage capital and operating costs,
5. used fuel bundles transported off-site,
6. transportation facility requirements (e.g., number of casks, trucks, etc),
7. annual transportation capital and operating costs,
8. used fuel bundles immobilized/disposed,
9. annual immobilization/disposal capital and operating costs, and
10. summary of all capital and operating costs.

All cost data is available in current and/or present worth dollars. The present worth discount rate as well as the discount year is variable and can be chosen by the user.

IV. TYPICAL USES

SCUFF has application in various used fuel management programs:

1. establishment, evaluation and optimization of the entire used fuel management strategy,

2. determination or verification of charges to be allotted for the used fuel management component of electrical rates,

3. tool for planning and optimizing additional fuel storage facility requirements and can provide a means for evaluating and comparing alternate storage concepts,

4. description of transportation requirements and schedules for use in optimization of the transport phase and evaluation and comparison of alternate transportation concepts,

5. evaluation and comparison of alternative disposal concepts including varying distances to the disposal facility, and

6. calculation of fuel flow and schedule data which will form the basis for a full environmental assessment of the disposal/immobilization phase.

The following example graphically illustrates how the SCUFF code can be applied. Figure 2 shows the effects of delaying the disposal/immobilization start date on used fuel management costs. Results are provided in both 1984 constant dollars and present worth costs discounted by a net discount rate (difference between borrowing rates and escalation) of 3% per year.

As illustrated in Figure 2a, disposal costs represent the dominant 1984 constant dollar cost component. All costs are not affected to a large extent by the disposal date except the storage component which increases when the disposal date is delayed. This is to be expected since delaying disposal will result in increased onsite storage space requirements leading to increased capital and operating costs.

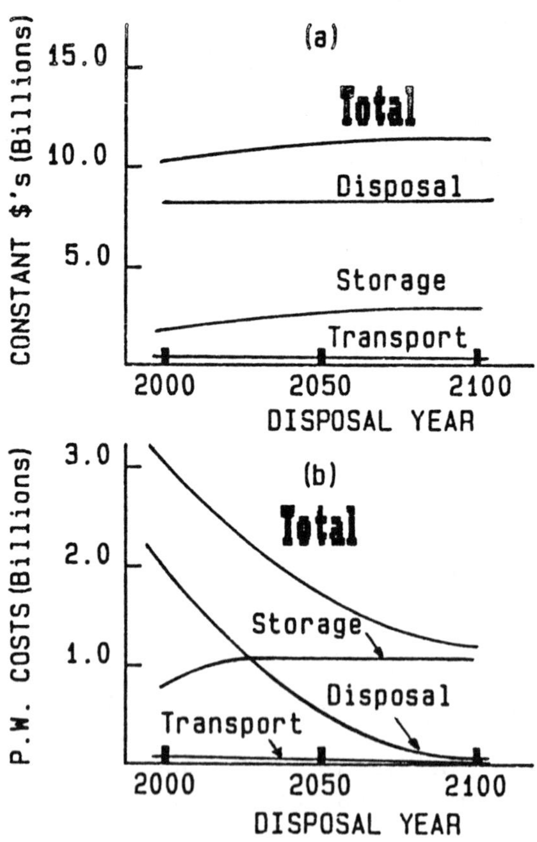

FIGURE 2. Costs Versus Disposal Year

The present worth breakdown of these same costs are shown in Figure 2b. In contrast to the constant dollar results, present worth costs decrease asymptotically as disposal is delayed. The storage cost component becomes dominant for a disposal start date beyond about the year 2025. Since storage costs are expended much earlier in time than disposal costs, their present worth value becomes more significant (i.e., less discounting effect). These results indicate that economically it may be desirable to delay disposal to beyond about the year 2025. In addition, although storage costs seem small in comparison to anticipated disposal and overall used fuel management costs, their present worth values are quite significant suggesting that emphasis should be placed on optimization of short-term storage costs as a means of minimizing overall long-term used fuel management costs. Further cost component breakdowns (by year and by generating station) can be made to identify detailed capital and operating costs.

V. FUTURE DEVELOPMENT AND SUMMARY

Several additional features are being incorporated into SCUFF to increase its usefulness and flexibility:

1. Subroutines are being added to allow modelling of alternate storage, transport, and disposal concepts. In the storage phase dry vault and modular canister systems may be incorporated into the code. Rail and barge/ship transport modes and disposal under sedimentary rock is also being considered for inclusion in the code.

2. Although originally intended to operate on the UNIVAC-1182 mainframe computer, SCUFF is being modified to run on an IBM-PC. This modification is nearing completion.

3. A subroutine is being developed which will enable automatic optimization of a preselected variable. This optimization will operate by minimizing with respect to a specified cost or quantity.

In its present form the SCUFF code can model the reference Ontario Hydro used fuel management program. It provides a tool to simulate and cost the entire program from an overall systems approach and provides a quick and consistent method of assessing, comparing, and optimizing used fuel management strategy and conducting sensitivity analysis.

REFERENCES

1. J. M. Cipolla and B. R.Reynolds, SCUFF: A Computer Code To Model Irradiated Fuel Management Costs and Logistics, Ontario Hydro Report 85145 (1985).

2. J. M. Cipolla and B. R. Reynolds, SCUFF Verification, Ontario Hydro (1985).

3. Atomic Energy of Canada Limited, A Disposal Centre For Irradiated Nuclear Fuel: Conceptual Design Study, AECL-6415 (1980).

Disposal in the Overall Waste Management System Context

COMPARISON OF DISPOSAL OPTIONS FOR DEFENSE HIGH-LEVEL RADIOACTIVE WASTE

Alan S. Goldfarb
Lester Ettlinger
Norman Lord
Brian Price
The MITRE Corporation
1820 Dolley Madison Blvd.
McLean, Virginia 22102

Arthur Follett
Office of Defense Waste and Byproducts Management
U.S. Department of Energy
Washington, D.C. 20545

ABSTRACT

Two disposal options for defense high-level radioactive waste are compared with respect to health and safety, regulation, transportation, public acceptability, and national security. None of these factors provided a reason to conclude that a separate defense-only repository is required. Comments received from the public on a draft report of the comparative analysis did not provide a basis for changing the conclusions of the study.

I. INTRODUCTION

The Nuclear Waste Policy Act of 1982[1] establishes a program for the siting, construction, operation, and decommissioning of a licensed geologic repository for the permanent disposal of spent nuclear fuel and high-level radioactive wastes from civilian nuclear power reactors and other sources. Section 8 of the Act required the President to evaluate the use of one or more repositories for permanent disposal of defense high-level radioactive waste. On April 30, 1985, President Reagan decided to accept the Department of Energy's (DOE) recommendation that defense and commercial radioactive wastes be commingled in a commercial repository on the basis of the projected cost advantage and the fact that no compelling requirement for a defense only repository was found. This paper reports on the results of a study performed for the DOE to provide input to the President's evaluation. The full study report is contained in DOE Report No. DOE/DP-0020/1.[2]

Two options for disposal of defense high-level waste were considered. They were:

- Defense high-level waste is disposed of in a commercial geologic repository.

- Defense high-level waste is disposed of in a defense only geologic repository.

Each option was examined in separate studies by DOE contractors with expertise in the specific factors that were to be considered in the evaluation according to the Act. The factors were cost efficiency, health and safety, regulation, transportation, public acceptability, and national security. The options were then compared to determine if a compelling requirement for a defense only repository could be identified. This paper discusses the findings for the factors other than cost efficiency. A paper on cost efficiency has been prepared separately for this conference.

Defense high-level waste is generated and stored at three Department of Energy sites: the Savannah River Plant, Aiken, South Carolina; the Idaho National Engineering Laboratory near Idaho Falls, Idaho; and the Hanford Reservation in Hanford, Washington.

Defense high-level waste is initially in the form of an acidic aqueous waste solution produced when irradiated nuclear fuel is reprocessed to recover uranium and plutonium. At Idaho National Engineering Laboratory, the acidic liquid is transformed directly to a dry granular solid in a calciner. At the Savannah River Plant and Hanford Reservation, the acidic waste is neutralized. When neutralized, a sludge containing most of the radionuclides, except cesium, precipitates. The cesium can be separated from the residual salt solution (supernate). The cesium free salt solution can be disposed of as low-level waste. At the Hanford Reservation, cesium and strontium were separated from waste produced before 1984 and converted to dry cesium chloride and strontium fluoride salts. These salts were sealed in double wall metal capsules and stored in water pools pending a decision on disposition.

The reference plan of the Department of Energy[3] is to immobilize new and readily retrievable old defense high-level waste and dispose of it in a geologic repository. Radioactive waste in 149 single shell tanks at Hanford, from which most of the strontium and cesium have been removed, will be stabilized in place if, after the requisite environmental documentation, it is determined that the short term risks and costs of retrieval and transportation outweigh the environmental benefits of disposal in a geologic repository.

A facility (the Defense Waste Processing Facility (DWPF)) is currently under construction at the Savannah River Plant in which Savannah River Plant high-level waste will be immobilized in borosilicate glass. This facility is being designed to produce 500 canisters of borosilicate glass per year beginning in 1989. A similar facility is planned for operation at Hanford beginning in the early 1990's. It will have a capacity of about 75 canisters per year. An immobilization facility is anticipated to be operational at Idaho beginning in 2007. Approximately 20,000 canisters of immobilized defense high-level waste from these three sites are projected to be available for repository disposal through the year 2020. The 20,000 canisters of immobilized defense high-level waste is considered equivalent to the reprocessed waste from 10,000 metric tons of commercial nuclear fuel.

II. HEALTH AND SAFETY

A comparative assessment was made by Kocher, Smith and Witherspoon[4] of the potential long- and short-term health and safety impacts of the two disposal options for defense high-level waste. The objective of this assessment was to determine the impact of various disposal scenarios on demonstrating that the repository performance objectives and criteria of the Nuclear Regulatory Commission (NRC) Regulation, 10 CFR Part 60,[5] will be met. The regulation is designed to assure that proposed Environmental Protection Agency (EPA) standards (47 FR 58195)[6] for radionuclide releases to the accessible environment from a high-level waste repository are met.

Release rates and levels were estimated, using a conservative based computer code, which, when combined with suitably conservative input data, produced corresponding estimates of radionuclide releases to the environment.

Radionuclide release rates from a repository were asssumed to depend on the temperature of the waste package when containment failure occurred. Two scenarios for containment failure were considered: (1) 300 years for defense waste packages without an overpack; and, (2) 1,000 years for packages with an overpack. All defense waste packages were assumed to fail simultaneously and completely. The temperature inside the commercial repository was based on the repository containing a 50/50 mix of commercial spent fuel and commercial high-level waste. Two repository media were considered: salt and hard rock.

The computer codes assumed that the groundwater travel time to the accessible environment could not be less than 1,000 years as required by the NRC in 10CFR60. Radionuclide retardation factors were chosen from conservative values recommended in the literature.

Because of the non-site-specific conservative assumptions used in the analysis, the calculated releases cannot be used to demonstrate compliance with the EPA standard, but only for comparing the relative performance of the disposal scenarios. The analysis showed that defense waste in a defense only repository can be expected to exhibit a lower release of radionuclides to the environment than defense waste in a commercial repository, although the differences are insignificant compared with the large uncertainties associated with realistic performance assessments of actual repositories.

Further, under less conservative assumptions based on available geologic data information from potential repository sites, the DOE has found that there should be no releases of radioactivity from a commercial radioactive waste repository in salt or hard rock during the first 10,000 years (the time period of the proposed EPA standard) following decommissioning of the repository. Therefore, there is no apparent difference between the disposal options with regard to long-term health and safety.

The short-term health and safety impacts of defense waste disposal are related to repository construction and operation. Estimates of impacts on workers and off-site populations from accidents and from radiological and nonradiological effluents were obtained from similar estimates for commercial repositories. Both disposal options should meet applicable standards and neither option appears to have a significant advantage over the other.

III. REGULATION

Laws and regulations applicable to the disposal of defense high-level waste were examined by Lord and Goldfarb[7] to identify their impacts on each disposal option. The Nuclear Waste Policy Act[1] establishes the legal framework for disposal of defense high-level waste. Section 8(b)(3) of the Act states that "Any repository for the disposal of high-level radioactive waste resulting from atomic energy defense activities only shall (A) be subject to licensing under section 202 of the Energy Reorganization Act of 1973 (42 U.S.C. 5842); and (B) comply with all requirements of the Commission for the siting, development, construction, and operation of a repository." The Act also specifies sections of the Act which are applicable to a defense only repository, providing that the remaining sections of the Act are not applicable to such a repository.

The provisions of the Act applicable to a defense-only repository specify the rights of participation and consultation of States and Indian tribes with respect to any proposed repository, and these are identical for a commercial repository and a defense only repository. The Act prescribes a sequence of actions which the Department of Energy must take to receive authorization to construct a repository and to receive a license to operate the repository. A defense only repository is not bound to these provisions of the Act; however, except for the requirement for issuance of guidelines for recommendation of sites for a repository and the procedures to nominate sites for site characterization, the NRC regulations (10 CFR 60),[5] which are applicable to both a commercial and a defense only repository, contain almost identical procedures leading to repository licensing. The NRC procedural regulations may be revised as necessary in light of the Nuclear Waste Policy Act (48 FR 28195, June 21, 1983) to eliminate any differences. Thus, there does not appear to be any significant regulatory procedural advantage to establishing a defense only repository which would make it possible to begin operation of such a repository earlier than a commercial repository.

The NRC regulations contain a technical subpart which sets forth repository performance objectives and site and design criteria which, if satisfied, would support a finding (by the Commission) of no unreasonable risk to the health and safety of the public. The central concept of the repository isolation system is the use of several different types of engineered and natural barriers to assure isolation from the accessible environment. Performance requirements for these barriers include some flexibility because it was recognized that the characteristics of the waste form, such as thermal release, specific radionuclide content, and surface radiation levels, and the immediate environment in which the waste is emplaced, would affect repository performance and could influence the selection of technical measures required to achieve performance objectives.

An important factor influencing specification of the performance requirements for the geologic repository system and corresponding technical measures required to meet those requirements is the repository temperature. High repository temperatures can accelerate corrosion of waste containers, increase solubility of the waste form, and reduce the sorptive capacity of the repository media, thereby hastening radionuclide releases. In a commercial repository, containing high-heat-release commercial waste, the defense waste, which is cooler than commercial waste, could be subjected to a higher temperature environment than would exist in a defense-only repository. Thus it is possible that the engineered barrier system for defense waste may be different in a defense-only repository than in a commercial repository. For example, in a defense-only repository, use of a lower cost overpack or no overpack on defense waste may be sufficient, whereas in a higher temperature environment a more costly overpack might be required to provide adequate isolation assurance.

It is clear, however, that by appropriate selection of waste package design and repository design, acceptable containment performance of defense high-level waste can be assured in the commercial repository. It was also shown in the cost analysis that even if an overpack were required for defense high-level waste in a commercial repository, but not in a defense-only repository, the cost savings was not sufficient to overcome the cost advantage of disposal of defense waste in a commercial repository.

IV. TRANSPORTATION

The costs and risks of truck and rail transportation of defense high-level waste to each of five potential repository regions were estimated in order to provide a reasonable indication of the range and magnitude of these factors. The five potential repository regions are the Hanford Reservation in Washington, the Nevada Test Site in Nevada, the Permian Basin, the Paradox Basin, and the Gulf Interior Region.

To perform the transportation analysis, Joy, Shappert, and Boyle[8] assumed that defense high-level waste would be shipped to the repository at the same rate that it is produced by the three DOE generators. The actual schedule had not been established at the time the study was performed and will be the subject of negotiation between DOE's Assistant Secretary for Defense Programs and the Office of Civilian Radioactive Waste Management. It is intended that the receipt of defense waste will not adversely affect the rate of receipt of commercial waste. Differences between the assumptions used in this study and the actual schedule are not expected to alter the qualitative conclusions of the study.

The total transportation costs for defense high-level waste include the capital and maintenance costs for the casks and carrier transportation charges.

Truck distances were calculated using a computerized routing model which is designed to simulate routes on the highway system in the U.S. under conditions of interest. Routes that might be used for general commerce were used.

No routing restrictions were assumed. (In actual practice, routes selected for transport of defense high-level waste to a specific repository site would have to conform with the Department of Transportation's final rule on highway routing of large quantity radioactive material shipments (DOT Docket HM-164)). The truck routes are symmetric, i.e., the return trip for the empty cask uses the same route as the loaded cask.

Rail distances were calculated, using a railroad routing model which is designed to simulate routing on the railroad system. All rail shipments were assumed to travel as general freight between the origin and destination. In general, rail routes are not symmetrical because the originating railroad tries to maximize the distance traveled on its own right of way.

The total cost for transportation of the 20,000 canisters of defense high-level waste from the three defense sites to each of the five potential repository sites is summarized in Table 1. The table shows that the mode of transport has a significant effect on the total transportation cost. The higher cost of rail transportation is mainly due to much slower rail speeds and more constraints on routing, e.g., more limited rail network, and maximization of use of carriers' rail line.

With respect to any designated repository, the cost for shipping defense high-level waste to that site does not depend on whether the site is a defense only or a commercial repository. Although it may be possible to locate a defense only repository such that transportation costs would be lower than at the commercial repository, the potential cost savings appear small in relation to the potential cost savings realized by disposing of defense waste in a commercial repository rather than a defense only repository.

There are two categories of risk associated with the transport of high-level waste: (1) the nonradiological risks which occur independent of the nature of the waste and include the health impacts to the general population from pollutant emissions, e.g., emissions from diesel engines, and impacts to both the public and the workers from transportation related accidents; and

TABLE 1. Summary of Costs of Transporting Defense High-Level Waste to a Repository Site

Destination	Cost, millions of 1984 dollars 20,000 Canisters	
	Truck	Rail
Hanford	162	257
Nevada Test Site	148	257
Paradox Basin	138	219
Permian Basin	105	223
Gulf Interior Region	110	203

(2) the radiological risks which are the health impacts to workers and the general population from potential exposure to radiation during transport or in the event of an accident.

An accident analysis indicated that there would be fewer than 10 nonradiological fatalities from transportation of defense high-level waste during the 25-year operating period of a repository compared to more than 94,000 truck and rail fatalities from other activities during the same period.

Radiological impacts to transportation workers and to the population along the transport route were calculated for routine operation and for accident scenarios. As in the case of costs, the risk associated with shipping defense high-level waste varies from one potential repository site to another but does not depend on whether a site is a defense only or a commercial repository. Because transportation casks are designed to survive extremely severe accidents without serious consequences, (i.e. release of radioactive substances into the environment) the probability of accidents resulting in adverse health effects is very small and independent of whether waste is going to a commercial repository or a defense-only repository.

V. PUBLIC ACCEPTABILITY

Nealy and coworkers[9] examined the possible positions that segments of the public, namely, Federal agencies, states and Indian tribes, local officials, nuclear utilities and pro-nuclear groups, nuclear critics, citizen groups, and the general public, will take with regard to each disposal option. When their study was initiated, there was a limited public record of discussion of the issue in the Congressional record and other public opinion sources. Thus, it was necessary to infer, from a knowledge of public interests and positions that would influence the acceptability of the disposal options, the potential reaction of the public. The actual stand that the public takes is likely to be influenced by how the Federal government deals with the institutions and public segments involved in the repository program. The program's cost, potential for delay in the repository schedule, and public health and safety were among the major issues that were seen to affect the acceptability of the disposal options.

Some Federal agencies would favor the low-cost option. Utilities may support placement of defense waste in the commercial repository under the expectation that it may lower the cost to utilitites for disposal of their waste. However, if codisposal were considered to be a complicating factor in licensing and operating a repository, then a defense only repository may be supported by utilities, pro-nuclear groups, and some Federal agencies in the belief that it would remove a potential obstacle to expeditious establishment of the commercial repository. Most groups are primarily interested in assurances that public health and safety will not be compromised. They will insist that their health and safety concerns be adequately addressed, regardless of the option chosen for disposal of defense waste. It was concluded that the differences in public acceptability between the options appear to be minor compared to gaining public acceptance for any nuclear waste repository.

A 1984 draft report of the study was made available to the public for review and comment to increase public awareness and develop a public record on the issue of disposal of defense high-level waste. Over 400 copies of the report were distributed. Thirty comment letters containing over 400 comments were received. A report containing copies of the letters and the DOE's response is in preparation.[10] In general, the comments received confirmed our expectations. None of the commenters disagreed with our conclusion that there was no basis for a finding that a defense only repository is required. Six commenters supported codisposal, some with the qualifications that it not delay the program or increase costs to utilities. Other commenters focused primarily on the technical details in the report. Some questioned some of the assumptions used in the analysis. Concern was also expressed about delay in the repository program, possible displacement of civilian waste by defense waste, and safety due to increased repository complexity and local transportation impacts.

VI. NATIONAL SECURITY

National Security was addressed by Hindman.[11] There are two key national security issues with respect to disposal of defense high-level waste:

1. There must be no interruption of or delay or NRC involvment in the defense material production process or nuclear weapons activities.

2. There must be no disclosure of classified information.

Interruption or shutdown of a defense production or utilization facility because of waste buildup problems could develop if the opening of the repository were delayed, if the repository accepted high-level waste at less than the expected rate, or if the repository were to be closed for regulatory or technical reasons. Shutdown or interruption of defense production or utilization facilities could also occur because of in-plant regulatory requirements and/or inspections imposed by geologic repository requirements on waste form preparation. There is also concern that NRC regulation of a disposal system at a geologic repository might reflect back into the production system to create an interruption or shutdown of production operations.

Steps being taken to avoid the possibility of interruption or a shutdown of defense nuclear production activities because of the above, include (1) provision for sufficient interim waste storage to permit continued operation of production or immobilization facilities in the event of shutdown or delays in the operation of the geologic repository; and (2) thorough technological exchange between the DOE and the NRC during all stages of development of both the waste form and the repository. The DOE has initiated and continues contact with the NRC staff to ensure that they have a full understanding of and opportunity to review plans to produce immobilized glass waste forms.

There is some classified defense waste in storage tanks, and there will be more in the future. This waste is handled separately and appropriately until mixed with other wastes. This results in a composition that is unclassified prior to vitrification. A time delay between the creation of waste materials and vitrification, and the act of mixing various waste streams from several points in the process into large waste storage tanks creates a mixture that is unclassified. Subsequent immobilization steps will be unclassified. Therefore, immobilized waste destined for a repository will not reveal classified information.

The Department of Energy foresees no reason for the licensing or regulation of a geologic repository to require access to classified defense information.

Because the repository licensing process is untried and the Nuclear Regulatory Commission's information requirements are largely unknown at this time, there is uncertainty about the extent to which the NRC may wish to inquire into the defense production activities, and what the national security implications may be. This concern exists equally for both disposal options. As a result, national security considerations do not form a basis for preference of either option.

VII. SUMMARY

The comparison of disposal options for defense high-level waste with respect to health and safety, regulation, transportation, public acceptability, and national security, did not provide a basis for suggesting that one option would be preferable to the other or that a defense-only repository is required.

ACKNOWLEDGMENT

The authors wish to acknowledge the invaluable assistance of Mr. Vic Trebules, of the DOE Office of Civilian Radioactive Waste Management, in providing the input of his Office to the final report and the comment response document which are the subject of this paper.

REFERENCES

1. Public Law 97-425, "Nuclear Waste Policy Act of 1982," 42 USC 10101, January 7, 1983.

2. U.S. Department of Energy, An Evaluation of Commercial Repository Capacity for the Disposal of Defense High-Level Waste, DOE/DP-0020/1, Assistant Secretary for Defense Programs, Washington, D.C. (1984).

3. U.S. Department of Energy, The Defense Waste Management Plan, DOE/DP-0015, Assistant Secretary for Defense Programs, Washington, D.C. (1983).

4. Kocher, D. C., E. D. Smith, and J. P. Witherspoon, Evaluation of Health and Safety Impacts of Defense High-Level Waste in Geologic Repositories, ORNL/NFW-83/43, Oak Ridge National Laboratory, Oak Ridge, TN (1984).

5. U.S. Nuclear Regulatory Commission, "Disposal of High-Level Radioactive Waste in Geologic Repositories: Technical Criteria," Final Rule 10 CFR 60, 48 FR 28194, U.S. Nuclear Regulatory Commission, Washington, D.C. (1983).

6. U.S. Environmental Protection Agency, "Environmental Standards for the Management and Disposal of Spent Nuclear Fuel, High-Level and Transuranic Radioactive Wastes," Proposed Rule 40 CFR Part 191, 47 FR 58195, U.S. Environmental Protection Agency, Washington, D.C. (1982).

7. N. Lord and A. Goldfarb, "Regulatory Differences Between a Defense Only and a Commercial Nuclear Waste Repository," MITRE Working Paper WP-83W00591, The MITRE Corporation, McLean, VA (1983).

8. D. S. Joy, L. B. Shappert, and J. W. Boyle, "The Impact of Transporting Defense High-Level Waste to a Geologic Repository," Draft Working Paper No. NFW-83/40, Oak Ridge National Laboratory, Oak Ridge, TN (1983).

9. S. M. Nealy et al, "Public Acceptability of Colocation of Defense and Commercial High-Level Radioactive Waste," Draft Working Paper, Battelle Human Affairs Research Center, Seattle, WA (1983).

10. U.S. Department of Energy, An Evaluation of Commercial Repository Capacity for the Disposal of Defense High-Level Waste: Comments and Responses, In preparation, Assistant Secretary for Defense Programs, Washington, D.C. (1985).

11. T. B. Hindman, Jr., "National Security for Disposal of Defense High-Level Waste in a Geologic Repository," Draft Working Paper, Savannah River Operations Office, Aiken, SC (1983).

A TRUCK CASK DESIGN FOR SHIPPING DEFENSE HIGH-LEVEL WASTE

Marcella M. Madsen
Sandia National Laboratories*
P.O. Box 5800
Albuquerque, New Mexico 87185

Alan Zimmer
GA Technologies Inc.**
P.O. Box 85608
San Diego, California 92138

ABSTRACT

The Defense High-Level Waste (DHLW) cask is a Type B packaging currently under development by the U.S. Department of Energy (DOE). This truck cask has been designed to initially transport borosilicate glass waste from the Defense Waste Processing Facility (DWPF) to the Waste Isolation Pilot Plant (WIPP). Specific program activities include designing, testing, certifying, and fabricating a prototype legal-weight truck cask system. The design includes such state-of-the-art features as integral impact limiters and remote handling features. A replaceable shielding liner provides the flexibility for shipping a wide range of waste types and activity levels.

I. INTRODUCTION

Large quantities of DHLW[1] produced at reprocessing sites of the DOE, exist at various locations throughout the country. In order to end interim storage, the waste must be immobilized and prepared for shipment to locations where it may be safely and permanently stored. Capabilities must be developed by the DOE to transport waste from sites located in South Carolina, Idaho, and Washington. The locations of these sites are shown in Figure 1. DHLW immobilization facilities are scheduled to begin operation in 1989 at the Savannah River Plant in South Carolina and in the 1990s at the Hanford Reservation in Washington. Shipment of high-level waste from both sites is scheduled to begin in 1998 to a permanent repository. The Idaho National Engineering Laboratory is scheduled to begin immobilization of waste and off-site transport in 2008.

Prior to the initiation of large-scale transportation operations in 1998, limited quantities of DHLW will be shipped to the WIPP in New Mexico, where research and development experiments will be performed on this waste. Solidified waste from the DWPF located at the Savannah River Plant is scheduled to be transported to the WIPP in 1989 for these experiments.

---

*Operated by Sandia Corporation under Contract DE-AC04-76DP00789 with the U.S. Department of Energy
**Work performed under contract DE-AC03-80SF10791 with the U.S. Department of Energy

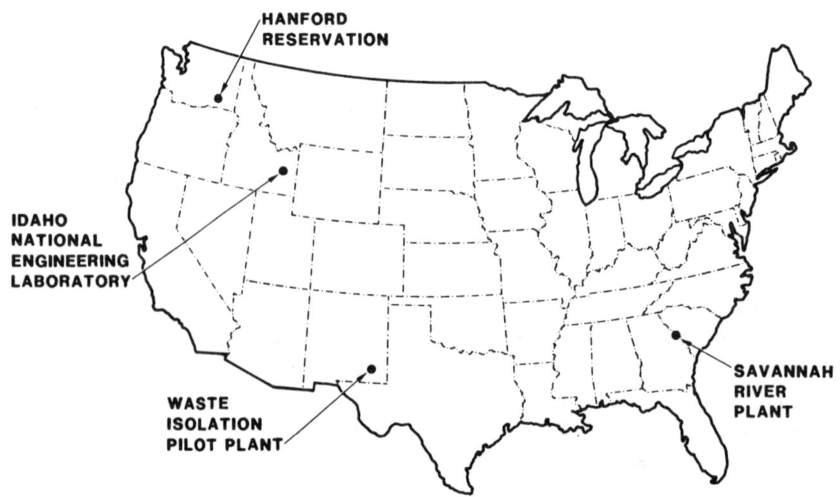

FIGURE 1. DHLW Generation and Storage Sites

In response to the need for a safe and efficient transportation system for DHLW, a program was initiated at Sandia National Laboratories by the DOE in 1979. Initial studies focused on a rail cask design which had the flexibility to ship up to eight canisters depending on activity level. Emphasis was changed to a truck cask design in 1982. GA Technologies, Inc. was selected to provide the detailed design of the shipping system. This shipping system will be certified by the DOE and will be designed, fabricated, tested, and operated in accordance with all applicable regulations.

II. TRANSPORTATION SYSTEM DESIGN REQUIREMENTS

A transportation system for truck transport includes the tractor, semi-trailer, packaging, mounting fixtures, tiedowns, and radioactive material contents. The packaging is the structure that shields the public from penetrating radiation, contains the radioactive material, and dissipates heat that is generated as a result of radioactive decay.

Transportation packagings are designed to stringent performance standards, are required to undergo an extensive quality assurance program, and are verified to meet or exceed the design standards through analysis or testing. The DHLW package is designed according to American National Standards Institute Standards, American Society for Testing and Materials specifications, Nuclear Regulatory Commission (NRC) regulatory guides, ASME Boiler and Pressure Vessel Code, and DOE, NRC, and Department of Transportation (DOT) regulations.

Packaging regulations describe design parameters for two physical environments that can exist during transport: normal and accident. A packaging for large quantities of radioactive materials is designed to contain and shield the contents when subjected to both environments. Normal conditions are defined by regulations[2] to bound expected normal conditions of transport such as heat, cold, vibration and shock normally incident to transport, rainfall, and a free drop through 0.3 m (1 ft). Federal regulations require that the packagings must survive the following hypothetical accident conditions:

1. impact of a fully loaded package on an unyielding target after a 9 m (30-ft) free fall,

2. impact of a fully loaded package on a 15-cm (6-in) diameter rigid steel pin following a 1-m (40-in) free fall,

3. exposure of a fully loaded package to a 800°C (1475°F) thermal source for 30 minutes, and

4. immersion of the package to a depth of 15 m (50 ft) in water for 8 hours.

The same packaging must undergo the first three tests sequentially and survive by meeting maximum leak rate requirements and external dose limits.

III. SYSTEM DESIGN

The conceptual design for this truck cask system was first reported in Reference 3. The DHLW shipping system is designed to meet a gross vehicle weight limit of 36,288 kilograms (80,000 lb). The cask is carried horizontally on a semitrailer as shown in Figure 2. The complete system includes the loaded cask, the semitrailer with supports, and the tractor. Design weight allowances for various system components are:

Unit	Weight
Tractor	7,938 kg (17,500 lb)
Trailer, Supports, and Tiedowns	5,443 kg (12,000 lb)
Cask and Contents	22,453 kg (49,500 lb)
Practical Limit	35,834 kg (79,000 lb)

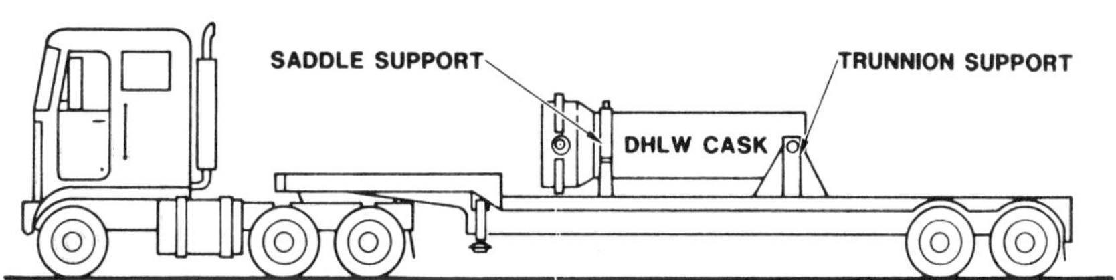

FIGURE 2. DHLW Transportation System-Truck

The support and tiedowns consist of the semitrailer mounted supports and the associated hardware required to secure the cask during transport. Use of a heavy duty tractor and a dedicated semitrailer provides for a flexible system.

Permanent storage repositories for radioactive waste will handle packages at frequencies far greater than those of today. Current handling methods would produce high radiation exposures and large personnel staffs would need to be maintained. To minimize the radiation dose to the facility operating personnel, the DHLW shipping cask transportation system is designed such that remote handling techniques can be used for all normal operations where personnel radiation exposure exists. A mock up of the cask was fabricated by the Westinghouse Hanford Company for a demonstration of remote handling techniques using robotics.[4]

Cask handling and loading operations are performed dry. The cask is top loaded while vertical. Loading operations may be performed by transferring the cask into a remote, shielded loading cell.

IV. PACKAGE CONTENTS

The DHLW shipping cask is designed to transport one canister of solidified DHLW produced by the DWPF.[5] The canister contains sludge (minimum age of 5 years) and supernate (minimum age of 15 years) in a borosilicate glass matrix. Subsequent to pouring the molten waste into the canister, the outer surface is cleaned and decontaminated. The canister is sealed by fusion welding a plug into the opening at the top of the canister.

The design basis canister shown in Figure 3 is 610 mm (24 in) in diameter by 2997-mm (118-in) long and has a loaded weight of 2356 kg (5195 lb). The canister material is type 304L stainless steel. Each canister contains approximately 282,000 curies of activity, has a surface radiation dose rate of 8900 rem/hr, and produces 750 watts of decay heat.

Gamma radiation is by far the predominant source of radiation outside the cask. The important gamma-emitting nuclides used for the source term analysis are Co-60, Zr-95, Rh-106, Nb-95, Sb-125, Ag-110m, Cs-134, Ba- 137m, Pr-144, Eu-152, Eu-154. Since the neutron source is small, special neutron shielding is not required.

V. PACKAGING DESIGN

The cask design relies on state-of-the-art technology, but the design is simple, since external cooling fins and special neutron shielding are not required. The cask components are designed to serve one or more of three functions: energy absorption, radiation shielding, and containment of the waste form. Those components that provide physical waste containment are referred to as primary containment boundaries. Each component will be discussed in the following paragraphs.

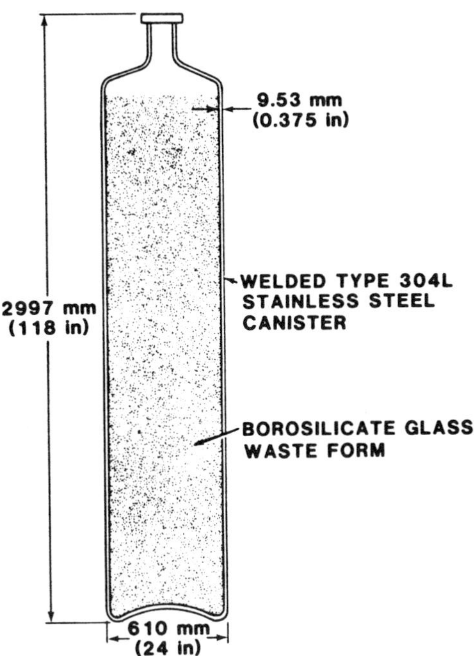

FIGURE 3. DWPF Canister

The packaging is 97.8 cm to 124.5 cm (38.5 in to 49.0 in) in diameter and 410.8 cm (151.75 in) in length. It is configured with a solid body of type 304 stainless steel and a gamma shielding liner of stainless steel jacketed, depleted uranium. This configuration was chosen to maximize the waste handling capacity of the cask while minimizing the external dose rates and staying within the legal weight truck limit.

A. Cask Body

The solid body cask, as shown in Figure 4, is of such high integrity that a large removable external impact limiter is not required, thereby improving the cask-handling characteristics and reducing the cask envelope dimensions. The 7.62-cm (3-in) thick stainless steel walls of the cask body are designed to ensure the structural integrity of the cask and to provide partial radiation shielding. The cask body in conjunction with the outer closure and double elastomer O-ring seals form a primary containment boundary. During impacts on the bottom of the cask, a ring impact limiter acts as an energy absorber. Although the impact limiter is an integral part of the cask body, it is not a primary containment boundary component. During impacts on the side of the cask, ring impact limiters located in line with the two upper lifting trunnions and the two lower tie-down trunnions act as energy absorbers.

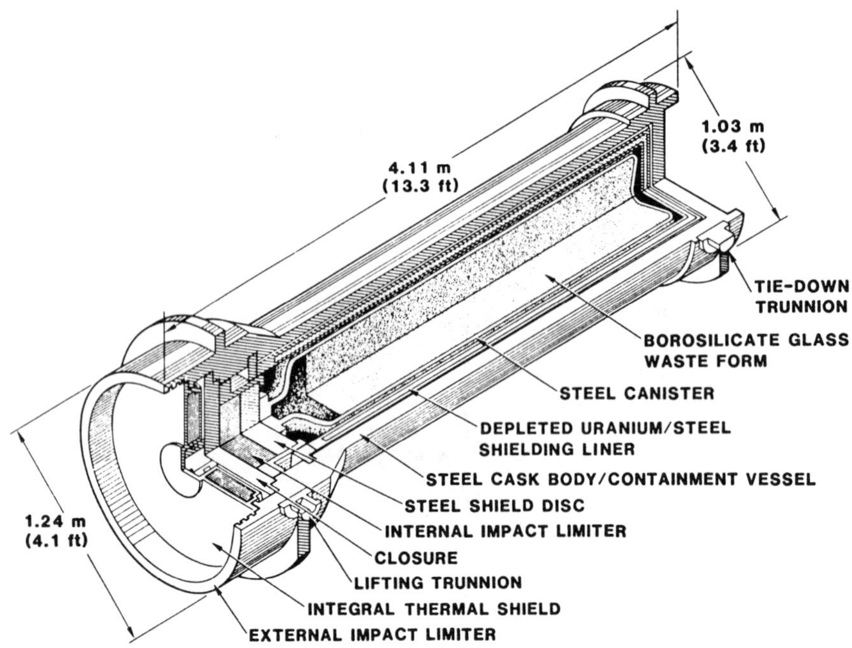

FIGURE 4. Defense High-Level Waste Truck Cask

B. Shielding Liner

Depleted uranium in the form of a removable shielding liner provides radiation shielding. It is restrained from axial movement by a segmented shear ring bolted to the top rim of the shielding liner. The shear ring extends into a circumferential groove machined into the inside wall of the cask body. The cask design allows this liner to be removed and replaced by different liners having other inside diameters. This feature allows the transport of larger or smaller sized waste forms having different shielding requirements than the DWPF waste. Other shielding liner designs could permit this cask to accommodate other waste forms such as commercial high-level waste or remote-handled transuranic waste.

C. External Impact Limiter

This component is a sculptured stainless steel ring attached to the closure end of the cask and absorbs energy by plastic deformation. Since the ring extends beyond the outer closure, it protects the closure bolts and seals from excessive deformation during the hypothetical accident condition 9-m (30-ft) free drop onto an unyielding surface. This ring, which is a separate part fitted to the cask body, protects the containment boundary near the closure from plastic deformation.

D. Closure Assembly

The closure assembly, which is part of the primary containment boundary, consists of a stainless steel plate that includes a double O-ring seal, a leak test port, and a gas sample port and is secured to the cask body by 24 Inconel 718 bolts. An internal aluminum honeycomb impact limiter assembly is bolted to the inner face of the closure. A plate distributes the load between the waste canister and impact limiter to limit the load transmitted to the closure and closure bolts during a 9-m free drop onto an unyielding surface. Both the closure plate and the internal impact limiter have radiation shielding functions.

E. Integral Thermal Shield

The function of the thermal shield is to protect the elastomer seal gaskets, the cavity gas sample port, and the seal leakage-rate test port built into the closure from excessive temperatures during the hypothetical accident condition thermal event. The thermal barrier is bolted to the outside face of the closure and must remain attached throughout the 9-m drop and the 1-m puncture hypothetical accident sequence. It consists of two thicknesses of stainless steel honeycomb contained between and bonded to stainless steel sheets and filled with vermiculite. The damage expected during those accident events will not substantially reduce its effectiveness in protecting the seals.

F. Trunnions

The trunnions are designed to safely support the DHLW shipping cask during handling and transport. They are also designed to deform or fail during the hypothetical accident conditions in a manner that will not breach or endanger the containment boundary.

VI. STRUCTURAL ANALYSIS

The structural design criteria are based on ASME Code, Section II, Appendix F, to define allowable inelastic limits. Two inelastic dynamic response finite element computer codes, HONDOII and DYNA3D, were used to analyze the impact events. An extensive study of the energy absorbing capabilities of the external impact limiter was performed to show that critical sections of the containment boundary are well below the allowable ASME limits. The critical sections analyzed for the 9 m closure end drop are shown in Figures 5 and 6. The cask neck junction and bottom centerline have the highest primary membrane stress at 13 ksi (25% of the allowable). Local membrane plus bending stresses are a maximum under the shear ring and the closure centerline (35 ksi and 47% of the allowable). A ring impact limiter is also provided at the lower end of the cask to protect the containment boundary. The available energy during the drop is absorbed through plastic deformation of the impact limiters and the cask body which is a part of the containment boundary. Different drop orientations provide the most severe loading to the closure and closure bolts. Only

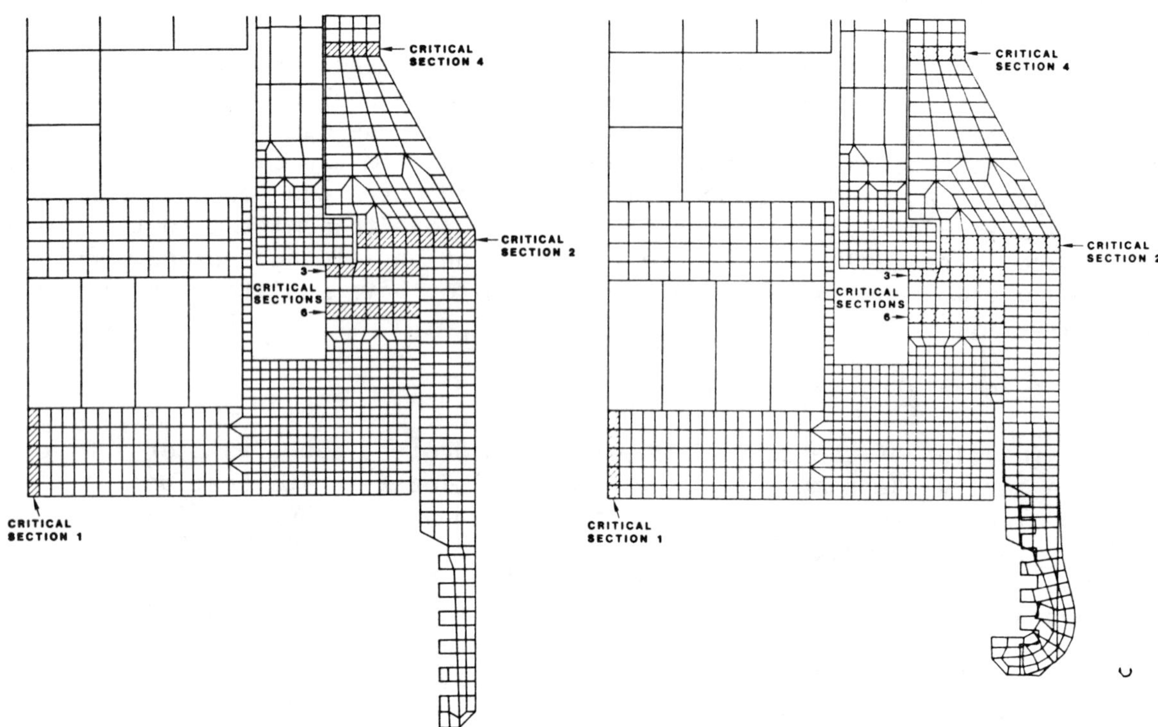

FIGURE 5. Undeformed and Deformed DHLW Structural Model for Closure End 9-m Drop

Critical Section	Maximum Primary Membrane P_m (ksi)	% of Allowable (a)	$P_{\ell+b}$ Maximum (ksi) Local Membrane + Bending Inner	Outer	Maximum % of Allowable[(a)]
1. Closure centerline	8	15	35	35	47
2. Cask body above	10	19	20	13	27
3. Cask body upper	11	21	35	29	47
4. Cask neck junction	13	25	16	30	40
5. Cask bottom centerline	13	25	32	35	47
6. Cask body lower section	10	19	14	24	32

[(a)]Allowables: P_m allowable = $S_y + 1/3\ (S_u - S_y)$ = 55 ksi
$P_{\ell+b}$ allowable = $0.7\ S_u$ = 79 ksi

FIGURE 6. Comparison of Containment Boundary Stresses to Design Criteria for Closure End 30-ft Drop

minimal plastic deformation of the closure occurs while the closure bolts remain well below yield. The center of gravity over the lower corner drop and the side drop provide the most severe loading for the remainder of the containment boundary. The DYNA3D analysis showed that the drop energy could easily be converted to strain energy by plastically deforming the impact limiter and the cask body while keeping within allowable stresses.

VII. FUTURE ACTIVITY

Fabrication of a half-scale model of the DHLW truck cask has been initiated with expected delivery during 1985. Testing on the model will be conducted at Sandia National Laboratories to verify the adequacy of the structural analysis and cask design. This series of destructive tests includes 9-m drops onto an essentially unyielding surface for several impact orientations. A 1-m drop onto a mild steel punch will follow the 9-m drop events. Following evaluation of the test results, a prototype will be fabricated and certified by the DOE. Initial shipments of immobilized waste from the DWPF are scheduled to be transported to the WIPP in 1989.

VIII. SUMMARY

The DHLW legal-weight truck cask has been designed to provide safe and efficient transport for defense high-level waste. All design and testing is in accordance with applicable regulations. Important design features of this state-of-the-art cask include:

- integral stainless steel impact limiters,
- removable depleted uranium shielding sleeve,
- remote handling capabilities,
- Type 304 stainless steel construction,
- bolted closure, and
- concentric elastomer O-ring seal

Testing of a half-scale model of the cask will commence during the fall of 1985. Certification is scheduled for completion in 1987. Initial shipments of immobilized waste from the Defense Waste Processing facility at the Savannah River Plant to the WIPP are scheduled for 1989.

ACKNOWLEDGMENTS

The authors wish to recognize that this paper describes the efforts of a great number of people who have significantly contributed to the technical details of the DHLW cask design. A list, which is not exhaustive, of the people deserving recognition are R. M. Burgoyne, N. W. Johanson, P. C. Rasmussen, and J. C. Reynaud of GA Technologies, Inc.: G. C. Allen, W. L. Uncapher, and R. G. Eakes of Sandia National Laboratories; R. G. Baxter and G. G. Chalfant of duPont Savannah River Plant; and K. G. Golliher and E. L. Wilmot of the DOE.

REFERENCES

1. N. W. Johanson et al., "HLDW Transportation system Design Basis and Concepts," Proceedings of the 6th International Symposium on Packaging and Transportation of Radioactive Materials, Berlin (West), Federal Republic of Germany (1980).

2. Title 1, Code of Federal Regulations, Part 71, Subpart F.

3. R. L. Moore et al., "Design Concept for a Defense High Level Waste Shipping Cask and Truck," Proceedings of the 7th International Symposium on Packaging and Transportation of Radioactive Materials, New Orleans, LA, (1983).

4. B. C. Gneiting et al., "Development Testing of a Nuclear Waste Cask Remote Handling System," Proceedings of Waste Management '85, Tucson, AZ (1985).

5. R. G. Baxter, Description of DWPF Reference Waste Form and Canister, E. I. duPont de Nemours & Co. (Savannah River Plant) Report DPSP-80-1033 (1983).

TRANSPORTATION SYSTEMS TO SUPPORT THE NUCLEAR WASTE POLICY ACT OF 1982

Edwin L. Wilmot
Robert E. Philpott
Office of Civilian Radioactive
Waste Management
U.S. Department of Energy
Washington, D.C. 20585

ABSTRACT

Late in 1982, the United States Congress enacted legislation for the disposal of spent nuclear fuel and high-level waste. The policy, embodied in Public Law 97-425 and referred to as the Nuclear Waste Policy Act of 1982 (NWPA), mandates that the Department of Energy (DOE) be responsible for the transport of commercial spent fuel and defense high-level waste from their points of origin to facilities constructed under provisions of the NWPA. It is the purpose of this paper to describe the preliminary transportation policies and plans developed by the Office of Civilian Radioactive Waste Management (OCRWM), within the DOE, to respond to the NWPA mandate.

I. TRANSPORTATION SYSTEM DESCRIPTION

Currently, about 13,000 MTU of spent fuel exist in reactor storage basins and at commercial storage facilities.[1] The ultimate goal of the OCRWM program is to transport this spent fuel and all newly generated spent fuel to a final geologic disposal site. The most current programmatic philosophy is to introduce an intermediate facility into the system in order to improve the performance of the overall system. This intermediate facility, which remains to be approved by Congress, is called an integrated monitored retrievable storage (MRS) facility. Its functions are delineated as follows: manage at-reactor spent fuel acceptance rates of the system; schedule and control transport; receive, inspect, and account for spent fuel for disposal; store fuel as needed; and perform any specialized packing, repairing or testing. If such a proposed facility is approved by Congress, most shipments from reactors would be to the MRS. The MRS would then perform its functions and ultimately ship consolidated spent fuel to the repository. Some shipments from reactors may proceed directly to the repository as may shipments from defense high-level waste (HLW) sources. Radioactive nonfuel components at the reactors could be shipped to an MRS or directly to a low-level disposal site or a repository, depending on an NRC decision on the destination for this waste form.

The candidate sites for the MRS are centrally located (all sites being considered are in Tennessee) to the inventory of spent fuel being stored so as to reduce the cask miles* required in the system. The cask miles required when shipments proceed directly from reactors to a repository are expected to be greater than the cask miles required if shipments proceed to the repository via the MRS.

One of the important benefits of the MRS is the uniform product that it will produce. It will enable subsequent repository shipments of uniform-sized canisters that potentially could have contents with uniform thermal and radiation levels. These canisters could be used for either final disposal or for interim handling.

The improved performance transportation system as described in Figure 1 has three discrete components: the front-end (i.e., shipments to the MRS), the back-end (i.e., shipments to the repository from the MRS); and direct shipments (i.e., shipments from reactors and HLW sources directly to the repository).

A. Front-End System

The front-end system provides a design challenge because of the diversity of shipping facilities interfacing with the MRS (i.e., approximately 100 reactors and HLW sources). The key to the design of this system will be to minimize the number of different casks required while reducing the number of shipments by maximizing cask capacities. All surface modes of transport will be considered. Legal and overweight truck cask designs will be pursued, but rail casks will be held to a 100-ton weight limit to ensure a gross vehicle weight that will allow unrestricted travel. Because not all reactors are accessible by rail or can accommodate rail casks, truck cask features facilitating intermodal transfer to railcars will be encouraged.

The front-end system will consist of a base fleet, referred to as from-reactor casks, that is capable of transporting a minimum of 80% of the fuel to be shipped as well as a supplementary fleet for the remainder. This latter fleet may be referred to as the nonstandard-fuel fleet. Currently, only the from-reactor fleet is being defined since it is hoped that the specifications for the nonstandard-fuel cask fleet will be defined by those contractors designing the base fleet. The expected maximum number of casks from-reactor is 120 if only the truck mode were considered and 70 if only rail were considered. The basis for these numbers is a capacity of 2 pressurized water reactor (PWR) spent fuel assemblies or 5 boiling water reactor (BWR) assemblies for truck casks and 14 PWR assemblies or 36 BWR assemblies for rail.

Figures 2, 3, and 4 are maximum design envelopes for the front-end system for truck, overweight truck, and rail modes, respectively. These envelopes are not meant to be representative of a particular system but to provide guidance for maximum sizes, total weights, and weight distribution. The transporter of

---

*Cask miles is the distance traveled by a cask regardless of the cask capacity. A mile traveled by a cask carrying two assemblies would be counted as one cask mile as would a mile traveled by a cask carrying 14 assemblies.

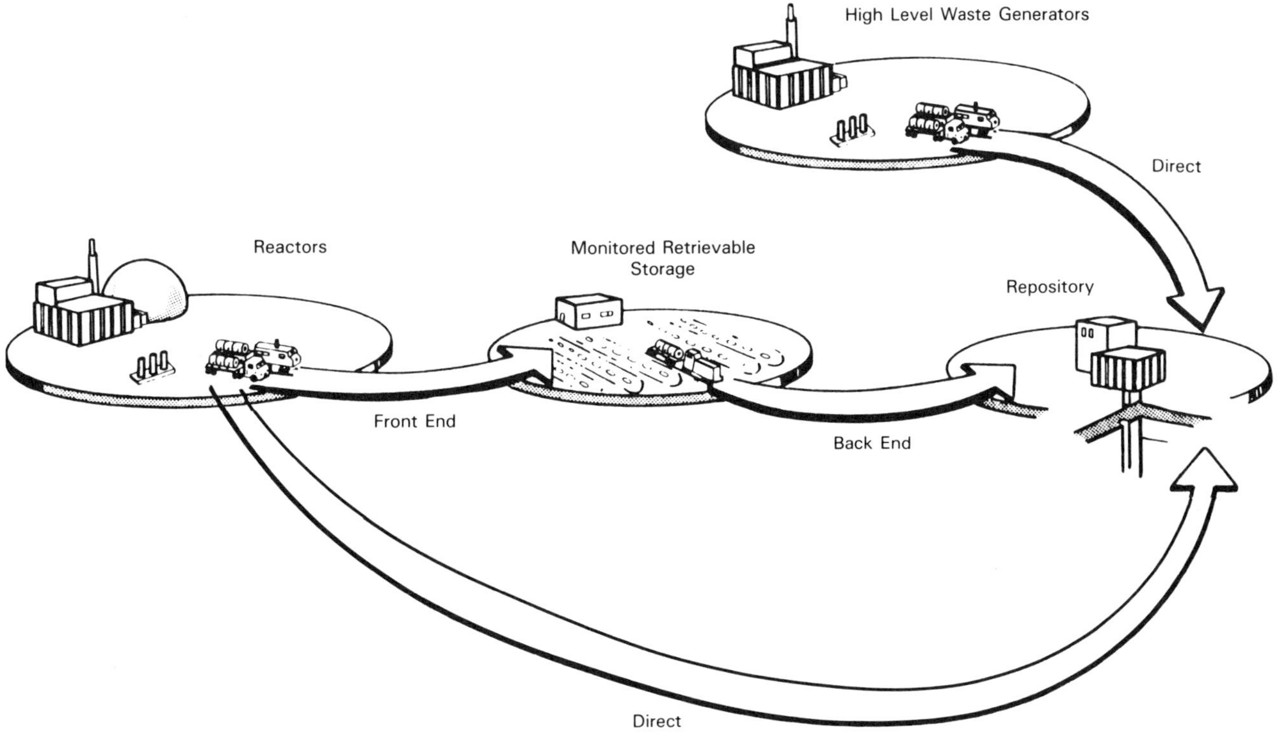

FIGURE 1. Transportation System

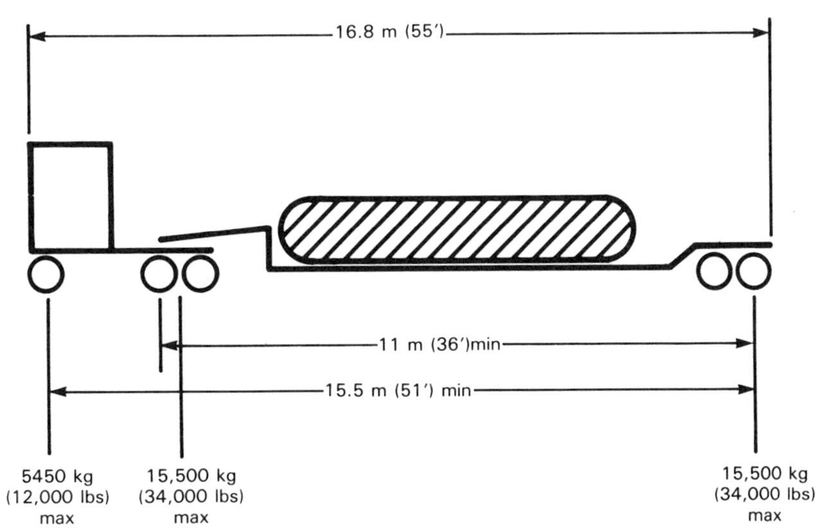

FIGURE 2. Truck Design Envelope

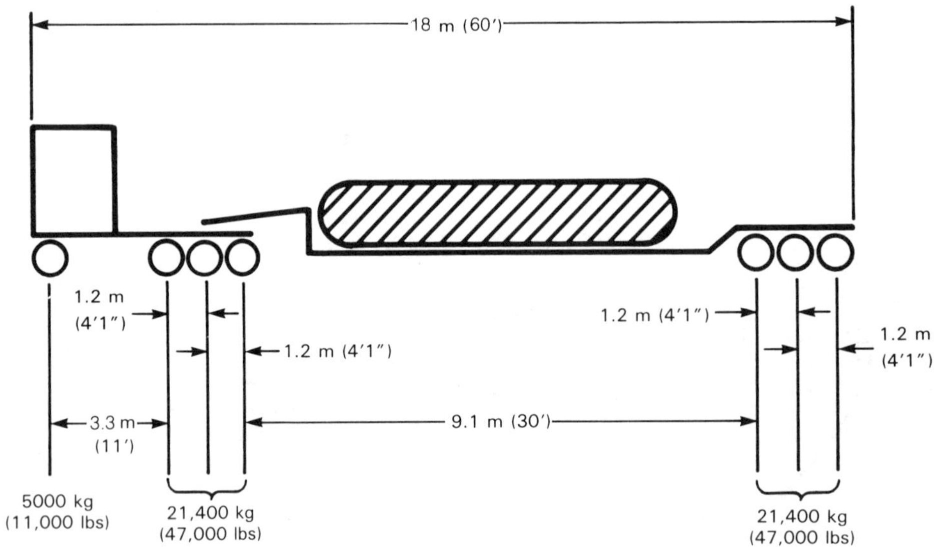

FIGURE 3. Overweight Truck Design Envelope

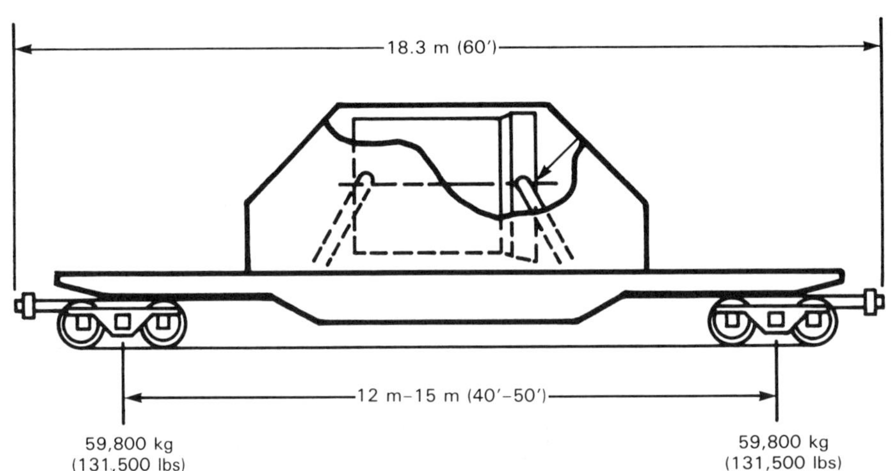

FIGURE 4. Rail Design Envelope

maximum size truck casks would require overweight permits but the weight restriction of 50,000-kg (110,000-lb) gross vehicle weight distribution over seven axles should allow overweight permitting in most states.[2] The rail system will be designed to allow transport in general commerce; however, there may be potential benefits in sequencing shipments or using marshalling yards in order to take advantage of dedicated trains. Some reactors are clustered so that, by proper sequencing of cask loading, a dedicated train could receive and accumulate cargo as it proceeds along a given route. Additionally, a specific marshalling yard could be designated to allow the assemblage of several rail casks in one location with a subsequent shipment via a single dedicated train.

B. Back-End System

The back-end system will most assuredly be the "showpiece" of the entire system. Much innovation and creativity as well as standardization will be possible. The back-end system (casks referred to as from-MRS) is so configured as to allow for identical interfaces at the origin and destination of the shipments since there is one source and one destination whose interfaces are still to be designed. The shipment contents will be uniform and the number of routes reduced.

A general description of the hardware might be as follows: cask weights approaching 136 T (150 t); the cask integrated into the rail car; capacities increased 25 MTU potentially); and all handling with automated remote equipment. As many as 30 casks may be required. An envelope of the physical limits of the rail system is shown in Figure 5. The gross vehicle weights could potentially approach 200,000 kg (440,000 lb) through the use of double trucks at each end of the railcar. With a reduced number of potential routes, size and weight restrictions can be negotiated more easily with the carriers. Such negotiations could potentially result in raising vehicle weight limits to accommodate heavier casks. The use of dedicated nuclear trains will enhance operational efficiencies. Although the primary focus of the from-MRS fleet will be the rail mode, other modes such as truck and barge will be developed to provide a back-up capability in the event of circumstances that temporarily preclude the use of rail.

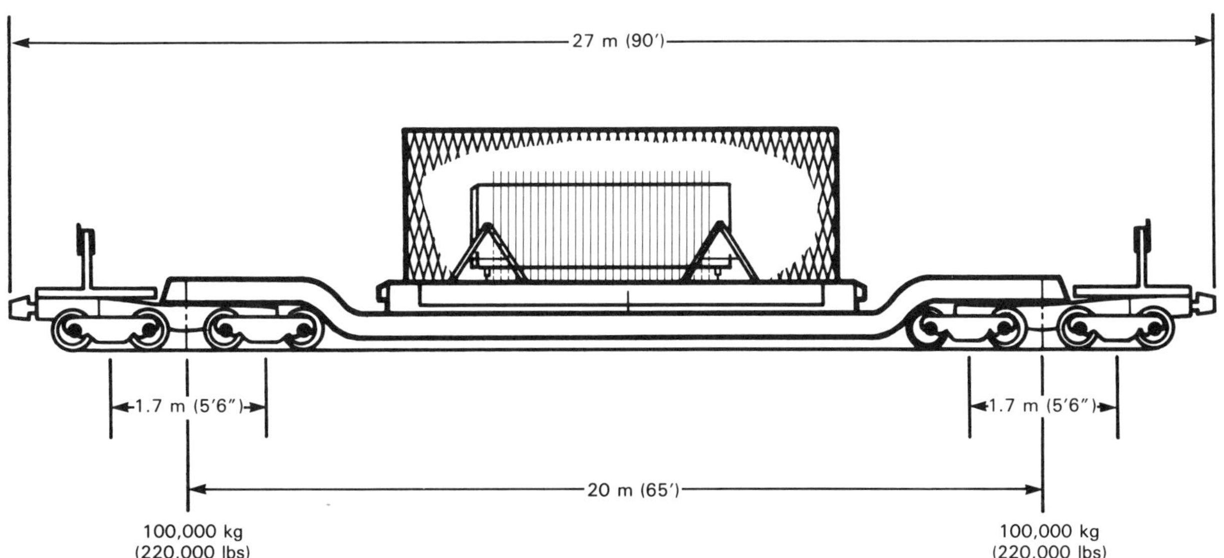

FIGURE 5. Heavy Railcar Design Envelope

Should the MRS be approved, other types of material will have to be shipped. These include skeletons and nonfuel assembly parts generated during rod consolidation. Current plans for hardware development have not embraced this need, but planning will be initiated if Congressional approval for the MRS concept is received.

C. Direct System

The system for shipping directly from the originating source to repository will accommodate two waste types: nonfuel components and high-level waste. The spent fuel components considered to be nonfuel are defined in Part 961 of Title 10 of the Code of Federal Regulations. The list includes such items as control spiders, burnable poisons, control rod elements, and neutron sources. The nonfuel component is currently not well characterized and cask development for them has a secondary priority until the front-end system has been initiated. Preliminary thoughts are to develop a flexible system that may be composed of a few cask bodies that have a number of interchangeable baskets to accommodate the assortment of miscellaneous items. The cask bodies would be expected to meet similar interface requirements to those imposed on the front-end fleet. If the MRS were not approved, the from-reactor fleet would also become part of the direct system.

Defense waste sources are DOE Savannah River Plant (SRP), the DOE Hanford Plant, and DOE Idaho National Engineering Laboratory. The waste form for SRP will be vitrified borosilicate glass in a 0.6 m x 3 m (2 ft x 10 ft) canister; other waste forms are being considered at the other sources. A truck cask is currently being designed to carry one canister. A rail cask is also being considered with a capacity of from five to seven 0.6 m x 3 m canisters of waste. Since HLW from West Valley is also a glass contained in the same size canister, the defense HLW cask may be used for West Valley shipments.

D. Role of Other Casks

Other casks exist or are being proposed that could have a significant role in the transportation system. Existing spent fuel casks that have been certified by the Nuclear Regulatory Commission (NRC), though not efficient for a major role in either the from-reactor or between-Federal-facilities fleet, may serve a useful purpose in carrying out nonstandard fuel, nonfuel components, or high-level waste. They will be given careful consideration for these types of applications.

As stated in Reference 3, the multipurpose cask, which is a cask designed for storage, transport, and disposal is unlikely to play a significant role in the transportation system primarily because of design inefficiencies and inherent uncertainties of verifying cask condition after extended storage periods. This cask is a viable alternative only if provisions of a one-time shipment can be granted.

II. DOE POLICY ON PACKAGE DESIGN FOR SHIPMENTS UNDER NWPA

Pursuant to U.S. Department of Transportation (DOT) regulations, DOE currently certifies its own casks using standards equivalent to those of the NRC; however, OCRWM plans that all package designations (casks) used in support of shipments under the NWPA be certified by the NRC prior to use. This policy is clearly delineated in a Procedural Agreement between the NRC and the DOE, published November 14, 1983 (48FR51875). DOE entered into the Agreement considering the intent of the NWPA [Section 137(a)] and taking into account the commercial source of the radioactive materials being transported under the NWPA. The terms of the Agreement apply to casks under the NWPA, including the repositories and potentially a monitored retrievable storage (MRS) facility, if such a facility were approved by Congress.

By coordinating with the NRC at the early stages of cask design, development, and testing, the OCRWM expects to reduce uncertainties and the potential for delays in certification. It will seek to resolve technical issues by preparing topical reports for submittal to the NRC for formal response. Their response will be expected to contain a written position on the issue in question. The NRC will also be asked to review and comment on cask performance specifications prior to their use in cask and vehicle design. The DOE anticipates NRC comments on such criteria as the hypothetical test conditions, which may reflect the results of their soon to be completed study evaluating regulatory standards against real accidents and the 1985 IAEA Safety Series 6 regulations.[4]

Private industry, under contract to the OCRWM, will submit final designs to the NRC independently after thorough review by OCRWM and OCRWM will not intercede unless general issues potentially affecting several designs arise.

III. DOE OBJECTIVES IN CASK DEVELOPMENT

Building upon 40 years of successful radioactive material shipping experience, OCRWM within the DOE is planning to embark on a major cask development program. OCRWM has defined three program objectives: safety, public acceptance, and efficiency.

A. Safety

Protection of the public health and safety will be the major objective of OCRWM in developing casks, but DOE will not assert its authority to certify cask designs for NWPA shipments. OCRWM will use its contractors to develop cask designs and will require them to submit their designs directly to the NRC.

OCRWM will minimize exposures by applying the principle of "as low as reasonably achievable" (ALARA). Prior to and during early stages of cask design, OCRWM will perform an ALARA study to determine the most cost-effective means of reducing radiation exposure. Such an evaluation will examine many facets of the transportation system including: loading, normal transport, expected values from exposure during accidents, and unloading. Intermodal transfers will be considered as appropriate. The ALARA study will also deal with the two general population contingents: the public and individuals exposed because of their occupation.

B. Efficiency

The second major objective that OCRWM seeks to attain with the cask development program is efficiency. An obvious step towards achieving this goal is to increase the capacity of existing cask designs. At a minimum, it is expected that existing capacities can be doubled by new designs because of the advanced age of the fuel to be shipped to the repository. New materials of construction will be explored and used where their application will result in efficiency and potential cost savings. Casks will be used that provide maximum flexibility in capacity of cavity size and dimension but not at the cost of safety and efficiency.

Another obvious step towards more efficient operations is to reduce the time required to load and unload. This will reduce the total number of casks required. Elimination of cask features that impede decontamination will be a high priority. Developing vehicle designs that make it difficult for a terrorist to conceal an explosive device will improve security inspection times. Low maintenance requirements and standard cask/facility interface design features will significantly increase time available for use and decrease turnaround times. Automated remote handling of casks has been shown to reduce handling times so features to facilitate robotic handling will be incorporated. Furthermore, advanced design features that could eliminate time consuming closure bolt installation and removal will also be considered, e.g., a breechlock closure may be a potential closure mechanism.

C. Public Acceptance

Despite an excellent record of safety and the firm OCRWM commitment to enhance levels of safety consistent with cost benefit analyses, concerned organizations and members of the public remain unconvinced about the high levels of protection afforded by the casks and administrative controls during operation. Hence, OCRWM has a prime objective to foster public acceptance through a cask development program conducted on an open and objective basis. For example, OCRWM will involve interested parties early in the cask development process by making design specifications and conceptual designs readily available for comment to anyone who is interested.

In addition to the "design in a fishbowl" approach, OCRWM will emphasize an active and rigorous quality assurance program throughout all phases of cask development and testing. OCRWM recognizes that quality assurance is critical to cask performance and integrity.

A further consideration in fostering public acceptance of cask designs is to physically test casks that will be used to transport waste under the provisions of NWPA. Although physical testing is not required to obtain an NRC certificate of compliance, OCRWM is currently formulating a testing program in which cask designs will, at a minimum, be tested by an independent testing laboratory using scale models. In addition, OCRWM will consider full-scale verification testing of prototype cask and vehicle systems.

By developing a safe and efficient system in an open environment with free exchange of information, OCRWM expects to increase public understanding and confidence.

IV. PARTICIPANTS AND THEIR ROLES

Several transportation system participants have been explicitly identified in the NWPA. OCRWM will take title to the spent fuel at the reactor sites and will be the shipper of record. As such, the OCRWM has ultimate responsibility for the safety and efficiency of transportation. OCRWM must perform its functions within the precepts of the two primary transportation regulators: the NRC and the DOT. Private industry will be utilized to the maximum extent to design and fabricate the system hardware as well as to provide carriage and operational services sites under contract to OCRWM. The utilities are the customer and, as such, are concerned about efficient use of utility ratepayer fees. The states and tribes, on the other hand, have the ultimate responsibility for the safety and welfare of their constituents and, therefore, may provide another independent surveillance, if they choose, on the system and operational safety. The interests of each of these parties must be consididred and satisfied before a smoothly operating transportation system can be realized.

V. SYSTEM PERTURBATIONS

Many parameters of the integrated system are, as yet, undefined and decisions remain to be made that could have significant impacts on planning and design. Decisions regarding the MRS, repository acceptance rates and spent fuel acceptance criteria could affect system configuration. Decisions by the NRC regarding a definition of HLW could have wide-reaching impacts. In addition, the NRC also is evaluating the hypothetical accident conditions used for cask design, and any change would have design implications. Extending the burnup of fuel has the potential to reduce cask capacities for some peak burnup projections as well as to reduce somewhat the quantities of fuel to be shipped. Anomalous source term parameters have been or could be discovered. For example, cobalt-60 in spent fuel assembly hardware not previously significant in shielding design for short-cooled fuel now has to be considered more carefully in long-cooled fuel casks. The above uncertainties are enumerated to indicate that all fleet designs and requirements are not currently available and cannot be for several years to come.

VI. SUMMARY

Early planning for a transportation system to support shipping and disposal of spent fuel and high-level waste is continuing and much of the necessary hardware is expected to be available in 1995. This schedule is aggressive despite the apparent abundance of time. By initiating design of the front-end fleet in 1986, it is expected that all objectives can be met by 1995. Safety can be assured through rigorous quality assurance, independent testing and prototype check-out. The public can be thoroughly familiarized with the designs, the testing, and capabilities of the fleet. Furthermore, efficiency can be assured by proper design and by allowing prototype cask handling at most facilities. The next 20 years promises to be an exciting and challenging period for significant innovations in transportation system development.

REFERENCES

1. C. M. Heeb et al., Reactor-Specific Spent Fuel Discharge Projections: 1984 to 2020, PNL-5396, Pacific Northwest Laboratory, Richland, Washington (1985).

2. J. C. Allen, S. Gupta and D. A. Ladd, Overweight Truck Shipments to Nuclear Waste Repositories: Legal, Political, Administrative and Operational Considerations (Draft), Battelle Project Management Division, Columbus, Ohio (1985).

3. G. C. Allen, "Advanced Transportation System Options for Spent Fuel and High-Level Waste," Packaging and Transportation of Radioactive Materials Seminar (1984).

4. International Atomic Energy Agency, "Regulations for the Safe Transport of Radioactive Materials," 1985 Edition, Safety Series No. 6, Vienna, Austria (1985).

MONITORED RETRIEVABLE STORAGE STATUS

J. H. Carlson
Office of Civilian Radioactive
Waste Management
U. S. Department of Energy
Washington, D.C. 20585

ABSTRACT

The Department of Energy (DOE) is preparing a proposal for Congressional consideration of the inclusion of a monitored retrievable storage (MRS) facility as part of the Federal Nuclear Waste Management System. The DOE plans to submit the proposal package to the Congress by January 15, 1986. The proposed preferred location of the MRS facility is the former Clinch River Breeder site in the state of Tennessee. If the Congress approves implementation of the MRS proposal, it is estimated that the facility could be licensed and operational in ten years.

I. INTRODUCTION

The Department of Energy (DOE), in response to the requirements of the Nuclear Waste Policy Act of 1982 (NWPA), is preparing a proposal for Congressional consideration of the inclusion of a monitored retrievable storage (MRS) facility as part of the Federal nuclear waste management system. The Nuclear Waste Policy Act directs the DOE to complete a study of the need for and feasibility of MRS, and to submit to Congress a proposal for the construction of one or more MRS facilities. The NWPA specifies that the proposal include a program for siting, development, construction and operation of an MRS facility; site-specific designs and cost estimates for the construction of the first facility; a plan for funding the construction and operation of such facilities; and a plan for integrating such facilities into the Federal waste management system. The NWPA is also very specific in stating that "...disposal of high-level radioactive waste and spent nuclear fuel in a repository developed under this NWPA should proceed regardless of any construction of a monitored retrievable storage facility...."

The NWPA directed submission of the MRS proposal and its accompanying documentation by June 1, 1985. Investigations of the need for, feasibility of, and integration of MRS resulted in the recognition late last year that an MRS facility deployed as an integral component of the Federal nuclear waste system, rather than a backup storage component, offered substantial benefits to the system. Development of this concept for the proposal, however, required substantial additional effort and precluded meeting the June 1985 completion date. The DOE is nearing completion of the proposal for this "improved performance" system including MRS, and plans to submit the proposal package to the Congress on or before January 15, 1986.

In this system the MRS facility would receive spent fuel by truck and rail from reactors, consolidate it, canister the consolidated rods and fuel hardware and then ship it by dedicated train in high capacity transportation casks to the repository. The MRS facility would have the capability to store up to 15,000 MTU of spent fuel, if necessary.

II. MRS PROGRAM ACTIVITIES COMPLETED TO DATE

The programmatic and technical accomplishments since enactment of the NWPA are summarized below.

A. Establishment of the MRS Program

The NWPA established the Office of Civilian Radioactive Waste Management (OCRWM) in DOE as the manager of the Nuclear Waste Program. The MRS activities are implemented through the DOE Richland Operation Office under OCRWM direction. When, as is discussed later, three candidate MRS sites were identified for evaluation in April 1985 the Oak Ridge Operations Office was assigned responsibility for outreach activities; that is, the interaction with the state and local officials and the public. The Pacific Northwest Laboratory (PNL) operated by Battelle Memorial Institute, is the prime contractor for the MRS effort. The Ralph M. Parsons Company of Pasadena, California, was selected by DOE through competitive procurement as the architect-engineer for the MRS designs to be included in the proposal. PNL selected several subcontractors to aid in preparation of documents under the program and for supporting studies.

B. MRS Research and Development Requirements

The NWPA directed DOE to submit to Congress, within six months after enactment of NWPA, a report detailing the research and development activities necessary to develop the MRS proposal. This report, designated as DOE/S-0021, Monitored Retrievable Storage Proposal Research and Development Report, was submitted in June 1983. This report concluded that passive dry storage technologies then under consideration for the MRS Program were the preferred technologies for this application, and that the technologies are sufficiently mature that the MRS proposal could be prepared using current engineering and design practice without additional research and development.

C. MRS Concept and Mission Definition

Before the enactment of the NWPA, DOE had evaluated concepts for the storage of spent fuel and radioactive wastes, and had identified eight storage concepts as those most promising for application to an MRS facility. These eight concepts were reexamined in greater detail for MRS use. Two of the concepts--the sealed storage cask (concrete cask) and the field drywell--were selected, respectively by DOE as the base and alternative concepts for storage at an MRS facility. This selection was documented in the DOE report

DOE/RL-82-2, Selection of Concepts for Monitored Retrievable Storage of Spent Nuclear Fuel and High-Level Radioactive Wastes, issued in April 1984.

Pending additional study of the need for and feasibility of MRS, DOE initially recognized benefits to having an MRS facility as a backup to the repository to be deployed in the event of repository delays. MRS designs and plans were consistent with this early view. However, later analyses showed that an MRS facility would be more beneficial when operated as an integral component of that system. Therefore, in the Fall of 1984, the DOE elected to base its proposal for MRS construction on this role. The concept selection was revisited and reconfirmed consistent with this role.

D. MRS Need and Feasibility Study

The NWPA directed the DOE to conduct "a detailed study of the need for and feasibility of...constructing one or more monitored retrievable storage facilities." A comprehensive study is underway to accomplish this. This study compares alternative waste system configurations with and without an MRS that could serve various roles. The analyses to date indicate that overall advantages to the system accrue from deployment of an MRS facility as an integral component of the Federal waste system, and that its construction is feasible and cost-effective in light of the benefits to be gained from its deployment. The system improvements include the following:

1. improved confidence in the DOE's ability to meet schedules, particularly in beginning to accept quantities of waste no later than January 31, 1998 (the integral MRS facility would be scheduled for initial operation as early as 1996),

2. increased reliability and flexibility in operating the system in an integrated cost-effective manner by incorporating an additional facility that can regulate the flow of waste to the repository,

3. ability to accept significantly larger quantities of waste in the early years of operation, substantially reducing the added cost of providing increased at-reactor storage capabilities, and

4. improved transportation efficiency because spent-fuel consolidation and packaging at the MRS facility, together with utilization of rail casks and dedicated trains, would reduce the number of shipments to the repository with concomitant reductions in potential transportation impacts to the public.

Analysis is continuing, and the final report considering the MRS as an integral part of the system, a system without an MRS and the MRS as a backup in the event of repository delays will be included in the environmental assessment.

E. Site Screening and Recommendation

Early efforts evaluated potential criteria for siting of an MRS facility and considered screening procedures based on the criteria. Factors considered included physical and meteorological conditions, sociological impacts, economics, and comparative risk assessment (although absolute risks were determined to be very low). It was determined that a large number of sites throughout the contiguous United States could qualify for the safe location of an MRS facility with little differentiation among those sites. Consideration of the transportation of spent fuel and radioactive wastes throughout the Federal system, on the other hand, disclosed a region of the Nation within which location of an MRS facility could reduce total shipment miles substantially, and thus could limit transportation impacts (e.g., risks) associated with such shipments. Further evaluation and screening of potential sites within that region resulted in the identification of three sites for further consideration. The former Clinch River Breeder Reactor site in the State of Tennessee is the preferred site. Two alternative sites, one on the DOE Reservation in Oak Ridge, Tennessee, and the other at the Tennessee Valley Authority's former Hartsville nuclear plant site near Nashville, Tennessee, have also been identified for evaluation. The DOE identification of those candidate sites was publicly announced on April 25, 1985. Identification of the sites is recorded in the report DOE/RW-0023, Screening and Identification of Sites for a Proposed Monitored Retrievable Storage Facility, dated April 1985. Congress has reserved the actual selection of the MRS site for itself.

F. MRS Design Activities

Initial design activities, in progress before the definition of the MRS mission as an integral component of the Federal waste management system, were based on reference, generic sites. An advanced conceptual design for an MRS facility using sealed storage casks of concrete construction and a conceptual design for an alternate concept employing field drywells for storage were completed in January 1985. These initial design efforts provided information which facilitated early need and feasibility analyses, the site screening and selection process, environmental studies of the MRS operations, and the development of other components of the MRS proposal. Following completion of these back-up facility designs, design efforts were directed to conceptual designs for the integral MRS facilities (using sealed casks and field drywells) for the recommended site and the two alternative sites. These designs, which are essentially complete, will be included in the MRS proposal. Figure 1 is an artist's rendition of an MRS facility on the candidate Clinch River site. The facility would receive, consolidate and canister spent fuel destined for the first repository. The canistered spent fuel would then be immediately shipped by dedicated train to the repository or stored temporarily until the repository was ready to receive it. The facility is designed for a maximum storage capacity of 15,000 MTU. Further detail on the design is presented in the next paper.

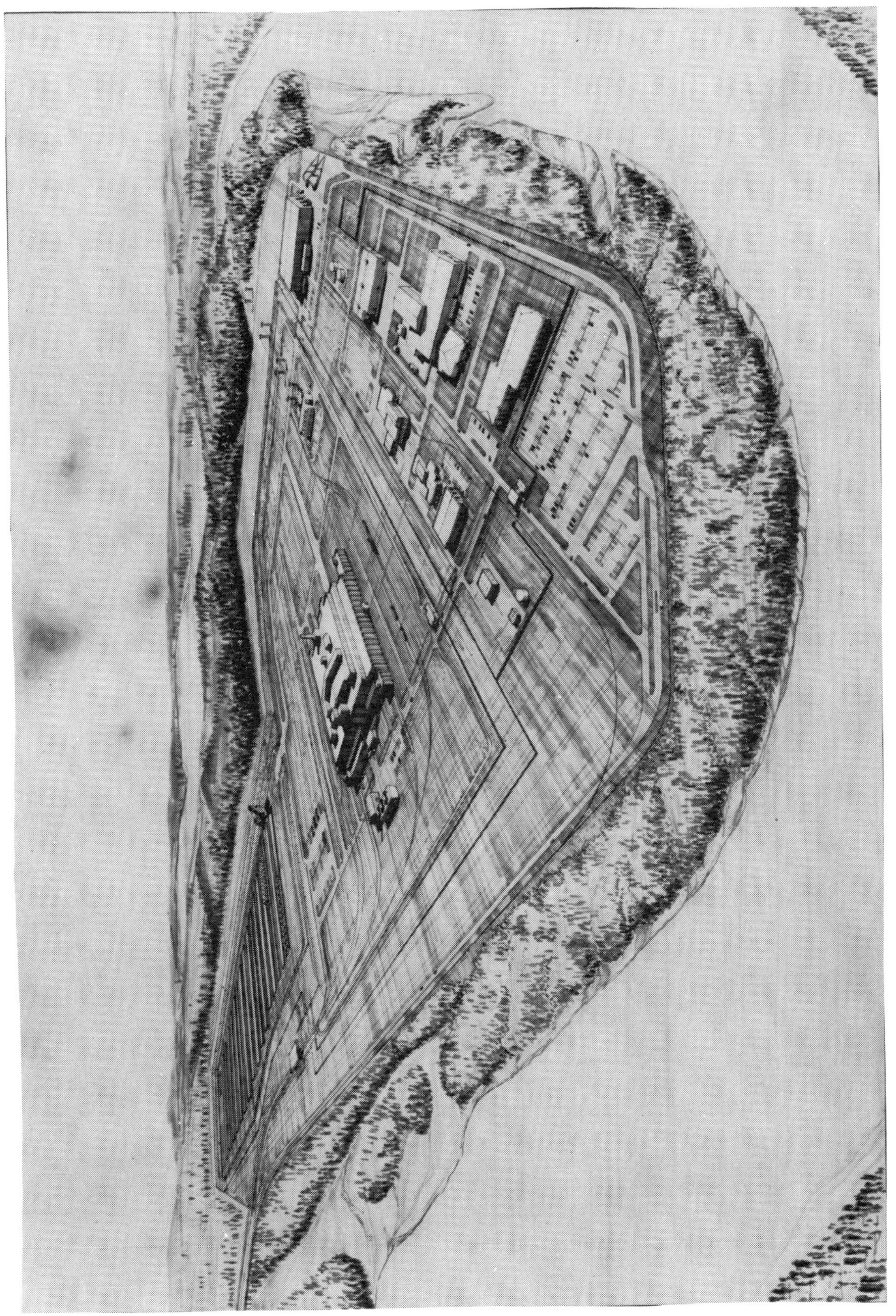

FIGURE 1. An Artist's Rendition of an MRS Facility on the Candidate Clinch River Site

G. MRS Program Plan

An MRS Program Plan document is in preparation, and will be submitted as part of the proposal package. This document will describe a plan and schedule for deployment and integration of the MRS facility to implement an "improved" performance waste management system, estimates of costs over the lifetime of the MRS facility, including a plan for funding MRS deployment, operation, and decommissioning. DOE intends to recommend in these plans measures that would permit the local governments to receive revenues equivalent to those that would be realized if this were a similar, private sector venture fully subject to normal taxing practices.

The plan will also contain estimated MRS facility costs and describe changes in costs of other waste management activities as a result of inclusion of MRS. DOE intends to recommend that costs of development, construction and operation of the facility be borne by the Waste Fund established by the NWPA.

H. MRS Environmental Assessment

The environmental assessment is currently undergoing internal review and will be submitted to Congress with the proposal. This assessment will include the final results of the need and feasibility analysis and a detailed analysis of the environmental impacts of the MRS designs at the identified sites. This site/design analysis is based on available data and will analyze the relative advantages and disadvantages of the alternate design and site combinations.

I. Site-Related Activities and Public Interactions

Beginning with the announcement of the MRS sites, interactions were initiated with the State of Tennessee, and through the State with local groups and government bodies, as well as with other States to explain and discuss DOE's current views on the siting and operation of the MRS facility. A $1.4 million grant has been made to Tennessee to aid them in understanding and evaluating the DOE proposal and its supporting bases. These activities will be supplemented by additional public information and interactions. These interactions are intended to provide public information and to cooperate with potentially affected governmental units on matters relating to the MRS facility. To facilitate local access to MRS information an MRS office has been set up at the DOE Oak Ridge Operations Office. MRS documents and information are being shared with Tennessee officials as they are developed. The complete set of supporting documents will be provided to Tennessee when they are provided to the NRC and EPA for review in December before submittal of the MRS proposal to Congress. The interactions with Tennessee will continue through submission of the proposal, and through congressional consideration of the proposal. If Congress authorizes MRS construction, an interaction program consistent with the NWPA will be initiated. State and local concerns expressed during this process will be provided to the Congress with the proposal.

III. IMPLEMENTATION OF MRS CONSTRUCTION AND OPERATION, IF APPROVED

If the Congress approves the implementation of the MRS proposal, it is estimated to take approximately ten years to complete those necessary activities leading to operation of an MRS facility. This would include the definitive design, preparation of an EIS in compliance with the National Environmental Policy Act, licensing by the Nuclear Regulatory Commission and facility construction. A schedule showing activities and elapsed time between Congressional approval and operation is shown in Figure 2.

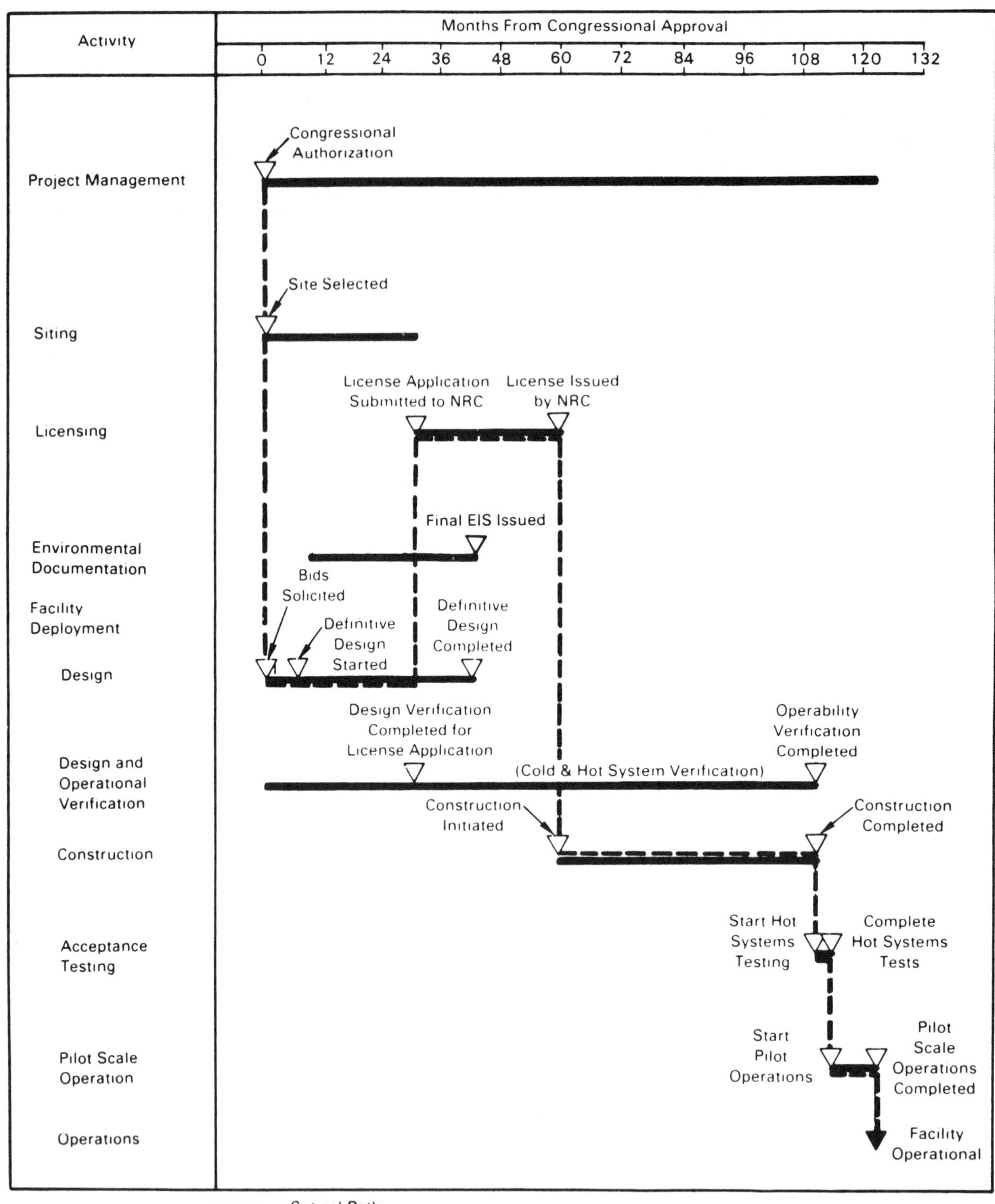

FIGURE 2. Proposed Schedule for an MRS Facility

MONITORED RETRIEVABLE STORAGE DESIGN

W. D. Woods
The Ralph M. Parsons Company
100 West Walnut Street
Pasadena, California 91124

ABSTRACT

The Nuclear Waste Policy Act of 1982 (NWPA) established a national policy for the safe storage and disposal of spent nuclear fuel and high-level radioactive waste. The NWPA requires that DOE . . . "submit a proposal to Congress on the need for and feasibility of one or more Monitored Retrievable Storage (MRS) Facilities" . . . In subsequent evaluations of the commercial nuclear waste management system, DOE has identified important advantages in providing an MRS Facility as an integral part of the total system. The integral MRS Facility serves as an independent, centralized spent nuclear fuel and high-level waste handling and packaging facility with a safe temporary storage capacity.

I. BACKGROUND

In accordance with NWPA, DOE has prepared a Mission Plan describing the information needs of the program being conducted by DOE to fulfill the requirements of this Act. The 1984 Draft Mission Plan assumed that the MRS Facility would serve as a backup facility to a repository. In this role, the MRS Facility would be built to provide safe temporary storage of waste only if a repository were significantly delayed. However, after subsequent evaluations of the nuclear waste management system, DOE has identified important advantages in providing an MRS Facility as an integral part of the total system.

If the Congress authorizes construction of an integral MRS Facility, it may serve as an independent, centralized spent nuclear fuel and high-level waste preparation facility with a safe temporary storage capability. All types of spent nuclear fuel and encapsulated high-level waste, from the nation's commercial reactors, will be shipped to this facility, in various cask configurations, for consolidation and packaging for temporary storage or final disposal. Thus, an increase in overall waste management flexibility will be provided. All waste packages prepared at the MRS Facility will be shipped in standardized casks by dedicated train to a geologic repository, thereby reducing total cask miles.

II. FUNCTIONAL DESIGN CRITERIA

A Functional Design Criteria (FDC) document was prepared by Pacific Northwest Laboratory (PNL) and approved by the Department of Energy (DOE).

It defines the project objectives and establishes minimum requirements for a safe MRS Facility that can be licensed, operated, maintained, and decommissioned.

The MRS Facility must include all infrastructure, facilities, and equipment required to routinely receive, unload, prepare for storage, and temporarily store spent fuel (SF), high-level waste (HLW), and transuranic waste (TRU). The facility is to be complete with all supporting facilities to make the MRS Facility a self-sufficient installation. The facility must be licensed by the Nuclear Regulatory Commission (NRC) in accordance with 10 CFR 72.

The base case MRS Facility must have the capability to receive, retrieve, and ship concurrently a minimum of 3,600 metric tons of uranium (MTU) per year, primarily as spent fuel, and a small amount (60 canisters) as HLW. It is to have in-building lag storage capacity for 1,000 metric tons of consolidated spent fuel in canisters plus a field storage capacity of 15,000 MTU of SF and a small amount (300 canisters) of HLW. The design shall assume a spent-fuel mix of 60% by weight from pressurized water reactors (PWRs) and 40% by weight from boiling water reactors (BWRs), based on 0.462 MTU per PWR assembly and 0.186 MTU per BWR assembly.

The spent fuel can be received at the MRS Facility fully assembled, either packaged or bare, or disassembled and consolidated in canisters. Fully assembled fuel received at the MRS will be sealed in canisters as one unit or disassembled, consolidated, and sealed in canisters. When consolidated, all fuel-bearing rods from an assembly will be placed in canisters. Nonfuel-bearing components of disassembled spent-fuel assemblies will require volume reduction and packaging at the MRS Facility in preparation for final storage.

The MRS Facility must be capable of receiving, unloading, and shipping all standard rail and truck SF, and HLW shipping casks. For design purposes, it may be assumed that the spent fuel will be received 70 wt% by rail and 30 wt% by truck. Nonstandard or off-normal cask shipments are to be routed to a special lag storage area while determination is made of the procedure to follow for routing through the MRS Facility handling and storage process.

Fuel and waste storage facilities shall be capable of attenuating the radiation and passively dissipating the heat generated primarily by 33,000 MWD/MTU spent fuel and/or equivalent HLW out of reactor 10 years. As much as 5% of the spent fuel received shall be considered fuel 5 years out of reactor and/or 10-yr-old spent fuel with burnups as high as 55,000 MWD/MTU. The storage configuration must be capable of maintaining spent-fuel cladding surface temperatures below 375°C in an inert-gas environment, and of maintaining HLW container surface temperatures below 375°C in air and borosilicate glass centerline temperature below 500°C.

The design life of the Receiving and Handling (R&H) Facility shall be 50 years, maintainable or replaceable to extend life to at least 100 years irrespective of licensing intervals. The design life of the storage facilities shall be the same. The design life of supporting facilities shall be 30 years, except for exterior finishes that may require periodic maintenance to satisfy the 30-yr requirement.

The dry and passive storage facilities shall be capable of confining packages of radioactive material within the storage containers (sealed storage cask or drywell) during the entire storage period, and shall have a monitoring system capable of detecting releases to the environs of radioactive material greater than one-tenth of that stated in 10 CFR 20, Appendix B, Table II. Stored material shall be protected against likely natural or man-created events, excluding acts of war. In addition, all packages of waste must be retrievable at any time for examination, repair, or shipment offsite.

For receipt and inventory verification, provisions shall be included to account for the quantity, type, and history of the material stored in the facility.

III. MRS SITES

DOE has identified three candidate sites to be included in its site-specific proposal to Congress. All three candidate sites are in the State of Tennessee, with the preferred one at the former Clinch River Breeder Reactor site. The two alternative candidate sites are the DOE Oak Ridge Reservation and the former Tennessee Valley Authority reactor site at Hartsville, Tennessee.

Using existing site data, conceptual designs have been developed for the three sites and the two storage concepts.

All sites are of sufficient area required for the MRS Facility. Based on existing site data, the design values for natural phenomena events, volume of excavation, and distance to existing infrastructure are essentially the same. The major design and cost differences among sites involve the need for demolition of existing facilities. Before termination of construction, a considerable amount of construction had been accomplished at the Hartsville site. Consequently, the Hartsville MRS arrangement will require demolition of some existing facilities. No existing facilities exist on either the Clinch River or Oak Ridge site; therefore, no demolition will be required at these locations.

IV. SITE ARRANGEMENT

Because of topography, existing site facilities, existing infrastructure locations, and access routes, MRS site arrangements vary at each of the three candidate sites. However, all arrangements require the same supporting, waste handling, and storage facilities, which are contained in either a limited or a protected access area. The relative locations of the limited and protected access areas were determined with regard to topography, security, accessibility, and the functional relationships between areas and their contained facility functions.

The limited-access area contains the facilities and services (administration, security, cold maintenance, utilities, emergency response, and industrial) required to support the waste handling and storage operations. It is enclosed within a single fence and provides controlled access, via a manned gatehouse, for all personnel and offsite vehicles and nonradioactive materials.

The protected area contains the Receiving and Handling (R&H) Building; the SF, HLW and HAW/RHTRU Storage Facility; and other safety-related support services. This area is enclosed within a double fence and provided with surveillance and detection devices in accordance with the requirements of 10 CFR 73 and DOE Order 5632, Chapter III.

Shipments containing radioactive materials enter the site through an inspection gatehouse, where the casks and carrier are inspected. Offsite tractors or engines are disconnected from the carriers, an onsite tractor or railcar mover is attached, and the carriers are transported through a fenced corridor to the protected area. Access control for personnel and nonradioactive shipments that enter the protected area from the limited area is provided via the Protected Area Gatehouse.

The SF, HLW, and HAW/RHTRU Storage Facility is located within the protected area, adjacent to the R&H Building. Access control into the storage facility is provided by a single fence with a locked gate.

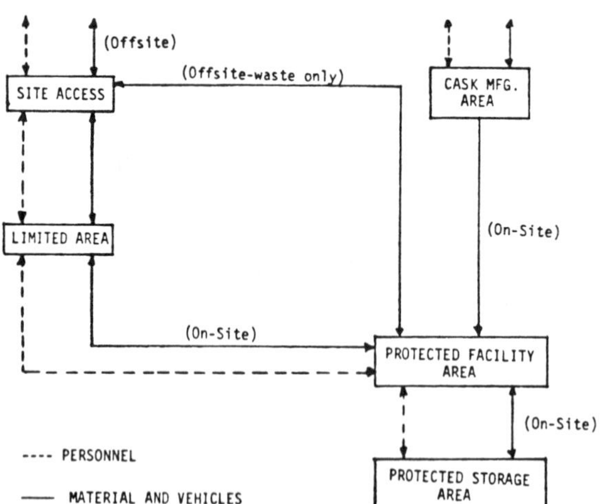

FIGURE 1. Site Flow Diagram

V. SUPPORT FACILITIES

To provide the necessary functions required for an independent MRS Facility, the design includes 14 separate support buildings or systems. The following is a summary of these buildings and their functions:

- Administration Building which contains offices and support areas for operations, quality assurance, health physics, accountability, State and Federal officials, and general site administration,
- Site Services Building which contains offices, shop areas for site (cold) maintenance, shielded process cell mockup, and redundant process computer,
- Warehouse for storage of nonradioactive bulk materials and equipment,
- Vehicle Maintenance Building which contains repair facilities for onsite vehicles (trucks, cars, cask/canister transportation, and rail equipment),
- Water Supply Facility which contains water supply and firewater pumps,
- Security Building which contains security fence administration and redundant hardened alarm monitoring station,
- Main Gate/Badgehouse which contains security station for personnel and offsite vehicle access control into the limited area,
- Inspection Gatehouse which contains security/inspection station for shipments of radioactive materials into or from the protected access area,
- Protected Area Gatehouse which contains security station for personnel and vehicle access to protected area, the primary alarm monitoring station, and the personnel monitoring station,
- Storage Facility Gate Station which contains access control to the SF and HLW Storage Facility,
- Fire Station which contains administration, central fire alarm and dispatch station, vehicle storage, and central first-aid station,
- Sewage Treatment Plant which contains sewage treatment system,
- Cask Manufacturing Facility which contains bulk cask material storage, concrete batch plant, and sealed storage cask casting and curing facilities, and
- Standby Generator Building which contains standby generator and normal and standby switchgear.

VI. RECEIVING AND HANDLING BUILDING

The R&H Building is the primary operating building of the MRS Facility. It is a 709,000-sq-ft, multilevel structure designed to physically contain and control all radioactive materials received and processed or generated by spent-fuel process operations. The facility includes spaces and equipment for dry remote receiving, monitoring, identifying, disassembling, consolidating, and packaging of spent fuel and processing and packaging of spent-fuel hardware for onsite storage or retrieving from storage and shipping to storage facilities offsite. The facility includes a lag storage vault with a capacity of 1,000 MTU of consolidated spent fuel.

The structural designs of the various areas composing the R&H Building are designed in accordance with their importance to safety. Areas required for confinement of radioactive materials, prevention of criticality, and safe shutdown are designed as Category I structures.

The building contains the following operational areas:

- two rail/truck receiving and unloading areas,
- four shielded processing cells with dedicated cask unloading areas,
- two HLW/RHTRU/repository overpack cells with two shipping cask unloading/loading areas and two onsite cask/transportation loading/unloading areas,
- two SF canister welding/decon/inspection stations,
- one compartmentized canistered consolidated fuel lag storage vault,
- one high-activity radwaste treatment area,
- one low-level radwaste treatment area,
- HVAC equipment areas,
- electrical equipment areas, and
- administration and personnel support areas.

Shipping casks arrive at the R&H Building in the receiving, inspection, and shipping area, where the carriers are washed, if required, and the casks removed onto cask carts. Casks containing canistered HLW are routed to a cask handling and decontamination room located beneath a canister handling area. Casks containing spent fuel are routed to a cask handling and decontamination rooms located beneath a shielded process cell. Here the cask's external surface is decontaminated, if required; the interior environment is sampled; gas-pressure measurements are taken; the outer cask lid is removed; the cask inner lid bolts are loosened; and special adapters are installed. The casks are then moved, by the cask cart, into a cask unloading room and mated to an unloading port located in the shielded process cell or canister handling area floor. The unloading port shield plug and the cask inner lid are removed and the spent-fuel assemblies or HLW waste canisters are removed by the area/cell crane. The spent-fuel assemblies are visually inspected, inventoried, and either placed into a lag storage pit for future processing or are routed directly to the disassembly/consolidation station. Upon visual inspection or from reactor records, any assembly that is judged to be damaged or would have a high probability of damage during consolidation is routed to a dedicated upender and canistered as a total intact assembly.

The consolidation equipment is composed of three basic elements: the laser cutter for severing the nozzle and bottom tie plate, the fuel disassembly system, and the consolidation system.

The disassembly and consolidation equipment are fitted with the appropriate module and tooling for a specific PWR or BWR assembly. The in-cell crane transfers three PWR or seven BWR assemblies to a vertically oriented clamping module. The laser cutter is positioned and activated to remove the nozzles by cutting the PWR guide tubes on the inside diameter or the BWR tie rods from the outside. The in-cell crane removes the freed nozzles and transfers them to a nonfuel bearing waste packaging station. After nozzle removal, the clamping module is rotated to the horizontal position. For the BWR assemblies, the laser cutter is repositioned to cut the lower ends of the tie rods to remove the lower nozzle. Removal of the center rod requires that it be rotated 45 deg about its axis to disengage its locking detents from the grids. Grippers securely grasp each fuel rod and simultaneously extract all rods through their support grids. During extraction, horizontal and vertical combs support the rods and maintain their arrangement. The disassembled fuel rods are consolidated into a circular bundle and pushed into a clean canister. The nonfuel-bearing structure is removed from the clamping module by a dedicated robot and transferred to a nonfuel bearing waste packaging station.

Spent-fuel packaging consists of canister positioning, inerting, welding, and decontamination.

The canister upender (located in the weld/test/decon area) is a mobile transportation and positioning cart for spent-fuel canisters, overpack canisters, or drums. The upender is mounted on rails to position the canister for loading of spent fuel from the consolidation station; to load it into the canister welding system, or to position it at either the cutting or the UT testing station.

An empty, clean canister is aligned with the axial centerline of the consolidated spent-fuel bundle retained in the spent-fuel consolidation station. The push mechanism on the consolidation station loads the rods into a clean storage canister. The loaded canister is transferred and positioned in a canister welding station. With the canister in place, air is evacuated from the canister and welding station, and both are backfilled with an argon-helium gas mixture. The positioned canister lid is welded to the canister by a resistance welding system. At the end of the welding cycle, the argon-helium mixture is evacuated from the welding chamber, a helium-leak-detection test is performed, and the exterior of the canister is decontaminated by a Freon liquid system. After decontamination, the canister is removed from the welding station and an ultrasonic test is performed on the canister weld.

The disassembly of spent-fuel assemblies and the consolidation of fuel rods leave behind nonfuel components, such as nozzles, grids, guide tubes, etc. The nozzles are placed directly into 55-gal drums and the other components are shredded before placement in 55-gal drums. All nonfuel scrap drums are sealed, decontaminated, and transferred to the overpack/weld/discharge area for shipment to storage.

Canistered spent fuel, HLW, and drummed high-activity waste are outloaded through the overpack/weld/discharge area. For onsite storage, the sealed storage cask or drywell transporter is positioned by its transporter in the loading/decon room, located beneath the overpack/weld/discharge area, and mated to a loading port. The loading port and the sealed storage cask shield plug are removed by the in-cell crane, and the canister/drum of waste is loaded into the cask. After loading, the cask shield plug and the loading port shield plug are replaced and the sealed storage cask is moved to the transfer/discharge corridor, where the sealed storage cask cover is welded in place.

The MRS Facility has also been designed to load waste into a repository container. This is accomplished by moving an empty container into the overpack/weld/discharge area and placing it into a weld pit. The waste canister/drums are placed in the container, by the in-cell crane, and the overpack cover is welded by the electron beam method.

Outloading of the repository overpack or canistered/drum waste for offsite shipment is accomplished by positioning, on a cask cart, an empty transportation cask in the shipping loadout room located beneath the overpack/weld/discharge area and mated to a loading port. The loading port shield plug and the cask inner lid are removed, and the canister/drums are loaded into the cask by the in-cell crane. After loading, the inner lid is replaced and the cask is moved into the cask loading room, where the outer lid is replaced and the cask exterior is decontaminated. The loaded cask is moved by the cask cart into the receiving, inspection, and shipping area, and loaded onto a rail carrier.

VII. STORAGE FACILITIES

The SF, HLW, and HAW/RHTRU Storage Facility is designed to store radioactive waste processed through the R&H Building. Two alternate dry passive storage concepts were designed: 1) a primary concept using an array of sealed storage casks (concrete casks) containing the waste canisters and stored in an open field above ground, and 2) an alternate concept using an array of sealed in-ground drywells containing the waste canisters. In the concrete cask concept, the concrete cask and steel liner are designed for radiation shielding and to remove the heat by convection and thermal radiation. The drywell concept is designed to use the surrounding soil to attenuate nuclear radiation and to dissipate the heat from the waste while in storage.

The concrete storage cask provides a sealed, self-shielded, dry storage container for intact spent-fuel assembly canisters, consolidated fuel rod canisters, and drums and canisters of HLW and HAW/RHTRU. The storage casks are cylindrical, reinforced concrete structures with a stepped, carbon-steel-lined cavity for storing waste containers. A cylindrical concrete shield plug fits into the open top of the cavity and a steel cover plate is seal-welded to the liner flange to close the cask. The casks are 22-ft high by 12-ft dia (outside diameter). The cask and the enclosed storage canisters provide double containment to withstand credible natural phenomena and man-induced events.

Monitoring of the concrete cask liner and cover integrity is performed on a periodic basis. Sampling of a cask internal atmosphere is performed by removing the access cover bolted to the gas sampling port housing. Gas samples are collected for analysis to determine the presence of any gaseous fission products or canister tag gas in the cask interior. Monitoring also includes temperature surveillance for a fixed number of casks. Temperature measurement is accomplished by using thermocouples installed in tubes, extending from the cask exterior to the liners of spent-fuel and commercial high-level waste casks.

The field drywell concept uses the surrounding geologic medium to attenuate radiation and to dissipate decay heat generated by the waste while it is sealed in the drywell. Four drywell sizes have been designed to accommodate the variations in waste drums and canisters configurations. Each drywell holds a single canister of waste. Each different waste type is stored in a unique array of drywells, sized according to the containers they must hold, and spaced according to the amount of waste heat that must be dissipated and the thermal conductivity of the soil at the proposed sites. Monitoring of the drywell liner and cover integrity is similar to that described for the sealed storage cask.

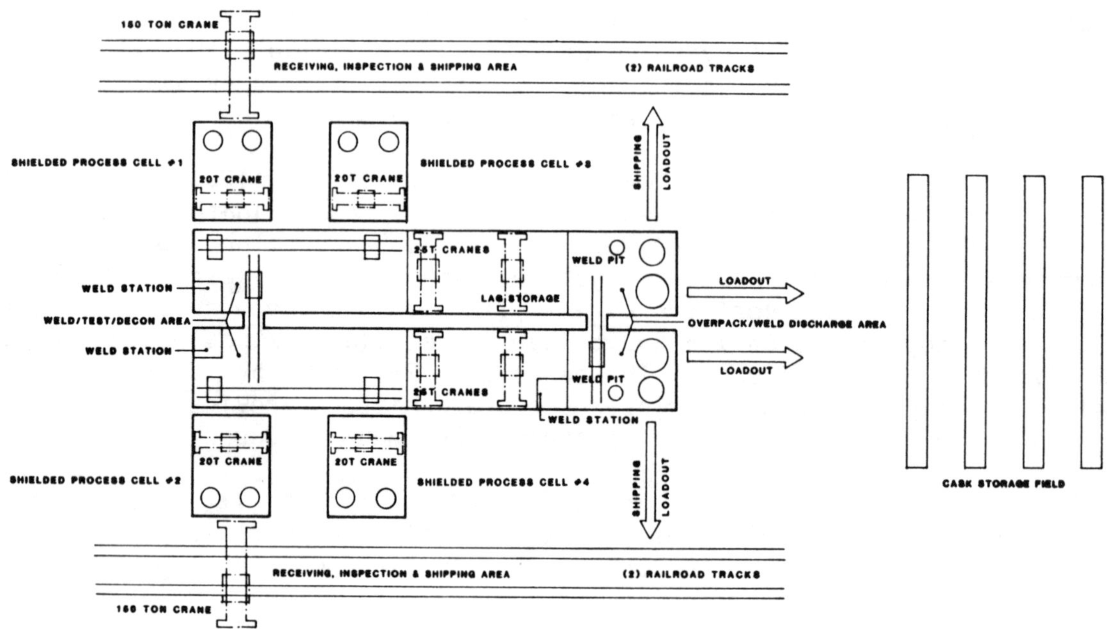

FIGURE 2. Receiving and Handling Building Material Flow

SUBSEABED DISPOSAL: A POTENTIAL ALTERNATIVE TO MINED REPOSITORIES

S. G. Bertram
D. R. Anderson
Seabed Programs Division
Sandia National Laboratories
P.O. Box 5800
Albuquerque, New Mexico 87185

ABSTRACT

The Subseabed Disposal Project at Sandia National Laboratories is investigating the disposal of high-level radioactive waste and spent fuel in the deep ocean floor. The project is closely integrated with international studies of subseabed disposal. The research program focuses on assessments in six major fields of study: site, barrier, environment, engineering, safety, and institutions. The international program will publish a Concept Assessment Report in 1987, presenting the research results available at that time. Subsequent publication of scientific, environmental and preliminary engineering results obtained in the various national programs is planned for about 1990.

I. INTRODUCTION

The Subseabed Disposal Project at Sandia National Laboratories is investigating the disposal of high-level radioactive waste (HLW) and spent fuel in the deep ocean floor. The SDP is managed by Sandia National Laboratories for the Department of Energy. The US project is part of an international program studying subseabed disposal. The international Seabed Working Group (SWG) is organized under the auspices of the Nuclear Energy Agency of the Organization for Economic Cooperation and Development. Member and observer nations during the 10-year history of the SWG have included Australia, Belgium, Canada, the Commission of European Communities, the Federal Republic of Germany, France, Italy, Japan, Netherlands, Switzerland, the United Kingdom, and the United States. This paper briefly describes the project's objectives and reviews some of the results. Few distinctions are made between national and international efforts because they are so closely linked.

II. THE CONCEPT OF SUBSEABED DISPOSAL

A subseabed disposal system would require facilities for handling, transportation, port operations, deployment, monitoring and salvage or retrieval. The nature of these facilities will depend on the subseabed emplacement method selected. Studies have not progressed sufficiently to support selection of an emplacement method, therefore no conceptual design has been developed. However, the SDP uses a preliminary description of the concept for reference purposes.

Spent fuel would be transported from the reactors to a reprocessing plant or a repackaging plant. The transportation and packaging systems developed for geologic repositories could be used for this step. After the appropriate processing and packaging, the waste would be transported to a port facility, where it would be inspected, fitted with some type of overpack, and loaded onto the ship. The transportation route would be selected to avoid shipping lanes and maritime activities and to minimize risks to the environment.

A special ship would be needed to transport and emplace the packaged waste at the disposal site. The ship would be designed so canisters could be maintained at a safe temperature. All handling facilities would be designed to minimize worker exposure. The waste canisters would be emplaced into the sediments by using a variation of either soil penetrometer or deep-sea drilling technology. Regardless of the emplacement method used, waste canisters would be fitted with locating devices, emplaced into a grid, precisely surveyed, and periodically monitored.

III. THE RESEARCH PROGRAM

The US SDP is conducting research primarily directed at answering the following questions:

Can the barrier system of waste form, canister, and sediment adequately inhibit the migration of hazardous quantities of radioactive materials while they decay to levels that can be effectively and safely dispersed by the ocean waters?

Can waste-filled canisters be reliably emplaced in the selected deep-ocean sedimentary formations without compromising the barrier properties of the containment system?

This work is being conducted in four consecutive phases. Proceeding from one phase to the next is contingent upon successful completion of the tasks in each phase. The phases are listed below.

Phase 1 - evaluation of the literature and historical data to determine whether field investigations and predictive analyses should be initiated (completed in 1976),

Phase 2 - modeling of the ocean and sediment processes and acquisition of new data to access the concept in more detail (scheduled for completion in 1990),

Phase 3 - determination of engineering feasibility and development of a conceptual subseabed disposal system, including a recommended site (identification of an institutional regime to establish criteria and procedures for an operational repository), and

Phase 4 - demonstration of disposal.

At present, the SDP is in Phase 2 of the research program, and is emphasizing modeling and data acquisition. Particular questions about emplacement and hole closure, which are relevant to determining engineering feasibility, are also being addressed during the current phase. This strategy was selected for several reasons, including the need to understand the close relationship between the sediment's barrier capability and disturbances caused by emplacement operations.

Efforts to develop a subseabed disposal system could not begin without a specific national policy decision. For this reason, the SDP is evaluating the institutional implications of subseabed disposal and is undertaking a comprehensive public communication program in order to establish a basis for the decision making process.

IV. PHASE 2 RESEARCH

A. Mathematical Models: Feasibility Assessment Tool Development

Because any releases from a subseabed repository would occur over millions of years, and because the radionuclides that could pose a threat to humans or to the environment have long half-lives, it is necessary to simulate isolation, containment, and dispersion processes in the deep ocean and the underlying sediments. The project places high priority on developing and applying appropriate mathematical models to predict long-term consequences of disposal. Most of the Phase 2 research is directed towards quantification of the appropriate processes, development of models, and acquisition of associated data.

Mathematical models are used to describe real processes, to manage large sets of data, to analyze the effects of uncertainty in the data, and to predict processes and phenomena that are difficult to observe. Models have been used in this way throughout the research effort. However, in deciding to use models to assess concept feasibility, the SDP recognizes that the project is responsible for establishing confidence in the model calculations and input parameters. Moreover, the project must understand and identify the limitations imposed by the assumptions incorporated in the models. Consequently, the SDP follows well-established procedures for developing mathematical models. The following steps summarize this procedure: 1) identify relevant processes and construct a model to describe those processes; 2) define and acquire the data needed to use the proposed model; 3) verify the mathematical accuracy of the model; 4) validate the model to ensure that it adequately represents the site-specific processes.[1]

B. Marine Geology: Seabed Location Studies

Geologists are investigating the geological, geophysical, and geochemical characteristics of seafloor sediments, which would comprise the primary barrier to the possible release of radionuclides. Site assessment requires detailed evaluations of potentially suitable deep-sea locations on the basis of stability and predictability guidelines developed by the SDP and the SWG. The geologic stability guideline requires evidence of a million years of continuous deposition, a minimal possibility of seismic and volcanic activity, and a minimal possibility of major climatic or oceanographic changes. The predictability guideline requires uniform, predictable sediment to ensure that the input parameters for sediment and barrier models would be representative of the entire location.[2]

Site assessment began with literature searches, which indicated that the most suitable deep-sea geological formations would be the clay deposits associated with abyssal hills and abyssal plains. Three potential locations in the North Pacific and two in the North Atlantic were identified on the basis of preliminary geologic guidelines. Reconnaissance surveys, detailed bathymetric mapping, deep-tow seismic surveys and long cores are used to study these locations in detail. The marine geologists have mapped seafloor topography and identified sediment structure at all three North Pacific locations and verified the sediment structure at two of the three locations. The international assessment now emphasizes characterization of locations in the North Atlantic, although work is continuing in the North Pacific. By 1986, the North Atlantic locations should be as well-characterized as the two in the North Pacific. The international task group has found that marked differences do exist among the study locations, but the significance of these differences has not yet been evaluated.

A number of current geologic issues must be clarified if substantative site evaluation is to continue. Resolving these issues will largely be an interdisciplinary effort. These issues include (1) obtaining sediment samples from the potential emplacement horizons; (2) determining if porewater advection is important at each location; (3) obtaining geochemical data and establishing limiting values; (4) determining the abundance and importance of bioturbation at each location; (5) establishing limits for erosion; (6) establishing the minimum period of years for geological stability; (7) evaluating the effects of waste disposal on slope stability; (8) defining the minimum dimensions of a disposal site; (9) evaluating the effects of small morphological features on in situ experiments and disposal; and (10) determining whether manganese nodules or other sediment surface features pose problems.[3]

C. Oceanography: Environmental Studies

Studies of environmental processes in the ocean are also being conducted in the North Pacific and North Atlantic. If radioactive material is released as a result of subseabed disposal, some will ultimately disperse in the ocean waters. Those chemical, physical, and biological mechanisms that affect the radionuclide's behavior after entering the water must be identified and modeled. At present, the objective of the environmental studies is to develop the ability to make realistic predictions of radionuclide concentrations in important marine biota, the ensuing human population dose, and the radio-sensitivity of marine organisms. These predictions will be made by interfacing physical dispersion models, which include the effects of geochemistry and biology, with radiological dose and pathways models.[1]

The physical dispersion (ocean transport simulation) model is composed of several submodels, which Sandia is developing or modifying: these include a three-dimensional bottom boundary layer model; a regional eddy-resolving model; and a large-scale ocean circulation model. The models will also include relevant geochemical and biological processes. Before results from the ocean transport model can be used in the concept assessment, model verification must be completed. This will require site specific as well as general oceanographic data.[1]

Biological pathways and the contribution of biological processes to geochemical cycles are of prime importance in the potential transport of radionuclides from the deep-sea sediments to the ocean surface. The critical interfaces for the biological-geochemical interaction are sedimentation and bioturbation. Thus, data are required on the rates and pathways by which organisms influence geochemical processes. The minimum set of biological observations identified by the SWG's Biological Oceanography Task Group for site assessment includes measurements of (1) primary production; (2) nekton biomass over the full water column; (3) sediment infauna biomass; (4) bioturbation rates; and (5) microbial activity to the emplacement depth.[4] The environmental task is also responsible for providing data on the estimated harvestable biomass and potential food web transfers through marine organisms to humans for eventual use in physical dispersion models.

D. Sediment Geochemistry and Mechanics: Barrier Studies

The primary barrier to radionuclide migration in a subseabed repository is the sediment in which the waste would be buried. The waste form and canister also delay the release of radionuclides. Thermal and radiation-induced impacts on these components are being evaluated. Preliminary analyses have shown that the waste-induced thermal period in the sediments near the canisters is essentially complete by 100 years after emplacement. Temperature-induced changes in the sediment could allow the canister to move downward or upward. However, preliminary predictions indicate that the canister, sediment, and porewater would have moved less than 1 meter after approximately 60 years. After that time, the movement would be effectively stopped.

The sediment barrier could also be affected if the radiation and the heat emitted by the buried canister altered the chemical structure of the sediment in the near-field. The near-field is defined as that region where the sediment temperature exceeds 100°C. Model predictions suggest that the near-field will never extend more than 0.8 m into the sediment surrounding the canister. The primary objective of the near-field chemical investigations has been to determine how hydrothermal processes and radiation affect the waste package and sediment. Extensive laboratory studies using deep-sea sediments are being conducted to identify and assess any chemical alterations that could affect the mobility of radionuclides. Several sets of laboratory experiments suggest that all the important seawater-sediment reactions that occur in the near-field of a subseabed repository probably have been identified. None of these known reactions would harm the canister or the effectiveness of the sediment barrier. Additional experiments will be required, however, to obtain data needed to develop models for long-term near-field behavior.[1]

The far-field is defined as the sediment between the near-field and the ocean floor. Reliable predictions of transport rates for radionuclides under far-field conditions are necessary to assess repository performance. Radionuclide mobility is influenced by the sediment type, the chemical composition of the waste, and the radionuclide complexes generated in the near-field. Preliminary results suggest that oxidized pelagic sediment would form a highly effective barrier to the migration of cationic radionuclides. Anions would be released slowly. Porewater advection is another important parameter to radionuclide transport through sediments; sensitivity studies have determined the maximum permissible porewater advection velocities for pertinent radionuclides in oxidized sediments. Studies of methods of measuring porewater movement and the barrier properties of reducing sediments are underway.[1]

The response of deep-sea sediment is the subject of a major experimental effort. The in-situ heat transfer experiment (ISHTE) was initially designed to provide data on the response of in-situ sediment to heating in order to verify laboratory experimental approaches and computer models. The experiment, which will remain on the seabed for one year, has been expanded to include experiments to measure the thermal field, determine the effective thermal conductivity of the sediment, measure pore pressure, evaluate radionuclide migration processes, sample porewater, study sediment chemistry, measure shear strength, and collect cores. ISHTE, scheduled to be deployed in the deep ocean in 1986, provides an opportunity to develop and demonstrate the technology necessary to conduct long-term experiments on and in the seafloor at depths of ~6 km. ISHTE will <u>not</u> simulate waste emplacement.[1]

E. Systems Analysis: Radionuclide Transport and Safety Studies

Preliminary radionuclide transport studies suggest that only a small subset of the total radionuclide composition of the waste or spent fuel would cross the sediment-seawater interface above the repository in quantities large enough to have any significant radiological effects on humans or biota. Certain sediments slow migration of most radionuclides, greatly delaying their release to the ocean. The burial depth is also an important delay factor. A case study[1] using specific parameters (oxidizing sediment, 20-m burial depth, high-level waste of defined form and composition, sorption coefficients of zero, a 100-year canister life, and a 1000-year leach rate) found that significant quantities of only four nuclides would enter the ocean in the first 10^4 years, with peak release rates occurring at 10^4 years. These radionuclides are technecium-99, carbon-14, iodine-129, and selenium-79. After 10^4 years, the release rates of four other nuclides would begin to peak at lower levels. These are neptunium-237, uranium-233, palladium-107, and thorium-229. Future studies will examine the long-term behavior of these and any other nuclides predicted to enter the ocean in significant quantities. The attenuation of released nuclides through dispersion and dilution in the ocean waters will be modeled.

Radionuclides may be released from an improperly emplaced or damaged waste canister, or from the expected gradual deterioration of a properly emplaced canister. Preliminary safety analyses have not identified any situations, given a properly emplaced canister, that would preclude disposing of radioactive wastes in the deep ocean sediments. Results from these studies suggest that the peak early dose to an individual would be a thousand times below the dose from natural background radiation. Safety analyses based on a variety of accident scenarios also are being conducted. These studies rely on models developed by the SDP to predict the consequences of accidents as well as of proper emplacement. Because regulatory standards for a subseabed waste repository are not available, the project compares dose calculations with different safety standards for perspective. Standards established by the US Environmental Protection Agency and the International Council for Radiation Protection are typically used.

F. Engineering: Emplacement and Hole Closure Studies

Phase two research includes efforts to resolve two questions: 1) Can waste canisters be reliably emplaced at the required depth in the sediments? 2) Can hole closure behind an emplaced waste canister be assured? The two emplacement technologies being considered are dynamic penetrators and deep-sea drilling. The first would rely on the hole closing dynamically behind the penetrator; the second would require the drilled hole to be filled.

A seabed penetrator, containing one waste canister, could descend at high velocities through the ocean water and bury itself at least twenty meters into the plastic sedimentary formations of the seafloor. An alternative penetrator type could be lowered to a given water depth and "boosted" into the sediment. Adapting drilling technology could result in a system in which canisters were stacked within the emplacement column to sediment depths of a hundred or more meters.

Modeling studies have predicted the emplacement depths achievable with penetrators, and field tests have confirmed the predictions. The dynamics of hole closure have been modeled; an international hole closure demonstration and validation experiment in the seabed is planned for 1986. Other participants in the international program have examined drilled emplacement, concluding that it is feasible; a British study includes preliminary conceptual designs for the ship, a drilling platform, and the emplacement process.[5] The US SDP has not evaluated this recent study.

G. Social Science: Institutional Studies

The goal of the institutional studies task is to maintain policy options for the subseabed concept until research results permit national and international authorities to make sound policy decisions. A comprehensive approach to meeting this goal involves legal, political, and social concerns. Distinguishing among these concerns is important for analytical purposes, but most institutional issues relevant to subseabed disposal have elements of all three. Moreover, in many cases, parallel legal, political, and social issues are applicable on both the national and international levels. In fact, these levels are so closely coupled that most issues must be addressed on both levels. The project's institutional plan is designed to address this complex of institutional issues.

The concept of a high-level waste repository in the international "global commons" is novel, with no clear precedent. The status of the concept and the research is not clearly defined in either US policy or international law. Within the US, the project is authorized as research on alternatives to geologic disposal by Section 222 of the Nuclear Waste Policy Act, while the Marine Protection, Research and Sanctuaries Act prohibits ocean dumping of HLW. In the international arena, it is not clear which of the possible bodies would regulate a subseabed repository. However, the London Dumping Convention appears to have assumed international regulatory jurisdiction over the concept. Federal and international laws probably would require amendments before a subseabed repository could be developed.

The SDP has sought to involve the public since its inception. The public communication activities are designed to reach the research community, public interest and environmental organizations, industry organizations, government decision makers, and other, unaffiliated citizens. The process includes presentations and workshops, publication of results in peer-reviewed literature, publication of overview materials in popular technical literature, and development of public information documents.

V. PRELIMINARY CONCLUSIONS

Although the feasibility assessment is far from complete, no evidence has yet been found that indicates the option is not feasible. The SWG plans to publish a preliminary international status report in 1987, in which the program will present an assessment of scientific, environmental and preliminary engineering feasibility of the subseabed concept. In 1990 each participating nation will prepare a national project concept assessment report based on the SWG results and on its own research.

REFERENCES

1. Seabed Programs Division, The Subseabed Disposal Program: 1983 Status Report, SAND83-1387, Sandia National Laboratories, Albuquerque, New Mexico (1983).

2. E. P. Laine, et al, Program Criteria for Subseabed Disposal of Radioactive Waste: Site Qualification Plan, SAND81-0709, Sandia National Laboratories, Albuquerque, New Mexico (1982).

3. D. R. Anderson, ed., "Site Assessment Task Group Report Addenda I," Tenth Annual Meeting of the Coordinated Program to Assess the Subseabed Disposal of Radioactive Waste, Halifax, Nova Scotia, April 29 to May 6, 1985, SAND85-1365, Sandia National Laboratories, Albuquerque, New Mexico (in prep).

4. D. R. Anderson, ed., "Biological Oceanography Task Group Report," Tenth Annual Meeting of the Coordinated Program to Assess the Subseabed Disposal of Radioactive Waste, Halifax, Nova Scotia, April 29 to May 6, 1985, SAND85-1365, Sandia National Laboratories, Albuquerque, New Mexico (in prep).

5. M. R. C. Bury, A Feasibility Study of the Offshore Disposal of Radioactive Waste by Drilled Emplacement, 2. Development of the Design, 014N/83/2539, Taylor Woodrow Construction Ltd., Southall, Middlesex, United Kingdom (1983).

THE USE OF MODEL FREE-FALL PENETRATORS IN INVESTIGATIONS ON THE FEASIBILITY OF EMPLACING HEAT GENERATING WASTES IN DEEP OCEAN SEDIMENTARY FORMATIONS

C. N. Murray
Commission of the European Communities
Joint Research Centre
21020 Ispra (VA) Italy

D. R. Anderson
Seabed Programs Division
Sandia National Laboratories
P.O. Box 5800
Albuquerque, New Mexico 87185

ABSTRACT

As a part of the International Seabed Working Group Coordinated Program on the study of the feasibility and safety of the disposal of heat generating wastes into deep ocean sedimentary formations, research is being undertaken by a number of countries of the OECD, including some members states of the European Community and the Joint Research Centre. In order to demonstrate the necessary engineering emplacement capability studies are being carried out on two main technologies: penetrators and deep ocean drilling. The present paper will describe the work done to date using large model free-fall penetrators to investigate the ability of abyssal plain sedimentary formations to act as potential disposal sites for long-lived nuclear waste.

I. INTRODUCTION

Since 1983 the International Seabed Working Group under the auspices of the OECD Nuclear Energy Agency, has been carrying out a coordinated research program to investigate the feasibility and safety of the disposal of heat generating wastes into deep ocean abyssal plain formations. The main objectives of the research are to assess the long-term safety of the option, to demonstrate the necessary engineering emplacement capability within oceanic geological formations, and to identify characteristic study zones in the North Atlantic and Pacific Oceans.

In order to show engineering capability, studies are being carried out on two main technologies: free-fall penetrators and deep ocean drilling. The investigations on the penetrator option require consideration of the processes of hydrodynamic stability in the water column, and penetration (for predictive purposes) and hole closure during the passage of the vehicle through the sediment column. To assess these processes data on the geotechnical characteristics of the sediments down to the emplacement depth are needed. To obtain these, two approaches are being followed: firstly, the measurement of sediment properties in the laboratory on samples obtained from sites of interest[2] and, secondly, the development of instrumented penetrators capable of measuring selected water and sediment column parameters in-situ.

At present consideration of the drilled option is being based on the experience gained in deep ocean drilling over the last 12-15 years and on feasibility studies identifying and analyzing the characteristics of ocean platforms needed for the handling of long-lived heat generating waste.[3] The present paper will briefly describe the work done to date using large model free-fall penetrators to investigate the ability of abyssal plain sedimentary geological formations to act as potential disposal sites for these hazardous wastes.

II. FREE-FALL PENETRATOR INVESTIGATIONS

The penetrator option is possibly the simplest of those which have been considered for the ocean disposal of radioactive waste.[4,5] The method basically consists of loading one or more canisters of heat generating waste into a torpedo like vehicle. These penetrators are then transported to an ocean disposal site, the subseabed repository area, and launched so that they come to rest within the sediment formation.

In order that this disposal method be developed it is necessary to show that it is feasible from an engineering standpoint, that is to say: the required embedment depth can be attained, the hole behind an embedded penetrator is closed, that the bulk sediment properties in the disturbed zone are such as to satisfy radiological safety requirements, and that repository design criteria can be satisfied.

A. Test Series I

The first series of deep-ocean penetrator experiments were performed in 1983 at a North Atlantic study site Great Meteor East (GME), in the southwestern distal part of the Madeira Abyssal Plain, as a collaborative experiment between the Building Research Establishment, UK, the Commission of the European Communities, Joint Research Centre, and the Institute of Oceanographic Sciences (IOS), UK. The GME area has been extensively studied by the Rijks Geologische Dienst of the Netherlands and the UK Institute of Oceanographic Sciences as well as the recent International Long Core Cruise (ESOPE). The main objectives of these first tests were to establish the technical feasibility of the method and to enable comparisons to be undertaken with the drilled option alternative.

In order to develop the free-fall penetrators used in these tests detailed hydrodynamic analyses were carried out on a number of different possible configurations, nose shape, length to diameter ratios and stabilizing surfaces.[6,7] These studies were aimed at designing a penetrator which would be stable during its passage through 5 km of water and would achieve maximum possible penetration on impact with the sediment. On the basis of these studies a design for a large model penetrator was proposed having the following characteristics, length 3.25 m, diameter 0.325 m, construction material mild steel of density 7.77 gm^{-3}, weight 1.8 tonnes in air.

In March 1983 four similar pentrators of the above dimensions were dropped in the GME area, each vehicle having been fitted before launch with an acoustic transmitter screwed into the tail section of the penetrator body, which emitted a constant 12 kHz signal on immersion in sea water.

As the unit falls through the water column the frequency of the signal received by the surface hydrophones decreases as the penetrator accelerates to its terminal velocity. The receiver on board ship measures the frequency shift and converts it to a D.C. voltage which is then recorded on F.M. tape recorder or chart recorder. By assuming values for the velocity of sound in seawater and sediment, the velocity of the penetrator can be calculated from the measured frequency shift (Doppler Principle) using the equation:

$$f' = f \frac{V_s}{V_s+V}$$

where

f' = the frequency received at hydrophone,
f = the frequency transmitted,
V_s = velocity of sound in the medium surrounding the transmitter
V = velocity of the penetrators.

The results of the four penetrator drops[8] are summarized in Table 1. The terminal velocities of the penetrators in the water column were found to be in good agreement with predicted values[6] and the impact speeds correspond to an average drag coefficient of 0.150 which compares well with that derived from theoretical calculations of 0.148.[5,6] Depths of penetration were derived from velocity-time profiles recorded from the moment of impact of the penetrator with the sediment to its rest value. For a detailed discussion of these results the reader should refer to reference 8.

Out of the four penetrators deployed, one (Number 4) failed to give a sufficiently clear acoustic signal upon entry into the sediment column and it was not possible to estimate its depth of penetration. The results obtained from the first three penetrators show that tail penetration was on average about 30 m. The slight differences in the terminal velocities are probably due to minor variations in manufacturing (surface finish, etc.).

B. Test Series II

The results of the first series of tests carried out in March 1983 demonstrated that a model penetrator can reach burial depths in excess of 30 m in suitable sediments. In order to extrapolate these results to large, variable density penetrators capable of transporting heat generating waste more detailed information was considered necessary to enable the development of a predictive capability[9,10] which would be capable of precisely describing the behavior of a full scale penetrator within the water column and during its impact and embedment within the sediment column.

TABLE 1. Penetrator Results from March 20, 1983

Drop N.°	Latitude (°N)	Longitude (°W)	Depth	Impact Velocity (ms^{-1})	Tail Penetration (m)
1.	31°25.5'	24°44.2'	5433	46.4	30.0
2.	31°25.6'	24°44.7'	5433	48.3	31.3
3.	31°23.6'	24°02.6'	5425	50.9	27
4.	31°23.5'	24°02.8'	5425	47.3	unknown

To successfully reach these objectives it was decided that an important engineering program would have to be undertaken having the following components:

1. the development of a data transfer system from the embeded penetrator to the sea surface,

2. the development of specialized sensors which would give information on the penetrator behavior and geotechnical information about the sediment column, and

3. the testing of penetrators of differing configurations in order to understand and optimize penetration capability and stability.

An initial start was made in February-March 1984 when a second series of penetrator tests were carried out in the Nares Abyssal Plain in the North Atlantic. The specific aims of this second series of tests were (1) to compare the successful acoustic Doppler-Swift Systems (ADSS) with two alternative data transmission systems which were under development, and (2) to investigate the effect of variations in design on penetration depth i.e., to test model prediction capability.

Two advanced data aquisition and transmission systems were investigated. The first is based on a low-frequency 3.5 kHz transponder system (LFTS) which is currently being developed at IOS and jointly sponsored by the UK Department of the Environment and the CEC Joint Research Centre, Ispra. The second, an Explosive Acoustic Telemetry System (EATS) is one developed by Sandia National Laboratories, USA which uses a number of small explosive charges. The detonation of the charges is controlled by an on-board microprocessor which encodes data from sensors as the time interval between explosions.

In order to investigate the effect of variations in the design of penetrators on depth of embedment, a parametric analysis of the performances of free-fall penetrators was undertaken[11,12] (Figure 1). This study resulted in the choice of six different designs (I-VI) which were selected to produce the greatest range of parameter variations possible (density, weight, length to diameter ratio and nose sharpness) within the constraints of budget and ship launching capability. Thirteen penetrators were constructed collaboratively by the Joint Research Centre and the Building Research Establishment U.K. Two further penetrator types (VII and VIII) were built by Sandia National

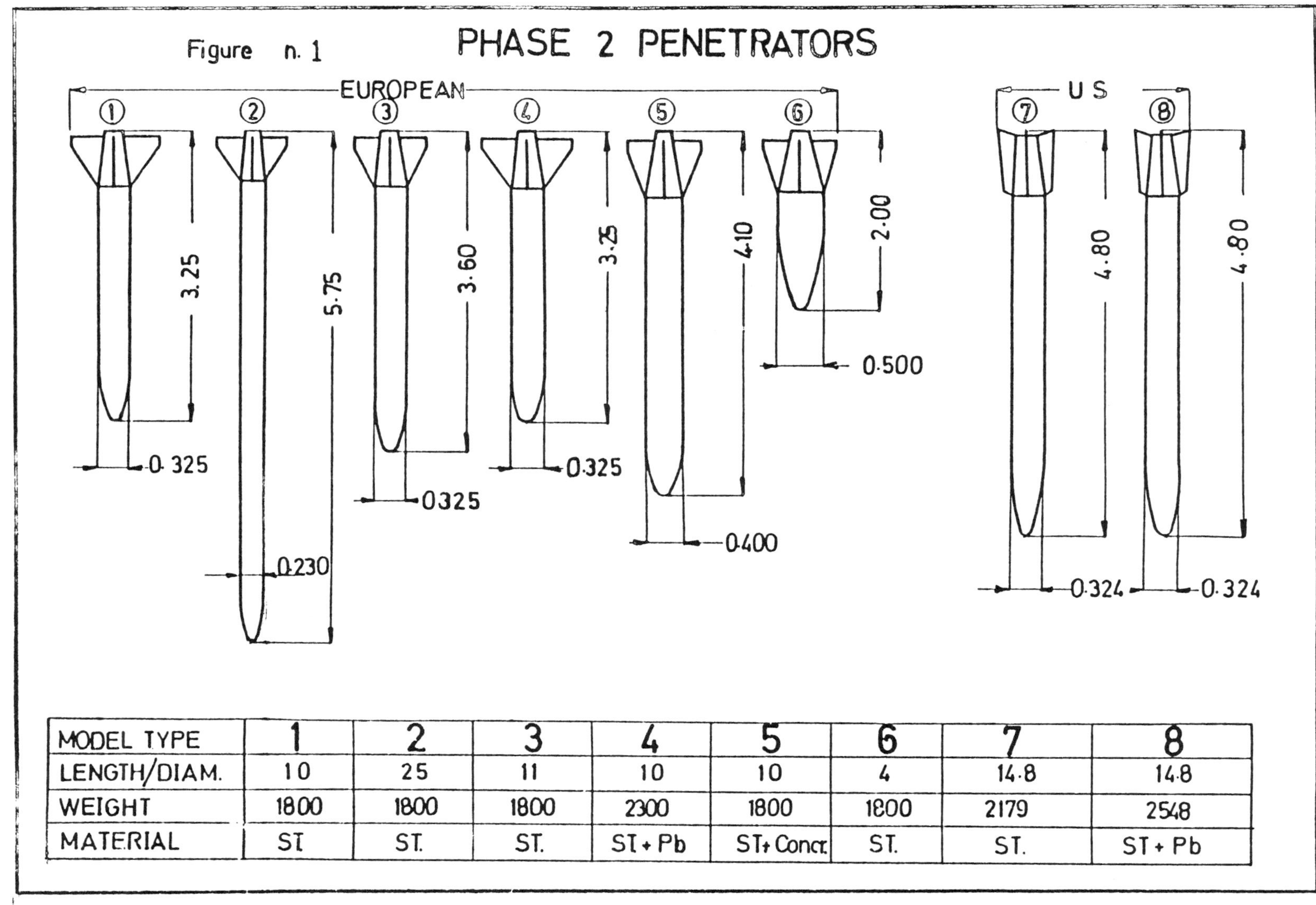

MODEL TYPE	1	2	3	4	5	6	7	8
LENGTH/DIAM.	10	25	11	10	10	4	14.8	14.8
WEIGHT	1800	1800	1800	2300	1800	1800	2179	2548
MATERIAL	ST	ST.	ST.	ST + Pb	ST + Concr.	ST.	ST.	ST + Pb

FIGURE 1. Type Instrumentation Construction

Laboratories bringing the total number of penetrators available for the test series to seventeen. Of these, 11 penetrators were instrumented with ADSS only, 4 with ADSS plus EATS and 2 with LFTS only.

The tests were carried out from the Dutch ship M.V. Tyro and involved the collaboration of the Building Research Establishment, UK, Joint Research Centre, CEC, Sandia National Laboratories, US and the Rijks Geologische Dienst of the Netherlands. The results of these series of tests are being prepared for publication.[13] Preliminary analysis of the ADSS recordings have been undertaken[14] and are shown in Table 2 along with data from LFTS and EATS instrumentation.

TABLE 2. Penetrator Results Test Series II, Nares Abyssal Plain

Test No	Penetrator	Instrumentation	Peak Velocity, ms^{-1}	Impact Velocity, ms^{-1}	Burial Depth, m
15	I	ADSS	57.5	56.5	30.5
10	IA	LFTS	--	44[(a)]	23[(a)]*
2	III	ADSS	56.5	56	30.5
16	IV	ADSS	64	58	38
14	VI	ADSS	50	46	22
7	VII	ADSS	55	55.5	33
		EATS	53.6[(b)]	53[(c)]	28[(c)]**
8	VIII	ADSS	53.5	55[(d)]	30[(d)]
		EATS	50.7[(b)]	53[(c)]	30[(c)]**

NB. *Personal communication LFTS data and notes 1-4; T. Freeman
** " " EATS data; D. Talbert

(a) The impact velocity was estimated by correlating the sub-bottom echoes from the transponder with those from shipboard 3.5 kHz profiles.

(b) Average velocity calculated from detonations in water column between 35 and 55 second after launch.

(c) These figures are provisional and as yet uncorrected for changes in the ship position during the measurements.

(d) The accuracy of these figures is questionable due to frequency instability of this particular ADSS unit. The peak velocity was recorded approximately 20-30 sec after launch; the penetrator then slowed slightly and apparently accelerated just before impact.

Comparison of the predicted embedment depths[11,12] have been generally in good agreement with the measured values when a shear strength profile of

$$Cu = 5 + 1.5z,$$

is used, where z is depth of tail penetration.

The broad agreement between the two independent measurements of burial depth on the same penetrator (EATS and ADSS) adds some confidence to the validity of both measurement systems. Both the advanced transmission systems (LFTS and EATS) were demonstrated to be capable of transmitting data from within the deep ocean bed although neither system was entirely reliable at this stage.

From the results of this test series II it may be concluded that the simple models for penetrator behavior have proved themselves useful in demonstrating penetrator feasibility. As more in situ data become available, this work should lead to a good predictive capability for both the penetration phase and sediment column behavior immediately after impact (i.e., hole closure).

C. Test Series III

The conclusions of the 1984 penetrator tests led to the planning of third series of penetrator tests which were recently carried out during the International Long Core Cruise by the M.V. Marion Dufresne to both the GME and NAP Atlantic Sites (June-July 1985). The objectives of these investigations in addition to the continuing development of more complex transmission systems (LETS and wire transmission) were:

1. to test, for the first time in the deep ocean, sensor instrumentation which could give information about the behavior of the penetrator during its passage in both the water and sediment column (acceleration-velocity, tilt, base pressure and temperature),

2. to fill in gaps in the data needed for the development of predictive models,

3. to compare penetrator results with geotechnical measurements on recovered long cores at the embedment depth, and

4. to test the feasibility of using a METEOSAT satellite communications systems for the re-transmission of data received from embeded penetrators to a land based European receiving Centre.

Although the data from the cruise are not as yet fully worked up, preliminary results show that both of the transmission systems tested are fully capable of sending back data through 5-6 Km water from an embeded penetrator, and that sensors are becoming available which function under the rugged conditions encountered. These sensors very successfully sent back data on the behavior of the penetrator in both the water and sediment columns. Further, it appears that the GME and the NAP sites are covered by satellite communications networks and that it is quite possible to use a low-frequency transponder system to encode data that are compatible with the Data Collection Platform (DCP) of the METEOSAT system.

The implication of this successful use of a deep ocean transmission link and advanced sensor data acquisition system is that autonomous sediment column experiments can be undertaken in real time at reasonable cost and that model validation can be envisaged to investigate critical aspects of the sub-seabed option even over long periods of time.

III. SUMMARY

Research into the feasibility of using large free-fall penetrators as practical means of emplacing solidified heat generating waste into deep ocean sedimentary geological formations has been actively carried out by the Engineering Studies Task Group of the International Seabed Working Group since 1982, the first in-situ tests being undertaken in 1983. From the studies carried out to date it may be concluded that:

- It is possible to design and construct penetrators that will enter into selected areas of deep ocean sediments and bury themselves at least 30 m.

- Models are continuing to be developed that allow prediction, in both the water and sediment columns, of penetrator behavior including hydrodynamic stability, depth of penetration and hole closure.

- Penetrator technology, as a tool for investigating selected deep ocean sediment formations, has advanced rapidly with the development of a number of transmission systems using on-board instrument packages. Such systems will allow much more detailed data to be collected on the geotechnical and eventually geochemical characteristics of these zones. These data will be specially important for investigating hole closure and nuclide migration phenomena.

- The demonstration of the possibility of transmitting data from an emplaced penetrator to the sea surface and thence via a data collection platform of the METLOSAT system to a land based receiving centre means that autonomous long-term investigations of the sediment column can be carried out in quasi real-time.

- Although more detailed engineering studies are needed (especially on hole closure) to complete a feasibility status report on the seabed option (to be published in 1989/90), the results to date have shown that there appear to be no major scientific, technical or economic reasons for not pursuing research into this option further, and that successful engineering techniques for safely emplacing heat generating nuclear wastes within deep ocean sedimentary formations can be demonstrated.

REFERENCES

1. S. G. Bertram and D. R. Anderson, "Subseabed Disposal: A Potential Alternative to Mined Repositories," in these proceedings (1986).

2. "International Long-Core Cruise," Preliminary Report (1985).

3. M. R. C. Bury, The Offshore Disposal of Radioactive Waste by Drilled Emplacement, Report prepared by Taylor Woodrow Construction Limited for the European Atomic Energy Community's Cost-sharing Research Program Published Graham & Trotman Ltd. (1985).

4. Atkins Planning, Concepts for the Disposal of High-Level Radioactive Waste: the Deep Ocean Bed, Work performed for the Department of the Environment, U.K. Report N. RW/82.015 (1982).

5. P. J. Valent, and H. J. Lee, Feasibility of Subseafloor Emplacement of Nuclear Waste, Marine Geotech. Vol. I, 267-293 (1976).

6. Aermacchi Spa, "Hydrodynamic Analysis and Design of High-Level Waste Disposal Model Penetrator," Rep. 419/STE/82, Commission of European Communities Joint Research Centre, Ispra (1982).

7. C. N. Murray, and L. Visentini "Hydrodynamic Aspects Concerning the Design of Free-Fall Penetrators for High-Level Radioactive Waste Disposal," Radioactive Waste Management and the Nuclear Fuel Cycle, 6(1),37-50 (1985).

8. T. J. Freeman, C. N. Murray, T. J. G. Francis, S. D. McPhail, and P. J. Schultheiss, "Modeling Radioactive Waste Disposal by Penetrator Experiments in the Abyssal Atlantic Ocean," Nature 5973, 130-133 (1984).

9. Owe Arup et al., Penetrator Engineering Study - Preliminary Feasibility, performed under contract 394-84-7-WAS-UK to the Commission of the European Communities and part-funded by the UK Department of the Environment, DOE Report N. RW/83.094 (1984).

10. J. V. Boisson, "Etudes des Condition d'Enfouissement de Penetreurs dans les Sediments Marins," Nuclear Science and Technology EUR 9667 (1985).

11. Aermacchi Spa, "Parametric Design of Penetrators for High-Level Radioactive Disposal into Deep Sea Sediments. Studies for Phase II Test Series" Rep. 210-099-81, Commission of European Communities, Joint Research Centre, Ispra (1983).

12. C. N. Murray, and L. Visentini, "Parametric Analysis of Performances of Free Fall Penetrators in Deep Ocean Sediments," Oceanic Engineer, 38-49 (1985).

13. T. J. Freeman, C. N. Murray, D. M. Talbert, "Penetrator Experiments in the Nares Abyssal Plain of the Atlantic Ocean" (in preparation).

14. Aermacchi Spa, "Analysis of Penetrator Data by a Digital Frequency Meter," Ref. 210-EUR-108, Commission of European Communities, Joint Research Centre, Ispra (1985).

OPTIMIZATION STUDIES OF RADIOACTIVE WASTE MANAGEMENT AT A PRE-INDUSTRIAL SCALE

F. Girardi
H. Dworschak
Commission of European Communities
Joint Research Centre
I-21020 Ispra (Varese)
Italy

ABSTRACT

Review of research conducted within the framework of the European Community R&D on waste management suggests that emphasis for the next five year period should be put on verification, demonstration and optimization of concepts already under study, more than on search of new concepts. Implementation of this policy requires a considerable upgrading of research facilities, which is accounted for in the program planning of both the Joint Research Centre (JRC) and the contractual shared-cost activities. In particular, the JRC is constructing a hot-cell facility (PETRA) which will be able to produce various types of fully active conditioned waste resulting from the operation of a PUREX reprocessing plant in standard or modified modes. The facility is proposed:

1. for studies of optimization of waste management and

2. for the production of typical and off-standard conditioned waste products for the study of quality control methods and for setting up large scale verification experiments in laboratories and underground facilities.

I. INTRODUCTION

The Commission of European Communities (CEC) has been engaged in radioactive waste management R&D for over one decade, with research carried out at its own Joint Research Centre (JRC) and through contractual activities with national laboratories and industries of member countries under shared-cost action programs. The second European Community Conference on Radioactive Waste Management and Disposal recently held in Luxembourg[1] has summarized and evaluated the knowledge obtained by the research of the last five year program (1980-1984), discussed the dominant technical orientations for the program 1985-1989 and discussed trends and policies for the successive period.

II. R&D TRENDS FOR THE COMING DECADE

The consensus expressed during the Conference at the various panels and discussion groups was that the research conducted up to now has set the basis for several disposal concepts which appear both feasible and safe even taking into account the many uncertainties connected with research carried out frequently at a limited laboratory scale, and in simulated conditions.

Although, of course, theoretical and laboratory activities should be continued, it appears that they are approaching in many areas the point of "diminishing return," and that some change in research trends is needed. Such changes may be summarized by three words: verification, demonstration, optimization.

Verification: safety assessment of waste disposal is largely based on theoretical evaluations and laboratory experiments. It is necessary in the next decade to "amplify" the scale, both in time and quantities, and verify that the scientific laws on which the safety case is based are indeed applicable to the "real" case. For this reason, in-situ studies should gradually replace laboratory activities, and real fully active waste should gradually replace simulated waste.

Demonstration: This word has frequently been used at the Conference, more or less with the same meaning of verification. It implies, however, a stronger interaction between scientists and decision makers (including the public) which will certainly occur during the verification phase, and which should not be neglected in setting up the lay-out of large-scale verification experiments.

Optimization: The opinion that the desire of assuring safety "at all costs" has led to disposal schemes which are perhaps unnecessarily redundant has frequently been expressed at the Conference. Although one may or may not agree with this opinion, it is undoubted that disposal concepts should be optimized to assure safety, avoiding unnecessary redundance and waste of resources: it is in essence the compliance to the ALARA principle (as low as reasonably achievable) which is at the base of the radiological protection, and which has been accepted by all countries.

III. TOWARDS IMPLEMENTATION AND OPTIMIZATION

A common feature of the three indicated trends is the need of fully active waste samples produced by semi-industrial plants in quantities and qualities tailored for the various needs, such as:

- Setting-up larger scale experiments, which show that the interactions between the various components of a waste disposal system are properly accounted for by the safety evaluations. "Larger scale" means not only larger and more realistic objects, but also implies longer experimental times. Verifying and demonstrating the validity of accelerated tests to simulate processes which occur in centuries will be a challenging task for scientists, and will require a well coordinated use of the experimental facilities existing or to be developed.

- Developing and establishing characterization methods which assure that the product which goes to disposal has the same characteristics as the product for which safety was assessed. Characterization and quality control of waste will require methods which are still largely to be developed. Such developments in turn will require not only the availability of "real" samples, but also the availability of off-standard products, to verify that the methods are indeed capable of assuring quality control.

Also, activities directed towards optimization of waste management will require a large use of semi-industrial facilities. It is well known among R&D policy makers than it is much easier to evaluate benefits of any new process or product that to estimate drawbacks. In the nuclear field particularly the passage from tracer laboratory experiments to fully active large-scale installations has always been a source of difficulties, which have frequently resulted in the abandonment of processes that appeared quite promising at the laboratory scale.

Optimization will also require a much stronger coordination between the various compartments of the full cycle, and particularly with reprocessing and refabrication. It is well known that the policy of the European Community favors more reprocessing of spent fuel and disposal of the resulting waste, than the "once-through" option, in which direct disposal of spent fuel is envisaged.

Table 1 shows the waste output of a reference reprocessing plant assumed by JRC for waste management optimization studies (throughput 1000 tons spent fuel/year). While conditioning of HAW as borosilicate glass is practically considered by all interested E.C. countries as the technique of choice, there is still considerable debate on techniques for immobilizing the other types of waste, and there is clearly an economic and safety incentive to design future reprocessing installations in which the waste output is minimized, or, better, which are optimized not only for the uranium and plutonium production, but also for the waste management.

Major changes in the reprocessing-refabrication schemes such as those required for the actinide separation and transmutation of minor actinides are not supported by C.E.C., since their cost appears overwhelming when compared with the potential benefit, at least for the present nuclear energy industry.[2,3] There are, however, minor changes for which the cost/benefit picture looks promising, at least at laboratory scale.

As examples, stream merging and oxalate precipitation are two schemes which have been studied by JRC in recent years, and which appear worthy of further investigations at a higher technological scale.

TABLE 1. Volumes of Conditioned Waste Assumed for a Reprocessing Plant (throughout 1000 ton/year spent fuel) in JRC Waste Management Optimization Studies

WASTE TYPE	VOLUME OF RAW WASTE m^3/1000 MTHM	CONDITIONED WASTES MATRIX TREATMENT	VOLUME m^3/y	No. OF CANISTERS/DRUMS/ FLASKS
HAW	500	GLASS	150	848/1949
MAW (i)aqueous	1500	CEMENT	2,500	12,500
(ii)phosphoric acids	17			
(iii) ^{14}C	0.1			
Tritiated water	2000	CEMENT	3,500	17,500
COMBUST. (TRU)	1845	COMPACTION	461	2,550
COMBUSTIBLE (non-TRU)	660	INCINERATION/ CEMENT	25	140
NON-COMBUSTIB.				
TRU	1071	PACKAGING	2,142	10,710
NON-TRU	154		308	1,540
HULLS/SPACERS etc.	350	CEMENT	600	1,000
DISSOLVER RESIDUES	50			
NOBLE GASES(Kr)	97	SEPARATION/ STORED UNDER PRESSURE		360-1000
COMBUSTIBLE	1320	COMPACTION	330	1,815
NON-COMBUSTIB.	260	CUTTING/COMPAC.	130	715
HEPA	100	COMPACTION	25	138
DECOMMISSION.	10	SIZE REDUCTION	10	55
SCRAP RECOVERY SOLUTIONS	117	CEMENT	230	1,280
SOLVENT(TBP)(1)	37			
ORGANIC(1) (oils,lubr.etc)	0.5			
EVAPORATOR(1) OVERHEADS	160			

(1) Treatment and conditioning to be decided

1. Stream merging.[4] Recent results on the degradation of vitrified waste in conditions of geological disposal[5] seem to indicate that the availability for migration towards the biosphere of long-lived risk-determining nuclides such as ^{99}Tc and the actinides is more dependent on the chemical characteristics of the geological environment than on the quality of the conditioned product: a low Eh-high pH environment, as frequently encountered in deep geological environments, tends to favour the tetravalent state for Tc, Np and Pu, giving rise to chemical forms with a very low mobility. Such conditions may also be artificially enhanced by the presence of large quantities of iron and concrete in backfilling or structural materials.

In such cases, some relaxation on the quality of the waste matrix is possible in principle, because the waste container could assure containment during the initial thermal peak period, while the local environment would assure the long-term containment.

A waste management concept in which all medium-level liquid waste streams (MLLW) of spent fuel reprocessing are merged into the high-activity waste (HAW) stream could be envisaged, leading to a conditioned waste with a higher sodium content and the worst leaching properties.

Table 2 shows the volume increase of the HAW for various types of vitrified waste, which largely depend on the final Na-concentration accepted in glass. It must, however, be considered that the cost of HAW disposal is probably more dependent on the heat dissipation requirements, which are not appreciably modified by the addition of MLLW, than by the vitrified waste volume. The advantage of not having to condition and dispose the medium level waste should be weighed against the additional chemical and technological complexity of stream merging.

2. OXAL process.[6] The OXAL process was originally conceived in the frame of the project on actinide separation from HAW and transmutation indicated above (2) and carried out by JRC in 1974-1980. It is based on actinide precipitation as oxalates in the presence of rare earth carriers. Successive studies[6] have shown that the scheme is applicable with minor modifications to MLLW streams of spent fuel reprocessing giving raise to conditioned MLLW products which could possibly be classified as "alpha-free." Merging of the actinide oxalate stream to the HAW stream would not cause any appreciable change to the conditioned HAW inventory.

A more ambitious scheme[7] has also been studied, in which the OXAL process would be applied to both HAW and MLLW stream, in order to split all reprocessing-refabrication waste into three streams:

1. an actinide - rare earth stream, to be conditioned and disposed for safety assurance for time periods longer than 1000 years,

TABLE 2. Volume Increases in HAW as a Result of Blending Different MLLW Stream Fractions

MLLW Stream Fractions	Kg Na_2O in waste	% Na_2O in glass	Volume increase with respect to reference HAW[(1)] in glass containing FP. 15 W %	10 W %
Proc. Specific Waste (PSW)	39	6	1.6	1.1
Proc. Specific Waste	39	10	1	1
Managem. Waste	122	6	5	3.3
Managem. Waste	122	10	3	2
Managem. Waste	122	20	1.5	1
PSW + Managem.	161	6	6.6	4.4
PSW + Managem.	161	10	4	2.7
PSW + Managem.	161	20	2	1.3
OXAL Process	161	10	1.04	

(1) Reference HAW, 10 W% Na_2O, volume of glass 0.15 m^3/MTHM, density of glass 2.7×10^3 kg/m^3.

2. a fission-product stream, to be conditioned for safety assurance limited to about 1000 years, and

3. an alpha-free MLLW stream, to be conditioned as low-level waste.

The scheme appears to give several advantages both in the upstream Purex operations and in the downstream side of waste management. It is recognized, however, that for such a scheme to be acceptable, a more detailed and quantified process scheme is required in addition to the experimental verification from a larger scale of operation. Also, extra costs incurred on implementing the scheme need to be assessed in detail in relation to the derived benefits as a result of its implementation. The combination of stream merging and oxalate precipitation, gives rise to a large variety of possible schemes, also depending on local conditions such as possibility of waste discharge at sea against a zero-release plant concept.[8]

In agreement with the trends indicated above, the financial resources of the C.E.C. for the next five years are largely devoted to scale up waste management towards the industrial scale, and to optimize waste management, more than to search for new and advanced disposal strategies. Under the shared-cost action program underground demonstration facilities will be constructed for fully active scale tests as reported in another part of the Conference.[9]

At the JRC a facility is being constructed, named PETRA (Project Evaluation of Treatments RAdioactive waste) which will be essentially devoted to the production of kg quantities of various types of waste, both for optimization studies, and for successive experiments of verification-demonstration. The facility, which has been fully described elsewhere,[10] is installed in a surface area of about 43 m^2 in three existing hot cells at the Ispra Center.

Several operations on the back-end of the fuel cycle can be studied by means of PETRA. It will, in fact, be possible to:

- characterize, chemically and physically, waste streams arising from PUREX type operations, and eventually follow their evolution during interim storage,

- explore the feasibility of variations in the reprocessing scheme, aiming at a reduction of waste generation and optimization of waste categories, such as the stream-merging and OXAL process mentioned above,

- study the performance of different waste treatment and conditioning techniques and processes, inclusive of particular matrix materials for waste fractions such as actinides, and

- prepare conditioned waste with the anticipated specific activity levels for characterization and behaviour testing, also in conditions relevant to interim storage and geological disposal.

The layout of the chemical process units allows the possibility to process LWR fuel material batches corresponding to about 6 kg of U. As a nominal maximum annual capacity the treatment of 10 such batches has been scheduled. Figure 1 shows the process units, the input streams and some basic data of the output products of PETRA, when operated in the reference mode.

The PETRA plant is expected to become operational at a fully active scale during 1987. The facility itself will be open to international cooperation so that management schemes developed in the laboratories of the European Community and other interested countries may be jointly tested at a fully active pre-industrial scale without major investments.

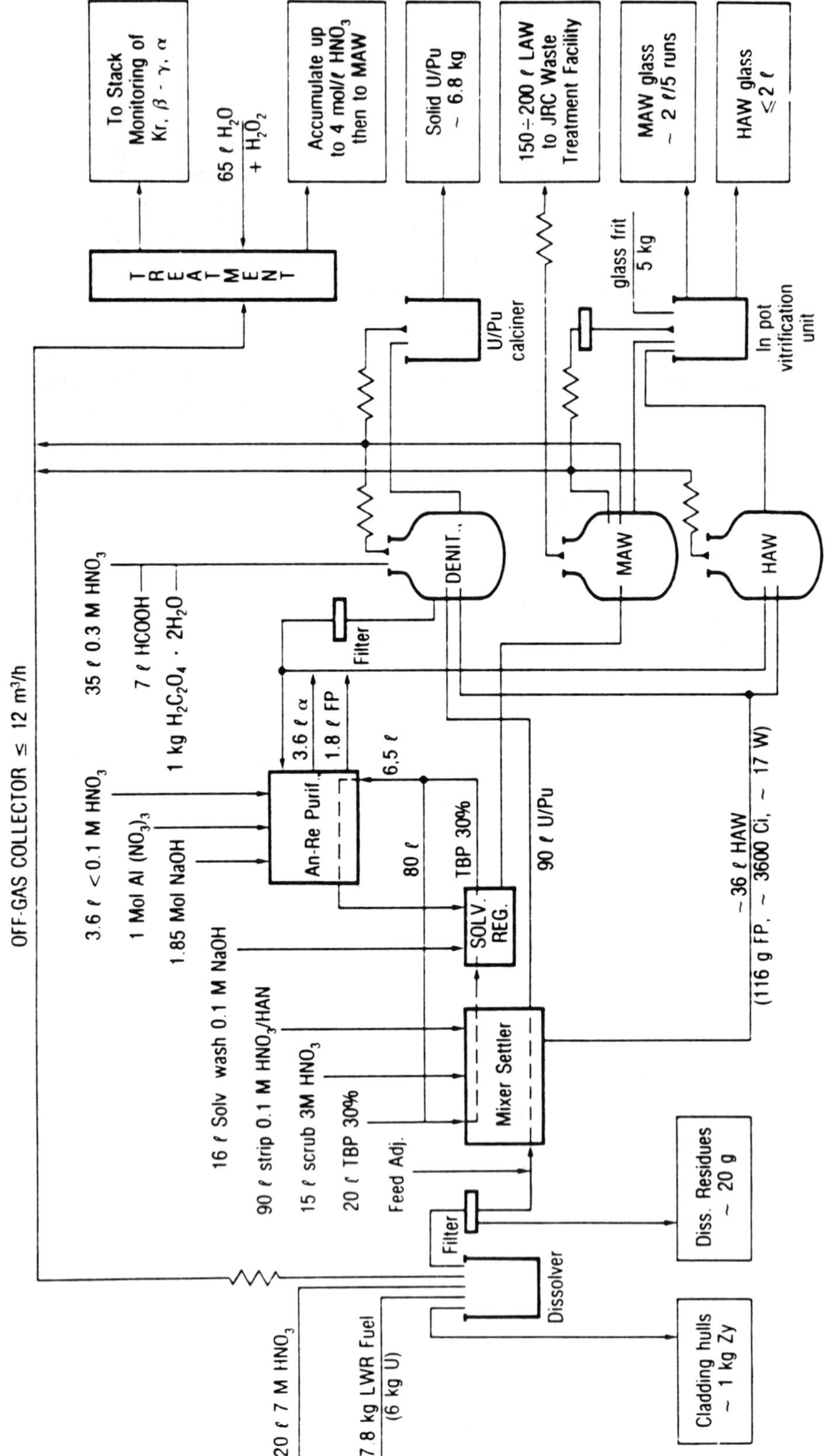

FIGURE 1. Process Units, Input Streams and Output Products of PETRA

REFERENCES

1. Proc. 2nd European Community Conference: Radioactive Waste Management and Disposal, Tech. Rep. EUR 10163 (in press).

2. F. Mannone, and H. Dworschak, "Chemical Separation of Actinides from High Activity Waste," J.R.C. Special Publication S.A. 1.07.03.84.02 (1984).

3. E. Schmidt, E. Zamorani, W. Hage, and S. Guardini, "Assessment Studies on Nuclear Transmutation of By-Product Actinides," J.R.C. Special Publication S.A., 1.05.03.83.13 (1983).

4. H. Dworschak, B. A. Hunt, and F. Mousty, "Study of Waste Management Strategies Which Minimize the Long-Term Risk: Waste Stream Merging," ANS Int. Topical Meeting on Fuel Reprocessing and Waste Management, Jackson Hole, Wyoming, USA (1984).

5. F. Lanza, G. Bidoglio, E. Zamorani, "Influence of Solubility and Insolubilization on the Release of Radioactive Products," Proc. 2nd Intern. Seminar on Rad. Waste Products, Julich (FRG), (June 1985) (in press).

6. F. Mousty, P. Barbero, G. Tanet, L'Acide Oxalique et le Traitement des Dechets Provenant du Traitement du Combustible Irradie, Tech. Rep. EUR 7975 (1982).

7. H. Dworschak et al., "Incentives for Integrating an Alpha Waste Management Strategy into the Fuel Cycle," Nuclear Technology, 61, 432 (1983).

8. B. A. Hunt, and H. Dworschak, "Alternative Treatment Modes for Medium Level Liquid Waste," Rad. Waste Manag. and the Fuel Cycle, 5, 1 (1984).

9. S. Orlowski, "Status of the Commission of European Communities Nuclear Waste Disposal Program," in these proceedings (1986).

10. H. Dworschak, F. Girardi, "PETRA, a Hot Cell Facility for Waste Management Studies," in Proc. 2nd European Community Conference: Radioactive Waste Management and Disposal, Techn. Rep. EUR 10163 (in press).

AUTHOR INDEX

SUBJECT INDEX

A

B

C

C (continued)

D

E

F

G

H

I

J

L

M

N

O

O (continued)

P

Q

R

R (continued)

S

T

U

V

V (continued)

W

Y

Z